The Periplasm

The Periplasm

EDITED BY

Michael Ehrmann

ASM PRESS   WASHINGTON, D.C.

Cover image: Isolated murein sacculus from *Caulobacter crescentus* (courtesy of W. Vollmer; see Chaper 11).

Copyright © 2007   ASM Press
American Society for Microbiology
1752 N St., N.W.
Washington, DC 20036-2904

Library of Congress Cataloging-in-Publication Data

The periplasm / edited by Michael Ehrmann.
    p. ; cm.
  Includes bibliographical references and index.
  ISBN-13: 978-1-55581-398-7 (alk. paper)
  ISBN-10: 1-55581-398-4 (alk. paper)
  1. Bacterial proteins. 2. Proteins—Synthesis. 3. Escherichia coli.
  4. Microbiology.   I. Ehrmann, Michael, Dr. II. American Society
  for Microbiology.
  [DNLM: 1. Periplasmic Proteins—physiology. 2. Bacterial Outer
  Membrane Proteins—physiology. 3. Escherichia coli—physiology.
  4. Membrane Transport Proteins—physiology. 5. Periplasm—physiology.   QW 52 P445 2006]

QR92.P75P47 2006
572'.69—dc22

                                                                    2006019687

10 9 8 7 6 5 4 3 2 1

Address editorial correspondence to ASM Press, 1752 N St., N.W., Washington, DC
20036-2904, USA

Send orders to ASM Press, P.O. Box 605, Herndon, VA 20172, USA
Phone: 800-546-2416; 703-661-1593
Fax: 703-661-1501
E-mail: books@asmusa.org
Online: estore.asm.org

CONTENTS

Contributors ix
Foreword Hiroshi Nikaido xiii
Preface xvii

I.   PROTEIN SECRETION   1

**1. Co- and Posttranslational Protein Targeting
to the SecYEG Translocon in *Escherichia coli***
Eitan Bibi
3

2. The Tat Protein Export Pathway
Tracy Palmer and Ben C. Berks
16

**3. Assembly of Integral Membrane Proteins from
the Periplasm into the Outer Membrane**
Jörg H. Kleinschmidt
30

**4. Structure, Function, and Transport of Lipoproteins
in *Escherichia coli***
*Hajime Tokuda, Shin-ichi Matsuyama,
and Kimie Tanaka-Masuda*
67

II.   PROTEIN FOLDING AND QUALITY
CONTROL   81

5. The Cpx Envelope Stress Response
Tracy L. Raivio
83

6. **Regulation and Function of the Envelope Stress Response Controlled by σ^E**
Carol A. Gross, Virgil A. Rhodius, and Irina L. Grigorova
107

7. **Disulfide Bond Formation in the Periplasm**
Mehmet Berkmen, Dana Boyd, and Jon Beckwith
122

8. **Periplasmic Chaperones and Peptidyl–Prolyl Isomerases**
Jean-Michel Betton
141

9. **Periplasmic Proteases and Protease Inhibitors**
Nicolette Kucz, Michael Meltzer, and Michael Ehrmann
150

III.   **KEY PHYSIOLOGICAL PROCESSES   171**

10. **Cell Division**
S. J. Ryan Arends, Kyle B. Williams, Ryan J. Kustusch, and David S. Weiss
173

11. **Structure and Biosynthesis of the Murein (Peptidoglycan) Sacculus**
Waldemar Vollmer
198

12. **Periplasmic Events in the Assembly of Bacterial Lipopolysaccharides**
Chris Whitfield, Emilisa Frirdich, and Anne N. Reid
214

13. **Electron Transport Activities in the Periplasm**
Stuart J. Ferguson
235

14. **Periplasmic Nitrate Reduction**
Jeff A. Cole
247

15. **The Biosynthesis of the Molybdenum Cofactor and Its Incorporation into Molybdoenzymes**
Silke Leimkühler
260

16. **Transfer of Energy and Information across the Periplasm in Iron Transport and Regulation**
Volkmar Braun and Susanne Mahren
276

17. Periplasmic ABC Transporters
Elie Dassa
287

18. Antimicrobial and Stress Resistance
Keith Poole
304

19. Osmoregulation in the Periplasm
Jean-Pierre Bohin and Jean-Marie Lacroix
325

IV. PRACTICAL IMPLICATIONS 343

**20. Practical Applications for Periplasmic
Protein Accumulation**
John C. Joly and Michael W. Laird
345

21. Periplasmic Expression of Antibody Fragments
David P. Humphreys
361

V. PROTEIN COMPOSITION 389

**22. Methods for the Computational Prediction
of Periplasmic Proteins**
Jennifer L. Gardy and Fiona S. L. Brinkman
391

Index 407

CONTRIBUTORS

S. J. Ryan Arends
Department of Microbiology, Roy J. and Lucille A. Carver College of Medicine,
University of Iowa, Iowa City, IA 52242

Jon Beckwith
Department of Microbiology and Molecular Genetics, Harvard Medical School,
Boston, MA 02115

Mehmet Berkmen
Department of Microbiology and Molecular Genetics, Harvard Medical School,
Boston, MA 02115

Ben C. Berks
Department of Biochemistry, University of Oxford, Oxford OX1 3QU,
United Kingdom

Jean-Michel Betton
Unité de Repliement et Modélisation des Protéines, Institut Pasteur–CNRS URA2185,
F-75724 Paris Cedex 15, France

Eitan Bibi
Department of Biological Chemistry, The Weizmann Institute of Science,
Rehovot 76100, Israel

Jean-Pierre Bohin
Unité de Glycobiologie Structurale et Fonctionnelle, UMR CNRS-USTL 8576, IFR118,
Université des Sciences et Technologies de Lille, F-59655 Villeneuve d'Ascq Cedex, France

Dana Boyd
Department of Microbiology and Molecular Genetics, Harvard Medical School,
Boston, MA 02115

Volkmar Braun
Mikrobiologie/Membranphysiologie, Auf der Morgenstelle 28,
D-72026 Tübingen, Germany

Fiona S. L. Brinkman
Department of Molecular Biology and Biochemistry, Simon Fraser University,
Burnaby, BC V5A 1S6, Canada

Jeff A. Cole
School of Biosciences, University of Birmingham, Birmingham B15 2TT,
United Kingdom

Elie Dassa
Unité des Membranes Bactériennes CNRS URA2172, Département de Microbiologie
Fondamentale et Médicale, Site Fernbach, Institut Pasteur, 25 Rue du Docteur Roux,
F-75724 Paris Cedex 15, France

Michael Ehrmann
Zentrum für Medizinische Biotechnologie, Universität Duisburg-Essen,
45117 Essen, Germany

Stuart J. Ferguson
Department of Biochemistry, University of Oxford, South Parks Road,
Oxford OX1 3QU, United Kingdom

Emilisa Frirdich
Department of Molecular and Cellular Biology, University of Guelph,
Guelph, Ontario N1G 2W1, Canada

Jennifer L. Gardy
Department of Molecular Biology and Biochemistry, Simon Fraser University,
Burnaby, BC V5A 1S6, Canada

Irina L. Grigorova
Graduate Group in Biophysics, University of California, San Francisco,
San Francisco, CA 94143-2200

Carol A. Gross
Department of Microbiology and Immunology, Genentech Hall, Room S376,
600 16th Street, Box 2200, University of California, San Francisco,
San Francisco, CA 94143-2200

David P. Humphreys
UCB-Celltech, 216 Bath Road, Slough SL1 4EN, United Kingdom

John C. Joly
Genentech, Inc., Early Stage Cell Culture Dept., 1 DNA Way,
South San Francisco, CA 94080

Jörg H. Kleinschmidt
Fachbereich Biologie, Fach M 694, Universität Konstanz,
D-78457 Konstanz, Germany

Nicolette Kucz
Zentrum für Medizinische Biotechnologie, Universität Duisburg-Essen,
45117 Essen, Germany

Ryan J. Kustusch
Department of Microbiology, Roy J. and Lucille A. Carver College of Medicine,
University of Iowa, Iowa City, IA 52242

Jean-Marie Lacroix
Unité de Glycobiologie Structurale et Fonctionnelle, UMR CNRS–USTL 8576,
IFR118, Université des Sciences et Technologies de Lille, F-59655
Villeneuve d'Ascq Cedex, France

Michael W. Laird
Late Stage Cell Culture Department, Genentech, Inc., 1 DNA Way,
South San Francisco, CA 94080

Silke Leimkühler
Universität Potsdam, Institut für Biochemie und Biologie,
AG Biochemie (Proteinanalytik), Karl-Liebknecht Str. 24–25, Haus 25,
D-14476 Potsdam, Germany

Susanne Mahren
Mikrobiologie/Membranphysiologie, Auf der Morgenstelle 28,
D-72026 Tübingen, Germany

Shin-ichi Matsuyama
Department of Life Science, Rikkyo University, 3-34-1 Nishi-ikebukuro,
Toshima-ku, Tokyo 171-8501, Japan

Michael Meltzer
Zentrum für Medizinische Biotechnologie, Universität Duisburg-Essen,
45117 Essen, Germany

Hiroshi Nikaido
Department of Molecular and Cell Biology, 426 Barker Hall, MC-3202,
University of California, Berkeley, CA 94720-3202

Tracy Palmer
Department of Molecular Microbiology, John Innes Centre, Colney Lane,
Norwich NR4 7UH, United Kingdom

Keith Poole
Department of Microbiology and Immunology, Queen's University,
Kingston, Ontario K7L 3N6, Canada

Tracy L. Raivio
Department of Biological Sciences, University of Alberta,
Edmonton, AB T6G 2E9, Canada

Anne N. Reid
Department of Molecular and Cellular Biology, University of Guelph,
Guelph, Ontario N1G 2W1, Canada

Virgil Rhodius
Department of Microbiology and Immunology, Genentech Hall, Room S376,
600 16th Street, Box 2200, University of California, San Francisco,
San Francisco, CA 94143-2200

Kimie Tanaka-Masuda
Institute of Molecular and Cellular Biosciences, University of Tokyo,
1-1-1 Yayoi, Bunkyo-ku, Tokyo 113-0032, Japan

Hajime Tokuda
Institute of Molecular and Cellular Biosciences, University of Tokyo,
1-1-1 Yayoi, Bunkyo-ku, Tokyo 113-0032, Japan

Waldemar Vollmer
Mikrobielle Genetik, Auf der Morgenstelle 28, Universität Tübingen,
D-72026 Tübingen, Germany

David S. Weiss
Department of Microbiology, Roy J. and Lucille A. Carver College of Medicine,
University of Iowa, Iowa City, IA 52242

Chris Whitfield
Department of Molecular and Cellular Biology, University of Guelph,
Guelph, Ontario N1G 2W1, Canada

Kyle B. Williams
Department of Microbiology, Roy J. and Lucille A. Carver College of Medicine,
University of Iowa, Iowa City, IA 52242

FOREWORD

We now take it for granted that the periplasm is a separate compartment within the gram-negative bacterial cell, with its characteristic components and physiological processes. However, this knowledge is relatively recent, and few scientists could foresee such a development in the early 1970s.

Perhaps the earliest study that showed the existence of characteristic periplasmic proteins was that of Harold Neu and Leon Heppel in 1964 (Neu and Heppel, 1964), showing that ribonuclease was released from *Escherichia coli* cells upon their conversion into spheroplasts by the EDTA-lysozyme treatment in sucrose. Before this study, we did not know how *E. coli* could survive with such "dangerous" enzymes as ribonucleases and phosphatases, which would destroy vital components of the living cell, and yet which were clearly present and could be released upon cell breakage. We learned, for the first time, that they are kept in the periplasmic space, away from the cytosol, in gram-negative bacteria. Heppel went on to show that a number of active transport systems in *E. coli* required the presence of specific binding proteins in the periplasm, thus broadening our knowledge of the physiological functions of the periplasmic components.

To deepen our understanding of the nature of the periplasm, it was necessary to know the properties of the barriers of this compartment. The cytoplasmic membrane, common to all cells, was of course well known. But the external barrier was commonly referred to as "cell wall." Some of us often argue that names are not important. I felt the same way in the late 1960s. Names are actually important, however, because they force us to think in certain ways. As far as I am aware, the term "outer membrane" was first used in the absolutely beautiful electron microscopic study of a periodontal pathogen *Veillonella* by Howard Bladen and Stephan Mergenhagen (Bladen and Mergenhagen, 1964) in the same year when Neu and Heppel discovered the release of periplasmic ribonuclease. This work even supplied some evidence for the localization of lipopolysaccharide (LPS) in the outer membrane. It is surprising that the notion of the outer membrane as a true bilayer membrane did not become popular immediately afterward. Possibly it was due to the bias of biochemists

and geneticists against morphological data, and also perhaps because of the use
of a bacterial species that was unfamiliar to most *E. coli/ Salmonella* biologists. As
regards the latter, *Veillonella*, together with *Selenomonas* and a few other gram-
negative genera, is now placed in the gram-positive subphylum of *Clostridium*,
on the basis of 16S rRNA-based taxonomy (Willems and Collins, 1995)! Even
the subsequent use of the outer membrane terminology in *E. coli* several years
later by Takashi Miura and Shoji Mizushima (Miura and Mizushima, 1968)
failed to make the word popular. I believe that the term came into general use
only when Mary Osborn emphasized the membranelike nature of this "cell
wall" layer through the analysis of its components (Osborn et al., 1972).
Indeed, once we thought of the outer membrane as a true biological membrane based on a lipid bilayer structure, many previous results became compre-
hensible. These include the observation that mutants producing very defective
LPS, a component of the outer membrane, became hypersusceptible to dyes,
detergents, and lipophilic antibiotics (studies from the Otto Westphal laboratory around 1969, cited in Nikaido and Vaara, 1985), as well as the finding by
Loretta Leive that the EDTA-induced partial release of LPS from intact *E. coli*
cells made them hypersusceptible to lipophilic antibiotics (Leive, 1968). I am
pleased to note that our laboratory also made contributions in this area by
showing that the lipid bilayer structure in the outer membrane is asymmetric,
with its outer leaflet containing only LPS (Kamio and Nikaido, 1976), and that
the outer membrane acts as a molecular sieve for hydrophilic solutes (Decad
and Nikaido, 1976), the latter finding leading to the discovery of porins.

The definition of the outer membrane as an effective permeability barrier
implied that the periplasm might be different from the environment also in
terms of the composition of small solutes. This was shown by the discovery of
Donnan potential across the outer membrane by Jeff Stock, Barbara Rauch,
and Saul Roseman in 1977 (Stock et al., 1977); the inside-negative potential
causes the accumulation of small cations in the periplasm. This was soon followed by the discovery by Howard Shulman and Eugene Kennedy of membrane-derived oligosaccharides (MDOs) in periplasm (Schulman and Kennedy,
1979), which were later shown to be a group of polyanionic oligosaccharides
too large to pass through the porins and were shown to be responsible for the
generation of Donnan potential.

As shown by the contents of this book, the amazingly complex and unique
nature of the periplasm as a separate cellular compartment has now been elucidated through the studies of many laboratories. Thus the pioneering studies
from the Jon Beckwith laboratory on the disulfide formation in periplasm led
to our current knowledge of the unique folding pathways of periplasmic and
outer membrane proteins. Laboratories of Carol Gross and Tom Silhavy showed
how the correct folding of these proteins was ensured by the unique chaperones and stress-sensing mechanisms in the periplasm. Tokuda's laboratory discovered how the lipoproteins are sorted and transported to the outer membrane. These are just a few examples, and many more fascinating developments
are covered in this book.

Finally, when we consider the many outer membrane proteins whose functions are affected by proteins in the cytoplasmic membrane, such as the TonB-

dependent transport systems or tripartite multidrug efflux pumps, it is now clear that the two barriers of the periplasm are not completely separated and independent from each other. In this context, we should recall the early studies by Manfred Bayer, whose discovery of the "adhesion sites" between the inner and outer membranes in 1968 (Bayer, 1968) was perhaps an example of "premature" discoveries in science. In subsequent years, this important result was often the recipient of skepticism and criticism rather than the accolades it deserved. We still do not know how the numerous and dynamic connections between the two membranes required by the tripartite efflux pumps, for example, are made through the peptidoglycan structure. Although we have come far in the past 35 years or so, many things remain to be discovered in this fascinating compartment, the periplasm.

REFERENCES

Bayer, M. E. 1968. Adsorption of bacteriophages to adhesions between wall and membrane of *Escherichia coli*. *J. Virol.* **2:**346–356.

Bladen, H. A., and S. E. Mergenhagen. 1964. Ultrastructure of *Veillonella* and morphological correlation of an outer membrane with particles associated with endotoxic activity. *J. Bacteriol.* **88:**1482–1492.

Decad, G. M., and H. Nikaido. 1976. Outer membrane of gram-negative bacteria. XII. Molecular-sieving function of cell wall. *J. Bacteriol.* **128:**325–336.

Kamio, Y., and H. Nikaido. 1976. Outer membrane of *Salmonella typhimurium*: accessibility of phospholipid head groups to phospholipase C and cyanogen bromide activated dextran in the external medium. *Biochemistry* **15:**2561–2570.

Leive, L. 1968. Studies on the permeability change produced in coliform bacteria by ethylenediaminetetraacetate. *J. Biol. Chem.* **243:**2373–2380.

Miura, T., and S. Mizushima. 1968. Separation by density gradient centrifugation of two types of membranes from spheroplast membrane of *Escherichia coli* K12. *Biochim. Biophys. Acta* **150:**159–161.

Neu, H. C., and L. A. Heppel. 1964. Some observations on the "latent" ribonuclease of *Escherichia coli*. *Proc. Natl. Acad. Sci. USA* **51:**1267–1274.

Nikaido, H., and M. Vaara. 1985. Molecular basis of bacterial outer membrane permeability. *Microbiol. Rev.* **49:**1–32.

Osborn, M. J., J. E. Gander, E. Parisi, and J. Carson. 1972. Mechanism of assembly of the outer membrane of *Salmonella typhimurium*. Isolation and characterization of cytoplasmic and outer membrane. *J. Biol. Chem.* **247:**3962–3972.

Schulman, H., and E. P. Kennedy. 1979. Localization of membrane-derived oligosaccharides in the outer envelope of *Escherichia coli* and their occurrence in other gramnegative bacteria. *J. Bacteriol.* **137:**686–688.

Stock, J. B., B. Rauch, and S. Roseman. 1977. Periplasmic space in *Salmonella typhimurium* and *Escherichia coli*. *J. Biol. Chem.* **252:**7850–7861.

Willems, A., and M. D. Collins. 1995. Phylogenetic placement of *Dialister pneumosintes* (formerly *Bacteroides pneumosintes*) within the *Sporomusa* subbranch of the *Clostridium* subphylum of the gram-positive bacteria. *Int. J. Syst. Bacteriol.* **45:**403–405.

HIROSHI NIKAIDO

PREFACE

The speed at which novel and elegant higher throughput and miniaturized technology develops, interdisciplinary activities spring up, and discoveries are made is quite incredible. This burst of recent activity resulted in a boost in publishing. It is my impression, however, that the greater the quantity of information that is made available, the more readily that journals spring up, and the more extensive that press coverage becomes, the less that the individual person reads anything thoroughly. This notion is supported by the fact that the half-life of citations continuously shrinks. Even the most elegant approaches and groundbreaking discoveries of the 1980s and 1990s are no longer in people's minds.

When I contacted a famous colleague about a potential contribution to this book, he responded, "My only worry is that, in today's times, no one is consulting (and citing!) a book anymore. . . . So, I am worried that the articles may end up in some obscurity." While I do appreciate the comment, I do not think that this analysis is realistic. Books have educated people over the centuries, and we will likely see a revival in book reading because, like no other medium, the book offers in-depth coverage of past and present scientific progress and provides a unique chance to study and contemplate in peace.

It was, therefore, a great pleasure to edit the first book dedicated to the periplasm, an extracytoplasmic compartment of gram-negative bacteria. While bacteria were identified hundreds of years ago, it took a bit of extra time for the periplasm to be recognized. Key events were Mitchell's work on transport; Heppel's osmotic shock technique, allowing the selective release of soluble periplasmic proteins; and advances in imaging using electron microscopy. In those early days, transport proteins, cell wall components, and various enzymes such as β-lactamase, ribonuclease, alkaline phosphatase, etc., were widely studied and provided general biological insights. The contributions to this volume clearly indicate that the periplasm is still at the center of a very active and expanding field of research that will continue to produce important results in the future.

It is quite surprising, however, that the periplasm rarely receives much attention in the relevant textbooks, most of which contain only a few, often trivial, sentences on this topic. Perhaps more amusing are the events leading to the production of this book. ASM's own Greg Payne was cruising the ASM General Meeting in New Orleans and happened to be unable to attend a session on the periplasm because of overcrowding. This sparked his interest, and downstream events that I do not understand in great detail ended, for me, in the honor of editing this fine book, which is meant to enthuse students and to provide the latest facts to the experts. The book aims to inform general biology students at the university undergraduate level and graduate students and post docs in microbiology, biochemistry, structural biology, and biotechnology as well as medical practitioners wanting to learn about the physiology of pathogenic microorganisms and some of their drug resistance strategies or lipopolysaccharide biosynthesis.

The periplasm was the first accessory cellular compartment that evolved to make cells more robust and efficient and thus competitive. At least with respect to protein transport, folding, and quality control, the periplasm shares many similarities with the endoplasmic reticulum of eukaryotic cells. Also, because of its relative simplicity it serves as a general model system to study protein biogenesis and function, the control of protein composition, bioenergetics, solute transport, and stress response on the molecular and mechanistic levels. Computational methods predicting cellular localization that are comprehensively reviewed in this book were first developed by using soluble periplasmic and cytoplasmic membrane proteins. Finally, pioneering studies on the production of recombinant proteins containing disulfide bonds led to wide applications, which are also reviewed in depth.

Gratitude is expressed to the participating scientists, all eminent figures in the field, who invested a good share of their nonexisting spare time to provide us with the facts and sometimes their own personal insights. Also, many thanks are due to Greg Payne and John Bell of ASM Press for their input, help, and patience during all stages of production.

M. EHRMANN

PROTEIN SECRETION

I

CO- AND POSTTRANSLATIONAL PROTEIN TARGETING TO THE SecYEG TRANSLOCON IN *ESCHERICHIA COLI*

Eitan Bibi

I

Targeting of membrane and secretory proteins to the *Escherichia coli* cytoplasmic membrane occurs either co- or posttranslationally. In vitro and in vivo studies have shown that the two targeting pathways converge at the SecYEG translocon, through which the proteins are either inserted into the membrane or translocated to the periplasm. For posttranslational targeting, presecretory proteins require chaperones, among which the translocation-specific chaperones SecB and SecA have been well characterized. SecA, which mediates the posttranslational transfer of presecretory proteins to the translocon, is also actively involved in the translocation process. The cotranslational pathway is mediated by the signal recognition particle (SRP) system, which plays an essential role in the biogenesis of membrane proteins and several secretory proteins. The SRP system is responsible for the targeting of ribosomes that translate SRP substrates to the translocon. Recent evidence suggests that the cotranslational targeting machinery physically interacts with the translocon, thus offering a possible clue as to how ribosomes are transferred from the SRP system to the translocon for proper cotranslational membrane insertion.

Selective targeting of proteins to and across biological membranes of various organelles and cellular compartments represents a basic phenomenon in cell biology (Schatz and Dobberstein, 1996; Wickner and Schekman, 2005). In gram-negative bacteria, such as *E. coli*, proteins of several groups engage in essential functions within the extracytoplasmic compartment, the periplasm. This includes integral cytoplasmic membrane proteins, which interact with the periplasm either through their periplasmic domains or their roles in the biogenesis or function of this compartment. Other groups include soluble periplasmic proteins or those peripherally associated with the periplasmic side of the inner or outer membranes, and also outer membrane proteins that protrude into the periplasmic space. Although many of these proteins utilize posttranslational secretory pathways (Mori and Ito, 2001; Driessen et al., 2001), integral membrane proteins and also some secretory proteins are inserted and/or translocated cotranslationally (Luirink and Sinning, 2004). The posttranslational pathway, which is mediated by general and specific chaperones, such as SecB and SecA, is relatively well understood and documented (for recent reviews, see Eichler and Duong, 2004; Veenendaal et al., 2004; Vrontou and Economou, 2004; Osborne et al., 2005). Therefore, this chapter focuses

Eitan Bibi, Department of Biological Chemistry, The Weizmann Institute of Science, Rehovot 76100, Israel.

The Periplasm
Edited by Michael Ehrmann © 2007 ASM Press, Washington, D.C.

more on the cotranslational pathway that is mediated by the SRP system of *E. coli* as a model for gram-negative bacteria. This pathway can be divided into three main steps: (i) the targeting step, through which ribosomes translating the relevant proteins are targeted to the membrane; (ii) the intermediate step, through which the ribosomes are transferred from the targeting system to the translocon; and (iii) the insertion/translocation step, through which the translating ribosomes extrude the nascent polypeptide chain into or across the cytoplasmic membrane. Several questions regarding step i were raised after in vivo studies and are discussed. Step ii is the least characterized element of the pathway, and it might constitute part of step i if the ribosomes are directly targeted to the translocon (Angelini et al., 2005). As mentioned, the cotranslational targeting pathway is catalyzed by the SRP system, and the pioneering work on the mammalian SRP system (reviewed in Walter and Johnson, 2004) provided the initial framework regarding how the SRP system in *E. coli* might mediate ribosomal targeting to the inner membrane. As discussed later in detail, the targeting is initiated by a signal recognition event that takes place immediately after a hydrophobic nascent peptide protrudes from the ribosome. In other words, according to the model, the translating ribosome interacts with the SRP, and the entire RNA-ribosome-nascent chain-SRP complex is then targeted to the membrane, where it associates with the SRP receptor. The *E. coli* SRP system includes a 4.5S RNA (reviewed in Brown, 1991; Bovia and Strub, 1996); Ffh, a homologue of the mammalian SRP54 protein; and FtsY, of which the 300-residue-long C-terminal domain resembles the C-terminal domain of the mammalian SRP receptor α-subunit (SRα) (Romisch et al., 1989; Bernstein et al., 1989). Notably, all the *E. coli* SRP components are essential for cell growth (Brown and Fournier, 1984; Gill and Salmond, 1987; Luirink et al., 1994; Phillips and Silhavy, 1992). The initial identification of an SRP-like system in *E. coli* and its implication in protein targeting was rather surprising

(Bassford et al., 1991; Beckwith, 1991; Poritz et al., 1991), because, unlike in the mammalian system, it was known that in *E. coli* secretory proteins can migrate to the membrane posttranslationally (Bassford et al., 1991). Subsequent genetic studies have provided an explanation for this discrepancy (Macfarlane and Muller, 1995; de Gier et al., 1996; Ulbrandt et al., 1997; Seluanov and Bibi, 1997; Tian et al., 2000) by suggesting that the *E. coli* SRP system may function preferentially in the biosynthetic pathway of cytoplasmic membrane proteins. A cotranslational pathway for targeting hydrophobic integral membrane proteins would be advantageous in preventing their synthesis and aggregation in the cytoplasm (Bernstein, 2000), but secretory proteins may also use a cotranslational pathway (Luirink et al., 1994; Phillips and Silhavy, 1992; Nakatogawa and Ito, 2001; Schierle et al., 2003; Shimohata et al., 2005). Proteins selected for SRP-mediated targeting in *E. coli* probably have a highly hydrophobic N-terminal polypeptide (signal peptide in presecretory proteins), as discussed in a recent review by Luirink and Sinning (2004). After having been targeted to the membrane, the ribosome-nascent chain complexes must be transferred to the translocon in a manner that is presently not fully understood, and then, via a second signal recognition event at the translocon (Jungnickel and Rapoport, 1995), the membrane insertion/secretion step is initiated. In the cotranslational pathway of membrane protein insertion, the translocon utilizes an additional protein, YidC (Samuelson et al., 2000), which is thought to mediate the assembly of energy-transducing membrane protein complexes (van der Laan et al., 2005).

POSTTRANSLATIONAL PROTEIN TARGETING

Despite their different mode of action, the co- and posttranslational pathways share one important element: recognition of a hydrophobic N-terminal sequence. All the polypeptides are targeted to the membrane by virtue of a hydrophobic sequence, a cleavable signal sequence, or a noncleaved transmembrane segment. In presecretory proteins the presence of an amino-terminal signal sequence is critical for recognition by the targeting system, and it also influences the folding of the fully translated preproteins in the cytosol (Park et al., 1988). Upon completion of their translocation across the inner membrane, signal sequences are usually removed from the preproteins by the translocation-specific signal peptidase (Paetzel et al., 2002). This signal consists of a short, positively charged N-terminal segment that is followed by a longer hydrophobic segment, and a short polar domain containing the signal peptidase cleavage site (von Heijne, 1990).

As previously described, most secretory proteins of *E. coli* are first synthesized in the cytoplasm and then targeted to the translocon by general and specific chaperones. SecB is a specific chaperone for certain preproteins (Randall and Hardy, 2002). By interacting with these polypeptides in the cytosol, SecB precludes their premature folding and also aggregation. In other words, SecB helps to maintain the noncompact, translocation-competent folding state of these precursors, a prerequisite for proper interaction with the translocon. Note, SecB is not essential for growth of *E. coli*, suggesting that many proteins are targeted to the translocon in its absence (Baars et al., 2005). Possibly, under these conditions other, general chaperones play an important role in maintaining the preproteins in a translocation-competent folding state (Baars et al., 2005). Signal sequences are not required for SecB recognition, and SecB interacts with portions of the mature part of preproteins (Randall et al., 1990). The question of how proteins are selected by SecB is still not fully understood (Ullers et al., 2004). The X-ray structure of the *E. coli* SecB (Dekker et al., 2003), combined with biochemical data, enabled this issue to be discussed, mainly suggesting that as a tetramer (dimer of dimers) (Topping et al., 2001), SecB has two, 70-amino-acid-long binding grooves, each on either side of the molecule, which is appropriate for the binding of two extended 20-residue-long peptides, as suggested (Xu et al., 2000). Based on its X-ray structure, the current view is that sub-strate-binding by SecB is initiated by electrostatic forces and that hydrophobic forces substantially increase the affinity, as suggested in the past (Randall and Hardy, 1995). For proper targeting to the translocon, the SecB-preprotein complex must interact with SecA. Part of this interaction has been observed in the X-ray structure of SecB with a C-terminal domain of SecA from *Haemophilus influenzae* (Zhou and Xu, 2003). Note, however, not only is SecB not essential for growth in *E. coli*, the C-terminal domain of SecA that mediates SecB interaction is also not essential (Karamanou et al., 2005). Nevertheless, the fact that the affinity between SecB and SecA is substantially increased in the context of the translocon compared with that measured in solution (Fekkes et al., 1997) reinforces the claim that SecB has a specific role in the targeting of presecretory proteins (Baars et al., 2005), in addition to its proposed role as a general chaperone that interacts with nonsecretory polypeptide proteins (Ullers et al., 2004).

Similarly to SecB, SecA has also been implicated in chaperoning newly synthesized nonsecretory proteins (Eser and Ehrmann, 2003), in addition to its well-characterized essential role in presecretory protein targeting and translocation. SecA is the membrane peripheral ATPase subunit of the translocon. In this capacity it functions both as a targeting chaperone and as an energy-driven polypeptide-translocating motor in the context of the SecYEG complex. Several X-ray structures of SecA have been resolved, including structures of the *Bacillus subtilis* protein (Hunt et al., 2002; Osborne et al., 2004) and that of *Mycobacterium tuberculosis* (Sharma et al., 2003). Structurally, SecA is a dimer, and each monomer comprises several functional domains. The domain organization of SecA was initially revealed by biochemical studies (e.g., Price et al., 1996), which were confirmed by the X-ray structures. It contains an N-terminal motor domain, homologous to DEAD box helicases (the DExH/D-box proteins compose a large family of ATPases that have been proposed to mediate RNA structural rearrangements in a variety of cellular

processes, including nuclear pre-mRNA splicing, ribosome assembly, protein synthesis, nuclear transport, and RNA degradation) (for
SecA, see Sianidis et al., 2001), which includes
the nucleotide binding site and an ATPase regulatory element. Additional important domains
in SecA are those specializing in substrate binding, including a recently proposed protein binding domain (termed PBD) (Papanikou et al.,
2005) protruding from within the DEAD motor domain and the C-terminal domain. The
C-terminal domain is also implicated in the
interaction with SecB and lipids (Breukink
et al., 1995). SecA forms a homodimer. Its
functional oligomeric state, however (Or et al.,
2005; de Keyzer et al., 2005; Jilaveanu et al.,
2005; Jilaveanu and Oliver, 2006), as well as the
explanation for the oligomerization in terms
of structure, has not yet been fully elucidated.
Previous studies suggested that, although SecA
is not required during the targeting of SRP
substrates (Scotti et al., 1999), it plays a role
during cotranslational membrane insertion
of periplasmic polypeptide domains (Qi and
Bernstein, 1999; Deitermann et al., 2005). This
observation raises the question of how ribosomes and SecA alternatively (or simultaneously) interact with the translocon during this
process.

COTRANSLATIONAL PROTEIN TARGETING

Although protein translocation can occur co-
or posttranslationally, the cotranslational pathway is more general and conserved among all
cells and organisms, where it is responsible for
the biosynthesis of most membrane proteins. As
mentioned, cotranslational targeting of proteins
to the membrane is mediated by the SRP system, which includes two essential proteins and
an essential RNA molecule as follows.

Ffh. Ffh is the 48-kDa SRP protein of
E. coli; it is homologous to the mammalian
SRP54 SRP subunit (Bernstein et al., 1989;
Romisch et al., 1989). It contains two major
domains: (i) The NG domain, which struc-
turally resembles the NG domain of the SRP

receptor, FtsY (see later). This domain contains
the GTP binding site of Ffh and it is involved
in the interaction of SRP with the ribosome
and with FtsY. The GTPase activity of Ffh
is required for proper functioning in vivo
(Samuelsson et al., 1995). (ii) The C-terminal
methionine-rich M domain is implicated in the
interaction with the hydrophobic nascent
polypeptide and with the *E. coli* 4.5S SRP-
RNA. Unlike the SRP receptor or the SRP-
RNA, the *E. coli* Ffh protein exhibits marked
similarity to all the known SRP54 proteins
throughout the entire sequence. Therefore, it is
likely that the interaction of Ffh with the ribosome, nascent polypeptide chains, SRP-
RNA, and FtsY, is remarkably conserved and
probably does not require additional cell type-
specific cofactors. This notion has received
some support from the observation that, in
"mixing" experiments, the Ffh protein from
one species can also form functional SRP in
remotely related systems both in vitro (Bernstein et al., 1993) and in vivo (Gutierrez et al.,
1999). High-resolution structural information
about the *E. coli* Ffh is rather limited (Batey
et al., 2000, 2001). However, the domains of
several other highly conserved Ffh heterologs
have been resolved; thus reasonable predictions
can be made. One important open question
deals with the possible interaction of Ffh between the M domain and NG, which might
clarify how NG-GTPase is controlled by the
4.5S RNA and the hydrophobic nascent chain
(Rosendal et al., 2003; Buskiewicz et al., 2005;
Spanggord et al., 2005).

4.5S RNA. Only a small portion of the
mammalian SRP-RNA (7SL RNA) is evolutionarily conserved. The 7SL RNA is more
than 300 nucleotides long and is schematically
divided into two major domains. The small
domain contains an Alu structure (stem and
loop composed of both termini of 7SL RNA)
and binds SRP proteins 9 and 14 (Strub et al.,
1991). This domain has been implicated in
translation arrest as a means of preventing non-
targeted secretory or membrane proteins from
being translated in the cytosol (for a review, see

Bui and Strub, 1999). *E. coli* 4.5S RNA (114 nucleotides long) does not contain an Alu domain (Poritz et al., 1988), suggesting that either the arrest function is absent (Raine et al., 2003) or is perhaps mediated by an alternative mechanism (Avdeeva et al., 2002; Bochkareva et al., unpublished data). The *E. coli* 4.5S RNA represents the minimal functional core of the SRP-RNA, which was termed domain IV and is relatively conserved in all cell types. This domain participates in the interaction with both the SRP54 protein (as proposed by Poritz et al., 1988; Struck et al., 1988) and the hydrophobic nascent chain (Batey et al., 2000), and therefore plays a pivotal role in the SRP pathway.

SRP. Initial experimental support for the notion that *E. coli* has an SRP complex came from the demonstration that an antiserum to Ffh precipitated from cell extract RNP complexes containing Ffh and the SRP-related 4.5S RNA (Poritz et al., 1990; Ribes et al., 1990). The formation of this complex was analyzed in detail and confirmed by in vitro studies also in other bacterial systems (Kurita et al., 1996; Lentzen et al., 1996; Samuelsson and Olsson, 1993; Wood et al., 1992). Indirect support for a physical association between 4.5S RNA and Ffh was also provided by experiments indicating that 4.5S RNA stabilizes Ffh both in vivo (Jensen and Pedersen, 1994) and in vitro (Zheng and Gierasch, 1997).

Many of the genetic and biochemical results have later been confirmed by structural studies that have provided an atomic view of the components of the SRP complex and their intermolecular interactions in several bacterial systems (Batey et al., 2000; Jovine et al., 2000; Keenan et al., 1998; Schmitz et al., 1999; Rosendal et al., 2003) (for a review, see Luirink and Sinning, 2004). As expected, the essential interactions of Ffh occur in the moiety that contains domain IV of SRP-RNA. Based on the structure of the M domain of Ffh bound to the 49-nucleotide segment of 4.5S RNA (Batey et al., 2000) and that of the M domain (Keenan et al., 1998), Bernstein (2000) proposed that a signal

sequence recognition surface exists, consisting of both protein and RNA, which interact with the signal sequence by a combination of hydrophobic and electrostatic interactions.

FtsY. Targeting of ribosomes to the cytoplasmic membrane in *E. coli* depends on the expression of the SRP receptor, FtsY (Herskovits and Bibi, 2000). In addition, when FtsY is depleted by genetic means, the expression of polytopic membrane proteins such as LacY (Seluanov and Bibi, 1997), SecY (Herskovits and Bibi, 2000), and MdfA (Bochkareva et al., unpublished) is repressed. Moreover, other studies implied that FtsY forms a complex with membrane-bound ribosomes also in the absence of SRP or the translocon (Herskovits et al., 2002), and recent studies demonstrated interactions between FtsY and the translocon (Angelini et al., 2005). These observations thus underscore the central role of FtsY in ribosome targeting and in the biogenesis of SRP substrates (Herskovits et al., 2000). FtsY contains three distinct domains: the C-terminal N and G domains (302 residues long; termed the NG domain), which constitute a universally conserved SRP-GTPase (Montoya et al., 1997), and an N-terminal A domain (195 residues long), which was shown to be dispensable under various growth conditions (Eitan and Bibi, 2004). Under steady-state conditions, FtsY is distributed between the cytoplasm and the membrane (Luirink et al., 1994), and it has no known membrane anchor partner homologous to the mammalian β-subunit of the SRP receptor, the third GTP-binding protein of eukaryotic SRP systems (Schlenker et al., 2006). Nevertheless, various studies have suggested that FtsY functions as a membrane-bound receptor (Zelazny et al., 1997; Bibi et al., 2001). A-domain-truncated FtsY versions exhibit a strong affinity for membranes (de Leeuw et al., 1997, 2000), possibly through the N domain (Millman and Andrews, 1999), an interaction that seems to be dominated by electrostatic forces (de Leeuw et al., 2000). In addition to its interaction with the inner membrane, the *E. coli* FtsY also functionally interacts with the

SRP protein Ffh in a nucleotide-dependent manner (reviewed in Luirink and Sinning, 2004). The structure of a complex between the two NG domains of Ffh and FtsY has recently been resolved by X-ray crystallography (Egea et al., 2004; Focia et al., 2004).

The *E. coli* SRP Pathway

As described (Luirink et al., 2005), the cotranslational targeting pathway is initiated by the interaction of SRP with ribosomes translating SRP substrates upon the emergence of a hydrophobic nascent peptide from the ribosome. In *E. coli*, this interaction was initially confirmed by using a heterologous in vitro system and cross-linking experiments demonstrating that Ffh binds functional signal peptides as they emerge from the ribosome but not if they have been fully released from the ribosome. This interaction requires 4.5S RNA (Luirink et al., 1992), and it is correlated with the hydrophobicity of the nascent peptide (Valent et al., 1995). These studies later received further support (Koch et al., 1999; Valent et al., 1998), when native *E. coli* in vitro translation systems were used to study the interactions between SRP and the nascent polypeptide chains. In the course of these studies, inner membrane proteins were identified as the preferred SRP substrates, in accordance with the earlier conclusions from genetic studies (de Gier et al., 1996; Macfarlane and Muller, 1995; Seluanov and Bibi, 1997; Ulbrandt et al., 1997). Moreover, cross-linking experiments showed that SRP and the ribosome-associated chaperone trigger factor exhibit different binding preferences to nascent polypeptide chains and might therefore direct different translation products to different destinations (Deuerling et al., 1999; Wegrzyn and Deuerling, 2005). Whereas SRP recognizes the sequence of a hydrophobic nascent chain of membrane protein, its possible interaction with a somewhat less hydrophobic signal sequence of a presecretory protein might be prevented by the trigger factor, which binds to the secretory nascent chain early during translation (Beck et al., 2000; Valent et al., 1995), where it exhibits an interesting interplay

with the SRP in the vicinity of the nascent polypeptide exit site of the ribosome (Eisner et al., 2006). The mode of SRP interaction with the ribosome has been studied by crosslinking approaches and identification of interacting ribosomal proteins (see, for example, Gu et al., 2003; Ullers et al., 2003). The ribosomal SRP-binding protein, L23, also interacts with the trigger factor and the emerging nascent polypeptide chain (Ullers et al., 2003; Raine et al., 2004; Buskiewicz et al., 2004; Houben et al., 2005).

As suggested, the next targeting step requires the ribosome-nascent chain-SRP complex to interact with the membrane-associated receptor, FtsY. In vitro studies have demonstrated that the *E. coli* SRP complex interacts with FtsY in a GTP-dependent manner, like their mammalian homologues. This was demonstrated using SRP reconstituted from purified Ffh and 4.5S RNA, as well as purified FtsY (Kusters et al., 1995; Miller et al., 1994; Jagath et al., 2000). In addition, indirect genetic observations have indicated that Ffh interacts with FtsY in vivo (Ulbrandt et al., 1997). The physiological involvement of 4.5S RNA in this interaction is not completely clear at present, however (Peluso et al., 2000, Jagath et al., 2001). Biochemical and structural studies confirmed that in several SRP systems, Ffh interacts with FtsY in the absence of SRP-RNA, and even in the absence of the Ffh M domain (Macao et al., 1997; Egea et al., 2004; Focia et al., 2004). As mentioned, both FtsY and Ffh contain homologous GTPase domains, and according to their X-ray structures (Freymann et al., 1997; Montoya et al., 1997), these domains also have similar tertiary structures. The enzymatic activity of each component in the SRP-FtsY complex was studied in a series of elegant experiments, which revealed their individual properties and mutual influence and response during GTP hydrolysis (see, for example, Powers and Walter, 1995; Moser et al., 1997; Shan et al., 2004; Shan and Walter, 2005). In brief, Ffh (reconstituted into SRP) and FtsY stimulate each other's GTPase activity. In addition, FtsY undergoes reversible conformational changes upon its GTP-dependent association with SRP. Finally, the GTP hydrolysis step is crucial for the dissociation of SRP from its receptor after the release of the ribosome nascent chain from the targeting complex.

As mentioned, the model presented earlier was developed predominantly based on in vitro experiments. However, several questions arose from the results of in vivo studies of the targeting step in *E. coli*. (i) Analysis of the amount of membrane-bound ribosomes in Ffh- or FtsY-depleted cells revealed that only FtsY depletion had a substantial inhibitory effect (Herskovits and Bibi, 2000). Further studies of this issue (Herskovits et al., 2002) indicated that an alternative, SRP-independent, FtsY-mediated ribosome-targeting pathway exists and that FtsY forms membrane-bound complexes with ribosomes also in the absence of SRP. The alternative ribosomal targeting model was described in detail elsewhere (Herskovits et al., 2000). (ii) Another important difference was observed in vivo between cells depleted of FtsY and those depleted of Ffh. Apparently, the expression of several membrane proteins was substantially decreased only in FtsY-depleted cells (Seluanov and Bibi, 1997; Herskovits et al., 2002; Bochkareva et al., unpublished data). This phenomenon is directly related to the ongoing debate of whether SRP-mediated translation control occurs also in *E. coli* (Raine et al., 2003; Avdeeva et al., 2002).

Finally, ribosomes translating SRP substrates are targeted to and assembled on the translocon (Valent et al., 1998; Prinz et al., 2000), such that the elongating polypeptide chain is transferred directly from the tunnel in the ribosome into the translocation channel. This is the least understood step in the *E. coli* cotranslational targeting pathway. Several possible mechanisms underlying ribosome transfer from the targeting system to the translocon may be considered. (i) Ribosomes are targeted directly to the translocon via the translocon-attached FtsY (Angelini et al., 2005). (ii) FtsY-ribosome-SRP complexes are initially anchored to the membrane through FtsY-lipid interactions (de Leeuw et al., 1997, 2000; Millman and Andrews, 1999)

and then diffuse and associate with the translocon. (iii) FtsY-ribosome-SRP complexes are bound to an unknown membrane protein that serves as a docking site (Herskovits et al., 2002) and mediates, together with FtsY (Angelini et al., 2005), the final transfer of the ribosome to the translocon.

THE POST- AND COTRANSLATIONAL PATHWAYS CONVERGE AT THE TRANSLOCON

The translocon of *E. coli* is formed from an evolutionarily conserved heterotrimeric membrane protein complex, called the SecYEG complex (reviewed in Mori and Ito, 2001; Driessen et al., 2001). In the cotranslational pathway ribosome binding to the *E. coli* translocon has not been investigated extensively, but several studies have suggested that a strong interaction occurs between the large ribosomal RNA and the translocon (Prinz et al., 2000). The recently determined X-ray structure of an archaebacterial core translocon (van den Berg et al., 2004) led to the hypothesis that the cytoplasmic loops in the C-terminal half of the SecY subunit of the translocon are responsible for the interaction. Surprisingly, ribosomes can also bind to the translocon when they lack nascent chains or carry nontranslocating polypeptides (Neuhof et al., 1998; Raden and Gilmore, 1998; Prinz et al, 2000). However, studies of eukaryotic systems showed that ribosomes carrying nascent chains with a signal sequence have a competitive advantage and can displace other ribosomes from the channels. The mechanism of competition is unclear at present, but it involves the function of the SRP and its receptor (Neuhof et al., 1998; Raden and Gilmore, 1998).

Posttranslationally, SecB-bound preproteins are targeted to the translocon-bound SecA protein (Hartle et al., 1990). The SecB-preprotein complex may initially weakly interact with SecA in the cytosol, after which the complex assembles on the translocon. Next, upon interaction of the signal sequence with SecA, the preprotein is released from SecB (Fekkes et al., 1997). Finally, binding of ATP by SecA terminates the targeting step, because it is accompanied with the release of SecB from the membrane and the initiation of the translocation step by the SecA motor.

Several questions arise from the dual involvement of the translocon in both the post- and cotranslational pathways. (i) SecA is also needed during the cotranslational insertion of hydrophilic polypeptide segments of SRP substrates across the membrane (Neumann-Haefelin et al., 2000; Qi and Bernstein, 1999; Deitermann et al., 2005). An open question is how, under these conditions, the translocon accommodates both the ribosomes and SecA. (ii) Is there competition between SecB–SecA–bound preprotein and ribosomes translating SRP substrates? And, how does each targeting system sense free translocons? (iii) Are there domains in the translocon that play distinct roles in either of the pathways? Studies of this question using SecY mutants indicated that SRP-dependent integration and SecA/SecB-mediated translocation are mechanistically distinct processes even at the level of the membrane, where they engage different domains of SecY and different components of the translocon (Newitt and Bernstein, 1998; Koch and Muller, 2000).

ACKNOWLEDGMENTS

Work in the author's laboratory is supported by GIF, the German-Israeli Foundation for Scientific Research and Development, and by the Israel Science Foundation.

REFERENCES

Angelini, S., S. Deitermann, and H. G. Koch. 2005. FtsY, the bacterial signal-recognition particle receptor, interacts functionally and physically with the SecYEG translocon. *EMBO Rep.* **6:**476–481.

Avdeeva, O. N., A. G. Myasnikov, P. V. Sergiev, A. A. Bogdanov, R. Brimacombe, and O. A. Dontsova. 2002. Construction of the 'minimal' SRP that interacts with the translating ribosome but not with specific membrane receptors in Escherichia coli. *FEBS Lett.* **514:**70–73.

Baars, L., A. J. Ytterberg, D. Drew, S. Wagner, C. Thilo, K. J. van Wijk, and J. W. de Gier. 2006. Defining the role of the E. coli chaperone SECB using comparative proteomics. *J. Biol. Chem.* **281:**10024–10034.

Bassford, P., J. Beckwith, K. Ito, C. Kumamoto, S. Mizushima, D. Oliver, L. Randall, T. Silhavy, P. C. Tai, and W. Wickner. 1991. The primary pathway of protein export in E. coli. *Cell* **65:**367–368.

Batey, R. T., R. P. Rambo, L. Lucast, B. Rha, and J. A. Doudna. 2000. Crystal structure of the ribonucleoprotein core of the signal recognition particle. *Science* **287:**1232–1239.

Batey, R. T., M. B. Sagar, and J. A. Doudna. 2001. Structural and energetic analysis of RNA recognition by a universally conserved protein from the signal recognition particle. *J. Mol. Biol.* **307:**229–246.

Beck, K., L. F. Wu, J. Brunner, and M. Muller. 2000. Discrimination between SRP- and SecA/SecB-dependent substrates involves selective recognition of nascent chains by SRP and trigger factor. *EMBO J.* **19:**134–143.

Beckwith, J. 1991. "Sequence-gazing?" *Science* **251:** 1161–1162.

Bernstein, H. D. 2000. The biogenesis and assembly of bacterial membrane proteins. *Curr. Opin. Microbiol.* **3:**203–209.

Bernstein, H. D., M. A. Poritz, K. Strub, P. J. Hoben, S. Brenner, and P. Walter. 1989. Model for signal sequence recognition from amino-acid sequence of 54K subunit of signal recognition particle. *Nature* **340:**482–486.

Bernstein, H. D., D. Zopf, D. M. Freymann, and P. Walter. 1993. Functional substitution of the signal recognition particle 54-kDa subunit by its Escherichia coli homolog. *Proc. Natl. Acad. Sci. USA* **90:**5229–5233.

Bibi, E., A. A. Herskovits, E. S. Bochkareva, and A. Zelazny. 2001. Putative integral membrane SRP receptors. *Trends Biochem. Sci.* **26:**15–16.

Bovia, F., and K. Strub. 1996. The signal recognition particle and related small cytoplasmic ribonucleoprotein particles. *J. Cell Sci.* **109:**2601–2608.

Breukink, E., N. Nouwen, A. van Raalte, S. Mizushima, J. Tommassen, and B. de Kruijff. 1995. The C terminus of SecA is involved in both lipid binding and SecB binding. *J. Biol. Chem.* **270:**7902–7907.

Brown, S. 1991. 4.5S RNA: does form predict function? *New Biol.* **3:**430–438.

Brown, S., and M. J. Fournier. 1984. The 4.5S RNA gene of Escherichia coli is essential for cell growth. *J. Mol. Biol.* **178:**533–550.

Bui, N., and K. Strub. 1999. New insights into signal recognition and elongation arrest activities of the signal recognition particle. *Biol. Chem.* **380:** 135–145.

Buskiewicz, I., E. Deuerling, S. Q. Gu, J. Jockel, M. V. Rodnina, B. Bukau, and W. Wintermeyer. 2004. Trigger factor binds to ribosome-signal-recognition particle (SRP) complexes and is excluded by binding of the SRP receptor. *Proc. Natl. Acad. Sci. USA.* **101**:7902–7906.

Buskiewicz, I., A. Kubarenko, F. Peske, M. V. Rodnina, and W. Wintermeyer. 2005. Domain rearrangement of SRP protein Ffh upon binding 4.5S RNA and the SRP receptor FtsY. *RNA* **11**:947–957.

Buskiewicz, I., F. Peske, H. J. Wieden, I. Gryczynski, M. V. Rodnina, and W. Wintermeyer. 2005. Conformations of the signal recognition particle protein Ffh from Escherichia coli as determined by FRET. *J. Mol. Biol.* **351**:417–430.

de Gier, J. W., P. Mansournia, Q. A. Valent, G. J. Phillips, J. Luirink, and G. von Heijne. 1996. Assembly of a cytoplasmic membrane protein in Escherichia coli is dependent on the signal recognition particle. *FEBS Lett.* **399**:307–309.

Deitermann, S., G. S. Sprie, and H. G. Koch. 2005. A dual function for SecA in the assembly of single spanning membrane proteins in Escherichia coli. *J. Biol. Chem.* **280**:39077–39085.

de Keyzer, J., E. O. van der Sluis, R. E. Spelbrink, N. Nijstad, B. de Kruijff, N. Nouwen, C. van der Does, and A. J. Driessen. 2005. Covalently dimerized SecA is functional in protein translocation. *J. Biol. Chem.* **280**:35255–35260.

Dekker, C., B. de Kruijff, and P. Gros. 2003. Crystal structure of SecB from Escherichia coli. *J. Struct. Biol.* **144**:313–319.

de Leeuw, E., D. Poland, O. Mol, I. Sinning, C. M. ten Hagen-Jongman, B. Oudega, and J. Luirink. 1997. Membrane association of FtsY, the E. coli SRP receptor. *FEBS Lett.* **416**:225–229.

de Leeuw, E., K. te Kaat, C. Moser, G. Menestrina, R. Demel, B. de Kruijff, B. Oudega, J. Luirink, and I. Sinning. 2000. Anionic phospholipids are involved in membrane association of FtsY and stimulate its GTPase activity. *EMBO J.* **19**:531–541.

Deuerling, E., A. Schulze-Specking, T. Tomoyasu, A. Mogk, and B. Bukau. 1999. Trigger factor and DnaK cooperate in folding of newly synthesized proteins. *Nature* **400**:693–696.

Driessen, A. J., E. H. Manting, and C. van der Does. 2001. The structural basis of protein targeting and translocation in bacteria. *Nat. Struct. Biol.* **8**:492–498.

Egea, P. F., S. O. Shan, J. Napetschnig, D. F. Savage, P. Walter, and R. M. Stroud. 2004. Substrate twinning activates the signal recognition particle and its receptor. *Nature* **427**:215–221.

Eichler, J., and F. Duong. 2004. Break on through to the other side—the Sec translocon. *Trends Biochem. Sci.* **29**:221–223.

Eisner, G., M. Moser, U. Schafer, K. Beck, and M. Muller. 2006. Alternate recruitment of signal recognition particle and trigger factor to the signal sequence of a growing nascent polypeptide. *J. Biol. Chem.* **281**:7172–7179.

Eitan, A., and E. Bibi. 2004. The core Escherichia coli signal recognition particle receptor contains only the N and G domains of FtsY. *J. Bacteriol.* **186**:2492–2494.

Eser, M., and M. Ehrmann. 2003. SecA-dependent quality control of intracellular protein localization. *Proc. Natl. Acad. Sci. USA* **100**:13231–13234.

Fekkes, P., C. van der Does, and A. J. Driessen. 1997. The molecular chaperone SecB is released from the carboxy-terminus of SecA during initiation of precursor protein translocation. *EMBO J.* **16**:6105–6113.

Focia, P. J., I. V. Shepotinovskaya, J. A. Seidler, and D. M. Freymann. 2004. Heterodimeric GTPase core of the SRP targeting complex. *Science* **303**:373–377.

Freymann, D. M., R. J. Keenan, R. M. Stroud, and P. Walter. 1997. Structure of the conserved GTPase domain of the signal recognition particle. *Nature* **385**:361–364.

Gill, D. R., and G. P. Salmond. 1987. The Escherichia coli cell division proteins FtsY, FtsE and FtsX are inner membrane-associated. *Mol. Gen. Genet.* **210**:504–508.

Gu, S. Q., F. Peske, H. J. Wieden, M. V. Rodnina, and W. Wintermeyer. 2003. The signal recognition particle binds to protein L23 at the peptide exit of the Escherichia coli ribosome. *RNA* **9**: 566–573.

Gutierrez, J. A., P. J. Crowley, D. G. Cvitkovitch, L. J. Brady, I. R. Hamilton, J. D. Hillman, and A. S. Bleiweis. 1999. Streptococcus mutans ffh, a gene encoding a homologue of the 54 kDa subunit of the signal recognition particle, is involved in resistance to acid stress. *Microbiology* **145**:357–366.

Hartl, F. U., S. Lecker, E. Schiebel, J. P. Hendrick, and W. Wickner. 1990. The binding cascade of SecB to SecA to SecY/E mediates preprotein targeting to the E. coli plasma membrane. *Cell* **63**: 269–279.

Herskovits, A. A., and E. Bibi. 2000. Association of Escherichia coli ribosomes with the inner membrane requires the signal recognition particle receptor but is independent of the signal recognition particle. *Proc. Natl. Acad. Sci. USA* **97**:4621–4626.

Herskovits, A. A., E. S. Bochkareva, and E. Bibi. 2000. New prospects in studying the bacterial signal recognition particle pathway. *Mol. Microbiol.* **38**:927–939.

Herskovits, A. A., E. Shimoni, A. Minsky, and E. Bibi. 2002. Accumulation of endoplasmic membranes and novel membrane-bound ribosome-signal recognition particle receptor complexes in Escherichia coli. *J. Cell Biol.* **159**:403–410.

Houben, E. N., R. Zarivach, B. Oudega, and J. Luirink. 2005. Early encounters of a nascent membrane protein: specificity and timing of contacts inside and outside the ribosome. *J. Cell Biol.* **170:**27–35.

Hunt, J. F., S. Weinkauf, L. Henry, J. J. Fak, P. McNicholas, D. B. Oliver, and J. Deisenhofer. 2002. Nucleotide control of interdomain interactions in the conformational reaction cycle of SecA. *Science* **297:**2018–2026.

Jagath, J. R., N. B. Matassova, E. de Leeuw, J. M. Warnecke, G. Lentzen, G. M. V. Rodnina, J. Luirink, and W. Wintermeyer. 2001. Important role of the tetraloop region of 4.5S RNA in SRP binding to its receptor FtsY. *RNA* **7:**293–301.

Jagath, J. R., M. V. Rodnina, and W. Wintermeyer. 2000. Conformational changes in the bacterial SRP receptor FtsY upon binding of guanine nucleotides and SRP. *J. Mol. Biol.* **295:**745–753.

Jensen, C. G., and S. Pedersen. 1994. Concentrations of 4.5S RNA and Ffh protein in Escherichia coli: the stability of Ffh protein is dependent on the concentration of 4.5S RNA. *J. Bacteriol.* **176:**7148–7154.

Jilaveanu, L. B., and D. Oliver. 2006. SecA dimer cross-linked at its subunit interface is functional for protein translocation. *J. Bacteriol.* **188:**335–338.

Jilaveanu, L. B., C. R. Zito, and D. Oliver. 2005. Dimeric SecA is essential for protein translocation. *Proc. Natl. Acad. Sci. USA.* **102:**7511–7516.

Jovine, L., T. Hainzl, C. Oubridge, W. G. Scott, J. Li, T. K. Sixma, A. Wonacott, T. Skarzynski, and Nagai, K. 2000. Crystal structure of the ffh and EF-G binding sites in the conserved domain IV of Escherichia coli 4.5S RNA. *Struct. Fold. Des.* **8:**527–540.

Jungnickel B., and T. A. Rapoport. 1995. A post-targeting signal sequence recognition event in the endoplasmic reticulum membrane. *Cell* **82:**261–270.

Karamanou, S., G. Sianidis, G. Gouridis, C. Pozidis, Y. Papanikolau, E. Papanikou, and A. Economou. 2005. Escherichia coli SecA truncated at its termini is functional and dimeric. *FEBS Lett.* **579:**1267–1271.

Keenan, R. J., D. M. Freymann, P. Walter, and R. M. Stroud. 1998. Crystal structure of the signal sequence binding subunit of the signal recognition particle. *Cell* **94:**181–191.

Koch, H. G., T. Hengelage, C. Neumann-Haefelin, J. MacFarlane, H. K. Hoffschulte, K. L. Schimz, B. Mechler, and M. Muller. 1999. In vitro studies with purified components reveal signal recognition particle (SRP) and SecA/SecB as constituents of two independent protein-targeting pathways of Escherichia coli. *Mol. Biol. Cell* **10:**2163–2173.

Koch, H. G., and M. Muller. 2000. Dissecting the translocase and integrase functions of the Escherichia coli SecYEG translocon. *J. Cell Biol.* **150:**689–694.

Kurita, K., K. Honda, S. Suzuma, H. Takamatsu, K. Nakamura, and K. Yamane. 1996. Identification of a region of Bacillus subtilis Ffh, a homologue of mammalian SRP54 protein, that is essential for binding to small cytoplasmic RNA. *J. Biol. Chem.* **271:**13140–13146.

Kusters, R., G. Lentzen, E. Eppens, A. van Geel, C. C. van der Weijden, W. Wintermeyer, and J. Luirink. 1995. The functioning of the SRP receptor FtsY in protein-targeting in E. coli is correlated with its ability to bind and hydrolyse GTP. *FEBS Lett.* **372:**253–258.

Lentzen, G., H. Moine, C. Ehresmann, B. Ehresmann, and W. Wintermeyer. 1996. Structure of 4.5S RNA in the signal recognition particle of Escherichia coli as studied by enzymatic and chemical probing. *RNA* **2:**244–253.

Luirink, J., and I. Sinning. 2004. SRP-mediated protein targeting: structure and function revisited. *Biochim. Biophys. Acta* **1694:**17–35.

Luirink, J., S. High, H. Wood, A. Giner, D. Tollervey, and B. Dobberstein. 1992. Signal-sequence recognition by an Escherichia coli ribonucleoprotein complex. *Nature* **359:**741–743.

Luirink, J., C. M. ten Hagen-Jongman, C. C. van der Weijden, B. Oudega, S. High, B. Dobberstein, and R. Kusters. 1994. An alternative protein targeting pathway in Escherichia coli: studies on the role of FtsY. *EMBO J.* **13:**2289–2296.

Luirink, J., G. von Heijne, E. Houben, and J. W. de Gier. 2005. Biogenesis of inner membrane proteins in Escherichia coli. *Annu. Rev. Microbiol.* **59:**329–355.

Macao, B., J. Luirink, and T. Samuelsson. 1997. Ffh and FtsY in a Mycoplasma mycoides signal-recognition particle pathway: SRP RNA and M domain of Ffh are not required for stimulation of GTPase activity in vitro. *Mol. Microbiol.* **24:**523–534.

Macfarlane, J., and M. Muller. 1995. The functional integration of a polytopic membrane protein of Escherichia coli is dependent on the bacterial signal-recognition particle. *Eur. J. Biochem.* **233:**766–771.

Miller, J. D., H. D. Bernstein, and P. Walter. 1994. Interaction of E. coli Ffh/4.5S ribonucleoprotein and FtsY mimics that of mammalian signal recognition particle and its receptor. *Nature* **367:**657–659.

Millman, J. S., and D. W. Andrews. 1999. A site-specific, membrane-dependent cleavage event defines the membrane binding domain of FtsY. *J. Biol. Chem.* **274:**33227–33234.

Montoya, G., C. Svensson, J. Luirink, and I. Sinning. 1997. Crystal structure of the NG domain from the signal-recognition particle receptor FtsY. *Nature* **385:**365–368.

Mori, H., and K. Ito. 2001. The Sec protein-translocation pathway. *Trends Microbiol.* **9:**494–500.

Moser, C., O. Mol, R. S. Goody, and I. Sinning. 1997. The signal recognition particle receptor of Escherichia coli (FtsY) has a nucleotide exchange factor built into the GTPase domain. *Proc. Natl. Acad. Sci. USA* **94:**11339–11344.

Nakatogawa H., and K. Ito. 2001. Secretion monitor, SecM, undergoes self-translation arrest in the cytosol. *Mol. Cell* **7:**185–192.

Neuhof, A., M. M. Rolls, B. Jungnickel, K. U. Kalies, and T. A. Rapoport. 1998. Binding of signal recognition particle gives ribosome/nascent chain complexes a competitive advantage in endoplasmic reticulum membrane interaction. *Mol. Biol. Cell.* **9:**103–115.

Neumann-Haefelin, C., U. Schafer, M. Muller, and H. G. Koch. 2000. SRP-dependent co-translational targeting and SecA-dependent translocation analyzed as individual steps in the export of a bacterial protein. *EMBO J.* **19:**6419–6426.

Newitt, J. A., and H. D. Bernstein. 1998. A mutation in the Escherichia coli secY gene that produces distinct effects on inner membrane protein insertion and protein export. *J. Biol. Chem.* **273:** 12451–12456.

Or, E., D. Boyd, S. Gon, J. Beckwith, and T. A. Rapoport. 2005. The bacterial ATPase SecA functions as a monomer in protein translocation. *J. Biol. Chem.* **280:**9097–9105.

Osborne, A. R., W. M. Clemons, Jr., and T. A. Rapoport. 2004. A large conformational change of the translocation ATPase SecA. *Proc. Natl. Acad. Sci. USA* **101:**10937–10942.

Osborne, A. R., T. A. Rapoport, and B. van den Berg. 2005. Protein translocation by the Sec61/SecY channel. *Annu. Rev. Cell Dev. Biol.* **21:**529–550.

Paetzel, M., A. Karla, N. C. Strynadka, and R. E. Dalbey. 2002. Signal peptidases. *Chem. Rev.* **102:** 4549–4580.

Papanikou, E., S. Karamanou, C. Baud, M. Frank, G. Sianidis, D. Keramisanou, C. G. Kalodimos, A. Kuhn, and A. Economou. 2005. Identification of the preprotein binding domain of SecA. *J. Biol. Chem.* **280:**43209–43217.

Park, S., G. Liu, T. B. Topping, W. H. Cover, and L. L. Randall. 1988. Modulation of folding pathways of exported proteins by the leader sequence. *Science* **239:**1033–1035.

Peluso, P., D. Herschlag, S. Nock, D. M. Freymann, A. E. Johnson, and P. Walter. 2000. Role of 4.5S RNA in assembly of the bacterial signal recognition particle with its receptor. *Science* **288:**1640–1643.

Phillips, G. J., and T. J. Silhavy. 1992. The E. coli ffh gene is necessary for viability and efficient protein export. *Nature* **359:**744–746.

Poritz, M. A., H. D. Bernstein, K. Strub, D. Zopf, H. Wilhelm, and P. Walter. 1990. An E. coli ribonucleoprotein containing 4.5S RNA resembles mammalian signal recognition particle. *Science* **250:**1111–1117.

Poritz, M. A., H. D. Bernstein, and P. Walter. 1991. Response to "Sequence-Gazing?" *Nature* **251:**1161–1162.

Poritz, M. A., K. Strub, and P. Walter. 1988. Human SRP RNA and E. coli 4.5S RNA contain a highly homologous structural domain. *Cell* **55:**4–6.

Powers, T., and P. Walter. 1995. Reciprocal stimulation of GTP hydrolysis by two directly interacting GTPases. *Science* **269:**1422–1424.

Price, A., A. Economou, F. Duong, and W. Wickner. 1996. Separable ATPase and membrane insertion domains of the SecA subunit of preprotein translocase. *J. Biol. Chem.* **271:**31580–31584.

Prinz, A., C. Behrens, T. A. Rapoport, E. Hartmann, and K. U. Kalies. 2000. Evolutionarily conserved binding of ribosomes to the translocation channel via the large ribosomal RNA. *EMBO J.* **19:**1900–1906.

Qi, H. Y., and H. D. Bernstein. 1999. SecA is required for the insertion of inner membrane proteins targeted by the Escherichia coli signal recognition particle. *J. Biol. Chem.* **274:**8993–8997.

Raden, D., and R. Gilmore. 1998. Signal recognition particle-dependent targeting of ribosomes to the rough endoplasmic reticulum in the absence and presence of the nascent polypeptide–associated complex. *Mol. Biol. Cell.* **9:**117–130.

Raine, A., N. Ivanova, J. E. Wikberg, and M. Ehrenberg. 2004. Simultaneous binding of trigger factor and signal recognition particle to the E. coli ribosome. *Biochimie* **86:**495–500.

Raine, A., R. Ullers, M. Pavlov, J. Luirink, J. E. Wikberg, and M. Ehrenberg. 2003. Targeting and insertion of heterologous membrane proteins in E. coli. *Biochimie* **85:**659–668.

Randall, L. L., and S. J. Hardy. 1995. High selectivity with low specificity: how SecB has solved the paradox of chaperone binding. *Trends Biochem. Sci.* **20:**65–69.

Randall, L. L., and S. J. Hardy. 2002. SecB, one small chaperone in the complex milieu of the cell. *Cell Mol. Life Sci.* **59:**1617–1623.

Randall, L. L., T. B. Topping, and J. S. Hardy. 1990. No specific recognition of leader peptide by SecB, a chaperone involved in protein export. *Science* **248:**860–863.

Ribes, V., K. Romisch, A. Giner, B. Dobberstein, and D. Tollervey. 1990. E. coli 4.5S RNA is part of a ribonucleoprotein particle that has properties related to signal recognition particle. *Cell* **63:**591–600.

Romisch, K., J. Webb, J. Herz, S. Prehn, R. Frank, M. Vingron, and B. Dobberstein. 1989. Homology of 54K protein of signal-recognition particle, docking protein and two E. coli proteins with putative GTP-binding domains. *Nature* **340:**478–482.

Rosendal, K. R., K. Wild, G. Montoya, and I. Sinning. 2003. Crystal structure of the complete core of archaeal signal recognition particle and implications for interdomain communication. *Proc. Natl. Acad. Sci. USA* **100:**14701–14706.

Samuelson, J. C., M. Chen, F. Jiang, I. Moller, M. Wiedmann, A. Kuhn, G. J. Phillips, and R. E. Dalbey. 2000. YidC mediates membrane protein insertion in bacteria. *Nature* **406:**637–641.

Samuelsson, T., and M. Olsson. 1993. GTPase activity of a bacterial SRP-like complex. *Nucleic Acids Res.* **21:**847–853.

Samuelsson, T., M. Olsson, P. M. Wikstrom, and B. R. Johansson. 1995. The GTPase activity of the Escherichia coli Ffh protein is important for normal growth. *Biochim. Biophys. Acta* **1267:**83–91.

Schatz, G., and B. Dobberstein. 1996. Common principles of protein translocation across membranes. *Science* **271:**1519–1526.

Schierle, C. F., M. Berkmen, D. Huber, C. Kumamoto, D. Boyd, and J. Beckwith. 2003. The DsbA signal sequence directs efficient, cotranslational export of passenger proteins to the Escherichia coli periplasm via the signal recognition particle pathway. *J. Bacteriol.* **185:**5706–5713.

Schlenker, O., A. Hendricks, I. Sinning, and K. Wild. 2006. The structure of the mammalian SRP receptor as prototype for the interaction of small GTPases with longin domains. *J. Biol. Chem.* **281:**8898–8906.

Schmitz, U., T. L. James, P. Lukavsky, and P. Walter. 1999. Structure of the most conserved internal loop in SRP RNA. *Nat. Struct. Biol.* **6:**634–638.

Scotti, P. A., Q. A. Valent, E. H. Manting, M. L. Urbanus, A. J. Driessen, B. Oudega, and J. Luirink. 1999. SecA is not required for signal recognition particle-mediated targeting and initial membrane insertion of a nascent inner membrane protein. *J. Biol. Chem.* **274:**29883–29888.

Seluanov, A., and E. Bibi. 1997. FtsY, the prokaryotic signal recognition particle receptor homologue, is essential for biogenesis of membrane proteins. *J. Biol. Chem.* **272:**2053–2055.

Shan, S. O., R. M. Stroud, and P. Walter. 2004. Mechanism of association and reciprocal activation of two GTPases. *PLoS Biol.* **2:**e320.

Shan, S. O., and P. Walter. 2005. Molecular crosstalk between the nucleotide specificity determinant of the SRP GTPase and the SRP receptor. *Biochemistry* **44:**6214–6222.

Sharma, V., A. Arockiasamy, D. R. Ronning, C. G. Savva, A. Holzenburg, M. Braunstein, W. R. Jacobs, Jr., and J. C. Sacchettini. 2003. Crystal structure of Mycobacterium tuberculosis SecA, a preprotein translocating ATPase. *Proc. Natl. Acad. Sci. USA* **100:**2243–2248.

Shimohata, N., Y. Akiyama, and K. Ito. 2005. Peculiar properties of DsbA in its export across the Escherichia coli cytoplasmic membrane. *J. Bacteriol.* **187:**3997–4004.

Sianidis, G., S. Karamanou, E. Vrontou, K. Boulias, K. Repanas, N. Kyrpides, A. S. Politou, and A. Economou. 2001. Cross-talk between catalytic and regulatory elements in a DEAD motor domain is essential for SecA function. *EMBO J.* **20:**961–970.

Spanggord, R. J., F. Siu, A. Ke, and J. A. Doudna. 2005. RNA-mediated interaction between the peptide-binding and GTPase domains of the signal recognition particle. *Nat. Struct. Mol. Biol.* **12:**1116–1122.

Strub, K., J. Moss, and P. Walter. 1991. Binding sites of the 9- and 14-kilodalton heterodimeric protein subunit of the signal recognition particle (SRP) are contained exclusively in the Alu domain of SRP RNA and contain a sequence motif that is conserved in evolution. *Mol. Cell Biol.* **11:**3949–3959.

Struck, J. C., H. Y. Toschka, T. Specht, and V. A. Erdman. 1988. Common structural features between eukaryotic 7SL RNAs, eubacterial 4.5S RNA and scRNA and archaebacterial 7S RNA. *Nucleic Acids Res.* **16:**7740.

Tian, H., D. Boyd, and J. Beckwith. 2000. A mutant hunt for defects in membrane protein assembly yields mutations affecting the bacterial signal recognition particle and Sec machinery. *Proc. Natl. Acad. Sci. USA* **97:**4730–4735.

Topping, T. B., R. L. Woodbury, D. L. Diamond, S. J. Hardy, and L. L. Randall. 2001. Direct demonstration that homotetrameric chaperone SecB undergoes a dynamic dimer-tetramer equilibrium. *J. Biol. Chem.* **276:**7437–7441.

Ulbrandt, N. D., J. A. Newitt, and H. D. Bernstein. 1997. The E. coli signal recognition particle is required for the insertion of a subset of inner membrane proteins. *Cell* **88:**187–196.

Ullers, R. S., E. N. Houben, A. Raine, C. M. ten Hagen-Jongman, M. Ehrenberg, J. Brunner, B. Oudega, N. Harms, and J. Luirink. 2003. Interplay of signal recognition particle and trigger factor at L23 near the nascent chain exit site on the Escherichia coli ribosome. *J. Cell Biol.* **161:**679–684.

Ullers, R. S., J. Luirink, N. Harms, F. Schwager, C. Georgopoulos, and P. Genevaux. 2004. SecB is a bona fide generalized chaperone in Escherichia coli. *Proc. Natl. Acad. Sci. USA* **101:**7583–7588.

Valent, Q. A., D. A. Kendall, S. High, R. Kusters, B. Oudega, and J. Luirink. 1995. Early events in preprotein recognition in E. coli: interaction of SRP and trigger factor with nascent polypeptides. *EMBO J.* **14:**5494–5505.

Valent, Q. A., P. A. Scotti, S. High, J. W. de Gier, G. von Heijne, G. Lentzen, W. Wintermeyer, B. Oudega, and J. Luirink. 1998. The Escherichia coli SRP and SecB targeting pathways converge at the translocon. *EMBO J.* **17:**2504–2512.

Van den Berg, B., W. M. Clemons, Jr., I. Collinson, Y. Modis, E. Hartmann, S. C. Harrison, and T. A. Rapoport. 2004. X-ray structure of a protein-conducting channel. *Nature* **427:** 36–44.

van der Laan, M., N. P. Nouwen, and A. J. Driessen. 2005. YidC, an evolutionary conserved device for the assembly of energy-transducing membrane protein complexes. *Curr. Opin. Microbiol.* **8:**182–187.

Veenendaal, A. K., C. van der Does, and A. J. Driessen. 2004. The protein-conducting channel SecYEG. *Biochim. Biophys. Acta* **1694:**81–95.

von Heijne, G. 1990. The signal peptide. *J. Membr. Biol.* **115:**195–201.

Vrontou, E., and A. Economou. 2004. Structure and function of SecA, the preprotein translocase nanomotor. *Biochim. Biophys. Acta* **1694:**67–80.

Walter, P., and A. E. Johnson. 1994. Signal sequence recognition and protein targeting to the endoplasmic reticulum membrane. *Annu. Rev. Cell Biol.* **10:**87–119.

Wegrzyn, R. D., and E. Deuerling. 2005. Molecular guardians for newborn proteins: ribosome-associated chaperones and their role in protein folding. *Cell Mol. Life Sci.* **62:**2727–2738.

Wickner, W., and R. Schekman. 2005. Protein translocation across biological membranes. *Science* **310:** 1452–1456.

Wood, H., J. Luirink, and D. Tollervey. 1992. Evolutionary conserved nucleotides within the E. coli 4.5S RNA are required for association with P48 in vitro and for optimal function in vivo. *Nucleic Acids Res.* **20:**5919–5925.

Xu, Z., J. D. Knafels, and K. Yoshino. 2000. Crystal structure of the bacterial protein export chaperone secB. *Nat. Struct. Biol.* **7:**1172–1177.

Zelazny, A., A. Seluanov, A. Cooper, and E. Bibi. 1997. The NG domain of the prokaryotic signal recognition particle receptor, FtsY, is fully functional when fused to an unrelated integral membrane polypeptide. *Proc. Natl. Acad. Sci. USA* **94:**6025–6029.

Zheng, N., and L. M. Gierasch. 1997. Domain interactions in E. coli SRP: stabilization of M domain by RNA is required for effective signal sequence modulation of NG domain. *Mol. Cell* **1:**79–87.

Zhou, J., and Z. Xu. 2003. Structural determinants of SecB recognition by SecA in bacterial protein translocation. *Nat. Struct. Biol.* **10:**942–947.

THE Tat PROTEIN EXPORT PATHWAY

Tracy Palmer and Ben C. Berks

2

All periplasmic functions depend on protein targeting into this cellular compartment. In *Escherichia coli*, it has been estimated that in excess of 450 proteins are exported across the cytoplasmic membrane (http://www.cf.ac.uk/biosi/staff/ehrmann/tools/ecce/ecce.htm). Most of these proteins are transported in an unfolded form by the Sec pathway (discussed in Chapter 1). However, a substantial subset of exported proteins is translocated by a radically different mechanism. These proteins are exported already folded, and sometimes as hetero-oligomers, by the twin arginine protein transport (Tat) pathway. Protein transport by the Tat pathway is powered solely by the proton electrochemical gradient. Below we describe what is known about the targeting and transport of proteins to the periplasm by the Tat pathway, focusing particularly on the model organism *E. coli*.

THE Tat-DEPENDENT PROTEOME
Usage of the Tat pathway for protein export varies greatly depending on the organism studied. Prediction programs such as TatFind and

TatP use the salient features of the Tat signal peptide (see "The Tat Targeting Signal" below) to identify candidate Tat substrates (Rose et al., 2002; Dilks et al., 2003; Bendtsen et al., 2005). Such bioinformatic approaches have suggested that most bacteria probably export relatively few (less than 50) proteins by the Tat pathway. Notable exceptions are the gram-positive organism *Streptomyces coelicolor*, which is likely to encode more than 150 Tat substrate proteins; the halophilic Archaeon *Halobacterium* sp. NRC-1 (64 Tat substrates from a secretome of probably not more than 80 proteins in total); and plant symbionts of the genus *Rhizobium* (estimated substrate number approaching a hundred) (Rose et al., 2002; Dilks et al., 2003). An indication of the accuracy of these substrate prediction programs can be obtained by comparing their predictions for the *E. coli* proteome with experimental data. So far 26 *E. coli* proteins have been shown to have authentic Tat signal peptides with another 5 Tat precursors assigned on the basis of known cofactor content or other expert analysis. TatFind predicts 27 of these substrates and 7 false positives, while TatP correctly predicts 23 proteins and misassigns another 7. If one also includes those proteins lacking signal peptides but known to associate with a Tat signal peptide-bearing partner, this brings the total number of proteins de-

Tracy Palmer, Department of Molecular Microbiology, John Innes Centre, Norwich NR4 7UH, and School of Biological Sciences, University of East Anglia, Norwich NR4 7TJ, United Kingdom. *Ben C. Berks*, Department of Biochemistry, University of Oxford, Oxford OX1 3QU, United Kingdom.

The Periplasm
Edited by Michael Ehrmann © 2007 ASM Press, Washington, D.C.

pendent on the Tat system for export in *E. coli* to 40. A list of all of the known or likely *E. coli* substrate proteins harboring Tat-targeting sequences and their probable cellular function is shown in Table 1.

The major class of proteins exported by the *E. coli* Tat system are those which contain redox cofactors and participate in the electron transport pathways discussed in detail in Chapter 13. Examples are proteins that bind iron-sulfur clusters, molybdopterin cofactors, or the complex nickel-iron cofactor found at the active site of many hydrogenase enzymes (for a comprehensive list of cofactor types see Berks et al., 2003). Tat substrates bind their cofactors in the cytoplasmic compartment prior to transport. Since this requires the proteins to fold, it is easy to see why these proteins cannot use the Sec pathway. It is important to realize, however, that not all cofactor-containing periplasmic proteins are targeted to Tat. For example, the *c*-type cytochromes, where hemes are attached to the unfolded apocytochrome in the periplasmic compartment (reviewed in Stevens et al., 2004), utilize the Sec pathway to transport the apo-cytochrome *c*.

Another significant class of proteins that are transported to the periplasm by the Tat pathway are the cell wall amidases. It is not intuitively obvious why these proteins should require the Tat pathway for their export, because they apparently do not contain cofactors. Moreover, of the three homologous amidase proteins in *E. coli*, AmiA and AmiC use the Tat pathway for their export, while AmiB is strictly Sec dependent (Ize et al., 2003; Bernhardt and de Boer, 2003). A similar case appertains to *Pseudomonas syringae*, which encodes two amidases (orthologs of the *E. coli* AmiB and AmiC), where only one of them (AmiC) is Tat dependent (Caldelari et al., 2006). Amidases play a key role in cell wall remodelling and in cleavage of the septal murein during cell wall division (discussed in more detail in Chapter 10). Many of the pleiotropic phenotypes of *E. coli tat* mutant strains, including formation of long chains of cells, hypersensitivity to some drugs and detergents, and inability to form biofilms, can be ascribed

to the mislocalization of AmiA and AmiC (Stanley et al., 2001; Bernhardt and de Boer, 2003; Ize et al., 2003, 2004). Tat substrate proteins also contribute to other areas of *E. coli* physiology, including response to osmotic shock (export of one of the enzymes involved in periplasmic glucans biosynthesis, OpgD, is Tat dependent (Lequette et al., 2004), copper homeostasis (through the Tat-dependent oxidase CueO, part of the Cu efflux regulon) (Outten et al., 2000; Sargent et al., 1999; Ize et al., 2004), and iron acquisition (via the Tat-exported periplasmic ferrichrome-binding protein, FhuD) (Ize et al., 2004).

THE Tat PROTEIN EXPORT MACHINERY

Two transcriptional units encode genes specifying components of the *E. coli* Tat protein transport machinery (Fig. 1A). A four-gene operon at minute 86 on the *E. coli* chromosome encodes genes designated *tatA* through *tatD*, while a monocistronic unit at minute 14 encodes *tatE*. A thorough deletion analysis indicates that the products of *tatA*, *B*, *C*, and *E* each contribute to protein transport by the Tat pathway (Sargent et al., 1998, 1999; Weiner et al., 1998; Bogsch et al., 1998). TatD, on the other hand, is a metal-dependent nuclease with no apparent role in protein export (Wexler et al., 2000). Tat transport occurs under all growth conditions. Consistent with this the *E. coli tat* genes are expressed constitutively (Jack et al., 2001).

A schematic representation of the topologies of each of the Tat components is shown in Fig. 1B. The TatA, TatB, and TatE proteins are all part of the same protein family (Settles et al., 1997; Chanal et al., 1998; Yen et al., 2002) and as such they share a common structure. They are anchored to the cytoplasmic membrane via an N-terminal transmembrane α-helix. This is almost immediately followed by a stretch of sequence that codes for an amphipathic α-helix and a highly charged C-terminal region of varying length (Settles et al., 1997; Sargent et al., 1998; Berks et al., 2000). The N_{out}-C_{in} topology of TatA and TatB predicted according

TABLE 1 The known or likely *E. coli* Tat-signal peptide–bearing substrate proteins[a]

Protein	Physiological role	Cofactors	Coexported partner?	Sequence of signal peptide
HyaA[b,c]	Hydrogen oxidation	3 × Fe-S clusters	HyaB	MNNEETFYQAM**RR**QGVTRRSFLKYCSLAATSLGLGAGMAPKIAWA
HybO[b,c]	Hydrogen oxidation	3 × Fe-S clusters	HyaB	MTGDNTLIHSHGIN**RR**DFMKLCAALAATMGLSSKAAA
HybA[b,c]	Hydrogen oxidation	4 × Fe-S clusters[d]	Unknown	MN**RR**NFIKAASCGALLTGALPSVSHA
NapG[b]	Nitrate reduction	4 × Fe-S clusters[d]	Unknown	MSRSAKPQNG**RRR**FLRDVVRTAGGLAAVGVALGLQQQTARA
NrfC[b,c]	Nitrite reduction	4 × Fe-S clusters[d]	Unknown	MTWS**RR**QFLTGVGVLAAVSGTAGRVVA
YagT[a]	Unknown	2 × Fe-S clusters[d]	YagR[d] YagS[d]	MSNQGEYPEDNRVGKHEPHDLSLT**RR**DLIKVSAATAVVYPHSTLAASVPA
YdhX[b,c]	Unknown	4 × Fe-S clusters[d]	Unknown	MSWIGWTVAATALGDNQMSFT**RR**KFVLGMGTVIFFTGSASSLLA
TorA[a,c]	TMAO reduction	MGD	None	MNNNDLFQAS**RR**RFLAQLGGLTVAGMLGPSLLTPRRATAAQA
TorZ[b,c]	TMAO reduction	MGD	None	MTLT**RR**EFIKHSGIAAGALVVTSAAPLPAWA
NapA[b,c]	Nitrate reduction	MGD, 1 × Fe-S cluster	None	MKLS**RR**SFMKANAVAAAAAAGLSVPGVA
DmsA[a]	DMSO reduction	MGD, 1 × Fe-S cluster	DmsB	MKTKIPDAVLAAEVS**RR**GLVKTTAIGGLAMASSALTLPFSRIAHA
YnfE[b,c]	DMSO reduction	MGD, 1 × Fe-S cluster[d]	YnfG[d]	MSKNERMVGIS**RR**TLVKSTAIGSLALAAGGFSLPFTLRNAAA
YnfF[b,c]	DMSO reduction	MGD, 1 × Fe-S cluster[d]	YnfG[d]	MMKIHTTEALMKAEISR**RR**SLMKTSALGSLALASSAFTLPFSQMVRA
FdnG[b,c]	Formate oxidation	MGD, 1 × Fe-S cluster[d]	FdnH	MDVS**RR**QFFKICAGGMAGTTVAALGFAPKQALA
FdoG[b,c]	Formate oxidation	MGD, 1 × Fe-S cluster[d]	FdoH[d]	MQVS**RR**QFFKICAGGMAGTTAAALGFAPSVALA
YedY[b,c]	TMAO/DMSO reduction	MPT	None	MK**RR**QVLKALGISATALSLPHAAHA
CueO[b,c]	Copper homeostasis	4 × Cu ions	None	MQ**RR**DFLKYSVALGVASALPLWSRAVFA
PcoA[b,e]	Copper resistance	4 × Cu ions[d,f]	None	MLLKTS**RR**TFLKGLTLSGVAGSLGVWSFNARSSLSLPVAA
SufI[b,c]	Possibly cell division	None	Unknown	MSLS**RR**QFIQASGIALCAGAVPLKASA
YahJ[a,c]	Unknown	1 × Fe ion[d]	Unknown	MKESNS**RR**EFLSQSGKMVTAAALFGTSVPLAHA
WcaM[b,c]	Colanic acid biosynthesis	Unknown	Unknown	MPFKKLS**RR**TFLTASSALAFLHTPFARA
MdoD[b,c]	Glucans biosynthesis	Unknown	Unknown	MD**RR**RFIKGSMAMAAVCGTSGIASLFSQAAFA

YcdB[a]	Unknown	Unknown	MQYKDENGVNEPS**RR**LLKVIGALALAGSCPVAHA
YcdO[c]	Unknown	Unknown	MTINF**RR**NALQLSVAALFSSAFMANA
YaeI[b,c]	Possible phosphodiesterase	Metal ions?	MIS**RR**FLQATAATIATSSGFGYMHYC
AmiA[b,c]	Cell wall amidase	Metal ions?	MSTFKPLKTLTS**RR**QVLKAGLAALTLSGMSQAIA
AmiC	Cell wall amidase	Metal ions?	MSGSNTAISR**RR**LLQGAGAMWLLSVSQVSLA
FhuD[a]	Ferrichrome binding	Fe(III) hydroxamates (reversibly)	MSGLPLISR**RR**LLTAMALSPLLWQMNTAHA
YcbK[b,c]	Unknown	Unknown	MDKFDANR**RK**LLALGGVALGAAILPTPAFA
Pac[g]	Penicillin amidase	None	MKN**RN**RMIVNCVTASLMYYWSLPALA
C3736[i]	Possible diene lactone hydrolase	Unknown	MPRLTAKDFPQELLDYYDYYAHGKISK**R**EFLNLAAKYAVGGMTALAFDLLKPNYALA
b3000[h,i]	Frame-shifted variant of C3736 in K-12 strains	Unknown	MTITLTGKSRNVSSSTL**RR**SAVGGMTALALFDLLKPNYALA

[a] The consecutive arginines of the twin-arginine motif (or Lys-Arg for C3736 and Arg-Asn for Pac) are shown in bold underline. For those proteins that bind Fe-S, MPT, or MGD cofactors, targeting necessarily must occur via the Tat pathway. Of the remaining proteins, the export of CueO (formerly YacK), SufI (both Sargent et al., 1999), MdoD (Lequette et al., 2004), AmiA, AmiC (both Ize et al., 2003; Bernhardt and de Boer, 2003), FhuD (Ize et al., 2004), Pac (Ignatova et al., 2002), and C3736 (B. Ize, B. C. Berks, and T. Palmer, unpublished work) has been demonstrated experimentally to require the Tat pathway. The signal peptides of YahJ, WcaM, YcdB, and YcbK have been shown to engage in a Tat pathway when fused to a Tat-specific reporter protein (D.Tullman-Ercek, M. DeLisa, Y. Kawarasaki, P. Iranpour, B. Ribnicky, T. Palmer, and G. Georgiou, unpublished data). Some of the leader peptidase cleavage sites have been determined experimentally (see Berks [1996] for some examples); others were predicted using SignalP (http://www.cbs.dtu.dk/services/SignalP/). Abbreviations: Fe-S, iron-sulfur cluster; MPT, molybdopterin-Mo; MGD, (molybdopterin guanine dinucleotide)₂-MO; DMSO, dimethyl sulfoxide.

[b] Predicted to be Tat substrates by TatFind (http://www.sas.upenn.edu/~pohlschr/Ectat.html).

[c] Predicted to be Tat substrates by TatP (http://www.cbs.dtu.dk/services/TatP-1.0/).

[d] Predictions inferred by homology, genetic linkage, or sequence analysis.

[e] PcoA is plasmid encoded.

[f] May bind additional copper ions through methionine-rich domains that are not present in CueO.

[g] Penicillin acylase is found in E. coli W strains.

[h] Ca²⁺ is not an active site cofactor but may be required for folding (Kasche et al., 2005).

[i] C3736 is encoded in the genomes of most strains of E. coli with the exception of K-12. In K-12 a recent frameshift has altered most of the signal peptide-coding region such that it has been replaced with an apparent twin-arginine signal peptide (protein b3000). Both signals are predicted to have the same cleavage site. A further frameshift has also truncated the open reading frame in K-12.

A

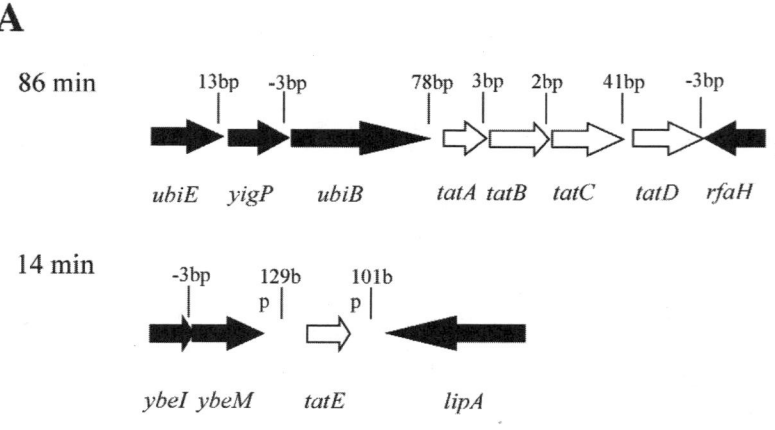

B

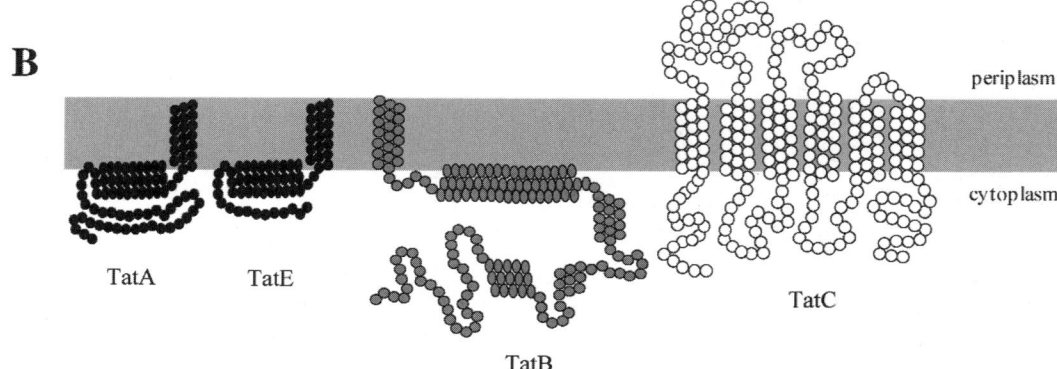

FIGURE 1 (A) Chromosomal location and organization of genes encoding components of the *E. coli* Tat pathway. Genes that are known or likely to be found in the same transcriptional unit (Jack et al., 2001) have the same fill. The spacing between each gene is shown in base pairs (bp) above the figure. (B) Schematic representation of the topological organization of the *E. coli* Tat components.

to the positive-inside rule (von Heijne, 1992) has now been confirmed experimentally (Bolhuis et al., 2001; Porcelli et al., 2002; P. A. Lee, G. Orriss, G. Buchanan, N. P. Greene, P. J. Bond, R. L. Jack, M. S. P. Sansom, B. C. Berks, and T. Palmer, submitted for publication).

The TatA and TatE proteins show high sequence similarity and share overlapping Tat transport functions. A strain lacking *tatA* shows a marked, but incomplete, block in Tat transport, while a strain additionally deleted for *tatE* completely lacks a functional Tat system. A strain deleted only for *tatE* is barely affected for Tat function (Sargent et al., 1998). Taken to-

gether with the observations that *tatE* is expressed more than 100-fold less than *tatA* and that many organisms lack a *tatE* gene, this suggests that *tatE* is a minor paralog of *tatA* that arose because of a late gene duplication event (Jack et al., 2001; Yen et al., 2002). The *tatB* gene is essential for a functional *E. coli* Tat system, and its inactivation results in a complete block in the transport of all native Tat substrates tested (Weiner et al., 1998; Sargent et al., 1999; Chanal et al., 1998). TatB shows only limited sequence similarity with TatA/E and cannot assume the function of TatA/E and vice versa (Sargent et al., 1999). Remarkably, some gram-positive organisms, for example, *Bacillus subtilis*, lack a TatB protein and therefore have functional Tat systems comprising only TatA and TatC (Pop et al., 2002; Jongbloed et al., 2004). This implies that TatB arose from a duplication of TatA that subsequently diverged in function. Sequence analysis suggests that this duplication and diversification event occurred independently at least twice during evolution (Yen et al., 2002). Two studies addressing the function of TatB in *E. coli* both showed that, although native Tat substrates were not detectably exported in a *tatB* deletion strain, there was some low-level export of sensitive Tat-dependent reporter proteins (Ize et al., 2002b; Blaudeck et al., 2005). Blaudeck et al. went on to isolate mutations that allowed the *E. coli* Tat system to work more efficiently in the absence of TatB. They isolated several mutations, each of which fell in the extreme N terminus of TatA, which permitted the export of the native Tat substrate trimethylamine-*N*-oxide (TMAO) reductase, TorA. The authors concluded that TatB is not essential for Tat transport but that it serves to greatly increase the efficiency of the transport process (Blaudeck et al., 2005).

TatC is an essential component of the Tat system (Bogsch et al., 1998). It is the largest and most hydrophobic of the Tat proteins. It has six transmembrane domains, with the N and C termini located in the cytoplasm (Drew et al., 2002; Behrendt et al., 2004; Ki et al., 2004) (see Fig. 1B). The TatC protein probably functions minimally as a dimer. Some of the TatC proteins encoded in the genomes of archaea are twice the size of eubacterial TatC proteins and consist of an internal duplication of TatC joined by a short linker sequence. Moreover, genetic experiments where coexpression of two inactive *tatC* alleles synthetically restored TatC function were also suggestive of an oligomeric function for the protein (Buchanan et al., 2002).

Tat COMPLEXES AND THE Tat PROTEIN TRANSPORT CYCLE

Under nontransporting, or resting, conditions, the Tat system of *E. coli* comprises two types of complexes (C. A. McDevitt, G. Buchanan,

F. Sargent, T. Palmer, and B. C. Berks, submitted for publication). One of these complexes, designated the TatBC complex, is made up of stoichiometric amounts of TatB and TatC (Bolhuis et al., 2001; McDevitt et al., submitted) (Fig. 2). When the TatABC components are overproduced, a low level of TatA is found to associate with the TatBC complex (Bolhuis et al., 2001; McDevitt et al., submitted). The TatBC complex varies in size from 370 kDa to greater than 650 kDa depending on the detergent used for membrane solubilization and the technique used to estimate the mass (Bolhuis et al., 2001; de Leeuw et al., 2002; Oates et al., 2005; McDevitt et al., submitted). Nonetheless, it clearly contains multiple copies of each of the two subunits. Low-resolution negative-stain electron microscopy of purified TatBC complexes from the closely related gram-negative organisms *E. coli*, *Salmonella enterica* serovar Typhimurium, and *Agrobacterium tumefaciens* shows that the complexes all have apparently similar ringlike geometries but probably comprise varying numbers of TatBC protomers, between 5 and 7 (Oates et al., 2003). A recent study using cysteine mutagenesis and cross-linking of TatB is consistent with this, since the pattern of cross-links is suggestive of the TatB component of the TatBC complex forming a ring of up to 5 TatB proteins. The greater size of TatC and the obligatory 1:1 stoichiometry of the complex would require that TatC be located on the outside of the TatB ring, presumably forming a larger, outer annulus (Lee et al., submitted).

The second type of Tat complex found in *E. coli* membranes under resting conditions contains only TatA (McDevitt et al., submitted). When the TatABC components are overproduced, the TatA complexes contain a very small amount of TatB (Sargent et al., 2001; Porcelli et al., 2002; de Leeuw et al., 2002; Oates et al., 2005). The TatA complexes vary in size from less than 100 kDa to greater than 700 kDa (Oates et al., 2005; Gohlke et al., 2005; McDevitt et al., submitted). Low-resolution 3D structures of TatA complexes have been produced by negative-stain electron microscopy

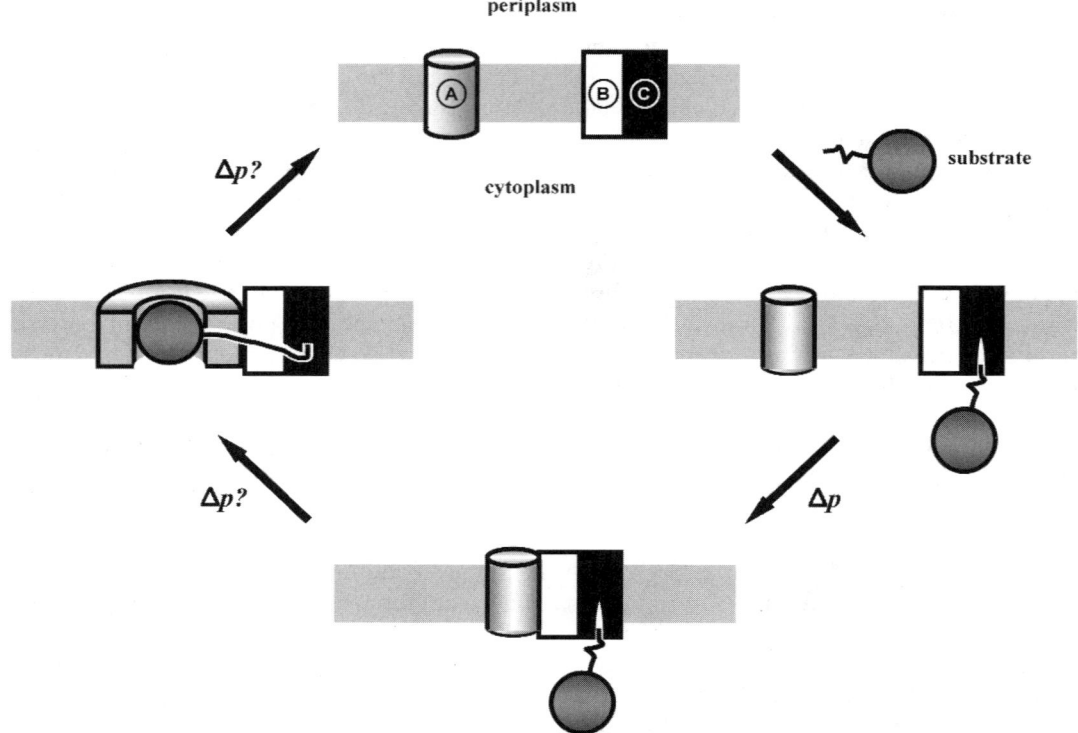

FIGURE 2 The Tat protein transport cycle. In the resting state, TatA and TatBC form separate, oligomeric complexes (top). A substrate docks at the TatBC complex, with the twin-arginine motif of the signal peptide interacting with TatC (right). This activated complex now interacts with TatA in an energy-dependent step (bottom). The substrate is transported through a channel made up of TatA. The energetic requirements for this are unknown (left). At some stage during the transport cycle the signal peptide is cleaved and the exported protein is released at the periplasmic face of the membrane (or integrated into the lipid bilayer for the membrane-anchored Tat substrates). The TatA and TatBC components dissociate and the system returns to the initial state (top). (Figure adapted from Ben C. Berks, Tracy Palmer, and Frank Sargent, Protein targeting by the bacterial twin-arginine translocation (Tat) pathway. *Curr. Opin. Microbiol.* **8:**174–181. Copyright [2005], with permission from Elsevier.)

and single-particle methods (Gohlke et al., 2005). As shown in Color Plate 1, the complexes are ring shaped with a lid structure covering one end of the central cavity. It is inferred that the complexes are oriented in the membrane in such a way that the central cavity could provide a transmembrane channel for protein transport. Individual TatA complexes differ in their ring circumferences presumably because different complexes contain different numbers of TatA protomers. The central channel varies in diameter between complexes. However, the channels in the biggest TatA complexes (ca. 70-

Å channel diameter) are capable of accommodating the largest known Tat substrates. Thus, the structures are entirely consistent with the idea that TatA forms the channel component of the Tat transporter. It has been proposed that the variation in size of the TatA complex may provide a mechanism to accommodate folded proteins of different size (Gohlke et al., 2005).

The first step in the translocation process is the engagement of a Tat substrate with the transport machinery (Fig. 2). Several studies have shown that the TatBC complex acts as the initial receptor for Tat substrate proteins. The first experiments in support of this came from work with the analogous Tat system present in the thylakoid membranes of plant chloroplasts. It was shown that substrate precursors bind to a complex comprising the plant orthologs of *E. coli* TatB and TatC (designated Hcf106 and cpTatC, respectively). This binding was shown not to require the proton motive force and is therefore energy independent (Cline and Mori, 2001; Ma and Cline, 2000; Musser and Theg, 2000). More recent studies using isolated *E. coli* Tat complexes (de Leeuw et al., 2002; McDevitt et al., submitted) or *E. coli* membrane vesicles containing elevated amounts of Tat proteins (Alami et al., 2003) confirmed that this was also true for the *E. coli* Tat system. The latter study made an elegant use of site-specific cross-linking reagents to identify TatC as the component that recognizes the twin-arginine motif of the signal peptide (Alami et al., 2003).

Once the substrate is bound to the TatBC complex, the next step is transport of the protein across the membrane. The structure of the TatA complexes reveals that they are likely to form the transport channel, and therefore for transport to occur, it requires interaction between the substrate-bound receptor complex and the TatA channel complexes. The main evidence for such an interaction again comes from studies of the thylakoid Tat system. Work from the Cline laboratory showed that, once a precursor was bound at TatBC, transport of the substrate required the proton motive force and the TatA ortholog, Tha4 (Cline and Mori, 2001). Indeed, in such circumstances, the TatA ortholog could be cross-linked to the TatBC complex, suggesting that the two complexes dynamically interact (Mori and Cline, 2002). More recent studies have indicated that this also appertains to the *E. coli* system (Alami et al., 2003). Such an interaction is presumably only transient since cross-links were not observed once transport was complete (Mori and Cline, 2002). Thus, it is inferred that the Tat transport process is a cycle that involves the assembly of a substrate-activated TatBC complex with a TatA channel complex to form a transient transport intermediate that returns to resting state after transport is complete.

Many questions are still outstanding as to how the Tat system operates. For example, it is currently not known whether the TatA complexes form rings of fixed size or whether individual rings adjust to accommodate varying sizes of substrates by gain or loss of TatA protomers. Likewise, the role of the essential amphipathic domains of TatA and TatB are still obscure. There is some suggestion that the amphipathic helix of TatA might, under some circumstances, be accessible from the periplasmic side of the membrane (Gouffi et al., 2004). This raises the attractive possibility that this region of the protein can fold back onto the transmembrane domain to provide a hydrophilic lining to the transport channel. Finally, it is also not known which of the Tat components is responsible for transducing the proton motive force nor whether the proton motive force is required for stages beyond formation of the TatABC complex. What is clear, however, is that protein transport by the Tat pathway is likely to be energetically expensive—a recent study of the thylakoid Tat pathway estimated that the transmembrane flux of almost 80,000 protons (equivalent to hydrolysis of about 10,000 molecules of ATP or 3% of the total energy output of the chloroplast) is required to transport a single substrate protein (Alder and Theg, 2003).

THE Tat TARGETING SIGNAL

Proteins are targeted to the Tat transport machinery by means of N-terminal signal peptides. These peptides are cleaved at an unknown point during the transport cycle by the membrane-bound, periplasmically facing protease leader peptidase (Yahr and Wickner, 2001). N-terminal signal peptides are common to proteins transported by both the Sec and Tat transport systems, and they share a number of similar features. They have an overall tripartite structure with an n-region (very short for Sec signal peptides but can be rather longer for Tat signals) that immediately precedes an h-region (an unbroken stretch of about 15 hydrophobic amino acids), which is followed by a short c-region that contains the recognition sequence for leader peptidase. However, despite their general similarities, there are several distinguishing features that are essential for targeting to the Tat pathway, and indeed to prevent mistargeting to the Sec machinery.

The most striking feature of Tat signal peptides is the twin-arginine motif. This is found at the C-terminal end of the n-region, immediately adjacent to the hydrophobic stretch, and can be defined as (S/T)-R-R-x-F-L-K (Berks, 1996). The critical feature of this motif is the consecutive, or twin, arginine residues that give rise to the name of the twin arginine translocation pathway. The twin arginines are almost always conserved in the signal peptides of proteins that are exported by the Tat pathway, and site-directed mutagenesis studies have shown that they are essential for efficient transport (e.g., Halbig et al., 1999; Cristóbal et al., 1999; Stanley et al., 2000). However, it is clear that in some circumstances substitution of either, but not both, of the two consensus arginines for lysine (or of the second arginine for glutamine or asparagine) may also be tolerated, although often with an associated reduction in export rate (Halbig et al., 1999; Cristóbal et al., 1999; Stanley et al., 2000; DeLisa et al., 2002). A small number of naturally occurring Tat substrates have a substitution at one of the two arginine residues, but such examples are still relatively rare (Hinsley et al., 2001; Ignatova et al., 2002; B. Ize, B.C. Berks, and T. Palmer, unpublished data) (see Table 1). Very few studies have addressed the role of the other consensus amino acids of the twin-arginine motif. However, it was shown that the hydrophobicity of the amino acid at the +2 position relative to the twin arginines (often phenylalanine) is an important determinant for export rate by the *E. coli* system (Stanley et al., 2000).

Although the twin-arginine motif is essential to target proteins to the Tat machinery, its presence does not, per se, render a protein incompatible for targeting and transport by the Sec pathway. There are numerous examples where fusing a Tat signal peptide to a Sec-compatible passenger protein allows transport to occur by the Sec pathway (e.g., Sanders et al.,

2001; DeLisa et al., 2003). However, for native Tat substrates, their folded state ensures that they are incompatible with the Sec pathway and that they are only competent for export via Tat. An additional possibility is that some or all Tat substrates may be guided to the Tat machinery by specific signal peptide-binding chaperone proteins (see below). The use of dual targeting reporter proteins has allowed further specific features of Tat signal peptides to be uncovered. For example, many Tat signal peptides contain a basic residue in their c-region that is almost never found in Sec-signal peptides. Several groups have shown that this is not required for targeting proteins to the Tat system, but it serves as a very effective "Sec-avoidance" signal, thus helping to prevent mistargeting between the two pathways (Bogsch et al., 1997; Cristóbal et al., 1999; Ize et al., 2002a; Blaudeck et al., 2003). A final difference between Sec- and Tat-targeting signal peptides lies in the hydrophobicity of the peptide h-region. Cristóbal et al. (1999) were the first to notice that the h-regions of Tat signal peptides are significantly less hydrophobic and contain proportionately more glycine and alanine residues than the h-regions of Sec-targeting signals. They were able to demonstrate that increasing the hydrophobicity of a Tat signal peptide was sufficient to reroute a dual Sec- and Tat-compatible reporter protein from the Tat pathway to the Sec pathway.

Finally, note that Tat signal peptides often have very extended n-regions prior to the twin-arginine motif (see, for example, some of the *E. coli* Tat signal peptides listed in Table 1). The function of these regions, which are often highly conserved between signal peptides of similar proteins from different bacteria, has not yet been fully addressed. One possibility is that they are primary binding sites for substrate-specific chaperone proteins. A very recent study has shown that the *E. coli* TorA signal peptide-specific chaperone TorD binds specifically at the h- and c-regions of the isolated TorA signal peptide rather than the n-region (Hatzixanthis et al., 2005). However, the n-region of the TorA signal peptide is rather short and is not conserved among TorA proteins from different species. This should be contrasted with the very long and highly conserved n-regions of the twin-arginine signal peptides of the hydrogenase small subunits, for example (Berks et al., 2000). Clearly, this is one area of research that merits further study.

COORDINATING ASSEMBLY AND EXPORT: Tat-DEPENDENT PROTEIN CHAPERONES

Clearly, for some Tat substrates at least, there is a complex chain of events that must happen prior to transport of the protein to the periplasm by the Tat pathway. For proteins that bind cofactors, the cell needs to coordinate cofactor biosynthesis and insertion into the apoprotein with the export process. Indeed, many of the cofactor-containing Tat substrates are extremely complex since they are exported as heterodimers, or even heterotrimers (see Table 1), yet the twin-arginine-containing signal peptide is found on only one of the subunits. In these cases the non-signal peptide-bearing partner protein(s) are exported by virtue of forming a complex with the signal-bearing subunit (the so-called hitchhiker mechanism; Rodrigue et al., 1999). Furthermore, these protein complexes are often anchored at the periplasmic side of the membrane by a C-terminal transmembrane α-helix (Jormakka et al., 2002; Hatzixanthis et al., 2003). This poses an additional problem to the cell since the Tat pathway is strictly posttranslational, and it therefore must prevent aggregation of the hydrophobic transmembrane segments in the cytoplasmic compartment. One way in which bacteria coordinate these processes is through the use of dedicated proofreading chaperones.

These dedicated proofreading chaperones bind to their substrate precursor proteins both through the cofactorless mature domain and through the twin-arginine signal peptide. When in complex with the chaperone the signal peptide of the substrate is assumed to be unable to engage the Tat apparatus. Thus, Tat transport can only occur when the chaperone is released.

Successful cofactor insertion into the mature domain causes release of the chaperone. Several proofreading chaperones have now been identified. Examples are DmsD, which binds to the signal peptide of the DmsA subunit of dimethylsulfoxide reductase (Oresnik et al., 2001); TorD, which interacts with the signal peptide of the trimethylamine-N-oxide reductase TorA (Jack et al., 2004); HyaE and HybE, which probably bind the signal peptides of the small subunits of hydrogenases 1 and 2, respectively (Dubini and Sargent, 2003; Jack et al., 2004). Remarkably, each chaperone is specific for its cognate signal peptide. Thus, if the signal peptide of the small subunit of hydrogenase 2 is replaced with that of TorA, hydrogenase 2 becomes completely dependent on TorD for coordinating its assembly (Jack et al., 2004). Hatzixanthis et al. (2005) have shown that TorD binds to the TorA signal peptide with moderate affinity (K_D ca. 2 μM). In addition, they demonstrated the presence of a hitherto unsuspected GTP-binding site on the TorD protein, which suggests that chaperone binding and release may be governed by cycles of GTP binding and hydrolysis. A schematic model for the steps involved in the preassembly of a Tat-dependent cofactor-containing protein is shown in Fig. 3.

The final stage in the chaperone-dependent assembly process for Tat substrates is the release of the substrate to allow its interaction with the Tat machinery. There is some suggestion that the chaperone protein DmsD can interact with the TatBC components of the Tat system, and thus it may also act as an "escortase," guiding the substrate to the transporter prior to disassembly (Papish et al., 2003). Although the proofreading events described in this section have currently only been described for cofactor-containing proteins, it is possible that such a mechanism may also operate to ensure correct folding (and possibly targeting) of all Tat substrates. Finally, it has been reported that the Tat system of both thylakoids and *E. coli* has an intrinsic "quality control" activity that allows it to distinguish between folded and unfolded substrates (Musser and Theg, 2000; DeLisa et al.,

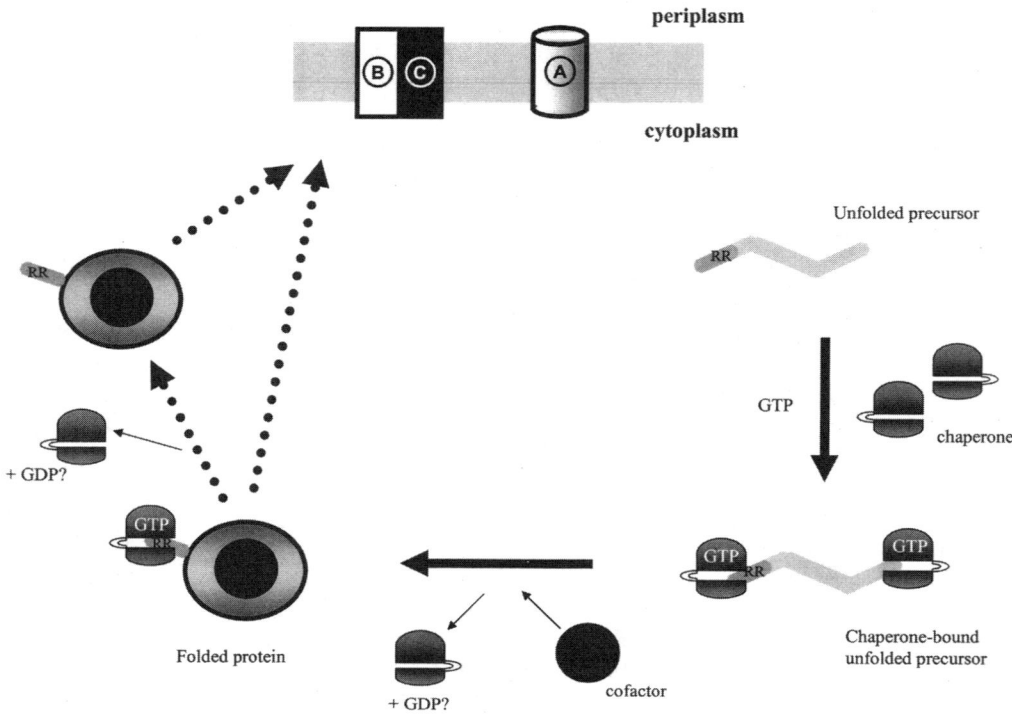

FIGURE 3 A model for chaperone-mediated assembly of a cofactor-containing Tat substrate based on studies with the paradigm TorD/TorA binary system. A de novo-synthesized, unfolded, precursor protein is initially recognized and bound by a specific chaperone. In some cases, the chaperone could bind to at least two sites on the protein, one being the signal peptide itself and the second site being elsewhere on the mature portion of the protein (Pommier et al., 1998; Jack et al., 2004). This is probably achieved by the independent binding of two chaperone molecules, but an alternative model would involve the binding of a single chaperone that contains two separate interaction sites. Association of the chaperone with the precursor leads to conformational changes in both. A binding site for GTP would be exposed on the chaperone (Hatzixanthis et al., 2005) (left), and the precursor would bind its cofactor(s). Once cofactor loading is complete, the mature portion of precursor would attain its final fully folded conformation and this would result in the release of at least one of the bound chaperones. Chaperone release may be triggered by GTP hydrolysis (bottom). If the Tat substrate precursor forms a cotranslocation complex with other proteins, they are probably associated at this stage. At this juncture, if the signal-bound chaperone also acts as an "escortase," the folded substrate will be targeted to the Tat machinery by the chaperone (left). Alternatively, release of the signal peptide-bound chaperone would allow the substrate to follow a generic route to the Tat machinery.

2003). Whether and how this quality control interrelates with the chaperone-mediated proof-reading process remains to be established.

ACKNOWLEDGMENTS

We thank our colleagues, past and present, with whom we have shared ideas about the Tat system. We are particularly indebted to our collaborators George Georgiou and Frank Sargent for continued and lively discussions. Tracy Palmer is an MRC Senior Non Clinical Research Fellow. Research in the authors' laboratories has been or is currently supported by the Biotechnology and Biological Sciences Research Council, the Medical Research Council, the Commission of the European Community, the Leverhulme Trust, the Wellcome Trust, the Royal Society, the John Innes Centre, the EPA Cephalosporin Fund, the University of East Anglia, and the University of Oxford.

REFERENCES

Alder, N. N., and S. M. Theg. 2003. Energetics of protein transport across biological membranes. A study of the thylakoid DeltapH-dependent/cpTat pathway. *Cell* **112:**231–242.

Alami, M., I. Luke, S. Deitermann, G. Eisner, H. G. Koch, J. Brunner, and M. Muller. 2003. Differential interactions between a twin-arginine signal peptide and its translocase in *Escherichia coli*. *Mol. Cell* **12:**937–946.

Behrendt, J., K. Standar, U. Lindenstrauss, and T. Brüser. 2004. Topological studies on the twin-arginine translocase component TatC. *FEMS Microbiol. Lett.* **234:**303–308.

Bendtsen, J. D., H. Nielsen, D. Widdick, T. Palmer, and S. Brunak. 2005. Prediction of twin-arginine signal peptides. *BMC Bioinformatics* **6:**167–175.

Berks, B. C. 1996. A common export pathway for proteins binding complex redox cofactors? *Mol. Microbiol.* **22:**393–404.

Berks, B. C., T. Palmer, and F. Sargent. 2003. The Tat protein translocation pathway and its role in microbial physiology. *Advances Microb. Physiolol.* **47:**187–254.

Berks, B. C., F. Sargent, and T. Palmer. 2000. The Tat protein export pathway. *Mol. Microbiol.* **35:**260–274.

Bernhardt, T. G., and P. A. de Boer. 2003. The *Escherichia coli* amidase AmiC is a periplasmic septal ring component exported via the twin-arginine transport pathway. *Mol. Microbiol.* **48:**1171–1182.

Blaudeck, N., P. Kreutzenbeck, R. Freudl, and G. A. Sprenger. 2003. Genetic analysis of pathway specificity during posttranslational protein translocation across the *Escherichia coli* plasma membrane. *J. Bacteriol.* **185:**2811–2819.

Blaudeck, N., P. Kreutzenbeck, M. Muller, G. A. Sprenger, and R. Freudl. 2005. Isolation and characterization of bifunctional *Escherichia coli* TatA mutant proteins that allow efficient Tat-dependent protein translocation in the absence of TatB. *J. Biol. Chem.* **280:**3426–3432.

Bogsch, E., S. Brink, and C. Robinson. 1997. Pathway specificity for a ΔpH-dependent precursor thylakoid lumen protein is governed by a 'Sec-avoidance' motif in the transfer peptide and a 'Sec-incompatible' mature protein. *EMBO J.* **16:**3851–3859.

Bogsch, E., F. Sargent, N. R. Stanley, B. C. Berks, C. Robinson, and T. Palmer. 1998. An essential component of a novel bacterial protein export system with homologues in plastids and mitochondria. *J. Biol. Chem.* **273:**18003–18006.

Bolhuis, A., J. E. Mathers, J. D. Thomas, C. M. Barrett, and C. Robinson. 2001. TatB and TatC form a functional and structural unit of the twin-arginine translocase from *Escherichia coli*. *J. Biol. Chem.* **276:**20213–20219.

Buchanan, G., E. de Leeuw, N. R. Stanley, M. Wexler, B. C. Berks, F. Sargent, and T. Palmer. 2002. Functional complexity of the twin-arginine translocase TatC component revealed by site-directed mutagenesis. *Mol. Microbiol.* **43:**1457–1470.

Caldelari, I., S. Mann, C. Crooks, and T. Palmer. 2006. The Tat pathway of the plant pathogen *Pseudomonas syringae* is required for optimal virulence. *Mol. Plant-Microbe Interact.* **19:**200–212.

Chanal, A., C.-L. Santini, and L.-F. Wu. 1998. Potential receptor function of three homologous components, TatA, TatB and TatE, of the twin-arginine signal sequence-dependent metalloenzyme translocation pathway in *Escherichia coli*. *Mol. Microbiol.* **30:**674–676.

Cline, K., and H. Mori. 2001. Thylakoid ΔpH-dependent precursor proteins bind to a cpTatC-Hcf106 complex before Tha4-dependent transport. *J. Cell. Biol.* **154:**719–729.

Cristóbal, S., J.-W. de Gier, H. Nielsen, and G. von Heijne. 1999. Competition between Sec- and Tat-dependent protein translocation in *Escherichia coli*. *EMBO J.* **18:**2982–2990.

de Leeuw, E., T. Granjon, I. Porcelli, M. Alami, S. B. Carr, M. Müller, F. Sargent, T. Palmer, and B. C. Berks. 2002 Oligomeric properties and signal peptide binding by *Escherichia coli* Tat protein transport complexes. *J. Mol. Biol.* **322:**1135–1146.

DeLisa, M. P., P. Samuelson, T. Palmer, and G. Georgiou. 2002. Genetic analysis of the twin arginine translocator secretion pathway in bacteria. *J. Biol. Chem.* **277:**29825–29831.

DeLisa, M. P., D. Tullman, and G. Georgiou. 2003. Folding quality control in the export of proteins by the bacterial twin-arginine translocation pathway. *Proc. Natl. Acad. Sci. USA* **100:**6115–6120.

Dilks, K., W. R. Rose, E. Hartmann, and M. Pohlschroder. 2003. Prokaryotic utilization of the twin-arginine translocation pathway: a genomic survey. *J. Bacteriol.* **185:**1478–1483.

Drew, D., D. Sjostrand, J. Nilsson, T. Urbig, C. N. Chin, J. W. de Gier, and G. von Heijne. 2002. Rapid topology mapping of *Escherichia coli* inner-membrane proteins by prediction and PhoA/GFP fusion analysis. *Proc. Natl. Acad. Sci. USA* **99:**2690–2695.

Dubini, A., and F. Sargent. 2003. Assembly of Tat-dependent NiFe hydrogenases: identification of precursor-binding accessory proteins. *FEBS Lett.* **549:**141–146.

Gohlke, U., L. Pullan, C. A. McDevitt, I. Porcelli, E. de Leeuw, T. Palmer, H. Saibil, and B. C. Berks. 2005. The TatA component of the twin-arginine protein transport system forms channel complexes of variable diameter. *Proc. Natl. Acad. Sci. USA* **102**:10482–10486.

Gouffi, K., F. Gerard, C. L. Santini, and L.-F. Wu. 2004. Dual topology of the *Escherichia coli* TatA protein. *J. Biol. Chem.* **279**:11608–11615.

Halbig, D., T. Wiegert, N. Blaudeck, R. Freudl, and G. A. Sprenger. 1999. The efficient export of NADP-containing glucose-fructose oxidoreductase to the periplasm of *Zymomonas mobilis* depends both on an intact twin-arginine motif in the signal peptide and on the generation of a structural export signal induced by cofactor binding. *Eur. J. Biochem.* **263**:543–551.

Hatzixanthis, K., T. A. Clarke, A. Oubrie, D. J. Richardson, R. J. Turner, and F. Sargent. 2005. Signal peptide-chaperone interactions on the twin-arginine protein transport pathway. *Proc. Natl. Acad. Sci. USA* **102**:8460–8465.

Hatzixanthis K., T. Palmer, and F. Sargent. 2003. A subset of bacterial inner membrane proteins integrated by the twin-arginine translocase. *Mol. Microbiol.* **49**:1377–1390.

Hinsley, A. P., N. R. Stanley, T. Palmer, and B. C. Berks. 2001. A naturally-occurring bacterial Tat signal peptide lacking one of the 'invariant' arginine residues of the consensus targeting motif. *FEBS Lett.* **49**:45–49.

Ignatova, Z., C. Hörnle, A. Nurk, and V. Kasche. 2002. Unusual signal peptide directs penicillin amidase from *Escherichia coli* to the Tat translocation machinery. *Biochem. Biophys. Res. Comm.* **291**:146–149.

Ize, B., F. Gérard, and L.-F. Wu. 2002a. *In vivo* assessment of the Tat signal peptide specificity in *Escherichia coli*. *Arch. Microbiol.* **178**:548–553.

Ize, B., F. Gérard, M. Zhang, A. Chanal, R. Volhoux, T. Palmer, A. Filloux, and L.-F. Wu. 2002b. *In vivo* dissection of the Tat translocation pathway in *Escherichia coli*. *J. Mol. Biol.* **317**:327–335.

Ize, B., I. Porcelli, S. Lucchini, J. C. Hinton, B. C. Berks, and T. Palmer. 2004. Novel phenotypes of *Escherichia coli tat* mutants revealed by global gene expression and phenotypic analysis. *J. Biol. Chem.* **279**:47543–47554.

Ize, B., N. R. Stanley, G. Buchanan, and T. Palmer. 2003. Role of the *Escherichia coli* Tat pathway in outer membrane integrity. *Mol. Microbiol.* **48**:1183–1193.

Jack, R. L., G. Buchanan, A. Dubini, K. Hatzixanthis, T. Palmer, and F. Sargent. 2004. Coordinating assembly and export of complex bacterial proteins. *EMBO J.* **23**:3962–3972.

Jack, R. L., F. Sargent, B. C. Berks, G. Sawers, and T. Palmer. 2001. Constitutive expression of *Escherichia coli tat* genes indicates an important role for the twin-arginine translocase during aerobic and anaerobic growth. *J. Bacteriol.* **183**:1801–1804.

Jongbloed, J. D., U. Grieger, H. Antelmann, M. Hecker, R. Nijland, S. Bron, and J.-M. van Dijl. 2004. Two minimal Tat translocases in *Bacillus*. *Mol. Microbiol.* **545**:1319–1325.

Jormakka, M., S. Törnroth, B. Byrne, and S. Iwata. 2002. Molecular basis of proton motive force generation: structure of formate dehydrogenase-N. *Science* **295**:1863–1868.

Kasche, V., Z. Ignatova, H. Markl, W. Plate, N. Punckt, D. Schmidt, K. Wiegandt, and B. Ernst. 2005. Ca^{2+} is a cofactor required for membrane transport and maturation and is a yield-determining factor in high cell density penicillin amidase production. *Biotechnol. Prog.* **21**:432–438.

Ki, J. J., Y. Kawarasaki, J. Gam, B. R. Harvey, B. L. Iverson, and G. Georgiou. 2004. A periplasmic fluorescent reporter protein and its application in high-throughput membrane protein topology analysis. *J. Mol. Biol.* **341**:901–909.

Lequette, Y., C. Ödberg-Ferragut, J.-P. Bohin, and J.-M. Lacroix. 2004. Identification of *mdoD*, and *mdoG* paralog which encodes a twin-arginine-dependent periplasmic protein that controls osmoregulated periplasmic glucan backbone structures. *J. Bacteriol.* **186**:3695–3702.

Ma, X., and K. Cline. 2000. Precursors bind to specific sites on thylakoid membranes prior to transport on the delta pH protein translocation system. *J. Biol. Chem.* **275**:10016–10022.

McDevitt, C. A., M. G. Hicks, T. Palmer, and B. C. Berks. 2005. Characterisation of Tat protein transport complexes carrying inactivating mutations. *Biochem. Biophys. Res. Commun.* **329**:693–698.

Mori, H., and K. Cline. 2002. A twin arginine signal peptide and the pH gradient trigger reversible assembly of the thylakoid ΔpH/Tat translocase. *J. Cell Biol.* **157**:205–210.

Musser, S. M., and S. M. Theg. 2000. Characterization of the early steps of OE17 precursor transport by the thylakoid ΔpH/Tat machinery. *Eur. J. Biochem.* **167**:2588–2597.

Oates, J., C. M. Barrett, J. P. Barnett, K. G. Byrne, A. Bolhuis, and C. Robinson. 2005. The *Escherichia coli* twin-arginine translocation apparatus incorporates a distinct form of TatABC complex, spectrum of modular TatA complexes and minor TatAB complex. *J. Mol. Biol.* **346**:295–305.

Oates, J., J. Mathers, D. Mangels, W. Kuhlbrandt, C. Robinson, and K. Model. 2003. Consensus structural features of purified bacterial TatABC complexes. *J. Mol. Biol.* **330**:277–286.

Oresnik, I. J., C. L. Ladner, and R. J. Turner. 2001. Identification of a twin-arginine leader-binding protein. *Mol. Microbiol.* **40**:323–331.

Outten, F. W., C. E. Outten, J. Hale, and T. V. O'Halloran. 2000. Transcriptional activation of the *Escherichia coli* copper efflux regulon by the chromosomal MerR homologue, CueR. *J. Biol. Chem.* **275**:31024–31029.

Papish, A. L., C. L. Ladner, and R. J. Turner. 2003. The twin-arginine leader-binding protein, DmsD, interacts with the TatB and TatC subunits of the *Escherichia coli* twin-arginine translocase. *J. Biol. Chem.* **278**:32501–32506.

Pommier, J., V. Mejean, G. Giordano, and C. Iobbi-Nivol. 1998. TorD, a cytoplasmic chaperone that interacts with the unfolded trimethylamine N-oxide reductase enzyme (TorA) in *Escherichia coli. J. Biol. Chem.* **273**:16615–16620.

Pop, O., U. Martin, C. Abel, and J. P. Muller. 2002. The twin-arginine signal peptide of PhoD and the TatAd/Cd proteins of *Bacillus subtilis* form an autonomous Tat translocation system. *J. Biol. Chem.* **277**:3268–3273.

Porcelli, I., E. de Leeuw, R. Wallis, E. van den Brink-van der Laan, B. de Kruijff, B. A. Wallace, T. Palmer, and B. C. Berks. 2002. Characterisation and membrane assembly of the TatA component of the *Escherichia coli* twin-arginine protein transport system. *Biochemistry* **41**:13690–13697.

Rodrigue, A., A. Chanal, K. Beck, M. Müller, and L.-F. Wu. 1999. Co-translocation of a periplasmic enzyme complex by a hitchhiker mechanism through the bacterial Tat pathway. *J. Biol. Chem.* **274**:13223–13228.

Rose, R. W., T. Brüser, J. C. Kissinger, and M. Pohlschröder. 2002. Adaptation of protein secretion to extremely high-salt conditions by extensive use of the twin-arginine translocation pathway. *Mol. Microbiol.* **45**:943–950.

Sanders, C., N. Wethkamp, and H. Lill. 2001. Transport of cytochrome *c* derivatives by the bacterial Tat protein translocation system. *Mol. Microbiol.* **41**:241–246.

Sargent, F., E. Bogsch, N. R. Stanley, M. Wexler, C. Robinson, B. C. Berks, and T. Palmer. 1998. Overlapping functions of components of a bacterial Sec-independent protein export pathway. *EMBO J.* **17**:3640–3650.

Sargent, F., U. Gohlke, E. de Leeuw, N. R. Stanley, T. Palmer, H. R. Saibil, and B. C. Berks. 2001. Purified components of the *Escherichia coli* Tat protein transport system form a double-layered ring structure. *Eur. J. Biochem.* **268**:3361–3367.

Sargent, F., N. R. Stanley, B. C. Berks, and T. Palmer. 1999. Sec-independent protein translocation in *Escherichia coli*: a distinct and pivotal role for the TatB protein. *J. Biol. Chem.* **274**:36073–36083.

Settles, A. M., A. Yonetani, A. Baron, D. R. Bush, K. Cline, and R. Martienssen. 1997. Sec-independent protein translocation by the maize Hcf106 protein. *Science* **278**:1467–1470.

Stanley, N. R., K. Findlay, B. C. Berks, and T. Palmer. 2001. *Escherichia coli* strains blocked in Tat-dependent protein export exhibit a defective cell separation morphology. *J. Bacteriol.* **183**:139–144.

Stanley, N. R., T. Palmer, and B. C. Berks. 2000. The twin arginine consensus motif of Tat signal peptides is involved in Sec-independent protein targeting in *Escherichia coli. J. Biol. Chem.* **257**:11591–11596.

Stevens, J. M., O. Daltrop, J. W. Allen, and S. J. Ferguson. 2004. *c*-type cytochrome formation: chemical and biological enigmas. *Acc. Chem. Res.* **37**:999–1007.

von Heijne, G. 1992. Membrane protein structure prediction. Hydrophobicity analysis and the positive-inside rule. *J. Mol. Biol.* **225**:487–494.

Weiner, J. H., P. T. Bilous, G. M. Shaw, S. P. Lubitz, L. Frost, G. H. Thomas, J. A. Cole, and R. J. Turner. 1998. A novel and ubiquitous system for membrane targeting and secretion of cofactor-containing proteins. *Cell* **93**:93–101.

Wexler, M., F. Sargent, R. L. Jack, N. R. Stanley, E. G. Bogsch, C. Robinson, B. C. Berks, and T. Palmer. 2000. TatD is a cytoplasmic protein with DNase activity. No requirement for TatD family proteins in sec-independent protein export. *J. Biol. Chem.* **275**:16717–16722.

Yahr, T. L., and W. T. Wickner. 2001. Functional reconstitution of bacterial Tat translocation *in vitro. EMBO J.* **20**:2472–2479.

Yen, M.-R., Y. H. Tseng, E. H. Nguyen, L.-F. Wu, and M. H. Saier, Jr. 2002. Sequence and phylogenetic analyses of the twin-arginine targeting (Tat) protein export system. *Arch. Microbiol.* **177**:441–450.

ASSEMBLY OF INTEGRAL MEMBRANE PROTEINS FROM THE PERIPLASM INTO THE OUTER MEMBRANE

Jörg H. Kleinschmidt

3

The outer membrane (OM) of gram-negative bacteria consists of a lipid bilayer, which is composed of phospholipids in the periplasmic leaflet and of lipopolysaccharide in the outer leaflet. Integral membrane proteins facilitate transport, for example, of nutrients, across this hydrophobic barrier. After their biosynthesis, outer membrane proteins (OMPs) are targeted to the cytoplasmic membrane in complex with SecB/SecA and they are then transported through the SecY/E/G translocon of the cytoplasmic membrane into the periplasm. In the periplasm, the signal sequences of the OMPs are cleaved by a leader peptidase and the proteins cross the periplasm in an unfolded form prior to their assembly into the OM. Misfolding of OMPs leads to their proteolysis and to the activation of extracytoplasmic stress response, which results in expression of periplasmic and outer membrane folding factors, such as chaperones, isomerases, and proteases. In the periplasm, the OMPs are kept soluble in largely unfolded form by periplasmic chaperones. A complex of several proteins in the OM is involved in targeting and assembly of the integral membrane proteins into the OM.

OMPs insert and fold into lipid bilayers by a highly concerted mechanism, in which secondary and tertiary structure formation is synchronized and coupled to lipid bilayer insertion. This chapter summarizes the current knowledge about the transport of OMPs through the periplasm and about their assembly into membranes.

Biological membranes are needed for maintaining the integrity of cells and cell organelles as well as their shape. All membranes including the OM of bacteria are composed of a lipid bilayer that has a hydrophobic core and polar surfaces. The lipid bilayer prevents the arbitrary passage of solutes across the membrane. Integral membrane proteins, also called transmembrane proteins (TMPs), are embedded into the lipid bilayers to perform many different functions, including the transport of solutes and signals across the hydrophobic barrier. Therefore, TMPs are essential for the survival of the cell. The biogenesis of membranes is of fundamental interest to basic research from the perspectives of cell biologists, structural biologists, biochemists, and biophysicists. Important for the understanding of membrane biogenesis is how integral membrane proteins are inserted and folded into the lipid bilayer host matrix. For insertion and folding into the OM, TMPs cross the cytoplasmic membrane in an unfolded

Jörg H. Kleinschmidt, Fachbereich Biologie, Fach M 694, Universität Konstanz, D-78457 Konstanz, Germany.

The Periplasm
Edited by Michael Ehrmann © 2007 ASM Press, Washington, D.C.

form. They must then traverse the periplasm. How insertion and folding of OMPs into the OM takes place is largely unknown, and this chapter aims to give an overview about our current knowledge on the insertion and folding of OMPs from the periplasm into the OM. To understand the insertion and folding process of integral membrane proteins, it is important to consider their transmembrane structure and biophysical properties. Therefore, some of the structures and properties of OMPs are described first in this chapter, followed by an overview of the currently known periplasmic folding factors of OMPs. Subsequently, recently discovered OMPs and lipoproteins that are associated with the OM and involved in OMP targeting, insertion, and/or folding are described. Finally, I summarize recently performed biochemical and biophysical investigations on OMP insertion and folding.

STRUCTURE OF OMPs

TMPs can be classified by their transmembrane secondary structure into the two categories α-helical and β-sheet membrane proteins. Within the hydrophobic core of the membrane, all polar amide hydrogen-bonding donors and carbonyl hydrogen-bonding acceptors of the polypeptide backbone form hydrogen bonds. Single helices can span the hydrophobic bilayer because, within the helix, each polar amide group of an amino acid forms a hydrogen bond with the carbonyl group of the fourth following amino acid in the helix. In a β-barrel, the hydrogen bonds between polar amide and carbonyl groups of the peptide bond are formed between neighboring strands. Single β-strands would expose the polar groups of the polypeptide chain toward the hydrophobic region of the membrane and therefore cannot cross the hydrophobic core of the lipid bilayer for energetic reasons. In TMPs, the nonpolar side chains face the hydrophobic acyl chains of the membrane lipids. Although the more abundant α-helical TMPs are found in the cytoplasmic (or inner) membranes, the TMPs with β-barrel structures are known from the OMs of bacteria, mitochondria, and chloroplasts.

In gram-negative bacteria, OMPs cross the periplasm before they assemble into the OM. All currently known OMPs from bacteria form transmembrane β-barrels. The β-barrel is characterized by the number of antiparallel β-strands and by the shear number, which is a measure for the inclination angle of the β-strands against the barrel axis. The OMPs of bacteria form transmembrane β-barrels with even numbers of β-strands ranging from 8 to 22 with shear numbers from 8 to 24 (Schulz, 2002). The strands are tilted by 36° to 44° relative to the barrel axis (Marsh and Páli, 2001; Schulz, 2002). Some examples for OMPs are given in Table 1, which also lists the molecular weight, the pI, the number of β-strands, the number of amino acid residues, the oligomeric state, and the function of the OMP. Structures of several OMPs are shown in Color Plate 2. β-Barrel membrane proteins of bacteria serve a wide range of different functions. Currently, they may be grouped into nine families: (i) general nonspecific diffusion pores (OmpC, OmpF, PhoE); (ii) passive, specific transporters, for example, for sugars (LamB, ScrY) or nucleosides (Tsx); (iii) active transporters for iron complexes (FhuA, FepA, FecA) or cobalamin (BtuB); (iv) enzymes such as proteases (OmpT), lipases (OmPlA), acyltransferases (PagP); (v) toxin binding defensive proteins (OmpX); (vi) structural proteins (OmpA, outer membrane protein A of *Escherichia coli*); (vii) adhesion proteins (NspA, OpcA); (viii) channels involved in solute efflux (TolC); and (ix) autotransporters (NalP, adhesin involved in diffuse adherence [AIDA]). Recent reports suggest that there may be OMPs with an even larger number of transmembrane β-strands than reported for the TonB-dependent active transporters. An example may be the OM usher PapC, for which a β-barrel with 26 TM β-strands has been proposed (Henderson et al., 2004; Thanassi et al., 2002). PapC is an OMP required for assembly and secretion of P pili by the chaperone/usher pathway in *E. coli* (Dodson et al., 1993; Norgren et al., 1987).

Recently developed screening algorithms for the genomic identification of β-barrel

TABLE 1 Examples of outer membrane proteins of known high-resolution structure

OMP	Organism	MW (kDa)	pI[a]	Residues	Residues in β-barrel domain	β-Strands in barrel domain	Oligomeric state	Function	PDB entry	Reference(s)
Outer membrane proteins with single-chain β-barrels										
OmpA	E. coli	35.2	5.6	325	171	8	Monomer	Structural	1QJP, 1BXW	Pautsch and Schulz, 1998, 2000
OmpA[b]	E. coli	35.2	5.6	325	171	8	Monomer	Structural	1G90	Arora et al., 2001
OmpX	E. coli	16.4	5.3	148	148	8	Monomer	Toxin binding	1QJ8	Vogt and Schulz, 1999
OmpX[b]	E. coli	16.4	5.3	148	148	8	Monomer	Toxin binding	1Q9F	Fernandez et al., 2004
NspA	N. meningitidis	16.6	9.5	153	153	8	Monomer	Cell adhesion	1P4T	Vandeputte-Rutten et al., 2003
PagP	E. coli	19.5	5.9	166	166	8	Monomer	Palmitoyltransferase	1HQT	Ahn et al., 2004
PagP[b]	E. coli	19.5	5.9	166	166	8	Monomer	Palmitoyltransferase	1MM4, 1MM5	Hwang et al., 2002
OmpT	N. meningitidis	33.5	5.4	297	297	10	Monomer	Protease	1I78	Vandeputte-Rutten et al., 2001
OpcA	N. meningitidis	28.1	9.5	254	254	10	Monomer	Adhesion protein	1K24	Prince et al., 2002
Tsx	E. coli	31.4	4.9	272	272	12	Monomer	Nucleoside uptake	1TLW, 1TLY	Ye and van den Berg, 2004
NalP[c]	N. meningitidis	111.5	6.7	1,063	267	12	Monomer	Autotransporter	1UYN	Oomen et al., 2004
OmPlA	E. coli	30.8	5.1	269	269	12	Monomer	Phospholipase	1QD6	Snijder et al., 1999
FadL	E. coli	45.9	4.9	421	378	14	Monomer	Fatty acid transporter	1T16, 1T1L	van den Berg et al., 2004a
Omp32	C. acidovorans	34.8	8.8	332	332	16	Trimer	Porin	1E54	Zeth et al., 2000
Porin	Rhodobacter capsulatus	31.5	4.0	301	301	16	Trimer	Porin	2POR	Weiss et al., 1991; Weiss and Schulz, 1992
Porin	R. blastica	30.6	3.8	290	290	16	Trimer	Porin	1PRN	Kreusch and Schulz, 1994
OmpF	E. coli	37.1	4.6	340	340	16	Trimer	Porin	2OMF	Cowan et al., 1992
PhoE	E. coli	36.8	4.8	330	330	16	Trimer	Porin	1PHO	Cowan et al., 1992
OmpK36	Klebsiella pneumoniae	37.6	4.4	342	342	16	Trimer	Porin	1OSM	Dutzler et al., 1999
LamB	E. coli	47.4	4.7	420	420	18	Trimer	Maltose-specific porin	1MAL, 1AF6	Schirmer et al., 1995; Wang et al., 1997

OMP	Organism	MW (kDa)	pI	Residues	Residues in β-barrel domain	β-Strands in barrel domain	Chains in the β-barrel	Function	PDB entry	References
Maltoporin	Serovar Typhimurium	48.0	4.7	427	427	18	Trimer	Maltose-specific porin	2MPR	Meyer et al., 1997
ScrY	Serovar Typhimurium	53.2	5.0	483	415	18	Trimer	Sucrose porin	1A0S, 1A0T	Forst et al, 1998
FhuA	E. coli	78.7	5.1	714	587	22	Monomer	Ferrichrome iron transporter	2FCP, 1BY3	Ferguson et al., 1998; Locher et al., 1998
FepA	E. coli	79.8	5.2	724	574	22	Monomer	Ferrienterobactin transporter	1FEP	Buchanan et al., 1999
FecA	E. coli	81.7	5.4	741	521	22	Monomer	Iron (III) dicitrate transporter	1KMO, 1PNZ	Ferguson et al., 2002; Yue et al., 2003
BtuB	E. coli	66.3	5.1	594	459	22	Monomer	Vitamin B_{12} transporter	1NQE, 1UJW	Chimento et al., 2003; Kurisu et al., 2003
FpvA	P. aeruginosa	86.5	5.1	772	538	22	Monomer	Ferripyoverdine transporter	1XKH	Cobessi et al., 2005
Outer membrane proteins with multichain β-barrels										
TolC	E. coli	51.5	5.2	471	285 (95 × 3)	12 (4 × 3)	Trimer	Export channel	1EK9	Koronakis et al., 2000
MspA	M. smegmatis	17.6	4.4	168	432 (32 × 8)	16 (2 × 8)	Octamer	Porin	1UUN	Faller et al., 2004
α-Hemolysin	S. aureus	33.2	7.9	293	378 (54 × 7)	14 (2 × 7)	Heptamer	Toxin	7AHL	Song et al., 1996

[a]Calculated by Protparam/SWISS-PROT.
[b]NMR structure.
[c]Translocator domain.

membrane proteins indicate that many en-coded OMPs are still not characterized, for ex-ample, in the genomes of *E. coli* and *Pseudomonas aeruginosa*, where their genes comprise about 2 to 3% of the entire genome (Wimley, 2002, 2003). Soluble bacterial toxins that can insert into membranes, such as α-hemolysine from *Staphylococcus aureus* (Song et al., 1996) and perfringolysine O from *Clostridium perfringens* (Heuck et al., 2000; Shepard et al., 1998), also form β-barrels, but these are oligomeric. Oligomeric β-barrels are also found in myco-bacteria, such as MspA from *Mycobacterium smegmatis* (Faller et al., 2004; for a review, see Niederweis, 2003). In this chapter, I focus on membrane insertion and assembly of the porins that form single-chain transmembrane β-barrels.

PERIPLASMIC PROTEINS INVOLVED IN OMP ASSEMBLY

After their biosynthesis, OMPs bind to the chaperone SecB in the cytoplasm and are then targeted in concert with the ATPase SecA to the cytoplasmic membrane (Driessen et al., 2001; Müller et al., 2001). Some secreted pro-teins instead use the signal recognition particle (SRP) pathway for targeting; for example, lipoproteins (Froderberg et al., 2004) or certain autotransporters (Sijbrandi et al., 2003). The OMPs are then translocated in an unfolded form across the cytoplasmic (inner) membrane via the SecYEG translocon (Breyton et al., 2002; Van den Berg et al., 2004b), requiring ATP and electrochemical energy. The pathway of OMPs from the cytoplasm toward the OM is illustrated in Fig. 1. After their translocation, a signal peptidase (SPase), which is bound to the cytoplasmic membrane, cleaves the N-ter-minal signal sequence of the OMP in the periplasmic space, recognizing the Ala-X-Ala motif at the end of the OMP signal sequence (Tuteja, 2005), which typically comprises the first 15 to 30 residues of the unprocessed OMP. After signal sequence cleavage, the mature OMP traverses the periplasm toward the OM for integration. Overproduction of OMPs or accumulation of unfolded OMPs in the peri-plasm activates the alternative stress σ-factor,

σ^E (RpoE) (Mecsas et al., 1993), in the cyto-plasm, which then causes production of peri-plasmic proteases and folding factors. Eσ^E RNA polymerase transcribes, for example, the genes of the periplasmic proteins Skp, SurA, DegP, and FkpA, which act as chaperones and affect the assembly of OMPs (Chen and Hen-ning, 1996; Lazar and Kolter, 1996; Missiakas et al., 1996; Rizzitello et al., 2001; Rouvière and Gross, 1996); the genes of periplasmic pro-teases such as DegP (HtrA); the genes of cer-tain outer membrane lipoproteins, such as YfiO; genes of enzymes involved in the biosynthesis of lipopolysaccharide (LPS), such as HtrM (RfaD), LpxD, and LpxA (Dartiga-longue et al., 2001; Rouvière et al., 1995); and the gene of the OMP Imp (OstA) (Dartiga-longue et al., 2001). Stress in the periplasm, such as heat, leads to misfolding of periplasmic proteins and OMPs, which is sensed by the trimeric protease DegS (Alba et al., 2001; Wilken et al., 2004) (for reviews, see, e.g., Alba and Gross, 2004; Ehrmann and Clausen, 2004) that is anchored to the periplasmic side of the inner membrane and binds the carboxy termi-nus of misfolded OMPs, recognizing the C-terminal peptide motif Tyr-X-Phe-COOH (Walsh et al., 2003). This motif is frequently found in the C terminus of the β-barrel do-main of OMPs (Struyve et al., 1991; Walsh et al., 2003). Upon binding of misfolded OMPs in a DegS PDZ-domain (a PDZ-domain is a protein domain involved in protein or peptide binding and named after three eukaryotic pro-teins in which it was observed: postsynaptic density protein, discs large protein, and zona occludens protein; see Sheng and Sala, 2001, for a recent review), DegS is activated and cleaves the anti-σ-factor RseA, which is in complex with RseB in the cytoplasmic mem-brane and with σ^E (RpoE) in the cytoplasm (Campbell et al., 2003; Collinet et al., 2000; De Las Peñas et al., 1997; Missiakas et al., 1997). Cleavage of RseA is then completed by the in-ner membrane metalloprotease YaeL (RseP) (Ades et al., 1999; Alba et al., 2002; Kanehara et al., 2003), leading to the release of σ^E into the cytoplasm (Ades et al., 2003). Under nor-

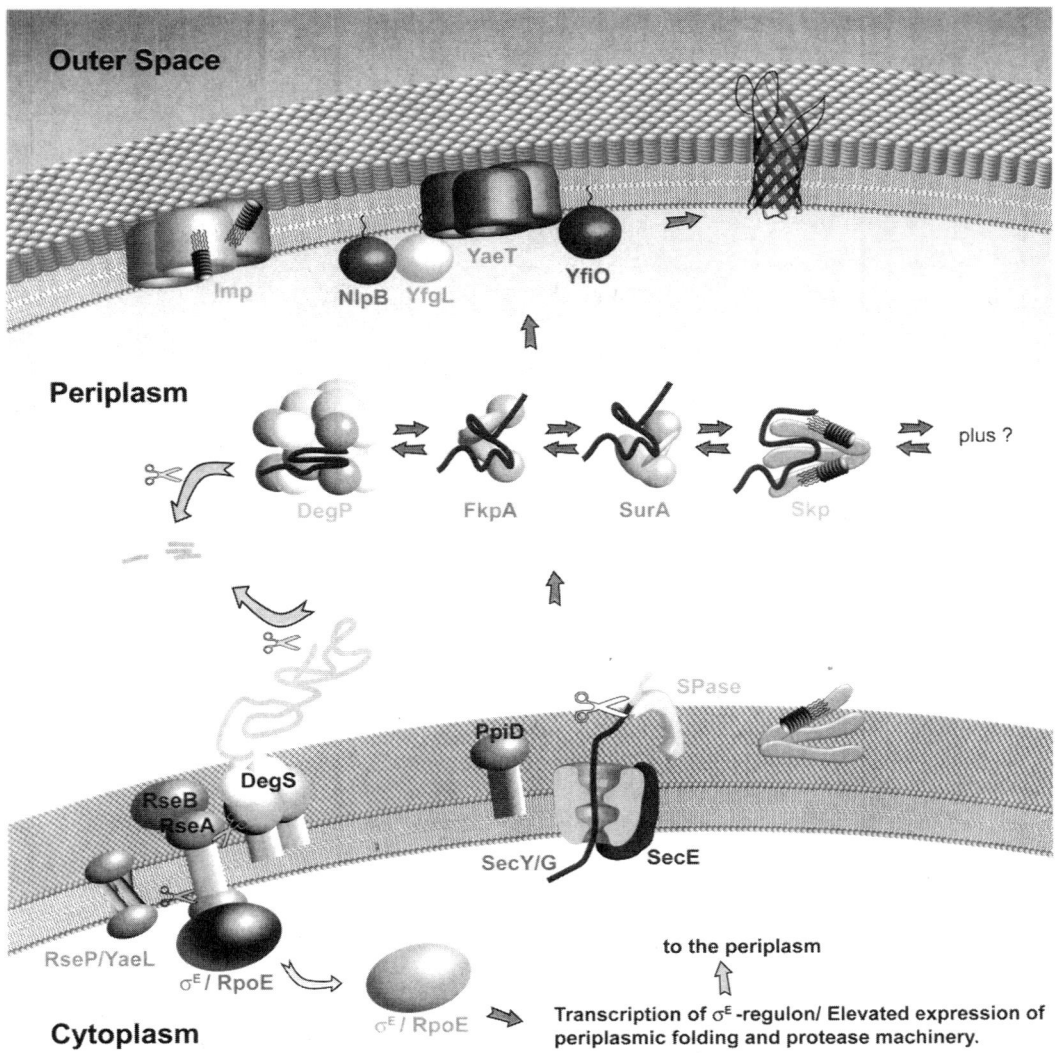

FIGURE 1 Export of integral membrane proteins through the periplasm to the OM. The OMP traverses the inner membrane through the SecY/G/E translocon in an unfolded form. A leader peptidase (SPase), anchored to the inner membrane, cleaves the signal sequence in the periplasm. The OMP then traverses the periplasm bound to a periplasmic chaperone. Among the periplasmic proteins that were either reported or likely to bind unfolded OMPs are Skp, SurA, DegP, and FkpA. All of the currently discovered soluble periplasmic folding factors are bifunctional. Skp binds unfolded OMPs and LPS; SurA and FkpA have chaperone function and—independent of the chaperone function—PPIase activity. DegP is a protease and a chaperone. From the periplasm, OMPs are targeted to or assembled into the outer membrane by membrane-bound proteins, namely YaeT (Omp85) and the lipoproteins YfiO, YfgL, and NlpB. This process has not been investigated yet. Misfolded proteins in the periplasm are degraded by proteases such as DegP and DegS. DegS is a sensor for misfolded OMPs and consequently cleaves the cytoplasmic RseA in the periplasm. In a second cleavage step, the cleaved RseA is degraded further by RseP (YaeL), leading to the release of σ^E (RpoE), which results in an elevated expression of periplasmic chaperones, isomerases, and proteases, and of OM-associated folding factors. See the text for further details.

mal conditions, the RseA-RseB complex binds about half of the σ^E content of the cell (Campbell et al., 2003; Collinet et al., 2000; De Las Peñas et al., 1997; Missiakas et al., 1997), and RseA degradation, initiated by DegS, upregulates the transcription of the heat shock genes by release of bound σ^E (Ades et al., 2003). The σ^E-dependent stress response is also activated in *rfa* mutants. The *htrM* (*rfaD*) gene product was shown to encode an ADP-L-glycero-D-mannoheptose-6-epimerase, an enzyme required for the biosynthesis of an LPS precursor (Pegues et al., 1990; Raina and Georgopoulos, 1991). Lack of the core heptose moiety in LPS was observed in such *htrM* (*rfaD*) mutants (Missiakas et al., 1996).

Searches for folding factors in the periplasm resulted in the discovery of several interesting proteins that function as chaperones or peptidyl-prolyl *cis*/*trans* isomerases (PPIases), which are discussed in the following sections. For example, the periplasmic seventeen kDa protein, Skp, was found to bind to unfolded OMPs on affinity columns, and the deletion of the *skp* gene led to reduced levels of OMPs in the *E. coli* OM (Chen and Henning, 1996). The concentrations of some OMPs in the OM of *E. coli* were also decreased, when one of the genes of the periplasmic PPIases, SurA (Lazar and Kolter, 1996; Rouvière and Gross, 1996) or PpiD (Dartigalongue and Raina, 1998), was deleted. There is no ATP in the periplasm (Wülfing and Plückthun, 1994) and therefore periplasmic chaperones are expected to function differently from cytoplasmic chaperones, which utilize ATP in their catalytic cycles (Craig, 1993).

Representatives of three different families of PPIases were found in the periplasm. These may assist the folding of OMPs, which traverse the periplasm in unfolded form. Examples are the parvulin type SurA (Behrens et al., 2001; Missiakas et al., 1996), the FKBP type FkpA (Bothmann and Plückthun, 2000; Missiakas et al., 1996; Ramm and Plückthun, 2000, 2001), and the cyclophilin type PpiA (RotA) (Liu and Walsh, 1990) (see Table 2).

Skp

The periplasmic seventeen-kilodalton protein, Skp (141 residues, 15.7 kDa), was identified as the major component of a mixture of periplasmic proteins that bound to sepharose-linked unfolded OMPs on affinity columns (Chen and Henning, 1996). *E. coli* cells lacking the *skp* gene display reduced levels of OmpA, OmpC, OmpF, and LamB in the OM (Chen and Henning, 1996; Missiakas et al., 1996), a phenotype which resembles that of *surA* mutants (Missiakas et al., 1996; Rouvière and Gross, 1996). Furthermore, Skp was found to improve the functional expression of a soluble fragment of the antibody 4-4-20 (Bedzyk et al., 1990; Whitlow et al., 1995) in the periplasm of *E. coli* (Bothmann and Plückthun, 1998). Skp almost completely prevents the aggregation of the soluble protein lysozyme at a molar ratio of 3:1 Skp/lysozyme (Walton and Sousa, 2004), consistent with previous observations on the 3:1 stoichiometry of Skp binding to OmpA (Bulieris et al., 2003).

Skp appears to be under control of both the σ^E (Alba and Gross, 2004; Ehrmann and Clausen, 2004) and the two-component CpxA/CpxR (Duguay and Silhavy, 2004) stress-response systems (Dartigalongue et al., 2001). Skp forms stable homotrimers in solution as determined by gel-filtration and cross-linking experiments (Schlapschy et al., 2004). The protein is highly basic with a calculated pI between 9.6 and 10.3 (depending on the algorithm used). The structure of the Skp trimer (Korndörfer et al., 2004; Walton and Sousa, 2004) (see Color Plate 3A) resembles a jellyfish with α-helical tentacles protruding about 60 Å from a β-barrel body and defining a central cavity. The entire Skp trimer is about 80 Å long and 50 Å wide. The Skp monomer has two domains. The small association domain (residues 1 to 21 and 113 to 141 of the mature sequence) is composed of three β-strands and two short α-helices, forms the limited hydrophobic core, and mediates the trimerization of Skp. The second, tentacle-shaped α-helical domain is formed by amino acids 22 to 112.

TABLE 2 Proteins suggested to be involved in assembly of outer membrane proteins

Protein	Organism	Residues[a]	Molecular mass (kDa)	Est. pI	Class	Oligomeric state	Essential	PDB entry	Reference(s)
Periplasmic proteins									
Skp[b]	E. coli	141	16	9.5	Chaperone	Trimer	No	1U2M, 1SG2	Korndörfer et al., 2004; Walton and Sousa, 2004
SurA[b]	E. coli	408	45	6.1	Chaperone/PPIase	Monomer	No	1M5Y	Bitto and McKay, 2002
DegP[b]	E. coli	448	47	7.9	Chaperone/Protease	Hexamer	No[c]	1KY9	Krojer et al., 2002
FkpA[b]	E. coli	245	26	6.7	Chaperone/PPIase	Dimer	No	1Q6U,1Q6H, 1Q6I	Saul et al., 2003, 2004
PpiA[b]	E. coli	166	18	8.2	PPIase	Monomer	No	1CLH, 1CSA, 1J2A	Clubb et al., 1994; Fejzo et al., 1994; Konno et al., 2004
PpiD[b]	E. coli	623	68	4.9	PPIase	Monomer	No		Dartigalongue and Raina, 1998
DegS	E. coli	327	35	5.0	Protease	Trimer	Yes	1SOT, 1SOZ, 1VCW, 1TE0	Wilken et al., 2004; Alba et al., 2001
Outer membrane proteins									
YfgL	E. coli	373	40	4.6	OM lipoprotein		No		Ruiz et al., 2005
YfiO	E. coli	226	26	5.5	OM lipoprotein		Yes		Wu et al., 2005
NlpB	E. coli	320	34	5.0	OM lipoprotein		No		Wu et al., 2005
YaeT	E. coli	790	88	4.9	Integral OMP		Yes		Doerrler and Raetz, 2005; Werner and Misra, 2005; Wu et al., 2005
Omp85	N. meningitidis	797	85	8.6	Integral OMP		Yes		Genevrois et al., 2003; Gentle et al., 2004; Voulhoux et al., 2003
HMW1B	H. influenzae	545	61	9.3	Integral OMP	Tetramer ?	No		Surana et al., 2004
Imp	N. meningitidis	802	89	8.8	Integral OMP				Bos et al., 2004
Imp	E. coli	760	87	4.9	Integral Omp		Yes		Braun and Silvahy, 2002

[a]With the exception of PpiD and the proteins from *N. meningitidis* and *H. influenzae*, parameters are given for the mature protein sequence. Number of residues, molecular mass, and pI were obtained from the SWISS-PROT database.

[b]While Skp, SurA, and DegP are individually not essential, double-null mutations in *skp* and *surA* (Rizzitello et al., 2001) and in *degP* and *surA* are (Rizzitello et al., 2001). Strains in which the four genes *surA*, *fkpA*, *ppiD*, and *ppiA* were deleted simultaneously were viable (Justice et al., 2005).

[c]DegP is essential for growth at elevated temperatures (Lipinska et al., 1990).

This domain is conformationally flexible. The charge distribution on the Skp surface gives the trimer an extreme dipole moment of ~3,700 Debye (770 eÅ) (Korndörfer et al., 2004), with positive charges all over the tentacle domain and, in particular, at the tips of the tentacle-like helices, while negative surface charge is found in the association domain. The surface of the tentacle-shaped domain contains hydrophobic patches inside the cavity formed by the tentacles. It may be that Skp binds its substrates in this central cavity (Korndörfer et al., 2004; Walton and Sousa, 2004). While the size of the cavity could be large enough to accommodate the transmembrane domain of OmpA in a folded form (Korndörfer et al., 2004), biochemical and spectroscopic data suggest that the OmpA barrel domain is largely unstructured when in complex with Skp (Bulieris et al., 2003). Also, the cavity would not be large enough for folded β-barrels of other OMPs to which Skp also binds, as shown for OmpF (Chen and Henning, 1996) and, in cross-linking experiments, for LamB and PhoE (Harms et al., 2001; Schäfer et al., 1999). As described later in this chapter, there is evidence for a connection between the function of Skp and LPS. Skp has a putative LPS binding site (Walton and Sousa, 2004) that was found by using a previously identified LPS binding motif (Ferguson et al., 2000). The binding site is formed on the surface of each Skp monomer by residues K77, R87, and R88, similar to the LPS binding motif in FhuA with residues K306, K351, and R382. Q99 in Skp may also form a hydrogen bond to an LPS phosphate, completing the four-residue LPS binding motif.

The structure of Skp resembles that of prefoldin (Pfd, GimC) of the archaeon *Methanobacterium thermotrophicum* (Siegert et al., 2000). However, in contrast to Skp, Pfd has a negative electrostatic surface potential and the pI is 4.6 (Devereux et al., 1984; Korndörfer et al., 2004). Pfd prevents aggregation of proteins in the cytosol and delivers them to class II cytosolic chaperonins (also called c-cpn, CCT, or TriC) independent of the presence of ATP or other nucleotides (Vainberg et al., 1998). Although prefoldin is a hexamer ($\alpha_2\beta_4$), it can be seen as

a dimer of trimers (Walton and Sousa, 2004), with the trimers resembling the Skp trimer. Electron microscopy reconstructions of Pfd bound to unfolded substrates have shown that the substrates are located within the cavity formed by the α-helical tentacles, where they appear to bind to the tips of the helices (Lundin et al., 2004; Martin-Benito et al., 2002). Eukaryotic prefoldin only binds stably to nascent actin chains if they are at least 145 residues long, suggesting a synergistic action of multiple weaker binding sites (Siegert et al., 2000). Pfd efficiently prevented aggregation of denatured rhodanese at a 1:1 molar ratio of the Pfd hexamer to rhodanese (Siegert et al., 2000), similar to the substrate binding stoichiometry of trimeric Skp.

Skp was found to insert into monolayers of negatively charged lipids (de Cock et al., 1999b). Consistent with this observation, two forms of Skp could be distinguished based on their sensitivity to proteolysis with trypsin or proteinase K: a free periplasmic form that is degraded and a form that is protected against digestion by association with membrane phospholipids (de Cock et al., 1999b). The presence of LPS in digestion experiments reduced the relative amount of protease-resistant Skp (de Cock et al., 1999b). Skp binds to the NH_2-terminal transmembrane β-barrel of OmpA in its unfolded form and is required for the release of OmpA into the periplasm (Schäfer et al., 1999). Skp binds neither to folded OmpA nor to the periplasmic domain (Chen and Henning, 1996), suggesting that Skp recognizes nonnative structures of OMPs. The *skp* gene maps at the 4-min region on the chromosome and is located upstream of genes that encode proteins involved in lipid A biosynthesis (Dicker and Seetharam, 1991; Roy and Coleman, 1994; Thome et al., 1990), an essential component of LPS of the OM. The gene *firA*, which codes for UDP-3-O-[3-hydroxymyristoyl]-glucosamine-*N*-acyltransferase starts only 4 bases downstream of the *skp* stop codon (Bothmann and Plückthun, 1998). The presence of a putative binding site for LPS in Skp (Walton and Sousa, 2004) could be related to the location of *skp* close to *firA*.

SurA

A third periplasmic protein, the survival factor A, SurA, has been demonstrated to affect OMP assembly. SurA was first identified as necessary for stationary-phase survival (Tormo et al., 1990). *E. coli* mutants in which the *surA* gene is deleted have reduced concentrations of OmpA and LamB in the OM (Lazar and Kolter, 1996; Rouvière and Gross, 1996). SurA⁻ strains are constitutively induced for the σ^E-dependent extracytoplasmic stress response (Missiakas et al., 1996; Rouvière and Gross, 1996), one of two signal transduction pathways known to communicate the folding state in the periplasm to the cytoplasm (Connolly et al., 1997; Danese and Silhavy, 1997; Mecsas et al., 1993).

The crystal structure of SurA is shown in Color Plate 3B. SurA consists of an N-terminal domain (N), which is composed of 148 amino acids and contains the α-helices H1 to H6. This domain is connected to the domain P1 (residues 149 to 260) and the domain P2 (residues 261 to 369). P2 connects the P1 domain to the C-terminal domain C (residues 370 to 428). Together, the N and C domains constitute a compact core with a broad deep crevice of about 50 Å in length. The P2 domain is tethered to this core by two extended peptide segments. The P1 and P2 domains have sequence similarity to parvulin, a cytoplasmic PPIase (Rahfeld et al., 1994).

The function of SurA in the transport and folding pathway to the OM was first assigned to its activity as a PPIase through its two parvulin-like domains, P1 and P2 (Color Plate 3B) (Lazar and Kolter, 1996; Missiakas et al., 1996; Rouvière and Gross, 1996). Only the P2 domain displayed PPIase activity in assays with reduced and carboxymethylated RNase T1 variant (RCM-T1), as demonstrated with SurA mutants lacking either the P1 or the P2 domain (Behrens et al., 2001). However, with a mutant form of SurA, from which the PPIase domains P1 and P2 were removed, it was shown that the N domain containing helices H1 to H6 functions as chaperone when linked with the C helix. A plasmid containing the gene for this mutant restored wt-SurA function and eliminated the extracytoplasmic stress response seen by activation of σ^E in *surA* deletion strains (Behrens et al., 2001). Since the two parvulin domains were obviously not necessary for SurA function in the maturation of OMPs, the SurA mutant lacking the parvulin domains was tested for function as a chaperone. In light-scattering assays on the aggregation of soluble citrate synthase during thermal stress, the presence of a 64-fold molar excess of SurA eliminated citrate synthase aggregation (Behrens et al., 2001).

The N-terminal amino acids 21 to 133 of SurA were able to bind peptides independent of the presence of proline (Webb et al., 2001). The SurA "core domain" has been proposed to bind the tripeptide motif aromatic-random-aromatic, which is prevalent in the aromatic girdles of β-barrel membrane proteins (Bitto and McKay, 2003). Isothermal titration calorimetry revealed that both the SurA and the SurA core domain bind a heptameric peptide with the sequence WEYIPNV with high affinity in the range of 1 to 14 μM (Bitto and McKay, 2003). Both forms of SurA exhibited affinity to the peptide consensus motif aromatic-polar-aromatic-nonpolar-proline (Ar-Pol-Ar-NonPol-Pro). Ar-X-Ar tripeptide motifs, where X can be any amino acid residue, are found with high frequency in OMPs, in particular, in two aromatic girdles close to the polar-apolar interfaces of the lipid bilayer. On average, the Ar-X-Ar motif is found about twice as often in membrane proteins as in cytoplasmic or periplasmic proteins (Bitto and McKay, 2003). For example, the number of Ar-X-Ar motifs in the β-barrel domains of OMPs is 7 for OmpF, 10 for LamB, 3 for OmpA, and 1 for TolC. Consistent with this, SurA bound unfolded OmpF and OmpG one order of magnitude more tightly than the soluble protein reduced carboxymethylated lactalbumin (Bitto and McKay, 2004). DegS binds the Ar-X-Ar-COOH motif with affinities in a range of 0.6 to 15 μM (Walsh et al., 2003), which is similar to the affinity of SurA to the WEYIPNV peptide and stronger than the affinity of SurA to the Ar-X-Ar motif (Bitto and McKay, 2003). Binding to SurA required a minimum of five amino acids, and a proline at position 5 increased the binding affinity (Bitto and McKay, 2003). A recent study indicated that, in addition to the presence of the Ar–X–Ar motif, the orientation of the amino acid side chains is important (Hennecke et al., 2005).

Although deletion of *surA* led to decreased levels of the OMPs LamB and OmpA in vivo (Lazar and Kolter, 1996; Rouvière and Gross, 1996), SurA had no effect on the assembly of TolC in vivo (Werner et al., 2003) or of AIDA in vitro (Mogensen et al., 2005), which contain only one and two Ar–X–Ar motifs, respectively, in their β-barrel domains. In preliminary experiments in the authors' laboratory, SurA facilitated the membrane insertion and folding of OmpA into preformed lipid bilayers (P. V. Bulieris, S. Behrens, and J. H. Kleinschmidt, manuscript in preparation). More recently, it was found that *surA* deletion severely diminished the expression of P and type 1 pili, which are adhesive cell surface organelles produced by uropathogenic strains of *E. coli* and assembled in the chaperone/usher pathway (Justice et al., 2005).

Genetic evidence suggests that SurA and Skp act as chaperones that are involved in parallel pathways of OMP targeting to the OM (Rizzitello et al., 2001). Null mutations in *skp* and *surA* as well as in *degP* and *surA* resulted in synthetic phenotypes. The *skp surA* null combination had a bacteriostatic effect and led to filamentation, while the *degP surA* null combination was bactericidal. It was suggested that the redundancy of Skp, SurA, and DegP is in the periplasmic chaperone activity, in which Skp and DegP are components of one pathway and SurA is a component of a parallel pathway. While the loss of either pathway was tolerated, the loss of both pathways was lethal (Rizzitello et al., 2001).

DegP

The widely conserved periplasmic DegP (also called HtrA, 448 residues, 47 kDa) is a heat shock protein and a member of the HtrA family of proteases (Lipinska et al., 1990; Spiess et al., 1999; Strauch et al., 1989), for which, in addition, a chaperone activity has been discovered (Spiess et al., 1999). Transcription of *degP* in response to periplasmic stress is under control of σ^E stress response system. Its double function as a protease and chaperone makes DegP an interesting quality control protein in the periplasm. In refolding experiments with soluble proteins as substrates, equimolar concentrations of DegP or DegP$_{S210A}$, which lacks protease activity due to substitution of serine by alanine in the active site, led to functional refolding of citrate synthase at 28°C (Spiess et al., 1999). Similarly, DegP facilitated refolding of periplasmic MalS, an α-amylase that hydrolyzes maltodextrins. Folding of MalS by DegP$_{S210A}$ was facilitated at moderate temperatures of 28°C or below, while experiments at 37°C or higher had only a very minor stimulating effect on the reactivation of MalS. By contrast, the proteolytic activity of wild-type DegP increased dramatically above 30°C (Spiess et al., 1999).

Two different conformational states of DegP, corresponding to the open and closed state of the homo-hexamer, were observed in the X-ray crystal structure of DegP (1KY9), which has been solved (Krojer et al., 2002). One DegP subunit is shown in Color Plate 4A. One DegP molecule consists of a protease domain and two protein binding domains, the PDZ domains. PDZ domains typically bind sequence specific a short C-terminal protein sequence, which folds into β-finger (Sheng and Sala, 2001). Six DegP molecules form a complex that consists of two loosely stacked rings of trimers (Krojer et al., 2002). The protease domains are located in a central cavity, while the 12 PDZ domains form the mobile side walls. These PDZ-domain side walls bind the protein substrates and mediate the opening and closing of the hexamer, allowing the exclusively lateral access of the substrate to the inner cavity formed by the two stacked rings of DegP trimers. The inner cavity is lined by hydrophobic residues, which are proposed to act as docking sites for unfolded proteins in the chaperone state of DegP. The protease domain is not active when DegP is in the chaperone conformation and further substrate binding is not possible. The structural organization of the DegP hexamer differs from other known cage-forming proteins, where access is allowed through narrow axial or lateral pores. The PDZ domains of the neighboring subunits interact with each other and form gatekeepers for the inner chamber (Krojer et al., 2002).

There are currently no experimental studies on OMP assembly that demonstrate a direct chaperone effect of DegP on the concentrations of TMPs in the OM. However, it was recently observed that DegP tightly binds to OmpC (M. Ehrmann, personal communication). Also, the chaperone function of OmpC was important to maintain viability of cells producing assembly-deficient mutants of OmpF, in which the conserved carboxy-terminal phenylalanine was nonconservatively replaced by dissimilar amino acids, leading to accumulation of monomeric, unfolded OmpF in the periplasm (Misra et al., 2000). In this study, the mutant $DegP_{S210A}$ (Spiess et al., 1999) was used. Overproduction of $DegP_{S210A}$ led to mislocalization of OmpF to the inner membrane. No effect of DegP or $DegP_{S210A}$ could be observed in folding studies with the autotransporter AIDA, which forms a transmembrane β-barrel domain in the OM. Instead, the presence of DegP resulted in proteolytic cleavage of AIDA (Mogensen et al., 2005). In contrast to observations with OmpA (Bulieris et al., 2003), the chaperone Skp also did not bind to AIDA and it did not affect the folding of the transmembrane β-barrel domain of AIDA (Mogensen et al., 2005). Therefore, further experiments are needed to clarify whether DegP directly affects the folding of OMPs into membranes.

FkpA

The structure of FkpA (245 residues, 26 kDa) has been determined in three different forms (Saul et al., 2004): for wild-type FkpA, for a truncated mutant lacking the last 21 residues, FkpAΔCT, and for this mutant (see Color Plate 4B) in complex with the macrolide FK506, which is an immunosuppressant. FkpA forms V-shaped dimers and the 245-residue subunit is divided into two domains. The N-terminal domain (residues 15 to 114) includes three α-helices that are interlaced with those of the other subunit. The C-terminal domains (residues 115 to 224) are located at the two ends of the V. The C domain, which is structurally similar to human FKBP12, belongs to the FKBP-type of PPIases, well-characterized PPIases that are inhibited by FK506. The two FKBP type C domains are held apart by the long α-helix 3 of the N domains, with their FK506 binding sites facing toward each other. In the FK506-bound FkpAΔCT crystal structure the centers of the ligands are about 49 Å apart from each other.

FkpA isomerase activity was measured in a protein-folding assay with Rnase T1 and determined to $k_{cat}/K_m = 4,000$ mM^{-1}·s^{-1} (Ramm and Plückthun, 2000). This underlines the efficiency of FkpA compared with other FKBPs, such as FKBP12 with $k_{cat}/K_m = 800$ mM^{-1}·s^{-1} (Dolinski et al., 1997) and trigger factor with $k_{cat}/K_m = 740$ mM^{-1}·s^{-1} (Stoller et al., 1995). In comparison, activities of the cyclophilin PpiA were determined to $k_{cat}/K_m = 6,000$ mM^{-1}·s^{-1} (Compton et al., 1992) and to 10,000 mM^{-1}·s^{-1} (Pogliano et al., 1997). The activity of the parvulin type SurA was $k_{cat}/K_m = 30$ to 40 mM^{-1}·s^{-1} (Behrens et al., 2001) and that of PpiD was $k_{cat}/K_m = 400$ to 3,400 mM^{-1}·s^{-1} (Dartigalongue and Raina, 1998).

A selection system for periplasmic chaperones (Bothmann and Plückthun, 1998), using a genomic library from an *skp* deletion strain (Bothmann and Plückthun, 2000), led to the discovery of FkpA as a folding promoter for the single-chain antibody fragment (scFv) of the antibody 4-4-20 (Bedzyk et al., 1990; Whitlow et al., 1995). Subsequently, coexpression of FkpA was also shown to improve functional expression of scFv (Bothmann and Plückthun, 2000), even for those antibody fragments that did not contain any *cis*-prolines, suggesting that the effect of FkpA as a folding facilitator is independent of its PPIase activity (Bothmann and Plückthun, 2000). In contrast, SurA or PpiA showed no increase in the expression levels of functional scFv, neither upon coexpression in the periplasm nor by the phage display method. The chaperone function of FkpA was also observed with the periplasmic protein MalE31, a maltose binding protein with a high tendency to aggregation (Arie et al., 2001), and could be assigned to the N-terminal domain, which exists in solution as a mixture of monomers and dimers (Saul et al., 2004), while the C-terminal FKBP domain contained the PPIase function (Saul et al., 2004). Light-scattering assays indicated that an 8-fold molar excess of FkpA reduced citrate synthase aggregation to a high extent and delayed it considerably (Ramm and Plückthun, 2001).

Homologues of FkpA are found in many pathogenic bacteria such as *Legionella pneumophila* (Horne and Young, 1995). FkpA is not essential, but *fkpA⁻* strains display an increased permeability of the OM to certain detergents and to antibiotics (Missiakas et al., 1996).

Periplasmic DegP is upregulated in *fkpA⁻* strains, suggesting that FkpA is involved in folding of proteins in the periplasm. Transcription of *fkpA* is under control of the alternative σ-factor σ^E (Danese and Silhavy, 1997; Dartigalongue et al., 2001; Missiakas et al., 1996). The elevated expression of the σ^E regulon that is induced in *htrM (rfaD)* deletion mutants, i.e., in strains producing a modified form of LPS, is turned off by overexpression of FkpA or SurA. This was monitored using *htrM* mutants, in which the gene for β-galactosidase, *lacZ*, was fused to σ^E-transcribed promoters (Missiakas et al., 1996; Raina et al., 1995). Such *htrM* mutants display increased activity of β-galactosidase because of increased σ^E stress response. The elevated activity of β-galactosidase was reduced upon overexpression of FkpA or SurA in these *htrM* mutants. A similar observation was made for *htrA dsbC* double-null mutants (Missiakas et al., 1996).

To date, a direct effect of FkpA on the assembly of OMPs has not been demonstrated, and FkpA will be an interesting target for future studies.

PpiA (Rotamase A, RotA)

PpiA (16 kDa, 166 residues), which is sometimes also called RotA, is a member of the cyclophilin class of PPIases (Hayano et al., 1991;

Liu and Walsh, 1990). Both the NMR solution structure of PpiA (Clubb et al., 1994; Fejzo et al., 1994) and the X-ray crystal structure (Konno et al., 2004) have been solved. PpiA is expressed under control of the two-component CpxR-CpxA stress response system (Pogliano et al., 1997). It is homologue to human cyclophilin (34% sequence identity) and has similar PPIase activity. The catalytic activity is close to the upper diffusional limit with $k_{cat}/K_m \sim 1.0 \times 10^7$ M⁻¹s⁻¹ at 10°C, but unlike human cyclophilin, PpiA is not inhibited by cyclosporin A (Pogliano et al., 1997), which binds to cyclophilin with high affinity ($K_d = 17$ nM), but to PpiD with low affinity ($K_d = 3.4$ μM) (Fejzo et al., 1994). PpiA is not essential (Kleerebezem et al., 1995) and deletion of *ppiA* did neither affect the levels of integral proteins in the OM nor the levels of soluble periplasmic proteins. Also, rates of OMP integration into the OM in vivo were not affected according to pulse-chase experiments, suggesting that PpiA does not play an important role in protein folding (Kleerebezem et al., 1995).

PpiD

PpiD belongs to the parvulin class of PPIases (Dartigalongue and Raina, 1998). Transcription is controlled by the CpxAR stress response system (Danese et al., 1995; Danese and Silhavy, 1997; Dartigalongue and Raina, 1998). PpiD (68 kDa, 623 residues) is anchored to the inner membrane by a transmembrane helix. Recently it was reported that inactivation of all four genes *ppiD*, *surA*, *ppiA*, and *fkpA* encoding the known periplasmic PPIases resulted in a viable strain. This strain had reduced levels of the OMPs LamB and OmpA, with OMP levels similar to that of the *surA* null mutant (Justice et al., 2005). This result suggests that neither PpiD, nor FkpA, nor PpiA has a direct role in OMP folding.

Other Periplasmic Chaperones

In addition to the chaperones discussed above, other classes of chaperones are in the periplasm. For example, the PapD-like superfamily of chaperones, such as PapD and FimC, are important in the chaperone/usher pathways for the assembly of bacterial surface organelles such as fimbria/pili, which are involved in cell-surface attachment and cell-cell contacts. The PapD-like chaperones act very specifically with nascent pilus (FGS chaperones) or nonpilus (FGL chaperones) subunits as they emerge from inner membrane translocon, and they subsequently deliver the subunits to the corresponding usher pore in the OM, at which the subunits are assembled into linear fibers before their secretion through the pore. A detailed discussion of these specialized chaperones is beyond the scope of this chapter. For a review, see Thanassi (2002). Another class of periplasmic chaperones is involved in transport of outer membrane lipoproteins across the periplasm to the OM. An example is the lipoprotein chaperone LolA (Taniguchi et al., 2005). Most of the 90 different species of lipoproteins are known to be associated with the periplasmic side of the OM, to which they are anchored by a lipoid linked to the N-terminal cysteine of the lipoprotein. For a recent review on the transport of outer membrane lipoproteins to the OM, see Tokuda and Matsuyama (2004).

OMPs INVOLVED IN OMP ASSEMBLY

Omp85, YaeT, and HMW1B

It was discovered recently that Omp85 of the OM of *Neisseria meningitidis* is essential for cell growth (Voulhoux et al., 2003). Omp85 was necessary for the integration of several OMPs, such as the trimeric porins PorA and PorB, the heterooligomeric complex of the siderophore receptor FrpB/RmpM, or the outer membrane phospholipase A (OmpLA). In strains depleted of Omp85, the passenger domain of the immunoglobulin A1 protease autotransporter (IgA1) was difficult to detect, while the full-length autotransporter accumulated, indicating that the passenger domain of IgA1 was not cleaved, possibly because the β-barrel domain of IgA1 did not insert into the OM to allow the translocation of the passenger domain. Oligomers of the secretin PilQ, which plays a role in type IV pili formation, were strongly re-

duced in the OM, while monomeric, likely unfolded PilQ was accumulated. Immunofluorescence microscopy with antibodies directed against PorB and PilQ and Alexa fluorochrome-conjugated secondary antibodies indicated reduced surface exposure of the OMPs PorB and PilQ in Omp85 deletion strains. A direct interaction of immobilized Omp85 on nitrocellulose membranes with denatured PorA was demonstrated with PorA antibodies. All these observations indicated that OMPs were not correctly inserted into the OM in *omp85*-deletion strains (Voulhoux et al., 2003). Omp85 is a highly conserved protein in gram-negative bacteria with homologues also in eukaryotic cells, such as Tob55 in mitochondria (Gentle et al., 2004; Paschen et al., 2003) and Toc75 in chloroplasts (Voulhoux et al., 2003).

Similar to depletion of Omp85 of *N. meningitidis*, depletion of homologue YaeT in *E. coli* also led to severe defects in the biogenesis of OMPs (Werner and Misra, 2005). OMPs were found to accumulate in the periplasm, suggesting that YaeT facilitates the insertion of soluble periplasmic intermediates into the OM, supporting the role of Omp85 and its homologues in other organisms in the targeting or assembly of TMPs into OMs (Werner and Misra, 2005).

Another homologue to Omp85 of *N. meningitidis* and YaeT of *E. coli* is HMW1B of *Haemophilus influenzae* (Surana et al., 2004). HMW1B has a pore size of about 2.7 nm and was found to form multimers, possibly tetramers. HMW1B is critical for secretion of the *H. influenzae* adhesin HMW1 and interacts with the N terminus of the adhesin.

It was also suggested that Omp85 may be involved in the transport of LPS and phospholipids to the OM (Genevrois et al., 2003) and that the effect of Omp85 on OMP assembly may be indirect. Omp85 is located in a cluster of genes involved in lipid A, fatty acid, and phospholipid biosynthesis, which would support a role in transport of LPS or phospholipid. However, the gene encoding for the periplasmic chaperone Skp (Bothmann and Plückthun, 1998; Chen and Henning, 1996) is located in the same cluster, and the Skp trimer

(Korndörfer et al., 2004; Walton and Sousa, 2004) has been shown to bind unfolded OMPs (Bulieris et al., 2003; Chen and Henning, 1996). It has been proposed that Skp also contains an LPS binding site (Walton and Sousa, 2004). Deletion of *omp85* in *N. meningitidis* led to a significant portion of cells showing signs of lysis in electron microscopy (Genevrois et al., 2003). These cells also showed accumulation of vesicular lipid structures and of electron-dense, proteinaceous material in the periplasm, suggesting that these components were no longer incorporated into the OM in the absence of Omp85. Electron microscopy also indicated some LPS accumulation at the outside of the OM, verified by a cytochemical reaction for polysaccharide, while a modified form of LPS was found to cofractionate with the inner membrane fraction in isopycnic sucrose gradient centrifugation. Radioactive labeling of phospholipids indicated that the phospholipid distribution between inner and outer membrane fractions was shifted toward the inner membrane in Omp85-depleted strains, suggesting a role of Omp85 in phospholipid transport toward the OM (Genevrois et al., 2003).

While a role of Omp85 of *N. meningitidis* in the transport of phospholipids and LPS toward the OM was reported, another study with a temperature-sensitive mutant of *E. coli*, carrying 9 amino acid substitutions in YaeT, displayed no alterations in the phospholipid and LPS compositions under permissive and nonpermissive temperatures for bacterial growth (Doerrler and Raetz, 2005). In this study, levels of OMPs in the OM of the mutant were low. Moderately overexpressed SecA, a cytoplasmic ATPase, was found to function as a multicopy suppressor of the temperature-sensitive phenotype, partially restoring the level of OMPs in the OM. One likely explanation for this was that SecA may be compensating for a decrease in secretion levels of mutant YaeT at elevated temperatures in the temperature-sensitive mutant. OMPs accumulated neither in the periplasm nor in the inner membrane, and it was suggested that most likely they were degraded (Doerrler and Raetz, 2005).

Imp

Besides *N. meningitidis* Omp85, another integral OMP, the increased membrane permeability protein (Imp), also called the oganic solvent tolerance protein A (OstA), was shown to play an important role in the assembly of the OM in *E. coli* (Braun and Silhavy, 2002) and later also in *N. meningitidis* (Bos et al., 2004). While Imp is essential in *E. coli*, it is not in *N. meningitidis*, which, in contrast to most other gram-negative bacteria, is viable even when LPS is not synthesized (Steeghs et al., 2001). Deletion of *imp* in *E. coli* resulted in the appearance of a novel high-density membrane fraction that contained properly folded OMPs in sodium dodecyl sulfate (SDS)-stable conformation (Braun and Silhavy, 2002). *E. coli* Imp cofractionated with OMPs such as LamB, but Imp was expressed at much lower levels. The *imp* gene encodes an 87-kDa protein composed of 784 amino acids and sequence analysis predicted that Imp has a high content of β-sheet secondary structure and is disulfide bonded in a high-molecular-weight complex (Braun and Silhavy, 2002). The *imp* gene is located upstream of the gene *surA*, which encodes a periplasmic PPIase with chaperone activity (Behrens et al., 2001), and both genes are cotranscribed. Imp depletion in cells led to filamentation, followed by membrane rupture and cell lysis (Braun and Silhavy, 2002). *Imp* deletion strains of *N. meningitidis* produced much lower amounts of LPS (Bos et al., 2004). Accessibility assays with LPS-modifying enzymes, neuraminidase and PagL, demonstrated that the residual LPS was not exposed to the cell surface, suggesting that Imp facilitates the integration of LPS into the outer leaflet of the OM. However, since *imp* deletion strains of neisseriae are viable (Bos et al., 2004), it is unlikely that Imp is involved in the transport of phospholipids, a role previously assigned to Omp85 (Genevrois et al., 2003).

YfgL, YfiO, and NlpB Outer Membrane Lipoproteins

YaeT of *E. coli* was found to be part of a larger complex, which contained the outer membrane lipoproteins YfgL, YfiO, and NlpB (Wu et al., 2005), suggesting that these lipoproteins are required for YaeT function. YfgL was identified by using chemical conditionality as a new genetic method to probe OM assembly (Ruiz et al., 2005). In this method, antibiotics were used in selection studies on strains with defects in OM permeability to create chemical conditions that demanded specific suppressor mutations to partially restore membrane impermeability against these antibiotics. The mutations were not in the target of the toxic molecule, but in the genes, which after mutation allowed survival of the cell by changing the composition and properties of the membrane to prevent the import of the antibiotics across the OM and therefore their action on the target molecules in the cell. The method was tested with an *imp* deletion strain of *E. coli* (Braun and Silhavy, 2002), with an increased permeability of the OM. These cells were sensitive to detergents and antibiotics (Sampson et al., 1989). Selections for resistance against the antibiotics chlorobiphenyl vancomycin and moenomycin, which both inhibit peptidoglycan biosynthesis, revealed only loss-of-function mutations in the gene *yfgL* (Eggert et al., 2001), which encodes a putative outer membrane lipoprotein. Lack of YfgL leads to changes in the OM that prevent the import of the antibiotics into the cell. YfgL deletion strains showed reduced levels of OMPs such as OmpA and LamB, suggesting a role of YfgL in OMP biogenesis (Ruiz et al., 2005). This effect is not as strong as the effect caused by depletion of the periplasmic chaperone and PPIase SurA. The reduction in LamB concentration is additive, because double mutants with deletions of *surA* and *yfgL* showed an even stronger reduction of LamB levels in the OM (Ruiz et al., 2005).

Coimmunoprecipitation experiments indicated that YfgL is present in a complex with YaeT and two other lipoproteins, NlpB and YfiO, which were identified by mass spectrometry. Essential for *E. coli* growth were YaeT (Werner and Misra, 2005; Wu et al., 2005) and YfiO (Onufryk et al., 2005), while YfgL and NlpB were not (Bouvier et al., 1991; Eggert et al., 2001; Ruiz et al., 2005).

IN VITRO STUDIES ON THE FOLDING OF OMPs

Folding of OMPs into Detergent Micelles

Many, but not all, outer membrane proteins that form transmembrane β-barrels can be successfully refolded in vitro from a urea-denatured state. Upon denaturation in urea at slightly elevated temperature, OMPs lose their quaternary, tertiary, and secondary structure, which was shown, for example, by circular dichroism (CD) spectroscopy. First, Henning and coworkers performed in vitro refolding studies of TMPs in 1978 and demonstrated that the 8-stranded β-barrel OmpA develops native structure when incubated with LPS and Triton X-100 after dilution of the denaturants SDS or urea (Schweizer et al., 1978). Similarly, it was later shown that after heat-induced unfolding in SDS micelles, OmpA refolds into micelles of the detergent octylglucoside even in absence of LPS (Dornmair et al., 1990). These results on the β-barrel OmpA, and the successful refolding of bacteriorhodopsin that consists of a bundle of seven transmembrane α-helices and was first refolded by Khorana and coworkers in 1981 (Huang et al., 1981), suggest that the information for the formation of native structure in TMPs is contained in their amino acid sequence, as previously described by the Anfinsen paradigm for soluble proteins (Anfinsen, 1973).

However, a difference in refolding experiments of α-helical and β-barrel membrane proteins is that the α-helical membrane proteins were first taken up in SDS-detergent micelles before they were refolded into mixed micelles of phosphatidylcholine lipid and the detergent 3-[(3-cholamidopropyl)-dimethylammonio]-1-propanesulfonate (CHAPS) or into lipid bilayers of phosphatidylcholine. Completely unfolded bacteriorhodopsin in organic solvent can be transferred into the denaturing detergent SDS, in which it develops a large degree of α-helical structure. It is not clear whether such an intermediate is an off-pathway product or a possible folding intermediate, since SDS is not found in cells.

Poor in vitro refolding of some OMPs upon denaturant dilution in the presence of pre-formed phospholipid bilayers appears to be a consequence of the fast aggregation of OMPs, which competes with bilayer insertion and folding. In vivo, molecular chaperones keep the OMPs soluble in the periplasm before they become part of the OM. The chaperones are likely more efficient at preventing OMP aggregation than the denaturant urea that has been used in folding studies in vitro and that must be diluted before OMPs can insert and fold into model membranes. In vivo, there must also be a targeting mechanism that prevents the insertion of OMPs from the periplasm into the cytoplasmic membrane and specifically directs them to the OM.

Folding of β-Barrel Membrane Proteins into Phospholipid Bilayers Is Oriented

Surrey and Jähnig (1992) showed that OmpA spontaneously inserts and folds into phospholipid bilayers upon denaturant dilution. Oriented insertion and folding of OmpA into lipid bilayers in the absence of detergent was observed when unfolded OmpA in 8 M urea was reacted with small unilamellar vesicles (SUVs) of 1,2-dimyristoyl-sn-glycero-3-phosphocholine ($diC_{14:0}PC$) under concurrent strong dilution of the urea. The insertion of OmpA into vesicles was oriented, because trypsin digestion was complete (100%) and resulted in a 24-kDa fragment, while the full-length OmpA (35 kDa) was no longer observed. Translocation of the periplasmic domain of OmpA across the lipid bilayer into the inside of the vesicle would have led to a full protection of OmpA from trypsin digestion. The 24-kDa fragment corresponded to the membrane-inserted β-barrel domain (19 kDa) and a smaller part of the periplasmic domain, which was largely digested by trypsin. By contrast, only 50% of detergent-refolded OmpA, which was reconstituted into $diC_{14:0}PC$ vesicles after refolding into micelles,

could be digested with trypsin, indicating random orientation of the periplasmic domain inside and outside of the phospholipid vesicles (Surrey and Jähnig, 1992).

For direct oriented insertion of OmpA into the bilayers, the preformed lipid vesicles had to be in the lamellar-disordered (liquid-crystalline) phase and the vesicles had to be sonicated (Rodionova et al., 1995; Surrey and Jähnig, 1995). By contrast, insertion and folding did not complete when the lipid bilayers were in the lamellar-ordered (gel) phase or when refolding attempts were made with $diC_{14:0}PC$ bilayers of large unilamellar vesicles (LUVs) that were prepared by extrusion through membranes with 100-nm pore size (Kleinschmidt and Tamm, 2002). Similarly, folding and trimerization of OmpF (Surrey et al., 1996) was observed after interaction of urea-unfolded OmpF with preformed lipid bilayers in the absence of detergent. Membrane-inserted dimers of OmpF were detected transiently. In vitro, the yields of folded OmpF in lipid bilayers are small (≲30 %), even under optimized conditions (Surrey et al., 1996) and when compared with OmpA, which quantitatively folds at pH 10. The pH dependence observed for the folding of OMPs in vitro is typically linked to an increased solubility of the OMPs, when acidic or basic amino acid side chains are present in their charged form. Refolding experiments in vitro therefore result in better folding yields when carried out at a pH that is sufficiently different from the pI of the protein. The pI of OmpA and of other OMPs of E. coli or Salmonella enterica serovar Typhimurium is in between 5 and 6, while OMPs of N. meningitidis or Comamonas acidovorans have a pI in the basic pH region (see Table 1). In vivo, solubility of OMPs is conferred by binding to a periplasmic chaperone.

β-Barrel Structure Formation Requires a Supramolecular Assembly of Amphiphiles

To determine basic principles for the folding of β-barrel TMPs, folding of OmpA was examined with a large set of different phospholipids and detergents at different concentrations (Kleinschmidt et al., 1999b). Folding of OmpA was successful with 64 different detergents and phospholipids that had very different compositions of the polar headgroup, did not carry a net charge, and had a hydrophobic carbon chain length ranging from 7 to 14 carbon atoms. In all cases, folding yields were near 100% at pH 10 (Kleinschmidt et al., 1999b), but folding kinetics of OmpA were different for the different detergents (unpublished results). For OmpA folding, the concentrations of these detergents or phospholipids must be above the critical micelle concentration (CMC), demonstrating that a supramolecular assembly (micelles or lipid bilayers) with a hydrophobic interior is the minimal requirement for the formation of a β-barrel transmembrane domain (Kleinschmidt et al., 1999b). In these experiments, OmpA folding was monitored by CD spectroscopy and by electrophoretic mobility measurements. Both methods indicated that after exposure to amphiphiles with short hydrophobic chains (with 14 or fewer carbons), OmpA assumes either both, secondary and tertiary structure (i.e., the native state) or no structure at all, dependent on the presence of supramolecular assemblies (micelles, bilayers). Thermodynamically, OmpA folding into micelles is a two-state process (Kleinschmidt et al., 1999b).

The necessary presence of amphiphiles (lipids, detergents) above the critical concentration for assembly (CCA) to induce the formation of native secondary and tertiary structure in OmpA also indicated that β-barrel structure does not develop while detergent or lipid monomers are adsorbed to a newly formed hydrophobic surface of the protein, i.e., absorption of single amphiphiles one after another will not help to form natively structured parts in the protein. To the contrary, a hydrophobic core of a micelle or bilayer must be present to allow folding of OmpA. (The term CCA is defined here to describe the amphiphile concentration at which a geometrically unique, water-soluble supramolecular assembly is formed, which can be a micelle, a lipid vesicle, or even an inverted or cubic lipid phase. The CCA is identical with the CMC in the special case of micelle-forming detergents. The CCA does not refer to the formation of random aggregates [for instance, of misfolded membrane proteins].) Conlan and Bayley (2003) reported later that another OMP, OmpG, folds into a range of detergents such as Genapol X-080, Triton X-100, n-dodecyl-β-D-maltoside, Tween 20, and octylglucoside. However, OmpG did neither fold into n-dodecylphosphocholine nor into the negatively charged detergents SDS and sodium cholate. Similar to OmpA, the detergent concentrations had to be above the CMC for OmpG folding (Conlan and Bayley, 2003). Different detergents have also been used for refolding of other β-barrel membrane proteins for subsequent membrane protein crystallization (for an overview, see, for example, Buchanan [1999]).

Electrophoresis as a Tool To Monitor Insertion and Folding of β-Barrel Membrane Proteins

SDS-polyacrylamide gel electrophoresis (SDS-PAGE) according to Laemmli (1970) has been very useful in monitoring the folding of OmpA into detergent micelles or lipid bilayers, provided that the samples are not boiled prior to electrophoresis (Bulieris et al., 2003; Dornmair et al., 1990; Kleinschmidt and Tamm, 1996, 2002; Kleinschmidt et al., 1999b; Kleinschmidt, 2003; Schweizer et al., 1978; Surrey and Jähnig, 1992, 1995). If samples are not heat denatured prior to electrophoresis, the folded and denatured OMPs migrate differently. For OmpA, Henning and coworkers described this property as heat modifiability (Schweizer et al., 1978). This has also been reported for other OMPs of bacteria such as FhuA (Locher and Rosenbusch, 1997), OmpG (Behlau et al., 2001; Conlan et al., 2000; Conlan and Bayley, 2003), and FomA (Kleivdal et al., 1995; Pocanschi et al., 2006; Puntervoll et al., 2002; C . L. Pocanschi, T. Dahmane, Y. Gohon, F. Rappaport, H.-J. Apell, J. H. Kleinschmidt, and J.-L. Popot, submitted for publication). Native OmpA, for example, migrates at 30 kDa, whereas unfolded

OmpA migrates at 35 kDa (Schweizer et al., 1978). See also Kleinschmidt (2006) for further examples.

Until now, all structural and functional experiments have shown a strict correlation between the 30-kDa form and structurally intact and fully functional OmpA. These previous studies included analysis of the OmpA structure by Raman, Fourier transform infrared (FT-IR), and CD spectroscopy (Dornmair et al., 1990; Kleinschmidt et al., 1999b; Rodionova et al., 1995; Surrey and Jähnig, 1992, 1995; Vogel and Jähnig, 1986), biochemical digestion experiments (Kleinschmidt and Tamm, 1996; Surrey and Jähnig, 1992), and functional assays such as phage inactivation (Schweizer et al., 1978) and single-channel conductivity measurements (Arora et al., 2000). A similar strict correlation was observed recently also for the major OMP of *Fusobacterium nucleatum*, FomA (Pocanschi et al., 2006; Pocanschi et al., submitted). There are, however, some β-barrel membrane proteins that do not exhibit a different migration on SDS-polyacrylamide gels. Examples are the nucleoside transporter Tsx and the mitochondrial hVDAC1, which unfolded in SDS even at room temperature, as can be shown by CD spectroscopy (B. Shanmugavadivu, H.-J. Apell, K. Zeth, and J. H. Kleinschmidt, submitted for publication).

It is possible to determine the kinetics of native structure formation in OmpA (Kleinschmidt and Tamm, 1996, 2002; Surrey and Jähnig, 1995), FomA (Pocanschi et al., 2006), OmpG (Conlan and Bayley, 2003), and probably also in other OMPs using the different electrophoretic mobility of folded and unfolded OMPs. Although SDS inhibits folding of OMPs, it often does not unfold them unless samples are boiled.

In an assay to determine, for example, the OmpA-folding kinetics, OmpA that was denatured in 8 M urea was reacted with preformed lipid bilayers (vesicles) under concurrent strong dilution of the denaturant (Kleinschmidt and Tamm, 1996, 2002). At defined times after initiation of folding, small volumes of the reaction mixture were taken out and an equal volume of SDS-treatment buffer, typically used in SDS-PAGE, was added. In these samples, SDS bound quickly to folded and unfolded OmpA and stopped further OmpA folding (Kleinschmidt and Tamm, 1996, 2002), while already folded OmpA was not unfolded at room temperature.

In the end of the kinetic refolding experiment, the fractions of folded OmpA in all samples were determined by cold SDS-PAGE (i.e., without heat denaturing the samples). The fractions of folded OmpA at each time were estimated by densitometric analyses of the bands of folded and of unfolded OmpA, thus monitoring the formation of tertiary structure in OmpA as a function of time (Kinetics of Tertiary Structure Formation by Electrophoresis, KTSE). This method was also applied successfully to studying the folding kinetics of FomA (Pocanschi et al., 2006) and OmpG (Conlan and Bayley, 2003).

Role of LPS

The OM contains mostly LPS in the outer leaflet. LPS has relatively short hydrocarbon chains, which are partially hydroxylated close to the glucosamine backbone at carbon 3, lowering the hydrophobic thickness of the OM. Several periplasmic proteins and LPS have been demonstrated to interact with OMPs in the periplasm, and initial studies suggested that LPS is required for efficient assembly of OMPs such as monomeric OmpA (Freudl et al., 1986; Schweizer et al., 1978) and trimeric PhoE (de Cock and Tommassen, 1996; de Cock et al., 1999a) into OMs.

Further evidence for a role of LPS came from genetic studies. In *rfa* mutants, the σ^E-dependent stress response was activated. The *htrM* (*rfaD*) gene product was shown to encode an ADP-L-glycero-D-mannoheptose-6-epimerase, an enzyme required for the biosynthesis of an LPS precursor (Pegues et al., 1990; Raina and Georgopoulos, 1991). Lack of the core heptose moiety in *htrM* mutants led to an altered LPS (Missiakas et al., 1996). In such mutants, the assembly of certain OMPs was affected because of the absence of proper LPS (Nikaido and Vaara, 1985; Schnaitman and Klena, 1993).

Also, the rate of OMP synthesis was decreased (Ried et al., 1990).

Together with divalent cations, LPS was reported to facilitate trimerization of PhoE in mixed micelles of Triton X-100 detergent in vitro (de Cock and Tommassen, 1996; de Cock et al., 1999a). However, in these studies, experiments were performed with micelles of LPS and Triton X-100 instead of phospholipid bilayers. It was later found that monomeric OmpA folds relatively fast into micelles but with rather slow kinetics into phospholipid bilayers (Surrey and Jähnig, 1995; Surrey et al., 1996). Proteins that are already folded in micelles may be easily inserted into OMs, especially when the membranes are present in large excess, but would end up with a random orientation in the bilayer, i.e., loops could be exposed to the periplasmic space instead of their normal orientation to the outer space when OMP-micelle complexes fuse with membranes. Since OmpA assumed a random orientation after micelle-bilayer fusion (Surrey and Jähnig, 1992), it is unlikely that OmpA would first fold into LPS micelles in the periplasm, which then fuse with the OM as first proposed for PhoE based on the appearance of a folded monomer in mixed micelles of LPS and Triton X-100 in vitro (de Cock and Tommassen, 1996). However, a PhoE mutant was later shown to fold in vivo and also in vitro into N-lauryl-N,N-dimethylamine-N-oxide (LDAO) micelles but not into mixed micelles of Triton X-100 and LPS, also leading to doubts about the existence of a folded monomeric intermediate of PhoE in LPS in vivo (Jansen et al., 2000).

Folding of the β-Barrel OmpA into Lipid Bilayers Is Assisted by Skp and LPS

Direct biochemical evidence for a chaperone-assisted three-step delivery pathway of an OMP to a model membrane was first given for OmpA (Bulieris et al., 2003). It was demonstrated that the periplasmic chaperone Skp keeps OmpA soluble in vitro at pH 7 in an unfolded form even when the denaturant urea was diluted out and not present at concentrations needed to keep OmpA soluble at pH 7. Skp was also shown to prevent folding of OmpA into LPS micelles and to inhibit the folding of OmpA into phospholipid bilayers composed of phosphatidylethanolamine, phosphatidylglycerol, and phosphatidylcholine (Bulieris et al., 2003). Only when Skp complexes with unfolded OmpA were reacted with LPS in a second stage, a folding-competent form of OmpA was formed that efficiently inserted and folded into phospholipid bilayers in a third stage. In this Skp/LPS-assisted folding pathway, faster folding kinetics and higher yields of folded OmpA were observed than with the direct folding of OmpA into the same lipid bilayers upon urea dilution in the absence of Skp and LPS (Bulieris et al., 2003). In the sole presence of either Skp or LPS, the kinetics of insertion and folding were inhibited (Fig. 2).

The higher folding yields of OmpA from the complex with Skp and LPS (in comparison with OmpA folding from the urea–denatured state) may be a consequence of faster Skp binding to unfolded OmpA in solution in comparison with the folding of OmpA into lipid bilayers. Faster rates of Skp binding in solution would result in relatively lower amounts of aggregated OmpA, thus increasing the amounts of OmpA available for folding. However, it was also shown that LPS is required for the efficient OmpA insertion from complexes with Skp into lipid bilayers (Bulieris et al., 2003). In this study, unfolded OmpA bound LPS or Skp or both. The binding stoichiometries were 25 molecules LPS with a binding constant of $K_{LPS} \sim 1.2 \pm 0.7$ mM^{-1} (i.e., with a free energy of binding $\Delta G = -8.3 \pm 0.3$ kcal/mol) and three molecules of Skp with a much larger binding constant of $K_{Skp} \sim 46 \pm 30$ mM^{-1} (i.e., with $\Delta G = -10.3 \pm 0.5$ kcal/mol) (Bulieris et al., 2003). The 8- to 150-fold greater OmpA binding constant of Skp explains that Skp prevents the folding of OmpA upon addition of LPS micelles. However, LPS was necessary to promote efficient folding of OmpA into preformed phospholipid membranes at optimal stoichiometries of 0.5 to 1.7 mol of LPS/mol of Skp and 3 mol of Skp/mol of unfolded OmpA. For fast kinetics and high yields

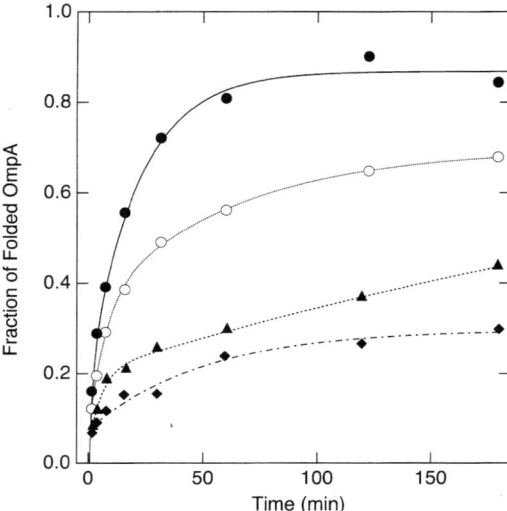

FIGURE 2 Folding of OmpA into lipid bilayers requires both Skp and LPS (adapted from Bulieris et al., 2003). Data shown correspond to folding experiments of urea-denatured OmpA into lipid bilayers, which were added 30 min after dilution of the denaturant urea in the absence of Skp and LPS (○), in the presence of Skp (♦), in the presence of LPS (▲), and in the presence of both Skp and LPS (●). The folding kinetics was fastest and folding yields were highest when both Skp and LPS were present. Folding was inhibited when either Skp or LPS was absent. The folding kinetics in the presence of Skp and LPS also compares favorably with the folding kinetics from the urea-denatured state in the absence of Skp and LPS, indicating that OmpA is insertion competent in vivo, in the absence of urea, when in complex with Skp and LPS. The data shown in Bulieris et al. (2003) also indicated that OmpA did not develop native structure in complex with Skp and LPS, but only in the presence of lipid bilayers.

of membrane insertion and folding of OmpA, about 1.5 to 5 mol of LPS bound to Skp-OmpA complexes (i.e., much lower amounts than observed in absence of Skp) (Bulieris et al., 2003). CD spectroscopy and KTSE assays indicated that large amounts of secondary and tertiary structure in OmpA only form in the third stage of the assembly pathway, upon addition of phospholipid bilayers (Bulieris et al., 2003), suggesting that Skp and LPS deliver OmpA to the membrane, which is absolutely necessary for the formation of complete secondary and tertiary structure in OmpA.

The role of LPS in this stage of membrane insertion and folding was questioned in a recent review (Tamm et al., 2004), arguing that the stimulation of folding and insertion of OmpA was only 19%. This number is the absolute, not the relative change (28%) in folding yield of OmpA, when comparing the results of two OmpA-folding experiments, which are performed either with denatured OmpA in 8 M urea or with OmpA in complex with Skp and LPS. This complex was formed after strong dilution of urea (see also Fig. 2). This comparison unfortunately is not very meaningful, because there is no urea in the periplasm. Instead, it is better to compare OmpA folding experiments into lipid bilayers with either only LPS or only Skp present and the same experiments with both Skp and LPS present (Fig. 2). Therefore, as indicated in Fig. 2, the effect of LPS on the folding of OmpA bound in complex with Skp is between 96% and 291%. The effect on the folding rate is even greater. Furthermore, in the same review Tamm et al. (2004) speculated that the role of LPS may just be to displace the basic Skp from OmpA. If that were the case, a part of OmpA would aggregate instead of leading to higher folding yields into lipid bilayers. Furthermore, binding of Skp to unfolded OmpA is stronger than binding of LPS (see above) as demonstrated in the reviewed paper, indicating that LPS is unable to completely displace Skp from OmpA. LPS is at least transiently associated with the periplasmic side of the inner membrane (Wang et al., 2004), a location also reported for Skp (de Cock et al., 1999b; Schäfer et al., 1999), suggesting an early interaction of Skp with LPS or of Skp/OmpA complexes with LPS.

The interaction of the OmpA/Skp/LPS complex with the lipid bilayer is apparently the most important event to initiate folding of OmpA in the presence of chaperones and LPS as folding catalysts. The described assisted folding pathway and discovered 3:1 stoichiometry for Skp binding to OmpA (Bulieris et al., 2003) was later supported by the observation that Skp is trimeric in solution (Schlapschy et al., 2004) and by the description of the crystal structure of Skp and a putative LPS binding site in Skp (Korndörfer et al., 2004; Walton and Sousa, 2004) (Color Plate 3A). One LPS binding site per Skp monomer is consistent with the observation of optimal folding kinetics of OmpA from an OmpA/Skp/LPS complex at 0.5 to 1.7 mol of LPS/Skp (Bulieris et al., 2003). In this case, a 1:1 mol/mol stoichiometry indicates that LPS only binds to the LPS binding site of Skp and OmpA is completely shielded from interactions with LPS, probably by binding in between the tentacles of the Skp trimer, while LPS is bound to the outer surface of Skp. A current folding model for this assisted OmpA-folding pathway is shown in Fig. 3.

MECHANISM OF INSERTION AND FOLDING OF β-BARREL MEMBRANE PROTEINS INTO LIPID BILAYERS

Lipid Acyl Chain Length Dependence and Rate Law for β-Barrel Membrane Protein Folding

The rate law of OmpA folding into a range of different phospholipid bilayers was determined using the method of initial rates (Kleinschmidt and Tamm, 2002). The folding kinetics of OmpA into LUVs of short-chain phospholipids, such as 1,2-lauroyl-*sn*-glycero-3-phosphocholine ($diC_{12:0}PC$), at 30°C, and also into

SUVs of phospholipids with longer chains, such as 1,2-dioleoyl-*sn*-glycero-3-phosphocholine ($diC_{18:1}PC$), at 40°C followed a single-step second-order rate law. The folding kinetics of OmpA into these phospholipids could also be approximated with a pseudo-first-order rate law, when the lipid concentration was high compared with the protein concentration (>90 mol of lipid per mol protein). With this approximation, a rate constant was observed that was identical with the product of the second-order rate constant and the lipid concentration. When fitted with a second-order rate law, the kinetic rate constants depended neither on the lipid nor on the protein concentration, if the lipid/protein ratio was above ~90 mol/mol, while the first-order rate constant depended on the lipid concentration. However, the second-order rate constants strongly depended on the acyl chain lengths of the lipids. When OmpA folding into bilayers of $diC_{12:0}PC$ was monitored by fluorescence spectroscopy, this rate constant was $k_{2.ord} \sim 0.4$ liter $mol^{-1} s^{-1}$, while it was $k_{2.ord} \sim 5.2$ liters $mol^{-1} s^{-1}$ for OmpA folding into bilayers of 1,2-diundecanoyl-*sn*-glycero-3-phosphocholine ($diC_{11:0}PC$) and $k_{2.ord} \sim 30$ liters $mol^{-1} s^{-1}$ for OmpA folding into 1,2-dicapryl-*sn*-glycero-3-phosphocholine ($diC_{10:0}PC$) bilayers (Klein-

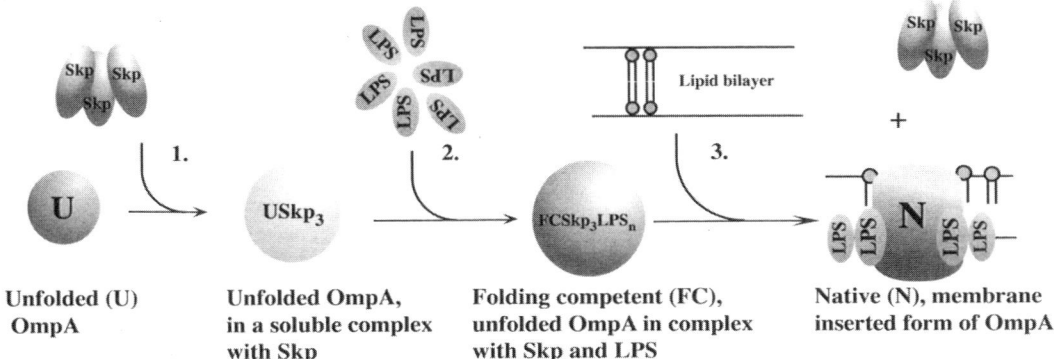

| Unfolded (U) OmpA | Unfolded OmpA, in a soluble complex with Skp | Folding competent (FC), unfolded OmpA in complex with Skp and LPS | Native (N), membrane inserted form of OmpA |

FIGURE 3 A model of the Skp/LPS-assisted folding pathway of the β-barrel protein OmpA of the OM of *E. coli* is depicted. After translocation across the cytoplasmic membrane by the SecY/E/G system in unfolded form (U), OmpA binds three molecules of the trimeric Skp, which is a periplasmic chaperone and keeps OmpA soluble in an unfolded state (USkp₃). The complex of unfolded OmpA and Skp interacts with LPS molecules to form a folding-competent intermediate of OmpA (FCSkp₃LPS*ₙ*). In the final step, folding-competent OmpA inserts and folds into the lipid bilayer. (Adapted from Kleinschmidt, 2003.)

schmidt and Tamm, 2002). Faster folding kinetics into thinner lipid bilayers were recently explained with an increased flexibility of the thinner bilayers (Marsh et al., 2006).

The Kinetics of Secondary and Tertiary Structure Formation in the β-Barrel Domain of OmpA Are Synchronized

The kinetics of membrane insertion and structure formation of OmpA initiated by denaturant dilution in the presence of preformed lipid bilayers may also be monitored by CD spectroscopy or by KTSE. When the kinetics of secondary structure formation were measured for OmpA insertion and folding into LUVs of saturated short-chain phospholipids, a similar dependence of the rate constants on the length of the hydrophobic acyl chains of the lipids was observed as by fluorescence spectroscopy. However, in general, the second-order rate constants were smaller than the corresponding rate constants of the fluorescence time courses (Kleinschmidt and Tamm, 2002). Secondary structure formation was fastest with $diC_{10:0}PC$ and slowest with $diC_{12:0}PC$ as determined from the CD kinetics at 204 nm. When OmpA was reacted with preformed lipid bilayers (LUVs) of $diC_{14:0}PC$ or $diC_{18:1}PC$, the CD signals did not change with time, indicating no changes in the secondary structure of OmpA upon incubation with these lipids.

Interaction of OmpA with the Lipid Bilayer Precedes Folding

When folding kinetic values were analyzed by using KTSE assays to determine the rate constants of tertiary structure formation, observations corresponded to those made by CD spectroscopy. The folding kinetics of OmpA depended on the length of the hydrophobic chains, but OmpA did not fold when the experiments were performed with $diC_{14:0}PC$ or $diC_{18:1}PC$. The OmpA folding kinetics into $diC_{12:0}PC$ bilayers at different concentrations were fitted to a second-order rate law, and second-order rate constants were determined. Over a range of different lipid concentrations,

the second-order rate constants obtained by KTSE were practically indistinguishable from the second-order rate constants of secondary structure formation. The rate constants of the secondary and tertiary structure formation of OmpA in $diC_{12:0}PC$ were both $^{s/t}k_{2.ord} \sim$ 0.090 liters $mol^{-1} s^{-1}$. By contrast, the second-order rate constant obtained from the fluorescence time courses of the OmpA folding kinetics into this lipid was about 4- to 5-fold higher ($^{pla}k_{2.ord} \sim 0.4$ liters $mol^{-1} s^{-1}$), indicating that the adsorption and insertion of the fluorescent tryptophan residues of OmpA into the hydrophobic core of the lipid bilayer were faster than the formation of the fully folded form of OmpA.

Four of the five tryptophans of OmpA are at the front end of the β-barrel and presumably interacted first with the hydrophobic core of the membrane, leading to fast fluorescence kinetics compared with the CD kinetics and kinetics of tertiary structure formation by electrophoresis. Together, these results indicated that the formation of the β-strands and the formation of the β-barrel of OmpA take place in parallel and are a consequence of the insertion of the membrane protein into the lipid bilayer. The previous observation that a preformed supramolecular amphiphile assembly is necessary for structure formation in OmpA was therefore further detailed by a kinetic characterization of the faster rates of interaction of OmpA with the lipid bilayer and by the slower rates of secondary and tertiary structure formation in OmpA.

Temperature Dependence and Multistep Folding Kinetics of the β-Barrel of OmpA into DOPC Bilayers

Early folding experiments with urea-unfolded OmpA and membranes of dimyristoyl-phosphatidylcholine ($diC_{14:0}PC$) indicated that OmpA folds into lipid bilayers of small unilamellar vesicles (SUVs) prepared by sonication, but not into bilayers of LUVs with a diameter of 100 nm prepared by extrusion (Surrey and Jähnig, 1992, 1995). Lipids with longer chains such as $diC_{14:0}PC$ and dioleoyl-phosphatidylcholine ($diC_{18:1}PC$) required the preparation of SUVs by ultrasonication and temperatures greater than ~25 to 28°C for successful OmpA insertion and folding (Kleinschmidt and Tamm, 1996; Surrey and Jähnig, 1992).

Lipid bilayers of SUVs have a high surface curvature and intrinsic curvature stress. OMP insertion increases the diameter of the vesicles and therefore reduces the curvature stress. The high curvature is also linked to an increased exposure of the hydrophobic surface to OmpA after it is adsorbed at the membrane-water interface, facilitating insertion of OmpA into SUVs compared with insertion of OmpA into bilayers of LUVs, where curvature stress is much lower and no insertion was observed.

The folding kinetics of OmpA into SUVs of $diC_{14:0}PC$ or $diC_{18:1}PC$ were slower than the folding kinetics into LUV short-chain phospholipids and strongly temperature dependent (Kleinschmidt and Tamm, 2002). The fluorescence kinetics of OmpA folding that could still be fitted to a single-step pseudo–first-order rate law at 40°C (Kleinschmidt and Tamm, 1996, 2002) were more complex when the temperature for folding was 30°C or less. A single-step rate law was not sufficient to describe the kinetics (Kleinschmidt and Tamm, 1996).

Insertion and folding of OmpA into bilayers of $diC_{18:1}PC$ (SUVs) were characterized by at least three kinetic phases, when experiments were performed at temperatures between 2 and 40°C. These phases could be approximated by pseudo-first-order kinetics at a lipid/protein ratio of 400. Two folding steps could be distinguished by monitoring the fluorescence time courses at 30°C. The first (faster) step was only weakly temperature dependent ($k_1 = 0.16$ min^{-1}, at 0.5 mM lipid). The second step was up to two orders of magnitude slower at low temperatures, but the rate constant approached the rate constant of the first step at higher temperatures (~0.0058 min^{-1} at 2°C and ~0.048 to 0.14 min^{-1} at 40°C, in the presence of 0.5 mM lipid). The activation energy for the slower process was 46 ± 4 kJ/mol (Kleinschmidt and Tamm, 1996). An even slower phase of OmpA folding was observed by

KTSE assays, indicating that tertiary structure formation was slowest with a rate constant of $k_3 = 0.9 \times 10^{-2}$ min^{-1} (at 3.6 mM lipid and at 40°C) (Kleinschmidt and Tamm, 1996). This is consistent with the smaller rate constants of secondary and tertiary structure formation in comparison with the rate constants of protein association with the lipid bilayer, which were later observed for OmpA folding into LUVs of short-chain phospholipids (Kleinschmidt and Tamm, 2002) (see "Interaction of OmpA with the Lipid Bilayer Precedes Folding," above).

The kinetic phases that were observed for OmpA folding into $diC_{18:1}PC$ bilayers (SUVs) suggest that at least two membrane-bound OmpA-folding intermediates exist when OmpA folds and inserts into lipid bilayers with 14 or more carbons in the hydrophobic acyl chains. These membrane-bound intermediates could be stabilized in fluid $diC_{18:1}PC$ bilayers at low temperatures between 2 and 25°C (the temperature for the phase transition of $diC_{18:1}PC$ from the lamellar-ordered to the lamellar-disordered, liquid crystalline phase is $T_c = -18$°C). The low-temperature intermediates could be rapidly converted to fully inserted, native OmpA, as demonstrated by temperature-jump experiments (Kleinschmidt and Tamm, 1996).

Characterization of Folding Intermediates by Fluorescence Quenching

Tryptophan fluorescence quenching by brominated phospholipids (see, e.g., Alvis et al., 2003; Bolen and Holloway, 1990; Everett et al., 1986; Ladokhin and Holloway, 1995; Ladokhin, 1999a, b; Markello et al., 1985; Williamson et al., 2002) or by lipid spin labels (see, e.g., Abrams and London, 1992, 1993; Cruz et al., 1998; Fastenberg et al., 2003; Piknova et al., 1997; Prieto et al., 1994) traditionally has been very valuable to determine characteristic elements of the transmembrane topology and lipid-protein interactions of TMPs.

The positions of fluorescent tryptophans with reference to the center of the phospholipid bilayer can be determined by using a set of membrane-integrated fluorescence quenchers that carry either two vicinal bromines or, alternatively, a doxyl group at the sn-2 acyl chain of the phospholipid. When in proximity to the fluorescent tryptophan residues of TMPs, these groups quench the tryptophan fluorescence. The positions of the bromines in 1-palmitoyl-2-(4,5-dibromo-)stearoyl-sn-glycero-3-phosphocholine (4,5-DiBrPC), in 6,7-DiBrPC, in 9,10-DiBrPC, and in 11,12-DiBrPC are known from X-ray diffraction to be 12.8, 11.0, 8.3, and 6.5 Å from the center of the lipid bilayer (McIntosh and Holloway, 1987; Wiener and White, 1991). From the extent of Trp fluorescence quenching by each of these membrane-inserted quenchers, the location of the Trp can be obtained by interpolation or extrapolation of the distance-dependent fluorescence-quenching profiles.

To further characterize the folding process of OmpA, we combined this method with the study of the folding kinetics of OmpA into bilayers (SUVs) of $diC_{18:1}PC$ (Kleinschmidt et al., 1999a; Kleinschmidt and Tamm, 1999). The average positions of the five fluorescent tryptophans of OmpA relative to the center of the lipid bilayer were measured as a function of time and therefore determined for the membrane-bound folding intermediates that were previously implicated by the discovery of multistep folding kinetics (Kleinschmidt and Tamm, 1996). A new method was developed by studying the kinetics of the refolding process in combination with the tryptophan fluorescence quenching at different depths in the lipid bilayer (Kleinschmidt and Tamm, 1999) using membrane-embedded quenchers.

The fluorescence intensity of the tryptophans of OmpA was measured as a function of time after initiation of OmpA folding by dilution of the denaturant in the presence of preformed lipid bilayers containing one of the brominated lipids as a fluorescence quencher. In a set of four equivalent folding experiments, bilayers were used that contained 30 mol% of one of the four brominated lipids and 70% $diC_{18:1}PC$. The fluorescence intensities in the four different time courses of OmpA folding in the presence of each of the four brominated lipids were subsequently normalized by division with fluorescence intensities obtained upon OmpA folding into bilayers of 100% $diC_{18:1}PC$ (i.e., in the absence of any quencher). Thus, depth-dependent quenching profiles were obtained at each time after initiation of OmpA folding. From these profiles, the vertical location of Trp in the membrane in projection to the bilayer normal was then determined by using the parallax method (Abrams and London, 1992; Chattopadhyay and London, 1987) or the distribution analysis (Ladokhin and Holloway, 1995; Ladokhin, 1999b).

A large set of experiments was performed in the temperature range between 2 and 40°C. At each selected temperature, the average distances of the tryptophans to the center of the lipid bilayer were determined as a function of time. Therefore, we called this method time-resolved distance determinations by tryptophan fluorescence quenching (TDFQ) (Kleinschmidt and Tamm, 1999). Previously unidentified folding intermediates on the pathway of OmpA insertion and folding into lipid bilayers were detected, trapped, and characterized. Three membrane-bound intermediates were described, in which the average distances of the Trps from the bilayer center were 14 to 16 Å, 10 to 11 Å, and 0 to 5 Å, respectively (Kleinschmidt and Tamm, 1999).

The first folding intermediate was stable at 2°C for at least 1 h. A second intermediate was characterized at temperatures between 7 and 20°C. The Trps moved 4 to 5 Å closer to the center of the bilayer at this stage. Subsequently, in an intermediate that was observed at 26 to 28°C, the Trps moved another 5 to 11 Å closer to the center of the bilayer. This intermediate appeared to be less stable. The distribution parameter, calculated from distribution analysis, was largest for the Trp distribution of this intermediate. This was a consequence of the mechanism of folding and of the structure of folded OmpA (Arora et al., 2001; Pautsch and Schulz, 1998, 2000). The large distribution parameter observed for this intermediate was consistent with experiments on single Trp mutants of OmpA (Kleinschmidt et al., 1999a) (see below). Trp-7 has to remain in the first leaflet of the lipid bilayer, while the other Trps must be translocated across the bilayer to the second leaflet. Therefore, with symmetrically incorporated brominated lipids as fluorescence quenchers, the largest distribution parameter was observed when the four translocating Trps were in the center of the lipid bilayer. Formation of the native structure of OmpA was observed at temperatures ≳28°C. In the end of these kinetic experiments, all 5 Trps were finally located on average about 9 to 10 Å from the bilayer center, Trp-7 in the periplasmic leaflet and the other 4 Trps in the outer leaflet of the OM.

When KTSE experiments were performed to monitor OmpA folding at 30°C, a band at 32 kDa was observed in the first few minutes of OmpA folding (Kleinschmidt and Tamm, 1996). The folding conditions for this experiment were nearly identical with those of the fluorescence-quenching experiments at 28 to 30°C. Therefore, this 32-kDa form is very likely identical with the third folding intermediate of OmpA, in which the average Trp location is 0 to 5 Å from the center of the lipid bilayer. The comparison indicated that in this intermediate, a significant part of the β-barrel had formed, which is resistant to treatment with SDS at room temperature.

The β-Barrel Domain of OmpA Folds and Inserts by a Concerted Mechanism

TDFQ experiments were subsequently performed with the five different single tryptophan mutants of OmpA. These mutants were prepared by site-directed mutagenesis (Kleinschmidt et al., 1999a), and each contained a single tryptophan and 4 phenylalanines in the 5 tryptophan positions of the wild-type protein. All mutants were isolated from the *E. coli* OM and refolded in vitro into lipid bilayers. TDFQ for each of the single Trp mutants of OmpA gave more structural detail on the folding mechanism of OmpA. These TDFQ experiments were carried out at selected temperatures between 2 and 40°C (Kleinschmidt et al., 1999a).

When kinetic experiments were performed below 30°C, each of the 5 tryptophans approached a distance of 10 to 11 Å from the bilayer center in the end of the fluorescence time course of OmpA folding. A distance decrease with time was observed even at 40°C for Trp-7 (Fig. 4A). The TDFQ results showed that Trp-7 did not migrate any closer to the bilayer center than ~10 Å independent of the experimental conditions. However, Trp-15, Trp-57, Trp-102, and Trp-143 were detected very close to the center of the lipid bilayer in the first minutes of refolding at temperatures of 30°C, 32°C, 35°C, and 40°C, respectively. This is shown for Trp-143 in Fig. 4B. TDFQ experiments performed at 40°C resolved the last two steps of OmpA refolding, and the translocation rate constants of the first phase of fast distance change were 0.55, 0.46, 0.26, and 0.43 min^{-1} for Trp-15, Trp-57, Trp-102, and Trp-143, respectively. The four Trps crossed the center of the bilayer and approached distances of ~10 Å from the bilayer center in the final folding step of OmpA. These experiments demonstrated that Trp-15, Trp-57, Trp-102, and Trp-143 are similarly located in three folding intermediates that were also observed previously for wild-type OmpA. The similar distances of these Trps from the membrane center in each of the membrane-bound folding intermediates indicate a simultaneous translocation of the transmembrane segments of OmpA, coupled to the formation of the β-barrel structure upon insertion.

The results of these kinetic studies on the folding mechanism of OmpA may be used to develop a tentative model of OmpA folding (Color Plate 5). The time courses of OmpA folding into phospholipid bilayers (LUVs) of $diC_{12:0}PC$ indicated that β-strand secondary and β-barrel tertiary structure formation are synchronized with the same rate constant (Kleinschmidt and Tamm, 2002), which is lower than the rate constant of the fluorescence time course of OmpA adsorption to the lipid bilayer.

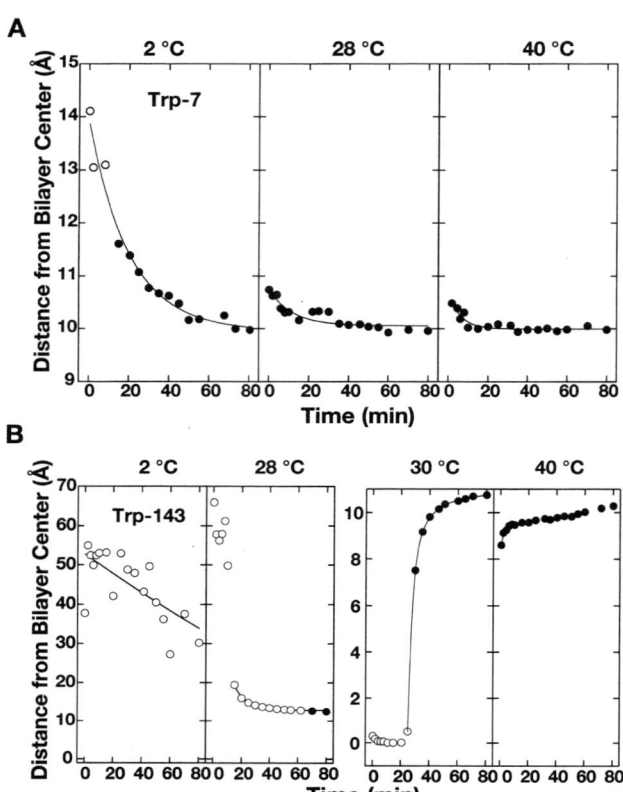

FIGURE 4 (A) Time courses of the movement of Trp-7 toward the bilayer center at 2, 28, and 40°C. Distances were obtained from curve fits to fluorescence-quenching profiles as described in the text. Data points represented by filled circles were the fitted quenching-profile minima, open circles denote extrapolated distances from the observed quenching profiles. The solid lines are fits of single- or double-exponential functions to the data. (B) Time courses of the movement of Trp-143 toward the bilayer center at 2 and 28°C and from the bilayer center at 30 and 40°C. At 2°C, the distances of Trp-143 could only be obtained by extrapolation (open circles). The solid lines are fits of the data to single- or double-exponential functions. (Adapted from Kleinschmidt et al., 1999a.)

Strongly temperature-dependent kinetic values were observed and several kinetic phases were distinguished when folding of OmpA was investigated with lipid bilayers of $diC_{18:1}PC$, which is a phospholipid with comparably long hydrophobic chains. OmpA first adsorbs to the water-membrane interface (intermediate A) and the intrinsic fluorescence of OmpA increases strongly due to the partitioning of the fluorescent Trps into the less polar environment at the membrane/water interface. Subsequently, the slower phase of the fluorescence changes reflects the migration of the Trps from the membrane/water interface into the hydrophobic core of the lipid bilayer.

The translocation of the Trps across the bilayer is best monitored with membrane-inserted fluorescence quenchers, because the intrinsic Trp fluorescence does not change much during Trp migration through the 30-Å hydrophobic core of $diC_{18:1}PC$. The average location of the Trps of 14 to 16 Å from the bilayer center after adsorption to the membrane-water interface was determined by TDFQ experiments at 2°C (Kleinschmidt and Tamm, 1999). At temperatures of 5 to 25°C, this initial phase of folding was fast and followed by a second, slower phase, in which the Trps move into more hydrophobic regions at a distance of about 10 Å from the bilayer center. The observed folding intermediate (B) is quite stable. A third membrane-bound intermediate (C) was identified at 27 to 29°C. In this intermediate, all Trps, except Trp-7, are detected a distance of 0 to 5 Å from the bilayer center in the first minutes of OmpA folding. Trp-7 remains at the same location as in intermediate B. Very likely, this intermediate is identical with the

32-kDa form of OmpA that was previously observed by KTSE experiments (Kleinschmidt and Tamm, 1996). Finally, at temperatures above 28 to 30°C, Trp-15, Trp-57, Trp-102, and Trp-143 move away from the center of the bilayer to a distance of about 10 Å. This distance of the Trp residues of OmpA compares well with the X-ray and NMR structures of OmpA (Arora et al., 2001; Pautsch and Schulz, 2000).

The basic elements of the model in Color Plate 5 are the synchronized kinetics of secondary and tertiary structure formation, the simultaneous migration of the tryptophans that cross the bilayer center, and the migration of Trp-7, which does not translocate. However, more structural information is needed to improve this preliminary model. For example, it is not known how the residues of the polar loops of OmpA cross the hydrophobic core of the lipid bilayer.

PERSPECTIVES

Although considerable progress has been made in recent years to understand the pathways and principles of targeting and assembly of OMPs into the OM, numerous unanswered questions remain. It is now clear that periplasmic chaperones, such as Skp and SurA, help to keep OMPs unfolded in the periplasm and prevent their aggregation without requiring ATP as an energy source. The identification of the integral OMP YaeT (Omp85 in *N. meningitidis*) and outer membrane lipoproteins as factors involved in targeting and/or insertion of OMPs into the OM on one side and the spontaneous assembly of OMPs into lipid bilayers in vitro on the other side raises several interesting questions. Obviously, there is no absolute requirement of proteinaceous machinery to refold OMPs to their functionally active state. Yet in cells, OMPs must be specifically sorted to the OM. Furthermore, the accumulation of misfolded OMPs in the periplasm upon deletion of *yaeT* (*omp85*) suggests that the properties of the OM lipid bilayer differ from the properties of the phospholipid bilayers into which OMPs successfully fold in vitro. In vitro, folding is successful with thin lipid bilayers (Kleinschmidt

and Tamm, 2002; Pocanschi et al., 2006; Marsh et al., 2006) that are flexible. Flexibility is required, because the lipids must make space for the OMP during insertion, a process that is much slower with thicker bilayers formed, for example, by dioleoyl phospholipids. In the OM, spontaneous protein insertion may not be a favored process for the cell, which must maintain its integrity. Therefore, some mechanism appears to be necessary in vivo to specifically recognize OMPs and to facilitate OM insertion only for them. It is likely that proteins mediate this process and that their role is to locally alter the properties of the membrane for the insertion of OMPs into the OM. Insertion, if not spontaneous, would require energy and it is unclear what the source for this energy may be. Future studies must address this problem. One possibility may be a mechanism of energy transduction across the periplasm, similar to the mechanism proposed for the active import of iron siderophores or cobalamin.

The basic physical mechanism of β-barrel formation in the OM is likely the same as the mechanism of β-barrel formation in lipid bilayers (Kleinschmidt et al., 1999a; Kleinschmidt and Tamm, 2002), but it will be interesting to examine how the complex of YaeT, YfiO, YfgL, and NlpB facilitates targeting and membrane insertion of OMPs. The recently discovered, highly concerted mechanism of secondary and tertiary structure formation in OMPs, which is coupled to the bilayer insertion process, must be examined in more detail in the absence and in the presence of proteinaceous folding factors. A number of biophysically interesting questions must be addressed in this context. For instance, how do the polar outer loops traverse the hydrophobic core of the OM? What are the topologies of complexes of OMPs and their chaperones? How are OMPs recognized by the chaperones and by OM-folding facilitators? How is YaeT inserted into the OM? Is this an autocatalytic process? Are the lipoproteins required? What exactly is the role of the lipoproteins in OMP insertion? It will be interesting to shed more light into these and additional problems of OM assembly in future studies.

ACKNOWLEDGMENT

I thank the Deutsche Forschungsgemeinschaft for supporting the work of J.H.K. with DFG-Grants KL 1024/2-2, 2-3, 2-5, and 2-6.

REFERENCES

Abrams, F. S., and E. London. 1992. Calibration of the parallax fluorescence quenching method for determination of membrane penetration depth: refinement and comparison of quenching by spin-labeled and brominated lipids. *Biochemistry* **31:** 5312–5322.

Abrams, F. S., and E. London. 1993. Extension of the parallax analysis of membrane penetration depth to the polar region of model membranes: use of fluorescence quenching by a spin-label attached to the phospholipid polar headgroup. *Biochemistry* **32:**10826–10831.

Ades, S. E., L. E. Connolly, B. M. Alba, and C. A. Gross. 1999. The *Escherichia coli* σ^E-dependent extracytoplasmic stress response is controlled by the regulated proteolysis of an anti-σ-factor. *Genes Dev.* **13:**2449–2461.

Ades, S. E., I. L. Grigorova, and C. A. Gross. 2003. Regulation of the alternative σ-factor σ^E during initiation, adaptation, and shutoff of the extracytoplasmic heat shock response in *Escherichia coli. J. Bacteriol.* **185:**2512–2519.

Ahn, V. E., E. I. Lo, C. K. Engel, L. Chen, P. M. Hwang, L. E. Kay, R. E. Bishop, and G. G. Prive. 2004. A hydrocarbon ruler measures palmitate in the enzymatic acylation of endotoxin. *EMBO J.* **23:**2931–2941.

Alba, B. M., and C. A. Gross. 2004. Regulation of the *Escherichia coli* σ^E-dependent envelope stress response. *Mol. Microbiol.* **52:**613–619.

Alba, B. M., J. A. Leeds, C. Onufryk, C. Z. Lu, and C. A. Gross. 2002. DegS and YaeL participate sequentially in the cleavage of RseA to activate the σ^E-dependent extracytoplasmic stress response. *Genes Dev.* **16:**2156–2168.

Alba, B. M., H. J. Zhong, J. C. Pelayo, and C. A. Gross. 2001. *degS* (*hhoB*) is an essential *Escherichia coli* gene whose indispensable function is to provide σ^E activity. *Mol. Microbiol.* **40:**1323–1333.

Alvis, S. J., I. M. Williamson, J. M. East, and A. G. Lee. 2003. Interactions of anionic phospholipids and phosphatidylethanolamine with the potassium channel KcsA. *Biophys. J.* **85:**3828–3838.

Anfinsen, C. B. 1973. Principles that govern the folding of protein chains. *Science* **181:**223–230.

Arie, J. P., N. Sassoon, and J. M. Betton. 2001. Chaperone function of FkpA, a heat shock prolyl isomerase, in the periplasm of *Escherichia coli. Mol. Microbiol.* **39:**199–210.

Arora, A., F. Abildgaard, J. H. Bushweller, and L. K. Tamm. 2001. Structure of outer membrane protein A transmembrane domain by NMR spectroscopy. *Nat. Struct. Biol.* **8:**334–338.

Arora, A., D. Rinehart, G. Szabo, and L. K. Tamm. 2000. Refolded outer membrane protein A of *Escherichia coli* forms ion channels with two conductance states in planar lipid bilayers. *J. Biol. Chem.* **275:**1594–1600.

Bedzyk, W. D., K. M. Weidner, L. K. Denzin, L. S. Johnson, K. D. Hardman, M. W. Pantoliano, E. D. Asel, and E. W. Voss, Jr. 1990. Immunological and structural characterization of a high affinity anti-fluorescein single-chain antibody. *J. Biol. Chem.* **265:**18615–18620.

Behlau, M., D. J. Mills, H. Quader, W. Kühlbrandt, and J. Vonck. 2001. Projection structure of the monomeric porin OmpG at 6 Å resolution. *J. Mol. Biol.* **305:**71–77.

Behrens, S., R. Maier, H. de Cock, F. X. Schmid, and C. A. Gross. 2001. The SurA periplasmic PPIase lacking its parvulin domains functions in vivo and has chaperone activity. *EMBO J.* **20:**285–294.

Bitto, E., and D. B. McKay. 2002. Crystallographic structure of SurA, a molecular chaperone that facilitates folding of outer membrane porins. *Structure (Camb)* **10:**1489–1498.

Bitto, E., and D.B. McKay. 2003. The periplasmic molecular chaperone protein SurA binds a peptide motif that is characteristic of integral outer membrane proteins. *J. Biol. Chem.* **278:**49316–49322.

Bitto, E., and D. B. McKay. 2004. Binding of phage-display-selected peptides to the periplasmic chaperone protein SurA mimics binding of unfolded outer membrane proteins. *FEBS Lett.* **568:** 94–98.

Bolen, E. J., and P. W. Holloway. 1990. Quenching of tryptophan fluorescence by brominated phospholipid. *Biochemistry* **29:**9638–9643.

Bos, M. P., B. Tefsen, J. Geurtsen, and J. Tommassen. 2004. Identification of an outer membrane protein required for the transport of lipopolysaccharide to the bacterial cell surface. *Proc. Natl. Acad. Sci. USA* **101:**9417–9422.

Bothmann, H., and A. Plückthun. 1998. Selection for a periplasmic factor improving phage display and functional periplasmic expression. *Nat. Biotechnol.* **16:**376–380.

Bothmann, H., and A. Plückthun. 2000. The periplasmic *Escherichia coli* peptidylprolyl *cis,trans*-isomerase FkpA. I. Increased functional expression of antibody fragments with and without *cis*-prolines. *J. Biol. Chem.* **275:**17100–17105.

Bouvier, J., A. P. Pugsley, and P. Stragier. 1991. A gene for a new lipoprotein in the *dapA–purC* interval of the *Escherichia coli* chromosome. *J. Bacteriol.* **173:**5523–5531.

Braun, M., and T. J. Silhavy. 2002. Imp/OstA is required for cell envelope biogenesis in *Escherichia coli. Mol. Microbiol.* **45:**1289–1302.

Breyton, C., W. Haase, T. A. Rapoport, W. Kühlbrandt, and I. Collinson. 2002. Three-dimensional structure of the bacterial protein-translocation complex SecYEG. *Nature* **418:**662–665.

Buchanan, S. K. 1999. β-barrel proteins from bacterial outer membranes: structure, function and refolding. *Curr. Opin. Struct. Biol.* **9:**455–461.

Buchanan, S. K., B. S. Smith, L. Venkatramani, D. Xia, L. Esser, M. Palnitkar, R. Chakraborty, D. van der Helm, and J. Deisenhofer. 1999. Crystal structure of the outer membrane active transporter FepA from *Escherichia coli. Nat. Struct. Biol.* **6:**56–63.

Bulieris, P. V., S. Behrens, O. Holst, and J. H. Kleinschmidt. 2003. Folding and insertion of the outer membrane protein OmpA is assisted by the chaperone Skp and by lipopolysaccharide. *J. Biol. Chem.* **278:**9092–9099.

Campbell, E. A., J. L. Tupy, T. M. Gruber, S. Wang, M. M. Sharp, C. A. Gross, and S. A. Darst. 2003. Crystal structure of *Escherichia coli* σ^E with the cytoplasmic domain of its anti-σ RseA. *Mol. Cell* **11:**1067–1078.

Chattopadhyay, A., and E. London. 1987. Parallax method for direct measurement of membrane penetration depth utilizing fluorescence quenching by spin-labeled phospholipids. *Biochemistry* **26:**39–45.

Chen, R., and U. Henning. 1996. A periplasmic protein (Skp) of *Escherichia coli* selectively binds a class of outer membrane proteins. *Mol. Microbiol.* **19:**1287–1294.

Chimento, D. P., A. K. Mohanty, R. J. Kadner, and M. C. Wiener. 2003a. Substrate-induced transmembrane signaling in the cobalamin transporter BtuB. *Nat. Struct. Biol.* **10:**394–401.

Chimento, D. P., A. K. Mohanty, R. J. Kadner, and M. C. Wiener. 2003b. Crystallization and initial X-ray diffraction of BtuB, the integral membrane cobalamin transporter of *Escherichia coli. Acta Crystallogr. D. Biol. Crystallogr.* **59:**509–511.

Clubb, R. T., S. B. Ferguson, C. T. Walsh, and G. Wagner. 1994. Three-dimensional solution structure of *Escherichia coli* periplasmic cyclophilin. *Biochemistry* **33:**2761–2772.

Cobessi, D., H. Celia, N. Folschweiller, I. J. Schalk, M. A. Abdallah, and F. Pattus. 2005. The crystal structure of the pyoverdine outer membrane receptor FpvA from *Pseudomonas aeruginosa* at 3.6 Å resolution. *J. Mol. Biol.* **347:**121–134.

Collinet, B., H. Yuzawa, T. Chen, C. Herrera, and D. Missiakas. 2000. RseB binding to the periplasmic domain of RseA modulates the RseA: σ^E interaction in the cytoplasm and the availability of σ^E-RNA polymerase. *J. Biol. Chem.* **275:**33898–33904.

Compton, L. A., J. M. Davis, J. R. Macdonald, and H. P. Bachinger. 1992. Structural and functional characterization of *Escherichia coli* peptidyl-prolyl *cis-trans* isomerases. *Eur. J. Biochem.* **206:**927–934.

Conlan, S., and H. Bayley. 2003. Folding of a monomeric porin, OmpG, in detergent solution. *Biochemistry* **42:**9453–9465.

Conlan, S., Y. Zhang, S. Cheley, and H. Bayley. 2000. Biochemical and biophysical characterization of OmpG: a monomeric porin. *Biochemistry* **39:**11845–11854.

Connolly, L., A. De Las Penas, B. M. Alba, and C. A. Gross. 1997. The response to extracytoplasmic stress in *Escherichia coli* is controlled by partially overlapping pathways. *Genes Dev.* **11:**2012–2021.

Cowan, S. W., T. Schirmer, G. Rummel, M. Steiert, R. Ghosh, R. A. Pauptit, J. N. Jansonius, and J. P. Rosenbusch. 1992. Crystal structures explain functional properties of two *E. coli* porins. *Nature* **358:**727–733.

Craig, E. A. 1993. Chaperones: helpers along the pathways to protein folding. *Science* **260:**1902–1903.

Cruz, A., C. Casals, I. Plasencia, D. Marsh, and J. Perez-Gil. 1998. Depth profiles of pulmonary surfactant protein B in phosphatidylcholine bilayers, studied by fluorescence and electron spin resonance spectroscopy. *Biochemistry* **37:**9488–9496.

Danese, P. N., and T. J. Silhavy. 1997. The σ^E and the Cpx signal transduction systems control the synthesis of periplasmic protein-folding enzymes in *Escherichia coli. Genes Dev.* **11:**1183–1193.

Danese, P. N., W. B. Snyder, C. L. Cosma, L. J. Davis, and T. J. Silhavy. 1995. The Cpx two-component signal transduction pathway of *Escherichia coli* regulates transcription of the gene specifying the stress-inducible periplasmic protease, DegP. *Genes Dev.* **9:**387–398.

Dartigalongue, C., D. Missiakas, and S. Raina. 2001. Characterization of the *Escherichia coli* σ^E-regulon. *J. Biol. Chem.* **276:**20866–20875.

Dartigalongue, C., and S. Raina. 1998. A new heat-shock gene, *ppiD*, encodes a peptidyl-prolyl isomerase required for folding of outer membrane proteins in *Escherichia coli. EMBO J.* **17:**3968–3980.

de Cock, H., K. Brandenburg, A. Wiese, O. Holst, and U. Seydel. 1999a. Non-lamellar structure and negative charges of lipopolysaccharides required for efficient folding of outer membrane protein PhoE of *Escherichia coli. J. Biol. Chem.* **274:**5114–5119.

de Cock, H., U. Schäfer, M. Potgeter, R. Demel, M. Müller, and J. Tommassen. 1999b. Affinity of the periplasmic chaperone Skp of *Escherichia coli* for phospholipids, lipopolysaccharides and non-native outer membrane proteins. Role of Skp in the biogenesis of outer membrane protein. *Eur. J. Biochem.* **259:**96–103.

de Cock, H., and J. Tommassen. 1996. Lipopolysaccharides and divalent cations are involved in the formation of an assembly-competent intermediate of outer-membrane protein PhoE of *E.coli. EMBO J.* **15**:5567–5573.

Delano, W. L. 2002. *The PyMOL Molecular Graphics System.* DeLano Scientific, San Carlos, Calif.

De Las Peñas, A., L. Connolly, and C. A. Gross. 1997. The σ^E-mediated response to extracytoplasmic stress in Escherichia coli is transduced by RseA and RseB, two negative regulators of σ^E. *Mol. Microbiol.* **24**:373–385.

Devereux, J., P. Haeberli, and O. Smithies. 1984. A comprehensive set of sequence analysis programs for the VAX. *Nucleic Acids Res.* **12**:387–395.

Dicker, I. B., and S. Seetharam. 1991. Cloning and nucleotide sequence of the *firA* gene and the *firA200*(Ts) allele from *Escherichia coli. J. Bacteriol.* **173**:334–344.

Dodson, K. W., F. Jacob-Dubuisson, R. T. Striker, and S. J. Hultgren. 1993. Outer-membrane PapC molecular usher discriminately recognizes periplasmic chaperone-pilus subunit complexes. *Proc. Natl. Acad. Sci. USA* **90**:3670–3674.

Doerrler, W. T., and C. R. Raetz. 2005. Loss of outer membrane proteins without inhibition of lipid export in an *Escherichia coli* YaeT mutant. *J. Biol. Chem.* **280**:27679–27687.

Dolinski, K., C. Scholz, R. S. Muir, S. Rospert, F. X. Schmid, M. E. Cardenas, and J. Heitman. 1997. Functions of FKBP12 and mitochondrial cyclophilin active site residues *in vitro* and *in vivo* in *Saccharomyces cerevisiae. Mol. Biol. Cell* **8**:2267–2280.

Dornmair, K., H. Kiefer, and F. Jähnig. 1990. Refolding of an integral membrane protein. OmpA of *Escherichia coli. J. Biol. Chem.* **265**:18907–18911.

Driessen, A. J., E. H. Manting, and C. van der Does. 2001. The structural basis of protein targeting and translocation in bacteria. *Nat. Struct. Biol.* **8**:492–498.

Duguay, A. R., and T. J. Silhavy. 2004. Quality control in the bacterial periplasm. *Biochim. Biophys. Acta* **1694**:121–134.

Dutzler, R., G. Rummel, S. Alberti, S. Hernandez-Alles, P. Phale, J. Rosenbusch, V. Benedi, and T. Schirmer. 1999. Crystal structure and functional characterization of OmpK36, the osmoporin of *Klebsiella pneumoniae. Structure Fold. Des.* **7**:425–434.

Eggert, U. S., N. Ruiz, B. V. Falcone, A. A. Branstrom, R. C. Goldman, T. J. Silhavy, and D. Kahne. 2001. Genetic basis for activity differences between vancomycin and glycolipid derivatives of vancomycin. *Science* **294**:361–364.

Ehrmann, M., and T. Clausen. 2004. Proteolysis as a regulatory mechanism. *Annu. Rev. Genet.* **38**:709–724.

Everett, J., A. Zlotnick, J. Tennyson, and P. W. Holloway. 1986. Fluorescence quenching of cytochrome b5 in vesicles with an asymmetric trans-bilayer distribution of brominated phosphatidylcholine. *J. Biol. Chem.* **261**:6725–6729.

Faller, M., M. Niederweis, and G. E. Schulz. 2004. The structure of a mycobacterial outer-membrane channel. *Science* **303**:1189–1192.

Fastenberg, M. E., H. Shogomori, X. Xu, D. A. Brown, and E. London. 2003. Exclusion of a transmembrane-type peptide from ordered-lipid domains (rafts) detected by fluorescence quenching: extension of quenching analysis to account for the effects of domain size and domain boundaries. *Biochemistry* **42**:12376–12390.

Fejzo, J., F. A. Etzkorn, R. T. Clubb, Y. Shi, C. T. Walsh, and G. Wagner. 1994. The mutant Escherichia coli F112W cyclophilin binds cyclosporin A in nearly identical conformation as human cyclophilin. *Biochemistry* **33**:5711–5720.

Ferguson, A. D., R. Chakraborty, B. S. Smith, L. Esser, D. van der Helm, and J. Deisenhofer. 2002. Structural basis of gating by the outer membrane transporter FecA. *Science* **295**:1715–1719.

Ferguson, A. D., E. Hofmann, J. W. Coulton, K. Diederichs, and W. Welte. 1998. Siderophore-mediated iron transport: crystal structure of FhuA with bound lipopolysaccharide. *Science* **282**:2215–2220.

Ferguson, A. D., W. Welte, E. Hofmann, B. Lindner, O. Holst, J. W. Coulton, and K. Diederichs. 2000. A conserved structural motif for lipopolysaccharide recognition by procaryotic and eucaryotic proteins. *Structure* **8**:585–592.

Fernandez, C., C. Hilty, G. Wider, P. Guntert, and K. Wüthrich. 2004. NMR structure of the integral membrane protein OmpX. *J. Mol. Biol.* **336**:1211–1221.

Forst, D., W. Welte, T. Wacker, and K. Diederichs. 1998. Structure of the sucrose-specific porin ScrY from *Salmonella typhimurium* and its complex with sucrose. *Nat. Struct. Biol.* **5**:37–46.

Freudl, R., H. Schwarz, Y. D. Stierhof, K. Gamon, I. Hindennach, and U. Henning. 1986. An outer membrane protein (OmpA) of *Escherichia coli* K-12 undergoes a conformational change during export. *J. Biol. Chem.* **261**:11355–11361.

Froderberg, L., E. N. Houben, L. Baars, J. Luirink, and J. W. de Gier. 2004. Targeting and translocation of two lipoproteins in *Escherichia coli* via the SRP/Sec/YidC pathway. *J. Biol. Chem.* **279**:31026–31032.

Genevrois, S., L. Steeghs, P. Roholl, J. J. Letesson, and P. van der Ley. 2003. The Omp85 protein of *Neisseria meningitidis* is required for lipid export to the outer membrane. *EMBO J.* **22**:1780–1789.

Gentle, I., K. Gabriel, P. Beech, R. Waller, and T. Lithgow. 2004. The Omp85 family of proteins is essential for outer membrane biogenesis in mitochondria and bacteria. *J. Cell Biol.* **164**:19–24.

Guex, N., and M. C. Peitsch. 1997. SWISS-MODEL and the Swiss-PdbViewer: an environment for comparative protein modeling. *Electrophoresis* **18:**2714–2723.

Harms, N., G. Koningstein, W. Dontje, M. Müller, B. Oudega, J. Luirink, and H. de Cock. 2001. The early interaction of the outer membrane protein phoe with the periplasmic chaperone Skp occurs at the cytoplasmic membrane. *J. Biol. Chem.* **276:**18804–18811.

Hayano, T., N. Takahashi, S. Kato, N. Maki, and M. Suzuki. 1991. Two distinct forms of peptidyl-prolyl-*cis-trans*-isomerase are expressed separately in periplasmic and cytoplasmic compartments of *Escherichia coli* cells. *Biochemistry* **30:**3041–3048.

Heins, L., H. Mentzel, A. Schmid, R. Benz, and U. K. Schmitz. 1994. Biochemical, molecular, and functional characterization of porin isoforms from potato mitochondria. *J. Biol. Chem.* **269:**26402–26410.

Henderson, N. S., S. S. So, C. Martin, R. Kulkarni, and D. G. Thanassi. 2004. Topology of the outer membrane usher PapC determined by site-directed fluorescence labeling. *J. Biol. Chem.* **279:**53747–53754.

Hennecke, G., J. Nolte, R. Volkmer-Engert, J. Schneider-Mergener, and S. Behrens. 2005. The periplasmic chaperone SurA exploits two features characteristic of integral outer membrane proteins for selective substrate recognition. *J. Biol. Chem.* **280:**23540–23548.

Heuck, A. P., E. M. Hotze, R. K. Tweten, and A. E. Johnson. 2000. Mechanism of membrane insertion of a multimeric β-barrel protein: perfringolysin O creates a pore using ordered and coupled conformational changes. *Mol. Cell* **6:**1233–1242.

Horne, S. M., and K. D. Young. 1995. *Escherichia coli* and other species of the Enterobacteriaceae encode a protein similar to the family of Mip-like FK506-binding proteins. *Arch. Microbiol.* **163:**357–365.

Huang, K. S., H. Bayley, M. J. Liao, E. London, and H. G. Khorana. 1981. Refolding of an integral membrane protein. Denaturation, renaturation, and reconstitution of intact bacteriorhodopsin and two proteolytic fragments. *J. Biol. Chem.* **256:**3802–3809.

Hwang, P. M., W. Y. Choy, E. I. Lo, L. Chen, J. D. Forman-Kay, C. R. Raetz, G. G. Prive, R. E. Bishop, and L. E. Kay. 2002. Solution structure and dynamics of the outer membrane enzyme PagP by NMR. *Proc. Natl. Acad. Sci. USA* **99:**13560–13565.

Jansen, C., M. Heutink, J. Tommassen, and H. de Cock. 2000. The assembly pathway of outer membrane protein PhoE of *Escherichia coli*. *Eur. J. Biochem.* **267:**3792–3800.

Justice, S. S., D. A. Hunstad, J. R. Harper, A. R. Duguay, J. S. Pinkner, J. Bann, C. Frieden, T. J. Silhavy, and S. J. Hultgren. 2005. Periplasmic peptidyl prolyl *cis-trans* isomerases are not essential for viability, but SurA is required for pilus biogenesis in *Escherichia coli. J. Bacteriol.* **187:**7680–7686.

Kanehara, K., K. Ito, and Y. Akiyama. 2003. YaeL proteolysis of RseA is controlled by the PDZ domain of YaeL and a Gln-rich region of RseA. *EMBO J.* **22:**6389–6398.

Kleerebezem, M., M. Heutink, and J. Tommassen. 1995. Characterization of an *Escherichia coli* rotA mutant, affected in periplasmic peptidyl-prolyl *cis/trans* isomerase. *Mol. Microbiol.* **18:**313–320.

Kleinschmidt, J. H. 2003. Membrane protein folding on the example of outer membrane protein A of *Escherichia coli. Cell Mol. Life Sci.* **60:**1547–1558.

Kleinschmidt, J. H. 2006. Folding kinetics of the outer membrane proteins OmpA and FomA into phospholipid bilayers. *Chem. Phys. Lipids* **141:**30–47.

Kleinschmidt, J. H., T. den Blaauwen, A. Driessen, and L. K. Tamm. 1999a. Outer membrane protein A of *E. coli* inserts and folds into lipid bilayers by a concerted mechanism. *Biochemistry* **38:**5006–5016.

Kleinschmidt, J. H., and L. K. Tamm. 1996. Folding intermediates of a β-barrel membrane protein. Kinetic evidence for a multi-step membrane insertion mechanism. *Biochemistry* **35:**12993–13000.

Kleinschmidt, J. H., and L. K. Tamm. 1999. Time-resolved distance determination by tryptophan fluorescence quenching: probing intermediates in membrane protein folding. *Biochemistry* **38:**4996–5005.

Kleinschmidt, J. H., and L. K. Tamm. 2002. Secondary and tertiary structure formation of the β-barrel membrane protein OmpA is synchronized and depends on membrane thickness. *J. Mol. Biol.* **324:**319–330.

Kleinschmidt, J. H., M. C. Wiener, and L. K. Tamm. 1999b. Outer membrane protein A of *E. coli* folds into detergent micelles, but not in the presence of monomeric detergent. *Protein Sci.* **8:**2065–2071.

Kleivdal, H., R. Benz, and H. B. Jensen. 1995. The *Fusobacterium nucleatum* major outer-membrane protein (FomA) forms trimeric, water-filled channels in lipid bilayer membranes. *Eur. J. Biochem.* **233:**310–316.

Konno, M., Y. Sano, K. Okudaira, Y. Kawaguchi, Y. Yamagishi-Ohmori, S. Fushinobu, and H. Matsuzawa. 2004. *Escherichia coli* cyclophilin B binds a highly distorted form of trans-prolyl peptide isomer. *Eur. J. Biochem.* **271:**3794–3803.

Koradi, R., M. Billeter, and K. Wüthrich. 1996. MOLMOL: a program for display and analysis of macromolecular structures. *J. Mol. Graph.* **14**:51–55, 29–32.

Korndörfer, I. P., M. K. Dommel, and A. Skerra. 2004. Structure of the periplasmic chaperone Skp suggests functional similarity with cytosolic chaperones despite differing architecture. *Nat. Struct. Mol. Biol.* **11**:1015–1020.

Koronakis, V., A. Sharff, E. Koronakis, B. Luisi, and C. Hughes. 2000. Crystal structure of the bacterial membrane protein TolC central to multidrug efflux and protein export. *Nature* **405**:914–919.

Kreusch, A., and G. E. Schulz. 1994. Refined structure of the porin from *Rhodopseudomonas blastica*. Comparison with the porin from *Rhodobacter capsulatus*. *J. Mol. Biol.* **243**:891–905.

Krojer, T., M. Garrido-Franco, R. Huber, M. Ehrmann, and T. Clausen. 2002. Crystal structure of DegP (HtrA) reveals a new protease-chaperone machine. *Nature* **416**:455–459.

Kurisu, G., S. D. Zakharov, M. V. Zhalnina, S. Bano, V. Y. Eroukova, T. I. Rokitskaya, Y. N. Antonenko, M. C. Wiener, and W. A. Cramer. 2003. The structure of BtuB with bound colicin E3 R-domain implies a translocon. *Nat. Struct. Biol.* **10**:948–954.

Ladokhin, A. S. 1999a. Evaluation of lipid exposure of tryptophan residues in membrane peptides and proteins. *Anal. Biochem.* **276**:65–71.

Ladokhin, A. S. 1999b. Analysis of protein and peptide penetration into membranes by depth-dependent fluorescence quenching: theoretical considerations. *Biophys. J.* **76**:946–955.

Ladokhin, A. S., and P. W. Holloway. 1995. Fluorescence of membrane-bound tryptophan octyl ester: a model for studying intrinsic fluorescence of protein-membrane interactions. *Biophys. J.* **69**:506–517.

Laemmli, U. K. 1970. Cleavage of structural proteins during the assembly of the head of bacteriophage T4. *Nature* **227**:680–685.

Lazar, S. W., and R. Kolter. 1996. SurA assists the folding of *Escherichia coli* outer membrane proteins. *J. Bacteriol.* **178**:1770–1773.

Lipinska, B., M. Zylicz, and C. Georgopoulos. 1990. The HtrA (DegP) protein, essential for *Escherichia coli* survival at high temperatures, is an endopeptidase. *J. Bacteriol.* **172**:1791–1797.

Liu, J., and C. T. Walsh. 1990. Peptidyl-prolyl *cis-trans*-isomerase from *Escherichia coli*: a periplasmic homolog of cyclophilin that is not inhibited by cyclosporin A. *Proc. Natl. Acad. Sci. USA* **87**:4028–4032.

Locher, K. P., B. Rees, R. Koebnik, A. Mitschler, L. Moulinier, J. P. Rosenbusch, and D. Moras. 1998. Transmembrane signaling across the ligand-gated FhuA receptor: crystal structures of free and

ferrichrome-bound states reveal allosteric changes. *Cell* **95**:771–778.

Locher, K. P., and J. P. Rosenbusch. 1997. Oligomeric states and siderophore binding of the ligand-gated FhuA protein that forms channels across *Escherichia coli* outer membranes. *Eur. J. Biochem.* **247**:770–775.

Lundin, V. F., P. C. Stirling, J. Gomez-Reino, J. C. Mwenifumbo, J. M. Obst, J. M. Valpuesta, and M. R. Leroux. 2004. Molecular clamp mechanism of substrate binding by hydrophobic coiled-coil residues of the archaeal chaperone prefoldin. *Proc. Natl. Acad. Sci. USA* **101**:4367–4372.

Markello, T., A. Zlotnick, J. Everett, J. Tennyson, and P. W. Holloway. 1985. Determination of the topography of cytochrome b_5 in lipid vesicles by fluorescence quenching. *Biochemistry* **24**:2895–2901.

Marsh, D., and T. Páli. 2001. Infrared dichroism from the X-ray structure of bacteriorhodopsin. *Biophys. J.* **80**:305–312.

Marsh, D., B. Shanmugavadivu, and J. H. Kleinschmidt. 2006. Membrane elastic fluctuations and the insertion and tilt of β-barrel proteins. *Biophys. J.* **91**:227–232.

Martin-Benito, J., J. Boskovic, P. Gomez-Puertas, J. L. Carrascosa, C. T. Simons, S. A. Lewis, F. Bartolini, N. J. Cowan, and J. M. Valpuesta. 2002. Structure of eukaryotic prefoldin and of its complexes with unfolded actin and the cytosolic chaperonin CCT. *EMBO J.* **21**:6377–6386.

McIntosh, T. J., and P. W. Holloway. 1987. Determination of the depth of bromine atoms in bilayers formed from bromolipid probes. *Biochemistry* **26**:1783–1788.

Mecsas, J., P. E. Rouviere, J. W. Erickson, T. J. Donohue, and C. A. Gross. 1993. The activity of σ^E, an *Escherichia coli* heat-inducible σ-factor, is modulated by expression of outer membrane proteins. *Genes Dev.* **7**:2618–2628.

Meyer, J. E., M. Hofnung, and G. E. Schulz. 1997. Structure of maltoporin from *Salmonella typhimurium* ligated with a nitrophenyl-maltotrioside. *J. Mol. Biol.* **266**:761–775.

Misra, R., M. Castillo-Keller, and M. Deng. 2000. Overexpression of protease-deficient DegP(S210A) rescues the lethal phenotype of *Escherichia coli* OmpF assembly mutants in a *degP* background. *J. Bacteriol.* **182**:4882–4888.

Missiakas, D., J. M. Betton, and S. Raina. 1996. New components of protein folding in extracytoplasmic compartments of *Escherichia coli* SurA, FkpA and Skp/OmpH. *Mol. Microbiol.* **21**:871–884.

Missiakas, D., M. P. Mayer, M. Lemaire, C. Georgopoulos, and S. Raina. 1997. Modulation of the *Escherichia coli* σ^E (RpoE) heat-shock transcription-factor activity by the RseA, RseB and RseC proteins. *Mol. Microbiol.* **24**:355–371.

Mogensen, J. E., J. H. Kleinschmidt, M. A. Schmidt, and D. E. Otzen. 2005. Misfolding of a bacterial autotransporter. *Protein Sci.* **14:**2814–2827.

Müller, M., H. G. Koch, K. Beck, and U. Schäfer, 2001. Protein traffic in bacteria: multiple routes from the ribosome to and across the membrane. *Prog. Nucleic Acid Res. Mol. Biol.* **66:**107–157.

Niederweis, M. 2003. Mycobacterial porins—new channel proteins in unique outer membranes. *Mol. Microbiol.* **49:**1167–1177.

Nikaido, H., and M. Vaara. 1985. Molecular basis of bacterial outer membrane permeability. *Microbiol. Rev.* **49:**1–32.

Norgren, M., M. Baga, J. M. Tennent, and S. Normark. 1987. Nucleotide sequence, regulation and functional analysis of the *papC* gene required for cell surface localization of Pap pili of uropathogenic *Escherichia coli. Mol. Microbiol.* **1:**169–178.

Onufryk, C., M. L. Crouch, F. C. Fang, and C. A. Gross. 2005. Characterization of six lipoproteins in the σ^E regulon. *J. Bacteriol.* **187:**4552–4561.

Oomen, C. J., P. Van Ulsen, P. Van Gelder, M. Feijen, J. Tommassen, and P. Gros. 2004. Structure of the translocator domain of a bacterial autotransporter. *EMBO J.* **23:**1257–1266.

Paschen, S. A., T. Waizenegger, T. Stan, M. Preuss, M. Cyrklaff, K. Hell, D. Rapaport, and W. Neupert. 2003. Evolutionary conservation of biogenesis of β-barrel membrane proteins. *Nature* **426:**862–866.

Pautsch, A., and G. E. Schulz. 1998. Structure of the outer membrane protein A transmembrane domain. *Nat. Struct. Biol.* **5:**1013–1017.

Pautsch, A., and G. E. Schulz. 2000. High-resolution structure of the OmpA membrane domain. *J. Mol. Biol.* **298:**273–282.

Pegues, J. C., L. S. Chen, A. W. Gordon, L. Ding, and W. G. Coleman, Jr. 1990: Cloning, expression, and characterization of the *Escherichia coli* K-12 *rfaD* gene. *J. Bacteriol.* **172:**4652–4660.

Piknova, B., D. Marsh, and T. E. Thompson. 1997. Fluorescence quenching and electron spin resonance study of percolation in a two-phase lipid bilayer containing bacteriorhodopsin. *Biophys. J.* **72:**2660–2668.

Pocanschi, C. L., H.-J. Apell, P. Puntervoll, B. T. Høgh, H. B. Jensen, W. Welte, and J. Kleinschmidt. 2006. The major outer membrane protein of *Fusobacterium nucleatum* (FomA) folds and inserts into lipid bilayers via parallel folding pathways. *J. Mol. Biol.* **355:**548–561.

Pogliano, J., A. S. Lynch, D. Belin, E. C. Lin, and J. Beckwith. 1997. Regulation of *Escherichia coli* cell envelope proteins involved in protein folding and degradation by the Cpx two-component system. *Genes Dev.* **11:**1169–1182.

Prieto, M. J., M. Castanho, A. Coutinho, A. Ortiz, F. J. Aranda, and J. C. Gomez-Fernandez. 1994. Fluorescence study of a derivatized diacylglycerol incorporated in model membranes. *Chem. Phys. Lipids* **69:**75–85.

Prince, S. M., M. Achtman, and J. P. Derrick. 2002. Crystal structure of the OpcA integral membrane adhesin from Neisseria meningitidis. *Proc. Natl. Acad. Sci. USA* **99:**3417–3421.

Puntervoll, P., M. Ruud, L. J. Bruseth, H. Kleivdal, B. T. Høgh, R. Benz, and H. B. Jensen. 2002. Structural characterization of the *fusobacterial* non-specific porin FomA suggests a 14-stranded topology, unlike the classical porins. *Microbiology* **148:**3395–3403.

Rahfeld, J. U., K. P. Rücknagel, B. Schelbert, B. Ludwig, J. Hacker, K. Mann, and G. Fischer. 1994. Confirmation of the existence of a third family among peptidyl-prolyl *cis/trans* isomerases. Amino acid sequence and recombinant production of parvulin. *FEBS Lett.* **352:**180–184.

Raina, S., and C. Georgopoulos. 1991. The *htrM* gene, whose product is essential for *Escherichia coli* viability only at elevated temperatures, is identical to the *rfaD* gene. *Nucleic Acids Res.* **19:**3811–3819.

Raina, S., D. Missiakas, and C. Georgopoulos. 1995. The *rpoE* gene encoding the σ^E (σ^{24}) heat shock σ-factor of *Escherichia coli. EMBO J.* **14:**1043–1055.

Ramm, K., and A. Plückthun. 2000. The periplasmic *Escherichia coli* peptidylprolyl *cis,trans*-isomerase FkpA. II. Isomerase-independent chaperone activity in vitro. *J. Biol. Chem.* **275:**17106–17113.

Ramm, K., and A. Plückthun. 2001. High enzymatic activity and chaperone function are mechanistically related features of the dimeric *E. coli* peptidyl-prolyl-isomerase FkpA. *J. Mol. Biol.* **310:**485–498.

Ried, G., I. Hindennach, and U. Henning. 1990. Role of lipopolysaccharide in assembly of *Escherichia coli* outer membrane proteins OmpA, OmpC, and OmpF. *J. Bacteriol.* **172:**6048–6053.

Rizzitello, A. E., J. R. Harper, and T. J. Silhavy. 2001. Genetic evidence for parallel pathways of chaperone activity in the periplasm of *Escherichia coli. J. Bacteriol.* **183:**6794–6800.

Rodionova, N. A., S. A. Tatulian, T. Surrey, F. Jähnig, and L. K. Tamm. 1995. Characterization of two membrane-bound forms of OmpA. *Biochemistry* **34:**1921–1929.

Rouvière, P. E., A. De Las Penas, J. Mecsas, C. Z. Lu, K. E. Rudd, and C. A. Gross. 1995. *rpoE*, the gene encoding the second heat-shock σ-factor, σ^E, in *Escherichia coli. EMBO J.* **14:**1032–1042.

Rouvière, P. E., and C. A. Gross. 1996. SurA, a periplasmic protein with peptidyl-prolyl isomerase activity, participates in the assembly of outer membrane porins. *Genes Dev.* **10:**3170–3182.

Roy, A. M., and J. Coleman. 1994. Mutations in *firA*, encoding the second acyltransferase in lipopolysaccharide biosynthesis, affect multiple steps in lipopolysaccharide biosynthesis. *J. Bacteriol.* **176:** 1639–1646.

Ruiz, N., B. Falcone, D. Kahne, and T. J. Silhavy. 2005. Chemical conditionality: a genetic strategy to probe organelle assembly. *Cell* **121:**307–317.

Sampson, B. A., R. Misra, and S. A. Benson. 1989. Identification and characterization of a new gene of *Escherichia coli* K-12 involved in outer membrane permeability. *Genetics* **122:**491–501.

Saul, F. A., J. P. Arie, B. Vulliez-le Normand, R. Kahn, J. M. Betton, and G. A. Bentley. 2004. Structural and functional studies of FkpA from *Escherichia coli*, a *cis/trans* peptidyl-prolyl isomerase with chaperone activity. *J. Mol. Biol.* **335:**595–608.

Saul, F. A., M. Mourez, B. Vulliez-Le Normand, N. Sassoon, G. A. Bentley, and J. M. Betton. 2003. Crystal structure of a defective folding protein. *Protein Sci.* **12:**577–585.

Schäfer, U., K. Beck, and M. Müller. 1999. Skp, a molecular chaperone of gram-negative bacteria, is required for the formation of soluble periplasmic intermediates of outer membrane proteins. *J. Biol. Chem.* **274:**24567–24574.

Schirmer, T., T. A. Keller, Y. F. Wang, and J. P. Rosenbusch. 1995. Structural basis for sugar translocation through maltoporin channels at 3.1 Å resolution. *Science* **267:**512–514.

Schlapschy, M., M. K. Dommel, K. Hadian, M. Fogarasi, I. P. Korndörfer, and A. Skerra. 2004. The periplasmic *E. coli* chaperone Skp is a trimer in solution: biophysical and preliminary crystallographic characterization. *Biol. Chem.* **385:**137–143.

Schnaitman, C. A., and J. D. Klena. 1993. Genetics of lipopolysaccharide biosynthesis in enteric bacteria. *Microbiol. Rev.* **57:**655–682.

Schulz, G. E. 2002. The structure of bacterial outer membrane proteins. *Biochim. Biophys. Acta* **1565:** 308–317.

Schwede, T., J. Kopp, N. Guex, and M. C. Peitsch. 2003. SWISS-MODEL: an automated protein homology-modeling server. *Nucleic Acids Res.* **31:** 3381–3385.

Schweizer, M., I. Hindennach, W. Garten, and U. Henning. 1978. Major proteins of the *Escherichia coli* outer cell envelope membrane. Interaction of protein II with lipopolysaccharide. *Eur. J. Biochem.* **82:**211–217.

Sheng, M., and C. Sala. 2001. PDZ domains and the organization of supramolecular complexes. *Annu. Rev. Neurosci.* **24:**1–29.

Shepard, L. A., A. P. Heuck, B. D. Hamman, J. Rossjohn, M. W. Parker, K. R. Ryan, A. E. Johnson, and R. K. Tweten. 1998. Identification of a membrane-spanning domain of the thiol-activated pore-forming toxin *Clostridium perfringens* perfringolysin O: an α-helical to β-sheet transition identified by fluorescence spectroscopy. *Biochemistry* **37:**14563–14574.

Siegert, R., M. R. Leroux, C. Scheufler, F. U. Hartl, and I. Moarefi. 2000. Structure of the molecular chaperone prefoldin: unique interaction of multiple coiled coil tentacles with unfolded proteins. *Cell* **103:**621–632.

Sijbrandi, R., M. L. Urbanus, C. M. ten Hagen-Jongman, H. D. Bernstein, B. Oudega, B. R. Otto, and J. Luirink. 2003. Signal recognition particle (SRP)-mediated targeting and Sec-dependent translocation of an extracellular *Escherichia coli* protein. *J. Biol. Chem.* **278:**4654–4659.

Snijder, H. J., I. Ubarretxena-Belandia, M. Blaauw, K. H. Kalk, H. M. Verheij, M. R. Egmond, N. Dekker, and B. W. Dijkstra. 1999. Structural evidence for dimerization-regulated activation of an integral membrane phospholipase. *Nature* **401:**717–721.

Song, L., M. R. Hobaugh, C. Shustak, S. Cheley, H. Bayley, and J. E. Gouaux. 1996. Structure of staphylococcal α-hemolysin, a heptameric transmembrane pore. *Science* **274:**1859–1866.

Spiess, C., A. Beil, and M. Ehrmann. 1999. A temperature-dependent switch from chaperone to protease in a widely conserved heat shock protein. *Cell* **97:**339–347.

Steeghs, L., H. de Cock, E. Evers, B. Zomer, J. Tommassen, and P. van der Ley. 2001. Outer membrane composition of a lipopolysaccharide-deficient *Neisseria meningitidis* mutant. *EMBO J.* **20:**6937–6945.

Stoller, G., K. P. Rücknagel, K. H. Nierhaus, F. X. Schmid, G. Fischer, and J. U. Rahfeld. 1995. A ribosome-associated peptidyl-prolyl *cis/trans* isomerase identified as the trigger factor. *EMBO J.* **14:**4939–4948.

Strauch, K. L., K. Johnson, and J. Beckwith. 1989. Characterization of *degP*, a gene required for proteolysis in the cell envelope and essential for growth of *Escherichia coli* at high temperature. *J. Bacteriol.* **171:**2689–2696.

Struyve, M., M. Moons, and J. Tommassen. 1991. Carboxy-terminal phenylalanine is essential for the correct assembly of a bacterial outer membrane protein. *J. Mol. Biol.* **218:**141–148.

Surana, N. K., S. Grass, G. G. Hardy, H. Li, D. G. Thanassi, and J. W. Geme III. 2004. Evidence for conservation of architecture and physical properties of Omp85-like proteins throughout evolution. *Proc. Natl. Acad. Sci. USA* **101:**14497–14502.

Surrey, T., and F. Jähnig. 1992. Refolding and oriented insertion of a membrane protein into a lipid bilayer. *Proc. Natl. Acad. Sci. USA* **89:**7457–7461.

Surrey, T., and F. Jähnig. 1995. Kinetics of folding and membrane insertion of a β-barrel membrane protein. *J. Biol. Chem.* **270:**28199–28203.

Surrey, T., A. Schmid, and F. Jähnig. 1996. Folding and membrane insertion of the trimeric β-barrel protein OmpF. *Biochemistry* **35:**2283–2288.

Tamm, L. K., H. Hong, and B. Liang. 2004. Folding and assembly of β-barrel membrane proteins. *Biochim. Biophys. Acta* **1666:**250–263.

Taniguchi, N., S. I. Matsuyama, and H. Tokuda. 2005. Mechanisms underlying energy-independent transfer of lipoproteins from LolA to LolB, which have similar unclosed β-barrel structures. *J. Biol. Chem.* **280:**34481–34488.

Thanassi, D. G. 2002. Ushers and secretins: channels for the secretion of folded proteins across the bacterial outer membrane. *J. Mol. Microbiol. Biotechnol.* **4:**11–20.

Thanassi, D. G., C. Stathopoulos, K. Dodson, D. Geiger, and S. J. Hultgren. 2002. Bacterial outer membrane ushers contain distinct targeting and assembly domains for pilus biogenesis. *J. Bacteriol.* **184:**6260–6269.

Thome, B. M., H. K. Hoffschulte, E. Schiltz, and M. Müller. 1990. A protein with sequence identity to Skp (FirA) supports protein translocation into plasma membrane vesicles of *Escherichia coli.* *FEBS Lett.* **269:**113–116.

Tokuda, H., and S. Matsuyama. 2004. Sorting of lipoproteins to the outer membrane in *E. coli.* *Biochim. Biophys. Acta* **1694:**IN1–IN9.

Tormo, A., M. Almiron, and R. Kolter. 1990. *surA*, an *Escherichia coli* gene essential for survival in stationary phase. *J. Bacteriol.* **172:**4339–4347.

Tuteja, R. 2005. Type I signal peptidase: an overview. *Arch. Biochem. Biophys.* **441:**107–111.

Vainberg, I. E., S. A. Lewis, H. Rommelaere, C. Ampe, J. Vandekerckhove, H. L. Klein, and N. J. Cowan. 1998. Prefoldin, a chaperone that delivers unfolded proteins to cytosolic chaperonin. *Cell* **93:**863–873.

van den Berg, B., P. N. Black, W. M. Clemons, Jr., and T. A. Rapoport. 2004a. Crystal structure of the long-chain fatty acid transporter FadL. *Science* **304:**1506–1509.

van den Berg, B., W. M. Clemons, Jr., I. Collinson, Y. Modis, E. Hartmann, S. C. Harrison, and T. A. Rapoport. 2004b. X-ray structure of a protein-conducting channel. *Nature* **427:**36–44.

Vandeputte-Rutten, L., M. P. Bos, J. Tommassen, and P. Gros. 2003. Crystal structure of *Neisserial* surface protein A (NspA), a conserved outer membrane protein with vaccine potential. *J. Biol. Chem.* **278:**24825–24830.

Vandeputte-Rutten, L., R. A. Kramer, J. Kroon, N. Dekker, M. R. Egmond, and P. Gros. 2001. Crystal structure of the outer membrane protease OmpT from *Escherichia coli* suggests a novel catalytic site. *EMBO J.* **20:**5033–5039.

Vogel, H., and F. Jähnig. 1986. Models for the structure of outer-membrane proteins of *Escherichia coli* derived from Raman spectroscopy and prediction methods. *J. Mol. Biol.* **190:**191–199.

Vogt, J., and G. E. Schulz. 1999. The structure of the outer membrane protein OmpX from *Escherichia coli* reveals possible mechanisms of virulence. *Structure Fold. Des.* **7:**1301–1309.

Voulhoux, R., M. P. Bos, J. Geurtsen, M. Mols, and J. Tommassen. 2003. Role of a highly conserved bacterial protein in outer membrane protein assembly. *Science* **299:**262–265.

Walsh, N. P., B. M. Alba, B. Bose, C. A. Gross, and R. T. Sauer. 2003. OMP peptide signals initiate the envelope-stress response by activating DegS protease via relief of inhibition mediated by its PDZ domain. *Cell* **113:**61–71.

Walton, T. A., and M. C. Sousa. 2004. Crystal structure of Skp, a prefoldin-like chaperone that protects soluble and membrane proteins from aggregation. *Mol. Cell* **15:**367–374.

Wang, X., M. J. Karbarz, S. C. McGrath, R. J. Cotter, and C. R. Raetz. 2004. MsbA transporter-dependent lipid A 1-dephosphorylation on the periplasmic surface of the inner membrane: topography of francisella novicida LpxE expressed in *Escherichia coli.* *J. Biol. Chem.* **279:**49470–49478.

Wang, Y. F., R. Dutzler, P. J. Rizkallah, J. P. Rosenbusch, and T. Schirmer. 1997. Channel specificity: structural basis for sugar discrimination and differential flux rates in maltoporin. *J. Mol. Biol.* **272:**56–63.

Webb, H. M., L. W. Ruddock, R. J. Marchant, K. Jonas, and P. Klappa. 2001. Interaction of the periplasmic peptidylprolyl *cis-trans* isomerase SurA with model peptides. The N-terminal region of SurA id essential and sufficient for peptide binding. *J. Biol. Chem.* **276:**45622–45627.

Weiss, M. S., A. Kreusch, E. Schiltz, U. Nestel, W. Welte, J. Weckesser, and G. E. Schulz. 1991. The structure of porin from *Rhodobacter capsulatus* at 1.8 Å resolution. *FEBS Lett.* **280:**379–382.

Weiss, M. S., and G. E. Schulz. 1992. Structure of porin refined at 1.8 Å resolution. *J. Mol. Biol.* **227:** 493–509.

Werner, J., and R. Misra. 2005. YaeT (Omp85) affects the assembly of lipid-dependent and lipid-independent outer membrane proteins of *Escherichia coli.* *Mol. Microbiol.* **57:**1450–1459.

Werner, J., A. M. Augustus, and R. Misra. 2003. Assembly of TolC, a structurally unique and multifunctional outer membrane protein of *Escherichia coli* K-12. *J. Bacteriol.* **185:**6540–6547.

Whitlow, M., A. J. Howard, J. F. Wood, E. W. Voss, Jr., and K. D. Hardman. 1995. 1.85 A structure of anti-fluorescein 4-4-20 Fab. *Protein Eng.* **8:**749–761.

Wiener, M. C., and S. H. White. 1991. Transbilayer distribution of bromine in fluid bilayers containing a specifically brominated analogue of dioleoylphosphatidylcholine. *Biochemistry* **30:**6997–7008.

Wilken, C., K. Kitzing, R. Kurzbauer, M. Ehr-mann, and T. Clausen. 2004. Crystal structure of the DegS stress sensor: How a PDZ domain recognizes misfolded protein and activates a protease. *Cell* **117:**483–494.

Williamson, I. M., S. J. Alvis, J. M. East, and A. G. Lee. 2002. Interactions of phospholipids with the potassium channel KcsA. *Biophys. J.* **83:** 2026–2038.

Wimley, W. C. 2002. Toward genomic identification of β-barrel membrane proteins: composition and architecture of known structures. *Protein Sci.* **11:** 301–312.

Wimley, W. C. 2003. The versatile β-barrel membrane protein. *Curr. Opin. Struct. Biol.* **13:**404–411.

Wu, T., J. Malinverni, N. Ruiz, S. Kim, T. J. Silhavy, and D. Kahne. 2005. Identification of a multi-component complex required for outer membrane biogenesis in *Escherichia coli. Cell* **121:**235–245.

Wülfing, C., and A. Plückthun. 1994. Protein folding in the periplasm of *Escherichia coli. Mol. Microbiol.* **12:**685–692.

Ye, J., and B. van den Berg. 2004. Crystal structure of the bacterial nucleoside transporter Tsx. *EMBO J.* **23:**3187–3195.

Yue, W. W., S. Grizot, and S. K. Buchanan. 2003. Structural evidence for iron-free citrate and ferric citrate binding to the TonB-dependent outer membrane transporter FecA. *J. Mol. Biol.* **332:**353–368.

Zeth, K., K. Diederichs, W. Welte, and H. Engel-hardt. 2000. Crystal structure of Omp32, the anion-selective porin from *Comamonas acidovorans*, in complex with a periplasmic peptide at 2.1 Å resolution. *Structure Fold. Des.* **8:**981–992.

STRUCTURE, FUNCTION, AND TRANSPORT OF LIPOPROTEINS IN *ESCHERICHIA COLI*

Hajime Tokuda, Shin-ichi Matsuyama, and Kimie Tanaka-Masuda

4

BIOGENESIS OF LIPOPROTEINS

The outer membrane of gram-negative bacteria such as *Escherichia coli* is composed of proteins, phospholipids, and lipopolysaccharide (LPS). LPS is present exclusively in the outer leaflet of this membrane (Muhlradt and Golecki, 1975), whereas phospholipids are mostly localized in its inner leaflet. This asymmetrical bilayer contains a few species of major proteins and several lipid-modified proteins, so-called lipoproteins. The major proteins span the outer membrane, whereas most lipoproteins are anchored to the outer membrane through the attached lipids. Neither the major outer membrane proteins nor the outer membrane-specific lipoproteins have hydrophobic stretches that form a transmembrane α-helix and function as stop transfer or signal anchor sequences (von Heijne, 1996). Instead, the major outer membrane proteins span the membrane with amphipathic β-strands possessing alternating

hydrophobic residues, which do not cause the retention of proteins in the inner membrane. Modification of lipoproteins with lipids occurs on the outer surface of the inner membrane (Pugsley, 1993) and therefore does not inhibit the translocation of protein moiety (Yakushi et al., 2000). Thus, both the β-structure and lipid modification are characteristic of outer membrane-associated proteins.

Lipoproteins are synthesized as a precursor with a signal peptide at the N terminus and then translocated across the inner membrane by Sec machinery (Hayashi and Wu, 1990; Pugsley, 1993). A lipoprotein precursor has a consensus sequence called a lipobox or lipoprotein box around the signal peptide cleavage site (Hayashi and Wu, 1990). Processing of the precursor to the mature form sequentially occurs on the periplasmic side of the inner membrane (Fig. 1): formation of a thioether linkage between the N-terminal Cys residue of the mature region and diacylglycerol by phosphatidylglycerol:prolipoprotein diacylglyceryl transferase (Lgt), cleavage of the signal peptide by prolipoprotein signal peptidase (LspA or signal peptidase II), and aminoacylation of the N-terminal Cys residue by phospholipid:apolipoprotein transacylase (Lnt). The mature lipoprotein thus formed has a lipid-attached Cys at the N terminus (Sankaran and Wu, 1994).

Hajime Tokuda, Institute of Molecular and Cellular Biosciences, University of Tokyo, 1-1-1 Yayoi, Bunkyo-ku, Tokyo 113-0032, Japan. *Shin-ichi Matsuyama,* Department of Life Science, Rikkyo University, 3-34-1 Nishi-ikebukuro, Toshima-ku, Tokyo 171-8501, Japan. *Kimie Tanaka-Masuda*, Institute of Molecular and Cellular Biosciences, University of Tokyo, 1-1-1 Yayoi, Bunkyo-ku, Tokyo 113-0032, Japan. Present address: Kyowa Hakko Kogyo Co., Ltd., 3-6-6 Asahi-machi, Machida-shi, Tokyo 194-8533, Japan.

The Periplasm
Edited by Michael Ehrmann © 2007 ASM Press, Washington, D.C.

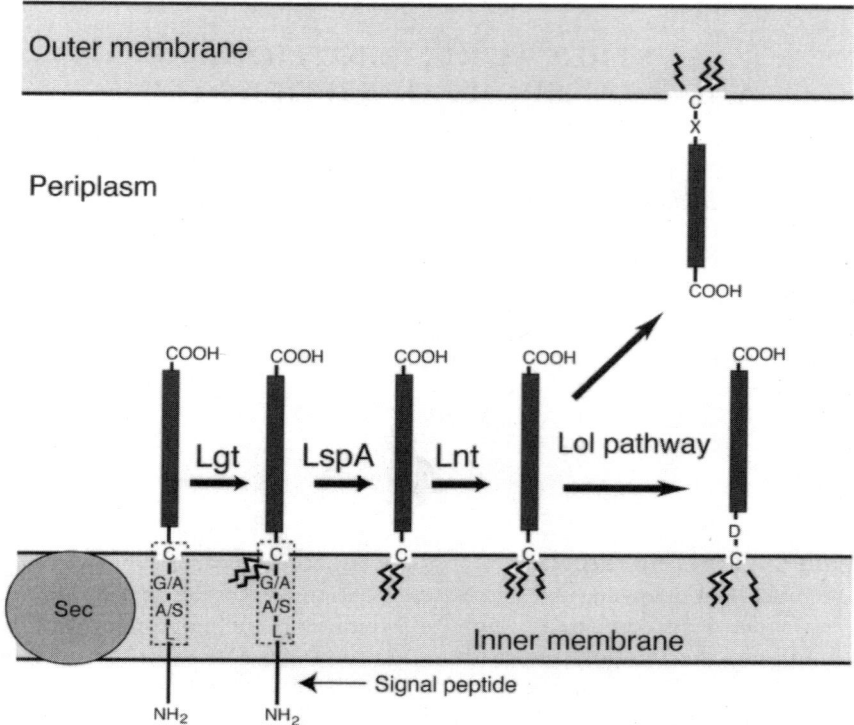

FIGURE 1 Biogenesis of lipoproteins. Lipoprotein precursors have a consensus sequence, -L-(A/S)-(G/A)-C-, called a lipobox (broken squares), around a signal cleavage site. X represents a residue other than Asp. Lgt, phosphatidylglycerol:prolipoprotein diacylglyceryl-transferase; LspA, prolipoprotein signal peptidase (also called Spase II); Lnt, phospholipid: apolipoprotein transacylase. A Lol pathway discussed later mediates the outer membrane localization of lipoproteins in a manner dependent on sorting signals.

Globomycin (Inukai et al., 1978) specifically inhibits LspA and causes the accumulation of diacylglyceryl prolipoproteins in the inner membrane (Mizushima, 1987). The three lipoprotein-processing enzymes Lgt, LspA, and Lnt are widely conserved in gram-negative bacteria, whereas Lnt homologs have not been found in gram-positive bacteria, although a result suggesting the aminoacylation of *Staphylococcus aureus* lipoproteins has been reported (Navarre et al., 1996).

In *E. coli*, lipoproteins are anchored to the periplasmic side of either the inner or outer membrane depending on the lipoprotein-sorting signal (Pugsley, 1993). Some gram-negative bacteria are known to possess lipoproteins on the outer surface of their outer membranes

(Haake, 2000). In gram-positive bacteria, lipoproteins are anchored to the outer leaflet of the cytoplasmic membrane.

ANALYSIS OF PUTATIVE LIPOPROTEIN GENE PRODUCTS

The complete genome sequence revealed many putative lipoprotein genes in various bacteria, for example, 114 are predicted for gram-positive *Bacillus subtilis* (Tjalsma et al., 1999), 105 for Lyme disease spirochete *Borrelia burgdorferi* (Fraser et al., 1997), and more than 100 for *E. coli* (Juncker et al., 2003; Brokx et al., 2004). Various membrane-associated activities presumably depend on lipoproteins. However, most lipoproteins, including even those of *E. coli*, have no known function and are only predicted to be lipoproteins. We recently cloned almost all the putative lipoprotein genes of *E. coli* and examined whether the proteins encoded by these genes could be labeled with radioactive palmitate. Some proteins were further examined as to their sensitivity to globomycin when palmitate labeling did not provide unequivocal results. Previously identified lipoproteins were also examined as controls. These analyses revealed that *E. coli* possesses at least 90 lipoproteins (Table 1). The protein moieties of most lipoproteins are predicted to be soluble and presumably exposed to the periplasm, suggesting that they play important roles in various activities in the periplasm.

The products of three putative lipoprotein genes, *yebF*, *yliB*, and *ymcA*, were neither labeled with palmitate nor inhibited by globomycin. We therefore determined their N-terminal sequences and found that these three proteins are processed to nonlipidated mature forms, indicating that the three genes do not encode lipoproteins. Three other gene products, YiaM, YjbH, and YpdI, were also resistant to globomycin and not labeled with palmitate. They are unlikely to be lipoproteins, although this has not been confirmed by N-terminal sequencing.

We disrupted each of the 90 lipoprotein genes and examined the growth of cells under various conditions. Only two lipoproteins, LolB (Tanaka et al., 2001) and YfiO (Onufryk et al., 2005), were essential as reported. Disruption of the lipoprotein genes encoding DcrB, NlpI, Pal, RlpA, RlpB, Spr, and YcfM caused

TABLE 1 Biochemically confirmed lipoproteins in *E. coli*[a]

Name	Synonym	A	B	C	Name	Synonym	A	B	C	Name	Synonym	A	B	C
AcrA		√	√		Slp		√	√		YehR			√	√
AcrE	EnvC	√	√		SlyB		√	√		YfeY			√	
ApbE		√	√		Spr			√		YfgH			√	
Blc		√	√		VacJ			√		YfgL			√	
CsgG		√	√		Wza			√		YfhG				√
CyoA		√	√		YaeC			√		YfiB				√
DcrB				√	YaeF			√		YfiL			√	
EcnA		√		√	YafT			√		YfiO	EcfD		√	
EcnB		√		√	YafY			√		YfjS			√	
FlgH		√	√		YaiW				√	YgdI			√	
HslJ			√		YajG			√		YgdR				√
LolB		√	√	√	YajI			√		YgeR			√	
Lpp		√	√	√	YbaY				√	YggG			√	
MltA		√	√		YbcU	BorD		√		YghG			√	
MltB		√	√		YbfN			√		YhfL			√	
MltC		√	√		YbfP			√	√	YhiU			√	
MltD			√		YbhC			√		YiaD			√	
MltE		√	√		YbjP			√		YidQ	EcfI			√
NlpA		√	√		YbjR			√		YiiG				√
NlpB		√	√		YcaL			√		YjbF			√	
NlpC				√	YccZ			√		YjeI			√	
NlpD		√	√		YcdR			√		YjfO			√	
NlpE	CutF	√	√		YceB			√		YlcB	CusC		√	
NlpI		√	√		YcfM			√		YmcC			√	
OsmB		√	√		YcjN			√		YnbE			√	
OsmE				√	YdhY			√		YnfC			√	
Pal	ExcC	√	√		YeaY			√		YoaF			√	
RcsF			√		YecR			√	√	YqhH			√	
RlpA		√	√		YedD			√		Yrak			√	
RlpB		√	√		YegR			√		YraP	EcfH		√	

[a]A, lipoproteins identified previously; B, labeling with radioactive palmitate; C, inhibition of precursor processing by globomycin.

temperature-sensitive growth, whereas that of the gene encoding YegR caused cold-sensitive growth in a strain-dependent manner. Disruption of the genes encoding AcrA, Lpp, NlpE, NlpI, Pal, Spr, YcfM, and YfgL made cells hypersensitive to one or many drugs. AcrA is a component of a multidrug export system (Ma et al., 1993). It has been proposed that YfgL and YfiO together with NlpB form a complex with YaeT, which is an *E. coli* homolog of Omp85 (Genevrois et al., 2003; Voulhoux et al., 2003), and are involved in outer membrane insertion of β-barrel proteins (Wu et al., 2005). Loss of the outer membrane integrity is one of the reasons for the drug hypersensitivity caused by the disruption of some lipoprotein genes.

Among previously identified lipoproteins, only NlpE is known upon overproduction to induce DegP, which is a periplasmic protease (Danese et al., 1995; Snyder et al., 1995). Overproduction of NlpE activates a two-component signal transduction system comprising CpxA and CpxR, and then stimulates the expression of *degP* (Danese et al., 1995; Snyder et al., 1995), which is also positively regulated by σ^E. This stress response system involving the Cpx two-component system and σ^E is thought to represent the quality control mechanism in the periplasm (Danese and Silhavy, 1997; Alba and Gross, 2004). When unfolded proteins are accumulated in the periplasm, DegP is induced to clean the periplasm. We overproduced each of the 90 lipoproteins and examined the level of *degP* expression. In addition to NlpE, new inner membrane lipoprotein YafY strongly induced *degP* expression in a manner dependent on the Cpx two-component system (Miyadai et al., 2004). The amino acid sequences of YafY and YfjS, another inner membrane lipoprotein, are almost identical, but overproduction of the latter did not induce *degP* expression. Construction of various YafY-YfjS chimeric lipoproteins revealed that only a few residues located in the N- and C-terminal regions were important for the induction of DegP. It is possible that NlpE and YafY are both components of a regulatory network, which also consists of the Cpx system, and that overproduction of one of the two lipoproteins disturbs the controlled expression of DegP.

IN VIVO ANALYSIS OF LIPOPROTEIN SORTING SIGNALS

Yamaguchi et al. (1988) first proposed the importance of the residue at position 2 of lipoproteins as to their destinations. They showed that replacement of Ser at position 2 of an outer membrane-specific lipoprotein by Asp caused the protein to remain in the inner membrane. Furthermore, replacement of Asp at position 2 of an inner membrane-specific lipoprotein by another residue caused outer membrane localization of the protein. These results strongly suggested that Asp at position 2 functions as an inner membrane retention signal for lipoproteins, whereas other residues cause outer membrane localization. However, they later found that lipoproteins having a certain residue at position 3 were localized in the outer membrane even though Asp was present at position 2 (Gennity and Inouye, 1991).

Pugsley and coworkers reported methods for comprehensive examination of the inner membrane retention signals of lipoproteins (Seydel et al., 1999). They constructed a maltose binding protein (MalE) derivative having a lipid-modified Cys at the N terminus (lipoMalE), and then expressed it in a chromosomal *malE* deletion mutant. When Asp was at the N-terminal second position, lipoMalE was localized in the inner membrane, where it functioned as a maltose binding protein, thereby enabling the mutant to grow in the presence of maltose. In contrast, lipoMalE localized in the outer membrane was not functional and did not support *malE* mutant growth. Systematic substitution of the residue at position 2 of lipoMalE revealed that in addition to Asp, five other residues (Phe, Trp, Tyr, Gly, and Pro) supported mutant growth, indicating that Asp at position 2 is not the sole inner membrane retention signal. However, *E. coli* native lipoproteins do not have Phe, Trp, Tyr, Gly, or Pro at position 2. Pugsley and coworkers (Robichon et al., 2003) then used a phage T5-encoded lipoprotein, Llp, which has Ser at position 2 and is localized in the outer membrane of *E. coli*. Llp localized in the outer membrane inhibits FhuA, an outer membrane receptor for phage T5, thereby protecting cells from phage T5 infection. They found that the Llp derivative having Asp at position 2 was more than 40% targeted to the outer membrane. However, the outer membrane targeting of this derivative was abolished on the insertion of a peptide spacer between Asp and the residue at position 3. From these results, they concluded that the conformation of the Llp derivative inhibited the recognition of the Asp residue at position 2.

To discuss the molecular mechanisms underlying lipoprotein sorting, we must understand the functions of the Lol pathway mediating the inner to outer membrane transport of lipoproteins.

THE LOL PATHWAY

Structure and Function of LolA and LolB

E. coli spheroplasts secrete various protein species that are destined for the periplasm or outer membrane. In contrast, the major outer membrane lipoprotein Lpp remained in the inner membrane of spheroplasts as a mature form unless periplasmic materials were added externally. The periplasmic Lpp-releasing factor (20 kDa) was purified and named LolA (Matsuyama et al., 1995). LolA was then shown to release other outer membrane lipoproteins such as Pal, NlpB, Slp, and RlpA, whereas inner membrane lipoproteins AcrA and NlpA were not released even in the presence of LolA (Yokota et al., 1999), indicating that LolA plays a critical role in the sorting of lipoproteins. Lipoproteins released from spheroplasts in the presence of LolA existed as a water-soluble complex with LolA. When this complex was incubated with the outer membrane, lipoproteins were incorporated into the outer membrane. Proteoliposomes were reconstituted from solubilized outer membrane proteins and *E. coli*

phospholipids to identify the outer membrane protein conferring this lipoprotein incorporation activity. The protein (23 kDa), named LolB, was found to be a novel outer membrane lipoprotein possessing receptor activity for lipoproteins (Matsuyama et al., 1997). A lipoprotein-LolB complex was formed upon incubation of the lipoprotein-LolA complex and LolB, indicating the transfer of the lipoprotein from LolA to LolB. The release and outer membrane localization of LolB per se also depend on LolA and LolB, respectively (Yokota et al., 1999).

LolA (Matsuyama et al., 1995) and LolB (Matsuyama et al., 1997) each function in a monomeric form. Although there is no apparent homology between the amino acid sequences of LolA and LolB, the crystal structures of LolA and LolB solved at 1.65 and 1.9 Å resolution, respectively, are strikingly similar to each other (Takeda et al., 2003). Both have a hydrophobic cavity consisting of an unclosed β-barrel and an α-helical lid (Color Plate 6). The inner surface of the β-sheet and three α-helices consists of hydrophobic residues. The hydrophobic cavity represents a possible binding site for the lipid moieties of lipoproteins. The side chain of Arg43 is oriented toward the interior of LolA and hydrogen bonded to the main chain carbonyls of residues in the α1- and α2-helices (not indicated), thereby fixing the helices to the β2-strand like a lid. Arg43 is the only residue that functions as a lock disconnecting the hydrophobic cavity of LolA from the solvent region. In contrast, the loops of LolB do not disconnect its hydrophobic cavity from the solvent. Indeed, one crystal form of LolB contained polyethylene glycol monomethyl ether, which was used in the crystallization process in its hydrophobic cavity.

The "lid" of LolA is expected to undergo opening and closing upon the accommodation and release of lipoproteins, respectively. It has been speculated that closing of the lid through hydrogen bonding between Arg at position 43 and loop residues is important for the efficient transfer of lipoproteins to LolB, whose lid is not closed (Takeda et al., 2003). To examine the functional importance of the LolA lid closing, Arg at position 43 was systematically mutagenized (Taniguchi et al., 2005). In contrast to the unique property of Arg at position 43, all derivatives except one having Leu instead of Arg (R43L) supported the growth of cells. All Arg43 derivatives retained the ability to accept lipoproteins from the inner membrane, whereas their abilities to transfer associated lipoproteins to LolB were variously reduced. It was then revealed that the hydrophobic interaction between LolA and lipoproteins should be weak for the efficient transfer of a lipoprotein from LolA to LolB, otherwise the LolA-lipoprotein complex accumulates in the periplasm (Taniguchi et al., 2005). The hydrophobic interaction between lipoproteins and the R43L derivative was as strong as that between lipoproteins and LolB. This seems to be the reason why the R43L derivative was most defective among all derivatives in the transfer of lipoproteins to LolB. It has been thought that a lack of hydrogen bonding between Arg at position 43 and loop residues causes growth inhibition because of the stabilization of liganded LolA relative to its free form (Takeda et al., 2003), which then leads to toxic mislocation of Lpp in the inner membrane (Yakushi et al., 1997). However, Val or Ile in place of Arg at position 43 did not inhibit growth (Taniguchi et al., 2005), suggesting that the formation of hydrogen bonding between Arg43 and residues in the loops is less important than previously speculated (Takeda et al., 2003). We speculate that other residues also contribute to the formation of the closed form of LolA (Taniguchi et al., 2005), which was found for the crystals of unliganded LolA.

Systematic mutagenesis of Phe at position 47 in the β3-strand of LolA revealed that replacement of Phe with polar residues severely inhibited its ability to accept lipoproteins (Taniguchi et al., 2005). It is strongly suggested that a polar residue at position 47 causes the formation of additional hydrogen bonds between the β3-strand and the lid helices, thereby inhibiting LolA lid opening.

Among five highly conserved Trp residues of LolB, replacement of the one at position 52 by Pro impaired the receptor activity of LolB and caused accumulation of the LolA-lipoprotein complex in the periplasm (Wada et al., 2004). In contrast, no defective mutant was obtained for Trp at position 183 despite its strong conservation.

A Novel ABC Transporter, LolCDE

The LolA-dependent release of lipoproteins from right-side-out membrane vesicles required ATP as well as an outer membrane-sorting signal (Yakushi et al., 1998). The inner membrane proteins conferring the release activity were purified by monitoring the activity reconstituted into proteoliposomes. The LolCDE complex thus identified has a subunit stoichiometry of $C_1D_2E_1$ with an expected molecular weight of 139,483 (Yakushi et al., 2000). The genes encoding the three proteins form an operon. LolD possesses Walker A and B motifs with the consensus sequence of the ABC (ATP binding cassette) transporter protein called the ABC signature motif (Higgins et al., 1986). Both LolC and LolE are predicted to span the membrane four times and to have a large domain exposed to the periplasm. The LolCDE complex is therefore an ABC transporter but mechanistically differs from all other ABC transporters in that it is not involved in the transmembrane transport of substrates. Mutations in the Walker A, B, and ABC signature motifs completely inhibit the release of lipoproteins (Yakushi et al., 2000; Masuda et al., 2002).

The amino acid sequences of membrane subunits LolC and LolE are similar to each other, the identity being 26%. Moreover, the N-terminal 60 residues of the two proteins exhibit 55% identity. Despite this sequence similarity, LolC and LolE are both essential for the release of lipoproteins (Narita et al., 2002). The Sec-dependent translocation and modification of lipoprotein precursors are completely independent of and not affected by Lol-mediated reactions since inhibition of the LolCDE function does not affect the translocation of a lipoprotein precursor across the inner membrane and subsequent processing to the mature lipoprotein (Yakushi et al., 2000; Narita et al., 2002).

LolD hydrolyzes ATP on the cytoplasmic side of the inner membrane, while LolC and/or LolE recognize and release lipoproteins from the periplasmic side of the inner membrane. Hence, communication between LolD and LolC/E to utilize the energy of ATP hydrolysis is essential for the lipoprotein release reaction. LolD contains a characteristic sequence called the LolD motif, which is highly conserved among LolD homologs but not other ABC transporters in *E. coli* (Yakushi et al., 2000). The LolD motif is located between the Walker A and ABC signature motifs and is suggested to be a contact site with LolC/E in the crystal structures of other ABC transporters (Locher et al., 2002; Smith et al., 2002). We isolated some dominant negative mutants having an altered residue in the LolD motif. Overexpression of these mutants arrested growth despite the chromosomal *lolD*⁺ background. Some mutations in the LolD motif decrease the ATPase activity of the LolCDE complex with little effect on the ATPase activity of the LolD subunit, suggesting that these mutations perturbed the communication between the membrane-spanning subunits and the ATPase subunit. We then selected suppressor mutations of the *lolC* and *lolE* genes that correct the growth defect caused by LolD mutants. Mutations of *lolC* suppressors were mainly located in the periplasmic loop, whereas those of *lolE* were mainly located in the cytoplasmic loop. These results strongly suggest that the mode of interaction with LolD differs between LolC and LolE (Y. Ito, H. Matsuzawa, S. Matsuyama, S. Narita, and H. Tokuda, unpublished observations).

E. coli is predicted to possess many ABC transporters (Linton and Higgins, 1998; Paulsen et al., 1998). As far as is known, LolCDE (Narita et al., 2002) and MsbA (Zhou et al., 1998) are the only essential ABC transporters of *E. coli*. MsbA is proposed to be involved in the transport of LPS from the inner to the outer membrane (Doerrler et al., 2001). Therefore, these two essential ABC transporters appear to be involved in the biogenesis of the *E. coli* envelope.

IN VITRO ANALYSIS OF LIPOPROTEIN-SORTING SIGNALS

Whether lipoproteins are specific to the inner or outer membrane, the mode of lipid modification at the N-terminal Cys is the same (Fukuda et al., 2002), indicating that the lipoprotein-sorting signal does not affect lipid modification but functions as a determinant of membrane localization.

To evaluate lipoprotein-sorting signals, the LolA-dependent release of lipoproteins from spheroplasts was examined. When the residue at position 3 was Ser, only Asp at position 2 caused the retention of lipoproteins in spheroplasts (Terada et al., 2001), confirming the importance of Asp at position 2 for the inner membrane retention of lipoproteins. However, residues at position 3 differentially affected the inner membrane retention of lipoproteins having Asp at position 2, indicating that Asp at position 2 alone is not sufficient for a strong inner membrane retention signal. The strongest inner membrane retention signals are Asp-Asp, Asp-Glu, and Asp-Gln. These signals have been found in native lipoproteins having Asp at position 2, whereas ambiguous sorting signals such as Asp-His, Asp-Lys, and Asp-Ile causing less efficient retention or release, and hence localization in both membranes (Gennity and Inouye, 1991; Seydel et al., 1999), have not been found in native lipoproteins. Since Asp-Asn is also a potent inner membrane retention signal, an acidic residue or its amide form at position 3 seems to make Asp at position 2 the strongest inner membrane retention signal. Judging from the results of analyses of lipoprotein-sorting signals, most of the *E. coli* lipoproteins listed in Table 1 seem to be localized in the outer membrane.

It was found that the LolA-dependent release of outer membrane-specific lipoproteins from proteoliposomes reconstituted with LolCDE was inhibited by other outer membrane-specific, but not inner membrane-specific, lipoproteins. Moreover, outer membrane-specific lipoproteins stimulated ATP hydrolysis by Lol-CDE, whereas inner membrane-specific ones did not. These results revealed a novel function of Asp at position 2, i.e., lipoproteins having this signal avoid being recognized by LolCDE, thereby remaining in the inner membrane (Masuda et al., 2002). A mutant that can release lipoproteins having Asp and Gln at positions 2 and 3, respectively, was isolated (Narita et al., 2003). The mutant carried an Ala to Pro mutation at position 40 of LolC. A significant portion of an inner membrane lipoprotein was localized to the outer membrane when this LolC mutant was expressed. LolA formed a complex with the released lipoprotein, which was subsequently incorporated into the outer membrane in a LolB-dependent manner, indicating that the inner membrane retention signal only functions with LolCDE.

Outer membrane-specific lipoproteins with Cys at position 2 were subjected to chemical modification followed by the release reaction in reconstituted proteoliposomes (Hara et al., 2003). SH-specific introduction of bulky non-protein molecules into Cys did not inhibit the LolCDE-dependent release, indicating that LolCDE releases outer membrane-specific lipoproteins without recognizing the second residue. Therefore, LolCDE only recognizes an N-terminal Cys possessing three acyl chains, the sole common structure of lipoproteins. Although SH-specific introduction of a negative charge to Cys did not cause the retention of lipoproteins, oxidation of Cys to cysteic acid resulted in generation of the Lol avoidance signal. Conversely, modification of the carboxylic acid of Asp at position 2 abolished its Lol avoidance function. Taken together, these results indicate that the Lol avoidance signal should have a negative charge that is within a certain distance from Cα of the second residue. Furthermore, phosphatidylethanolamine (PE) was required for the Lol avoidance function of Asp at position 2, whereas the Lol avoidance signal also functioned in proteoliposomes reconstituted with phosphatidylcholine, which *E. coli* does not contain. These results strongly suggest that the electrostatic interaction between Asp at position 2 and phospholipids

having a positive charge is responsible for the Lol avoidance mechanism (Hara et al., 2003). It seems likely that the third residue should not disturb the interaction between Asp at position 2 and PE for the Lol avoidance mechanism. Possible hydrogen bonds are formed between Cys at position 1 and the PE molecule interacting with Asp at position 2. This may strengthen the interaction between Asp at position 2 and the PE molecule. As a result, a tight lipoprotein-PE complex that has five acyl chains is formed, which cannot be accommodated in LolCDE. When Glu is present at position 2, the PE molecules involved in the electrostatic interaction with Glu and the hydrogen bond formation with Cys at position 1 would be different because of its longer side chain. This seems to be the reason why Glu at position 2 does not have the Lol avoidance function.

It has been reported that not only Asp, but also residues such as Phe, Pro, and Trp at position 2, followed by Asn at position 3, cause the inner membrane retention of lipoproteins (Seydel et al., 1999), although *E. coli* native lipoproteins do not have these residues at position 2. It seems likely that the mechanism of inner membrane retention caused by these hydrophobic residues at position 2 is different from the mechanism by which native lipoproteins remain in the inner membrane.

Phospholipid compositions affect not only the Lol avoidance function of Asp at position 2 but also the activity of the LolCDE complex. PE is known as a nonbilayer lipid (Gruner, 1985), which has been proposed to be important for the functions of various membrane proteins (Curnow et al., 2004; Booth et al., 2001; van der Does et al., 2000; Mikhail et al., 1996). Cardiolipin also forms non-bilayer structures in the presence of a high concentration of magnesium ions (Rietveld et al., 1993). Both ATP hydrolysis and the release of lipoproteins by LolCDE reconstituted into cardiolipin proteoliposomes significantly increased with an increase in the concentration of Mg^+. However, the Lol avoidance function of Asp was abolished in the cardiolipin proteoliposomes and inner membrane-specific lipoproteins were efficiently released even in the presence of high

Mg$^+$ (S. Miyamoto and H. Tokuda, unpublished observations).

OVERALL MECHANISMS FOR LIPOPROTEIN SORTING TO OUTER MEMBRANES

Five Lol proteins, A to E, are highly conserved
in various gram-negative bacteria. The sorting
and localization of lipoproteins in many, if not
all, gram-negative bacteria therefore seem to be
mediated by a Lol pathway constituting the
LolCDE complex in the inner membrane,
LolA in the periplasm and LolB in the outer
membrane (Fig. 2). All Lol proteins are essential
for *E. coli* growth (Tanaka et al., 2001; Narita
et al., 2002; Tajima et al., 1998). In contrast, a
LolA homolog of *Helicobacter pylori* has been reported to be nonessential (Chalker et al., 2001).

When the LolCDE complex interacts with
outer membrane-specific lipoproteins in the
outer leaflet of the inner membrane, the LolD
subunit hydrolyzes ATP on the cytoplasmic
side of the membrane. This energy is transferred from LolD to LolC/LolE, and then utilized to release lipoproteins from the outer
leaflet of the membrane, leading to the formation of a LolA-lipoprotein complex in the
periplasm. This requires the opening of the

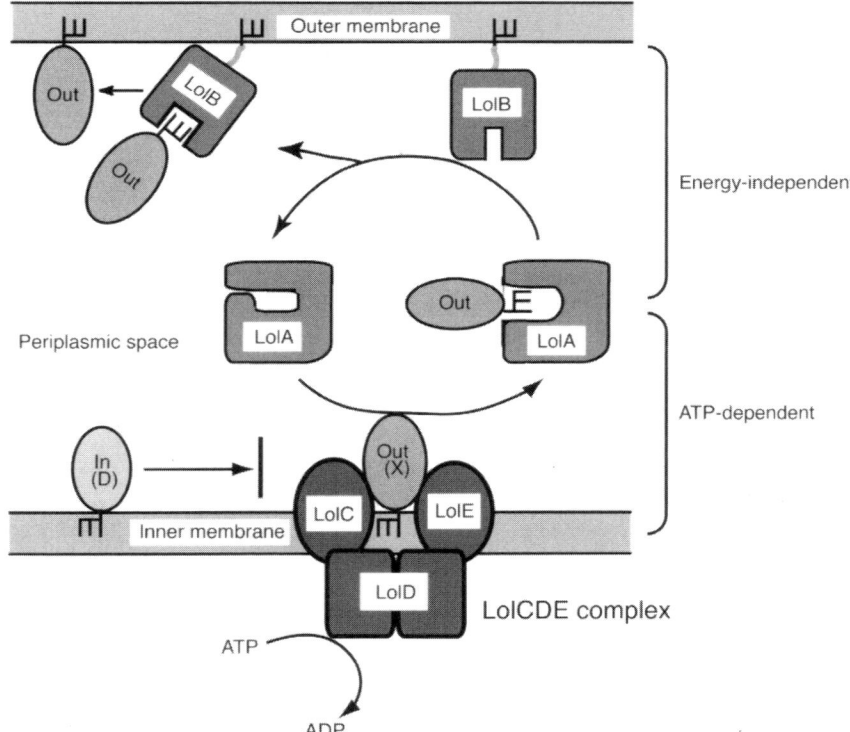

FIGURE 2 Sorting and outer membrane localization of lipoproteins through the Lol
pathway. "In" and "Out" represent inner membrane-specific and outer membrane-specific
lipoproteins, respectively. An ABC transporter, LolCDE, releases outer membrane-specific
lipoproteins from the inner membrane, causing the formation of a complex between the
released lipoproteins and the periplasmic molecular chaperone LolA. When this complex
interacts with outer membrane receptor LolB, the lipoproteins are transferred from LolA to
LolB and then localized to the outer membrane. The inner membrane retention signal Asp at
position 2 inhibits the recognition of lipoproteins by LolCDE, thereby causing their retention in the inner membrane. Phosphatidylethanolamine plays an important role in the Lol
avoidance function of Asp at position 2.

LolA lid through disruption of the hydrogen bonds between Arg43 and the lid-helices of LolA, and presumably hydrophobic interactions between other residues. The ATP energy released on the cytoplasmic side of membranes is thus utilized on the periplasmic side of the membrane to form a LolA-lipoprotein complex. The lipoprotein transfer from LolA to LolB, which have similar unclosed β-barrel structures, is unidirectional and very efficient, but requires no energy input. Arg at position 43 of LolA is important to weaken the hydrophobic interaction between LolA and lipoproteins (Taniguchi et al., 2005). LolB anchored to the outer membrane then transfers the associated lipoproteins to the inner leaflet of the outer membrane to which lipoproteins are most stably anchored through three acyl chains. A LolB derivative, which lacks the N-terminal lipid anchor, functions to incorporate associated lipoproteins into the lipid bilayer. However, the derivative does not distinguish the inner and outer membranes. Only the outer membrane anchoring of LolB determines the membrane specificity of lipoprotein localization (J. Tsukahara, S. Narita, and H. Tokuda, unpublished observations).

LIPOPROTEIN SORTING IN OTHER GRAM-NEGATIVE BACTERIA

Asp at position 2 may not always be the Lol avoidance signal in other bacteria, although Lol proteins are conserved in various gram-negative bacteria. MexA in *Pseudomonas aeruginosa* is an inner membrane lipoprotein possessing Gly and Lys at positions 2 and 3, respectively. Since the ionic interaction between PE and the second residue is not involved in the Lol avoidance mechanism in this particular case, the inner membrane retention signal of MexA was systematically examined. We found that Lys at position 3 and Ser at position 4 are important for the inner membrane retention of MexA (S. Narita and H. Tokuda, unpublished observations). On the other hand, lipoproteins with Asp at position 2 were not released from proteoliposomes reconstituted with LolCDE of *P. aeruginosa*, indicating the same Lol avoidance

mechanism (K. Tanaka, S. Narita, and H. Tokuda, unpublished observations). Since membrane localization and sorting signals have been biochemically determined for only a few lipoproteins in other bacteria, it remains to be determined how widely the Lol avoidance mechanism mentioned above is applicable to the inner membrane-specific lipoproteins in bacteria.

The Lol avoidance mechanism is required for the localization of PulA of *Klebsiella oxitoca* on the outer surface of the outer membrane. Pugsley and collaborators examined the membrane localization of PulA in *E. coli* with or without a subset of *pul* genes, which comprise the Type II secretion pathway (Pugsley, 1993). Wild-type PulA having Asp at position 2 is localized on the outer surface of the outer membrane when expressed with the Type II secretion pathway. In contrast, PulA is exclusively localized on the periplasmic surface of the inner membrane in the absence of the Type II pathway, indicating that Asp at position 2 functions as a Lol avoidance signal in the absence of the Type II pathway. Substitution of Asp with another residue results in the localization of PulA on both the periplasmic surface and the outer surface of the outer membrane when the Type II secretion pathway is present. However, this PulA derivative is exclusively localized on the periplasmic surface of the outer membrane in the absence of the Type II secretion pathway. Therefore, PulA expressed in *E. coli* and, presumably, *K. oxitoca* should have a Lol avoidance signal to be efficiently translocated to the outer surface of the outer membrane through the Type II pathway; otherwise the Lol pathway causes the localization of PulA on the periplasmic surface of the outer membrane.

B. burgdorferi, the Lyme disease spirochete, has been reported to possess more than 100 lipoproteins, some of which are on the outer surface of the outer membrane (Haake, 2000) and cause an immunoresponse of host cells (Scragg et al., 2000). This bacterium does not seem to have a complete Lol pathway because of an apparent lack of LolB homologs. It is important to clarify the mechanism underlying the sorting of more than 100 lipoproteins in *B. burgdorferi*.

ACKNOWLEDGMENTS

We thank Shoji Watanabe for preparation of Color Plate 6 and Rika Ishihara for help in the preparation of this review.

This work was supported by grants to H. T. from the Ministry of Education, Science, Sports and Culture of Japan.

REFERENCES

Alba, B. M., and C. A. Gross. 2004. Regulation of the *Escherichia coli* sigma-dependent envelope stress response. *Mol. Microbiol.* **52:**613–619.

Booth, P. J., R. H. Templer, W. Meijberg, S. J. Allen, A. R. Curran, and M. Lorch. 2001. *In vitro* studies of membrane protein folding. *Crit. Rev. Biochem. Mol. Biol.* **36:**501–603.

Brokx, S. J., M. Ellison, T. Locke, D. Bottorff, L. Frost, and J. H. Weiner. 2004. Genome-wide analysis of lipoprotein expression in *Escherichia coli* MG1655. *J. Bacteriol.* **186:**3254–3258.

Chalker, A. F., H. W. Minehart, N. J. Hughes, K. K. Koretke, M. A. Lonetto, K. K. Brinkman, P. V. Warren, A. Lupas, M. J. Stanhope, J. R. Brown, and P. S. Hoffman. 2001. Systematic identification of selective essential genes in *Helicobacter pylori* by genome prioritization and allelic replacement mutagenesis. *J. Bacteriol.* **183:**1259–1268.

Curnow, P., M. Lorch, K. Charalambous, and P. J. Booth. 2004. The reconstitution and activity of the small multidrug transporter EmrE is modulated by non-bilayer lipid composition. *J. Mol. Biol.* **343:** 213–222.

Danese, P. N., and T. J. Silhavy. 1997. The sigma(E) and the Cpx signal transduction systems control the synthesis of periplasmic protein-folding enzymes in *Escherichia coli*. *Genes Dev.* **11:**1183–1193.

Danese, P. N., W. B. Snyder, C. L. Cosma, L. J. B. Davis, and T. J. Silhavy. 1995. The Cpx two-component signal transduction pathway of *Escherichia coli* regulates transcription of the gene specifying the stress-inducible periplasmic protease, DegP. *Genes Dev.* **9:**387–398.

Doerrler, W. T., M. C. Reedy, and C. R. Raetz. 2001. An *Escherichia coli* mutant defective in lipid export. *J. Biol. Chem.* **276:**11461–11464.

Fraser, C. M., S. Casjens, W. M. Huang, G. G. Sutton, R. Clayton, R. Lathigra, O. White, K. A. Ketchum, R. Dodson, E. K. Hickey, M. Gwinn, B. Dougherty, J. F. Tomb, R. D. Fleischmann, D. Richardson, J. Peterson, A. R. Kerlavage, J. Quackenbush, S. Salzberg, M. Hanson, R. van Vugt, N. Palmer, M. D. Adams, J. Gocayne, J. Weidman, et al. 1997.

Genomic sequence of a Lyme disease spirochaete, *Borrelia burgdorferi*. *Nature* **390:**580–586.

Fukuda, A., S. Matsuyama, T. Hara, J. Nakayama, H. Nagasawa, and H. Tokuda. 2002. Aminoacylation of the N-terminal cysteine is essential for Lol-dependent release of lipoproteins from membranes but does not depend on lipoprotein sorting signals. *J. Biol. Chem.* **277:**43512–43518.

Genevrois, S., L. Steeghs, P. Roholl, J.-J. Letesson, and P. van der Ley. 2003. The Omp85 protein of *Neisseria meningitidis* is required for lipid export to the outer membrane. *EMBO J.* **22:**1780–1789.

Gennity, J. M., and M. Inouye. 1991. The protein sequence responsible for lipoprotein membrane localization in *Escherichia coli* exhibits remarkable specificity. *J. Biol. Chem.* **266:**16458–16464.

Gruner, S. 1985. Intrinsic curvature hypothesis for biomembrane composition: a role for nonbilayer lipids. *Proc. Natl. Acad. Sci. USA* **82:**3665–3669.

Haake, D. A. 2000. Spirochaetal lipoproteins and pathogenesis. *Microbiology* **146**(Pt 7):1491–1504.

Hara, T., S. Matsuyama, and H. Tokuda. 2003. Mechanism underlying the inner membrane retention of *E. coli* lipoproteins caused by Lol avoidance signals. *J. Biol. Chem.* **278:**40408–40414.

Hayashi, S., and H. C. Wu. 1990. Lipoproteins in bacteria. *J. Bioenerg. Biomembr.* **22:**451–471.

Higgins, C. F., I. D. Hiles, G. P. Salmond, D. R. Gill, J. A. Downie, I. J. Evans, I. B. Holland, L. Gray, S. D. Buckel, A. W. Bell, and M. A. Hermodsen. 1986. A family of related ATP-binding subunits coupled to many distinct biological processes in bacteria. *Nature* **323:**448–450.

Inukai, M., R. Enokita, A. Torikata, M. Nakahara, S. Iwado, and M. Arai. 1978. Globomycin, a new peptide antibiotic with spheroplast-forming activity. I. Taxonomy of producing organisms and fermentation. *J. Antibiot. (Tokyo)* **31:**410–420.

Juncker, A. S., H. Willenbrock, von G. Heijne, S. Brunak, H. Nielsen, and A. Krogh. 2003. Prediction of lipoprotein signal peptides in Gram-negative bacteria. *Protein Sci.* **12:**1652–1662.

Linton, K. J., and C. F. Higgins. 1998. The *Escherichia coli* ATP-binding cassette (ABC) proteins. *Mol. Microbiol.* **28:**5–13.

Locher, K. P., A. T. Lee, and D. C. Rees. 2002. The *E. coli* BtuCD structure: a framework for ABC transporter architecture and mechanism. *Science* **296:**1091–1098.

Ma, D., D. N. Cook, M. Alberti, N. G. Pon, H. Nikaido, and J. E. Hearst. 1993. Molecular cloning and characterization of *acrA* and *acrE* genes of *Escherichia coli*. *J. Bacteriol.* **175:**6299–6313.

Masuda, K., S. Matsuyama, and H. Tokuda. 2002. Elucidation of the function of lipoprotein-sorting signals that determine membrane localization. *Proc. Natl. Acad. Sci. USA* **99:**7390–7395.

Matsuyama, S., T. Tajima, and H. Tokuda. 1995. A novel periplasmic carrier protein involved in the sorting and transport of *Escherichia coli* lipoproteins destined for the outer membrane. *EMBO J.* **14:** 3365–3372.

Matsuyama, S., N. Yokota, and H. Tokuda. 1997. A novel outer membrane lipoprotein, LolB (HemM), involved in the LolA (p20)-dependent localization of lipoproteins to the outer membrane of *Escherichia coli. EMBO J.* **16:**6947–6955.

Mikhail, B., J. Sun, H. R. Kaback, and W. Dowhan. 1996. A phospholipid acts as a chaperone in assembly of a membrane transport protein. *J. Biol. Chem.* **271:**11615–11618.

Miyadai, H., K. Tanaka-Masuda, S. Matsuyama, and H. Tokuda. 2004. Effects of lipoprotein overproduction on the induction of DegP (HtrA) involved in the quality control of *Escherichia coli* periplasm. *J. Biol. Chem.* **279:**39807–39813.

Mizushima, S. 1987. Assembly of membrane proteins, p. 163–185. *In* M. Inouye (ed.), *Bacterial Outer Membranes as Model Systems.* John Wiley & Sons, New York, N.Y.

Muhlradt, P. F., and J. R. Golecki. 1975. Asymmetrical distribution and artifactual reorientation of lipopolysaccharide in the outer membrane bilayer of *Salmonella typhimurium. Eur. J. Biochem.* **51:** 343–352.

Narita, S., K. Kanamaru, S. Matsuyama, and H. Tokuda. 2003. A mutation in the membrane subunit of an ABC transporter LolCDE complex causing outer membrane localization of lipoproteins against their inner membrane-specific signals. *Mol. Microbiol.* **49:**167–177.

Narita, S., K. Tanaka, S. Matsuyama, and H. Tokuda. 2002. Disruption of *lolCDE* encoding an ATP-binding-cassette transporter is lethal for *Escherichia coli* and prevents the release of lipoproteins from the inner membrane. *J. Bacteriol.* **184:**1417–1422.

Navarre, W. W., S. Daefler, and O. Schneewind. 1996. Cell wall sorting of lipoproteins in *Staphylococcus aureus. J. Bacteriol.* **178:**441–446.

Onufryk, C., M. L. Crouch, F. C. Fang, and C. A. Gross. 2005. Characterization of six lipoproteins in the sigmaE regulon. *J. Bacteriol.* **187:**4552–4561.

Paulsen, I. T., M. K. Sliwinski, and M. H. Saier, Jr. 1998. Microbial genome analyses: global comparisons of transport capabilities based on phylogenies, bioenergetics and substrate specificities. *J. Mol. Biol.* **277:**573–592.

Pugsley, A. P. 1993. The complete general secretory pathway in Gram-negative bacteria. *Microbiol. Rev.* **57:**50–108.

Rietveld, A. G., J. A. Killian, W. Dowhan, and B. de Kruijff. 1993. Polymorphic regulation of membrane phospholipid composition in *Escherichia coli. J. Biol. Chem.* **268:**12427–12433.

Robichon, C., M. Bonhivers, and A. P. Pugsley. 2003. An intramolecular disulphide bond reduces the efficacy of a lipoprotein plasma membrane sorting signal. *Mol. Microbiol.* **49:**1145–1154.

Sankaran, K., and H. C. Wu. 1994. Lipid modification of bacterial prolipoprotein. Transfer of diacylglyceryl moiety from phosphatidylglycerol. *J. Biol. Chem.* **269:**19701–19706.

Scragg, I. G., D. Kwiatkowski, V. Vidal, A. Reason, T. Paxton, M. Panico, A. Dell, and H. Morris. 2000. Structural characterization of the inflammatory moiety of a variable major lipoprotein of *Borrelia recurrentis. J. Biol. Chem.* **275:** 937–941.

Seydel, A., P. Gounon, and A. P. Pugsley. 1999. Testing the '+2 rule' for lipoprotein sorting in the *Escherichia coli* cell envelope with a new genetic selection. *Mol. Microbiol.* **34:**810–821.

Smith, P. C., N. Karpowich, L. Millen, J. E. Moody, J. Rosen, P. J. Thomas, and J. F. Hunt. 2002. ATP binding to the motor domain from an ABC transporter drives formation of a nucleotide sandwich dimer. *Mol. Cell.* **10:**139–149.

Snyder, W. B., L. J. B. Davis, P. N. Danese, C. L. Cosma, and T. J. Silhavy. 1995. Overproduction of NlpE, a new outer membrane lipoprotein, suppresses the toxicity of periplasmic LacZ by activation of the Cpx signal transduction pathway. *J. Bacteriol.* **177:**4216–4223.

Tajima, T., N. Yokota, S. Matsuyama, and H. Tokuda. 1998. Genetic analyses of the *in vivo* function of LolA, a periplasmic chaperone involved in the outer membrane localization of *Escherichia coli* lipoproteins. *FEBS Lett.* **439:**51–54.

Takeda, K., H. Miyatake, N. Yokota, S. Matsuyama, H. Tokuda, and K. Miki. 2003. Crystal structures of bacterial lipoprotein localization factors, LolA and LolB. *EMBO J.* **22:**3199–3209.

Tanaka, K., S. Matsuyama, and H. Tokuda. 2001. Deletion of *lolB* encoding an outer membrane lipoprotein is lethal for *Escherichia coli* and causes the accumulation of lipoprotein localization intermediates in the periplasm. *J. Bacteriol.* **183:**6538–6542.

Taniguchi, N., S. Matsuyama, and H. Tokuda. 2005. Mechanisms underlying energy-independent transfer of lipoproteins from LolA to LolB, which have similar unclosed β-barrel structures. *J. Biol. Chem.* **280:**34481–34488.

Terada, M., T. Kuroda, S. Matsuyama, and H. Tokuda. 2001. Lipoprotein sorting signals evaluated as the LolA-dependent release of lipoproteins from the cytoplasmic membrane of *Escherichia coli. J. Biol. Chem.* **276:**47690–47694.

Tjalsma, H., V. P. Kontinen, Z. Pragai, H. Wu, R. Meima, G. Venema, S. Bron, M. Sarvas, and J. M. van Dijl. 1999. The role of lipoprotein processing by signal peptidase II in the Gram-positive eubacterium *Bacillus subtilis. J. Biol. Chem.* **274:** 1698–1707.

van der Does, C., J. Swaving, van W. Klompenburg, and A. J. Driessen. 2000. Non-bilayer lipids stimulate the activity of the reconstituted bacterial protein translocase. *J. Biol. Chem.* **275:** 2472–2478.

von Heijne, G. 1996. Principles of membrane protein assembly and structure. *Prog. Biophys. Mol. Biol.* **66:**113–139.

Voulhoux, R., M. P. Bos, J. Geurtsen, M. Mols, and J. Tommassen. 2003. Role of a highly conserved bacterial protein in outer membrane protein assembly. *Science* **299:**262–265.

Wada, R., S. Matsuyama, and H. Tokuda. 2004. Targeted mutagenesis of five conserved tryptophan residues of LolB involved in membrane localization of *Escherichia coli* lipoproteins. *Biochem. Biophys. Res. Commun.* **323:**1069–1074.

Wu, T., J. Malinverni, N. Ruiz, S. Kim, T. J. Silhavy, and D. Kahne. 2005. Identification of a multicomponent complex required for outer membrane biogenesis in *Escherichia coli. Cell* **121:** 235–245.

Yakushi, T., K. Masuda, S. Narita, S. Matsuyama, and H. Tokuda. 2000. A new ABC transporter mediating the detachment of lipid-modified proteins from membranes. *Nat. Cell Biol.* **2:**212–218.

Yakushi, T., T. Tajima, S. Matsuyama, and H. Tokuda. 1997. Lethality of the covalent linkage between mislocalized major outer membrane lipoprotein and the peptidoglycan of *Escherichia coli. J. Bacteriol.* **179:**2857–2862.

Yakushi, T., N. Yokota, S. Matsuyama, and H. Tokuda. 1998. LolA-dependent release of a lipid-modified protein from the inner membrane of *Escherichia coli* requires nucleotide triphosphate. *J. Biol. Chem.* **273:**32576–32581.

Yamaguchi, K., F. Yu, and M. Inouye. 1988. A single amino acid determinant of the membrane localization of lipoproteins in *E. coli. Cell* **53:** 423–432.

Yokota, N., T. Kuroda, S. Matsuyama, and H. Tokuda. 1999. Characterization of the LolA-LolB system as the general lipoprotein localization mechanism on *Escherichia coli. J. Biol. Chem.* **274:**30995–30999.

Zhou, Z., K. A. White, A. Polissi, C. Georgopoulos, and C. R. Raetz. 1998. Function of *Escherichia coli* MsbA, an essential ABC family transporter, in lipid A and phospholipid biosynthesis. *J. Biol. Chem.* **273:**12466–12475.

PROTEIN FOLDING AND QUALITY CONTROL

THE Cpx ENVELOPE STRESS RESPONSE

Tracy L. Raivio

5

Over 20 years ago Silverman and colleagues isolated mutations in a locus termed *cpx* (conjugative pilus expression) that had adverse effects on F plasmid conjugation in *Escherichia coli*. The same *cpx* mutations altered a number of diverse cellular phenotypes and were localized in an operon that encoded the membrane-localized CpxA sensor protein. More than a decade later, the Silhavy lab isolated new *cpx* mutations in response to genetic selections for *E. coli* mutants able to withstand the expression of toxic, misfolded, mislocalized envelope proteins. These mutants turned out to be functionally similar to those originally isolated by Silverman and their characterization ultimately led to a description of the Cpx envelope stress response, an adaptation to envelope insults that is controlled by the membrane-bound sensor kinase CpxA and the cytoplasmic response regulator CpxR. CpxA recognizes envelope perturbations and communicates these to CpxR, which results in increased expression of genes encoding factors involved in envelope protein folding and degradation. The study of the Cpx response, together with the σ^E response (see Chapter 6), resulted in the characterization of a

suite of envelope protein-folding factors that reside in the periplasm and greatly increased our knowledge of how envelope proteins are matured in this cellular compartment. Curiously, just as Silverman initially noted that *cpx* mutants affected diverse, apparently unrelated phenotypes, recent study of the Cpx response has yielded surprising downstream targets and functions that do not appear to be involved in envelope protein biogenesis. In this chapter, the studies that led to identification and characterization of the Cpx envelope stress response are summarized and recent work that hints at a diverse range of Cpx-influenced cellular phenotypes is highlighted.

Cpx—CONJUGATIVE PILUS EXPRESSION

In 1980 McEwen and Silverman reported the isolation of chromosomal *cpx* mutations in *E. coli* that diminished F plasmid conjugal DNA transfer (McEwen and Silverman, 1980a). They used nitrosoguanidine mutagenesis followed by selection for mutants of *E. coli* carrying the F plasmid that were resistant to the Qβ bacteriophage, which gains entry to the cell through adsorption to the F pilus (Fig. 1). These mutants produced no detectable F pili and had lost conjugal donor activity. Further, the *cpx* mutants expressed reduced levels of at least one of

Tracy L. Raivio, Department of Biological Sciences, University of Alberta, Edmonton, Alberta T6G 2E9, Canada.

The Periplasm
Edited by Michael Ehrmann © 2007 ASM Press, Washington, D.C.

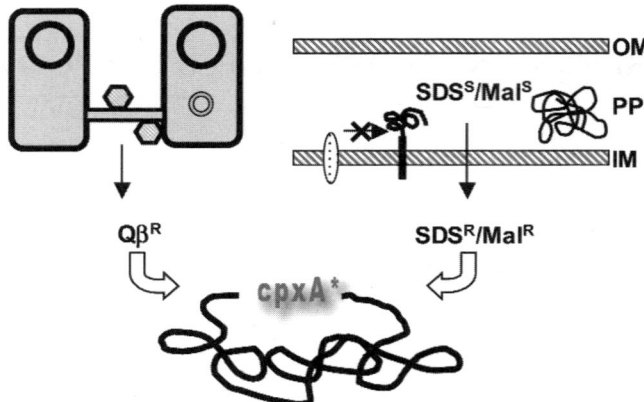

FIGURE 1 Independent genetic selections identified the Cpx locus. (Left) Silverman and colleagues selected conjugal F plasmid donors (right rectangle) that were resistant to the Qβ phage (hexagon), which adsorbs to the F pilus (thin grey line connecting donor and recipient cells). These mutants were defective in conjugal DNA transfer to F-recipient cells (left rectangle) and mapped to the *cpxA* locus. (Right) Silhavy and colleagues selected for mutants resistant to the expression of the toxic, misfolded, mislocalized proteins LamBA23D (rectangle joined to squiggly line) or a tribrid fusion protein LamB-LacZ-PhoA (squiggly line). LamBA23D contains a signal-sequence mutation that prevents cleavage of the signal sequence by leader peptidase (oval) and causes slowed processing, altered localization, and sensitivity to the inducer maltose (Mal^S) and SDS (SDS^S). LamB-LacZ-PhoA leads to the production of disulfide-bonded aggregates of β-galactosidase in the periplasm (squiggly line), which is manifest as sensitivity to the inducer maltose (Mal^S). SDS^R or Mal^R mutants mapped to the *cpxA* locus. OM, outer membrane; PP, periplasm; IM, inner membrane.

the F plasmid transfer proteins, TraT. The mutations localized to two regions of the chromosome that were designated *cpxA* (88′) and *cpxB* (41′) (McEwen and Silverman, 1980b). It was later shown that the *cpxB* mutation was cryptic in some strains and served only to potentiate the effects of the *cpxA* mutation. The *cpxB* mutation has never been described and will not be discussed further here. Silverman and colleagues went on to show that the *cpxA* mutations were associated with a variety of other, diverse, apparently unrelated phenotypes, including temperature sensitivity, auxotrophy for isoleucine and valine, aminoglycoside resistance, inability to grow on nonfermentable carbon sources, the ability to use L-serine as a carbon source, diminished ion-driven transport of lactose and proline, and altered inner and outer membrane protein composition (McEwen and Silverman, 1980b, 1980c, 1982; McEwen et al., 1983; Rainwater and Silverman, 1990; Sutton et al., 1982). While some of these phe-

notypes appeared to be due to effects on protein activity or localization, others seemed to be caused by a decrease in transcription. For example, the isoleucine/valine auxotrophy was linked to a decrease in activity of the biosynthetic enzyme acetohydroxyacid synthase isozyme I (Sutton et al., 1982), while diminished levels of OmpF were shown to be due in part to reduced transcription (McEwen et al., 1983). Further, while some phenotypes could be directly tied to alterations in envelope-localized proteins (Lpp, OmpF, F pilus), others appeared to be the result of effects on cytoplasmic resident proteins (acetohydroxyacid synthase isozyme I) (McEwen and Silverman, 1980a, 1982; McEwen et al., 1983; Sutton et al., 1982). Thus, from the beginning, it has been clear that the Cpx response influences numerous cellular processes in assorted ways. Because all of the phenotypes of the *cpxA* mutants had some association with the envelope, Silverman and colleagues concluded that the "primary effect of the mutations is to alter selectively the synthesis of certain envelope proteins" (McEwen and Silverman, 1982).

Later experiments focused on genetically characterizing the *cpxA* mutations as well as identifying the CpxA protein product. It was observed that the *cpxA* mutations were easily revertible and this, together with their temperature sensitivity, suggested that they were point mutations that altered activity of the wild-type protein (McEwen and Silverman, 1980b). Cloning and sequencing of the *cpxA* gene predicted it to encode an integral inner membrane protein with a single periplasmic domain and a cytoplasmic domain that shared homology with the EnvZ sensor histidine kinase, predictions that were borne out by studies with various fusion proteins and antisera (Albin and Silverman, 1984; Albin et al., 1986; Weber and Silverman, 1988). These observations were made just at the time that two-component regulatory systems were being recognized as a major means of regulating bacterial gene expression in response to the environment, and Silverman and colleagues proposed that CpxA likely functions as a transmembrane sensor protein (Weber and Silverman, 1988). They also concluded, based on sequence analysis, that *cpxA* was the 3′ gene of an operon (Weber and Silverman, 1988). This finding was later confirmed; *cpxR*, the gene encoding the cognate response regulator for CpxA, is encoded upstream of *cpxA* in an operon (Dong et al., 1993). Finally, the Silverman *cpxA* mutations were shown to be gain-of-function mutations, since *cpxA* deletion mutations had little effect on F plasmid conjugation (Gubbins et al., 2002; Rainwater and Silverman, 1990).

Thus, Silverman and colleagues provided the kernels for our current understanding of the Cpx envelope stress response. They demonstrated that gain-of-function mutations in the gene encoding the two-component sensor CpxA could alter a wide variety of cellular phenotypes, many of which were envelope associated, and ultimately hypothesized that it was the deregulation of CpxA kinase activity that led to the observed phenotypes.

THE Cpx ENVELOPE STRESS RESPONSE

Over a decade after Silverman's first report of *cpxA* mutations, the Silhavy lab identified new *cpxA* mutations in completely independent genetic selections. They set out to discover genes involved in the biogenesis of secreted envelope proteins by selecting for mutants that could survive in the presence of misfolded and/or mislocalized, toxic envelope proteins (Fig. 1). The *lamBA23D* gene encodes a variant of the outer membrane protein LamB in which the signal sequence processing site that is cleaved upon translocation across the inner membrane has been altered (Carlson and Silhavy, 1993). The slowed processing and altered cellular location of this mutant protein are thought to lead to toxicity, which is manifest as sensitivity to the inducer maltose as well as the detergent sodium dodecyl sulfate (SDS). Similarly, a LamB-LacZ-PhoA tribrid protein that is secreted to the periplasm is also toxic and sensitive to the inducer maltose (Snyder and Silhavy, 1995). The toxicity of this protein is thought to be due to the formation of high-molecular-weight, disulfide-bonded aggregates. Selections for either maltose (LamB-LacZ-PhoA)- or SDS (LamBA23D)-resistant mutants of *E. coli* strains expressing either of these toxic proteins yielded mutations that mapped to the *cpxA* locus and were shown to confer the same pleiotropic phenotypes as Silverman's *cpxA* alleles (Fig. 1) (Cosma et al., 1995). The Silhavy group termed these mutations *cpxA★* to denote the fact that they were not null mutations. In a parallel approach, multicopy suppressors of LamB-LacZ-PhoA toxicity were sought. NlpE, a novel outer membrane lipoprotein, answered this selection and was shown to require the *cpxRA* locus to alleviate tribrid toxicity (Snyder et al., 1995). An important part of the suppression mechanism in all three genetic selections was mediated through the periplasmic protease DegP (Cosma et al., 1995; Snyder et al., 1995). *cpxA★* mutations and NlpE overexpression conferred suppression of LamB-LacZ-PhoA toxicity through DegP-dependent degradation of the hybrid protein.

DegP also appeared to be important for suppression of LamBA23D toxicity, since its mutation diminished the degree of suppression conferred by the mutant *cpxA** alleles. These observations suggested that *cpxA** mutations and NlpE overexpression might both exert suppression by activating the *cpxRA* two-component system, leading to up-regulated *degP* transcription. To test this idea, β-galactosidase expression from a *degP-lacZ* reporter gene was examined in strains carrying *cpxA**, *cpxR⁻*, and *cpxA⁻* mutations, as well as in response to NlpE overexpression (Danese et al., 1995). It was found that NlpE overexpression did indeed induce *degP* transcription in a *cpxRA*-dependent fashion, and similar increases in *degP* transcription could be conferred by *cpxA** alleles (Table 1). The transcription of *degP* had previously been shown to be controlled by σ^E, an alternative sigma factor proposed to regulate an extracytoplasmic heat shock response (see Chapter 6) (Erickson and Gross, 1989; Lipinska et al., 1988). This link, more than any other, led to the hypothesis that the Cpx proteins might also play an integral role in regulating expression of protein-folding factors in the periplasm in response to insults to the envelope.

The idea that the Cpx two-component system might regulate periplasmic factors, besides DegP, involved in envelope protein trafficking was supported by the observation that *cpxA** mutations still conferred some suppression of LamBA23D toxicity in the absence of the *degP* gene (Cosma et al., 1995). In a search for these factors, Danese and Silhavy examined periplasmic protein profiles in strains overexpressing NlpE. N-terminal sequencing of a 23 kDa band that was up-regulated by NlpE overexpression in wild-type strains, but not in a *cpxR* mutant, identified DsbA, the major disulfide oxidase in the periplasm that is responsible for disulfide bond formation in secreted proteins (Danese and Silhavy, 1997) (Table 1). Promoter mapping experiments and the analysis of a *lacZ* operon fusion revealed that NlpE overexpression caused a CpxR-dependent 5- to 6-fold increase in transcription from a promoter found upstream of the *yihE/orfA* gene, which is situated 5′ to, and in an operon with,

dsbA. Simultaneously, the Beckwith lab investigated DsbA and DegP synthesis via pulse-chase immunoprecipitation in various *cpx* mutant backgrounds and in response to NlpE overexpression (Pogliano et al., 1997). They observed similar Cpx-dependent changes in DsbA expression, which were reflected in levels of transcription from the promoter upstream of *yihE/orfA*. Beckwith and coworkers went on to define a potential CpxR~P binding site by comparing DNase I footprints upstream of the *dsbA* and *degP* genes. They noted the presence of the 5′GTAAA(N)₆GTAA-3′ consensus binding site upstream of the promoter for the *ppiA* gene, encoding a periplasmic peptidyl-prolyl-isomerase, and showed that *ppiA* transcription was elevated by NlpE overexpression or in the presence of *cpxA** mutations (Table 1). Further, DNase I footprinting showed that CpxR~P bound upstream of *ppiA*. Thus, these experiments showed that the Cpx two-component system regulated expression of the periplasmic protease/chaperone DegP, the disulfide oxidase DsbA, and the peptidyl-prolyl-isomerase PpiA. Since all of these factors are involved, or predicted to be involved, in protein folding in the periplasm, together these studies definitively proved that a major function of the Cpx response is in the biogenesis of proteins secreted across the inner membrane. What remained mysterious were the conditions under which Cpx-regulated expression of these functions might be important.

Cpx SIGNAL TRANSDUCTION

CpxA and CpxR Make Up a Two-Component Regulatory System

The Cpx envelope stress response is regulated by a typical two-component regulatory system (Fig. 2). Silverman's group first noted the similarity of CpxA to other sensor proteins and they showed that it was localized to the inner membrane (Weber and Silverman, 1988). Alkaline phosphatase fusions demonstrated that CpxA contained an N-terminal periplasmic domain flanked by transmembrane segments and a C-terminal cytoplasmic region (Weber and Silverman, 1988). The DNA sequence sug-

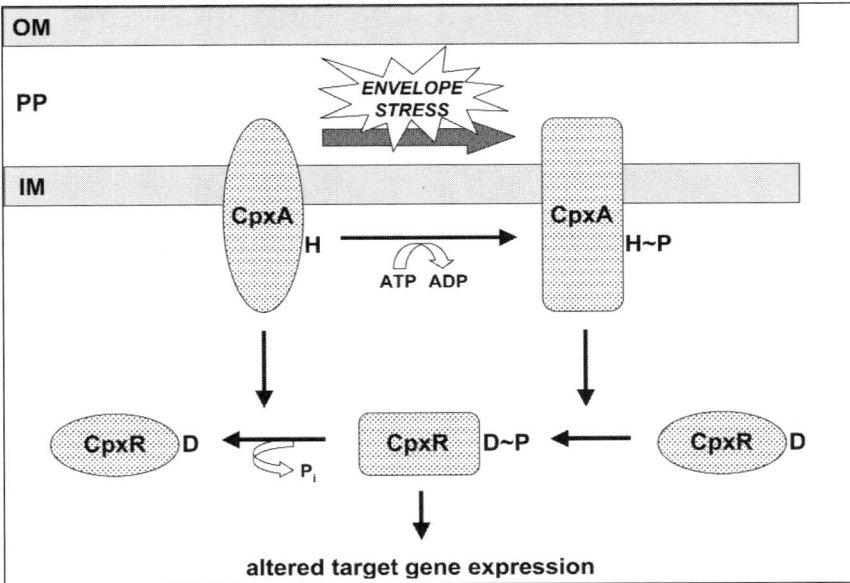

FIGURE 2 Cpx signal transduction is mediated by a two-component regulatory system. Envelope stresses are sensed by an inner membrane-localized sensor histidine kinase, CpxA. In the absence of envelope stress, CpxA functions as a CpxR~P (light shaded rectangle) phosphatase (dark shaded oval). In the presence of envelope stress, CpxA is thought to undergo a conformational change (dark shaded rectangle), which causes it to take on autokinase and CpxR (light shaded oval) kinase activities. Phosphorylation of CpxR converts it to a transcription factor able to bind with increased affinity to the promoters of target genes. OM, outer membrane; PP, periplasm; IM, inner membrane; P$_i$, inorganic phosphate; H, histidine; D, aspartate; P, phosphate.

gested that *cpxA* was the 3′ gene in an operon. Lin and coworkers showed that the region upstream of *cpxA* encoded a potential protein with homology to response regulator elements of two-component signal transduction systems and designated the gene *cpxR* (Dong et al., 1993). Sequencing of *cpxA⋆* alleles showed that all of these gain-of-function mutations were localized to the *cpxA* gene (Raivio and Silhavy, 1997). The *cpxA⋆* mutations clustered in areas encoding the periplasmic, transmembrane, or predicted kinase domains of the CpxA protein. Since these domains had been shown to foster signal sensing, transduction, and enzymatic activity, respectively, in other two-component sensor kinases, these observations intimated that CpxA might function in an analogous manner. Indeed, biochemical assays showed that CpxA, like other two-component sensor proteins, possessed autokinase, CpxR kinase,

and CpxR~P phosphatase activities (Raivio and Silhavy, 1997). CpxA⋆ mutant proteins exhibited a deficit in phosphatase activity, which indicated that the "on" or activated state of the signal transduction pathway occurred when CpxA functioned predominantly as a CpxR kinase (Raivio and Silhavy, 1997). Further, it was demonstrated that CpxR~P had enhanced affinity for the *degP*, *yihE/dsbA*, and *ppiA* promoters (Pogliano et al., 1997; Raivio and Silhavy, 1997). Thus, a model for Cpx signal transduction emerged in which inducing cues would elevate the CpxA kinase:phosphatase ratio, leading to an accumulation of CpxR~P and increased transcription of target genes (Fig. 2).

What Is the Cpx-Inducing Signal?

But what signal promotes CpxA kinase activity? There is a great deal of interest in the answer to this question, since the sensing mechanisms for other stress response pathways (i.e., σ^E, σ^H) have yielded significant insight into the processes of protein biogenesis in the envelope (σ^E) and the cytoplasm (σ^H). Although several studies provide good clues, the answer to this question remains enigmatic. The Cpx pathway is activated by a number of general insults to the bacterial envelope, by the overexpression of a variety of envelope proteins, and by attachment to abiotic, hydrophobic surfaces (Fig. 3).

Cpx RESPONSE ACTIVATION BY GENERAL ENVELOPE PERTURBATIONS

Early studies indicated that CpxA might be involved in regulating gene expression in response to pH. Nakayama and Watanabe identified cpxA in a screen for E. coli transposon mutants that caused deregulated expression of the Shigella spp. virF regulatory gene (Nakayama and Watanabe, 1995). Transposon insertions in cpxA caused virF to be expressed at elevated levels at the normally repressive pH of 6.0, suggesting that CpxA might somehow be involved in pH sensing. Danese and Silhavy subsequently confirmed this hypothesis, showing that elevated pH caused induction of the Cpx response (Danese and Silhavy, 1998). Alterations to the inner or outer membranes also trigger Cpx signal transduction. The Cpx response is activated in mutants lacking phosphatidylethanolamine, which leads to a disruption in both inner and outer membrane structure (Mileykovskaya and Dowhan, 1997). Similarly, in mutants that accumulate lipid II, an inner membrane-localized precursor of the outer membrane glycolipid enterobacterial common antigen (ECA), the Cpx response is induced (Danese et al., 1998). These mutants display phenotypes indicative of outer membrane disruption; thus, it is likely that both the inner membrane (through accumulation of lipid II) and the outer membrane are perturbed. It has also been reported that EDTA serves to induce the Cpx response, another condition that disrupts the outer membrane (DiGiuseppe and Silhavy, 2003). Similarly, high osmolarity has been shown to elevate expression of a cpxP-lacZ reporter (Prigent-Combaret et al., 2001). Considering that these inducers (pH, lipid II accumulation, PE deficiency) are likely to cause many pleiotropic effects in the envelope, it is difficult to infer anything about

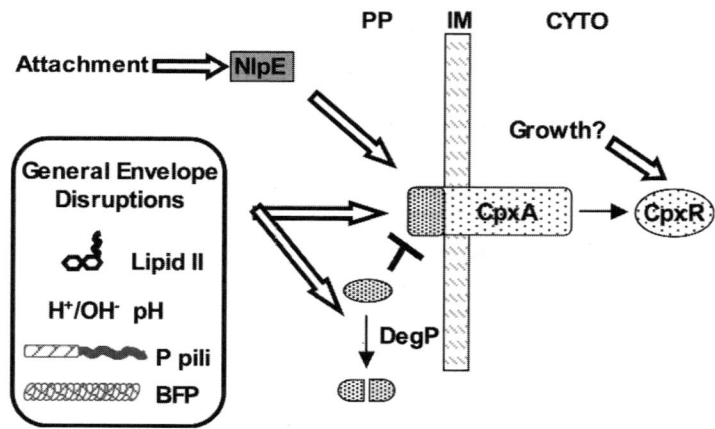

FIGURE 3 Cpx-inducing cues and -signaling proteins. All Cpx-inducing cues are sensed via the periplasmic sensing domain of CpxA (small shaded rectangle). Attachment to abiotic surfaces is signaled first through the outer membrane lipoprotein NlpE (shaded rectangle). Most other Cpx-inducing cues (boxed at bottom left) do not require NlpE for signaling. Some of these (pH, H^+/OH^-; P pilus subunit overexpression) lead to DegP-dependent degradation of the inhibitory signaling protein CpxP (light shaded oval in periplasm). Additional growth-related cues likely enter the signaling pathway downstream of CpxA but upstream of CpxR. PP, periplasm; IM, inner membrane; CYTO, cytoplasm; BFP, bundle-forming pilus.

the molecular nature of the signal(s) that might promote the Cpx response. Indeed, this collection of activating signals leaves open the possibility that the CpxA sensor might detect multiple types of cues.

Cpx RESPONSE ACTIVATION BY OVEREXPRESSION OF ENVELOPE PROTEINS

The Cpx response can also be initiated by the overexpression of a variety of proteins (Fig. 3). The first of these to be identified was NlpE. As discussed above, NlpE was identified by the Silhavy group in a screen for multicopy suppressors of the toxicity exerted by overexpression of the toxic tribrid protein LamB-LacZ-PhoA (see Fig. 1) (Snyder et al., 1995). Sequence analysis and biochemical characterization revealed that *nlpE* encoded a novel outer membrane lipoprotein. It is still not known how NlpE overexpression activates the Cpx response, although this is not likely to be a general feature of lipoprotein overproduction since the vast majority of *E. coli* lipoproteins do not induce the Cpx response when their expression is elevated (Miyadai et al., 2004). It has been proposed, based on studies of Cpx induction by pilus subunits (see below), that NlpE overexpression may induce the Cpx response because it leads to the production of misfolded protein in the envelope (Raivio and Silhavy, 1999, 2001). Alternatively, NlpE overexpression may mimic bacterial attachment to abiotic, hydrophobic surfaces, which has been shown to activate the Cpx response in an NlpE-dependent fashion (see below) (Otto and Silhavy, 2002).

The Cpx response can also be induced through overexpression of pilus subunits. Uropathogenic isolates of *E. coli* synthesize a multisubunit P pilus on the cell surface, which is required for attachment to kidney cells. The P pilus is the paradigm for the large class of chaperone-usher-type pili that are likely all assembled in a similar fashion (Thanassi and Hultgren, 2000). The multiple subunits of the pilus are secreted across the inner membrane and complex with a pilus-specific chaperone in the periplasm, which facilitates their folding and assembly into the pilus at the outer membrane usher platform. Enteropathogenic strains of *E. coli* elaborate a bundle-forming pilus (BFP) that is thought to be involved in initial attachment to intestinal epithelial cells upon infection. This pilus is a member of the type IV pilus family, whose members exhibit markedly different structures and assembly pathways from the chaperone-usher pili (Wolfgang et al., 2000). They consist of a helically arranged repeat of a single subunit, which is assembled at the inner membrane and extruded across the outer membrane. Overexpression of the P pilus subunits PapE or PapG in the absence of the PapD chaperone or overexpression of the BFP subunit BfpA was shown to foster CpxR-dependent increases in expression of *degP-lacZ*, *cpxP-lacZ*, or *spy-lacZ* reporter genes (Jones et al., 1997; Nevesinjac and Raivio, 2005). Since, in all of these cases, the end result is the production of aggregates of misfolded protein associated with the inner membrane, these data strongly argue that misfolded envelope protein is a Cpx inducer. In support of this, overexpression of a misfolded variant of the periplasmic maltose binding protein MalE, MalE31, which forms periplasmic inclusion bodies, also induces the Cpx response (Hunke and Betton, 2003). It does not appear to be aggregation per se that induces the Cpx pathway, since mutant, misfolded MalE variants that fail to aggregate still promote Cpx pathway activity (Hunke and Betton, 2003). Further, it was found that PapE aggregation and Cpx pathway activity were not correlated (Lee et al., 2004). The Pap subunits are assembled by means of a lock-and-key-type mechanism wherein the N-terminal extension of one subunit completes the immunoglobulin-like fold of the neighboring subunit (Sauer et al., 1999). In the absence of the periplasmic chaperone, this N-terminal extension fosters the formation of periplasmic aggregates of misfolded protein. Genetically engineered variants of PapE revealed that, while the N-terminal extension was necessary for Cpx pathway activation, it was not sufficient. Further, since activation of the Cpx response by the various engineered PapE derivatives was not associated with their aggregative properties, it appears that some inherent property of PapE folding leads to Cpx induction (Lee et al., 2004). So, while it remains unclear what aspect of envelope protein overexpression serves to induce the Cpx response, it seems probable that the answer lies in understanding the folding pathways and intermediates of those proteins that, upon overexpression, lead to Cpx response activation.

BACTERIAL ATTACHMENT TO SURFACES INDUCES THE Cpx PATHWAY

It is not just pilus subunit overexpression that triggers Cpx pathway activity. Rather, upregulation of both P pilus and BFP expression also serves to induce the Cpx response (Jones et al., 1997; Nevesinjac and Raivio, 2005). These observations raised the hypothesis that perhaps CpxA was specifically attuned to changes in pilus assembly. One attractive idea that springs from this tenet is that the Cpx pathway could sense changes in pilus assembly that might be incurred upon attachment. This is a particularly tantalizing idea with respect to the type IV pili, since many of them have been shown to be capable of retraction. Thus, upon host cell attachment, one could envisage changes in the levels of pilus subunit intermediates in the envelope by means of slowed assembly and/or pilus retraction. Although there are currently no data to support this idea, it was recently shown that the Cpx response does in fact sense bacterial attachment. Silhavy and colleagues demonstrated that the Cpx pathway is induced upon attachment to abiotic, hydrophobic surfaces (Otto and Silhavy, 2002). They compared the expression of *lacZ* reporters with various Cpx-regulated genes in planktonic cultures and populations of bacteria attached to hydrophobic glass beads. They showed not only that attachment activates expression of Cpx-regulated genes, but also that this activation requires NlpE. Further, by using a quartz crystal microbalance to monitor cell-surface interactions, they demonstrated that *cpx* and *nlpE* mu-

tants displayed altered cell-surface interactions. Thus, NlpE appears to be a part of the signal transduction pathway that permits Cpx sensing of surface attachment and also may play a physical role in the actual attachment process (Fig. 3). Since the laboratory strain of *E. coli* used in these studies was nonpiliated, it is unclear whether pilus-mediated attachment might also induce the Cpx response, and if so, whether NlpE is also a part of this signaling pathway.

How Does CpxA Sense Envelope Stress?

Most of what we know of how CpxA senses envelope stress signals derives from studies of *cpxA★* mutations, which cause the Cpx signal transduction pathway to be constitutively active. Raivio and Silhavy showed by sequence analysis that one class of these mutations, including one of the original Silverman mutations, *cpxA2*, altered amino acids in the periplasmic domain of CpxA (Raivio and Silhavy, 1997). These mutations clustered roughly in the center of the periplasmic domain and included both point mutations as well as an in-frame deletion, *cpxA24*, which removed 32 amino acids. Further, these mutations conferred a signal-blind phenotype. The Cpx pathway could no longer be activated by inducing cues in strains bearing this class of *cpxA★* mutation (Raivio and Silhavy, 1997). These observations suggested that the periplasmic *cpxA★* mutations altered a CpxA sensory domain.

Given the diversity of Cpx-activating signals, it seemed possible that CpxA might sense multiple activating cues, perhaps using different mechanisms. DiGiuseppe and Silhavy investigated this question by examining the expression of Cpx-regulated *lacZ* reporters in wild-type and *cpxA24* mutants in response to a variety of inducers. They showed that the *cpxA24* mutant was blind to all envelope stress signals tested as well as to surface attachment (DiGiuseppe and Silhavy, 2003). This suggests that all extracytoplasmic signaling events are sensed by CpxA using a common mechanism and also, possibly, that all Cpx-inducing cues generate the same molecular signal in the envelope (Fig. 3). NlpE was found to be required only for sensing surface attachment and so does not seem to be part of the core signal transduction apparatus. The Cpx response can also be activated by growth, and this activation requires CpxR but is only partially dependent on CpxA (Fig. 3). Thus, it is possible that a subset of inducers enter the Cpx-signaling cascade independently of CpxA. Beyond the identification of a putative sensing domain within CpxA, little work has been done to address the molecular mechanism behind envelope stress sensation and so this question remains largely open.

An additional signaling molecule involved in controlling Cpx pathway activity has been identified (Fig. 3). CpxP is a novel periplasmic protein with no informative homologues. It was identified in a genetic screen that sought random *lacZ* operon fusions that exhibited enhanced expression in the presence of NlpE overexpression (Danese and Silhavy, 1998) (Table 1). The *cpxP* open reading frame was found to be directly upstream of, and in the opposite orientation from, the *cpxRA* operon. Its sequence predicted CpxP to be a secreted, periplasmic protein, and this was confirmed by alkaline phosphatase fusions. Subsequent studies showed that overexpression of CpxP led to a decrease in the expression of Cpx-regulated *lacZ* fusions, suggesting that CpxP might play a role in signaling (Raivio et al., 1999). Inhibition required the same domain in CpxA previously identified as necessary for sensing envelope stresses, since the Cpx pathway could not be inhibited by CpxP overexpression in *cpxA** mutants bearing mutations in this sensing domain. These observations led to the proposal that CpxP inhibition might be exerted by a direct interaction between CpxP and the sensing domain of CpxA. This model is supported by experiments with spheroplasts (Raivio et al., 2000). Spheroplast formation of wild-type *E. coli* cells leads to a potent induction of the Cpx response, as measured by Western blot analysis of β-galactosidase expression in a strain bearing a Cpx-regulated *spy-lacZ* fusion. In strains overexpressing a MBP-CpxP fusion protein, which inhibits Cpx-regulated gene expression, spheroplast formation still led to pathway activation. However, if the MBP-CpxP fusion was tethered to the inner membrane through a signal sequence mutation, activation of the Cpx response by spheroplast formation was negated. The simplest explanation for this result is that a direct interaction between CpxA and CpxP is required to inhibit the pathway, and when the fusion protein is not tethered to the inner membrane, this interaction is lost during spheroplast formation when the periplasmic contents are leaked out of the cell (Fig. 3). Notably, this model has become dogma, although there is still no definitive proof that CpxA and CpxP interact.

CpxP does not appear to be required for signaling, since, when *cpxP* is either deleted or overexpressed, the Cpx response is still activated by inducing cues (Raivio et al., 1999). What, then, might its role be? Recent experiments show that *cpxP* is the most responsive Cpx regulon member identified to date (DiGiuseppe and Silhavy, 2003). CpxP expression is increased more than any other Cpx-regulated gene in response to activating cues. Further, Buelow and Raivio recently demonstrated that a CpxP-Bla fusion protein is rapidly degraded in a DegP-dependent fashion in the presence of Cpx inducers (Buelow and Raivio, 2005). Two hypotheses have been put forward to explain these data. First, it may be that CpxP is required to rapidly reset the Cpx signaling pathway to "unstressed" levels upon the relief of envelope stress. Hence, it would be expressed at high levels upon induction and degraded by the DegP protease as long as envelope stress existed. Upon the relief of envelope stress, CpxP would accumulate and shut down the Cpx response. Alternatively, or in addition, CpxP may function to prevent gratuitous activation of the Cpx pathway. This model argues that maximal Cpx response induction requires the removal of CpxP first. Thus, CpxP would act as a safeguard, preventing maximal induction of the response until DegP-mediated CpxP proteolysis occurred. An unanswered question in this field

TABLE 1 The Cpx regulon

Functional group	Gene(s)[a]	Protein(s) function	Cpx effect[b]	Evidence[c]	Coregulation[d]
Envelope protein fate	degP	Periplasmic serine protease/chaperone	Activation	• Induced by NlpE, cpxA★ • CpxR binds upstream	σ^E
	yihEdsbA	Unknown/periplasmic disulfide oxidase	Activation	• Induced by NlpE, cpxA★ • CpxR binds upstream	
	ppiA	Peptidyl-prolyl-isomerase	Activation	• Induced by NlpE, cpxA★ • CpxR binds upstream	
	cpxP	Periplasmic inhibitor protein	Activation	• Induced by NlpE, cpxA★	
	cpxRA	Two-component system	Activation	• Induced by NlpE, cpxA★	σ^S
	skp	Periplasmic chaperone	Activation	• Expression reduced in cpxR^− background • CpxR consensus binding site upstream	
	rpoErseABC	σ^E envelope stress regulator and signal transduction proteins	Repression	• CpxR consensus binding site in promoter • cpxA★ elevates and cpxR^− represses transcription	σ^E
Attachment	pap	P pilus and regulators	Repression	• cpxA★ and NlpE overexpression inhibit phase variation and papBA transcription • CpxR binds to phase variation switch region • CpxR binds at and upstream of promoters	
	csgBA csgDEFG	Curli adhesin, regulators, assembly factors	Repression	• csgD induction by high salt requires CpxR	OmpR
Pathogenesis	virF^pSS	Virulence regulatory protein	?	• E. coli CpxA mutants deregulate expression in response to pH • CpxR binds upstream	
	hilA^ST	Virulence regulatory protein	?	• cpxA mutations reduce expression	
	icm-dot^LP	Virulence genes	Activation	• icm gene expression reduced in LP cpxR mutants • CpxR^LP binds icmR promoter region	
	mviM	Homologue of MviA virulence protein	Activation	• CpxR consensus binding site upstream • cpxA★ elevates and cpxR^− represses transcription	

Envelope functions	motABcheAW	Motility and chemotaxis	Repression	• CpxR binds promoter • $cpxA\star$ inhibits transcription	
	aer	Aerotaxis	Repression	• CpxR consensus binding site in promoter • $cpxA\star$ inhibits, $cpxR^-$ activates transcription	
	spy	?	Activation	• NlpE and $cpxA\star$ activate	
	ompC	Outer membrane porin C	Activation	• NlpE and $cpxA\star$ activate, $cpxR^-$ inhibits transcription	OmpR
	ompF	Outer membrane porin F	Repression	• CpxR binds upstream of promoter • NlpE and $cpxA\star$ inhibit, $cpxR^-$ activates transcription	
	psd	Phosphatidylserine decarboxylase	Activation	• CpxR consensus binding site upstream of promoter • $cpxA\star$ activates, $cpxR^-$ inhibits transcription	
	secA	Secretion			
	acrD mdtABC	Multidrug efflux components	Activation	• $cpxA$ and $cpxR$ mutations reduce indole-induced transcription • CpxR binds upstream of promoters	BaeR
Cytoplasmic functions	htpX	Plasma membrane protease	Activation	• Expression induced by $cpxA\star$ and reduced by $cpxR$ mutation	σ^{H}
	ung	Uracil-N-glycosylase	Repression	• CpxR binds to promoter and inhibits in vitro transcription	
	aroK	Shikimate kinase	Activation	• NlpE represses and $cpxR^-$ elevates transcription • CpxR consensus binding site upstream of promoter • $cpxA\star$ activates, $cpxR^-$ inhibits transcription	

[a]SS, S. sonnei; ST, serovar Typhimurium; LP, L. pneumophila.
[b]Effect of Cpx on transcription of the indicated gene(s).
[c]See text for references; LP, L. pneumophila.
[d]Other envelope stress or global regulatory responses besides CpxAR and/or other regulatory proteins that have been shown to interact with CpxR to influence transcription of the gene(s) in question.

is how CpxP degradation is initiated. Simple overexpression or deletion of *degP* has little effect on Cpx-mediated gene expression and so it is clear that additional signals must be generated in the presence of Cpx inducers to facilitate CpxP degradation and increased Cpx pathway activity. Silhavy and colleagues have hypothesized that CpxP functions as a chaperone to bind misfolded proteins and target them for degradation by DegP and that it is this function that would allow its preferential removal in the presence of envelope stresses. In support of this idea, they demonstrated that CpxP is required for proteolysis of misfolded P pilus subunits and that CpxP is degraded in the presence of these substrates (Isaac et al., 2005). Another possibility is that CpxP itself becomes misfolded in the presence of envelope stressors and this causes it to become a substrate for DegP. Periplasmic molecules that seem to perform similar accessory roles in other signal transduction pathways have recently been described and so it seems CpxP may be a member of a new class of regulatory molecules that function in adjusting the level of signal transduction to a stressor-appropriate level (Beck et al., 2004; Grigorova et al., 2004; Matson and DiRita, 2005).

Control of Transcription by CpxR~P

The ultimate effect of Cpx-inducing cues is to up-regulate phosphotransfer between CpxA and CpxR, leading to an accumulation of CpxR~P and consequent changes in gene expression. There has been very little investigation of the mechanism(s) used by CpxR~P to effect changes in transcription. Mutagenesis, DNA binding studies, and the analysis of in vivo gene expression in various *cpx* mutant backgrounds or in response to Cpx inducers show that CpxR~P exerts both positive and negative effects on transcription. Positive effects, by analogy with the highly related response regulator OmpR, are thought to be due to stimulatory interactions with RNA polymerase (RNAP). Binding site analyses predict that repressive effects result from precluding RNAP and/or activator binding. In addition,

recent experiments suggest that CpxR~P can also function as a potentiator of other transcription factors.

CpxR AS AN ACTIVATOR

Early footprinting experiments in the Beckwith lab showed that phosphorylation of CpxR increased its affinity for DNA sequences upstream of the *ppiA*, *degP*, and *yihE-dsbA* genes (Pogliano et al., 1997) (Table 1). Comparison of the footprinted regions of these three genes identified a CpxR~P consensus binding site of 5′ GTAAA − (N)$_5$ − GTAA 3′. The direct repeat nature of this motif suggests that CpxR~P, like other response regulators, may bind DNA as a dimer. Using the upstream regions of these genes, together with those upstream of 7 other genes or operons that had been shown to be affected in vivo by the Cpx response (*motABcheAW*, *tsr*, *ygjT*, *ppiD*, *cpxRA*, *cpxP*, *csgBA*), Lin and colleagues used a motif-finding algorithm to derive a weighted matrix for CpxR~P recognition (De Wulf et al., 2002). This matrix was essentially the same as the binding site identified by the Beckwith lab, except extended by one base pair, 5′ GTAAA − (N)$_5$ − GTAAA 3′. They used this consensus to screen the *E. coli* genome for potential new Cpx regulon members. Northern analysis of 11 of these predicted targets showed that 8 displayed altered levels of RNA in *cpx* mutant backgrounds, indicating that they may indeed be true Cpx regulon members. Together these studies convincingly demonstrated that the 5′ GTAAA − (N)$_5$ − GTAAA 3′ sequence element makes up a core CpxR~P binding motif.

CpxR is highly homologous to OmpR, the defining member of the OmpR class of response regulators. OmpR has been heavily studied using both genetic and biochemical approaches. Further, the crystal structure of the C-terminal DNA binding domain has been solved (Martinez-Hackert and Stock, 1997). It is a member of the winged helix class of transcription factors and consists of a helix-turn-helix with an extended turn that is flanked by loops or wings. Genetic and biochemical studies indicate that OmpR and other members of this class of transcription factors bind DNA via the wings and recognition helix. The extended turn or alpha loop in the helix–turn–helix has been shown genetically to facilitate interaction between OmpR and the C-terminal domain of the alpha subunit of RNAP (RNAPαCTD) (Pratt and Silhavy, 1994). Thus, it is predicted that phosphorylation of OmpR stimulates DNA binding and facilitates stimulatory interactions with RNAPαCTD that activate transcription. By analogy, CpxR is expected to function in a similar manner. In support of this conjecture, DiGiuseppe and Silhavy mutated the predicted conserved site of phosphorylation, D51, and a conserved residue in the recognition helix, M199, and showed that these mutations prevented or severely attenuated pH-mediated induction of *cpxP-lacZ* expression (DiGiuseppe and Silhavy, 2003). Thus, although little evidence exists thus far, it seems likely that CpxR may function in a manner similar to OmpR to activate transcription.

CpxR AS A REPRESSOR

CpxR~P has also been shown to repress gene expression. In genomic searches using a consensus sequence, Lin and colleagues identified putative CpxR~P binding sites upstream of the *motABcheAW*, *tsr*, *ung*, *aer*, and *ompF* operons/genes (De Wulf et al., 2002) (Table 1). For all of these sites, both in vivo studies of gene expression and in vitro studies of DNA binding indicate that CpxR~P inhibits transcription of these genes. Motility, as well as transcription of the *motABcheAW*, *tsr*, and *aer* genes, is inhibited in strains carrying constitutively activated *cpxA★* alleles (De Wulf et al., 1999, 2002). Further, mobility shift assays demonstrate that CpxR~P binds to sequences upstream of the *motABcheAW* and *tsr* transcription start sites (De Wulf et al., 1999). In all cases, the CpxR~P consensus binding site is predicted to overlap a promoter region. Activation of the Cpx response by mutation or NlpE overexpression leads to inhibition of *ompF* transcription and CpxR~P protects sites upstream of *ompF* that overlap, or are found upstream of, those of the

OmpR transcription factor (Batchelor et al., 2005; McEwen et al., 1983). Similarly, overexpression of CpxR inhibited uracil-DNA glycosylase activity, the product of the *ung* gene (Ogasawara et al., 2004). Overexpression of CpxR or NlpE leads to diminished transcription of the *ung* gene as measured by S1 nuclease or in vitro transcription assays and CpxR~P bound to sites upstream of the *ung* gene (Ogasawara et al., 2004). In this case, the predicted CpxR~P binding site is located upstream of the promoter, but is oriented in the opposite direction of the promoter. Finally, Dorel and colleagues demonstrated that activation of the Cpx response, either by mutation or NlpE overexpression, inhibited transcription of the *csgAB* operon that encodes the curlin subunit necessary for production of the curli adhesin (Dorel et al., 1999) (Table 1). Mobility shift experiments demonstrated that CpxR bound upstream of both operons and sequence analysis identified several putative CpxR~P binding sites, most of which overlap binding sites for OmpR~P, which is an activator of these operons, or a promoter (Jubelin et al., 2005; Prigent-Combaret et al., 2001). Based on these data, it appears that CpxR~P can specifically bind to, and repress, transcription from several promoters. The location of CpxR~P consensus binding sites suggests that in most cases, the mechanism of repression is to block RNAP and/or activator binding. In the case of the *ung* gene, it is not clear how repression would be exerted. Since the CpxR box also contains a putative UP element, which is predicted to enhance transcription through interaction with the RNAPαCTD, it has been hypothesized that CpxR~P binding would exert repression by preventing this interaction. In any case, it is clear that CpxR~P is capable of directly repressing the expression of numerous genes. Clarification of the mechanism(s) awaits further experimentation.

CpxR AS A POTENTIATOR

In addition to functioning as an activator and a repressor, a new role for CpxR~P in transcription has recently been put forth, that of potentiator. Recent experiments showed that the Cpx response affected expression of three different operons primarily through another regulatory pathway. Further, CpxR has been shown to bind to sites upstream of these operons that abut those bound by a "primary" regulatory protein, suggesting that it may function as an enhancer of these transcription factors. Increasing Cpx pathway activity led to elevated transcription of the *ompC* gene, and this effect was dependent on the porin regulatory protein, OmpR (Batchelor et al., 2005) (Table 1). DNA-footprinting experiments showed that CpxR~P bound to sites that overlap, or are upstream from, OmpR binding sites. Similarly, overexpression of CpxR caused a 4.1- and 3.4-fold increase in transcripts from the *acrD* and *mdtA* genes, both of which encode components of multidrug efflux pumps (Hirakawa et al., 2003). Analysis of *lacZ* fusions to the promoters of these genes showed that inducible expression was attributable largely to the BaeSR two-component system, but that the Cpx pathway facilitated maximal induction (Hirakawa et al., 2005) (Table 1). Mobility shift assays demonstrated that CpxR bound to the *acrD* and *mdtA* promoters and footprinting analysis showed that the CpxR binding sites were located upstream of, and partially overlapping, the BaeR binding site (Hirakawa et al., 2005). Together, these observations suggest that CpxR~P can enhance or potentiate transcription activation that is fostered by other regulatory proteins. The mechanism involved has not been elucidated. Perhaps CpxR~P serves to attract or anchor other regulatory proteins via direct protein:protein interactions. Alternatively, maybe CpxR~P binding alters DNA structure, facilitating regulator binding, RNAP binding, and/or open complex formation. Another possibility is the formation of higher order protein complexes that facilitate multiple contacts between RNAP and several transcription factors.

NEW TARGETS, NEW ROLES?
Over 20 years ago, it was noted that mutations in the *cpx* locus conferred a number of pleio-

tropic phenotypes that were not easily explained. Similarly, recent studies aimed at elucidating the Cpx regulon indicate that the Cpx response influences a diverse range of physiological processes (Table 1). To date, these include, but are not limited to, attachment to abiotic surfaces, pilus elaboration, biofilm formation, regulation of virulence, conjugation, control of cytoplasmic membrane stability, regulation of porin synthesis, and modulation of mutation rate. The effect of the Cpx response on many of these processes can be explained by its regulation of envelope protein-folding and degrading factors. Other Cpx-influenced processes, however, such as the direct transcriptional control of genes encoding virulence regulators and cytoplasmic proteins, suggest that the Cpx response has been coopted to induce processes unrelated to envelope protein folding that might benefit the bacterium under conditions where envelope protein folding is disrupted.

Microbial Surface Interaction
The Cpx response has been demonstrated to play roles in the elaboration of pili, attachment to abiotic surfaces, and the formation of biofilms (Table 1). While many of these effects are attributable to Cpx regulation of envelope protein-folding factors, the Cpx response has also been shown to directly influence the transcription of genes encoding a variety of adhesins. Thus, it appears that one major role of the Cpx response is to regulate interactions with both living and abiotic surfaces.

THE Cpx RESPONSE REGULATES PILUS ASSEMBLY AND EXPRESSION
The Cpx response is induced by up-regulation of both P pilus and BFP expression (see above) (Jones et al., 1997; Nevesinjac and Raivio, 2005). The sensitivity of the Cpx response to these processes, and their requirement for Cpx-regulated folding factors, suggest that one role of the Cpx response may be to facilitate pilus assembly. The Silhavy and Hultgren labs tested this idea by examining P-pilus assembly in

wild-type and *cpxR* mutant laboratory strains of *E. coli* overexpressing all of the *pap* genes required for P-pilus assembly. Using electron microscopy and two different pilus isolation techniques, they showed that mutant bacteria lacking the Cpx signal transduction pathway expressed pili with an aberrant, short morphology (Hung et al., 2001). Since this phenotype occurred in strains expressing the *pap* gene cluster from an exogenous promoter, these effects were proven to be posttranscriptional. Similarly, Nevesinjac and Raivio used electron microscopy, immunoblotting, and epithelial cell binding assays to show that enteropathogenic *E. coli* (EPEC) *cpxR* mutants produced diminished amounts of BFP (Nevesinjac and Raivio, 2005). Further, constitutive activation of the Cpx response was required in laboratory strains of *E. coli* for BFP elaboration to occur from a plasmid expressing all of the required *bfp* genes from an exogenous promoter. Thus, the Cpx response also appears to facilitate BFP elaboration at a posttranscriptional level. While the assembly pathways and structures of P pili and BFP are dramatically different (see above), they both share a requirement for the Cpx-regulated folding factor DsbA. DsbA is required for disulfide bond formation and proper folding of the PapD periplasmic chaperone and the major BFP subunit BfpA (Jacob-Dubuisson et al., 1994; Zhang and Donnenberg, 1996). Further, DegP is required for the proteolysis of toxic, misfolded Pap subunits that arise during P-pilus biogenesis (Jones et al., 1997). Although it is unknown whether other Cpx-regulated folding factors might also be required for pilus elaboration, cumulatively, these observations support the conclusion that the Cpx response is attuned to pilus assembly because some of its regulon members are required for proper biogenesis of these adhesins.

Curiously, the Cpx response also seems to be involved in controlling transcription of the *pap* gene cluster, although in a negative fashion. Low and colleagues used *lac* fusions to the *pap* promoter region to show that activation of the Cpx response by mutation or NlpE overexpression led to diminished, phase-variable expression of the *pap* gene cluster and reduced piliation (Hernday et al., 2004). Mobility shift and DNA footprinting experiments suggested that CpxR~P inhibited expression of the *pap* gene cluster by competing with the positive regulator Lrp for binding to sites that are required for phase switching. Although it is difficult to reconcile this result with the positive role of Cpx in P-pilus assembly demonstrated by the Hultgren and Silhavy labs, it may be that, in the presence of envelope stresses (including P-pilus overexpression or up-regulation), Cpx response activation serves both to efficiently assemble the multitude of subunits already expressed as well as to shut down pilus synthesis at the level of transcription to prevent any further trauma to the cell envelope. It is unknown whether the Cpx response also influences expression of the *bfp* gene cluster. Nonetheless, together, these studies show that a major role of the Cpx response is to monitor and modulate the assembly and possibly the transcription of pili.

ATTACHMENT TO ABIOTIC SURFACES REQUIRES THE Cpx RESPONSE

Pilus studies showed that the Cpx response could influence microbial attachment through its effects on adhesin expression. The Silhavy lab has demonstrated that the Cpx response also mediates attachment of *E. coli* to abiotic, hydrophobic surfaces in a pilus-independent manner. Otto and Silhavy studied attachment of wild-type and mutant strains of the non-piliated laboratory strain MC4100 to hydrophobic quartz crystals using acridine orange to count the numbers of adhered bacteria and a quartz crystal microbalance to monitor changes in surface mass and viscoelastic properties of the attached bacteria (Otto and Silhavy, 2002). They found that CpxR and NlpE were both required for wild-type attachment and that, in the absence of either of these proteins, the adhered population of cells exhibited dramatically different viscoelastic properties. Thus, the Cpx response is required to sense attachment to abiotic surfaces (see above), and some Cpx-regulated gene product(s) must be required for this process to occur in a wild-type fashion.

THE Cpx RESPONSE INFLUENCES BIOFILM FORMATION

Further support for a role for the Cpx response in surface attachment derives from studies of the involvement of the Cpx response in biofilm formation. Lejeune and colleagues identified mutations in *cpxA* that led to increased CpxR~P levels in a screen for mutants defective in biofilm formation in 96-well polystyrene plates (Dorel et al., 1999). Activation of the Cpx response by *cpxA** mutations or NlpE overexpression had a similar negative effect on biofilm formation and they showed that this correlated with diminished expression of the *csg* genes specifying the curli adhesin. Subsequent studies identified CpxR binding sites in the promoters of the *csgBA* and *csgDEFG* operons, which suggested direct repression of *csg* transcription by CpxR~P (see above) (Jubelin et al., 2005; Prigent-Combaret et al., 2001) (Table 1). Finally, Ghigo and colleagues showed by microarray analysis of biofilm and planktonic populations of cells that the Cpx-regulated genes *cpxP* and *spy* were highly up-regulated in biofilms and that *cpxP* and *cpxR* mutants formed biofilms with reduced mass and substrate coverage (Beloin et al., 2004). To accommodate these observations, it has been proposed that Cpx induction within biofilms serves to down-regulate adhesin synthesis (i.e., curli), since they are no longer needed, and also to up-regulate expression of other factors that facilitate efficient biofilm formation (i.e., CpxP, Spy?). Thus, although the mechanism is unclear, the evidence suggests that the Cpx response is involved at some level in biofilm formation.

Pathogenesis

Since P pili and BFP are required to attach to kidney and epithelial cells, respectively, to initiate urinary tract or intestinal infections, the documented effects of the Cpx response on pilus expression and assembly (see above) suggested a part for the Cpx pathway in controlling the virulence of bacterial pathogens and, in particular, the early stages of infection involving host cell attachment. Although work in this area is just beginning, it has been shown that mutation of the *cpx* locus alters infection in animal models and the Cpx response has been shown to influence transcription of a variety of key virulence genes (Table 1).

In *Salmonella enterica* serovar Typhimurium, Roberts and colleagues showed that *cpxA** mutations caused reduced attachment to and invasion of eukaryotic cells and both gain (*cpxA**) and loss- (*cpxA*)-of-function mutations in *cpxA* led to a growth defect in mice (Humphreys et al., 2004). Another study showed that insertional inactivation of *cpxA* in serovar Typhimurium caused reduced expression of HilA, a key regulator of invasion genes, at certain pH values (Nakayama et al., 2003). Curiously, *cpxR* mutants were unaffected in a mouse model of serovar Typhimurium infection (Humphreys et al., 2004). Since, in the absence of CpxA phosphatase activity, CpxR~P levels can build up in the cell due to phosphorylation by low-molecular-weight phosphodonors such as acetylphosphate (Danese et al., 1995), one explanation for these apparently contradictory observations is that it is activation of the Cpx response, and not its elimination, that reduces serovar Typhimurium virulence. Thus, in both *cpxA** and *cpxA*- bacteria, HilA expression would be reduced and virulence functions, including attachment and invasion, would be adversely affected. These studies, combined with the observation that Cpx pathway activation can repress transcription of the P pilus encoding *pap* operon, suggest that perhaps the main function of the Cpx response in pathogenesis is to inhibit virulence-determinant expression from occurring at inappropriate times.

The Cpx response is also implicated in the regulation of virulence gene expression in other pathogens, including *Shigella sonnei*, *Legionella pneumophila*, and *Yersinia enterocolitica*. In *S. sonnei*, the Cpx response is thought to control expression of the central regulatory gene *virF*, which encodes a regulator of invasion functions (Nakayama and Watanabe, 1995, 1998). *E. coli cpxA* mutants exhibit deregulated expression of an exogenous *virF* promoter in response to pH, and CpxR~P has been shown to bind to the *virF* promoter region. Further, *cpxA* mutants reduce the levels of a second regulatory protein, InvE, at a posttranscriptional level (Mitobe et al., 2005). Since the phenotypes of *cpxR* mutants are not known, it is possible that in *Shigella* species, as in serovar Typhimurium, the *cpxA* mutant actually mimics Cpx pathway activation and that this is the signal that leads to altered VirF and InvE expression. Finally, it has been shown that CpxR homologues in *L. pneumophila* and *Y. enterocolitica* are likely involved in transcription of the *icm-dot* genes encoding the type IV secretion system and *degP*, respectively, both of which are necessary for virulence (Gal-Mor and Segal, 2003; Heusipp et al., 2004). CpxR was identified in a complementation screen designed to identify *L. pneumophila* genes, which when cloned in multiple copies, could restore the low levels of expression of an *icmR-lacZ* fusion observed in *E. coli*. *L. pneumophila cpxR* mutants exhibited diminished expression of the gene encoding the IcmR chaperone as well as several other *icm-lacZ* fusions. The *L. pneumophila* CpxR was shown to bind to the *icmR* promoter, and deletion of a putative CpxR binding site led to a reduction in *icmR-lacZ* expression equivalent to that seen when *cpxR* was mutated. In *Y. enterocolitica*, a deletion of *cpxR* was shown to ablate induction of *degP* expression by overexpression of the outer membrane protein (OMP) Ail. Curiously, OMP overexpression is a trigger for the σ^E response in *E. coli*. Thus, these data suggest that the Cpx signaling pathway may function differently in *Y. enterocolitica*. Cumulatively, these experiments implicate the Cpx response in regulating the transcription of virulence genes in various pathogens. Some data suggest that this effect may be predominantly negative and it is unclear how these findings can be reconciled with those that indicate a stimulatory role for the Cpx response in pilus assembly in *E. coli*. One possibility is that the Cpx response facili-

tates early steps in infection, such as pilus biogenesis, while simultaneously inhibiting later steps (invasion) that would be ineffective until after host cell attachment takes place.

Conjugation/Regulation of Cytoplasmic Protein Turnover

The earliest studies of the *cpx* locus identified a role for the Cpx response in controlling conjugal transfer of the *E. coli* F plasmid. More recent studies following up on this observation, together with experiments designed to identify Cpx regulon members, suggest that this effect may be predominantly due to Cpx control of genes encoding factors involved in cytoplasmic protein elaboration and/or turnover.

Silverman's early experiments identified gain-of-function *cpxA** mutations that constitutively activated the Cpx response and caused F plasmid–containing donor cells to be insensitive to Qβ phage, which adsorbs to cells via the F pilus (McEwen and Silverman, 1980a). His group showed that these mutants failed to make the F pilus. Further, analysis of *lacZ* reporters indicated that expression of *tra* genes, which are required for conjugal DNA transfer, was diminished, and this correlated with a decrease in TraJ protein levels (Silverman et al., 1993). Since TraJ is a positive regulator of *tra* operon expression, it was concluded that the effect of the *cpxA** mutations on conjugation and F pilus expression was due to a decrease in TraJ levels. Almost 10 years after this initial observation, the Frost and Raivio labs demonstrated by immunoblot analysis that activation of the Cpx response by mutation (*cpxA**) or NlpE overexpression caused a decrease in the levels of the TraJ, TraY, and TraM regulatory proteins (Gubbins et al., 2002). Northern blot and *traJ-lacZ* expression analyses showed that Cpx pathway activation had little effect on *traJ* transcription. Analysis of TraJ, TraY, and TraM protein half-lives in wild-type and Cpx-induced strains indicated that TraJ stability was specifically decreased when the Cpx response was activated. This study was the first to indicate that the Cpx response might influence the fates of intracellular proteins, in addition to those found in the envelope.

Although the mechanism of Cpx-mediated TraJ instability is not known, some tantalizing observations support the hypothesis that the Cpx response controls the fates of proteins not found in the envelope. The Beckwith and Lin groups both noted the presence of a putative CpxR~P consensus binding site upstream of *rpoH* (De Wulf et al., 2002; Pogliano et al., 1997) (Table 1). Since *rpoH* encodes σ^H, which regulates a cotillion of cytoplasmic proteases and chaperones, this observation suggested that the Cpx pathway might indeed influence cytoplasmic protein fates. However, there are no published data to support Cpx-dependent regulation of *rpoH*, so this link remains tentative. Lin and colleagues used a bioinformatic approach to identify genes containing putative CpxR~P binding sites in their upstream regions and predicted that the *tig*, *dnaK*, *hslTS*, and *rpsP* genes might be Cpx regulated (De Wulf et al., 2002). These genes encode trigger factor (*tig*), which is involved in folding newly translated proteins, σ^H regulated heat shock chaperones (*dnaK*) or proteins (*hslTS*), and a 30S ribosomal subunit protein. Further, Akiyama and colleagues showed that, under conditions where proteolysis of plasma membranes is prevented, overexpression of this class of proteins induced the Cpx response (Shimohata et al., 2002). Additionally, the Cpx response was shown to regulate transcription of HtpX, a plasma membrane zinc metalloproteinase in which the protease domain is localized to the cytoplasm (Shimohata et al., 2002) (Table 1). Altogether, these experiments argue that, in addition to regulating the fates of envelope-localized proteins, the Cpx response also influences the stability of intracellular proteins. Perhaps the Cpx response has evolved to coordinate protein folding and degrading activities in the cytoplasm and envelope.

Other Roles?

Lin and colleagues developed a 15-bp weighted matrix for CpxR recognition based on the sequences of known Cpx-regulated genes and used it to scan the *E. coli* genome in 15-bp segments for related sites (De Wulf et al., 2002). They identified more than 100 genes/operons with putative CpxR binding sites located upstream of the transcriptional start site. The genes identified encode or are predicted to encode proteins with diverse functions. Some of these include motility and chemotaxis, transport, and metabolism. Only eight of these potential targets were studied further. Northern analysis showed that transcription of the *ung* (uracil-N-glycosylase), *ompC* (outer membrane porin C), *psd* (phosphatidyl serine decarboxylase), *mviM* (virulence factor), *aroK* (shikimate kinase I), *rpoErseABC* (σ^E and σ^E regulatory proteins), *secA* (secretion), and *aer* (aerotaxis) genes was altered by *cpxR* deletion and/or a *cpxA** mutation (Table 1). Subsequent studies have shown that *ung* is negatively regulated by the Cpx response, and this leads to an increased mutation rate in the presence of Cpx induction (Ogasawara et al., 2004). Similarly, Cpx pathway activation has been shown to activate *ompC* expression in a direct manner (Batchelor et al., 2005). Presumably, Cpx regulation of these gene targets endows some physiological adaptation to signals that induce the Cpx response (envelope stress), but at this time it is not clear how many of the genes cited by DeWulf et al. represent bona fide Cpx targets, and what their function in responding to envelope stresses might be.

OVERLAP WITH OTHER STRESS RESPONSES

The Cpx envelope stress response exhibits significant connectivity to other adaptive responses. The Cpx pathway shares common inducers and downstream targets with the σ^E and Bae envelope stress responses, as well as the EnvZ-OmpR response to osmolarity (Table 1). In addition, the σ^S-regulated stationary-phase response and the σ^H-mediated heat shock response are tied to the Cpx pathway (Table 1). Although the logic for the observed overlap between the Cpx response and other adaptive mechanisms is not completely clear, presumably integration of stress responses allows for optimal physiological adaptation in the face of a given environment that may contain a mixture of complex conditions.

The Cpx and σ^E Envelope Stress Responses

Significant stress response overlap was first noted for the Cpx and σ^E envelope stress responses. The analysis of Cpx- and σ^E-regulated *lacZ* fusions indicated that some inducing cues, such as overexpression of the adhesin PapG in the absence of its cognate periplasmic chaperone, and accumulation of ECA lipid II precursor in the inner membrane, activated both pathways (Danese et al., 1998; Jones et al., 1997). Similarly, some genes were activated by both responses (Table 1). The *degP* gene, specifying a periplasmic protease; the *skp* gene, encoding a periplasmic chaperone; and the *rpoErseABC* operon, encoding σ^E and its regulators, are thought to be jointly regulated by the Cpx and σ^E responses (Danese et al., 1995; Dartigalongue et al., 2001; De Wulf et al., 2002). Studies of the stress-inducible expression of *degP* led to the identification of σ^E, and *cpxA★* alleles were initially shown to alleviate toxic envelope stress through increased transcription of *degP* (Cosma et al., 1995; Danese et al., 1995; Erickson and Gross, 1989; Lipinska et al., 1988; Snyder et al., 1995). Dartigalongue et al. have shown that expression of β-galactosidase from a *skp-lacZ* fusion responds to both the σ^E and Cpx responses (Dartigalongue et al., 2001). Finally, σ^E autoactivates its own operon and De Wulf and colleagues identified a putative CpxR~P binding site upstream of the *rpoErseABC* operon (De Wulf et al., 2002; Raina et al., 1995; Rouviere et al., 1995). They showed that deletion of *cpxR* led to elevated expression of this operon, while constitutive activation of the Cpx pathway by *cpxA★* mutation led to diminished *rpoErseABC* transcription, suggesting that Cpx response activation may inhibit σ^E pathway activity (De Wulf et al., 2002). Thus, it seems that some inducing cues must generate activating signals for both responses and that some aspects of adaptation to the specific cues that trigger each response must require common protein-folding and -degrading factors. It is not clear why the Cpx response would inhibit the σ^E pathway. Perhaps, in the face of Cpx-specific envelope stresses that disrupt envelope protein folding, it is desirable to inhibit or halt OMP biogenesis, in which the σ^E response is centrally involved (see Chapter 6).

The Cpx and Bae Envelope Stress Responses

The Cpx and Bae envelope stress responses are also interrelated. The Bae pathway was initially identified based on its ability to induce expression of the Cpx regulon member *spy* in response to Cpx-inducing cues in a *cpxRA* mutant (Raffa and Raivio, 2002). Thus, it has been clear from the start that these two pathways must respond to signals that are closely related and that Bae- and Cpx-mediated adaptation must share some features. Recently, it was shown that one component of this shared adaptation includes induction of genes encoding components of multidrug efflux pumps (Table 1). Yamaguchi and colleagues demonstrated that indole-induced increases in expression of the *mdtABC* and *acrD* genes, encoding multidrug efflux pump components, was jointly mediated by the Cpx and Bae envelope stress responses (Hirakawa et al., 2005). Further, they showed via DNA-footprinting experiments that CpxR bound upstream of BaeR to multiple sites at both the *mdtABC* and *acrD* promoters. Since deleting BaeSR led to a greater reduction in the ability to induce *mdtABC* or *acrD* expression than deletion of CpxAR did, and since the BaeR binding site is located proximal to the promoter, it has been proposed that the Cpx response plays an accessory role in efflux pump induction, enhancing BaeR-mediated activation of these genes.

Cpx Overlap with the EnvZ/OmpR Regulon

De Wulf et al. suggested the presence of CpxR~P binding sites upstream of the *ompC* and *ompF* genes based on bioinformatic analysis and demonstrated that *ompC* transcript levels were elevated in strains bearing *cpxA*★ mutations and diminished in those carrying *cpxR* deletion alleles (De Wulf et al., 2002). Silverman's group showed in 1983 that OmpF levels were reduced in *cpxA*★ mutants (McEwen et al., 1983). Recently, Goulian and coworkers identified mutations in *cpxA* in a transposon mutagenesis screen for genes affecting expression of the OmpC and OmpF porins (Batchelor et al., 2005). They showed that elevated Cpx pathway activity caused increased *ompC* transcription and reduced *ompF* expression. DNase I footprinting showed that CpxR bound to sites upstream of, and overlapping, those of the OmpR response regulator, which, together with the EnvZ histidine kinase, functions to regulate porin expression in response to osmolarity. It is hypothesized that CpxR binding at the *ompF* promoter interferes with either OmpR or RNAP binding to mediate repression. In contrast, since the Cpx-mediated activation of the *ompC* promoter required OmpR, it was proposed that CpxR must bind upstream of *ompC* in conjunction with OmpR to affect activation. Thus, at the *ompC* promoter, CpxR appears to fulfill an accessory role, as it does at the *mdtABC* and *acrD* promoters. These data suggest that the EnvZ/OmpR and CpxA/CpxR responses are linked (Table 1). In support of this idea, the Cpx response has also been shown to influence the *csgBA* and *csgDEFG* promoters, which specify the curli adhesin and are also EnvZ/OmpR regulated (Jubelin et al., 2005; Prigent-Combaret et al., 2001). Activation of the Cpx pathway led to inhibition of expression of the *csg* operons and CpxR was shown to bind upstream of both *csgD* and *csgB* to sites that overlapped those of OmpR. Since OmpR is required for the activation of curli expression in response to increased osmolarity, it has been hypothesized that CpxR exerts repression by competing with OmpR for binding upstream of the *csgD* and/or *csgB* promoters. Thus, it appears that the CpxAR and EnvZ/OmpR two-component signal transduction pathways func-

tion together at multiple promoters to set the level of transcription. Presumably this signaling balances adaptive requirements in response to the signals sensed by EnvZ and CpxA.

The Cpx Response and Stationary Phase

Finally, the Cpx response also interfaces with the starvation/stationary response regulated by σ^S (Table 1). Lin's group showed that, in some strains of *E. coli*, σ^S contributes to the elevated activity of the Cpx pathway in stationary phase by mediating an increase in *cpxRA* expression upon entry to stationary phase (De Wulf et al., 1999). DiGiuseppe and Silhavy have also shown that Cpx pathway activity increases in stationary phase (DiGiuseppe and Silhavy, 2003). They found that some of this increased activity was independent of the sensor kinase CpxA, but dependent on CpxR, suggesting that stationary-phase-specific signals might enter the Cpx signal transduction cascade at points beyond CpxA, in the cytoplasm. In their laboratory strain, σ^S did not contribute to stationary-phase regulation of Cpx gene expression, and so there may be additional signaling pathways that mesh with the Cpx response upon entry to stationary phase or under starvation conditions. Clearly, though, there is some coordination in activation of the Cpx and σ^S stress responses.

SUMMARY AND FUTURE PERSPECTIVES

Characterization of the original Cpx mutants revealed that the *cpx* locus affected multiple, apparently unrelated phenotypes. Studies over the past 20 years have confirmed this early observation and given us some insight into why this might be. One of the major functions of the Cpx response is to alter the fates of envelope proteins in response to challenges to the bacterial envelope. Undoubtedly, many of the diverse phenotypes observed in *cpxA*★ mutants bearing a constitutively active Cpx signal transduction pathway are due to this well-defined role in the regulation of envelope protein-folding factors. Recent bioinformatic, genetic, and molecular analyses, however, indicate that the picture is more complicated. The Cpx response also influences the transcription of many genes with no apparent role in envelope protein folding. Further, the Cpx response appears able to impinge on and alter the outcomes of other stress responses. Only through a careful cataloging of Cpx regulon members will we begin to understand other functions this response might have, outside envelope protein biogenesis. Other outstanding questions remain. How does CpxR function as a transcription factor? CpxR~P can activate, repress, or enhance transcription mediated by other regulators. While binding site locations often provide insight into potential mechanisms, in other cases it is unclear how CpxR~P binding might exert the observed effects on transcription. This is especially true at promoters where CpxR seems to exert its effects by potentiating the activity of another regulatory protein. Finally, although it has become dogma that CpxA senses and is activated by misfolded envelope proteins, the mechanism(s) involved remain elusive. Hopefully, an analysis of the folding pathways of proteins that induce the Cpx response upon overexpression in conjunction with more detailed studies of the CpxA sensing domain will prove enlightening. CpxP clearly has a role in setting the activity level of the CpxA kinase, but how this occurs is not known. It is commonly accepted that CpxP and CpxA interact directly, although no evidence for this exists. Undoubtedly, future experiments will be directed toward understanding the extent of the Cpx regulon and the molecular mechanisms used to sense and transduce envelope stress signals into a specific pattern of altered gene expression.

REFERENCES

Albin, R., and P. M. Silverman. 1984. Identification of the *Escherichia coli* K-12 *cpxA* locus as a single gene: construction and analysis of biologically-active *cpxA* gene fusions. *Mol. Gen. Genet.* **197:**272–279.

Albin, R., R. Weber, and P. M. Silverman. 1986. The Cpx proteins of *Escherichia coli* K12. Immuno-logic detection of the chromosomal *cpxA* gene product. *J. Biol. Chem.* **261:**4698–4705.

Batchelor, E., D. Walthers, L. J. Kenney, and M. Goulian. 2005. The *Escherichia coli* CpxA-CpxR envelope stress response system regulates expression of the porins *ompF* and *ompC. J. Bacteriol.* **187:**5723–5731.

Beck, N. A., E. S. Krukonis, and V. J. DiRita. 2004. TcpH influences virulence gene expression in *Vibrio cholerae* by inhibiting degradation of the transcription activator TcpP. *J. Bacteriol.* **186:**8309–8316.

Beloin, C., J. Valle, P. Latour-Lambert, P. Faure, M. Kzreminski, D. Balestrino, J. A. Haagensen, S. Molin, G. Prensier, B. Arbeille, and J. M. Ghigo. 2004. Global impact of mature biofilm lifestyle on *Escherichia coli* K-12 gene expression. *Mol. Microbiol.* **51:**659–674.

Buelow, D. R., and T. L. Raivio. 2005. Cpx signal transduction is influenced by a conserved N-terminal domain in the novel inhibitor CpxP and the periplasmic protease DegP. *J. Bacteriol.* **187:**6622–6630.

Carlson, J. H., and T. J. Silhavy. 1993. Signal sequence processing is required for the assembly of LamB trimers in the outer membrane of *Escherichia coli. J. Bacteriol.* **175:**3327–3334.

Cosma, C. L., P. N. Danese, J. H. Carlson, T. J. Silhavy, and W. B. Snyder. 1995. Activation of the Cpx two-component signal transduction pathway in *Escherichia coli* suppresses envelope associated stresses. *Mol. Microbiol.* **18:**491–505.

Danese, P. N., G. R. Oliver, K. Barr, G. D. Bowman, P. D. Rick, and T. J. Silhavy. 1998. Accumulation of the enterobacterial common antigen lipid II biosynthetic intermediate stimulates *degP* transcription in *Escherichia coli. J. Bacteriol.* **180:**5875–5884.

Danese, P. N., and T. J. Silhavy. 1997. The σᴱ and the Cpx signal transduction systems control the synthesis of periplasmic protein-folding enzymes in *Escherichia coli. Genes Dev.* **11:**1183–1193.

Danese, P. N., and T. J. Silhavy. 1998. CpxP, a stress-combative member of the Cpx regulon. *J. Bacteriol.* **180:**831–839.

Danese, P. N., W. B. Snyder, C. L. Cosma, L. J. Davis, and T. J. Silhavy. 1995. The Cpx two-component signal transduction pathway of *Escherichia coli* regulates transcription of the gene specifying the stress-inducible periplasmic protease, DegP. *Genes Dev.* **9:**387–398.

Dartigalongue, C., D. Missiakas, and S. Raina. 2001. Characterization of the *Escherichia coli* sigma E regulon. *J. Biol. Chem.* **276:**20866–20875.

De Wulf, P., O. Kwon, and E. C. C. Lin. 1999. The CpxRA signal transduction system of *Escherichia coli*: growth-related autoactivation and control of unanticipated target operons. *J. Bacteriol.* **181**:6552–6778.

De Wulf, P., A. M. McGuire, X. Liu, and E. C. Lin. 2002. Genome-wide profiling of promoter recognition by the two-component response regulator CpxR-P in *Escherichia coli. J. Biol. Chem.* **277**:26652–26661.

DiGiuseppe, P. A., and T. J. Silhavy. 2003. Signal detection and target gene induction by the CpxRA two-component system. *J. Bacteriol.* **185**:2432–2440.

Dong, J. S., S. Iuchi, S. Kwan, Z. Lu, and E. C. C. Lin. 1993. The deduced amino-acid sequence of the cloned *cpxR* gene suggests the protein is the cognate regulator for the membrane sensor, CpxA, in a two-component signal transduction system of *Escherichia coli. Gene* **136**:227–230.

Dorel, C., O. Vidal, C. Prigent-Combaret, I. Vallet, and P. Lejeune. 1999. Involvement of the Cpx signal transduction pathway of *E. coli* in biofilm formation. *FEMS Microbiol. Lett.* **178**:169–175.

Erickson, J. W., and C. A. Gross. 1989. Identification of the σ^E subunit of *Escherichia coli* RNA polymerase: a second alternate σ factor involved in high-temperature gene expression. *Genes Dev.* **3**:1462–1471.

Gal-Mor, O., and G. Segal. 2003. Identification of CpxR as a positive regulator of *icm* and *dot* virulence genes of *Legionella pneumophila. J. Bacteriol.* **185**:4908–4919.

Grigorova, I. L., R. Chaba, H. J. Zhong, B. M. Alba, V. Rhodius, C. Herman, and C. A. Gross. 2004. Fine-tuning of the *Escherichia coli* sigmaE envelope stress response relies on multiple mechanisms to inhibit signal-independent proteolysis of the transmembrane anti-sigma factor, RseA. *Genes Dev.* **18**:2686–2697.

Gubbins, M. J., I. Lau, W. R. Will, J. M. Manchak, T. L. Raivio, and L. S. Frost. 2002. The positive regulator, TraJ, of the *Escherichia coli* F plasmid is unstable in a *cpxA*★ background. *J. Bacteriol.* **184**:5781–5788.

Hernday, A. D., B. A. Braaten, G. Broitman-Maduro, P. Engelberts, and D. A. Low. 2004. Regulation of the *pap* epigenetic switch by CpxAR: phosphorylated CpxR inhibits transition to the phase ON state by competition with Lrp. *Mol. Cell* **16**:537–547.

Heusipp, G., K. M. Nelson, M. A. Schmidt, and V. L. Miller. 2004. Regulation of htrA expression in *Yersinia enterocolitica. FEMS Microbiol. Lett.* **231**:227–235.

Hirakawa, H., Y. Inazumi, T. Masaki, T. Hirata, and A. Yamaguchi. 2005. Indole induces the expression of multidrug exporter genes in *Escherichia coli. Mol. Microbiol.* **55**:1113–1126.

Hirakawa, H., K. Nishino, T. Hirata, and A. Yamaguchi. 2003. Comprehensive studies of drug resistance mediated by overexpression of response regulators of two-component signal transduction systems in *Escherichia coli. J. Bacteriol.* **185**:1851–1856.

Humphreys, S., G. Rowley, A. Stevenson, M. F. Anjum, M. J. Woodward, S. Gilbert, J. Kormanec, and M. Roberts. 2004. Role of the two-component regulator CpxAR in the virulence of *Salmonella enterica* serotype Typhimurium. *Infect. Immun.* **72**:4654–4661.

Hung, D. L., T. L. Raivio, C. H. Jones, T. J. Silhavy, and S. J. Hultgren. 2001. Cpx signaling pathway monitors biogenesis and affects assembly and expression of P pili. *EMBO J.* **20**:1508–1518.

Hunke, S., and J. M. Betton. 2003. Temperature effect on inclusion body formation and stress response in the periplasm of *Escherichia coli. Mol. Microbiol.* **50**:1579–1589.

Isaac, D. D., J. S. Pinkner, S. J. Hultgren, and T. J. Silhavy. 2005. The extracytoplasmic adaptor protein CpxP is degraded with substrate by DegP. *Proc. Natl. Acad. Sci. USA* **102**:17775–17779.

Jacob-Dubuisson, F., J. Pinkner, X. Xu, R. Striker, A. Padmanhaban, and S. J. Hultgren. 1994. PapD chaperone function in pilus biogenesis depends on oxidant and chaperone-like activities of DsbA. *Proc. Natl. Acad. Sci. USA* **91**:11552–11556.

Jones, C. H., P. N. Danese, J. S. Pinkner, T. J. Silhavy, and S. J. Hultgren. 1997. The chaperone-assisted membrane release and folding pathway is sensed by two signal transduction systems. *EMBO J.* **21**:6394–6406.

Jubelin, G., A. Vianney, C. Beloin, J. M. Ghigo, J. C. Lazzaroni, P. Lejeune, and C. Dorel. 2005. CpxR/OmpR interplay regulates curli gene expression in response to osmolarity in *Escherichia coli. J. Bacteriol.* **187**:2038–2049.

Lee, Y. M., P. A. DiGiuseppe, T. J. Silhavy, and S. J. Hultgren. 2004. P pilus assembly motif necessary for activation of the CpxRA pathway by PapE in *Escherichia coli. J. Bacteriol.* **186**:4326–4337.

Lipinska, B., S. Sharma, and C. Georgopoulos. 1988. Sequence analysis and regulation of the *htrA* gene of *Escherichia coli*: a σ^{32}-independent mechanism of heat-inducible transcription. *Nucleic Acids Res.* **16**:10053–10067.

Martinez-Hackert, E., and A. M. Stock. 1997. The DNA-binding domain of OmpR: crystal structures of a winged helix transcription factor. *Structure* **5**:109–124.

Matson, J. S., and V. J. DiRita. 2005. Degradation of the membrane-localized virulence activator TcpP by the YaeL protease in *Vibrio cholerae. Proc. Natl. Acad. Sci. USA* **102**:16403–16408.

McEwen, J., L. Sambucetti, and P. M. Silverman. 1983. Synthesis of outer membrane proteins in *cpxA cpxB* mutants of *Escherichia coli* K-12. *J. Bacteriol.* **154**:375–382.

McEwen, J., and P. M. Silverman. 1980a. Chromosomal mutations of *Escherichia coli* that alter expression of conjugative plasmid functions. *Proc. Natl. Acad. Sci. USA* **77:**513–517.

McEwen, J., and P. M. Silverman. 1980b. Genetic analysis of *Escherichia coli* K-12 chromosomal mutants defective in expression of F-plasmid functions: identification of genes *cpxA* and *cpxB. J. Bacteriol.* **144:**60–67.

McEwen, J., and P. M. Silverman. 1980c. Mutations in genes *cpxA* and *cpxB* of *Escherichia coli* K-12 cause a defect in isoleucine and valine syntheses. *J. Bacteriol.* **144:**68–73.

McEwen, J., and P. M. Silverman. 1982. Mutations in genes *cpxA* and *cpxB* alter the protein composition of *Escherichia coli* inner and outer membranes. *J. Bacteriol.* **151:**1553–1559.

Mileykovskaya, E., and W. Dowhan. 1997. The Cpx two-component signal transduction pathway is activated in *Escherichia coli* mutant strains lacking phosphatidylethanolamine. *J. Bacteriol.* **179:**1029–1034.

Mitobe, J., E. Arakawa, and H. Watanabe. 2005. A sensor of the two-component system CpxA affects expression of the type III secretion system through posttranscriptional processing of InvE. *J. Bacteriol.* **187:**107–113.

Miyadai, H., K. Tanaka-Masuda, S. Matsuyama, and H. Tokuda. 2004. Effects of lipoprotein overproduction on the induction of DegP (HtrA) involved in quality control in the *Escherichia coli* periplasm. *J. Biol. Chem.* **279:**39807–39813.

Nakayama, S., A. Kushiro, T. Asahara, R. Tanaka, L. Hu, D. J. Kopecko, and H. Watanabe. 2003. Activation of *hilA* expression at low pH requires the signal sensor CpxA, but not the cognate response regulator CpxR, in *Salmonella enterica* serovar Typhimurium. *Microbiology* **149:**2809–2817.

Nakayama, S.-I., and H. Watanabe. 1995. Involvement of *cpxA*, a sensor of a two-component regulatory system, in the pH-dependent regulation of expression of *Shigella sonnei virF* gene. *J. Bacteriol.* **177:**5062–5069.

Nakayama, S.-I., and H. Watanabe. 1998. Identification of *cpxR* as a positive regulator essential for expression of the *Shigella sonnei virF* gene. *J. Bacteriol.* **180:**3522–3528.

Nevesinjac, A. Z., and T. L. Raivio. 2005. The Cpx envelope stress response affects expression of the type IV bundle-forming pili of enteropathogenic Escherichia coli. *J. Bacteriol.* **187:**672–686.

Ogasawara, H., J. Teramoto, K. Hirao, K. Yamamoto, A. Ishihama, and R. Utsumi. 2004. Negative regulation of DNA repair gene (*ung*) expression by the CpxR/CpxA two-component system in *Escherichia coli* K-12 and induction of mutations by increased expression of CpxR. *J. Bacteriol.* **186:**8317–8325.

Otto, K., and T. J. Silhavy. 2002. Surface sensing and adhesion of *Escherichia coli* controlled by the Cpx-signaling pathway. *Proc. Natl. Acad. Sci. USA* **99:**2287–2292.

Pogliano, J. A., S. Lynch, D. Belin, E. C. C. Lin, and J. Beckwith. 1997. Regulation of *Escherichia coli* cell envelope proteins involved in protein folding and degradation by the Cpx two-component system. *Genes Dev.* **11:**1169–1182.

Pratt, L. A., and T. J. Silhavy. 1994. OmpR mutants specifically defective for transcriptional activation. *J. Mol. Biol.* **243:**579–594.

Prigent-Combaret, C., E. Brombacher, O. Vidal, A. Ambert, P. Lejeune, P. Landini, and C. Dorel. 2001. Complex regulatory network controls initial adhesion and biofilm formation in *Escherichia coli* via regulation of the *csgD* gene. *J. Bacteriol.* **183:**7213–7223.

Raffa, R. G., and T. L. Raivio. 2002. A third envelope stress signal transduction pathway in *Escherichia coli. Mol. Microbiol.* **45:**1599–1611.

Raina, S., D. Missiakas, and C. Georgopoulos. 1995. The *rpoE* gene encoding the σ^E (σ^{24}) heat shock sigma factor of *Escherichia coli. EMBO J.* **14:**1043–1055.

Rainwater, S., and P. M. Silverman. 1990. The Cpx proteins of *Escherichia coli* K-12: evidence that *cpxA*, *ecfB*, *ssd*, and *eup* mutations all identify the same gene. *J. Bacteriol.* **172:**2456–2461.

Raivio, T. L., M. W. Laird, J. C. Joly, and T. J. Silhavy. (2000). Tethering of CpxP to the inner membrane prevents spheroplast induction of the Cpx envelope stress response. *Mol. Microbiol.* **37:**1186–1197.

Raivio, T. L., D. L. Popkin, and T. J. Silhavy. 1999. The Cpx envelope stress response is controlled by amplification and feedback inhibition. *J. Bacteriol.* **181:**5263–5272.

Raivio, T. L., and T. J. Silhavy. 1997. Transduction of envelope stress in *Escherichia coli* by the Cpx two-component system. *J. Bacteriol.* **179:**7724–7733.

Raivio, T. L., and T. J. Silhavy. 1999. The σ^E and Cpx regulatory pathways: overlapping but distinct envelope stress responses. *Curr. Opin. Microbiol.* **2:**159–165.

Raivio, T. L., and T. J. Silhavy. 2001. Periplasmic stress and ECF sigma factors. *Annu. Rev. Microbiol.* **55:**591–524.

Rouviere, P. E., A. De Las Penas, J. Mecsas, C. Z. Lu, K. E. Rudd, and C. A. Gross. 1995. *rpoE*, the gene encoding the second heat-shock sigma factor, σ^E, in *Escherichia coli. EMBO J.* **14:**1032–1042.

Sauer, F. G., K. Futterer, J. S. Pinkner, K. W. Dodson, S. J. Hultgren, and G. Waksman. 1999. Structural basis of chaperone function and pilus biogenesis. *Science* **285:**1058–1061.

Shimohata, N., S. Chiba, N. Saikawa, K. Ito, and Y. Akiyama. 2002. The Cpx stress response system of *Escherichia coli* senses plasma membrane proteins and controls HtpX, a membrane protease with a cytosolic active site. *Genes Cells* **7**:653–662.

Silverman, P. M., L. Tran, R. Harris, and H. M. Gaudin. 1993. Accumulation of the F plasmid TraJ protein in *cpx* mutants of *Escherichia coli*. *J. Bacteriol.* **175**:921–925.

Snyder, W. B., L. J. B. Davis, P. N. Danese, C. L. Cosma, and T. J. Silhavy. 1995. Overproduction of NlpE, a new outer membrane lipoprotein, suppresses the toxicity of periplasmic LacZ by activation of the Cpx signal transduction pathway. *J. Bacteriol.* **177**:4216–4223.

Snyder, W. B., and T. J. Silhavy. 1995. β-galactosidase is inactivated by intermolecular disulfide bonds and is toxic when secreted to the periplasm of *Escherichia coli*. *J. Bacteriol.* **177**:953–963.

Sutton, A., T. Newman, J. McEwen, P. M. Silverman, and M. Freundlich. 1982. Mutations in genes *cpxA* and *cpxB* of *Escherichia coli* K-12 cause a defect in acetohydroxyacid synthase I function in vivo. *J. Bacteriol.* **151**:976–982.

Thanassi, D. G., and S. J. Hultgren. 2000. Assembly of complex organelles: pilus biogenesis in gram-negative bacteria as a model system. *Methods* **20**: 111–126.

Weber, R. F., and P. J. Silverman. 1988. The Cpx proteins of *Escherichia coli* K-12: structure of the CpxA polypeptide as an inner membrane component. *J. Mol. Biol.* **203**:467–476.

Wolfgang, M., J. P. van Putten, S. F. Hayes, D. Dorward, and M. Koomey. 2000. Components and dynamics of fiber formation define a ubiquitous biogenesis pathway for bacterial pili. *EMBO J.* **19**:6408–6418.

Zhang, H. Z., and M. S. Donnenberg. 1996. DsbA is required for stability of the type IV pilin of enteropathogenic *Escherichia coli*. *Mol. Microbiol.* **21**: 787–797.

REGULATION AND FUNCTION OF THE ENVELOPE STRESS RESPONSE CONTROLLED BY σ^E

Carol A. Gross, Virgil A. Rhodius, and Irina L. Grigorova

6

When cells experience heat or other stresses that unfold proteins, adaptive responses that counteract cellular damage are induced. Distinct stress responses cope with damage in each compartment of the gram-negative bacterial cell. In *Escherichia coli*, the cytoplasmic heat shock response is mediated primarily by σ^{32}, an alternative σ-factor whose regulon members promote protein, RNA, DNA, and inner membrane homeostasis (reviewed in Gross, 1996; Yura and Nakahigashi, 1999; and Nonaka et al., 2006). In addition, σ^S is slowly induced by heat stress and contributes additional protective functions for the cell (reviewed in Hengge-Aronis, 2002). σ^E and the two-component CpxRA pathways are the primary responders to envelope stress (reviewed in Raivio and Silhavy, 2001) and these two responses are linked. σ^E is essential for viability under nonstressed conditions; cells lacking σ^E survive because of unmapped suppressor mutations that permit growth at low but not high temperature (Con-

nolly et al., 1997; De Las Penas et al., 1997a). However, overexpression of the CpxRA regulon permits growth at high temperature, indicating that some stress functions are shared by both regulons (Connolly et al., 1997). Both stress pathways are considered in this volume. Here, we consider regulation and function of the σ^E envelope stress response. The reader is also referred to several excellent recent reviews of the σ^E pathway (Raivio and Silhavy, 2001; Ades, 2004; Alba and Gross, 2004).

THE DISCOVERY OF σ^E AND ITS STRESS RESPONSE

σ^E was discovered in a quest for the factor that regulates σ^{32} expression at lethal temperatures. Mapping the promoters controlling σ^{32} expression indicated that *rpoH*, the gene encoding σ^{32}, was transcribed from three major promoters (Erickson et al., 1987). Transcription from the middle promoter, P2, increased dramatically at lethal temperature and when cells were subjected to lethal concentrations of ethanol (Erickson et al., 1987). However, the P2 promoter was transcribed neither by σ^{70}, the housekeeping σ-factor, nor by σ^{32}. Purification of the factor responsible for transcription of this promoter identified σ^E (Erickson and Gross, 1989). σ^E was also identified from a heat-resistant fraction of RNA polymerase

Carol A. Gross, Departments of Microbiology and Immunology and Cell and Tissue Biology, University of California, San Francisco, San Francisco, CA 94143. *Virgil A. Rhodius*, Department of Microbiology and Immunology, University of California, San Francisco, San Francisco, California 94143. *Irina L. Grigorova*, Graduate Group in Biophysics, University of California, San Francisco, San Francisco, CA 94143.

The Periplasm
Edited by Michael Ehrmann © 2007 ASM Press, Washington, D.C.

holoenyzme that was able to transcribe *rpoH*. The gene encoding σ^E, *rpoE*, was identified both by searching for a gene with homology to other σ's and as an insertion mutation that decreased σ^E activity (Raina et al., 1995; Rouviere et al., 1995).

Because σ^E was identified as the factor required to transcribe σ^{32} at lethal temperatures, it was initially considered to be the extreme stress σ (Erickson et al., 1987). However, accumulating evidence that σ^E responded to conditions in the envelope soon challenged this idea. First, DegP, a periplasmic chaperone/protease, was shown to be transcribed by σ^E, indicating that the σ^E regulon included envelope proteins (Lipinska et al., 1988; Erickson and Gross, 1989). Then, an overexpression screen identified outer membrane porins, including OmpC, OmpF, and OmpX, as potent inducers of σ^E activity (Mecsas et al., 1993). Significantly, overexpressed porins induced σ^E activity only when the export pathway was intact, suggesting that the inducing signal was generated in the envelope compartment. Mutants having aberrant lipopolysaccharide (LPS) or lacking periplasmic folding agents also induced σ^E activity (Missiakas et al., 1996; Rouviere and Gross, 1996; Tam and Missiakas, 2005). Finally, reducing the amounts of OmpC and OmpF was sufficient to lower basal σ^E activity, indicating that the signal must be intimately related to the expression level of porins (Mecsas et al., 1993). Taken together, these studies suggested that aberrant conditions in the cell envelope induced σ^E activity and that the signal transduction pathway that conveyed this information might monitor porin status.

COMPONENTS OF THE SIGNAL TRANSDUCTION PATHWAY REGULATING σ^E

Overview

σ^E is regulated by its antisigma factor, RseA; when complexed with RseA, σ^E is transcriptionally inactive. Stressed cells degrade RseA via a proteolytic cascade. RseA is cleaved first by DegS on its periplasmic face and then by RseP (formerly known as YaeL) on its cytoplasmic face to release the RseA/σ^E complex from the membrane (Alba et al., 2002; Kanehara et al., 2002). The cytoplasmic fragment of RseA is then degraded by ClpXP, an ATP-dependent cytoplasmic protease, and active σ^E is released (Flynn et al., 2004). This proteolytic cascade is activated by unassembled porins, which accumulate during appropriate envelope stress. The steps of RseA degradation and σ^E release are illustrated in Color Plate 7.

The RseA and RseB Negative Regulators

RseA, a 24-kDa protein, is the primary negative regulator of σ^E; deletion of *rseA* increases σ^E activity about 25-fold (De Las Penas et al., 1997; Missiakas et al., 1997). RseA is ideally suited to transmit a signal from the envelope to the cytoplasm because it is a membrane-spanning protein, approximately evenly split between the periplasm and cytoplasm (De Las Penas et al., 1997b; Missiakas et al., 1997; Alba et al., 2002; Kanehara et al., 2002; Walsh et al., 2003). The cytoplasmic face of RseA binds σ^E (De Las Penas et al., 1997b; Missiakas et al., 1997), thereby occluding its RNA polymerase-binding determinants and rendering σ^E inactive for transcription (Campbell et al., 2003). Degradation of RseA by appropriate envelope stress releases σ^E that is active for transcription (Ades et al., 1999).

RseB, a 36-kDa protein, is a secondary regulator. RseB binds to the periplasmic face of RseA, but deletion of *rseB* only increases basal σ^E activity 1.5- to 2-fold (De Las Penas et al., 1997b; Missiakas et al., 1997). However, as described below, RseB plays an important role in preventing inappropriate degradation of RseA (Grigorova et al., 2004).

DegS AND ITS ACTIVATION

DegS is an ATP-independent, trimeric serine protease (Wilken et al., 2004). Insertion of DegS into the inner membrane is mediated by its N terminus with the remainder of the protein protruding into the periplasm (Alba et al., 2001). Thus, DegS is positioned to cleave periplasmic proteins. DegS-type proteases are distinguished by having a protease domain followed by a single PDZ domain and are a branch of the larger HtrA/DegP protease family (Clausen et al., 2002; Ehrmann and Clausen, 2004). DegS initiates RseA degradation by cleaving RseA about 30 aa C-terminal to its membrane-spanning segment (Walsh et al., 2003) (Color Plate 7A and B).

DegS was first implicated in RseA degradation because Δ*degS* strains have very low σ^E activity and are unable to degrade RseA, even under inducing conditions that normally lead to rapid RseA degradation (Ades et al., 1999). DegS is thought to be unable to cleave RseA on its own; its latent activity is unmasked when the C termini of porins bind to the DegS PDZ domain (Walsh et al., 2003). This view is supported by the following observations: (i) porin C-terminal peptides bind selectively to the DegS PDZ domain in vitro; tripeptides with a YXF consensus motif are good binders with the YYF motif binding the most strongly; (ii) overexpression of these same peptides in vivo induces σ^E activity; stronger binders more highly induce σ^E activity; (iii) induction by C-terminal porin peptides requires the PDZ domain of DegS; cells with a DegS variant lacking the PDZ domain are not induced by overexpression of these peptides; and (iv) purified DegS alone is unable to cleave RseA; addition of porin C-terminal peptides activates cleavage in proportion to its strength of binding to the PDZ domain. Based on these observations, Walsh et al. (2003) proposed that unbound PDZ domains sterically occluded the protease chamber; binding of the inducing peptide was thought to reposition the PDZ domains to allow access to the proteolytic chamber.

The recent crystal structures of apo (free) and peptide-bound DegS reveal that the proteolytic chamber is fully accessible in both forms (Wilken et al., 2004), eliminating the idea that the inducing porin peptide relieves steric occlusion mediated by the PDZ domains (Walsh et al., 2003). Instead, these structures suggest a novel mechanism of induction by the porin peptides. The catalytic triad is in the active conformation in the peptide-bound state but in an inactive conformation in the apo state of DegS. Moreover, the inducing tripeptide contacts the protease domain of DegS. Together, these observations led to the provocative idea that binding of the tripeptide to the DegS protease domain initiates an allosteric transition that repositions the catalytic triad, thereby activating Deg (Wilken et al., 2004).

Two issues remain to be resolved concerning the proposed mechanism of induction. First, the critical contact between DegS and the porin is made by the −1 position of the peptide, an amino acid that is not a critical determinant of induction ability (Wilken et al., 2004). Second, whereas wild-type (wt) DegS is active only in the presence of added C-terminal peptides, DegSΔPDZ cleaves RseA on its own (B. Cezairliyan and R. Sauer, unpublished data). The inducing model does not explain these observations in a straightforward way, although they can certainly be reconciled with the proposed model. The role of the C-terminal extension of the DegSPDZ domain may be critical to resolving this issue. This extension forms a two-stranded antiparallel β-sheet that forms multiple contacts with the protease domain (Wilken et al., 2004). Such contacts might induce an allosteric transition that mispositions the catalytic triad; peptide binding might then antagonize this autoinhibitory cascade to activate cleavage.

The idea that porin C termini are the physiological activators of DegS is consistent both with known porin structure and with physiological analysis of σ^E activity. Porins are trimeric β-barrel proteins that form channels in the outer membrane so that solutes can enter the cell (reviewed in Nikaido, 1996, 2003). Published structures of porins indicate that their C termini are buried in the trimer interface (Cowan et al., 1995); thus accumulation of free C termini indicates that the complex pathway that trimerizes porins and inserts them in the outer membrane is unable to do so efficiently. Cells with reduced porin expression would be expected to have only a few free C termini available to activate DegS, thereby accounting for the low σ^E activity of strains lacking OmpR, the transcriptional activator of OmpC, and OmpF (Mecsas et al., 1993). Conversely, cells overexpressing porins would have more free C termini and would activate DegS robustly, thereby accounting for their high σ^E activity (Mecsas et al., 1993).

RseP and Its Cleavage of RseA

RseP (YaeL) is a member of the regulated intramembrane proteolysis (RIP) family of proteases (Kanehara et al., 2001) that are found in organisms from bacteria to humans (reviewed in Brown et al., 2000). These membrane-spanning, Zn^{2+} metalloproteases have an active face that cleaves membrane proteins within or close to the cytoplasmic side of the membrane (Brown et al., 2000). RseP also has a periplasmic PDZ domain, which is characteristic of prokaryotic RIP orthologues but absent in their eukaryotic counterparts (Kanehara et al., 2001). RIP proteases usually participate in signal transduction cascades that involve transmembrane signaling (Brown et al., 2000). RIP proteases almost always propagate the signal rather than initiating the signaling cascade (Brown et al., 2000). This is true in the σ^E pathway as well: cleavage by RseP follows that by DegS (Color Plate 7, A and B).

RseP was first implicated in RseA degradation because RseP deletion strains have very low σ^E activity, even after induction (Alba et al., 2002; Kanehara et al., 2002). RseP deletion strains accumulate a membrane-localized fragment of RseA, which consists of a small portion of the periplasmic domain and the entire cytoplasmic domain of RseA (Alba et al., 2002; Kanehara et al., 2002). Production of this fragment requires DegS (Alba et al., 2002; Kanehara et al., 2002), and its size is consistent with the known periplasmic DegS cleavage site (Walsh et al., 2003). Thus, this fragment is likely to result from DegS cleavage. Purified RseP cleaves RseA at a cysteine residue within the membrane-spanning segment of RseA, confirming that RseP functions to release the

RseA/σ^E complex from the membrane (Akiyama et al., 2004). Cleavage at this cysteine was predicted from the known cleavage site of its mammalian orthologue, site II protease, attesting to the high conservation of function in this protease group (Alba et al., 2002).

DegS cleavage is prerequisite for cleavage by RseP (Alba et al., 2002; Kanehara et al., 2002, 2003); RseA is completely stable in strains lacking DegS even though RseP is present. RseP cleavage of intact RseA is inhibited by at least two independent pathways (Color Plate 7A). In the first pathway, the PDZ domain of RseP inhibits its own function in a reaction that requires both RseB and the Gln-rich region of the RseA periplasmic domain (Kanehara et al., 2003; Grigorova et al., 2004). This inhibitory pathway is relieved when DegS cleaves RseA because most of its inhibitory periplasmic domain is removed. In a second inhibitory pathway, DegS itself inhibits the RseP cleavage reaction. This inhibition occurs even when RseP lacks its PDZ domain, thereby distinguishing it from the first pathway (Kanehara et al., 2003; Grigorova et al., 2004). The molecular mechanisms underlying these inhibitory mechanisms are unknown.

Studies of RseP in vitro indicate that this protease is rather promiscuous: it is able to cleave transmembrane segments with no similarity to the RseA membrane-spanning segment provided that they contain helix-destabilizing residues (Akiyama et al., 2004). Tight control over the activity of this protease may ensure its selectivity toward appropriate substrates.

Release of Free σ^E

The cytoplasmic domain of RseA (RseA-cyto) is sufficient to prevent σ^E binding to RNA polymerase, necessitating its degradation (De Las Penas et al., 1997b; Campbell et al., 2003). Upon cleavage by RseP, the sequence of the RseA C terminus becomes VAA, which is an attractive substrate for cytoplasmic proteases. ClpXP is the major cytoplasmic protease that degrades RseA-cyto: σ^E/RseA-cyto binds to ClpX, in an SspB facilitated reaction; RseA-cyto is degraded by ClpXP; and active σ^E is released (Flynn et al., 2004) (Color Plate 7C). Even in the absence of ClpX, RseA-cyto is rapidly degraded, implicating several other ATP-dependent cytoplasmic proteases in this reaction (R. Chaba, unpublished data).

PROPERTIES OF THE SIGNAL TRANSDUCTION PATHWAY REGULATING σ^E

Overview

RseA regulates the transcriptional activity of σ^E. Because the complex of RseA-cyto and σ^E is extremely stable (I. Grigorova, unpublished data) and the number of RseA molecules is likely to be somewhat in excess of σ^E (Collinet et al., 2000; Grigorova, unpublished) the anti-sigma factor can drastically downregulate σ^E activity. Accumulation of the porin signal triggers the regulated degradation of RseA, thereby increasing σ^E activity. This signal transduction system directly translates changes in the rate of RseA degradation to changes in σ^E activity. In addition, it exhibits a graded response to unassembled porins over at least a 40-fold range.

Transducing the Degradation Signal

When RseA is degraded, the transcriptional activity of σ^E increases. For many systems governed by negative regulators, increased activity is governed by the change in the ratio of the transcription factor to its negative regulator, and this is true for σ^E as well (Ades et al., 1999). However, σ^E and its negative regulators are in the same operon, which is transcribed predominantly by σ^E (Raina et al., 1995; Rouviere et al., 1995). Thus, in the short term, changes in RseA degradation may be compensated by changes in its synthesis rate. For example, even though RseA is stabilized upon temperature downshift, the ratio of RseA to σ^E does not show the expected increase because the synthesis of RseA also decreases.

The properties of the RseA/σ^E system allow σ^E activity to be regulated by changes in the rate of RseA degradation, even when the relative levels of σ^E and RseA remain unchanged. σ^E dissociates from RseA-cyto extremely slowly in vitro ($k_{off} \ll 10^{-4}$ s^{-1}; Grigorova, unpublished) and it is very likely that this occurs very slowly in vivo as well. As a consequence, RseA degradation is the major mechanism for σ^E release into the cytoplasm. Therefore, changing the rate of RseA degradation changes the rate of σ^E release, which is immediately translated into changes in σ^E activity. This method of regulation is very rapid and is not dependent on changes in the relative levels of RseA and σ^E. This form of regulation may be important in other systems governed by an unstable negative regulator.

Graded Response to Porin Status

A graded response to a wide range of porin signals is ensured by three critical properties of the proteolytic cascade. These are described below, along with their consequences for the signal transduction pathway.

1. DegS, the only identified sensor of the unassembled porin signal (Grigorova et al., 2004), obligatorily initiates the proteolytic pathway (Alba et al., 2002; Kanehara et al., 2002). When obligatory cleavage by DegS is circumvented, the signal transduction pathway is sensitive to a decreased range of porin signal (Grigorova et al., 2004). This point is illustrated by comparing σ^E activity in $\Delta ompR$ cells with that in $\Delta ompR \Delta rseB$ cells, which are partially defective in inhibiting RseP from cleaving intact RseA. Whereas $\Delta ompR$ cells show a 20-fold decrease in basal σ^E activity, $\Delta ompR \Delta rseB$ cells show less than a 4-fold decrease in basal σ^E activity as compared with that in wt cells. Higher basal σ^E activity in $\Delta ompR \Delta rseB$ cells results from constitutive RseP cleavage of RseA in the absence of the inducing porin signal (Grigorova et al., 2004).

2. DegS is in excess over the physiologically relevant range of unassembled porins that bind to its PDZ domain. This was inferred from the finding that σ^E activity is very sensitive to variations in porin expression but insensitive to variations in DegS levels (Mecsas et al., 1993; Alba et al., 2002; Grigorova et al., 2004). Together these data suggest that DegS activation is limited by the porin signal, rather than by the cellular concentration of DegS. This feature is necessary so that the frequency of the initial cleavage event will be signal sensitive over a wide range of concentrations of unassembled porins.

3. Cleavage by DegS is the rate-limiting step of the proteolytic pathway. That DegS cleavage is rate limiting is inferred from the fact that no cleavage intermediates can be detected in wt cells (Alba et al., 2002). In addition, the rate of degradation of the cytoplasmic fragment of RseA has been directly measured in vivo and shown to be almost an order of magnitude faster than DegS cleavage under inducing conditions (R. Chaba, unpublished data). This feature allows the rate of RseA degradation and thus σ^E activity to vary in accordance with the magnitude of inducing signal, thereby ensuring that the proteolytic cascade faithfully transmits the signal.

FUNCTIONS OF THE σ^E REGULON

Overview

A variety of global approaches have been used to identify members of the σ^E regulon in *E. coli*. These include genetic screens for σ^E promoters (Dartigalongue et al., 2001; Rezuchova et al., 2003), whole genome expression analysis (Rhodius et al., 2006), and a computational model that allows prediction of σ^E promoters with high precision (Rhodius et al., 2006). The current validated σ^E regulon in *E. coli* K-12 is presented in Table 1.

The σ^E promoter prediction model was used to predict promoters in *E. coli* K-12 and also in closely related organisms, chosen because their σ^E promoters were expected to be virtually identical with those of *E. coli* K-12 (Rhodius et al., 2006). This allows distinction between those genes controlled by σ^E in many organisms and those controlled by σ^E in only

one or a few organisms. The small portion of the regulon that is relatively conserved across the nine organisms examined (the core regulon; 19 transcription units encoding 23 proteins) is functionally coherent, ensuring the synthesis, assembly, and homeostasis of LPS and porins, two critical components of the outer membrane. The predicted larger, variable portion of the extended regulon (~70 transcription units) contains many pathogenesis-related functions. The functions of the core and extended components of the σ^E regulon are summarized in Fig. 1.

Conserved Functions of the Regulon

Most genes with a known function in the core regulon ensure the synthesis or assembly of LPS or porins, or encode the transcriptional circuitry that maintains homeostasis of these key molecules (Table 2; Fig. 1). Together, LPS and porins constitute almost the entire external surface of the bacterium and are responsible for the high resistance of gram-negative bacteria to lipophilic solutes (reviewed in Nikaido, 1996). Thus, the core function of the σ^E regulon is to maintain the integrity of the outer membrane and the barrier function of the cell.

Five members of the core regulon affect synthesis of Lipid A (PlsB, LpxA, LpxD, LpxB, and LpxP); a sixth contributes to LPS assembly (BacA; Table 1). The high resistance of gram-negative bacteria to hydrophobic compounds is in large part due to Lipid A, the hydrophobic anchor of the LPS (Nikaido, 1996, 2003). Diffusion of hydrophobic compounds through the outer membrane is slowed dramatically by the high density of saturated fatty acid chains in Lipid A and by its potential for many lateral interactions (Nikaido, 1996, 2003). Lipid A is assembled from R-3-hydroxymyristate, a 14-carbon fatty acid, rather than from the longer fatty acids in standard glycerophospholipids (Cronan and Rock, 1996; Raetz, 1996). PlsB, LpxA, and LpxD each favor synthesis of Lipid A over longer fatty acids. An increase in PlsB, which transfers fatty acids to glycerol 3-phosphate, decreases the average chain length of fatty acids, thereby increasing

TABLE 1 Functional classification of the σ^E regulon members in *E. coli* K-12

Functional category	Regulon members[a]
Envelope	
Envelope proteases	AnsB DegP (PtrA) YfgC★
Periplasmic chaperones, folding catalysts	DsbC FkpA (Imp) Skp DegP (SurA) YaeT
OM biosynthesis	BacA Ddg FadD LpxA LpxB LpxD MdoG MdoH PlsB Psd
Lipid detoxification	AhpF
Lipoproteins	Lpp (NlpB) SmpA★ YeaY★ YfeY★ YfgL YfiO (YidQ★) YraP
OMP/channels/receptors	(OmpA) (OmpC) (OmpF) OmpX (Tsx) (YbiL★)
Transport proteins	GspA YicJ SbmA★ PtsN★ YhbG★
Other known/predicted envelope	NarW NarV RseA RseB RseC YaiW★ (YcbK) Ydhl★ YdhJ★ YdhK★ YfeK★ YfgD★ YggN★ YgiM★ (YhcN★) (YhjW★) YjeP★
Capsule	Wzb Wzc
Cytoplasmic	
Transcription	GreA (FabR) (RpoD) RpoE RpoH RpoN RseA SixA YcdC★
Translation	FusA KsgA PrfB YhbH★
DNA recombination/repair	(RecB) (RecD) RecJ RecR
DNA/RNA modification	(CafA) Cca DnaE Lhr RnhB
Cytoplasmic proteases	ClpX Lon YhjJ★
Cytoplasmic chaperones	(DnaK DnaJ)
Fatty acid biosynthesis	(FabR) FabZ
Leucine biosynthesis	(LeuA) (LeuB) (LeuC)
Pyridoxine biosynthesis	(PdxA)
Miscellaneous	
Carbon utilization	MalQ YiaK YiaL★ YicI★
Cell structure/division	(MreB) (MreC) (MreD) FtsZ PioO (YhdE★)
Metal	(CutC) (YbiL★) YecI★
Nitrate/nitrite respiration	NarV NarW YbjV YbjW
Prophage	YbcR★ YbcS★ YbcT★
Stress adaptation	YiiT★
Unknown function	(ApaG) (PqiA) (PqiB) YbaB YbfG (YbiX) YfeS YfeX YfjO Yhbj (YidR) YieE YiiS (YmbA)

[a]Proteins with no identified σ^E promoter as described in Rhodius et al. (2006) are in parentheses. ★ denotes proteins in which the function of σ^E regulon members is predicted from amino-acid-sequence BLAST analysis for related proteins of known/predicted function. Proteins that have no significant sequence homology to any protein of known/predicted function are labeled *unknown function*. Note that some proteins are in more than one functional group. OMP, outer membrane proteins; OM, outer membrane.

the concentration of R-3 hydroxymyristate, the Lipid A precursor (Cronan and Rock, 1996; Raetz, 1996). LpxA and LpxD transfer the first and second myristoyl moieties, respectively, to their polysaccharide (Cronan, 1996; Raetz, 1996). LpxP is responsible for a cold-resistant form of Lipid A (Carty et al., 1999; Vorachek-Warren et al., 2002), and BacA regenerates the lipid carrier protein necessary to transport the carbohydrate moieties of LPS through the outer membrane so that they can be assembled (Raetz and Whitfield, 2002; El Ghachi et al., 2004).

Six members of the core regulon are involved in porin assembly (YfiO, YraP, Skp, FkpA, DegP, and YaeT; Table 2; Fig. 1), a complex process that requires maintaining intermediates in a folding competent state after secretion to the periplasm, transport of these intermediates to the outer membrane, and their insertion into the outer membrane. The FkpA and DegP general chaperones are likely to maintain folding competence of porins (Missiakas et al., 1996; Rizzitello et al., 2001), and YraP, a lipoprotein, also plays a role in this process (Onufryk et al., 2005). Skp has a special

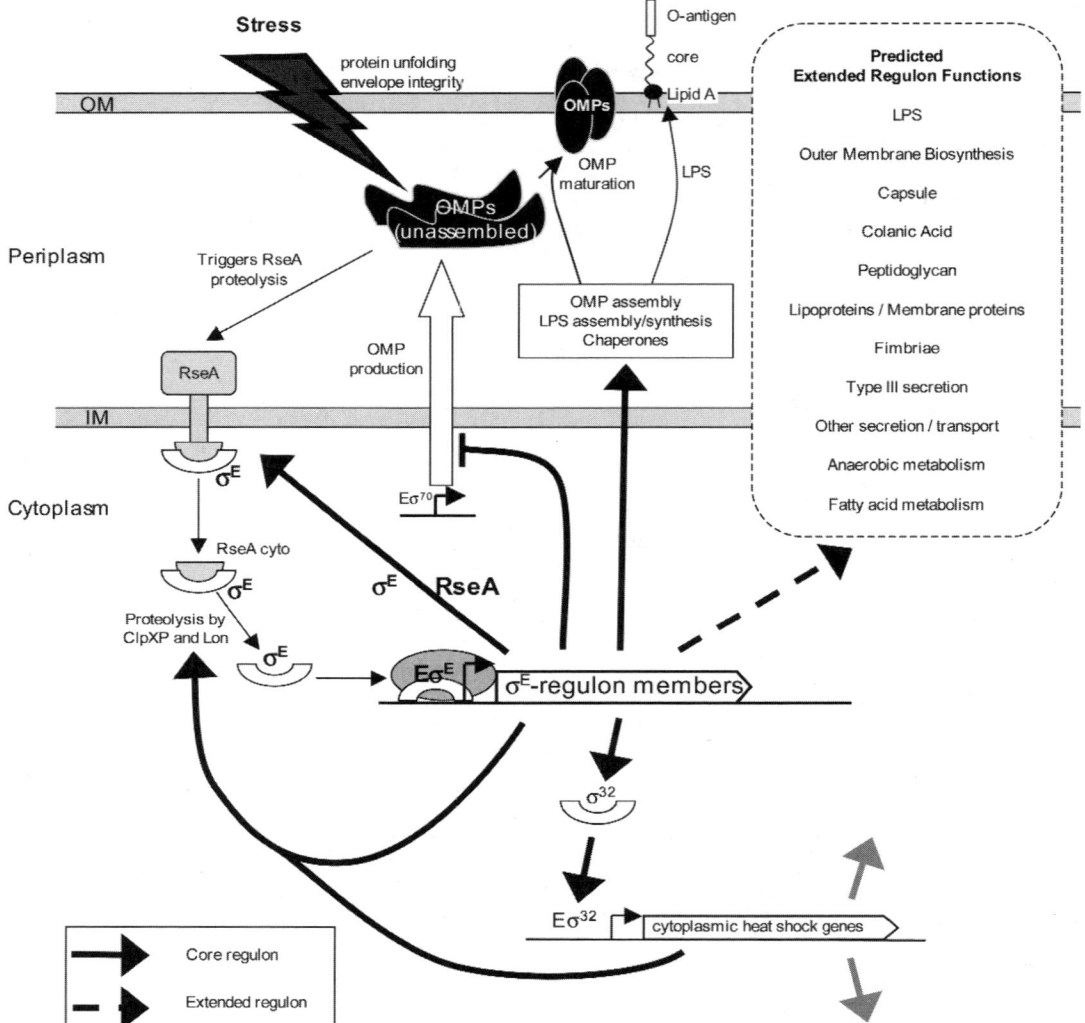

FIGURE 1 Regulon functions of the core and extended σ^E regulons. Stresses such as heat lead to the accumulation of unassembled OMPs in the periplasm; this leads to the sequential proteolysis of RseA, releasing free σ^E into the cytoplasm. σ^E then binds to RNA polymerase (E) and regulates the expression of target core and extended regulon members (Rhodius et al., 2006). Core regulon members are highly conserved across 9 genomes and are primarily involved in the regulation of the σ^E response. σ^E upregulates functions required for the synthesis, assembly, and/or the insertion of both OMPs and LPS, as well as envelope-folding catalysts and chaperones. σ^E also upregulates expression of itself; its negative regulator, RseA; the cytoplasmic proteases ClpX and Lon, thereby ensuring there is sufficient protease degradation of RseA-cyto; and σ^{32}. Note that ClpXP protease is also expressed from the σ^{32} regulon. σ^E downregulates OMP production from σ^{70} promoters, thereby reducing the accumulation of unassembled OMPs, which presumably limits the duration of the response. The extended regulon is less well conserved across related genomes and encodes many functions involved in pathogenesis.

TABLE 2 Predicted core σ^E regulon members from Rhodius et al. (2006)[a]

Gene	Function/description
Lipoproteins	
yfiO★★	Lipoprotein (essential); OMP assembly
yeaY★★	Lipoprotein
yraP★	Lipoprotein; OMP assembly
OM protein modification	
yaeT★	OMP assembly
skp★	OMP chaperone
fkpA★	Peptidyl-prolyl isomerase
degP	Periplasmic chaperone and serine protease
Cell envelope structure	
plsB★	Phospholipid biosynthesis
bacA★	Peptidoglycan, LPS, and teichoic acid biosynthesis
lpxA/B/D/P	Lipid A biosynthesis
ahpF	Lipid modification
Other cell envelope proteins	
ygiM	Putative membrane protein
yggN	Putative periplasmic protein
Transcriptional circuitry	
rpoE★★	Sigma E
rpoH★	Sigma H
rseA★	Negative regulator of sigma E
greA★	Transcription elongation factor
ompX	OMP (reverse promoter)
Cell division	
ftsZ	Cell division
Other	
yecI	Fe^{2+} acquisition

[a]Orthologous genes predicted to be regulated by σ^E in six or more genomes. Those predicted in seven genomes are indicated by ★; those in eight genomes are indicated by ★★. Genomes examined were: *E. coli* K-12 MG1655; *E. coli* CFT073; *E. coli* O157:H7 EDL933; *Shigella flexneri* 2a str. 2457T; *Salmonella enterica* subsp. *enterica* serovar Typhi str. CT18; serovar Typhimurium LT2; *Photorhabdus luminescens* subsp. *laumondii* TTO1; *Erwinia carotovora* subsp. *atroseptica* SCR11043; *Yersinia pestis* CO92.

role in porin maturation and may deliver porins to the outer membrane (reviewed in Gentle et al., 2005). At the outer membrane, porins must interact with YaeT, an outer membrane protein that functions to insert all β-barrel proteins into the outer membrane (Wu et al., 2005; reviewed in Gentle et al., 2005). YaeT, an Omp85 orthologue, works in concert with three lipoproteins: YfiO, YfgL, and NlpB (Wu et al. 2005). YfiO is a member of the core regulon and, like YaeT, is essential in *E. coli* (Onufryk et al., 2005; Wu et al., 2005). YfgL is transcribed by σ^E in *E. coli* and in four other organisms, so it represents a reasonably conserved function of σ^E; NlpB is indirectly upregulated when σ^E is overexpressed. These data indicate that the complex of proteins necessary for membrane insertion of β-barrel proteins is under σ^E control in *E. coli*.

An additional feature of the response is that when the cell is under stress, production of new porins is downregulated (Fig. 1). Although the exact mechanism must be determined, the current thinking is that downregulation is mediated by σ^E-controlled small RNAs, which may destabilize porin mRNAs and are transcribed divergently from the genes they downregulate. At least one candidate, the σ^E promoter divergent from OmpX, is a member of the core regulon (Table 2), suggesting that OMP downregulation may be a characteristic of the response in many organisms (Rhodius et al., 2006).

The transcriptional circuitry that governs σ^E is also encoded in the core regulon (Table 2, Fig. 1). The *rpoErseArseB* operon is driven by two σ^E promoters: one is located upstream of *rpoE* that drives the entire operon, and a second is located upstream of *rseA* that drives only the negative regulators of σ^E (Rhodius et al., 2006).

Upregulation of σ^{32}, which mediates the cytoplasmic heat shock response, is also a conserved feature of the transcriptional circuitry (Table 2, Fig. 2) (Rhodius et al. 2006). These two responses may be linked simply because it is advantageous to partially induce the cytoplasmic response during envelope stress, but there may be a more direct connection between the two responses. We have recently shown that numerous proteins localized to the cytoplasmic membrane are controlled by σ^{32}, including two proteins essential for lipoprotein maturation: signal peptidase II and CutE (Nonaka et al., 2006). Thus, upregulation of σ^{32} may also be necessary to maintain adequate production of the σ^E-controlled lipoproteins that act in concert with YaeT (Omp85) to facilitate insertion of β-barrel proteins in the outer membrane.

The Variable Portion of the Extended σ^E Regulon

About 70 of the σ^E-controlled transcription units identified by the promoter-prediction model are predicted to be present in fewer than six of the nine organisms examined; some are predicted to occur in only a single organism (Rhodius et al., 2006). However, the majority of these have functions with a coherent pathogenesis-related theme, i.e., they are known to be important during interaction with host cells. Two functions present in the core regulon also fit in this category: an iron acquisition system (YecI; Andrews, 1998) that facilitates growth in the iron-poor host environment and a component of a peroxide-detoxifying system (AhpF; Poole, 2005) that enhances survival during the respiratory burst encountered on interaction with macrophages.

The functional categories of all transcription units predicted to be in the extended σ^E regulon are summarized in Fig. 1; a complete listing of the predicted transcription units is found in Rhodius et al., 2006. Among the pathogenesis-related functions whose predictions have been validated in at least one organism are (i) components for the synthesis of capsule, which is important for adhesion to host cells and for protection against the innate immune system; (ii) a recombination system that could be important for DNA damage engendered by the respiratory burst; and (iii) components to allow respiration by nitrate reduction important for survival in microaerophilic or anaerobic environments. The nitrate reduction system is one of the few non-envelope-related functions encoded in the σ^E regulon. Additional pathogenesis-related functions predicted but not yet validated include production of chorismate and colanic acid, type III secretion, and LPS modification.

These predicted functions fit well with the finding that σ^E is an important pathogenesis

determinant in many organisms, including *Salmonella enterica* serovar Typhimurium (Humphreys et al., 1999; Testerman et al., 2002) and *E. coli* CFT073 (Redford et al., 2003) for which σ^E regulon members have been predicted, and also *Haemophilus influenzae* (Craig et al., 2002), *Vibrio cholerae* (Kovacikova and Skorupski, 2002), and the σ^E orthologue, AlgU, in *Pseudomonas aeruginosa* (Martin et al., 1994; Yu et al., 1995). Of these, serovar Typhimurium has been studied extensively and σ^E has been shown to be required for stationary-phase survival and virulence in mice (Humphreys et al., 1999; Kenyon et al., 2002; Testerman et al., 2002), oxidative stress resistance (Bang et al., 2005), and also resistance to antimicrobial peptides (Crouch et al., 2005). The exact role of σ^E regulon members in virulence has not been determined, though some mechanisms are now being elucidated in serovar Typhimurium. For example, the σ^E-regulated periplasmic peptidylprolyl isomerases such as FkpA, HtrA, and SurA are important for survival in mice (Humphreys et al., 1999, 2003) and a cascade of sigma factor expression in which σ^E enhances expression of σ^{32} and in turn the RNA-binding protein, Hfq, which then leads to an increased expression of the stationary-phase factor, σ^S (Bang et al., 2005). Expression of σ^S-dependent genes is required for resistance to oxidative stress. However, host-pathogen interactions are multifaceted and complex, and it will be worthwhile to examine some of the new predicted σ^E targets for their role in virulence.

RELATIONSHIP BETWEEN REGULATION OF THE σ^E REGULON AND ITS CORE FUNCTIONS

Overview

All regulatory systems evolve in concert with the functions they regulate. From this perspective, it is instructive to examine the control of the σ^E regulon in relationship to its core function of maintaining appropriate LPS and porin content of the outer membrane so that the barrier function of the outer membrane is intact. Both the signal and the signal transduction system are readily understood in light of this core regulon function. Moreover, the transcriptional circuitry encoded in the core regulon collaborates with the signal transduction system to maintain homeostasis of core functions.

The Signal

An important consideration is whether the activating signal reflects the status of the core functions of the regulon. The unassembled porin signal that activates σ^E does so. The assembly and function of porins and LPS are intertwined: LPS intermediates facilitate porin assembly and vice versa (Pages et al., 1990; Ried et al., 1990; Kloser et al., 1998; de Cock et al., 2001; Wu et al., 2005); and the porin assembly pathway is also involved in LPS assembly, e.g., Imp and Skp interact with LPS precursors (De Cock et al., 1999; Braun and Silhavy, 2002; Bos et al., 2004). As a consequence of these interrelationships, the porin signal monitors both LPS and porin status, and also ensures that these molecules occur in the proper ratio in the membrane, which is necessary for proper functioning of the membrane as a barrier for the cell.

The porin signal allows the cell to use the σ^E response to respond to general envelope stress as well as specific problems in porin/LPS status. This follows from the fact that the porin assembly pathway utilizes general chaperones and protein-folding catalysts like FkpA and DegP, in addition to committed agents like YaeT (Omp85). Any conditions that perturb periplasmic protein folding will titrate the general chaperones with unfolded proteins. As a result, porin assembly will be less efficient and unassembled porins will accumulate, thereby activating the response and increasing the general folding capacity of the envelope.

Thus far, we have discussed how the porin signal is used to maintain envelope homeostasis in response to environmental stress. However, the cell also requires extensive envelope remodeling during septation. Is σ^E activity important for cell cycle progression? There is currently no direct evidence for this proposition, but FtsZ, which is involved in initiating cell division, is a member of the core regulon (Rhodius et al., 2006), inviting speculation that σ^E activity is also upregulated during septation.

The Signal Transduction Pathway/Core Transcription Functions

As we have described in earlier sections, the signal is transduced through a proteolytic cascade that degrades RseA. This proteolytic cascade is organized so that cellular σ^E activity accurately reflects steady-state porin status and can change rapidly in response to both small and large changes in porin status over a wide range. Several features of the transcriptional circuitry improve the cellular response to changes in porin status. First, RseA is driven by two σ^E promoters, the operon promoter and a σ^E promoter upstream of RseA. This high expression of RseA is likely to be important both to maintain the appropriate RseA/σ^E ratio and to provide a sufficient excess of RseA to ensure rapid downregulation when the signal disappears. Second, in *E. coli* K-12 and several other organisms, ClpX and Lon (Flynn et al., 2004), the two major proteases that degrade the cytoplasmic fragment of RseA, are encoded in the regulon, ensuring that their amounts increase with demand. Finally, induction of σ^E downregulates new porin synthesis. Decreasing the flow of unassembled porins to the periplasm until existing unassembled porins are either folded or degraded increases the efficacy of the response.

In summary, the identified signal for the σ^E response, the signal transduction pathway, and the conserved transcriptional circuitry function together to ensure that the core functions of the regulon are maintained. From this point of view, it is interesting to consider why the variable pathogenesis-related functions are part of the regulon. Interaction with host cells may be a stress that triggers accumulation of unassembled porins, thereby inducing the σ^E response. The addition of pathogenesis-related functions to regulon ensures their expression immediately on contact with the host. This "early-warning" system may improve survival in the host organism.

SUMMARY AND PROSPECTS

We now know a great deal about the function and regulation of the σ^E regulon in *E. coli* and closely related organisms. The core function of the regulon is to maintain porin and LPS homeostasis so that the barrier function of the cell is intact. The signal transduction system and the transcriptional circuitry of the core regulon are set up to ensure that this function is performed efficiently. The unassembled porin signal reports on both LPS and porin status and activates a proteolytic cascade that degrades the negative regulator, RseA. This system is designed so that the rate of degradation is translated directly into σ^E activity over a wide range of signals.

Our knowledge thus far raises a host of new questions, which will undoubtedly be the subject of future investigations. These questions are of several types. First, we need a great deal more understanding at the molecular level. For example, we do not understand how RseP is regulated, or how this class of proteases cleaves its substrates; we know little about the mechanism for inserting β-barrel proteins into the outer membrane and even less about LPS insertion and synthesis. Second, although we have speculated about how the various aspects of the circuitry alter the response, these speculations must be subject to experimental test, both at the population and single-cell level. Third, there may be other inputs to the signal transduction pathway that activate σ^E. For example, some inducers, such as LamB, do not contain the canonical C-terminal sequence that activates DegS (Rouviere and Gross, 1996, Kenyon et al., 2005). If such inducers are not processed to expose the inducing signal, there may be an alternative mechanism of σ^E activation. Finally, examining how this response is modified to meet the needs of host organisms that are distantly related to *E. coli* K-12 seems certain to provide insights into the regulation of prokaryotic transcriptional response networks.

REFERENCES

Ades, S. E. 2004. Control of the alternative sigma factor sigmaE in Escherichia coli. *Curr. Opin. Microbiol.* **7**:157–162.

Ades, S. E., L. E. Connolly, B. M. Alba, and C. A. Gross. 1999. The Escherichia coli sigma(E)-dependent extracytoplasmic stress response is controlled by the regulated proteolysis of an anti-sigma factor. *Genes Dev.* **13**:2449–2461.

Akiyama, Y., K. Kanehara, and K. Ito. 2004. RseP (YaeL), an Escherichia coli RIP protease, cleaves transmembrane sequences. *EMBO J.* **23**:4434–4442.

Alba, B. M., and C. A. Gross. 2004. Regulation of the Escherichia colisigma-dependent envelope stress response. *Mol. Microbiol.* **52**:613–619.

Alba, B. M., J. A. Leeds, C. Onufryk, C. Z. Lu, and C. A. Gross. 2002. DegS and YaeL participate sequentially in the cleavage of RseA to activate the sigma(E)-dependent extracytoplasmic stress response. *Genes Dev.* **16**:2156–2168.

Alba, B. M., H. J. Zhong, J. C. Pelayo, and C. A. Gross. 2001. degS (hhoB) is an essential Escherichia coli gene whose indispensable function is to provide sigma (E) activity. *Mol. Microbiol.* **40**:1323–1333.

Andrews, S. C. 1998. Iron storage in bacteria. *Adv. Microb. Physiol.* **40**:281–351.

Bang, I. S., J. G. Frye, M. McClelland, J. Velayudhan, and F. C. Fang. 2005. Alternative sigma factor interactions in Salmonella: sigma and sigma promote antioxidant defences by enhancing sigma levels. *Mol. Microbiol.* **56**:811–823.

Bos, M. P., B. Tefsen, J. Geurtsen, and J. Tommassen. 2004. Identification of an outer membrane protein required for the transport of lipopolysaccharide to the bacterial cell surface. *Proc. Natl. Acad. Sci. USA* **101**:9417–9422.

Braun, M., and T. J. Silhavy. 2002. Imp/OstA is required for cell envelope biogenesis in Escherichia coli. *Mol. Microbiol.* **45**:1289–1302.

Brown, M. S., J. Ye, R. B. Rawson, and J. L. Goldstein. 2000. Regulated intramembrane proteolysis: a control mechanism conserved from bacteria to humans. *Cell* **100**:391–398.

Campbell, E. A., J. L. Tupy, T. M. Gruber, S. Wang, M. M. Sharp, C. A. Gross, and S. A. Darst. 2003. Crystal structure of Escherichia coli sigmaE with the cytoplasmic domain of its anti-sigma RseA. *Mol. Cell* **11**:1067–1078.

Carty, S. M., K. R. Sreekumar, and C. R. Raetz. 1999. Effect of cold shock on lipid A biosynthesis in Escherichia coli. Induction At 12 degrees C of an acyltransferase specific for palmitoleoyl-acyl carrier protein. *J. Biol. Chem.* **274**:9677–9685.

Clausen, T., C. Southan, and M. Ehrmann. 2002. The HtrA family of proteases: implications for protein composition and cell fate. *Mol. Cell* **10**:443–455.

Collinet, B., H. Yuzawa, T. Chen, C. Herrera, and D. Missiakas. 2000. RseB binding to the periplasmic domain of RseA modulates the RseA:sigmaE interaction in the cytoplasm and the availability of sigma E. RNA polymerase. *J. Biol. Chem.* **275:** 33898–33904.

Connolly, L., A. De Las Penas, B. M. Alba, and C. A. Gross. 1997. The response to extracytoplasmic stress in Escherichia coli is controlled by partially overlapping pathways. *Genes Dev.* **11:** 2012–2021.

Cowan, S. W., R. M. Garavito, J. N. Jansonius, J. A. Jenkins, R. Karlsson, N. Konig, E. F. Pai, R. A. Pauptit, P. J. Rizkallah, J. P. Rosenbusch, G. Rummel, and T. Schirmer. 1995. The structure of OmpF porin in a tetragonal crystal form. *Structure* **3:**1041–1050.

Craig, J. E., A. Nobbs, and N. J. High. 2002. The extracytoplasmic sigma factor, final sigma(E), is required for intracellular survival of nontypeable Haemophilus influenzae in J774 macrophages. *Infect. Immun.* **70:**708–715.

Cronan, J. E., and C. O. Rock. 1996. Biosynthesis of membrane lipids, p. 612–636. *In* F. C. Neidhardt, J. L. Ingraham, E. C. C. Lin, K. B. Low, B. Magasanik, W. S. Reznikoff, M. Riley, M. Schaechter, and H. E. Umbarger (ed.), Escherichia coli *and* Salmonella: *Cellular and Molecular Biology*, vol. 1. ASM Press, Washington, D.C..

Crouch, M. L., L. A. Becker, I. S. Bang, H. Tanabe, A. J. Ouellette, and F. C. Fang. 2005. The alternative sigma factor sigma is required for resistance of Salmonella enterica serovar Typhimurium to anti-microbial peptides. *Mol. Microbiol.* **56:** 789–799.

Dartigalongue, C., D. Missiakas, and S. Raina. 2001. Characterization of the Escherichia coli sigma E regulon. *J. Biol. Chem.* **276:**20866–20875.

de Cock, H., M. Pasveer, J. Tommassen, and E. Bouveret. 2001. Identification of phospholipids as new components that assist in the in vitro trimerization of a bacterial pore protein. *Eur. J. Biochem.* **268:**865–875.

De Cock, H., U. Schafer, M. Potgeter, R. Demel, M. Muller, and J. Tommassen. 1999. Affinity of the periplasmic chaperone Skp of Escherichia coli for phospholipids, lipopolysaccharides and non-native outer membrane proteins. Role of Skp in the biogenesis of outer membrane protein. *Eur. J. Biochem.* **259:**96–103.

De Las Penas, A., L. Connolly, and C. A. Gross. 1997a. SigmaE is an essential sigma factor in Escherichia coli. *J. Bacteriol.* **179:**6862–6864.

De Las Penas, A., L. Connolly, and C. A. Gross. 1997b. The sigmaE-mediated response to extracytoplasmic stress in Escherichia coli is transduced by RseA and RseB, two negative regulators of sigmaE. *Mol. Microbiol.* **24:**373–385.

Ehrmann, M., and T. Clausen. 2004. Proteolysis as a regulatory mechanism. *Annu. Rev. Genet.* **38:** 709–724.

El Ghachi, M., A. Bouhss, D. Blanot, and D. Mengin-Lecreulx. 2004. The bacA gene of Escherichia coli encodes an undecaprenyl pyrophosphate phosphatase activity. *J. Biol. Chem.* **279:** 30106–30113.

Erickson, J. W., and C. A. Gross. 1989. Identification of the sigma E subunit of Escherichia coli RNA polymerase: a second alternate sigma factor involved in high-temperature gene expression. *Genes Dev.* **3:**1462–1471.

Erickson, J. W., V. Vaughn, W. A. Walter, F. C. Neidhardt, and C. A. Gross. 1987. Regulation of the promoters and transcripts of rpoH, the Escherichia coli heat shock regulatory gene. *Genes Dev.* **1:**419–432.

Flynn, J. M., I. Levchenko, R. T. Sauer, and T. A. Baker. 2004. Modulating substrate choice: the SspB adaptor delivers a regulator of the extracytoplasmic-stress response to the AAA+ protease ClpXP for degradation. *Genes Dev.* **18:**2292–2301.

Gentle, I. E., L. Burri, and T. Lithgow. 2005. Molecular architecture and function of the Omp85 family of proteins. *Mol. Microbiol.* **58:**1216–1225.

Grigorova, I. L., R. Chaba, H. J. Zhong, B. M. Alba, V. Rhodius, C. Herman, and C. A. Gross. 2004. Fine-tuning of the Escherichia coli sigmaE envelope stress response relies on multiple mechanisms to inhibit signal-independent proteolysis of the transmembrane anti-sigma factor, RseA. *Genes Dev.* **18:**2686–2697.

Gross, C. A. 1996. Function and regulation of the heat shock proteins, p. 1382–1399. *In* F. C. Neidhardt, J. L. Ingraham, E. C. C. Lin, K. B. Low, B. Magasanik, W. S. Reznikoff, M. Riley, M. Schaechter, and H. E. Umbarger (ed.), Escherichia coli *and* Salmonella: *Cellular and Molecular Biology*, vol. 1. ASM Press, Washington, D.C.

Hengge-Aronis, R. 2002. Signal transduction and regulatory mechanisms involved in control of the sigma(S) (RpoS) subunit of RNA polymerase. *Microbiol. Mol. Biol. Rev.* **66:**373–395, table of contents.

Humphreys, S., G. Rowley, A. Stevenson, W. J. Kenyon, M. P. Spector, and M. Roberts. 2003. Role of periplasmic peptidylprolyl isomerases in Salmonella enterica serovar Typhimurium virulence. *Infect. Immun.* **71:**5386–5388.

Humphreys, S., A. Stevenson, A. Bacon, A. B. Weinhardt, and M. Roberts. 1999. The alternative sigma factor, sigmaE, is critically important for the virulence of Salmonella typhimurium. *Infect. Immun.* **67:**1560–1568.

Kanehara, K., Y. Akiyama, and K. Ito. 2001. Characterization of the yaeL gene product and its S2P-protease motifs in Escherichia coli. *Gene* **281:**71–79.

Kanehara, K., K. Ito, and Y. Akiyama. 2002. YaeL (EcfE) activates the sigma(E) pathway of stress response through a site-2 cleavage of anti-sigma(E), RseA. *Genes Dev.* **16:**2147–2155.

Kanehara, K., K. Ito, and Y. Akiyama. 2003. YaeL proteolysis of RseA is controlled by the PDZ domain of YaeL and a Gln-rich region of RseA. *EMBO J.* **22:**6389–6398.

Kenyon, W. J., D. G. Sayers, S. Humphreys, M. Roberts, and M. P. Spector. 2002. The starvation-stress response of Salmonella enterica serovar Typhimurium requires sigma(E)-, but not CpxR-regulated extracytoplasmic functions. *Microbiology* **148**(Pt 1)**:**113–122.

Kenyon, W. J., S. M. Thomas, E. Johnson, M. J. Pallen, and M. P. Spector. 2005. Shifts from glucose to certain secondary carbon-sources result in activation of the extracytoplasmic function sigma factor sigmaE in Salmonella enterica serovar Typhimurium. *Microbiology* **151**(Pt 7)**:**2373–2383.

Kloser, A., M. Laird, M. Deng, and R. Misra. 1998. Modulations in lipid A and phospholipid biosynthesis pathways influence outer membrane protein assembly in Escherichia coli K-12. *Mol. Microbiol.* **27:**1003–1008.

Kovacikova, G., and K. Skorupski. 2002. The alternative sigma factor sigma(E) plays an important role in intestinal survival and virulence in Vibrio cholerae. *Infect. Immun.* **70:**5355–5362.

Lipinska, B., S. Sharma, and C. Georgopoulos. 1988. Sequence analysis and regulation of the htrA gene of Escherichia coli: a sigma 32-independent mechanism of heat-inducible transcription. *Nucleic Acids Res.* **16:**10053–10067.

Martin, D. W., M. J. Schurr, H. Yu, and V. Deretic. 1994. Analysis of promoters controlled by the putative sigma factor AlgU regulating conversion to mucoidy in Pseudomonas aeruginosa: relationship to sigma E and stress response. *J. Bacteriol.* **176:**6688–6696.

Mecsas, J., P. E. Rouviere, J. W. Erickson, T. J. Donohue, and C. A. Gross. 1993. The activity of sigma E, an Escherichia coli heat-inducible sigma-factor, is modulated by expression of outer membrane proteins. *Genes Dev.* **12B:**2618–2628.

Missiakas, D., J. M. Betton, and S. Raina. 1996. New components of protein folding in extracytoplasmic compartments of Escherichia coli SurA, FkpA and Skp/OmpH. *Mol. Microbiol.* **21:**871–884.

Missiakas, D., M. P. Mayer, M. Lemaire, C. Georgopoulos, and S. Raina. 1997. Modulation of the Escherichia coli sigmaE (RpoE) heat-shock transcription-factor activity by the RseA, RseB and RseC proteins. *Mol. Microbiol.* **24:**355–371.

Nikaido, H. 1996. Outer membrane, p. 29–47. In F. C. Neidhardt, J. L. Ingraham, E. C. C. Lin, K. B. Low, B. Magasanik, W. S. Reznikoff, M. Riley, M. Schaechter, and H. E. Umbarger (ed.), Escherichia coli and Salmonella: *Cellular and Molecular Biology*, vol. 1. ASM Press, Washington, D.C.

Nikaido, H. 2003. Molecular basis of bacterial outer membrane permeability revisited. *Microbiol. Mol. Biol. Rev.* **67:**593–656.

Nonaka, G., M. Blankenschien, C. Herman, C. A. Gross, and V. A. Rhodius. 2006. Regulon and promoter analysis of the E. coli heat-shock factor, σ^{32}, reveals a multifaceted cellular response to heat stress. *Genes Dev.* **20:**1776–1789.

Onufryk, C., M. L. Crouch, F. C. Fang, and C. A. Gross. 2005. Characterization of six lipoproteins in the sigmaE regulon. *J. Bacteriol.* **187:**4552–4561.

Pages, J. M., J. M. Bolla, A. Bernadac, and D. Fourel. 1990. Immunological approach of assembly and topology of OmpF, an outer membrane protein of Escherichia coli. *Biochimie* **72:**169–176.

Poole, L. B. 2005. Bacterial defenses against oxidants: mechanistic features of cysteine-based peroxidases and their flavoprotein reductases. *Arch. Biochem. Biophys.* **433:**240–254.

Raetz, C. R., and C. Whitfield. 2002. Lipopolysaccharide endotoxins. *Annu. Rev. Biochem.* **71:** 635–700.

Raetz, C. R. H. 1996. Structure and biosynthesis of lipid A in *Escherichia coli*, p. 1035–1063. In F. C. Neidhardt, J. L. Ingraham, E. C. C. Lin, K. B. Low, B. Magasanik, W. S. Reznikoff, M. Riley, M. Schaechter, and H. E. Umbarger (ed.), Escherichia coli and Salmonella: *Cellular and Molecular Biology*, vol. 1. ASM Press, Washington, D.C.

Raina, S., D. Missiakas, and C. Georgopoulos. 1995. The rpoE gene encoding the sigma E (sigma 24) heat shock sigma factor of Escherichia coli. *EMBO J.* **14:**1043–1055.

Raivio, T. L., and T. J. Silhavy. 2001. Periplasmic stress and ECF sigma factors. *Annu. Rev. Microbiol.* **55:**591–624.

Redford, P., P. L. Roesch, and R. A. Welch. 2003. DegS is necessary for virulence and is among extraintestinal Escherichia coli genes induced in murine peritonitis. *Infect. Immun.* **71:** 3088–3096.

Rezuchova, B., H. Miticka, D. Homerova, M. Roberts, and J. Kormanec. 2003. New members of the Escherichia coli sigmaE regulon identified by a two-plasmid system. *FEMS Microbiol. Lett.* **225:**1–7.

Rhodius, V. A., W. C. Suh, G. Nonaka, J. West, and C. A. Gross. 2006. Conserved and variable functions of the sigma(E) stress response in related genomes. *PLoS Biol.* **4:**e2 43–59.

Ried, G., I. Hindennach, and U. Henning. 1990. Role of lipopolysaccharide in assembly of Escherichia coli outer membrane proteins OmpA, OmpC, and OmpF. *J. Bacteriol.* **172:**6048–6053.

Rizzitello, A. E., J. R. Harper, and T. J. Silhavy. 2001. Genetic evidence for parallel pathways of chaperone activity in the periplasm of Escherichia coli. *J. Bacteriol.* **183:**6794–6800.

Rouviere, P. E., A. De Las Penas, J. Mecsas, C. Z. Lu, K. E. Rudd, and C. A. Gross. 1995. rpoE, the gene encoding the second heat-shock sigma factor, sigma E, in Escherichia coli. *EMBO J.* **14:**1032–1042.

Rouviere, P. E., and C. A. Gross. 1996. SurA, a periplasmic protein with peptidyl-prolyl isomerase activity, participates in the assembly of outer membrane porins. *Genes Dev.* **10:**3170–3182.

Tam, C., and D. Missiakas. 2005. Changes in lipopolysaccharide structure induce the sigma(E)-dependent response of Escherichia coli. *Mol. Microbiol.* **55:**1403–1412.

Testerman, T. L., A. Vazquez-Torres, Y. Xu, J. Jones-Carson, S. J. Libby, and F. C. Fang. 2002. The alternative sigma factor sigmaE controls antioxidant defences required for Salmonella virulence and stationary-phase survival. *Mol. Microbiol.* **43:**771–782.

Vorachek-Warren, M. K., S. M. Carty, S. Lin, R. J. Cotter, and C. R. Raetz. 2002. An Escherichia coli mutant lacking the cold shock-induced palmitoleoyltransferase of lipid A biosynthesis: absence of unsaturated acyl chains and antibiotic hypersensitivity at 12 degrees C. *J. Biol. Chem.* **277:**14186–14193.

Walsh, N. P., B. M. Alba, B. Bose, C. A. Gross, and R. T. Sauer. 2003. OMP peptide signals initiate the envelope-stress response by activating DegS protease via relief of inhibition mediated by its PDZ domain. *Cell* **113:**61–71.

Wilken, C., K. Kitzing, R. Kurzbauer, M. Ehrmann, and T. Clausen. 2004. Crystal structure of the DegS stress sensor: how a PDZ domain recognizes misfolded protein and activates a protease. *Cell* **117:**483–494.

Wu, T., J. Malinverni, N. Ruiz, S. Kim, T. J. Silhavy, and D. Kahne. 2005. Identification of a multicomponent complex required for outer membrane biogenesis in Escherichia coli. *Cell* **121:**235–245.

Yu, H., M. J. Schurr, and V. Deretic. 1995. Functional equivalence of Escherichia coli sigma E and Pseudomonas aeruginosa AlgU: E. coli rpoE restores mucoidy and reduces sensitivity to reactive oxygen intermediates in algU mutants of P. aeruginosa. *J. Bacteriol.* **177:**3259–3268.

Yura, T., and K. Nakahigashi. 1999. Regulation of the heat-shock response. *Curr. Opin. Microbiol.* **2:**153–158.

DISULFIDE BOND FORMATION IN THE PERIPLASM

Mehmet Berkmen, Dana Boyd, and Jon Beckwith

7

The concept that active proteins have a specific three-dimensional shape that is lost upon denaturation was proposed by Mirsky and Pauling some 70 years ago (Mirsky and Pauling, 1936). In the passing decades, remarkable progress has been made toward appreciating the numerous biological mechanisms that modify and assist the folding of a protein into its native form in a living cell. However, many of the biological principles that govern a protein acquiring its final active folded state remain elusive. While approaches have been developed that do allow the prediction of some native structures based on amino acid sequences, there are still no general methodologies for such predictions. For some proteins, this difficulty could be due to our lack of knowledge of the posttranslational biological processes that assist and regulate protein folding in vivo. The discoveries of posttranslational covalent modifications, cofactor associations, proteolytic processing, compartmentalization, and folding assistance by chaperones have dramatically altered our naïve conception that all proteins fold into their biologically active state unassisted. One such important posttranslational modification is the formation of disulfide bonds.

Protein disulfide bonds are formed by an oxidative process in which the thiol (SH) groups of two cysteines are joined by covalent linkage, resulting in the loss of two hydrogens from the protein. In certain proteins (see below), disulfide bonds can be eliminated by a reductive process which restores two free cysteines. Finally, in some proteins, disulfide bonds can be rearranged—that is, the array of disulfide bonds is altered—in a process called disulfide bond isomerization. These different reactions are catalyzed by a group of enzymes termed disulfide oxidoreductases and isomerases which are present in all domains of life. Disulfide bonds usually contribute to the three-dimensional structure of a protein and oftentimes are required for the full activity of a protein. The formation of a disulfide bond can be viewed as a structural determinant that decreases the conformational entropy, resulting in increased stability of a polypeptide.

In many organisms, disulfide bond formation and isomerization occur in specialized oxidizing compartments. In gram-negative bacteria, proteins that require disulfide bonds for their final folded state are translocated across the cytoplasmic membrane into the periplasm where they interact with enzymes that catalyze

Mehmet Berkmen, Dana Boyd, and Jon Beckwith, Department of Microbiology and Molecular Genetics, Harvard Medical School, Boston, MA 02115.

The Periplasm
Edited by Michael Ehrmann © 2007 ASM Press, Washington, D.C.

the formation and isomerization of disulfide bonds. Although proteins with disulfide bonds are found in the cytoplasm, these proteins are almost always ones in which the cysteines involved are in the active site of the protein. For example, the cytoplasmic enzyme ribonucleotide reductase carries out its reduction of ribonucleotides using electrons derived from its active site cysteines. Thus, the protein becomes oxidized and must itself be reduced by cytoplasmic proteins such as the thioredoxins to be regenerated as an active enzyme (Prinz et al., 1997). In other words, in general, cytoplasmic disulfide bonds are ephemeral, being formed and then reduced in the process of participating in electron transfer reactions.

This chapter reviews the process of disulfide bond formation in the periplasm by following the life of a protein, from the "birth" of a protein at the ribosome to its entrance into the periplasm as an unfolded infant ("The Infant Protein—Entering the Periplasm," below), to the oxidation of the maturing protein ("The Maturing Protein—Disulfide Bond Formation," next page). The oxidative life of the aging protein is not infallible and sometimes succumbs to mistakes of oxidation ("Making Mistakes in the Adolescent Protein," p. 129). Those proteins that are misoxidized are corrected by the isomerization system ("Getting It Right—Correction by Disulfide Bond Isomerization," p. 130).

HISTORY

In trying to understand the relationship between the amino acid sequence of a protein and its final folded state, Anfinsen and colleagues studied the refolding of reduced bovine pancreatic ribonuclease A. Ribonuclease A (RNase A) contains four disulfide bonds at positions [26–84], [40–95], [58–110], and [65–72], all of which are essential for RNase A to be fully active. Anfinsen and colleagues showed that denatured and fully reduced RNase A could refold spontaneously to its active oxidized form in the presence of molecular oxygen (Anfinsen and Haber, 1961; Anfinsen et al.,

1961; White, 1960). Under optimal conditions, the physical properties of the in vitro oxidized RNase A were indistinguishable from the in vivo purified RNase A. Anfinsen and colleagues concluded that ". . . no special genetic information, beyond that contained in the amino acid sequence, is required for the proper folding of the molecule and for the formation of 'correct' disulfide bonds (Goldberger et al., 1963)." However, the spontaneous rate of disulfide bond formation in vitro ($\sim$20 min under optimized conditions) was significantly slower than the rate in living cells in vivo (<1 min). Furthermore, the yield of correctly folded protein was low, with much of the protein containing the incorrect disulfide bonds.

Because of the significant yield of incorrectly folded protein, Anfinsen and coworkers postulated that there might be a biological system that corrected the wrong disulfide bonds and led to the formation of native disulfide bonds. The subsequent search for a biological catalyst for correct disulfide bond formation led to the discovery of the eukaryotic endoplasmic reticulum enzyme, protein disulfide isomerase (PDI) (De Lorenzo et al., 1966; Goldberger et al., 1963). During the years after the discovery of PDI as an enzyme that isomerizes already existing disulfide bonds, the actual process of disulfide bond formation was incorrectly believed to be spontaneous. Cells were thought to require only the isomerase activity of PDI for rapid *rearrangement* of disulfide bonds. Ironically, PDI, which was recently shown to be a catalyst of disulfide bond formation in vivo, has not yet been shown to be an isomerase in vivo. It was not until the discovery of the periplasmic DsbA protein in *Escherichia coli* that it became clear that disulfide bond forming catalysts (Table 1) were necessary for efficient formation of disulfide bonds (Akiyama et al., 1992; Bardwell et al., 1991; Peek and Taylor, 1992).

THE INFANT PROTEIN—
ENTERING THE PERIPLASM

The formation of structural disulfide bonds between cysteines in proteins is catalyzed by the

TABLE 1 Properties of disulfide bond-forming enzymes in the periplasm

Protein	Localization	Function	Primary phenotype of mutant	C-X-X-C	Redox (mV)
DsbA	Periplasm	Oxidase	Defective in protein oxidization, flagella and F pilus assembly Sensitive to DTT, Hg^{2+}, and Cd^{2+} Resistant to M13 infection Slow growth rate and mucoidy in minimal media	C_{30}-PH-C_{33}	-120
DsbB	Inner membrane	Oxidize DsbA	Defective in DsbA oxidation	C_{41}-VL-C_{44} C_{104}-C_{130}	-210 -225
DsbC	Periplasm	Isomerase	Misoxidized nonconsecutive disulfide-bonded proteins Sensitive to DTT and Cu^{2+}	C_{98}-GY-C_{101}	-130
DsbD	Inner membrane	Reduce CcmG, DsbC, DsbG	Defective in CcmG, DsbC, and DsbG reduction	C_{103}-C_{109} α C_{163}-C_{285} β C_{461}-VA-C_{464} γ	-229 α ? β -241 γ
DsbG	Periplasm	Isomerase	None observed	C_{109}-PY-C_{112}	?

DsbA protein, located in the *E. coli* periplasm (Color Plate 8). Thus, proteins destined to contain structural disulfide bonds must be translocated from their cytoplasmic birthplace across the cytoplasmic membrane into the periplasm. All of the identified proteins known to acquire disulfide bonds in the periplasm are exported via the SecYEG membrane-embedded translocon. Both cotranslational and posttranslational mechanisms employ SecYEG for passage of their substrate proteins across the membrane (see Chapter 1). Numerous studies have shown that to successfully pass through the SecYEG pore proteins must remain in an unfolded state. In the cotranslational mechanism, export of a protein begins shortly after initiation of its synthesis, thus avoiding cytoplasmic folding. In posttranslational export, various mechanisms have been uncovered (antifolding factors, etc.) for maintaining the proteins in an unfolded state so that they remain export competent. This requirement for an unfolded state is also strongly suggested by the recently determined structure of the SecYEG complex, which appears to have a pore size too small to encompass anything of dimensions larger than a linear polypeptide chain, or perhaps an α-helical structure. Thus, nascent polypeptide chains, as they appear initially in the periplasm, are likely to have no or very little structure.

THE MATURING PROTEIN— DISULFIDE BOND FORMATION

DsbA is a periplasmic enzyme that interacts directly with cysteines of substrate proteins to introduce disulfide bonds. The unfolded nature of nascent polypeptide chains that are being translocated into the periplasm may have important implications for the mechanism by which DsbA introduces disulfide bonds into its substrates. This would be the case if disulfide bonds were formed during the translocation process. Direct evidence that disulfide bonds are formed on nascent chains in eukaryotic cells comes from studies on the translocation of immunoglobulins into the endoplasmic reticulum (Bergman and Kuehl, 1979). Both indirect and some direct evidence, presented later in this review, suggests that the cotranslocational formation of disulfide bonds also occurs in *E. coli*. Thus, the interaction between DsbA and substrate may take place between a completely unfolded protein, a protein that has assumed some secondary structure, or one that has folded even more extensively in the periplasm. Determining at which stage of folding disulfide bond formation takes place has important implications for the mechanism of recognition of substrate cysteines by DsbA.

Because of the significant role disulfide bonds play in the folding and/or stability of numerous periplasmic proteins, mutations in DsbA and its homologues in other gram-negative bacteria result in a myriad of phenotypes. Bacteria lacking a functional copy of *dsbA* have increased sensitivity to reduced dithiothreitol (DTT); benzylpenicillin (Missiakas et al., 1993); Cd^{2+}, Hg^{2+} (Stafford et al., 1999), and Zn^{2+} (Hayashi et al., 2000); they are defective in intracellular survival (Yu et al., 2000); they have increased sensitivity to infection (Gonzalez et al., 2001; Peek and Taylor, 1992); they affect competence (Tomb, 1992), biofilm formation (Genevaux et al., 1999), toxin (Stenson and Weiss, 2002), fimbria (Bouwman et al., 2003), and pilus production (Jacob-Dubuisson et al., 1994). Thus far there are more than two dozen periplasmic proteins in *E. coli* that have been shown to depend on DsbA for their correct folding (Agp, AppA, ArtJ, Bla, DegP, DppA, FlgI, GltI, GltX, HisJ, Imp, LivJ, LivK, MepA, OmpA, OppA, PhoA, RcsF, Rna, STa, STb, UgpB, YbeJ, YbjP, YcdO, YedD, YggN, YodA, and ZnuA).

Although disulfide bonds play a central role in many biological processes, none of the genes involved in their formation, including *dsbA* and other *dsb* genes (see subsequent sections), are essential for growth in laboratory conditions. This could mean that there are no essential proteins that contain structural disulfide bonds or that, if there are, those proteins still retain some activity in the absence of their disulfide bonds. Alternatively, the survival of a *dsbA* null mutant may be due to the weak-background, disulfide bond-forming activity found in such a strain, which may result from the inefficient oxidation of cysteines by oxygen itself (Bardwell et al., 1991, Leichert and Jakob, 2004). This background activity may be sufficient to yield enough activity of any essential protein that requires disulfide bonds for its proper functioning.

DsbA is a monomeric soluble protein with 189 amino acids (21 kDa) in its mature form. When discovered, DsbA was found to be a member of the thioredoxin family of proteins because of its characteristic Cys-X-X-Cys motif (where X is any amino acid) and homology to thioredoxin itself. The structure of DsbA has been determined by both NMR and X-ray crystallography of the oxidized and reduced forms (Guddat et al., 1998; Martin, 1995; Martin et al., 1993a, 1993b; Schirra et al., 1998). These studies have revealed a classic thioredoxin fold along with an extra 76-amino-acid helical domain (Color Plate 9). The helical domain folds around the thioredoxin fold, constructing a hydrophobic patch around the active site. The residues that form the hydrophobic patch around the active site are conserved and are suggested to assist DsbA in preferentially binding to unfolded proteins. A proposed uncharged groove running along the active site of DsbA may promote, through hydrophobic interactions, the binding of DsbA to unfolded substrates (Frech et al., 1996; Guddat et al., 1997; Vinci et al., 2002). Studies of the X-ray crystal structure of $DsbA_{C33A}$ dimer supported the presence of the proposed hydrophobic peptide-binding groove. In this first structural protein:protein interaction study of DsbA, the dimer interface of $DsbA_{C33A}$ overlapped with the peptide-binding groove. The peptide-binding groove lacked any specific interaction supporting DsbA's ability to oxidize a large spectrum of substrates. Further support for the lack of specificity of the peptide-binding groove was observed by a variety of hydrophobic compounds present in the crystallization media found occupying the groove (Ondo-Mbele et al., 2005).

Thioredoxin family members utilize their redox active cysteines either to reduce disulfide bonds in proteins (see discussion of ribonucleotide reductase in the introduction) or to catalyze the formation of disulfide bonds. Thioredoxin 1 of *E. coli*, the prototype of thioredoxin proteins, is the most reducing of this family of proteins, with a redox potential of -270 mV, while DsbA is the most oxidizing member with a redox potential of -120 mV. However, remarkably, genetic alterations that cause cytoplasmic thioredoxin 1 to accumulate in the oxidized form in the cytoplasm allow it to relatively efficiently catalyze the *formation* of disulfide bonds (Stewart et al., 1998). Conversely, the initial characterization of DsbA as a member of the thioredoxin family involved the demonstration that reduced DsbA could effectively *reduce* the disulfide bonds of insulin (Bardwell et al., 1991), an assay often used for thioredoxin activity.

DsbA is directed into the periplasm by its 19-amino-acid signal peptide. The mechanism by which DsbA is exported to the periplasm is unusual compared with most other periplasmic proteins. The translocation of DsbA is co-translational, dependent on SRP for its interaction with the SecYEG translocon (Schierle et al., 2003; Huber et al., 2005), whereas most periplasmic proteins are translocated posttranslationally. This cotranslational export of DsbA appears to be necessary because the protein folds rapidly in the cytoplasm (Huber et al. 2005). DsbA appears in the periplasm with its active site cysteines reduced. However, to act as a donor of disulfide bonds, DsbA itself must be in the oxidized form. This oxidation step is carried out by the membrane protein DsbB, thus converting DsbA to an active enzyme (see below for details on DsbB). The disulfide bond is formed between the two cysteines of the thioredoxin-like motif of DsbA, Cys_{30}–Pro_{31}–His_{32}–Cys_{33}.

The active site of DsbA is the crucial feature of the protein that permits it to be an efficient oxidant of substrates (Fig. 1). Exchanging the two amino acids lying between the two cysteines of DsbA with those of other thioredoxin family members results in striking effects on the redox potential of the proteins. The changes correlate with the redox potential of the source of the Cys-X-X-Cys sequence (Mössner et al., 1999). In addition, mutations that result in a change in a proline and a histidine residue within the Cys-X-X-Cys motif also result in significant changes in the oxidizing activity of DsbA (Grauschopf et al., 1995). These results suggest that the two amino acids between the two cysteines play a major role in distinguishing the redox potential of thioredoxin-like proteins.

Other studies show that the biochemical properties of the amino acids surrounding the active site also influence the redox properties of the enzyme (Kortemme and Creighton, 1995). The N-terminal $cysteine_{30}$ in the active site of DsbA has an unusually low pK_a of 3.5, favoring its protonated thiolate form under physiological conditions (Nelson and Creighton, 1994). $Histidine_{32}$ along with $glutmate_{97}$ appears to

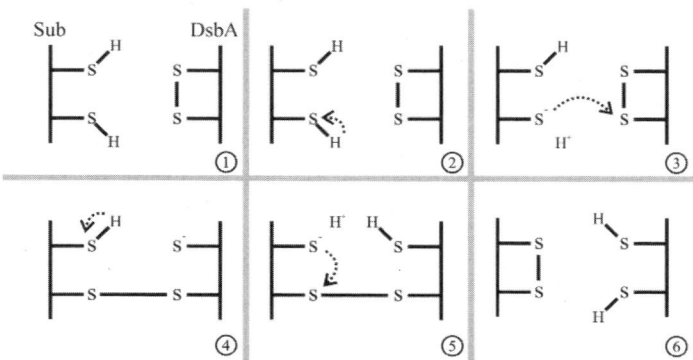

FIGURE 1 The mechanism of disulfide bond formation. DsbA catalyzes the formation of disulfide bonds in a polypeptide with reduced cysteines. The cysteines within the Cys-X-X-Cys active site of DsbA are oxidized (S—S) and the thiol side groups of cysteine residues in the substrate are reduced (SH) ①. Disulfide bond formation is initiated by deprotonation of a thiol group in the substrate ②. The resulting thiolate anion can initiate a nucleophilic attack on the disulfide bond of DsbA ③. The resolution of the mix-disulfide-bonded complex could occur by deprotonation of another thiol group ④, which can attack the substrate-DsbA disulfide bond ⑤. The result of this reaction is the oxidation of the substrate and the reduction of DsbA ⑥.

further stabilize the thiolate ion of cysteine$_{30}$ (Gane et al., 1995; Warwicker, 1998). The sulfur atom of cysteine$_{30}$ is fully exposed in the oxidized crystal structure of DsbA, making it highly reactive with reduced cysteines in substrate proteins. The C-terminal cysteine$_{33}$ in the active site of DsbA is buried with a high pK_a of approximately 9.5 and therefore is unlikely to be involved in the initial mixed-disulfide-bonded complex with a substrate (Nelson and Creighton, 1994). Furthermore, oxidized DsbA is more flexible and less stable than when it is reduced (Wunderlich and Glockshuber, 1993; Zapun et al., 1993). Taken together, the instability of the active site disulfide bond drives DsbA thermodynamically toward reacting with newly synthesized reduced substrates, resulting in their oxidation.

Another key residue important in the function of DsbA is *cis*-proline$_{151}$, a proline that is found in a similar position in the structure of most thioredoxin family members. In crystallized thioredoxins, these *cis*-prolines lie very close to the active site cysteines. Mutations in proline$_{151}$ to alanine destabilize DsbA significantly (Moutiez et al., 1999), resulting in alterations in the structure of the active site (Charbonnier et al., 1999). Mutations resulting in a change of proline$_{151}$ to serine, histidine, asparagine, tryptophan, and glycine result in accumulation of DsbA$_{P151X}$-DsbB mixed disulfide complexes (Kadokura et al., 2005), while a change to threonine results in accumulation of DsbA$_{P151T}$-substrate mixed disulfide complexes (Kadokura et al., 2004). Purification of these complexes resulted in the identification of numerous substrates of DsbA in *E. coli*. The proline$_{151}$ residue appears to be critical in the resolving of disulfide-bonded DsbA complexes.

The *dsbA* gene is part of a two-gene operon with an uncharacterized upstream gene, *rdoA* (*yihE*). There appear to be two promoters: a distal promoter upstream of the *rdoA* gene and a proximal promoter within the 3′ end of *rdoA*. Transcription from the proximal promoter is constitutive at low levels (Belin and Boquet, 1994). Under normal laboratory growth conditions, *dsbA* is expressed constitutively in the

periplasm at a level of about 850 molecules per cell (Akiyama et al., 1992). The expression of *dsbA* is increased in response to cell envelope protein-folding defects. This increase depends on positive regulation by the Cpx two-component system (Danese and Silhavy, 1997; Pogliano et al., 1997) (for further details see Chapter 5). DsbA is also induced at high pH under both aerobic and anaerobic conditions (Yohannes et al., 2004).

DsbB

The oxidation of proteins by DsbA results in the reduction of the cysteines in its active site (Fig. 1). DsbA is restored to its oxidized active state by the inner membrane protein, DsbB. DsbB is a 176-amino-acid (20-kDa) protein, and based on alkaline phosphatase (PhoA) fusion studies, DsbB has been shown to have four transmembrane domains and two periplasmic loops (Jander et al., 1994) (Fig. 2). Purified DsbB protein contains bound quinones, which play an important role in this electron transfer process. Electrons from reduced DsbA are relayed to membrane-associated quinones via the cysteines in DsbB.

Of the six cysteines present in DsbB, only the four periplasmic cysteines (Cys$_{41}$-Cys$_{44}$, Cys$_{104}$-Cys$_{130}$) are necessary for its activity (Jander et al., 1994). The first step in the reoxidation of DsbA by DsbB involves the attack of cysteine$_{30}$ of DsbA on the Cys$_{104}$-Cys$_{130}$ disulfide bond in the second periplasmic loop of DsbB. This reaction results in the formation of a mixed disulfide bond between cysteine$_{104}$ of DsbB and cysteine$_{30}$ of DsbA (Guilhot et al., 1995; Kishigami et al., 1995). For regeneration of oxidized and active DsbA, the mixed disulfide must be resolved by the attack of cysteine$_{33}$ on the DsbB-DsbA mixed disulfide.

Two models, both based on experimental results, have been proposed to explain how this reaction takes place (Color Plate 10). In the first concerted reaction model (Color Plate 10, pathway A), based mainly on in vivo studies, before the DsbA-DsbB mixed disulfide is resolved, a second reaction takes place between cysteine$_{130}$ of DsbB and the Cys$_{41}$-Cys$_{44}$ disul-

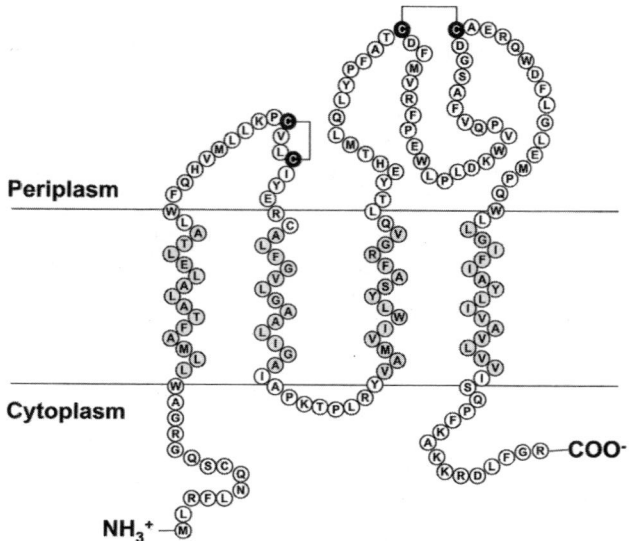

Periplasm

Cytoplasm

FIGURE 2 Topology of DsbB. The topology of DsbB based on alkaline phosphatase fusion studies (Jander et al., 1994). The active site cysteines are shown in their oxidized state, and the putative transmembrane domain amino acids are highlighted.

fide bond of DsbB (Kadokura and Beckwith, 2002). This reaction results in a protein complex containing both the DsbA-DsbB disulfide and an interdomain disulfide (Cys_{41}-Cys_{130}) between the first and second periplasmic domains of DsbB. The C-terminal cysteine pair of DsbB is less oxidizing (~225 mV [Inaba et al., 2004]) than the cysteine pair of DsbA (~120 mV). Therefore, the formation of the interdomain disulfide bond (Cys_{130}-Cys_{41}) in DsbB may be required to prevent a "backward" reaction (Color Plate 10, Stage 0). This backward reaction could be catalyzed by $cysteine_{130}$ attacking the disulfide bond formed between $cysteine_{30}$ of DsbA and $cysteine_{104}$ DsbB (Color Plate 10, Stage II).

According to the second sequential reaction model (Color Plate 10, pathway B), based on in vitro studies, the first step is the resolution of the DsbA-DsbB mixed disulfide complex by attack of DsbA's $cysteine_{33}$, leaving both $cysteine_{104}$ and $cysteine_{130}$ in the reduced form (Grauschopf et al., 2003). Subsequent steps transfer electrons to the first periplasmic domain and then to quinones. The results of these reactions, according to both models, are the reoxidation of DsbA and the reduction of quinones. Additional in vitro studies report ev-

idence for both mechanisms (Inaba et al., 2005). Further experimental evidence will be necessary to determine the nature of the process in vivo.

The quinone used by DsbB for transferring electrons varies depending on the growth conditions. In aerobiosis, reduced DsbB is oxidized by ubiquinone, which in turn is reduced to ubiquinol. Ubiquinol shuttles its electrons to either of the two membrane-embedded electron transfer cytochrome complexes *bo/bd* to be reoxidized back to ubiquinone. The terminal electron acceptor in this pathway is oxygen. In anaerobic conditions, the electron acceptor from DsbB is menaquinone, which is reoxidized by donating its electrons to terminal electron acceptors such as fumarate reductase or nitrate reductase (Bader et al., 1999, 2000; Kobayashi et al., 1997; Takahashi et al., 2004). A potential quinone-binding site has been identified in the second periplasmic loop of DsbB (Xie et al., 2002). In vitro studies of purified DsbB revealed that DsbB is purple (Regeimbal et al., 2003). Reduction, unfolding, and cysteine mutants of DsbB resulted in the loss of the purple color. The authors postulated that the color was due to a quinhydrone form of quinone associated with DsbB. It remains to be shown whether a quinhydrone plays a role in DsbB function in vivo. In addition, a mutation that results in a replacement of arginine$_{48}$ with histidine in DsbB (the only conserved amino acid besides the redox active cysteines) resulted in poor disulfide formation under anaerobic conditions. Purified DsbB$_{R 48H}$ has an increased K_m for ubiquinone and little or no activity with menaquinone in an in vitro assay for its activity toward DsbA (Kadokura et al., 2000).

In eukaryotic cells the protein disulfide isomerase (PDI) represents the counterpart of DsbA, being responsible for the formation of disulfide bonds in the endoplasmic reticulum. Analogously to DsbB, two ER proteins, Ero1p and Erv2p, regenerate PDI by oxidizing its active site cysteines. While DsbB is an integral membrane protein, Ero1p and Erv2p are peripheral membrane proteins, containing no hydrophobic transmembrane segments. In addition, the eukaryotic proteins use FAD as an electron receptor, while DsbB uses quinones. Nevertheless, a comparison of the crystal structures of Ero1 and Erv2 and modeling the α-helices of those proteins onto the transmembrane α-helices of DsbB suggest a model in which all three proteins share mechanistic features (Sevier et al., 2005). The role of the tryptophan residues in stacking of the redox active cofactor FAD in Ero1p and Erv2p was suggested to be similar for amino acid residue interactions with quinones in DsbB. The results of mutant analysis with DsbB are consistent with this proposal.

MAKING MISTAKES IN THE ADOLESCENT PROTEIN

We have pointed out that DsbA may act to generate disulfide bonds in proteins as nascent polypeptide chains emerge into the periplasm. Furthermore, we suggest that these nascent substrates for DsbA may show very little structure. If that is the case, we can ask how DsbA recognizes the appropriate pairs of cysteines to join into disulfide bonds—the disulfide bonds that are found in the final folded structure of the substrate protein. The answer is that it

probably does not. In fact, there is evidence that DsbA may be relatively indiscriminate in forming disulfide bonds in substrate proteins. This evidence comes from two sources. First, cysteine-containing proteins (β-galactosidase and a *Bacteriodes* β-lactamase) that are not ordinarily translocated into the *E. coli* periplasm, when they are forced to do so by the attachment of a signal sequence, acquire disulfide bonds that interfere with their activity (Alksne et al., 1995; Bardwell et al., 1991). Thus, DsbA can make disulfide bonds in proteins that were not meant to be oxidized at all. Second, *E. coli* and other bacteria express a protein disulfide isomerase (DsbC) that is necessary to correct the "errors" that are made by DsbA. DsbC acts by promoting rearrangement of disulfide bonds, allowing generation of the proper array in the final folded product. In *E. coli* mutants lacking DsbC, some disulfide-bonded proteins are reduced in activity or degraded (Berkmen et al., 2005; Hiniker and Bardwell, 2004; Joly and Swartz, 1997; Rietsch et al., 1997).

Why incorrect disulfide bonds are formed can be explained by assuming that cotranslational/cotranslocational disulfide bond formation occurs on unstructured polypeptide chains. DsbA may simply react with the first cysteine of the polypeptide chain to cross into the periplasm and join it in a disulfide bond with the next one to appear. Assuming this mechanism, consider a protein with four cysteines and its interaction with DsbA. According to this "consecutive" model of disulfide bond formation, the product of interaction of DsbA with this protein would contain disulfide bonds between cysteines 1 and 2 in the sequence and cysteines 3 and 4. If these are the disulfide bonds found in the mature protein, the product would contain the appropriate disulfide bonds and would need no correction (isomerization). In contrast, if the mature folded protein should contain disulfide bonds between cysteines 1 and 3 and cysteines 2 and 4, correction by isomerization would be necessary. This explanation would imply that only proteins that contain nonconsecutive disulfide bonds in their final folded structure would be subject to incorrect disulfide bond formation, thus necessitating the action of an isomerase.

Experimental evidence is consistent with this distinction between proteins containing consecutive or nonconsecutive disulfide bonds. First, several disulfide-bonded proteins, both prokaryotic and eukaryotic, have been expressed in *E. coli* and their dependence on DsbC has been tested. A fairly strict correlation exists between native disulfide bond patterns and DsbC dependence; proteins that contain nonconsecutive disulfide bonds in their final structure depend on DsbC for proper folding while proteins that should be made with consecutive disulfide bonds show very little or no dependence on DsbC (Berkmen et al., 2005; Hiniker and Bardwell, 2004; Joly and Swartz, 1997; Rietsch et al., 1997). Second, two nearly identical proteins of *E. coli*, phytase and glucose-1-phosphatase, differ mainly because phytase contains one nonconsecutive disulfide bond and glucose-1-phosphatase contains only consecutive disulfide bonds. Phytase depends on DsbC for assembly into an active enzyme and glucose-1-phosphatase does not. Introduction into glucose-1-phosphatase of the additional pair of cysteines placed comparably to those found in phytase led to the conversion of the protein from one that is DsbC independent to a DsbC-dependent enzyme.

A bioinformatic analysis of *E. coli* proteins that contain disulfide bonds reveals some surprising findings. Of those periplasmic proteins whose structure has been determined, approximately 66% contain disulfide bonds. Most of these contain only one or two disulfide bonds and these bonds are all formed between cysteines following consecutively in the sequence starting at the amino terminus. So far, only three native *E. coli* proteins whose structure is known show nonconsecutive disulfide bonds—phytase (AppA, four disulfide bonds), a murein endopeptidase (MepA, three disulfide bonds), and RNase I (four disulfide bonds). Each of these shows strong dependence on DsbC (Berkmen et al., 2005; Hiniker and Bardwell, 2004). A caveat to these conclusions is that the crystallized proteins may have acquired disul-

fide bonds by air oxidation during purification so that the disulfide bond pattern may not accurately represent the native state.

GETTING IT RIGHT— CORRECTION BY DISULFIDE BOND ISOMERIZATION

As described above, the action of DsbA on substrate proteins sometimes leads to the formation of incorrect disulfide bonds. This property of DsbA leads to the requirement for an enzyme, a protein disulfide isomerase, to promote rearrangement of the disulfide bonds in the protein so that it can achieve its native folded state. For a catalyst to be an effective disulfide bond isomerase, it must (i) recognize substrates that are misoxidized and misfolded, (ii) break the wrongly formed disulfide bonds, and (iii) allow the misoxidized protein to be rearranged to achieve its native disulfide-bonded state. For the known misoxidized proteins in *E. coli*, the disulfide bond isomerization pathway is performed by the periplasmic disulfide bond isomerase DsbC (Missiakas et al., 1994; Shevchik et al., 1994).

DsbC is a 236-amino-acid (25-kDa) homodimer directed to the periplasm by its 20-amino-acid-long amino-terminal signal peptide. Each monomer of DsbC contains four cysteines. Cysteine$_{141}$ and cysteine$_{163}$ are joined in a disulfide bond which is essential for the folding and structural stability of DsbC (Liu and Wang, 2001). The more amino-terminal cysteines (C_{98}-GY-C_{101}), the redox-active cysteines of DsbC, are normally found in a reduced state (Rietsch et al., 1997). These two cysteines are in a domain of DsbC that forms a thioredoxin fold and they correspond to the Cys-X-X-Cys site of other thioredoxin family members. The crystal structures of DsbC from *E. coli* (McCarthy et al., 2000) and from *Haemophilus influenzae* (Zhang et al., 2004) show a dual-domain structure consisting of a dimerization domain and the thioredoxin domain. The dimerization domain lies at the amino terminus of DsbC and is linked to the thioredoxin domain by a short α-helix (Color Plate 11A) (McCarthy et al., 2000; Sun and Wang, 2000).

The dimerization of the two monomers of DsbC results in the formation of an uncharged hydrophobic pocket (Color Plate 11 B and C). This property of DsbC likely explains another property of the protein, its ability to act as a protein-folding chaperone. DsbC can assist the refolding of denatured proteins such as lysozyme and glyceraldehyde 3-phosphate in vitro (Chen et al., 1999). Since misoxidized proteins will most likely be misfolded, chaperone properties would allow DsbC to recognize and bind to those proteins that contain incorrect disulfide bonds (Darby et al., 1998). The chaperone activity of DsbC is independent of its redox-active cysteines but depends on the dimerization domain.

Although there is evidence that DsbA also possesses some chaperone activity (Zheng et al., 1997), the presumed peptide-binding sites of each protein differ significantly. The DsbA peptide-binding site is a long cleft, perhaps allowing it to interact with extended nonfolded substrates, whereas the peptide-binding site of DsbC is a large pocket with room to accommodate misfolded proteins. These properties are consistent with a DsbA that interacts with nascent chains emerging from the secretion apparatus while DsbC interacts with fully translocated yet misfolded proteins. This view is supported by the fact that the chaperone activity of DsbC is not hindered by the addition of short peptides or by native folded proteins (Zheng et al., 1997). According to this picture of DsbC, it would not interact with properly folded proteins and attack their disulfide bonds.

While it seems likely that the chaperone properties of DsbC allow it to interact with misfolded proteins, the mechanism by which the subsequent steps in disulfide bond isomerization occur is not clear. Two mechanisms have been considered (Fig. 3). In both, the first step is the attack by one of the cysteines of DsbC on an incorrect disulfide bond exposed by the misfolding of the protein (Fig. 3, ①). The result is the formation of a mixed disulfide between active site cysteines of DsbC and a substrate cysteine that was formerly covalently bound to another substrate cysteine. After the

formation of this mixed disulfide, the two mechanisms diverge.

In the reducing model, the DsbC–substrate mixed disulfide is resolved by attack of the remaining free cysteine of DsbC (Fig. 3, ②). This reaction generates a substrate with a reduced disulfide bond in the substrate and an oxidized DsbC (Fig. 3, ③). Another attack on the remaining disulfide bond is catalyzed by DsbC (Fig. 3, ④). Further steps to generate the correct disulfide bonds would require that DsbA interact with the reduced substrate and form disulfide bonds between the appropriate cysteines to generate the final folded structure (Fig. 3, ⑤).

In the isomerase model (Fig. 3, ⑥), it is argued that the mixed disulfide between DsbC and substrate is attacked by another cysteine of the substrate (not the cysteine originally involved in the substrate disulfide bond), generating a new disulfide bond which is more likely to be a correct one (see below). Disulfide bond isomerization of a substrate by DsbC does not result in any net gain or loss of electrons. Thus, at the end of the isomerization reaction, DsbC remains in its reduced form (Fig. 3, ⑦).

Several lines of evidence are consistent with the former reducing model. First, $E.\ coli$ expresses another Dsb protein, DsbD, which is essential to maintain the function of DsbC. DsbD provides electrons to keep the Cys-X-X-Cys motif of DsbC in the reduced state. In mutants lacking DsbD, DsbC accumulates in the oxidized state. If DsbC acted by the first mechanism, it would always be returned to the reduced state at the end of its reaction with substrate and would not appear to require a protein such as DsbD to keep it reduced. However, this argument may not be relevant if it is simply the oxidizing environment of the periplasm that results in formation of the DsbC disulfide bond, and DsbD is present to reverse this unwanted oxidation (Collet et al., 2002; Rietsch et al., 1997; Walker and Gilbert, 1997).

Both of the mechanisms proposed for "isomerization" face the problem of how the protein finally ends up with the correct disulfide bonds. For the first mechanism, true isomeriza-

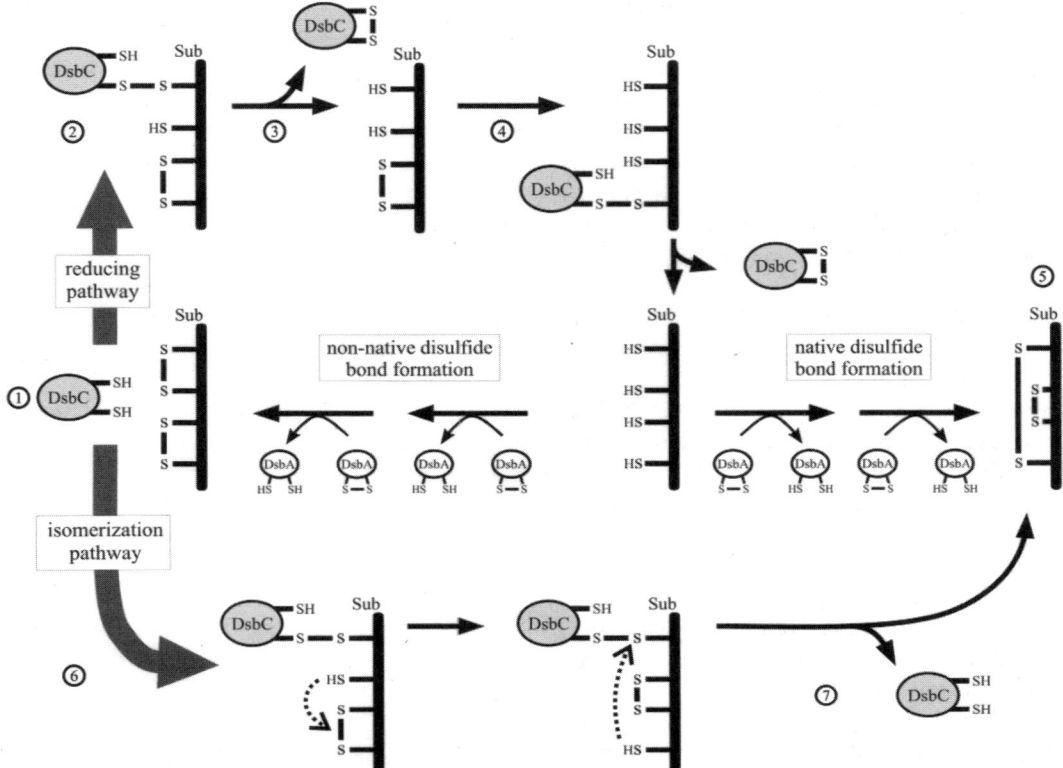

FIGURE 3 Proposed mechanism of isomerization by DsbC. For the purpose of clarity only a monomer of DsbC is shown. Reduced active DsbC recognizes misoxidized substrate ① and forms a mix-disulfide-bonded complex. This complex ② could be resolved by the reduction of the disulfide bond in the substrate, resulting in the oxidation of DsbC ③. A secondary cycle of reduction is necessary for the substrate to be fully reduced ④, allowing DsbA to reoxidize the substrate ⑤. Alternatively, the disulfide bonds in the complex could be shuffled ⑥ by the isomerase action of DsbC, resulting in native disulfide-bonded substrate and reduced DsbC ⑦.

tion, a shifting of disulfide bonds without any further oxidation steps, would require that the shift lead to a correct disulfide bond. In the second mechanism, reduction by DsbC and reoxidation by DsbA would require that DsbA, in its second pass at the protein, recognize the appropriate cysteines to join. Either of these concerns would not be as much of a problem, if, by the time these steps have occurred, the substrate protein has assumed significant aspects of its tertiary structure. As a result, the cysteines that are meant to be joined in the native structure may be proximal to each other. Then, the most likely cysteines to be joined by DsbA

would be those that are closest in the folded structure.

Based on in vitro reduction assays of insulin, DsbC displayed similar reducing capacity as the cytoplasmic reductase TrxA (Zhao et al., 2003). This property of DsbC is believed to be essential in breaking wrongly formed disulfide bonds in misoxidized substrates. While DsbD helps maintain DsbC in the reduced state, the dimeric structure of DsbC ensures that its redox active cysteines are sequestered from oxidation by DsbB. Since the in vitro properties of the active site cysteines of DsbC are similar to that of DsbA (Zapun et al., 1995), DsbC is capable of acting as an oxidase when the interaction with the reducing DsbD is removed (Rietsch et al., 1997). In addition, mutations that alter the dimerization domain of DsbC result in a monomeric form of the protein, which is now capable of interacting with DsbB to become oxidized. Oxidized monomers of DsbC can functionally replace DsbA and oxidize reduced substrates (Bader et al., 2001).

Recent studies suggest that DsbC may have evolved by some strikingly simple steps. In particular, it appears that the fusion of a dimerization domain to thioredoxin-like proteins generally leads to a protein with DsbC-like features. Chimeras of DsbC were created where the thioredoxin domain of DsbC was replaced with that of the oxidizing DsbA, the reducing TrxA, or the thioredoxin active domain "a" of the eukaryotic isomerase PDI. These chimeras all promoted isomerization of misoxidized substrates in vivo and displayed chaperone activity in vitro at varying levels (Segatori et al., 2004; Zhao et al., 2003).

The *dsbC* gene is in a bicistronic operon with the downstream gene *recJ* that encodes a $5' \rightarrow 3'$ exonuclease. Promoter analysis of *dsbC* shows putative CpxR-binding sites indicating regulation by the Cpx two-component stress response, similar to that of DsbA. DsbC transcription is also under the control of the σ^E pathway (Dartigalongue et al., 2001). The σ^E pathway responds to cell envelope stress and relays the signal via a regulated cascade of intermembrane proteolysis (for further details see Chapter 6). The cytoplasmic RNase E also regulates *dsbC* mRNA transcript levels. RNase E is an essential endoribonuclease that degrades mRNAs and assists the maturation of a variety of rRNAs and tRNAs in *E. coli*. Mutations in *rne*, which encodes RNase E, result in an increased half-life of *dsbC* mRNA, leading to increases in disulfide bond isomerase activity in *E. coli* (Zhan et al., 2004).

DsbG

DsbG is a second disulfide bond isomerase (24% amino acid sequence identity to DsbC) present in the periplasm of *E. coli*. DsbG was discovered in 1997 by screening for mutants that conferred DTT sensitivity or for genes that conferred DTT resistance when overexpressed (Andersen et al., 1997). DsbG is a soluble homodimeric protein with 231 amino acids (27 kDa) in its mature form and is expressed at approximately 25% the levels of DsbC. The crystal structure of DsbG dimer reveals a similar structure to DsbC with some differences (Heras et al., 2004). The α-helix linker of DsbG is 2.5 turns longer, making the hydrophobic pocket of DsbG significantly larger. Furthermore, unlike the uncharged pocket of DsbC there are seven negatively charged amino acids within the pocket of DsbG and three in the dimerization domain. These subtle biochemical and structural differences in the peptide-binding site could provide a different substrate specificity and activity of DsbG. In vitro, DsbG cannot reduce insulin but does promote refolding of substrate proteins, presumably by the same mechanism as DsbC (Shao et al., 2000). Multicopy expression of DsbG complements a *dsbC⁻* mutant and promotes isomerization and proper assembly of bovine pancreatic trypsin inhibitor in vivo. Its in vivo function depends on DsbD, which reduces the oxidized form as it does with DsbC (Bessette et al., 1999). However, DsbG's role in *E. coli* is not known as no substrate for it has been identified.

The genes for DsbG and AhpC (a cytoplasmic alkyl hydroperoxide reductase) are located adjacent to each other on the *E. coli* chromosome. However, they are transcribed in opposite directions from promoter regions located between the *dsbG* gene and *ahpCF* operon. Both are regulated by the oxidative stress regulator, OxyR. OxyR is a cytoplasmic Lys-R-type transcriptional regulator which is activated upon having its cysteines oxidized to form a disulfide bond. Footprinting analyses have confirmed the binding of oxidized OxyR 54 bp upstream of the *dsbG* gene (Zheng et al., 2001). However, in vitro analysis showed that it was the *ahpCF*-proximal binding site 238 bp upstream of *dsbG* that was responsible for the up-regulation of both *dsbG* and *ahpCF* transcripts. The evidence that only oxidized OxyR was capable of binding upstream of *dsbG*, along with the fact that *dsbG* transcripts were detected with primer extension analysis only after the induction of oxidative stress, indicates that DsbG is induced under conditions of oxidative stress.

DsbD

To be active as isomerases, DsbC (Rietsch et al., 1997) and DsbG (Andersen et al., 1997; Bessette et al., 1999) require their redox active cysteines to be maintained in their reduced state. Similarly, the membrane-anchored protein CcmG, which contains a thioredoxin fold involved in cytochrome *c* maturation, also requires its active site cysteines to be maintained in a reduced state (Beck et al., 1994; Crooke and Cole, 1995). The reduction of the oxidized forms of these proteins is performed by the protein DsbD.

DsbD is a 546-amino-acid (59-kDa) cytoplasmic membrane protein whose primary function is to shuttle electrons from the cytoplasmic thioredoxins to its substrates in the periplasm. This intriguing electron relay mechanism occurs via three distinguishable domains of DsbD: α, β and γ (Color Plate 12). The amino-terminal α-domain assumes an immunoglobulin-like fold and is targeted to the periplasm via the 19-amino-acid signal peptide of DsbD (Goulding et al., 2002). Topological studies of the central β-domain predict eight transmembrane domains (Chung et al., 2000). The carboxyl-terminal γ-domain assumes a thioredoxin fold (Rozhkova et al., 2004). Each domain has a pair of cysteines essential for the electron transfer activity of DsbD (Katzen and Beckwith, 2000; Krupp et al., 2001). The electron relay pathway is:

$$TrxA \rightarrow DsbD\ (\beta \rightarrow \gamma \rightarrow \alpha \rightarrow substrates)$$

Although not all of the intermediate disulfide-bonded complexes of the electron relay pathway of DsbD have been captured, based on the cumulative in vivo and in vitro data, a cascade of disulfide bond reduction has emerged. The first step in this electron relay occurs between the central β-membrane-embedded domain of DsbD and the cytoplasmic reductase TrxA. This initial step is catalyzed between cysteine$_{163}$ of the β-domain of DsbD and cysteine$_{32}$ of TrxA. The disulfide-bonded complex between the two proteins utilizing these two cysteines has been demonstrated (Katzen and Beckwith, 2000). The subsequent transfer of electrons to the γ-domain occurs via cysteine$_{464}$, resulting in the oxidation of TrxA and the reduction of the γ-domain. Electrons are then transferred to the α-domain, presumably via cysteine$_{103}$. Once the α-domain is reduced, it can interact with its oxidized substrates (DsbC, DsbG, and CcmG) via cysteine$_{109}$ (Haebel et al., 2002).

Crystal structures of DsbD α- (Goulding et al., 2002) and γ- (Kim et al., 2003) domains have been solved. The DsbD α-domain in complex with its substrates has revealed gross structural flexibility in DsbD. The DsbD α-domain has been crystallized in a disulfide-bonded complex with CcmG (Stirnimann et al., 2005), DsbC (Haebel et al., 2002), and DsbD γ-domain (Rozhkova et al., 2004). These studies have revealed structural and kinetic properties that hinder interactions between the oxidative pathway (DsbA-DsbB) and the reductive pathway (DsbD-DsbC) (Rozhkova et al., 2004). Additional models have proposed a "funnel"-like opening and closing of DsbD, allowing for the interaction of the γ-domain with the redox active membrane embedded cysteines in the β-domain (Porat et al., 2004). The mechanism whereby DsbD translocates electrons from the cytoplasm to the periplasm via the pair of cysteines in its membrane-embedded domain is unclear. This unusual electron transfer process remains one of the fascinating problems in the study of pathways leading to protein disulfide bond formation.

THE EXPANDING DSB FAMILY AND FUTURE DIRECTIONS AND UNANSWERED QUESTIONS

Since the discovery of the Dsb system in *E. coli*, variations on the theme of oxidoreductases are starting to emerge. In *Neisseria meningitidis*, two membrane-embedded DsbA homologues have been recently characterized (Tinsley et al., 2004). In *Staphylococcus aureus* lipidated membrane-anchored DsbA has been characterized (Dumoulin et al., 2005). Another thiol-disulfide oxidoreductase system in the cold-adapted bacterium *Pseudoalteromonas haloplanktis* has also been discovered recently. In *Campylobacter jejuni* proteins with DsbB-like activity and structure with extra C-terminal domains have recently been characterized (Raczko et al., 2005).

While these new findings open the field of disulfide bond formation to new horizons, many fundamental questions remain. The exact timing of disulfide bond formation remains unknown. At what stage of protein translocation are disulfide bonds formed? What is the mechanism of disulfide bond isomerization? How are electrons shuffled across the membrane? Have all the partners of redox regulation been discovered? Could redox active compounds such as glutathione or cysteine be involved in disulfide bond formation in the periplasm?

CONCLUDING REMARKS

The study of disulfide bond formation in proteins has a long history. In vitro studies on this process began in the 1960s with the important finding by Anfinsen and coworkers that proteins with disulfide bonds could be reduced and denatured and yet still reassemble into the proper structures without added enzyme catalysts. Since that time, numerous laboratories have studied the folding of a variety of disulfide-bonded proteins by denaturation and refolding in vitro. Here, we have shown that this folding process must be considerably more complicated in vivo than in vitro studies would have revealed. In particular, several features of disulfide bond formation discovered in recent years illustrate this complexity. First, disulfide bond formation likely occurs on nascent polypeptide chains as they are being translocated into the periplasm in bacteria and into the endoplasmic reticulum in eukaryotic cells. Second, for efficient formation of disulfide bonds, proteins require enzyme catalysts whose proto-

types are DsbA in bacterial cells and protein disulfide isomerase in eukaryotic cells. Third, the formation of the correct disulfide bonds requires enzyme catalysts such as DsbC. (It is still not entirely clear whether PDI also carries out the isomerization step in eukaryotic cells as proposed.)

We have suggested that it is the cotranslocational nature of disulfide bond formation that results in incorrect disulfide bonds being formed in DsbA's substrates. We have presented a picture of DsbA as quite indiscriminate in its interaction with cysteines in substrate proteins, showing little or no specificity in the cysteines it interacts with. However, this may not be entirely true. Elsewhere we have suggested that features of the surrounding amino acids or of secondary structure may also determine which cysteines of substrates are preferred in the attack on the disulfide bond of DsbA. Additional in vivo experiments could help to identify these features (Kadokura et al., 2004). Furthermore, even the distance between two cysteines in the primary sequence may play some role in which disulfide bonds are formed, as the greater the distance, the greater the possibility that formation of structure between two cysteines will influence whether that disulfide bond is formed. In other words, the sequential versus nonsequential correlation with DsbC dependence may not be as strict as we have presented.

Bioinformatic analysis conducted in our lab revealed that at least 90% of the periplasmic proteins with more than one cysteine contain disulfide bonds. This calculation excludes those proteins that use their cysteines to coordinate heme molecules, metal ions, or lipidation. For those periplasmic proteins with only a single free cysteine, their crystal structures showed these cysteines to be buried within the protein structure (AlsB, AraF, FhuD, GlpQ, and Tpx). These findings are consistent with the proposal that any protein that appears in the periplasm with more than one cysteine will be acted on by DsbA. The only exception is a predicted periplasmic putative L-asparaginase (AsgX) that apparently contains four (Protein Data Bank

[PDB] IDs: 1JN9 and 1K2X) or five (PDB ID: 1T3M) cysteines in reduced states. Thus, proteins that have no need for disulfide bonds may generally avoid cysteine codons in their genes because of the likely indiscriminate oxidizing activity of DsbA. In contrast, there are few if any cases of proteins with disulfide bonds found in the cytoplasm, either in prokaryotes or eukaryotes. Why are periplasmic proteins predominantly disulfide bonded? The most evident explanation is that (i) the periplasm is a more destructive environment than the cytoplasm, being much more subject to external toxic agents and perhaps richer in proteases than the cytoplasm, and (ii) disulfide bonds add stability to proteins, making them much more resistant to these insults.

ACKNOWLEDGMENTS

We thank Hiroshi Kadokura for insightful comments, Markus Eser and Seung-Hyun Cho for figures, and Melanie Berkmen for helpful editing.

REFERENCES

Akiyama, Y., S. Kamitani, N. Kusukawa, and K. Ito. 1992. In vitro catalysis of oxidative folding of disulfide-bonded proteins by the *Escherichia coli* *dsbA* (*ppfA*) gene product. *J. Biol. Chem.* **267:** 22440–22445.

Alksne, L. E., D. Keeney, and B. A. Rasmussen. 1995. A mutation in either *dsbA* or *dsbB*, a gene encoding a component of a periplasmic disulfide bond-catalyzing system, is required for high-level expression of the *Bacteroides fragilis* metallo-beta-lactamase, CcrA, in *Escherichia coli*. *J. Bacteriol.* **177:** 462–464.

Andersen, C. L., A. Matthey-Dupraz, D. Missiakas, and S. Raina. 1997. A new *Escherichia coli* gene, *dsbG*, encodes a periplasmic protein involved in disulphide bond formation, required for recycling DsbA/DsbB and DsbC redox proteins. *Mol. Microbiol.* **26:** 121–132.

Anfinsen, C. B., and E. Haber. 1961. Studies on the reduction and re-formation of protein disulfide bonds. *J. Biol. Chem.* **236:** 1361–1363.

Anfinsen, C. B., E. Haber, M. Sela, and F. H. White, Jr. 1961. The kinetics of formation of native ribonuclease during oxidation of the reduced polypeptide chain. *Proc. Natl. Acad. Sci. USA* **47:** 1309–1314.

Bader, M., W. Muse, D. P. Ballou, C. Gassner, and J. C. Bardwell. 1999. Oxidative protein folding is driven by the electron transport system. *Cell* **98:** 217–227.

Bader, M. W., A. Hiniker, J. Regeimbal, D. Goldstone, P. W. Haebel, J. Riemer, P. Metcalf, and J. C. Bardwell. 2001. Turning a disulfide isomerase into an oxidase: DsbC mutants that imitate DsbA. *EMBO J.* **20:** 1555–1562.

Bader, M. W., T. Xie, C. A. Yu, and J. C. Bardwell. 2000. Disulfide bonds are generated by quinone reduction. *J. Biol. Chem.* **275:** 26082–26088.

Bardwell, J. C., K. McGovern, and J. Beckwith. 1991. Identification of a protein required for disulfide bond formation *in vivo*. *Cell* **67:** 581–589.

Beck, R., H. Crooke, M. Jarsch, J. Cole, and H. Burtscher. 1994. Mutation in *dipZ* leads to reduced production of active human placental alkaline phosphatase in *Escherichia coli*. *FEMS. Microbiol. Lett.* **124:** 209–214.

Belin, P., and P. L. Boquet. 1994. The *Escherichia coli dsbA* gene is partly transcribed from the promoter of a weakly expressed upstream gene. *Microbiology* **140** (Pt 12): 3337–3348.

Bergman, L. W., and W. M. Kuehl. 1979. Co-translational modification of nascent immunoglobulin heavy and light chains. *J. Supramol. Struct.* **11:** 9–24.

Berkmen, M., D. Boyd, and J. Beckwith. 2005. The nonconsecutive disulfide bond of *Escherichia coli* phytase (AppA) renders it dependent on the protein-disulfide isomerase, DsbC. *J. Biol. Chem.* **280:** 11387–11394.

Bessette, P. H., J. J. Cotto, H. F. Gilbert, and G. Georgiou. 1999. *In vivo* and *in vitro* function of the *Escherichia coli* periplasmic cysteine oxidoreductase DsbG. *J. Biol. Chem.* **274:** 7784–7792.

Bouwman, C. W., M. Kohli, A. Killoran, G. A. Touchie, R. J. Kadner, and N. L. Martin. 2003. Characterization of SrgA, a *Salmonella enterica* serovar Typhimurium virulence plasmid-encoded paralogue of the disulfide oxidoreductase DsbA, essential for biogenesis of plasmid-encoded fimbriae. *J. Bacteriol.* **185:** 991–1000.

Charbonnier, J. B., P. Belin, M. Moutiez, E. A. Stura, and E. Quemeneur. 1999. On the role of the cis-proline residue in the active site of DsbA. *Protein Sci.* **8:** 96–105.

Chen, J., J. L. Song, S. Zhang, Y. Wang, D. F. Cui, and C. C. Wang. 1999. Chaperone activity of DsbC. *J. Biol. Chem.* **274:** 19601–19605.

Chung, J., T. Chen, and D. Missiakas. 2000. Transfer of electrons across the cytoplasmic membrane by DsbD, a membrane protein involved in thiol-disulphide exchange and protein folding in the bacterial periplasm. *Mol. Microbiol.* **35:** 1099–1109.

Collet, J. F., J. Riemer, M. W. Bader, and J. C. Bardwell. 2002. Reconstitution of a disulfide isomerization system. *J. Biol. Chem.* **277:** 26886–26892.

Crooke, H., and J. Cole. 1995. The biogenesis of c-type cytochromes in *Escherichia coli* requires a membrane-bound protein, DipZ, with a protein disulphide isomerase-like domain. *Mol. Microbiol.* **15**:1139–1150.

Danese, P. N., and T. J. Silhavy. 1997. The sigma(E) and the Cpx signal transduction systems control the synthesis of periplasmic protein-folding enzymes in *Escherichia coli. Genes Dev.* **11**:1183–1193.

Darby, N. J., S. Raina, and T. E. Creighton. 1998. Contributions of substrate binding to the catalytic activity of DsbC. *Biochemistry* **37**:783–791.

Dartigalongue, C., D. Missiakas, and S. Raina. 2001. Characterization of the *Escherichia coli* sigma E regulon. *J. Biol. Chem.* **276**:20866–20875.

De Lorenzo, F., R. F. Goldberger, E. Steers, Jr., D. Givol, and B. Anfinsen. 1966. Purification and properties of an enzyme from beef liver which catalyzes sulfhydryl-disulfide interchange in proteins. *J. Biol. Chem.* **241**:1562–1567.

Dumoulin, A., U. Grauschopf, M. Bischoff, L. Thony-Meyer, and B. Berger-Bachi. 2005. *Staphylococcus aureus* DsbA is a membrane-bound lipoprotein with thiol-disulfide oxidoreductase activity. *Arch. Microbiol.* **184**:117–128.

Frech, C., M. Wunderlich, R. Glocksuber, and F. X. Schmid. 1996. Preferential binding of an unfolded protein to DsbA. *EMBO J.* **15**:392–398.

Gane, P. J., R. B. Freedman, and J. Warwicker. 1995. A molecular model for the redox potential difference between thioredoxin and DsbA, based on electrostatics calculations. *J. Mol. Biol.* **249**:376–387.

Genevaux, P., P. Bauda, M. S. DuBow, and B. Oudega. 1999. Identification of Tn10 insertions in the *dsbA* gene affecting *Escherichia coli* biofilm formation. *FEMS Microbiol. Lett.* **173**:403–409.

Goldberger, R. F., C. J. Epstein, and C. B. Anfinsen. 1963. Acceleration of reactivation of reduced bovine pancreatic ribonuclease by a microsomal system from rat liver. *J. Biol. Chem.* **238**:628–635.

Gonzalez, M. D., C. A. Lichtensteiger, and E. R. Vimr. 2001. Adaptation of signature-tagged mutagenesis to *Escherichia coli* K1 and the infant-rat model of invasive disease. *FEMS Microbiol. Lett.* **198**:125–128.

Goulding, C. W., M. R. Sawaya, A. Parseghian, V. Lim, D. Eisenberg, and D. Missiakas. 2002. Thiol-disulfide exchange in an immunoglobulin-like fold: structure of the N–terminal domain of DsbD. *Biochemistry* **41**:6920–6927.

Grauschopf, U., A. Fritz, and R. Glocksuber. 2003. Mechanism of the electron transfer catalyst DsbB from *Escherichia coli. EMBO J.* **22**:3503–3513.

Grauschopf, U., J. R. Winther, P. Korber, T. Zander, P. Dallinger, and J. C. Bardwell. 1995. Why is DsbA such an oxidizing disulfide catalyst? *Cell* **83**:947–955.

Guddat, L. W., J. C. Bardwell, and J. L. Martin. 1998. Crystal structures of reduced and oxidized DsbA: investigation of domain motion and thiolate stabilization. *Structure* **6**:757–767.

Guddat, L. W., J. C. Bardwell, T. Zander, and J. L. Martin. 1997. The uncharged surface features surrounding the active site of *Escherichia coli* DsbA are conserved and are implicated in peptide binding. *Protein Sci.* **6**:1148–1156.

Guilhot, C., G. Jander, N. L. Martin, and J. Beckwith. 1995. Evidence that the pathway of disulfide bond formation in *Escherichia coli* involves interactions between the cysteines of DsbB and DsbA. *Proc. Natl. Acad. Sci. USA* **92**:9895–9899.

Haebel, P. W., D. Goldstone, F. Katzen, J. Beckwith, and P. Metcalf. 2002. The disulfide bond isomerase DsbC is activated by an immunoglobulin-fold thiol oxidoreductase: crystal structure of the DsbC-Dsb-Dalpha complex. *EMBO J.* **21**:4774–4784.

Hayashi, S., M. Abe, M. Kimoto, S. Furukawa, and T. Nakazawa. 2000. The *dsbA-dsbB* disulfide bond formation system of *Burkholderia cepacia* is involved in the production of protease and alkaline phosphatase, motility, metal resistance, and multidrug resistance. *Microbiol. Immunol.* **44**:41–50.

Heras, B., M. A. Edeling, H. J. Schirra, S. Raina, and J. L. Martin. 2004. Crystal structures of the DsbG disulfide isomerase reveal an unstable disulfide. *Proc. Natl. Acad. Sci. USA* **101**:8876–8881.

Hiniker, A., and J. C. Bardwell. 2004. *In vivo* substrate specificity of periplasmic disulfide oxidoreductases. *J. Biol. Chem.* **279**:12967–12973.

Huber, D., D. Boyd, Y. Xia, M. H. Olma, M. Gerstein, and J. Beckwith. 2005. Use of thioredoxin as a reporter to identify a subset of *Escherichia coli* signal sequences that promote signal recognition particle-dependent translocation. *J. Bacteriol.* **187**:2983–2991.

Inaba, K., Y. H. Takahashi, N. Fujieda, K. Kano, H. Miyoshi, and K. Ito. 2004. DsbB elicits a redshift of bound ubiquinone during the catalysis of DsbA oxidation. *J. Biol. Chem.* **279**:6761–6768.

Inaba, K., Y. H. Takahashi, and K. Ito. 2005. Reactivities of quinone-free DsbB from *Escherichia coli. J. Biol. Chem.* **280**:33035–33044.

Jacob-Dubuisson, F., J. Pinkner, Z. Xu, R. Striker, A. Padmanhaban, and S. J. Hultgren. 1994. PapD chaperone function in pilus biogenesis depends on oxidant and chaperone-like activities of DsbA. *Proc. Natl. Acad. Sci. USA* **91**:11552–11556.

Jander, G., N. L. Martin, and J. Beckwith. 1994. Two cysteines in each periplasmic domain of the membrane protein DsbB are required for its function in protein disulfide bond formation. *EMBO. J.* **13**:5121–5127.

Joly, J. C., and J. R. Swartz. 1997. *In vitro* and *in vivo* redox states of the *Escherichia coli* periplasmic oxidoreductases DsbA and DsbC. *Biochemistry* **36:**10067–10072.

Kadokura, H., M. Bader, H. Tian, J. C. Bardwell, and J. Beckwith. 2000. Roles of a conserved arginine residue of DsbB in linking protein disulfide-bond-formation pathway to the respiratory chain of *Escherichia coli. Proc. Natl. Acad. Sci. USA* **97:**10884–10889.

Kadokura, H., and J. Beckwith. 2002. Four cysteines of the membrane protein DsbB act in concert to oxidize its substrate DsbA. *EMBO J.* **21:**2354–2363.

Kadokura, H., L. Nichols II, and J. Beckwith. 2005. Mutational alterations of the key cis proline residue that cause accumulation of enzymatic reaction intermediates of DsbA, a member of the thioredoxin superfamily. *J. Bacteriol.* **187:**1519–1522.

Kadokura, H., H. Tian, T. Zander, J. C. Bardwell, and J. Beckwith. 2004. Snapshots of DsbA in action: detection of proteins in the process of oxidative folding. *Science* **303:**534–537.

Katzen, F., and J. Beckwith. 2000. Transmembrane electron transfer by the membrane protein DsbD occurs via a disulfide bond cascade. *Cell* **103:**769–779.

Kim, J. H., S. J. Kim, D. G. Jeong, J. H. Son, and S. E. Ryu. 2003. Crystal structure of DsbDgamma reveals the mechanism of redox potential shift and substrate specificity(1). *FEBS Lett.* **543:**164–169.

Kishigami, S., Y. Akiyama, and K. Ito. 1995. Redox states of DsbA in the periplasm of *Escherichia coli. FEBS Lett.* **364:**55–58.

Kobayashi, T., S. Kishigami, M. Sone, H. Inokuchi, T. Mogi, and K. Ito. 1997. Respiratory chain is required to maintain oxidized states of the DsbA–DsbB disulfide bond formation system in aerobically growing *Escherichia coli* cells. *Proc. Natl. Acad. Sci. USA* **94:**11857–11862.

Kortemme, T., and T. E. Creighton. 1995. Ionisation of cysteine residues at the termini of model alpha-helical peptides. Relevance to unusual thiol pKa values in proteins of the thioredoxin family. *J. Mol. Biol.* **253:**799–812.

Krupp, R., C. Chan, and D. Missiakas. 2001. DsbD-catalyzed transport of electrons across the membrane of *Escherichia coli. J. Biol. Chem.* **276:**3696–3701.

Leichert, L. I., and U. Jakob. 2004. Protein thiol modifications visualized in vivo. *PLoS. Biol.* **2:**e333.

Liu, X., and C. C. Wang. 2001. Disulfide-dependent folding and export of *Escherichia coli* DsbC. *J. Biol. Chem.* **276:**1146–1151.

Martin, J. L. 1995. Thioredoxin—a fold for all reasons. *Structure* **3:**245–250.

Martin, J. L., J. C. Bardwell, and J. Kuriyan. 1993. Crystal structure of the DsbA protein required for disulphide bond formation *in vivo. Nature* **365:**464–468.

Martin, J. L., G. Waksman, J. C. Bardwell, J. Beckwith, and J. Kuriyan. 1993. Crystallization of DsbA, an *Escherichia coli* protein required for disulphide bond formation *in vivo. J. Mol. Biol.* **230:**1097–1100.

McCarthy, A. A., P. W. Haebel, A. Torronen, V. Rybin, E. N. Baker, and P. Metcalf. 2000. Crystal structure of the protein disulfide bond isomerase, DsbC, from *Escherichia coli. Nat. Struct. Biol.* **7:**196–199.

Mirsky, A. E., and L. Pauling. 1936. On the structure of native, denatured and coagulated proteins. *Proc. Natl. Acad. Sci. USA* **22:**439–447.

Missiakas, D., C. Georgopoulos, and S. Raina. 1993. Identification and characterization of the *Escherichia coli* gene *dsbB*, whose product is involved in the formation of disulfide bonds in vivo. *Proc. Natl. Acad. Sci. USA* **90:**7084–7088.

Missiakas, D., C. Georgopoulos, and S. Raina. 1994. The *Escherichia coli dsbC* (*xprA*) gene encodes a periplasmic protein involved in disulfide bond formation. *EMBO J.* **13:**2013–2020.

Mössner, E., M. Huber-Wunderlich, A. Rietsch, J. Beckwith, R. Glockshuber, and F. Aslund. 1999. Importance of redox potential for the *in vivo* function of the cytoplasmic disulfide reductant thioredoxin from *Escherichia coli. J. Biol. Chem.* **274:**25254–25259.

Moutiez, M., T. V. Burova, T. Haertle, and E. Quemeneur. 1999. On the non-respect of the thermodynamic cycle by DsbA variants. *Protein Sci.* **8:**106–112.

Nelson, J. W., and T. E. Creighton. 1994. Reactivity and ionization of the active site cysteine residues of DsbA, a protein required for disulfide bond formation *in vivo. Biochemistry* **33:**5974–5983.

Ondo-Mbele, E., C. Vives, A. Kone, and L. Serre. 2005. Intriguing conformation changes associated with the trans/cis isomerization of a prolyl residue in the active site of the DsbA C33A mutant. *J. Mol. Biol.* **347:**555–563.

Peek, J. A., and R. K. Taylor. 1992. Characterization of a periplasmic thiol:disulfide interchange protein required for the functional maturation of secreted virulence factors of *Vibrio cholerae. Proc. Natl. Acad. Sci. USA* **89:**6210–6214.

Pogliano, J., A. S. Lynch, D. Belin, E. C. Lin, and J. Beckwith. 1997. Regulation of *Escherichia coli* cell envelope proteins involved in protein folding and degradation by the Cpx two-component system. *Genes Dev.* **11:**1169–1182.

Porat, A., S. H. Cho, and J. Beckwith. 2004. The unusual transmembrane electron transporter DsbD and its homologues: a bacterial family of disulfide reductases. *Res. Microbiol.* **155:**617–622.

Prinz, W. A., F. Aslund, A. Holmgren, and J. Beckwith. 1997. The role of the thioredoxin and glutaredoxin pathways in reducing protein disulfide bonds in the *Escherichia coli* cytoplasm. *J. Biol. Chem.* **272:**15661–15667.

Raczko, A. M., J. M. Bujnicki, M. Pawlowski, R. Godlewska, M. Lewandowska, and E. K. Jagusztyn-Krynicka. 2005. Characterization of new DsbB-like thiol-oxidoreductases of *Campylobacter jejuni* and *Helicobacter pylori* and classification of the DsbB family based on phylogenomic, structural and functional criteria. *Microbiology* **151:** 219–231.

Regeimbal, J., S. Gleiter, B. L. Trumpower, C. A. Yu, M. Diwakar, D. P. Ballou, and J. C. Bardwell. 2003. Disulfide bond formation involves a quinhydrone-type charge-transfer complex. *Proc. Natl. Acad. Sci. USA* **100:**13779–13784.

Rietsch, A., P. Bessette, G. Georgiou, and J. Beckwith. 1997. Reduction of the periplasmic disulfide bond isomerase, DsbC, occurs by passage of electrons from cytoplasmic thioredoxin. *J. Bacteriol.* **179:**6602–6608.

Rozhkova, A., C. U. Stirnimann, P. Frei, U. Grauschopf, R. Brunisholz, M. G. Grutter, G. Capitani, and R. Glockshuber. 2004. Structural basis and kinetics of inter- and intramolecular disulfide exchange in the redox catalyst DsbD. *EMBO. J.* **23:**1709–1719.

Schierle, C. F., M. Berkmen, D. Huber, C. Kumamoto, D. Boyd, and J. Beckwith. 2003. The DsbA signal sequence directs efficient, cotranslational export of passenger proteins to the *Escherichia coli* periplasm via the signal recognition particle pathway. *J. Bacteriol.* **185:**5706–5713.

Schirra, H. J., C. Renner, M. Czisch, M. Huber-Wunderlich, T. A. Holak, and R. Glockshuber. 1998. Structure of reduced DsbA from *Escherichia coli* in solution. *Biochemistry* **37:**6263–6276.

Segatori, L., P. J. Paukstelis, H. F. Gilbert, and G. Georgiou. 2004. Engineered DsbC chimeras catalyze both protein oxidation and disulfide-bond isomerization in *Escherichia coli*: reconciling two competing pathways. *Proc. Natl. Acad. Sci. USA* **101:**10018–10023.

Sevier, C. S., H. Kadokura, V. C. Tam, J. Beckwith, D. Fass, and C. A. Kaiser. 2005. The prokaryotic enzyme DsbB may share key structural features with eukaryotic disulfide bond forming oxidoreductases. *Protein Sci.* **14:**1630–1642.

Shao, F., M. W. Bader, U. Jakob, and J. C. Bardwell. 2000. DsbG, a protein disulfide isomerase with chaperone activity. *J. Biol. Chem.* **275:**13349–13352.

Shevchik, V. E., G. Condemine, and J. Robert-Baudouy. 1994. Characterization of DsbC, a periplasmic protein of *Erwinia chrysanthemi* and *Escherichia coli* with disulfide isomerase activity. *EMBO J.* **13:**2007–2012.

Stafford, S. J., D. P. Humphreys, and P. A. Lund. 1999. Mutations in *dsbA* and *dsbB*, but not *dsbC*, lead to an enhanced sensitivity of *Escherichia coli* to Hg^{2+} and Cd^{2+}. *FEMS Microbiol. Lett.* **174:**179–184.

Stenson, T. H., and A. A. Weiss. 2002. DsbA and DsbC are required for secretion of pertussis toxin by *Bordetella pertussis*. *Infect. Immun.* **70:**2297–2303.

Stewart, E. J., F. Aslund, and J. Beckwith. 1998. Disulfide bond formation in the *Escherichia coli* cytoplasm: an in vivo role reversal for the thioredoxins. *EMBO J.* **17:**5543–5550.

Stirnimann, C. U., A. Rozhkova, U. Grauschopf, M. G. Grutter, R. Glockshuber, and G. Capitani. 2005. Structural basis and kinetics of DsbD-dependent cytochrome c maturation. *Structure (Camb.)* **13:**985–993.

Sun, X. X., and C. C. Wang. 2000. The N-terminal sequence (residues 1-65) is essential for dimerization, activities, and peptide binding of *Escherichia coli* DsbC. *J. Biol. Chem.* **275:**22743–22749.

Takahashi, Y. H., K. Inaba, and K. Ito. 2004. Characterization of the menaquinone-dependent disulfide bond formation pathway of *Escherichia coli*. *J. Biol. Chem.* **279:**47057–47065.

Tinsley, C. R., R. Voulhoux, J. L. Beretti, J. Tommassen, and X. Nassif. 2004. Three homologues, including two membrane-bound proteins, of the disulfide oxidoreductase DsbA in *Neisseria meningitidis*: effects on bacterial growth and biogenesis of functional type IV pili. *J. Biol. Chem.* **279:**27078–27087.

Tomb, J. F. 1992. A periplasmic protein disulfide oxidoreductase is required for transformation of *Haemophilus influenzae* Rd. *Proc. Natl. Acad. Sci. USA* **89:**10252–10256.

Vinci, F., J. Couprie, P. Pucci, E. Quemeneur, and M. Moutiez. 2002. Description of the topographical changes associated to the different stages of the DsbA catalytic cycle. *Protein Sci.* **11:**1600–1612.

Walker, K. W., and H. F. Gilbert. 1997. Scanning and escape during protein–disulfide isomerase-assisted protein folding. *J. Biol. Chem.* **272:**8845–8848.

Warwicker, J. 1998. Modeling charge interactions and redox properties in DsbA. *J. Biol. Chem.* **273:**2501–2504.

White, F. H., Jr. 1960. Regeneration of enzymatic activity by airoxidation of reduced ribonuclease with observations on thiolation during reduction with thioglycolate. *J. Biol. Chem.* **235:**383–389.

Wunderlich, M., and R. Glockshuber. 1993. Redox properties of protein disulfide isomerase (DsbA) from *Escherichia coli*. *Protein Sci.* **2:**717–726.

Xie, T., L. Yu, M. W. Bader, J. C. Bardwell, and C. A. Yu. 2002. Identification of the ubiquinone-binding domain in the disulfide catalyst disulfide bond protein B. *J. Biol. Chem.* **277:**1649–1652.

Yohannes, E., D. M. Barnhart, and J. L. Slonczewski. 2004. pH-dependent catabolic protein expression during anaerobic growth of *Escherichia coli* K-12. *J. Bacteriol.* **186:**192–199.

Yu, J., B. Edwards-Jones, O. Neyrolles, and J. S. Kroll. 2000. Key role for DsbA in cell-to-cell spread of *Shigella flexneri*, permitting secretion of Ipa proteins into interepithelial protrusions. *Infect. Immun.* **68:**6449–6456.

Zapun, A., J. C. Bardwell, and T. E. Creighton. 1993. The reactive and destabilizing disulfide bond of DsbA, a protein required for protein disulfide bond formation *in vivo*. *Biochemistry* **32:**5083–5092.

Zapun, A., D. Missiakas, S. Raina, and T. E. Creighton. 1995. Structural and functional characterization of DsbC, a protein involved in disulfide bond formation in *Escherichia coli*. *Biochemistry* **34:** 5075–5089.

Zhan, X., J. Gao, C. Jain, M. J. Cieslewicz, J. R. Swartz, and G. Georgiou. 2004. Genetic analysis of disulfide isomerization in *Escherichia coli*: expression of DsbC is modulated by RNase E-dependent mRNA processing. *J. Bacteriol.* **186:**654–660.

Zhang, M., A. F. Monzingo, L. Segatori, G. Georgiou, and J. D. Robertus. 2004. Structure of DsbC from *Haemophilus influenzae. Acta. Crystallogr. D Biol. Crystallogr.* **60:**1512–1518.

Zhao, Z., Y. Peng, S. F. Hao, Z. H. Zeng, and C. C. Wang. 2003. Dimerization by domain hybridization bestows chaperone and isomerase activities. *J. Biol. Chem.* **278:**43292–43298.

Zheng, M., X. Wang, B. Doan, K. A. Lewis, T. D. Schneider, and G. Storz. 2001. Computation-directed identification of OxyR DNA binding sites in *Escherichia coli. J. Bacteriol.* **183:**4571–4579.

Zheng, W. D., H. Quan, J. L. Song, S. L. Yang, and C. C. Wang. 1997. Does DsbA have chaperone-like activity? *Arch. Biochem. Biophys.* **337:**326–331.

PERIPLASMIC CHAPERONES AND PEPTIDYL-PROLYL ISOMERASES

Jean-Michel Betton

8

Although the native structure of proteins is encoded in their amino acid sequences, the biogenesis of many proteins requires the participation of molecular chaperones to ensure their efficient folding on a biologically relevant timescale (Ellis and Hartl, 1999). The cellular functions of molecular chaperones include protecting newly synthesized polypeptides from misfolding or aggregation, and promoting disaggregation and refolding of stress-denaturated proteins (Bukau, 1999). The levels of these ubiquitous proteins are regulated according to the conformational states of their protein substrates in the respective cellular compartments (Morimoto, 1998; Alba and Gross, 2004).

BIOGENESIS OF EXTRACYTOPLASMIC PROTEINS

In gram-negative bacteria, the extracytoplasmic periplasmic and outer membrane proteins are synthesized in the cytoplasm as precursors with a cleavable N-terminal signal sequence and exported to their final destination in the cell envelope. Although, depending on the nature of these precursors, different pathways exist for their transport across the inner membrane (see Chapters 1 and 2), most are translocated in unstructured states through a channel formed by the heterotrimeric membrane complex SecYEG (van den Berg et al., 2004). After translocation and signal sequence cleavage, the newly exported extracytoplasmic proteins are sorted inside the periplasm via different folding pathways. While periplasmic localization is generally thought to be the default destination for proteins carrying cleavable signal sequences (Danese and Silhavy, 1998), folding intermediates of outer membrane proteins must transit through the periplasm in a conformational state compatible with their insertion into the lipid bilayer (Mogensen and Otzen, 2005). The outer membrane of *Escherichia coli* contains two types of proteins, lipoproteins and β-barrel membrane proteins. Since the biogenesis of lipoproteins is dealt with separately (see Chapter 4), the term *outer membrane protein* will only refer, throughout this chapter, to the later integral membrane proteins (Schulz, 2003).

The periplasm of gram-negative bacteria provides a special environment for protein folding. First, as a naturally oxidative compartment it favors the formation of disulfide bonds, and second, it lacks ATP that is essential for the activity of most cytoplasmic molecular chaperones (Young et al., 2004). Furthermore, the periplasm is separated from the extracellular

Jean-Michel Betton, Unité de Repliement et Modélisation des Protéines, Institut Pasteur–CNRS URA2185, 75724 Paris cedex 15, France.

The Periplasm
Edited by Michael Ehrmann © 2007 ASM Press, Washington, D.C.

milieu only by the porous outer membrane and is thus more susceptible to changes in the external environment than the cytoplasm. Therefore, it was debated whether periplasmic chaperones might exist that could associate with newly exported polypeptides emerging at the translocation sites to assist their folding process (Wülfing and Plückthun, 1994), as cytoplasmic chaperones do with newly synthesized polypeptides emerging from the ribosomes (Ferbitz et al., 2004). While some favor the idea that periplasmic proteins have evolved to resist the formation of aggregates under stress conditions (Liu et al., 2004), the transit of outer membrane proteins through the periplasm and peptidoglycan layer presents a different challenge for these β-barrel membrane proteins (Mogensen and Otzen, 2005). Several periplasmic folding factors that contribute rather differently to extracytoplasmic protein folding have been characterized, and recent studies of envelope stress response have revealed how they are regulated (Duguay and Silhavy, 2004).

PERIPLASMIC FOLDING PROTEINS

Many molecular chaperones, evolutionarily conserved from humans to bacteria, were initially identified because they are induced in response to stress conditions leading to the accumulation of intracellular nonnative proteins. Similarly, genetic studies of E. coli extracytoplasmic stress response induced by increased amounts of nonnative envelope proteins led to the discovery of both the signaling sigma E (σ^E) (see Chapter 6) and Cpx two-component systems (see Chapter 5), and most of the peri-

plasmic folding factors, including molecular chaperones and folding catalysts (Table 1).

Independently of their ability to hydrolyze ATP, the term *molecular chaperone* was originally restricted to cellular factors that assist protein folding by transiently interacting with nonnative proteins. Molecular chaperones do not accelerate folding reactions, but rather increase the number of intermediate species on a productive folding pathway (Ranson et al., 1998). Thus, the only known periplasmic folding factor that fulfills this definition of *holding* chaperones (Stirling et al., 2003) is Skp, a protein involved in the first folding steps of outer membrane proteins. Several periplasmic binding proteins of ABC transporters involved in nutrient uptake (see Chapter 17) have the ability to suppress the aggregation of a variety of nonnative proteins upon renaturation (Richarme and Caldas, 1997; Matsuzaki et al., 1998). However, these results were obtained from in vitro experiments and there are no data available suggesting that these binding proteins are true periplasmic chaperones in vivo.

Although not historically classified as molecular chaperones, enzymes that catalyze two potentially rate-limiting folding reactions, or folding catalysts, are abundant in the periplasm. These enzymes include the protein disulfide isomerase (PDI) or thiol/disulfide bond oxidoreductases (the Dsb proteins in E. coli) (see Chapter 7) and the peptidyl-prolyl isomerases (PPIases) that catalyze, respectively, the formation of disulfide bonds and the *cis-trans* isomerization of prolyl peptide bonds (Fischer and Aumüller, 2003). The classification of periplasmic folding factors into these two groups is not strict, as many of the folding catalysts display

TABLE 1 Periplasmic chaperones and peptidyl-prolyl isomerases

Protein	Chaperone activity	PPIase family	Oligomeric state	Transcriptional control
Skp (OmpH)	+		Trimer	Cpx, σ^E
PpiA (RotA)	−	Cyclophilin	Monomer	Cpx
FkpA	+	FKBP	Dimer	σ^E
SurA	+	Parvulin	Monomer	σ^E
PpiD	−	Parvulin	Monomer	Cpx, σ^{32}

secondary, noncatalytic chaperone activities by binding and stabilizing nonnative proteins. Thus, these periplasmic enzymes might also be considered as molecular chaperones (Miot and Betton, 2004). Similarly, the periplasmic protease DegP, involved in the degradation of misfolded extracytoplasmic proteins, may switch between protease and chaperone activities in a temperature-dependent manner (see Chapter 9).

Aside from folding factors with apparent global specificity for nonnative proteins, there are specific periplasmic chaperones like FimC and PapD that stabilize nascent pilus subunits by structural complementation (Choudhury et al., 1999; Sauer et al., 1999). The structures of pilus subunits contain an immunoglobin fold lacking a β-strand, which, in the assembled pilus structure, is provided by the neighboring subunit. Before pilus assembly, these specific molecular chaperones complement, and thus stabilize, a pilus subunit by donating a β-strand in a similar manner (Sauer et al., 2002). Most interestingly, these specific periplasmic chaperones represent a new type of protein-folding catalyst, which acts by accelerating the folding of pilus subunits (Vetsch et al., 2004).

Skp, A TRUE PERIPLASMIC CHAPERONE

Among the periplasmic folding proteins, Skp (or OmpH) is certainly the best-studied factor. Although this small (17-kDa) protein from *E. coli* was originally copurified with chromosomal DNA (Holk and Kleppe, 1988), Skp is a periplasmic protein synthesized with a cleavable N-terminal signal sequence (Thome and Müller, 1991). Two different global searches for periplasmic chaperones suggested an important role of Skp in the folding pathway of outer membrane proteins. First, Skp was retained on an affinity column with Sepharose-bound OmpF (Chen and Henning, 1996), and second, by genetic screening, mutations in the *skp* gene were isolated based on their increased σ^E activity (Missiakas et al., 1996). Although *skp* null mutants exhibit a significantly reduced production of outer membrane proteins, Skp is a nonessential periplasmic protein (Chen and

Henning, 1996). However, double mutants lacking both Skp and the periplasmic protease DegP are not viable at 37°C, probably due to accumulation of protein aggregates in the periplasm (Schäfer et al., 1999). Skp was further identified from a multicopy library of *E. coli* genes by selecting for increased production of a functional antibody fragment displayed on filamentous phage (Bothmann and Plückthun, 1998). Consistent with its periplasmic chaperone activity, further biochemical experiments showed that Skp selectively binds the newly synthesized outer membrane protein PhoE immediately after its translocation across the inner membrane (Harms et al., 2001).

In vitro, Skp forms stable complexes with refolded OmpA, preventing its aggregation. The addition of lipopolysaccharides (LPS) to this complex facilitates its insertion into lipid vesicles (Bulieris et al., 2003). These results demonstrate that Skp functions in the periplasm as a molecular chaperone for several β-barrel proteins by maintaining soluble conformations before their insertion and folding into the outer membrane bilayer.

The crystal structure of this highly basic protein (Color Plate 13A) reveals three long α-helical hairpins protruding from a central β-barrel structure that defines a trimerization domain (Korndörfer et al., 2004; Walton and Sousa, 2004). This homotrimeric complex has the overall appearance of a jellyfish with flexible tentacles defining an internal cavity which could constitute a substrate binding site. Indeed, it has been suggested that partialy folded or unfolded outer membrane proteins bind inside this central, predominantly hydrophobic cavity by surface-surface interactions. From structural comparisons with the outer membrane protein FhuA in complex with LPS, it was proposed that Skp contains a similar LPS-binding site (Walton and Sousa, 2004). Furthermore, the asymmetric electrostatic outer surface of Skp suggests a specific molecular orientation of the protein relative to the lipid bilayer (Korndörfer et al., 2004). Finally, the overall shape of Skp bears a remarkable resemblance to prefoldin, a molecular chaperone found in the cytoplasm of most eukaryotes and archaea, but not in bacteria (Siegert et al., 2000). Although these two molecular chaperones function in different protein-folding pathways, they probably share a common strategy for binding and stabilizing nonnative protein substrates.

PERIPLASMIC PEPTIDYL-PROLYL ISOMERASES

PPIases are ubiquitous and highly conserved enzymes that catalyze prolyl *cis-trans* isomerization, an intrinsically slow and potentially rate-limiting reaction during in vitro protein refolding (Schmid, 2001). Although PPIases were originally identified as protein targets of cyclosporin A and FK506, two immunosuppresive drugs, their physiological function remains elusive. However, *cis-trans* isomerizations of peptidyl-prolyl bonds in native proteins can be used as conformational switches to regulate cellular processes such as protein phosphorylation (Liou et al., 2003). Recently, *cis-trans* proline isomerization of a filamentous phage protein was found to function as a molecular switch for timing the infection of *E. coli* (Eckert et al., 2005). Unlike the PDI and Dsb proteins that function in naturally oxidative compartments, PPIases are found in all cellular compartments. PPIases are classified into three distinct families, comprising the cyclophilins, the FK506 binding proteins (FKBP), and the parvulins (Fischer and Aumüller, 2003). These families are unrelated to each other in terms of primary and tertiary structures, have distinct substrate specificities toward proline-containing peptides, and are sensitive to different types of inhibitors.

Representatives of the three PPIase families have been identified in the periplasm: PpiA (or RotA) is related to the cyclophilin family, FkpA belongs to the FKBP family, and SurA and PpiD are two parvulin homologs. In vitro, these purified proteins catalyze the *cis/trans* isomerization of peptidyl-prolyl bonds in oligopeptide substrates (Fischer et al., 1984). With the exception of *surA*, the absence of a significant phenotype for null mutants encoding these proteins indicates that the four periplas-

mic PPIases individually are not essential for viability and have collectively redundant physiological functions.

PpiA

The first report of PPIase activity in the periplasm of *E. coli* resulted from the biochemical characterizations of a gene product sharing sequence homology with human cyclophilins (Liu and Walsh, 1990). The corresponding periplasmic protein was designated RotA, and despite sequence and structural (Color Plate 13B) homologies to cyclophilins, its PPIase activity is only marginally inhibited by cyclosporin A. When compared with different structures of the human cyclophilin A, the cyclosporin-binding site of the periplasmic *E. coli* protein differs substantially in four regions (Clubb et al., 1994). These structural differences may determine protein substrate specificity and sensitivity to cyclosporin A. The observation that the Cpx pathway modulates expression of the *ppiA* gene suggests a significant role for this folding catalyst in the envelope of *E. coli* (Pogliano et al., 1997). However, folding and steady-state levels of periplasmic and outer membrane proteins are unaffected in a *ppiA* mutant (Kleerebezem et al., 1995). Although PpiA contributes mainly to the level of periplasmic PPIase activity in the standard oligopeptide assay, its exact function is still unknown.

FkpA

Like PpiA, the *fkpA* gene was originally identified in *E. coli* based on sequence homology with other known FK506-binding proteins (Horne and Young, 1995). However, the first evidence that FkpA plays a significant role as a periplasmic folding factor came from genetic screening based on extracytoplasmic stress response. In *E. coli* strains producing constitutively nonnative envelope proteins, *fkpA* and *surA* (see below), were identified as multicopy suppressors of the σ^E-dependent stress response (Missiakas et al., 1996). The nature of inducing signals in these strains suggested that FkpA participates in envelope protein folding. Moreover, null mutations in the *fkpA* gene induce the σ^E

regulon, but the corresponding mutants have no observable phenotype. Additional evidence for the function of FkpA in periplasmic folding came from genetic selections to improve a functional antibody fragment displayed on filamentous phages (Bothmann and Plückthun, 2000). This cellular effect of FkpA, consistent with chaperone activity, was further confirmed by Arié et al. (2001), who showed that overproduction of FkpA suppresses the formation of periplasmic inclusion bodies from MalE31, a defective folding variant of the maltose-binding protein (Saul et al., 2003). These studies found that the overproduction of other periplasmic PPIases, PpiA and SurA, did not increase the yield of native periplasmic proteins, either for functional antibody fragments or MalE31 substrates. However, overproduction of predicted active site FkpA variants displaying no PPIase activity prevented the aggregation of MalE31 (Arié et al., 2001). Therefore, it appeared that the catalytic PPIase activity of FkpA is not essential for its chaperone activity. The exact role of its noncatalytic activity in vivo remains to be established, although FkpA has been shown to improve the overproduction of correctly folded recombinant proteins either upon coexpression (Zhang et al., 2003) or as a fusion protein partner (Scholz et al., 2005).

In vitro, FkpA displays strong PPIase activity, as assessed either by the ribonuclease T1 refolding assay (Ramm and Plückthun, 2000; 2001) or by the standard oligopeptide assay. Furthermore, its enzyme activity is inhibited by FK506 and by rapamycin, confirming that FkpA is indeed a full member of the FKBP family. As mentioned previously, the chaperone activity of FkpA is independent of its PPIase activity (Ramm and Plückthun, 2000; Arié et al., 2001). However, since the interactions between FkpA and protein-folding intermediates are transient, no stable complexes between FkpA and its substrates could be detected either in vivo or in vitro. Finally, FkpA was proposed to act on early unfolding intermediates in the thermal aggregation of citrate synthase, a widely used chaperone substrate (Ramm and Plückthun, 2001).

The crystal structure of FkpA (Color Plate 13C) shows that this homodimeric protein is divided into two domains connected by a long α-helix (Saul et al., 2004). The N-terminal domain includes three α-helices that are interlaced with those of the other subunit to provide all intersubunit contacts maintaining the dimeric structure. The C-terminal domain, which shares the topology of the FKBPs, binds the FK506 ligand. The overall structure of the FkpA dimer is V-shaped with two C-terminal FKBP domains at the extremities. A deletion variant comprising the N-domain only exists in solution as a mixture of monomeric and dimeric species and displays chaperone activity (Saul et al., 2004). By contrast, a deletion variant comprising the C-domain only is monomeric, and although it shows full PPIase activity, it is devoid of chaperone function (Arié et al., 2001). These observations suggest that the chaperone and catalytic activities of FkpA reside, respectively, in the N- and C-terminal domains. Finally, the FkpA structure reveals the presence of a hydrophobic pocket at the dimeric interface, which could constitute a binding site for nonnative protein substrates.

SurA

As mentioned previously, global genetic searches for mutations activating the σ^E stress response identified the *surA* gene, which was previously isolated among genes required for survival of *E. coli* in the stationary phase (Tormo et al., 1990). The physiological defects of *surA* mutants are indicative of outer membrane perturbations, but are more severe than those of *skp* mutants (Lazar and Kolter, 1996; Missiakas et al., 1996; Rouviere and Gross, 1996). These defects include hypersensitivity to bile salts, detergents and hydrophobic antibiotics, and outer membrane protein compositions with reduced levels of porins. In *surA* mutants, unfolded monomeric species accumulate in the folding pathway of the maltoporin LamB (Rouviere and Gross, 1996). Like *skp* mutants, a combination of *surA* and *degP* null mutations exhibits a synthetic lethal phenotype, suggesting redundant periplasmic chaperone activity for these proteins (Rizzitello et al., 2001).

The primary structure of SurA is made up of four separate regions: an N-terminal region of approximately 150 residues, two PPIase domains belonging to the parvulin family, and a short C-terminal extension. Although SurA exhibits low PPIase activity in vitro, PPIase domains are not required for its function in vivo, whereas both the N- and C-terminal regions are essential for chaperone activity, its principal periplasmic function (Behrens et al., 2001). The crystal structure of SurA (Color Plate 13D) reveals two distinct structural domains of different sizes (Bitto and McKay, 2002). The larger domain forms a core module including the N-terminal region, the first PPIase domain and the short C-terminal extension. The smaller domain is composed of the second PPIase domain, tethered about 30 Å distant from the core structure. This core structural domain has a deep elongated crevice that could accommodate unfolded polypeptides, suggesting a binding site for unfolded outer membrane proteins. Consistent with this observation, SurA is known to bind several peptides with amino acid motifs frequently found in the sequences of outer membrane proteins (Bitto and McKay, 2003, 2004; Hennecke et al., 2005).

PpiD

Finally, PpiD, the second periplasmic parvulin homolog, was identified as a multicopy suppressor of a null mutation in surA (Dartigalongue and Raina, 1998). Like surA mutants, ppiD null mutants are viable but exhibit sensitivity to hydrophobic antibiotics and detergents, and reduced production of outer membrane proteins. However, null mutations of ppiD confer synthetic lethality in surA mutants. Although these phenotypes suggest that PpiD, like the other parvulin SurA, is involved in outer membrane protein folding, no direct evidence of this role has been obtained. The primary structure and topology of PpiD indicate that it is anchored in the inner membrane by one transmembrane α-helix and its parvulin PPIase domain extends into the periplasm. Its periplasmic localization suggests that PpiD could act on newly exported outer membrane proteins as they emerge from the SecYEG translocon.

CONCLUSIONS

In addition to protein disulfide isomerases and proteases, the periplasm contains at least five distinct protein-folding factors, including the nonessential molecular chaperone Skp, and four PPIases (PpiA, FkpA, SurA, and PpiD) that are individually, but not collectively, dispensable for viability in E. coli. Although Skp, SurA, and probably PpiD assist the folding of various outer membrane proteins, their precise contribution to overall periplasmic protein folding remains to be established. However, these periplasmic folding factors have been shown to be important both in physiological protein folding and under stress conditions. Many periplasmic folding catalysts play an auxiliary role as molecular chaperones, and it is clear that this function is not exclusively related to their enzymatic activity. In addition to the periplasmic PPIases FkpA and SurA, two periplasmic oxydoreductases, DsbC and DsbG, also exhibit biochemical chaperone activity (Chen et al., 1999; Shao et al., 2000). Thus, it is tempting to speculate that, because of their unique active sites, periplasmic folding catalysts can also perform chaperone activity to compensate for the apparent deficiency of molecular chaperone machineries in the periplasm. Indeed, it appears from their crystal structures that these periplasmic catalysts have evolved specialized surfaces and cavities to bind nonnative proteins. For example, the cleft formed in the DsbC dimer by association of the N-domains was proposed to interact with unfolded proteins (McCarthy et al., 2000). In FkpA, the cleft formed by association of the N-domains has approximately the same dimensions as that of DsbC. Therefore, an attractive idea is that V-shaped periplasmic folding catalysts cradle the unfolded polypeptide substrates while giving access to the catalytic sites (Zhao et al., 2003). In contrast to the hydrophobic nature of the DsbC cleft, however, that of FkpA has an intermediate negative potential, probably reflecting a difference in substrate specificity.

Based on genetic experiments, it has been proposed that several folding factors function in two distinct parallel folding pathways in the periplasm (Rizzitello et al., 2001). These data also suggest that periplasmic folding factors have preferred substrates, although biochemical experiments will be necessary to confirm these hypotheses. Therefore, to what extent are the periplasmic substrates dependent on a specific folding pathway, and is such dependence linked to structural properties? Further studies combining genetic and biochemical approaches will be necessary to address how the periplasmic folding factors achieve a balance between specificity and redundancy.

REFERENCES

Alba, B. M., and C. A. Gross. 2004. Regulation of the *Escherichia coli* σ^E-dependent envelope stress response. *Mol. Microbiol.* **52:**613–619.

Arié, J.-P., N. Sassoon, and J.-M. Betton. 2001. Chaperone function of FkpA, a heat shock prolyl isomerase, in the periplasm of *Escherichia coli*. *Mol. Microbiol.* **39:**199–210.

Behrens, S., R. Maier, H. de Cock, F. X. Schmid, and C. A. Gross. 2001. The SurA periplasmic PPIase lacking its parvulin domains functions *in vivo* and has chaperone activity. *EMBO J.* **20:** 285–294.

Bitto, E., and D. B. McKay. 2002. Crystallographic structure of SurA, a molecular chaperone that facilitates folding of outer membrane porins. *Structure* **10:**1489–1498.

Bitto, E., and D. B. McKay. 2003. The periplasmic molecular chaperone protein SurA binds a peptide motif that is characteristic of integral outer membrane proteins. *J. Biol. Chem.* **278:**49316–49322.

Bitto, E., and D. B. McKay. 2004. Binding of phage-display selected peptides to the periplasmic chaperone protein SurA mimics binding of unfolded outer membrane proteins. *FEBS Lett.* **568:** 94–98.

Bothmann, H., and A. Plückthun. 1998. Selection for a periplasmic factor improving phage display and functional periplasmic expression. *Nat. Biotechnol.* **16:**376–380.

Bothmann, H., and A. Plückthun. 2000. The periplasmic *Escherichia coli* peptidyl-prolyl *cis/trans* isomerase FkpA. I. Increased functional expression of antibody fragments with and without *cis*-prolines. *J. Biol. Chem.* **275:**17100–17105.

Bukau, B. 1999. *Molecular Chaperones and Folding Catalysts: Regulation, Cellular Function and Mechanisms.* Harwood Academic Publishers, Amsterdam, The Netherlands.

Bulieris, P. V., S. Behrens, O. Holst, and J. H. Kleinschmidt. 2003. Folding and insertion of the outer membrane protein OmpA is assisted by the chaperone Skp and by lipopolysaccharide. *J. Biol. Chem.* **278:**9092–9099.

Chen, J., J.-L. Song, S. Zhang, Y. Wang, D.-F. Cui, and C. C. Wang. 1999. Chaperone activity of DsbC. *J. Biol. Chem.* **274:**19601–19605.

Chen, R., and U. Henning. 1996. A periplasmic protein (Skp) of *Escherichia coli* selectively binds a class of outer membrane proteins. *Mol. Microbiol.* **19:**1287–1294.

Choudhury, D., A. Thompson, V. Stojanoff, S. Langermann, J. S. Pinkner, S. J. Hultgren, and G. Waksman. 1999. X-ray structure of the FimC-FimH chaperone-adhesin complex from uropathogenic *Escherichia coli*. *Science* **285:**1061–1066.

Clubb, R. T., S. B. Ferguson, C. T. Walsh, and G. Wagner. 1994. Three-dimensional solution structure of *Escherichia coli* periplasmic cyclophilin. *Biochemistry* **33:**2761–2772.

Danese, P. N., and T. J. Silhavy. 1998. Targeting and assembly of periplasmic and outer-membrane proteins in *Escherichia coli*. *Annu. Rev. Genet.* **32:**59–94.

Dartigalongue, C., and S. Raina. 1998. A new heat-shock gene, *ppiD*, encodes a peptidyl-prolyl isomerase required for folding of outer membrane proteins in *Escherichia coli*. *EMBO J.* **17:**3968–3980.

Duguay, A. R., and T. J. Silhavy. 2004. Quality control in the bacterial periplasm. *Biochim. Biophys. Acta* **1694:**121–134.

Eckert, B., A. Martin, J. Balbach, and F. X. Schmid. 2005. Prolyl isomerization as a molecular timer in phage infection. *Nat. Struct. Mol. Biol.* **12:**619–623.

Ellis, R. J., and F.-U. Hartl. 1999. Principles of protein folding in the cellular environment. *Curr. Opin. Struct. Biol.* **9:**102–110.

Ferbitz, L., T. Maier, H. Patzelt, B. Bukau, E. Deuerling, and N. Ban. 2004. Trigger factor in complex with the ribosome forms a molecular cradle for nascent proteins. *Nature* **431:**590–596.

Fischer, G., and T. Aumüller. 2003. Regulation of peptide bond *cis/trans* isomerization by enzyme catalysis and its implication in physiological processes. *Rev. Physiol. Biochem. Pharmacol.* **148:**105–150.

Fischer, G., H. Bang, and C. Mech. 1984. Determination of enzymatic catalysis for the *cis-trans* isomerization of peptide binding in proline-containing peptides. *Biomed. Biochim. Acta* **43:**1101–1111.

Harms, N., G. Koningstein, W. Dontje, M. Muller, B. Oudega, J. Luirink, and H. de Cock. 2001. The early interaction of the outer membrane protein PhoE with the periplasmic chaperone Skp occurs at the cytoplasmic membrane. *J. Biol. Chem.* **276:**18804–18811.

Hennecke, G., J. Nolte, R. Volkmer-Engert, J. Schneider-Mergener, and S. Behrens. 2005. The periplasmic chaperone SurA exploits two features characteristics of integral outer membrane proteins for selective substrate recognition. *J. Biol. Chem.* **280:**23540–23548.

Holk, A., and K. Kleppe. 1988. Cloning and sequencing of the gene for the DNA-binding 17K protein of *Escherichia coli. Gene* **67:**117–124.

Horne, S. M., and K. D. Young. 1995. *Escherichia coli* and other species of the *Enterobacteriaceae* encode a protein similar to the Mip-like FK506-binding protein. *Arch. Microbiol.* **163:**357–365.

Kleerebezem, M., M. Heutink, and J. Tommassen. 1995. Characterization of an *Escherichia coli rotA* mutant, affected in periplasmic peptidyl-prolyl *cis/trans* isomerase. *Mol. Microbiol.* **18:**313–320.

Korndörfer, I. P., M. K. Dommel, and A. Skerra. 2004. Structure of the periplasmic chaperone Skp suggests functional similarity with cytosolic chaperones despite differing architecture. *Nat. Struct. Mol. Biol.* **11:**1015–1020.

Lazar, S. W., and R. Kolter. 1996. SurA assists the folding of *Escherichia coli* outer membrane protein. *J. Bacteriol.* **178:**1770–1773.

Liou, Y.-C., A. Sun, A. Ryo, X. Z. Zhou, Z.-X. Yu, H.-K. Huang, T. Uchida, R. Bronson, G. Bing, X. Li, T. Hunter, and K. P. Lu. 2003. Role of prolyl isomerase Pin1 in protecting against age-dependent neurodegeneration. *Nature* **424:**556–561.

Liu, J., and C. T. Walsh. 1990. Peptidyl-prolyl *cis-trans* isomerase from *Escherichia coli*: a periplasmic homolog of cyclophilin that is not inhibited by cyclosporin A. *Proc. Natl. Acad. Sci. USA* **87:**4028–4032.

Liu, Y., X. Fu, J. Shen, H. Zhang, H. W., and Z. Chang. 2004. Periplasmic proteins of *Escherichia coli* are highly resistant to aggregation: reappraisal for roles of molecular chaperones in periplasm. *Biochem. Biophys. Res. Commun.* **316:**795–801.

Matsuzaki, M., Y. Kiso, I. Yamamoto, and T. Satoh. 1998. Isolation of a periplasmic molecular chaperone-like protein of *Rhodobacter sphaeroides* f. sp. *denitrificans* that is homologous to the dipeptide transport protein DppA of *Escherichia coli. J. Bacteriol.* **180:**2718–2722.

McCarthy, A. A., P. W. Haebel, A. Törrönen, V. Rybin, E. N. Baker, and P. Metcalf. 2000. Crystal structure of the protein disulfide bond isomerase, DsbC, from *Escherichia coli. Nat. Struct. Biol.* **7:**196–199.

Miot, M., and J.-M. Betton. 2004. Protein quality control in the bacterial periplasm. *Microb. Cell Fact.* **3:**4.

Missiakas, D., J.-M. Betton, and S. Raina. 1996. New components of protein folding in extracytoplasmic compartments of *Escherichia coli* SurA, FkpA and Skp/OmpH. *Mol. Microbiol.* **21:**871–884.

Mogensen, J. E., and D. E. Otzen. 2005. Interactions between folding factors and bacterial outer membrane proteins. *Mol. Microbiol.* **57:**326–346.

Morimoto, R. I. 1998. Regulation of the heat shock transcriptional response: cross talk between a family of heat shock factors, molecular chaperones, and negative regulators. *Genes Dev.* **12:**3788–3896.

Pogliano, J., A. S. Lynch, D. Belin, E. C. C. Lin, and J. Beckwith. 1997. Regulation of *Escherichia coli* cell envelope proteins involved in protein folding and degradation by the Cpx two-component system. *Genes Dev.* **11:**1169–1182.

Ramm, K., and A. Plückthun. 2000. The periplasmic *Escherichia coli* peptidylprolyl *cis,trans*-isomerase FkpA. II. Isomerase-independent chaperone activity *in vitro. J. Biol. Chem.* **275:**17106–17113.

Ramm, K., and A. Plückthun. 2001. High enzymatic activity and chaperone function are mechanistically related features of the dimeric *E. coli* peptidyl-prolyl isomerase FkpA. *J. Mol. Biol.* **310:**485–498.

Ranson, N. A., H. E. White, and H. R. Saibil. 1998. Chaperonins. *Biochem. J.* **333:**233–242.

Richarme, G., and T. D. Caldas. 1997. Chaperone properties of the bacterial periplasmic substrate-binding proteins. *J. Biol. Chem.* **272:**15607–15612.

Rizzitello, A. E., J. R. Harper, and T. J. Silhavy. 2001. Genetic evidence for parallel pathways of chaperone activity in the periplasm of *Escherichia coli. J. Bacteriol.* **183:**6794–6800.

Rouviere, P. E., and C. A. Gross. 1996. SurA, a periplasmic protein with peptidyl-prolyl isomerase activity, participates in the assembly of outer membrane porins. *Genes Dev.* **10:**3170–3182.

Sauer, F. G., K. Fütterer, J. S. Pinkner, K. W. Dodson, S. J. Hultgren, and G. Waksman. 1999. Structural basis of chaperone function and pilus biogenesis. *Science* **285:**1058–1061.

Sauer, F. G., J. S. Pinkner, G. Waksman, and S. J. Hultgren. 2002. Chaperone priming of pilus subunits facilitates a topological transition that drives fiber formation. *Cell* **111:**543–551.

Saul, F. A., J.-P. Arié, B. Vulliez-le Normand, R. Kahn, J.-M. Betton, and G. A. Bentley. 2004. Structural and functional studies of FkpA from Escherichia coli, a cis/trans peptidyl-prolyl isomerase with chaperone activity. *J. Mol. Biol.* **335:**595–608.

Saul, F. A., M. Mourez, B. Vulliez-Le Normand, N. Sassoon, G. A. Bentley, and J.-M. Betton. 2003. Crystal structure of a defective folding protein. *Protein Sci.* **12:**577–585.

Schäfer, U., K. Beck, and M. Müller. 1999. Skp, a molecular chaperone of gram-negative bacteria, is required for the formation of soluble periplasmic intermediates of outer membrane proteins. *J. Biol. Chem.* **274:**24567–24574.

Schmid, F. X. 2001. Prolyl isomerases. *Adv. Protein Chem.* **59:**243–282.

Scholz, C., P. Schaarschmidt, A. M. Engel, H. Andres, U. Schmitt, E. Faatz, J. Balbach, and F. X. Schmid. 2005. Functional solubilization of aggregation-prone HIV envelope proteins by covalent fusion with chaperone modules. *J. Mol. Biol.* **345:**1229–1241.

Schulz, G. E. 2003. Transmembrane β-barrel proteins. *Adv. Protein Chem.* **63:**47–70.

Shao, F., M. W. Bader, U. Jakob, and J. C. A. Bardwell. 2000. DsbG, a protein disulfide isomerase with chaperone activity. *J. Biol. Chem.* **275:**13349–13352.

Siegert, R., M. R. Leroux, C. Scheufler, F.-U. Hartl, and I. Moarefi. 2000. Structure of the molecular chaperone prefoldin: unique interaction of multiple coiled coil tentacles with unfolded proteins. *Cell* **103:**621–632.

Stirling, P. C., V. F. Lundin, and M. R. Leroux. 2003. Getting a grip on non-native proteins. *EMBO Rep.* **4:**565–570.

Thome, B. M., and M. Müller. 1991. Skp is a periplasmic *Escherichia coli* protein requiring SecA and SecY for export. *Mol. Microbiol.* **5:**2815–2821.

Tormo, A., M. Almiron, and R. Kolter. 1990. *surA*, an *Escherichia coli* gene essential for survival in stationary phase. *J. Bacteriol.* **172:**4339–4347.

van den Berg, B., W. M. Clemons, Jr., I. Collinson, Y. Modis, E. Hartmann, S. C. Harrison, and T. A. Rapoport. 2004. X-ray structure of a protein-conducting channel. *Nature* **427:**36–44.

Vetsch, M., C. Puorger, T. Spirig, U. Grauschopf, E. U. Weber-Ban, and R. Glockshuber. 2004. Pilus chaperones represent a new type of protein-folding catalyst. *Nature* **431:**329–332.

Walton, T. A., and M. C. Sousa. 2004. Crystal structure of Skp, a prefoldin-like chaperone that protects soluble and membrane proteins from aggregation. *Mol. Cell* **15:**367–374.

Wülfing, C., and A. Plückthun. 1994. Protein folding in the periplasm of *Escherichia coli. Mol. Microbiol.* **12:**685–692.

Young, J. C., V. R. Agashe, K. Siegers, and F.-U. Hartl. 2004. Pathways of chaperone-mediated protein folding in the cytosol. *Nat. Rev. Mol. Cell Biol.* **5:**781–791.

Zhang, Z., L.-P. Song, M. Fang, F. Wang, D. He, R. Zhao, J. Liu, Z.-Y. Zhou, C.-C. Yin, Q. Lin, and H.-L. Huang. 2003. Production of soluble and functional engineered antibodies in *Escherichia coli* improved by FkpA. *Biotechniques* **35:**1032–1042.

Zhao, Z., Y. Peng, S.-F. Hao, Z.-H. Zeng, and C. C. Wang. 2003. Dimerization by domain hybridization bestows chaperone and isomerase activities. *J. Biol. Chem.* **278:**43292–43298.

PERIPLASMIC PROTEASES AND PROTEASE INHIBITORS

Nicolette Kucz, Michael Meltzer, and Michael Ehrmann

9

Proteases and peptidases are catalytic enzymes that hydrolyze peptide bonds. They are classified according to the active site residue or ion involved in catalysis. The molecular mechanisms of the main classes, that is, serine, cysteine, aspartic, and metalloproteases, have been widely studied and are well understood (for review see Barrett et al., 2004). In general, proteases are involved in diverse functions; the most notable include digestive, protective, and regulatory processes. Digestive proteases are involved in protein degradation for nutritional purposes. Protective proteases degrade misfolded and damaged proteins or protein fragments that might aggregate and thus interfere with vital cellular functions. Regulatory proteases play roles in protein maturation and in activation or inactivation of proteins that are part of signal transduction cascades (Ehrmann and Clausen, 2004; Gottesman, 2003). Given the vast number of physiological events that involve proteolytic mechanisms, it is not surprising that each cellular compartment has its own and large set of proteolytic enzymes. The periplasm of *Escherichia coli* contains 20 proteases, 9 of these are serine, 6 are metallo, and 3 are (potential) cysteine proteases (Table 1). While there are no periplasmic aspartate proteases, two proteins belong to a class of unclassified proteases of unknown mechanism.

SERINE PROTEASES

About 35% of all entries in the MEROPS database are classified as serine proteases (Rawlings and Barrett, 1994). Serine peptidases are named after the nucleophilic Ser residue in the active site. They are found in bacteria, eukaryotes, and viruses and act as exopeptidases, endopeptidases, oligopeptidases, and omega peptidases (Rawlings and Barrett, 1994). Approximately 50 families of serine peptidases can be distinguished based on their amino acid sequences and most of these families were grouped into about nine clans (Rawlings and Barrett, 2004b).

In addition to the nucleophilic Ser residue, the catalytic mechanism of serine peptidases generally includes a proton donor. In clans PA(S), SB, SC, SH, SK, and SN, this proton donor is a histidine residue. Furthermore, a third residue, usually an aspartate (except in clan SH), is required probably for the proper orientation of the imidazolium ring of the histidine (Hedstrom, 2002; Rawlings and Barrett, 2004b). In clans SE, SF, and SM, a lysine residue

Nicolette Kucz, Michael Meltzer, and Michael Ehrmann, Zentrum für Medizinische Biotechnologie, Universität Duisburg-Essen, 45117 Essen, Germany.

The Periplasm
Edited by Michael Ehrmann © 2007 ASM Press, Washington, D.C.

TABLE 1 List of periplasmic proteases and a protease inhibitor

Serine Proteases

PBP-5, D-alanyl-D-alanine carboxypeptidase fraction A (*dacA*)

Peptidase family S11
Function . . . . Cell wall formation
Structure[a] 1NJ4
MW[b] . . . . 41,337 Da
                                                  403 aa (1–29 signal peptide)

PBP-4, D-alanyl-D-alanine carboxypeptidase (*dacB*)

Peptidase family S13
Function . . . . Cell wall formation
Structure . . . . Unknown
MW[b] . . . . 49,569 Da
                                                  477 aa (1–20 signal peptide)

PBP-6, D-alanyl-D-alanine carboxypeptidase fraction C (*dacC*)

Peptidase family S11
Function . . . . Cell wall formation
Structure . . . . Unknown
MW[b] . . . . 40,775 Da
                                                  400 aa (1–27 signal peptide)

PBP-6B, D-alanyl-D-alanine carboxypeptidase (*dacD*)

Peptidase family S11
Function . . . . Cell wall formation
Structure . . . . Unknown
MW[b] . . . . 41,022 Da
                                                  388 aa (1–21 signal peptide)

Protease Do (*degP/htrA*)

Peptidase family S2C
Function . . . . Heat shock protein, serine protease and chaperone activities, required at high
                                                  temperature, virulence factor, involved in refolding and degradation of
                                                  damaged proteins
Structure[a] 1KY9
                                                  Hexamer, 2 PDZ domains/monomer
MW[b] . . . . 46,829 Da
                                                  474 aa (1–26 signal peptide)

Protease DegQ (*degQ*)

Peptidase family S2C
Function . . . . Unknown
Structure . . . . Unknown, 2 PDZ domains/monomer
MW[b] . . . . 44,446 Da
                                                  455 aa (1–27 signal peptide)

PBP-7, D-alanyl-D-alanine-endopeptidase (*pbpG*)

Peptidase family S11
Function . . . . Cell wall formation
Structure . . . . Unknown
MW . . . . 31,200 Da
                                                  313 aa (1–28 signal peptide)

(Continued)

TABLE 1 *Continued*

Tail-specific protease Prc/Tsp (prc/*tsp*)
Peptidase family S41
Function . . . . Degrades proteins containing an 11-residue C-terminal ssrA degradation tag
Structure . . . . Unknown, 1 PDZ domain
MW[b] . . . . 74,323 Da
              682 aa (1–22 signal peptide)

Putative protease YdgD (*ydgD*)
Peptidase family S2B
Function . . . . Unknown
Structure . . . . Unknown
MW[b] . . . . 27,124 Da
              273 aa (1–21 signal peptide)

Metalloproteases
Alkaline phosphatase isozyme conversion protein IAP (*iap*)
Peptidase family M28C
Function . . . . Processing of alkaline phosphatase
Structure . . . . Unknown
MW[b] . . . . 35,343 Da
              345 aa (1–24 signal peptide)

Penicillin-insensitive D-alanyl-D-alanine-endopeptidase (*mepA*)
Peptidase family M74
Function . . . . Murein endopeptidase
Structure[a] 1U10
MW[b] . . . . 28,295 Da
              274 aa (1–19 signal peptide)

Pitrilysin/Protease III (*ptrA*)
Peptidase family M16
Function . . . . Endopeptidase
Structure[a] 1Q2L
MW[b] . . . . 105,113 Da
              962 aa (1–23 signal peptide)

Hypothetical protein YebA (*yebA*)
Peptidase family M37
Function . . . . Unknown
Structure . . . . Unknown
MW[b] . . . . 45,384 Da
              440 aa (1–34 signal peptide)

Hypothetical protein YfgC (*yfgC*)
Peptidase family M48
Function . . . . Unknown
Structure . . . . Unknown
MW[b] . . . . 51,018 Da
              487 aa (1–27 signal peptide)

Protein YhjJ (*yhjJ*)
Peptidase family M16
Function . . . . Unknown
Structure . . . . Unknown
MW[b] . . . . 53,050 Da
              498 aa (1–24 signal peptide)

(Continued)

TABLE 1 *Continued*

Cysteine Proteases
Putative Cys protease NlpC (*nlpC*)
Peptidase family C40
Function . . . . Unknown
Structure . . . . Unknown, lipoprotein
MW[b] . . . . 15,653 Da
                      154 aa (1–15 signal peptide)

Putative Cys protease YafL (*yafL*)
Peptidase family C40
Function . . . . Unknown
Structure . . . . Unknown
MW[b] . . . . 27,016 Da
                      249 aa (1–17 signal peptide)

Putative Cys protease YdhO (*ydhO*)
Peptidase family C40
Function . . . . Unknown
Structure . . . . Unknown
MW[b] . . . . 26,965 Da
                      271 aa (1–27 signal peptide)

Aspartic Proteases
None

Unknown Classification
Putative protease YdcP (*ydcP*)
Peptidase family U32
Function . . . . Unknown
Structure . . . . Unknown
MW[b] . . . . 70,582 Da
                      653 aa (1–20 signal peptide)

Putative protease YhbU (*yhbU*)
Peptidase family U32
Function . . . . Unknown
Structure . . . . Unknown
MW[b] . . . . 35,524 Da
                      331 aa (1–15 signal peptide)

Protease Inhibitors
Ecotin (*eco*)
Inhibitor family I11
Function . . . . Inhibits serine proteases including chymotrypsin, trypsin, elastases, factor X,
                      kallikrein, and a variety of other proteases
Structure[a] . . . . 1ECY contains one S—S bond
MW[b] . . . . 16,100 Da
                      162 aa (1–20 signal peptide)

[a]The PDB (Protein Data Bank) identification number.
[b]MW is the molecular mass of the mature form. The number of amino acid (aa) residues of the precursor form is given. The number of residues of the signal peptide is listed in parentheses.

takes over the role of the proton donor and a third residue is not required. Besides, some peptidases in clans SF and SM have a Ser/His catalytic dyad (Dodson and Wlodawer, 1998; Rawlings and Barrett, 2004b).

The catalytic mechanisms of serine peptidases differ between the clans and have been covered in several reviews (e.g., Hedstrom, 2002; Perona and Craik, 1995). Apart from structural differences, the mechanism of catalysis for the three clans PA(S), SB, and SC is the same. In the more recently discovered clans SE, SF, SH, and PB the catalytic residues and the mechanisms are different (Polgár, 2004). Most of the knowledge about the mechanistic features of serine peptidases is based on investigations on chymotrypsin (clan PA(S)).

The catalytic triad of chymotrypsin is part of an extensive hydrogen-bonding network and comprehends residues His57, Asp102, and Ser195 (for review see Hedstrom, 2002). Hydrogen bonds exist between the Nδ1-H of His57 and Oδ1 of Asp102 and between the OH of Ser195 and the Nϵ2-H of His57. Furthermore, the OH of Ser214 forms a hydrogen bond with Oδ1 of Asp102 and hydrogen bonds are also observed between the Oδ2 of Asp102 and the main-chain NHs of Ala56 and His57 that are believed to orient Asp102 and His57. An oxyanion hole is formed by the backbone NHs of Gly193 and Ser195. These atoms form a pocket of positive charge that activates the carbonyl of the peptide bond and stabilizes the negatively charged oxyanion of a tetrahedral intermediate. The generally accepted mechanism of action of serine proteases, especially for chymotrypsin-like proteases, has been described in several reviews (Hedstrom, 2002; Kraut, 1977; Steitz and Shulman, 1982). Ser195 attacks the carbonyl of the peptide substrate, assisted by His57 acting as a general base, generating a tetrahedral intermediate. While the resulting His57-H$^+$ is stabilized by the hydrogen bond to Asp102, the oxyanion of the tetrahedral intermediate is stabilized by interaction with the main-chain NHs of the oxyanion hole. Subsequently, the tetrahedral intermediate degrades with the re-

lease of leaving group, assisted by His57-H$^+$ acting as a general acid, to form the acyl-enzyme intermediate. In the deacylation reaction, water attacks the acyl-enzyme intermediate yielding a second tetrahedral intermediate. Ultimately, this intermediate breaks down and extrudes Ser195 and carboxylic acid product.

The periplasm of E. coli hosts 9 serine proteases. These enzymes (DegQ, DegP, Tsp, YdgD, PBP4, PBP5, PBP6, PBP6B, and PBP7) belong either to clan PA(S) or to clans SE and SK.

Clan PA(S)

Serine peptidases of clan PA(S) occur in prokaryotes, eukaryotes, and RNA viruses. All enzymes of this clan are endopeptidases. Among these, chymotrypsin and trypsin are the most widely studied of all serine proteases (Rawlings and Barrett, 2004b). PA(S) proteases adopt a two-domain structure, each domain containing a six-stranded β-barrel. The active site cleft is located at the interface between the two domains and the order of the catalytic triad is His, Asp, Ser (Clausen et al., 2002; Rawlings and Barrett, 2004b). The active site is arranged by six loops located at the C-terminal side of both domains. While loops LA, LB, and LC are located at the N-terminal β-barrel, those of the C-terminal barrel are termed L1, L2, and L3 (Perona and Craik, 1995). Many of the PA(S) proteases are expressed as an inactive propeptide with an N-terminal inhibitory peptide and proteolytic maturation is required for activation (Clausen et al., 2002). This process is also known as zymogen activation. The clan PA(S) includes several families such as S1, S3, S6, and S7.

FAMILY S1

Most proteins of family S1 are soluble, secreted enzymes, while a few are membrane bound. The soluble enzymes are mostly synthesized as so-called pre-proforms. An N-terminal signal peptide targets them to the secretory pathway. After translocation and initial processing by leader peptidases, the proforms contain an N-terminal extension acting as an inhibitor. Thus, a second proteolytic maturation is required to generate the active enzyme (Rawlings and Barrett, 1994, 2004b). The S1 family is divided into five subfamilies, named S1A to S1E. While subfamily S1A is mainly found in animals, the other subfamilies include basically bacterial endopeptidases (Rawlings and Barrett, 2004b).

DegP

E. coli DegP, also termed protease Do, was first identified and characterized by the laboratory of Fred Goldberg (Swamy et al., 1983). The periplasmic serine peptidase belongs to clan PA(S), family S1C, and like all PA(S) proteases DegP contains a catalytic triad (His-105, Asp-135, and Ser-210). It is synthesized as a 51-kDa precursor protein with an N-terminal signal peptide (26 amino acids long) that targets DegP to the periplasm. After cleavage of this signal peptide mature DegP has a molecular mass of 48 kDa.

The gene *degP*, also termed *htrA* (for high temperature requirement), was first identified by two different groups, who observed two distinct phenotypes of null mutants. While Lipinska and coworkers found that null mutants lack the ability to grow at elevated temperatures (Lipinska et al., 1988, 1989), Strauch and Beckwith observed that null mutants are unable to degrade misfolded and hybrid proteins (Strauch and Beckwith, 1988; Strauch et al., 1989). It has also been observed that *degP* is a heat shock gene (Lipinska et al., 1988), and subsequent studies revealed that the activity of the *degP* promoter is regulated by two signal transduction pathways, sigmaE and CpxAR, that are involved in the unfolded protein response of the cell envelope (see Chapters 5 and 6).

DegP can have two functions. It acts as a chaperone at low temperatures and as a protease at elevated temperatures (Spiess et al., 1999). In the chaperone conformation, DegP binds to hydrophobic surfaces of unfolded substrates assisting in refolding of various chemically denatured proteins such as MalS or citrate synthase. With increasing temperature protease activity is switched on and misfolded substrates are degraded. Only in the proteolytically inactive DegPS210A mutant is chaper-

one activity present at low and elevated temperatures (Spiess et al., 1999). The observed switch in activity is possible because DegP, and probably all members of the HtrA family, are not synthesized as pre-proforms. In contrast to classical serine proteases such as trypsin, these proteins do not depend on processing for activation of the enzymatic function. Rather, the mechanism of reversible zymogen activation allows these proteins to reversibly switch from the proteolytically inactive to the active conformation and vice versa. As a protease with housecleaning function DegP has a broad spectrum of substrates, including misfolded or denatured MBP, MalS, and PhoA (Baneyx and Georgiou, 1991; Betton et al., 1998; Snyder and Silhavy, 1995; Spiess et al., 1999; Strauch and Beckwith, 1988). In fact, all known substrates are at least partially unfolded.

The crystal structure of DegP sheds light on the architecture of this novel type of cage-forming protease-chaperone machine (Krojer et al., 2002). DegP consists of three functionally distinct domains: a conserved N-terminal protease domain resembling trypsin and two C-terminal PDZ domains (Fanning and Anderson, 1996; Krojer et al., 2002; Songyang et al., 1997). In general, PDZ domains mediate protein-protein interactions by binding C-terminal residues of their target proteins (Fanning and Anderson, 1996; Ponting, 1997). DegP forms a hexameric particle consisting of two trimeric rings (Krojer et al., 2002). The trimers exhibit a funnel-like architecture with the rather rigid protease domains on the top and the flexible PDZ domains protruding outward. The PDZ domains are believed to capture substrate to subsequently hand them over to the active sites within the hexameric particle. The ability to differentiate between folded and unfolded substrates and its broad substrate spectrum indicates that DegP represents an important protein quality control factor of the periplasm. So far, DegP has been considered to function as a protective protease that prevents the formation of protein aggregates. However, recent evidence suggests that it also has a regulatory function in the activation of the

CpxAR stress response system. In this system, the activity of the (stress) sensor kinase CpxA is modulated by a periplasmic protein CpxP. When a misfolded protein binds to CpxP, this complex is then degraded by DegP, leading to elevated CpxA activity. Therefore, the proteolytic function of DegP is not only to clear the cell of misfolded protein but also to regulate stress response (Buelow and Raivio, 2005; Isaac et al., 2005).

DegQ

E. coli DegQ (HhoA), which is homologous to DegP, was first identified and characterized by Bass et al. (1996) and Waller and Sauer (1996). The proteins DegP and DegQ have a similar size, consisting of 455 (DegQ) and 474 residues (DegP), and exhibit about 60% sequence identity. DegQ is synthesized with a signal sequence of 27 amino acids. Like DegP it belongs to the peptidase family S1C and contains a classical His Asp Ser catalytic triad. DegQ has two C-terminal PDZ domains but differs from DegP in lacking a Gly/Gln-rich sequence (residues 65 to 90). It was demonstrated that overproduction of DegQ rescues the temperature-sensitive growth defect of a *degP* null strain and that the two proteins DegP and DegQ are undistinguishable in specificity toward several substrates in vitro (Kolmar et al., 1996; Waller and Sauer, 1996). Therefore, the similarities of DegP and DegQ suggest that the two enzymes might have similar functions in vivo.

The *degQ* gene is located directly upstream of *degS*. DegS is a regulatory protease that is also homologous to DegP and Q. It is tethered to the cytoplasmic membrane via one transmembrane segment and is involved in activation of the sigma E pathway (Ehrmann and Clausen, 2004; Walsh et al., 2003; Wilken et al., 2004). The *degQ* and *S* genes are transcribed in the same direction, but each gene appears to be regulated independently. In contrast to *degP*, neither of them is heat inducible (Waller and Sauer, 1996). The *degQ* gene is not essential for normal growth and *degQ* mutants do not show an obvious phenotype under a variety of

growth conditions, including growth at 44°C (Waller and Sauer, 1996).

It was discovered that several DegP/DegQ homologs are implied in the pathogenic virulence of bacteria, for example, the DegP homolog in *Salmonella enterica* serovar Typhimurium (Johnson et al., 1991) as well as the *Brucella abortus* DegP homolog (Tatum et al., 1994). These findings suggest that the DegQ protein potentially plays a protective role when *E. coli* enters a host organism.

Tail-Specific Protease Prc/Tsp

Tail-specific protease (Tsp) is also known as C-terminal processing protease-1 Prc (processing involving C-terminal cleavage) and protease Re. It belongs to the clan SK, family S41. The SK clan also contains the Clp (Hsp100) proteases as well as tricorn and signal sequence peptidase A. Mature Tsp has 660 residues and a molecular mass of 74 kDa and the signal peptide of the precursor consists of 22 residues. The active site consists of a Ser, Lys catalytic dyad. The protein was first purified by Fred Goldberg's laboratory (Park et al., 1988); they recognized that Tsp is an endoprotease. The gene was identified and mapped by Hara et al. (1991). The substrates so far identified include FtsI (penicillin-binding protein 3) (Nagasawa et al., 1989), SecM (Nakatogawa and Ito, 2001), the lipoprotein NlpI (Tadokoro et al., 2004), and recombinant and unfolded proteins (Chen et al., 2004). The laboratory of Bob Sauer established that Tsp prefers substrates with non-polar C termini with little primary sequence specificity and postulated a role for Tsp in the removal of incompletely synthesized protein fragments resulting from truncated mRNAs (Keiler and Sauer, 1996; Keiler, et al., 1995, 1996; Silber et al., 1992).

tsp mutants are temperature sensitive and have a leaky phenotype, which explains the susceptibility of the mutant against multiple antibiotics (Seoane et al., 1992). Bass et al. (1996) isolated multicopy suppressors of *prc* mutants. Among those were the HtrA family members DegP, Q, and S, indicating a functional redundancy and thus perhaps a role for *prc* in the tolerance against protein-folding stress (Bass et al., 1996). Furthermore, its PDZ domain classifies it within the PDZ protease superfamily. Two other periplasmic serine proteases containing PDZ domains, DegP and DegQ, are also part of this family. However, the PDZ domain of Tsp is located N-terminal to the protease domain while the PDZ domains of DegP,Q are located C-terminal to the protease domains.

Putative Protease YdgD

YdgD has 252 residues corresponding to a molecular mass of 27 kDa and its signal sequence consists of 21 residues. It is classified in MEROPS as part of the S1A family whereas UniProt aligns it to the S1B family. There are no data on this protease except that its gene is nonessential.

Clan SE with the families S11, S12, and S13

Families S11, S12, and S13, which contain the catalytic dyad Ser, Lys, belong to clan SE (Granier et al., 1992; Palomeque-Messia et al., 1991). The peptidases of this clan are mostly involved in the biosynthesis, turnover, and lysis of bacterial cell walls. Many of the enzymes are termed penicillin-binding proteins, which usually act on D-alanyl bonds. While family S11 contains D–Ala–D–Ala peptidases and a transpeptidase of *Streptomyces* K15, family S12 includes enzymes with diverse specificities, e.g., the *Streptomyces* R61 D–Ala–D–Ala peptidase and class C β-lactamases. Family S13 includes D–Ala–D–Ala peptidases from *Actinomadura* and *E. coli* (Korat et al., 1991).

Penicillin–Binding Proteins

During the final stages of bacterial cell wall synthesis enzymes called penicillin-binding proteins (PBPs) catalyze the cross-linking of peptide chains from contiguous glycan strands of generated peptidoglycan (reviewed in Georgopapadakou et al., 1983; Ghuysen, 1991; Goffin and Ghuysen, 1998; Waxman and Strominger,

1983). Peptidoglycan, which is a macromolecule of interlinked glycan chains and peptide bridges, represents the rigid component of the eubacterial cell wall that contributes to the osmotic stability and shape of the cells. PBPs bear their name because of the covalent binding of β-lactam (Henderson et al., 1997). *E. coli* maintains 12 classically known PBPs as well as recent additions including PBP6B (Baquero et al., 1996), which can be divided into two groups: the high-molecular-weight (HMW) PBPs and the low-molecular-weight (LMW) PBPs. While the HMW PBPs, including PBPs 1a, 1b, 1c, 2, and 3, are essential for cell viability (Spratt, 1975), the LMW PBPs, consisting of PBPs 4, 5, 6, 7, DacD, AmpC, and AmpH, are nonessential (Baquero et al., 1996; Broome-Smith, 1985; Henderson et al., 1995, 1997; Matsuhashi et al., 1977, 1978; Spratt, 1980). HMW PBPs are bifunctional transglycosylase/transpeptidases or monofunctional transpeptidases. These enzymes synthesize and integrate individual peptidoglycan strands into the murein sacculus (Baquero et al., 1996; Schiffer and Holtje, 1999). LMW PBPs can be split into four subclasses: monofunctional DD–carboxypeptidases, bifunctional DD–carboxypeptidase/DD–endopeptidases, DD–endopeptidases, and β-lactamases (Baquero et al., 1996; Henderson et al., 1995, 1997; Holtje, 1998; Korat et al., 1991; Romeis and Holtje, 1994).

PENICILLIN-BINDING PROTEIN 4

PBP4 from *E. coli* is the product of the *dacB* gene, which is located between *greA* encoding a transcription elongation factor and *yhbZ*, a gene of unknown function. PBP4 has a molecular mass of 52 kDa and belongs to the family S13, clan SE. The peptidases of this family have a greater molecular mass than peptidases of the family S11 including PBP5 and PBP6 (Thunnissen et al., 1995). Family S13 contains D–Ala–D–Ala peptidases from *Actinomadura* and *E. coli*. Although only few crystallographic data for this family are available, conservation of sequences around the catalytic residues indicates comparable tertiary structures and a distant evolutionary relationship to other members of the clan SE.

PBP4 is a D-Ala-D-Ala peptidase that exhibits endo- as well as carboxypeptidase activity (Korat et al., 1991). During peptidoglycan synthesis it hydrolyzes cross-linked side chains previously formed by HMW PBPs. Furthermore, it performs the transpeptidation reaction of adjacent cross-linked peptides. As other D-Ala-D-Ala peptidases, it is synthesized with a signal peptide of 20 amino acids for targeting to the cell membrane. After translocation and removal of the signal sequence D-Ala-D-Ala peptidases are retained at the membrane by a C-terminal membrane anchor. In contrast to PBP5 and PBP6, the C-terminal region of PBP4 only forms weakly amphiphilic α-helices (Roberts et al., 1997; Harris et al., 1998). While the C-terminal regions of PBP5 and PBP6 tend to penetrate the hydrophobic membrane, the C-terminal sequence of PBP4 associates only weakly with the membrane lipid headgroup region by mainly electrostatic interactions (Harris et al., 2002).

Experiments to examine the physiological function of PBP4 indicated that shape abnormalities of a PBP5 mutant were increased by deletion of PBP4 (Meberg et al., 2004). This effect indicates that PBP4 represents an accessory enzyme for moderating cell shape. It has also been determined that neither PBP4 overexpression nor the loss of PBP4 is lethal (Baquero et al., 1996; Denome et al., 1999; Korat et al., 1991; van der Linden et al., 1992).

PENICILLIN-BINDING PROTEIN 5

PBP5 from *E. coli* is one of the best-studied PBPs and was first identified by Spratt (1975, 1977). It is the product of the *dacA* gene, which is located on the chromosome between the gene *ybeD*, whose function is unknown, and *rlpA*, encoding the rare lipoprotein A. PBP5 is a DD-carboxypeptidase that cleaves the terminal D-alanine residue from the pentapeptide side chains of murein components (Baquero et al., 1996; Matsuhashi et al., 1979). It is synthesized with a 29-amino-acid leader peptide, which

targets the protein to the cell membrane. After removal of the leader peptide, PBP5 is retained at the membrane by an amphiphilic C-terminal α-helix anchor (21 residues), resulting in a localization at the outer leaflet of the cytoplasmic membrane (Jackson and Pratt, 1987, 1988; Pratt et al., 1986).

PBP5 has a molecular mass of 44 kDa and belongs to the peptidase family S11, clan SE. The larger part of PBP5 comprises two domains orientated approximately 90° to each other (Davies et al., 2001). Domain I, the DD-carboxypeptidase domain, is highly conserved in the group of LMW PBPs, whereas the C-terminal domain II, which is rich in β-sheets is less conserved and no further homologues in published databases are known. It is assumed that domain II mediates protein-protein interaction between PBP5 and other components of the murein biosynthetic apparatus or that the domain serves as an linker to position the active site near its peptidoglycan substrate in the periplasm (Davies et al., 2001). Like in all PBPs, three conserved sequence motifs can be identified within the N-terminal 275 residues of PBP5. These conserved motifs are the SXXK tetrad, which contains the active site serine residue (amino acid 73), the SXN triad, and the KTG triad (Ghuysen, 1994; Joris et al., 1988).

Although PBP5 is the most abundant of the PBPs, its loss is not lethal to cells (Matsuhashi et al., 1978; Spratt, 1980). Nelson and Young examined several mutants lacking various PBPs (Nelson and Young, 2000). They found that PBP5 mutants were producing cells with the most abnormal morphologies. Mutants lacking all LMW PBPs still formed cells with nearly wild-type morphology as long as PBP5 remained active. Also, no other PBP could substitute PBP5 in vivo (Nelson and Young, 2000). These results confirm that this protein plays a central role in shape determination in *E. coli* (Meberg et al., 2004). The overexpression of PBP5 leads to a spherical growth of *E. coli* and ultimately to cell lysis (Markiewicz et al., 1982; Stoker et al., 1983).

PENICILLIN-BINDING PROTEIN 6

E. coli PBP6 is one of the main penicillin-binding components in the cell. Together with PBP5 it constitutes 85% of the cellular penicillin-binding capacity (Spratt, 1977; Spratt and Stro-minger, 1976). The protein is encoded by the *dacC* gene, which is surrounded by the genes *deoR* and *yliJ* on the chromosome. While *deoR* is a regulatory gene for the *deo* operon, the function of the *yliJ* is unknown. The sequences of *dacC* and *dacA*, the gene encoding for the PBP5, show a homology of 62% (Broome-Smith et al., 1988).

Like PBP5, PBP6 is a D-Ala-D-Ala-carboxy-peptidase and belongs to the family S11, clan SE. It cleaves the terminal D-alanine from the pentapeptide side chains of murein compo-nents during cell wall synthesis (Baquero et al., 1996). The specific activity of PBP6 was mea-sured at three to four times lower than the ac-tivity of PBP5 (Amanuma and Strominger, 1980). PBP6 is synthesized as a precursor pro-tein and translocated across the cytoplasmic membrane by an N-terminal signal sequence (27 amino acids). The mature protein consists of 373 residues corresponding to a molecular mass of 41 kDa (Pratt et al., 1981). PBP6 is as-sociated to the membrane by a C-terminal membrane anchor and thereby exposes the N-terminal part to the periplasmic space. The membrane anchor consists of 22 residues form-ing an amphiphilic helix (Jackson and Pratt, 1987).

Buchanan and Sowell recognized a consid-erable rise in expression level of PBP6 in sta-tionary-phase cells compared with exponen-tially growing cells (Buchanan and Sowell, 1982). These data indicate that PBP6 is in-volved in maintenance of the peptidoglycan during stationary phase. It was also shown that overexpression of PBP6 did not affect growth of *E. coli* (Markiewicz et al., 1982; Spratt, 1975; van der Linden et al., 1992) and that the loss of PBP6 is not lethal (Broome-Smith, 1985). Although PBP5 and PBP6 display similar sub-strate specificity, a certain difference between PBP5 and PBP6 mutants was observed. While the loss of PBP5 leads to an accumulation of muramyl pentapeptides, the loss of PBP6 does not (van der Linden et al., 1992). Together, these data support the view that PBP5 and PBP6 play a nonidentical physiological role in the cell.

PENICILLIN-BINDING PROTEIN 6B

Penicillin-binding protein 6B from *E. coli* (PBP6B) was first identified by Baquero et al. (1996). Its gene *dacD* is located between the genes *sbcB* and *sbmC*. While *sbmC* is a station-ary-phase-induced *E. coli* SOS gene involved in MccB17 susceptibility (Baquero et al., 1995), *sbcB* encodes for exonuclease I. PBP6b has a molecular mass of 43 kDa and was classified as belonging to the family S11 (clan SE) because it shares several molecular features of this group. It was demonstrated that PBP6b binds penicillin and has DD-carboxypeptidase activity. Moreover it exhibits the defined amino acid motifs of PBPs, which are an SXXK tetrad, an SXN triad, and the KTG triad (Ghuysen, 1991). In the case of PBP6B these boxes were identified as SLTK, containing the active site serine at position 63, SGN and KTG. Because of the high similarity of the amino acid se-quence to PBP5 and PBP6, it was suggested that PBP6B is also associated with the inner membrane (Baquero et al., 1996). Subsequent analyses indicated that the C-terminal region of PBP6B is forming weak amphiphilic α-he-lices. Possibly like PBP4, PBP6B binds to the membrane by weak association of its C termi-nus with the membrane lipid headgroups via mainly electrostatic interactions (Harris et al., 2002).

PENICILLIN-BINDING PROTEIN 7

PBP7 from *E. coli* was first identified by Spratt (1975). The corresponding gene *pbpG* is lo-cated between *dld* and *yohC* on the bacterial chromosome. *dld* encodes a lactate dehydroge-nase, while the *yohC* gene product is a con-served multispanning cytoplasmic membrane protein of unknown function. With a molecu-lar mass of 34 kDa, PBP7 is smaller than the other known LMW PBPs. Like PBP5 and PBP6, it belongs to the peptidase family S11 (clan SE) and contains the typical consensus-like sequences SXXK, SXN, and KTG, which were identified as SISK, containing the active site serine at position 70, SEN, and KTG. Furthermore, the amino acid sequence of PBP7 is very similar to the sequences of PBP5 and PBP6 (Henderson et al., 1995). PBP7 specifically hydrolyzes DD-diaminopimelate-alanine bonds in HMW murein sacculi. It is therefore classified as a DD-endopeptidase. Different from other LMW PBPs, PBP7 is a soluble periplasmic protein that does not have a defined lipid anchor (Henderson et al., 1995). Osmotic shock treatment consistently releases PBP7 from whole cells of *E. coli*. Tuomanen and Schwartz (1987) discovered that PBP7 is bound by antibiotics that lyse nongrowing *E. coli* cells, which is atypical for β-lactams.

Analyses concerning the physiological function of PBP7 revealed that shape aberrations from a PBP5 mutant were aggravated by deleting PBP7 (Meberg et al., 2004). This result suggests that PBP7 represents an accessory enzyme moderating cell shape. Because of its unique substrate requirements it may play a specialized role in remodelling of the cell wall. Furthermore, it has been reported that neither PBP7 overexpression nor the loss of PBP7 is lethal (Baquero et al., 1996; Edwards and Donachie, 1993; Korat et al., 1991). Recently, it was elucidated that penicillin-binding protein 8 (PBP8), which also was identified by Spratt (1975), is a proteolytic product of PBP7 (Henderson et al., 1994). PBP7 and PBP8 have a common N terminus and vary only in an OmpT-mediated truncation of the C terminus of PBP7. Like PBP7, PBP 8 is a soluble periplasmic protein which has meso-diaminopimelate-D-alanine DD-endopeptidase activity.

METALLOPROTEASES

The 50 families and 16 clans of metalloproteases recognized to date indicate that they are the most diverse of the four main types of proteases. Usually one divalent cation, in most cases zinc, is directly involved in catalysis, but in some families there are two metal ions that act together. The metal ion is usually held in place by three amino acyl residues, His, Glu, Asp, or Lys, but at least one additional residue is required for catalysis, playing an electrophilic role.

About half of the families of metallopeptidases contain the active site consensus sequence HEXXE, which forms the metal-binding site (Rawlings and Barrett, 1995). This sequence is quite common in proteins, but less than 20% of the motifs found in the SwissProt database are in known metallopeptidases. The motif can be more stringently defined for metalloproteases as abXHEbbHbc, where b is an uncharged residue, c is hydrophobic, and a is most commonly Val or Thr and forms part of the S1′ subsite at least in thermolysin (family M4) and neprilysin (M13) (Vijayaraghavan et al., 1990). X can be nearly any amino acid with the exception of proline, which is never found in this region. A proline residue might not be tolerated because the abXHEbbHbc motif adopts a helical conformation in all metallopeptidases of known structure (Rawlings and Barrett, 1995).

For the peptidases discussed in this chapter there are two relevant mechanisms of catalysis. The first mechanism involves the catalytic His His Glu/Asp sites for zinc binding. It is best studied in carboxypeptidase A (CPD A). Therefore, the nomenclature is standardized according to this peptidase. Structural and kinetic studies allow proposing the following mechanism for CPD A (Auld, 1987; Christianson and Lipscomb, 1989; Matthews, 1988). The first step aims at allowing the metal-bound hydroxide to attack the peptide carbonyl. Here, Glu270 catalyzes the removal of a proton from the metal-bound water. The negative charge in the transition state is stabilized by Arg127 (Phillips et al., 1992) and the substrate is fixed by other active site residues (e.g., Asn144, Arg145, and Tyr248) (Christianson and Lipscomb, 1989). In the second step Glu270 catalyzes the donation of a proton to the leaving amine. The third step follows immediately. The metal-bound tetrahedral intermediate collapses into products. Subsequently, the N-terminal product leaves and water returns to the metal (fourth step). Because of the still existing bond between the C-terminal product to Glu270 the enzyme complex is poised for the reverse reaction, that is, the synthesis of a peptide bond. The last two steps were derived from in situ X-ray absorption fine structure spectroscopy (XAFS) studies (Auld, 1997).

The second mechanism of catalysis involves cocatalytic zinc sites as postulated by Rawlings and Barrett (2004a) for *Aeromonas proteolytica* aminopeptidase (AAP) and *Streptomyces griseus* aminopeptidase (SGAP). In the first step zinc(1) interacts with the scissile peptide carbonyl. The amino acids Glu151/Glu131 (AAP/SGAP numbering) could act as a general base first by removing a proton from the zinc(2)-bound water. In the next step zinc(2)-bound hydroxide could attack the carbonyl of the scissile peptide bond. Thereby the tetrahedral intermediate is formed. In the last step this intermediate could collapse to the product complex, provided the protonated form of Glu131/Glu151 acts as a general acid catalyst by donating a proton to the scissile amide nitrogen.

Nonselective inhibitors affecting the metalprotease complex often achieve inhibition of metallopeptidases. There are three groups of such inhibitors. The first group includes chelating agents like ethylenediamine-N,N,N',N'-tetraacetic acid (EDTA), ethylene glycol bis(β-aminoethyl ether)-N,N,N',N'-tetraacetic acid (EGTA), 8-hydroxyquinoline-5-sulfonic acid (HQSA), pyridine-2,6-dicarboxylic acid (DPA), and 1,10-phenanthroline. The second type of inhibition involves displacement of the native metal by a second metal in the metalloenzyme. The third way is an excess of metal ions, like zinc.

There are at least four members of three clans (ME, MH, MO) and three families (M16, M28, M23) of metalloproteases found in the periplasm of *E. coli*. The following sections will provide a summary of the information available for pitrilysin, MepA, alkaline phosphatase isozyme conversion protein, YebA, YfgC, and YhjJ.

Pitrilysin

E. coli pitrilysin is the product of the *ptr* gene previously known as protease III or protease Pi (Cheng and Zipser, 1979). The gene encoding this enzyme is located on the chromosome between the *recB* and *recC* genes, which encode subunits of endonuclease V (Dykstra et al., 1984). The mature form of pitrilysin consists of 939 residues with a theoretical molecular mass of 105 kDa. The signal peptide of the preform consists of 23 amino acid residues. The MEROPS database aligns pitrilysin to the peptidase subfamily M16A, clan ME. Although the members of the M16 family are distributed through all kingdoms of life, pitrilysin itself has so far only been found in γ-proteobacteria. The most prominent members of M16 family are the mammalian insulysins (insulin-degrading enzymes, IDEs) (Affholter et al., 1988; Baumeister et al., 1993; Kuo et al., 1990), which are capable of cleaving peptides like insulin and amyloid β-protein, the latter being involved in Alzheimer disease (Kurochkin, 2001).

The active site consensus sequence of the members of this family is $HXXEX(X_n)E$, where n is 76 in most cases and X can be any amino acyl residue. In contrast to the active site sequences of the other metallopeptidases (see above), the initial five amino acids of the motif are inverted. It has been shown that two histidine residues and the second glutamate within the consensus sequence are necessary for binding the zinc ion and the first glutamate is required for catalysis (Becker and Roth, 1992; Gehm et al., 1993; Perlman et al., 1993; Perlman and Rosner, 1994).

The crystal structure of pitrilysin revealed a classical α/β-fold and two bowl-like halves participating in forming the central cavity (Maskos, 2004). Each bowl can be further divided into two units of about 240 residues. These four domains exhibit a similar fold even though the sequence conservation is very low. The crystal structure suggests that the catalytic mechanism is similar to thermolysin (Bode et al., 1993; Maskos, 2004).

The physiological role of pitrilysin in the cell is unknown, since no natural substrate has been identified so far. It was suggested, however, that the enzyme might degrade misfolded or aggregated protein and signal peptides (in analogy to insulysins and other members of the M16 family, respectively) (Authier et al., 1995; Betton et al., 1998). This notion might be supported by the fact that pitrilysin degrades protein fragments such as amyloid β-protein (Cornista et al., 2004). In contrast, however, several substrates with tertiary structure are also cleaved, i.e., oxidized insulin B chain and β-galactosidase (for review, see Maskos, 2004). In the absence of natural substrates, two methods for assaying pitrilysin activity have been developed. One method, the Insulin Degradation Assay, employs radioactively labeled insulin as substrate. The second assay is the Quenched Fluorescence Assay, in which degradation of the quenched fluorescence substrate QF27 (designed on the basis of the fragment 16–28 of the vasoactive intestinal peptide) results in the increase of the fluorescence (Anastasi and Barrett, 1995). No solid information on sequence specificity of pitrilysin is available. In addition, no selective inhibitor has been described to date. Unspecific inhibitors include chelating agents like EDTA (Anastasi et al., 1993) and an excess of metal ions (especially zinc) (Larsen and Auld, 1989). Also, zinc-bacitracin, an antibiotic produced by *Bacillus subtilis*, acts as a potent inhibitor (Anastasi et al., 1993; Ding et al., 1992).

MepA

Mature MepA consists of 255 amino acids and has a molecular mass of 28 kDa; its signal peptide consists of 19 residues. The protein is a penicillin-insensitive peptidoglycan amidase (Marcyjaniak et al., 2004) belonging to the M74 family and the LAS (lysostaphin, D-<u>A</u>la-D-Ala carboxypeptidase and <u>s</u>onic hedgehog) group. MepA is found in many gram-negative bacteria and as a murein endopeptidase it cleaves the D-alanyl-meso-2,6-diamino-pimelyl amide bond in *E. coli* peptidoglycan. The biological function of MepA is believed to be the removal of murein and/or the integration of additional strands into the murein sacculus.

The enzyme is sensitive to metal chelators and mutational analyses verified that the predicted Zn^{2+} ligands are His113, Asp120, and His211. The MepA fold consists of a central, six-stranded, mixed β-sheet. The four central strands of this sheet are antiparallel exhibiting the strand order 1, 2, 4, 3 (in LAS nomenclature) and the active site residues are present in this core motif. The protein is a dimer and each monomer has 3 disulfide bonds connecting residues 44–265, 187–235, and 216–223. These disulfide bonds are believed to hold loosely connected C-terminal residues in place. The cysteines involved in disulfide bond formation are nonconsecutive, suggesting a requirement of disulfide isomerase DsbC for proper folding. Consistently, MepA has been shown to be less stable in *dsbC* mutants (Hiniker and Bardwell, 2004).

Alkaline Phosphatase Isozyme Conversion Protein (IAP)

The mature form of IAP consists of 321 amino acids and has a predicted molecular mass of 35 kDa. IAP was initially described as a potential aminopeptidase. Alkaline phosphatase (AP) of *E. coli* exists as a dimer. The proteolytic removal of the N-terminal Arg residue from monomeric AP is responsible for the formation of three isoenzymes (isoenzyme 1 containing full-length monomers, isoenzyme 2 containing mixed monomers, and isoenzyme 3 containing truncated monomers). The posttranslational removal of the Arg from alkaline phosphatase isozyme 1 is assumed to be accomplished by IAP (Ishino et al., 1987).

The MEROPS database assigns IAP to the peptidase subfamily M28C, clan MH. Species containing IAP, as well as other species in which subfamily M28C is represented, are all members of the group *Enterobacteriaceae*. The M28 family and the MH clan are distributed through all kingdoms of life, except for viruses. It is unusual, but not unprecedented, that the family possesses aminopeptidases (like IAP) and carboxypeptidases. All peptidases of the M28 family require, in addition to the first zinc ion, a second one as a cocatalyst. Active site consensus sequence for the members of this family is HDDE(E/D)H. Furthermore, an aspartate and a glutamate are believed to also be required for catalysis (Fundoiano-Hershcovitz et al., 2004). As with all metalloproteases, chelating agents like EDTA can inhibit IAP. Other inhibitors are antipain, leupeptin, bestatin, and amastatin (Nakata et al., 1979). Because no further work has been done since its initial characterization, there is no knowledge about physiological role or mechanism of IAP.

Hypothetical Metalloprotease YebA

The hypothetical metalloprotease YebA consists of 406 amino acid residues with a predicted molecular mass of 45 kDa. The precursor has an unusually long signal peptide, consisting of 34 amino acids. YebA is assigned to the peptidase subfamily M23B, clan MO. This subfamily has so far only been found in bacteria, viruses, and one insect while the family M23 and the clan MO are distributed among bacteria, archaea, animals, and viruses. No experimental data are available for YebA, but in analogy to other members of the M23 family an involvement in peptidoglycan processing could be postulated. Other peptidases of the M23 family cleave either an N-acylmuramoyl-Ala bond between cell wall peptidoglycan and cross-linking peptide or a bond within the cross-linking peptide (Kessler, 2004; Kessler and Ohmann, 2004; Park and Mecham, 2004). Like all members of family M23, YebA is expected to bind water by a single zinc ion ligated to the active site consensus sequence His, Asp, and His, the second His being the catalytic residue (Odintsov et al., 2004). The active site residues occur in the motifs HXXXD and HXH, which is uncommon for metalloproteases.

Hypothetical Metalloprotease YfgC

Mature YfgC consists of 460 residues with a theoretical molecular mass of 51 kDa. Its signal peptide consists of 27 amino acids. YfgC belongs to the peptidase family M48 awaiting further classification. The only data available for this protein are from Carol Gross and coworkers (see Chapter 6) indicating that *yfgC* expression is under sigmaE control, suggesting that it might be part of the unfolded protein response. As a protease, YfgC could perhaps have some function in protection versus stress.

Hypothetical Metalloprotease YhjJ

Mature YhjJ consists of 474 residues and has a theoretical molecular mass of 53 kDa. Its signal peptide consists of 24 amino acids. YhjJ belongs to the peptidase subfamily M16B, clan ME; however, it is classified as a nonpeptidase as it seems to have lost its active site residues. No experimental data are available for this protein.

CYSTEINE PROTEASES

In recent years the information on cysteine proteases multiplied; therefore, the number of families has more than doubled since 1994. Today there are more than 45 families known. These families are divided into seven clans, the most prominent of which is clan CA. Members of this clan are described as "papainlike." Surprisingly, only members of family C1, which belongs to clan CA, are found in all kingdoms of life. Moreover, nearly half of the known families are represented exclusively in viruses. Up to now, it has been shown that the active site consensus sequence of most cysteine peptidases depends on a catalytic dyad of cysteine and histidine. The order of the cysteine and histidine residues in the consensus sequence differs between the families. For some families evidence suggests the requirement of a third residue for proper orientation of the imidazolium ring of histidine.

The catalytic mechanism of cysteine peptidases is similar to that of serine-type peptidases. A nucleophile and a proton donor base are required. As in the majority of serine peptidases, a histidine residue acts as the proton donor. The most widely studied cysteine protease is papain. The single polypeptide chain of papain forms two domains with a large cleft between them (Drenth et al., 1971a, 1971b; Kamphuis et al., 1984). The catalytic dyad of papain consists of Cys25 and His159. The two residues are separated by the cleft. Within the plane of the imidazole ring of His159 the sulfur atom of Cys25 is found. By hydrogen bonding of the imidazole ring to the side chain of Asn175 a Cys-His-Asn triad is generated (Vernet et al., 1995). Furthermore, the hydrogen bond is shielded from solvent by the indole ring of Trp177. Three-dimensional structure analyses show that the side-chain amide of Gln19 and the backbone NH group of Cys25 constitute the oxyanion of the tetrahedral intermediate (Asboth et al., 1985). The periplasm of *E. coli* hosts three potential cysteine proteases. They all belong to the clan CA, family C40.

Putative Cysteine Protease NlpC

Mature NlpC consists of 139 residues with a theoretical molecular mass of 16 kDa. Its signal peptide of 15 amino acid residues is rather short. Anantharaman and Aravind performed a bioinformatic analysis of the C40 family resulting in the description of the "NlpC/P60" superfamily (Anantharaman and Aravind, 2003). The most prominent eukaryotic members of this superfamily are nematode developmental regulator Egl-26, lecithin retinal acyltransferase (LRAT), and a candidate tumor suppressor H-rev107.

Not much is known about the putative cysteine protease NlpC, except that it is a lipoprotein (see Chapter 4). Its catalytic domain contains a conserved amino-terminal cysteine and a carboxyl-terminal histidine residue. The best-studied member of C40 family is dipeptidylpeptidase VI (DPP VI) that hydrolyzes substrates at L-Ala-γ-D-Glu. It is activated by 5 mM EDTA and 1 mM DTT and is inhibited by *N*-ethylmaleimide, iodoacetamide, and *p*-hydroxymercuribenzoate (Guinand, 2004; Vacheron et al., 1979; Valentin et al., 1983). DPP VI is expressed during sporulation and is responsible for the degradation of bacterial cell wall components (Guinand, 2004; Guinand et al., 1974).

Putative Cysteine Protease YafL

The putative cysteine protease YafL has 232 amino acid residues corresponding to a molecular mass of 27 kDa. The precursor form includes a signal peptide consisting of 17 amino acids. Based on the homology of YafL to other cysteine proteases, the MEROPS database aligns the enzyme to the family C40, clan CA. Like NlpC, it is a lipoprotein and classified as a member of the superfamily NlpC/P60 (see above). No experimental data are available for YafL.

Putative Cysteine Protease YdhO

The putative cysteine protease YdhO has 244 amino acid residues and a molecular mass of 27 kDa. The precursor protein has a signal peptide of 27 amino acids. Like the putative cysteine proteases NlpC and YafL, YdhO belongs to the C40 family and to the superfamily NlpC/P60. Again there is no information concerning the structure and the function of this protease.

PROTEASES OF UNKNOWN CLASSIFICATION

Proteases of unknown classification have an unknown catalytic mechanism, indicating that the tertiary structure of the active site catalytic residues has not been determined. Both potential proteases listed below belong to the U32 family. Some of these family members including PrtC from *Porphyromonas gingivalis* have been reported to degrade type I collagen (Kato et al., 1992).

Putative Protease YdcP

YdcP has 633 residues corresponding to a molecular mass of 71 kDa and its signal sequence consists of 20 residues.

Putative Protease YhbU

YhbU has 316 residues corresponding to a molecular mass of 36 kDa and its signal sequence consists of 15 residues.

PROTEASE INHIBITORS

The only known protease inhibitor in the periplasm is ecotin (for review, see McGrath et al., 1995; Rawlings et al., 2004). The mature form has 142 residues and a molecular mass of 16 kDa; the ecotin precursor contains a signal peptide of 20 residues. The dimeric protein is a general inhibitor of serine proteases. It potently inhibits, for example, chymotrypsin, trypsin, elastases, factor X, cathepsin G, and kallikrein among others in picomolar to micromolar concentrations. A crystal structure of ecotin in complex with trypsin revealed that ecotin binds to the active site of trypsin like a substrate (McGrath et al., 1994). While no native periplasmic substrates of ecotin are known, it is generally believed that ecotins protect bacteria from proteolytic attack occurring in host organisms (Eggers et al., 2004).

ACKNOWLEDGMENTS

We gratefully acknowledge funding from the British Biotechnology and Biological Sciences Research Council and the Deutsche Forschungsgemeinschaft.

ADDENDUM IN PROOF

While this chapter was being typeset, two additional potential periplasmic metalloproteases, YggG and YcaL, were found. Both should belong to the M48 family. YggG and YcaL share sequence homology of about 50%. No experimental data are available for these proteins.

REFERENCES

Affholter, J. A., V. A. Fried, and R. A. Roth. 1988. Human insulin-degrading enzyme shares structural and functional homologies with *E. coli* protease III. *Science* **242:**1415–1418.

Amanuma, H., and J. L. Strominger. 1980. Purification and properties of penicillin-binding proteins 5 and 6 from *Escherichia coli* membranes. *J. Biol. Chem.* **255:**11173–11180.

Anantharaman, V., and L. Aravind. 2003. Evolutionary history, structural features and biochemical diversity of the NlpC/P60 superfamily of enzymes. *Genome Biol.* **4:**R11.

Anastasi, A., and A. J. Barrett. 1995. Pitrilysin. *Methods Enzymol.* **248:**684–692.

Anastasi, A., C. G. Knight, and A. J. Barrett. 1993. Characterization of the bacterial metalloendopeptidase pitrilysin by use of a continuous fluorescence assay. *Biochem. J.* **290:**601–607.

Asboth, B., E. Stokum, I. U. Khan, and L. Polgar. 1985. Mechanism of action of cysteine proteinases: oxyanion binding site is not essential in the hydrolysis of specific substrates. *Biochemistry* **24:**606–609.

Auld, D. S. 1987. Acyl group transfer—metalloproteinases, p. 241–258. *In* M. I. Page and A. Williams (ed.), *Enzyme Mechanisms.* Royal Society of Chemistry, London, United Kingdom.

Auld, D. S. 1997. Zinc catalysis in metalloproteases. *Struct. Bonding (Berlin)* **89:**29–50.

Authier, F., J. J. Bergeron, W. J. Ou, R. A. Rachubinski, B. I. Posner, and P. A. Walton. 1995. Degradation of the cleaved leader peptide of thiolase by a peroxisomal proteinase. *Proc. Natl. Acad. Sci. USA* **92:**3859–3863.

Baneyx, F., and G. Georgiou. 1991. Construction and characterization of *Escherichia coli* strains deficient in multiple secreted proteases: protease III degrades high-molecular-weight substrates in vivo. *J. Bacteriol.* **173:**2696–2703.

Baquero, M. R., M. Bouzon, J. C. Quintela, J. A. Ayala, and F. Moreno. 1996. *dacD*, an *Escherichia coli* gene encoding a novel penicillin-binding protein (PBP6b) with DD-carboxypeptidase activity. *J. Bacteriol.* **178:**7106–7111.

Baquero, M. R., M. Bouzon, J. Varea, and F. Moreno. 1995. *sbmC*, a stationary-phase induced SOS *Escherichia coli* gene, whose product protects cells from the DNA replication inhibitor microcin B17. *Mol. Microbiol.* **18:**301–311.

Barrett, A. J., N. D. Rawlings, and J. F. Woessner. 2004. *Handbook of Proteolytic Enzymes.* Elsevier, London, United Kingdom.

Bass, S., Q. Gu, and A. Christen. 1996. Multicopy suppressors of *prc* mutant *Escherichia coli* include two HtrA (DegP) protease homologs (HhoAB), DksA, and a truncated R1pA. *J. Bacteriol.* **178:**1154–1161.

Baumeister, H., D. Muller, M. Rehbein, and D. Richter. 1993. The rat insulin-degrading enzyme. Molecular cloning and characterization of tissue-specific transcripts. *FEBS Lett.* **317:**250–254.

Becker, A. B., and R. A. Roth. 1992. An unusual active site identified in a family of zinc metalloendopeptidases. *Proc. Natl. Acad. Sci. USA* **89:**3835–3839.

Betton, J.-M., N. Sassoon, M. Hofnung, and M. Laurent. 1998. Degradation versus aggregation of misfolded maltose-binding protein in the periplasm of *Escherichia coli. J. Biol. Chem.* **273:**8897–8902.

Bode, W., F. X. Gomis-Rüth, and W. Stöcker. 1993. Astacins, serralysins, snake venom and matrix metalloproteinases exhibit identical zinc-binding environments (HEXXHXXGXXH and Met-turn) and topologies and should be grouped into a common family, the 'metzincins'. *FEBS Lett.* **331:**134–140.

Broome-Smith, J. K. 1985. Construction of a mutant of *Escherichia coli* that has deletions of both the penicillin-binding protein 5 and 6 genes. *J. Gen. Microbiol.* **131:**2115–2118.

Broome-Smith, J. K., I. Ioannidis, A. Edelman, and B. G. Spratt. 1988. Nucleotide sequences of the penicillin-binding protein 5 and 6 genes of *Escherichia coli. Nucleic Acids Res.* **16:**1617.

Buchanan, C. E., and M. O. Sowell. 1982. Synthesis of penicillin-binding protein 6 by stationary-phase *Escherichia coli. J. Bacteriol.* **151**:491–494.

Buelow, D. R., and T. L. Raivio. 2005. Cpx signal transduction is influenced by a conserved N-terminal domain in the novel inhibitor CpxP and the periplasmic protease DegP. *J. Bacteriol.* **187**:6622–6630.

Chen, C., B. Snedecor, J. C. Nishihara, J. C. Joly, N. McFarland, D. C. Andersen, J. E. Battersby, and K. M. Champion. 2004. High-level accumulation of a recombinant antibody fragment in the periplasm of *Escherichia coli* requires a triple-mutant (*degP prc spr*) host strain. *Biotechnol. Bioeng.* **85**:463–474.

Cheng, Y. S., and D. Zipser. 1979. Purification and characterization of protease III from *Escherichia coli. J. Biol. Chem.* **254**:4698–4706.

Christianson, D. W., and W. N. Lipscomb. 1989. Carboxypeptidase A. *Acc. Chem. Res.* **22**:62–69.

Clausen, T., C. Southan, and M. Ehrmann. 2002. The HtrA family of proteases. Implications for protein composition and cell fate. *Mol. Cell* **10**:443–455.

Cornista, J., S. Ikeuchi, M. Haruki, A. Kohara, K. Takano, M. Morikawa, and S. Kanaya. 2004. Cleavage of various peptides with pitrilysin from *Escherichia coli*: kinetic analyses using beta-endorphin and its derivatives. *Biosci. Biotechnol. Biochem.* **68**:2128–2137.

Davies, C., S. W. White, and R. A. Nicholas. 2001. Crystal structure of a deacylation-defective mutant of penicillin-binding protein 5 at 2.3-A resolution. *J. Biol. Chem.* **276**:616–623.

Denome, S. A., P. K. Elf, T. A. Henderson, D. E. Nelson, and K. D. Young. 1999. *Escherichia coli* mutants lacking all possible combinations of eight penicillin binding proteins: viability, characteristics, and implications for peptidoglycan synthesis. *J. Bacteriol.* **181**:3981–3993.

Ding, L., A. B. Becker, A. Suzuki, and R. A. Roth. 1992. Comparison of the enzymatic and biochemical properties of human insulin-degrading enzyme and *Escherichia coli* protease III. *J. Biol. Chem.* **267**:2414–2420.

Dodson, G., and A. Wlodawer. 1998. Catalytic triads and their relatives. *Trends Biochem. Sci.* **23**:347–352.

Drenth, J., J. N. Jansonius, R. Koekoek, and B. G. Wolthers. 1971a. Papain X-ray structure, p. 485–499. *In* P. D. Boyer (ed.), *The Enzymes*, vol. 3. Academic Press, New York, N.Y.

Drenth, J., J. N. Jansonius, R. Koekoek, and B. G. Wolthers. 1971b. The structure of papain. *Adv. Protein Chem.* **25**:79–115.

Dykstra, C. C., D. Prasher, and S. R. Kushner. 1984. Physical and biochemical analysis of the cloned *recB* and *recC* genes of *Escherichia coli* K-12. *J. Bacteriol.* **157**:21–27.

Edwards, D., and W. Donachie. 1993. Construction of a triple deletion of penicillin-binding proteins 4, 5 and 6 in *Escherichia coli*, p. 369–374. *In* M. de Pedro, J.-V. Höltje, and W. Löffelhardt (ed.), *Bacterial Growth and Lysis.* Plenum Press, New York, N.Y.

Eggers, C. T., I. A. Murray, V. A. Delmar, A. G. Day, and C. S. Craik. 2004. The periplasmic serine protease inhibitor ecotin protects bacteria against neutrophil elastase. *Biochem. J.* **379**:107–118.

Ehrmann, M., and T. Clausen. 2004. Proteolysis as a regulatory mechanism. *Annu. Rev. Genet.* **38**:709–724.

Fanning, A. S., and J. M. Anderson. 1996. Protein-protein interactions: PDZ domain networks. *Curr. Biol.* **6**:1385–1388.

Fundoiano-Hershcovitz, Y., L. Rabinovitch, Y. Langut, V. Reiland, G. Shoham, and Y. Shoham. 2004. Identification of the catalytic residues in the double-zinc aminopeptidase from *Streptomyces griseus. FEBS Lett.* **571**:192–196.

Gehm, B. D., W. L. Kuo, R. K. Perlman, and M. R. Rosner. 1993. Mutations in a zinc-binding domain of human insulin-degrading enzyme eliminate catalytic activity but not insulin binding. *J. Biol. Chem.* **268**:7943–7948.

Georgopapadakou, N. H., S. A. Smith, C. M. Cimarusti, and R. B. Sykes. 1983. Binding of monobactams to penicillin-binding proteins of *Escherichia coli* and *Staphylococcus aureus*: relation to antibacterial activity. *Antimicrob. Agents Chemother.* **23**:98–104.

Ghuysen, J. M. 1991. Serine beta-lactamases and penicillin-binding proteins. *Annu. Rev. Microbiol.* **45**:37–67.

Ghuysen, J. M. 1994. Molecular structures of penicillin-binding proteins and beta-lactamases. *Trends Microbiol.* **2**:372–380.

Goffin, C., and J. M. Ghuysen. 1998. Multimodular penicillin-binding proteins: an enigmatic family of orthologs and paralogs. *Microbiol. Mol. Biol. Rev.* **62**:1079–1093.

Gottesman, S. 2003. Proteolysis in bacterial regulatory circuits. *Annu. Rev. Cell. Dev. Biol.* **19**:565–587.

Granier, B., C. Duez, S. Lepage, S. Englebert, J. Dusart, O. Dideberg, J. Van Beeumen, J. M. Frere, and J. M. Ghuysen. 1992. Primary and predicted secondary structures of the Actinomadura R39 extracellular DD-peptidase, a penicillin-binding protein (PBP) related to the Escherichia coli PBP4. *Biochem. J.* **282**(Pt 3):781–788.

Guinand, M. 2004. Dipeptidyl-peptidase VI, p. 1399–1401. *In* A. J. Barrett, N. D. Rawlings, and J. F. Woessner (ed.), *Handbook of Proteolytic Enzymes.* Elsevier, London, United Kingdom.

Guinand, M., G. Michel, and D. J. Tipper. 1974. Appearance of gamma-D-glutamyl-(L) meso-diaminopimealate peptidoglycan hydrolase during sporulation in Bacillus sphaericus. *J. Bacteriol.* **120:** 173–184.

Hara, H., Y. Yamamoto, A. Higashitani, H. Suzuki, and Y. Nishimura. 1991. Cloning, mapping, and characterization of the *Escherichia coli prc* gene, which is involved in C-terminal processing of penicillin-binding protein 3. *J. Bacteriol.* **173:**4799–4813.

Harris, F., K. Brandenburg, U. Seydel, and D. Phoenix. 2002. Investigations into the mechanisms used by the C-terminal anchors of *Escherichia coli* penicillin-binding proteins 4, 5, 6 and 6b for membrane interaction. *Eur. J. Biochem.* **269:**5821–5829.

Harris, F., R. Demel, B. de Kruijff, and D. A. Phoenix. 1998. An investigation into the lipid interactions of peptides corresponding to the C-terminal anchoring domains of *Escherichia coli* penicillin-binding proteins 4, 5 and 6. *Biochim. Biophys. Acta* **1415:**10–22.

Hedstrom, L. 2002. Serine protease mechanism and specificity. *Chem. Rev.* **102:**4501–4524.

Henderson, T. A., P. M. Dombrosky, and K. D. Young. 1994. Artifactual processing of penicillin-binding proteins 7 and 1b by the OmpT protease of *Escherichia coli*. *J. Bacteriol.* **176:**256–259.

Henderson, T. A., M. Templin, and K. D. Young. 1995. Identification and cloning of the gene encoding penicillin-binding protein 7 of *Escherichia coli*. *J. Bacteriol.* **177:**2074–2079.

Henderson, T. A., K. D. Young, S. A. Denome, and P. K. Elf. 1997. AmpC and AmpH, proteins related to the class C beta-lactamases, bind penicillin and contribute to the normal morphology of *Escherichia coli*. *J. Bacteriol.* **179:**6112–6121.

Hiniker, A., and J. C. Bardwell. 2004. In vivo substrate specificity of periplasmic disulfide oxidoreductases. *J. Biol. Chem.* **279:**12967–12973.

Holtje, J. V. 1998. Growth of the stress-bearing and shape-maintaining murein sacculus of *Escherichia coli*. *Microbiol. Mol. Biol. Rev.* **62:**181–203.

Isaac, D. D., J. S. Pinkner, S. J. Hultgren, and T. J. Silhavy. 2005. The extracytoplasmic adaptor protein CpxP is degraded with substrate by DegP. *Proc. Natl. Acad. Sci. USA* **102:**17775–17779.

Ishino, Y., H. Shinagawa, K. Makino, M. Amemura, and A. Nakata. 1987. Nucleotide sequence of the iap gene, responsible for alkaline phosphatase isozyme conversion in Escherichia coli, and identification of the gene product. *J. Bacteriol.* **169:**5429–5433.

Jackson, M. E., and J. M. Pratt. 1987. An 18 amino acid amphiphilic helix forms the membrane-anchoring domain of the *Escherichia coli* penicillin-binding protein 5. *Mol. Microbiol.* **1:**23–28.

Jackson, M. E., and J. M. Pratt. 1988. Analysis of the membrane-binding domain of penicillin-binding protein 5 of *Escherichia coli*. *Mol. Microbiol.* **2:**563–568.

Johnson, K., I. Charles, G. Dougan, D. Pickard, P. O'Gaora, G. Costa, T. Ali, I. Miller, and C. Hormaeche. 1991. The role of a stress-response protein in *Salmonella typhimurium* virulence. *Mol. Microbiol.* **5:**401–407.

Joris, B., J. M. Ghuysen, G. Dive, A. Renard, O. Dideberg, P. Charlier, J. M. Frere, J. A. Kelly, J. C. Boyington, and P. C. Moews. 1988. The active-site-serine penicillin-recognizing enzymes as members of the Streptomyces R61 DD-peptidase family. *Biochem. J.* **250:**313–324.

Kamphuis, I. G., K. H. Kalk, M. B. Swarte, and J. Drenth. 1984. Structure of papain refined at 1.65 A resolution. *J. Mol. Biol.* **179:**233–256.

Kato, T., N. Takahashi, and H. K. Kuramitsu. 1992. Sequence analysis and characterization of the *Porphyromonas gingivalis prtC* gene, which expresses a novel collagenase activity. *J. Bacteriol.* **174:**3889–3895.

Keiler, K. C., and R. T. Sauer. 1996. Sequence determinants of C-terminal substrate recognition by the Tsp protease. *J. Biol. Chem.* **271:**2589–2593.

Keiler, K. C., K. R. Silber, K. M. Downard, I. A. Papayannopoulos, K. Biemann, and R. T. Sauer. 1995. C-terminal specific protein degradation: activity and substrate specificity of the Tsp protease. *Protein Sci.* **4:**1507–1515.

Keiler, K. C., P. R. Waller, and R. T. Sauer. 1996. Role of a peptide tagging system in degradation of proteins synthesized from damaged messenger RNA. *Science* **271:**990–993.

Kessler, E. 2004. beta-Lytic metalloendopeptidase, p. 998–1000. In A. J. Barrett, N. D. Rawlings, and J. F. Woessner (ed.), *Handbook of Proteolytic Enzymes*. Elsevier, London, United Kingdom.

Kessler, E., and D. E. Ohmann. 2004. Staphylolysin, p. 1001–1003. In A. J. Barrett, N. D. Rawlings, and J. F. Woessner (ed.), *Handbook of Proteolytic Enzymes*. Elsevier, London, United Kingdom.

Kolmar, H., P. Waller, and R. Sauer. 1996. The DegP and DegQ periplasmic endoproteases of *Escherichia coli*: specificity for cleavage sites and substrate conformation. *J. Bacteriol.* **178:**5925–5929.

Korat, B., H. Mottl, and W. Keck. 1991. Penicillin-binding protein 4 of *Escherichia coli*: molecular cloning of the dacB gene, controlled overexpression, and alterations in murein composition. *Mol. Microbiol.* **5:**675–684.

Kraut, J. 1977. Serine proteases: structure and mechanism of catalysis. *Annu. Rev. Biochem.* **46:**331–358.

Krojer, T., M. Garrido-Franco, R. Huber, M. Ehrmann, and T. Clausen. 2002. Crystal structure of DegP (HtrA) reveals a new protease-chaperone machine. *Nature* **416:**455–459.

Kuo, W. L., B. D. Gehm, and M. R. Rosner. 1990. Cloning and expression of the cDNA for a *Drosophila* insulin-degrading enzyme. *Mol. Endocrinol.* **4:**1580–1591.

Kurochkin, I. V. 2001. Insulin-degrading enzyme: embarking on amyloid destruction. *Trends Biochem. Sci.* **26**:421–425.

Larsen, K. S., and D. S. Auld. 1989. Carboxypeptidase A: mechanism of zinc inhibition. *Biochemistry* **28**:9620–9625.

Lipinska, B., O. Fayet, L. Baird, and C. Georgopoulos. 1989. Identification, characterization, and mapping of the *Escherichia coli htrA* gene, whose product is essential for bacterial growth only at elevated temperatures. *J. Bacteriol.* **171**:1574–1584.

Lipinska, B., S. Sharma, and C. Georgopoulos. 1988. Sequence analysis and regulation of the *htrA* gene of *Escherichia coli*: a sigma 32-independent mechanism of heat-inducible transcription. *Nucleic Acids Res.* **16**:10053–10067.

Marcyjaniak, M., S. G. Odintsov, I. Sabala, and M. Bochtler. 2004. Peptidoglycan amidase MepA is a LAS metallopeptidase. *J. Biol. Chem.* **279**:43982–43989.

Markiewicz, Z., J. K. Broome-Smith, U. Schwarz, and B. G. Spratt. 1982. Spherical *E. coli* due to elevated levels of D-alanine carboxypeptidase. *Nature* **297**:702–704.

Maskos, K. 2004. Pitrilysins/inverzincins. *In* A. Messerschmidt, M. Cygler, and W. Bode (ed.), *Handbook of Metalloproteins*, vol. 3. John Wiley & Sons, Ltd., Chichester, England.

Matsuhashi, M., I. N. Maruyama, Y. Takagaki, S. Tamaki, Y. Nishimura, and Y. Hirota. 1978. Isolation of a mutant of *Escherichia coli* lacking penicillin-sensitive D-alanine carboxypeptidase IA. *Proc. Natl. Acad. Sci. USA* **75**:2631–2635.

Matsuhashi, M., Y. Takagaki, I. N. Maruyama, S. Tamaki, Y. Nishimura, H. Suzuki, U. Ogino, and Y. Hirota. 1977. Mutants of *Escherichia coli* lacking in highly penicillin-sensitive D-alanine carboxypeptidase activity. *Proc. Natl. Acad. Sci. USA* **74**:2976–2979.

Matsuhashi, M., S. Tamaki, S. J. Curtis, and J. L. Strominger. 1979. Mutational evidence for identity of penicillin-binding protein 5 in *Escherichia coli* with the major D-alanine carboxypeptidase IA activity. *J. Bacteriol.* **137**:644–647.

Matthews, B. W. 1988. Structural basis of the action of thermolysin and related proteases. *Acc. Chem. Res.* **21**:333–340.

McGrath, M. E., T. Erpel, C. Bystroff, and R. J. Fletterick. 1994. Macromolecular chelation as an improved mechanism of protease inhibition: structure of the ecotin-trypsin complex. *EMBO J.* **13**:1502–1507.

McGrath, M. E., S. A. Gillmor, and R. J. Fletterick. 1995. Ecotin: lessons on survival in a protease-filled world. *Protein Sci.* **4**:141–148.

Meberg, B. M., A. L. Paulson, R. Priyadarshini, and K. D. Young. 2004. Endopeptidase penicillin-binding proteins 4 and 7 play auxiliary roles in determining uniform morphology of *Escherichia coli*. *J. Bacteriol.* **186**:8326–8336.

Nagasawa, H., Y. Sakagami, A. Suzuki, H. Suzuki, H. Hara, and Y. Hirota. 1989. Determination of the cleavage site involved in C-terminal processing of penicillin-binding protein 3 of *Escherichia coli*. *J. Bacteriol.* **171**:5890–5893.

Nakata, A., H. Shinagawa, and J. Kawamata. 1979. Inhibition of alkaline phosphatase isozyme conversion by protease inhibitors in *Escherichia coli* K-12. *FEBS Lett.* **105**:147–150.

Nakatogawa, H., and K. Ito. 2001. Secretion monitor, SecM, undergoes self-translation arrest in the cytosol. *Mol. Cell* **7**:185–192.

Nelson, D. E., and K. D. Young. 2000. Penicillin binding protein 5 affects cell diameter, contour, and morphology of *Escherichia coli*. *J. Bacteriol.* **182**:1714–1721.

Odintsov, S. G., I. Sabala, M. Marcyjaniak, and M. Bochtler. 2004. Latent LytM at 1.3A resolution. *J. Mol. Biol.* **335**:775–785.

Palomeque-Messia, P., S. Englebert, M. Leyh-Bouille, M. Nguyen-Disteche, C. Duez, S. Houba, O. Dideberg, J. Van Beeumen, and J. M. Ghuysen. 1991. Amino acid sequence of the penicillin-binding protein/DD-peptidase of *Streptomyces K15*. Predicted secondary structures of the low Mr penicillin-binding proteins of class A. *Biochem. J.* **279**:223–230.

Park, J. H., Y. S. Lee, C. H. Chung, and A. L. Goldberg. 1988. Purification and characterization of protease Re, a cytoplasmic endoprotease in *Escherichia coli*. *J. Bacteriol.* **170**:921–926.

Park, P. W., and R. P. Mecham. 2004. Lysostaphin, p. 1004–1005. *In* A. J. Barrett, N. D. Rawlings, and J. F. Woessner (ed.), *Handbook of Proteolytic Enzymes*. Elsevier, London, United Kingdom.

Perlman, R. K., B. D. Gehm, W. L. Kuo, and M. R. Rosner. 1993. Functional analysis of conserved residues in the active site of insulin-degrading enzyme. *J. Biol. Chem.* **268**:21538–21544.

Perlman, R. K., and M. R. Rosner. 1994. Identification of zinc ligands of the insulin-degrading enzyme. *J. Biol. Chem.* **269**:33140–33145.

Perona, J. J., and C. S. Craik. 1995. Structural basis of substrate specificity in the serine proteases. *Protein Sci.* **4**:337–360.

Phillips, M. A., L. Hedstrom, and W. J. Rutter. 1992. Guanidine derivatives restore activity to carboxypeptidase lacking arginine-127. *Protein Sci.* **1**:517–521.

Polgár, L. 2004. Catalytic mechanism of serine and threonine peptidases, p. 1440–1448. *In* A. J. Barrett, N. D. Rawlings, and J. F. Woessner (ed.), *Handbook of Proteolytic Enzymes*, 2nd ed., vol. 2. Elsevier, London, United Kingdom.

Ponting, C. P. 1997. Evidence for PDZ domains in bacteria, yeast, and plants. *Protein Sci.* **6**:464–468.

Pratt, J. M., I. B. Holland, and B. G. Spratt. 1981. Precursor forms of penicillin-binding proteins 5 and 6 of *E. coli* cytoplasmic membrane. *Nature* **293:** 307–309.

Pratt, J. M., M. E. Jackson, and I. B. Holland. 1986. The C terminus of penicillin-binding protein 5 is essential for localisation to the *E. coli* inner membrane. *EMBO J.* **5:**2399–2405.

Rawlings, N., and A. Barrett. 1994. Families of serine peptidases. *Methods Enzymol.* **244:**19–61.

Rawlings, N. D., and A. J. Barrett. 1995. Evolutionary families of metallopeptidases. *Methods Enzymol.* **248:**183–228.

Rawlings, N. D., and A. J. Barrett. 2004a. Catalytic mechanisms for metalloproteases, p. 283–284. *In* A. J. Barrett, N. D. Rawlings, and J. F. Woessner (ed.), *Handbook of Proteolytic Enzymes*, vol. 1. Elsevier, London, United Kingdom.

Rawlings, N. D., and A. J. Barrett. 2004b. Introduction: serine peptidases and their clans, p. 1417–1439. *In* A. J. Barrett, N. D. Rawlings, and J. F. Woessner (ed.), *Handbook of Proteolytic Enzymes*, 2nd ed., vol. 2. Elsevier, London, United Kingdom.

Rawlings, N. D., D. P. Tolle, and A. J. Barrett. 2004. Evolutionary families of peptidase inhibitors. *Biochem. J.* **378:**705–716.

Roberts, M. G., D. A. Phoenix, and A. R. Pewsey. 1997. An algorithm for the detection of surface-active alpha helices with the potential to anchor proteins at the membrane interface. *Comput. Appl. Biosci.* **13:**99–106.

Romeis, T., and J. V. Holtje. 1994. Penicillin-binding protein 7/8 of Escherichia coli is a DD-endopeptidase. *Eur. J. Biochem.* **224:**597–604.

Schiffer, G., and J. V. Holtje. 1999. Cloning and characterization of PBP 1C, a third member of the multimodular class A penicillin-binding proteins of *Escherichia coli. J. Biol. Chem.* **274:**32031–32039.

Seoane, A., A. Sabbaj, L. M. McMurry, and S. B. Levy. 1992. Multiple antibiotic susceptibility associated with inactivation of the *prc* gene. *J. Bacteriol.* **174:**7844–7847.

Silber, K. R., K. C. Keiler, and R. T. Sauer. 1992. Tsp: a tail-specific protease that selectively degrades proteins with nonpolar C termini. *Proc. Natl. Acad. Sci. USA* **89:**295–299.

Snyder, W., and T. Silhavy. 1995. Beta-galactosidase is inactivated by intermolecular disulfide bonds and is toxic when secreted to the periplasm of *Escherichia coli. J. Bacteriol.* **177:**953–963.

Songyang, Z., A. S. Fanning, C. Fu, J. Xu, S. M. Marfatia, A. H. Chishti, A. Crompton, A. C. Chan, J. M. Anderson, and L. C. Cantley. 1997. Recognition of unique carboxyl-terminal motifs by distinct PDZ domains. *Science* **275:**73–77.

Spiess, C., A. Beil, and M. Ehrmann. 1999. A temperature-dependent switch from chaperone to protease in a widely conserved heat shock protein. *Cell* **97:**339–347.

Spratt, B. G. 1975. Distinct penicillin binding proteins involved in the division, elongation, and shape of *Escherichia coli* K12. *Proc. Natl. Acad. Sci. USA* **72:**2999–3003.

Spratt, B. G. 1977. Properties of the penicillin-binding proteins of *Escherichia coli* K12. *Eur. J. Biochem.* **72:**341–352.

Spratt, B. G. 1980. Deletion of the penicillin-binding protein 5 gene of *Escherichia coli. J. Bacteriol.* **144:**1190–1192.

Spratt, B. G., and J. L. Strominger. 1976. Identification of the major penicillin-binding proteins of *Escherichia coli* as D-alanine carboxypeptidase IA. *J. Bacteriol.* **127:**660–663.

Steitz, T. A., and R. G. Shulman. 1982. Crystallographic and NMR studies of the serine proteases. *Annu. Rev. Biophys. Bioeng.* **11:**419–444.

Stoker, N. G., J. K. Broome-Smith, A. Edelman, and B. G. Spratt. 1983. Organization and subcloning of the *dacA-rodA-pbpA* cluster of cell shape genes in *Escherichia coli. J. Bacteriol.* **155:**847–853.

Strauch, K. L., and J. Beckwith. 1988. An *Escherichia coli* mutation preventing degradation of abnormal periplasmic proteins. *Proc. Natl. Acad. Sci. USA* **85:**1576–1580.

Strauch, K. L., K. Johnson, and J. Beckwith. 1989. Characterization of degP, a gene required for proteolysis in the cell envelope and essential for growth of Escherichia coli at high temperature. *J. Bacteriol.* **171:**2689–2696.

Swamy, K. H., C. H. Chung, and A. L. Goldberg. 1983. Isolation and characterization of protease do from *Escherichia coli*, a large serine protease containing multiple subunits. *Arch. Biochem. Biophys.* **224:** 543–554.

Tadokoro, A., H. Hayashi, T. Kishimoto, Y. Makino, S. Fujisaki, and Y. Nishimura. 2004. Interaction of the *Escherichia coli* lipoprotein NlpI with periplasmic Prc (Tsp) protease. *J. Biochem. (Tokyo)* **135:**185–191.

Tatum, F. M., N. F. Cheville, and D. Morfitt. 1994. Cloning, characterization and construction of *htrA* and *htrA*-like mutants of *Brucella abortus* and their survival in BALB/c mice. *Microb. Pathog.* **17:**23–36.

Thunnissen, M. M., F. Fusetti, B. de Boer, and B. W. Dijkstra. 1995. Purification, crystallisation and preliminary X-ray analysis of penicillin binding protein 4 from Escherichia coli, a protein related to class A beta-lactamases. *J. Mol. Biol.* **247:**149–153.

Tuomanen, E., and J. Schwartz. 1987. Penicillin-binding protein 7 and its relationship to lysis of nongrowing *Escherichia coli. J. Bacteriol.* **169:**4912–4915.

Vacheron, M. J., M. Guinand, A. Francon, and G. Michel. 1979. [Characterisation of a new endopeptidase from sporulating Bacillus sphaericus which is specific for the gamma-D-glutamyl-L-lysine and gamma-D-glutamyl-(L)meso-diaminopimelate linkages of peptidoglycan substrates (author's transl)]. *Eur. J. Biochem.* **100**:189–196.

Valentin, C., M. J. Vacheron, C. Martinez, M. Guinand, and G. Michel. 1983. Action of *Bacillus sphaericus* endopeptidases on bacterial peptidoglycans and peptidoglycan fragments. *Biochimie* **65**:239–245.

van der Linden, M. P., L. de Haan, M. A. Hoyer, and W. Keck. 1992. Possible role of *Escherichia coli* penicillin-binding protein 6 in stabilization of stationary-phase peptidoglycan. *J. Bacteriol.* **174**:7572–7578.

Vernet, T., D. C. Tessier, J. Chatellier, C. Plouffe, T. S. Lee, D. Y. Thomas, A. C. Storer, and R. Menard. 1995. Structural and functional roles of asparagine 175 in the cysteine protease papain. *J. Biol. Chem.* **270**:16645–16652.

Vijayaraghavan, J., Y. A. Kim, D. Jackson, M. Orlowski, and L. B. Hersh. 1990. Use of site-directed mutagenesis to identify valine-573 in the S'1 binding site of rat neutral endopeptidase 24.11 (enkephalinase). *Biochemistry* **29**:8052–8056.

Waller, P., and R. Sauer. 1996. Characterization of *degQ* and *degS*, *Escherichia coli* genes encoding homologs of the DegP protease. *J. Bacteriol.* **178**:1146–1153.

Walsh, N. P., B. M. Alba, B. Bose, C. A. Gross, and R. T. Sauer. 2003. OMP peptide signals initiate the envelope-stress response by activating DegS protease via relief of inhibition mediated by its PDZ domain. *Cell* **113**:61–71.

Waxman, D. J., and J. L. Strominger. 1983. Penicillin-binding proteins and the mechanism of action of beta-lactam antibiotics. *Annu. Rev. Biochem.* **52**:825–869.

Wilken, C., K. Kitzing, R. Kurzbauer, M. Ehrmann, and T. Clausen. 2004. Crystal structure of the DegS stress sensor: how a PDZ domain recognizes misfolded protein and activates a protease domain. *Cell* **117**:483–494.

KEY PHYSIOLOGICAL PROCESSES

CELL DIVISION

*S. J. Ryan Arends, Kyle B. Williams, Ryan J. Kustusch,
and David S. Weiss*

10

The cell cycle culminates in the formation of a septum that separates the mother cell into two daughters. In *Escherichia coli*, septum assembly occurs at the midcell and involves coordinated inward growth of all three layers of the cell envelope—the cytoplasmic membrane, the peptidoglycan wall, and the outer membrane. How the division septum is assembled and how its assembly is coordinated with other events in the cell cycle, such as chromosome segregation, are the topics of this chapter. Genetic analysis of cell division in *E. coli* began in the 1960s with the isolation of "filamentation thermosensitive mutants" (*fts*) that exhibited normal morphology at 30°C, but grew as long filaments with regularly spaced nucleoids upon shift to 42°C. By now, at least 15 proteins that participate directly in cell division have been identified in *E. coli* (Color Plate 14, Table 1). These proteins all localize to the midcell, where they appear to form a multiprotein complex called the "septal ring" or "divisome" that orchestrates assembly of the division septum. Besides these proteins, another five are known to regulate septal ring assembly, and thus cell division, but are not constituents of the septal ring (Fig. 1, Table 2). Bacterial cell division has been the subject of several recent reviews (Errington et al., 2003; Romberg and Levin, 2003; Weiss, 2004; Goehring et al., 2005; Rothfield et al., 2005; Vicente et al., 2006).

The first recognized event in cell division is assembly of a tubulin-like protein named FtsZ into a contractile ring at the division site. The Z ring serves as a landing pad for recruitment of other proteins to the division site and probably uses energy from GTP hydrolysis to drive cytokinesis. The remaining proteins that comprise the septal ring can be divided into several functional classes. (i) A collection of FtsZ-binding proteins (FtsA, ZipA, and ZapA) have various roles in promoting Z-ring assembly, linking the Z ring to the cytoplasmic membrane, and recruiting additional proteins into the septal ring. (ii) FtsK facilitates chromosome segregation. (iii) FtsI is a septum-specific peptidoglycan synthase. (iv) Several murein hydrolases, especially AmiC and EnvC, help separate daughter cells. (v) Many of the division proteins are of essentially unknown function, including FtsEX, FtsW, FtsQ, FtsB, FtsL, and FtsN. FtsEX belongs to the ABC transporter family. The remaining proteins of unknown function likely support peptidoglycan synthesis. (vi) Current models for outer membrane

S. J. Ryan Arends, Kyle B. Williams, Ryan J. Kustusch, and David S. Weiss, Department of Microbiology, Roy J. and Lucille A. Carver College of Medicine, University of Iowa, Iowa City, IA 52242.

The Periplasm
Edited by Michael Ehrmann © 2007 ASM Press, Washington, D.C.

TABLE 1 Proteins found in the septal ring of *E. coli*

Protein	Primary functions[a]	Remarks
Early Localization		
FtsZ	Form Z ring: recruit all other proteins, drive cytokinesis	Tubulin-like GTPase
FtsA	Stabilize Z ring, link it to CM, recruit late proteins	Member of actin superfamily
ZipA	Stabilize Z ring, link it to CM, recruit late proteins	Restricted to γ-*proteobacteria*
ZapA	Stabilize Z ring	Widely conserved but nonessential
Late Localization		
FtsEX	ABC transporter?	Widely conserved but only essential in low salt
FtsK	Assembly factor and chromosome segregation	AAA ATPase family
FtsQ	Probably only an assembly factor	Forms complex with FtsBL
FtsBL	Probably only an assembly factor	Associate via periplasmic coiled-coil motif
FtsW	Transporter?	SEEDS family member
FtsI (PBP3)	PG synthesis (transpeptidase)	Substrates not known
FtsN	Unknown; binds PG, but this function not essential	Restricted to γ-*proteobacteria*, but PG-binding domain widely conserved
AmiC	PG hydrolase (amidase), separation of daughter cells	Many murein hydrolases contribute to separation, but only AmiC and EnvC are known to localize to septal ring
EnvC	PG hydrolase, separation of daughter cells	

[a]All late proteins from FtsEX through FtsN are involved in recruitment, but in some cases this function is probably incidental. CM, cytoplasmic membrane; PG, peptidoglycan.

invagination involve Braun's lipoprotein and OmpA, which mediate attachment of the outer membrane to the peptidoglycan layer. Surprisingly, no outer membrane proteins are known to be specifically required for cell division.

MODELS FOR STRUCTURE AND CONSTRICTION OF THE Z-RING

Purified FtsZ assembles in the presence of GTP into protofilaments in which FtsZ molecules are stacked end-to-end with the GTP sandwiched at the subunit interface (reviewed in Romberg and Levin, 2003). Because protofilaments are observed under a variety of conditions, there is a general consensus that protofilaments are the fundamental building block of the Z ring. The Z ring is estimated to be about six to eight protofilaments wide in *E. coli* and probably consists of a collection of many short protofilaments rather than a few long ones such that each protofilament circumscribes a short arc, not a loop around the cell's waist (Stricker

et al., 2002; Anderson et al., 2004). An important question is how the protofilaments are arranged with respect to one another. The structure of the Z ring has implications for how it constricts and how (or whether) constriction provides energy that drives cytokinesis against the outward-directed force of turgor pressure. FtsZ protofilaments have been observed in vitro to associate side-to-side to form a variety of higher order structures—pairs of intertwined protofilaments, sheets of parallel and antiparallel protofilaments, and bundles of protofilaments arranged in complex ways have all been reported. Among the conditions that promote association of protofilaments into higher-order structures are cations such as Ca^{2+}, Mg^{2+}, and DEAE-dextran (Romberg and Levin, 2003). The division proteins ZipA and ZapA also promote bundling (RayChaudhuri, 1999; Hale et al., 2000; Gueiros-Filho and Losick, 2002; Low et al., 2004). Because association of protofilaments is so dependent on in

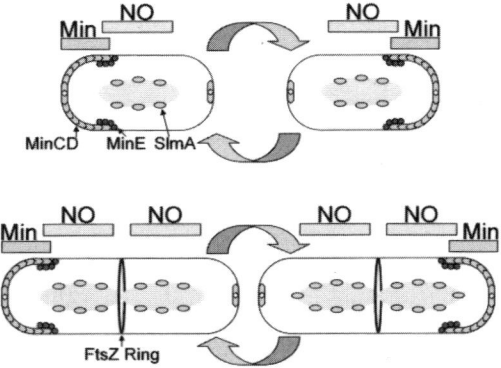

FIGURE 1 Regulation of Z-ring assembly. The Min system and nucleoid occlusion (NO) work together to ensure that the only permissive site for Z-ring assembly is the gap that opens up at the midcell as nucleoids segregate late in the cell cycle. (Top) In a newborn cell, nucleoid occlusion prevents the Z ring from assembling at the midcell, while the Min system prevents Z rings from forming in the DNA-free region near either pole. Nucleoid occlusion in *E. coli* is mediated by a DNA-binding protein named SlmA. The MinCD inhibitor complex oscillates from pole to pole under control of MinE, which is concentrated near the midcell. See text for details. (Bottom) Chromosome segregation relieves the midcell of nucleoid occlusion. MinCD continues to oscillate, but its inhibitory effects do not reach as far as the midcell because MinE limits growth of MinD polymers.

vitro conditions, it is difficult to say whether any of the higher-order structures reported to date are physiologically relevant.

Observations of GFP-tagged FtsZ in live cells have revealed a few surprises. One is the discovery of a weak, helical fluorescence signal that oscillates from pole to pole (Thanedar and Margolin, 2004). Does FtsZ play a role in elongation? Simple *ftsZ* mutants elongate normally, but under some circumstances inhibition of FtsZ function can produce cells with strikingly aberrant morphologies (Varma and Young, 2004). The second discovery is that FtsZ rings are extremely dynamic, undergoing constant remodeling even when simple time-lapse photography suggests the rings are static (Stricker et al., 2002). The most recent estimate puts the half-time for complete turnover of the ring at about 9 s (Anderson et al., 2004). These observations raise some interesting questions. How can such a dynamic structure maintain coherence during constriction? How can it generate force? What is the function of rapid remodeling?

Rapid remodeling of the Z ring could have several functions, none of which are mutually exclusive. First, as elaborated below, remodeling could be the basis of a net depolymerization reaction that drives cytokinesis against the outward force of turgor pressure. Second, turnover could allow the structure to constantly adapt to perturbations and remain at the leading edge of the invagination, even if constriction starts to go awry. Finally, rapid remodeling has implications for the regulation of cell division because a cell must constantly recommit itself to division by providing a continuous supply of fresh FtsZ·GTP to maintain the Z ring even after it has been assembled. This makes it easy for a cell to abort the division process, as occurs during the SOS response to DNA damage, or to redeploy FtsZ to another site, as occurs during the switch from medial to polar division during sporulation in *Bacillus subtilis*.

TABLE 2 Proteins that regulate cell division in *E. coli*[a]

Protein	Primary function	Remarks
SlmA	Nucleoid occlusion	DNA-binding protein, blocks Z-ring assembly in vivo but promotes bundling of protofilaments in vitro
MinC	The Min proteins work together to block division near the cell poles	Interacts directly with FtsZ to inhibit assembly
MinD		Membrane-associated ATPase, forms inhibitor complex with MinC
MinE		Topological specificity factor, restricts MinCD complex to poles
SulA	Binds to and sequesters FtsZ	Induced during SOS response

[a]All these proteins are negative regulators. Some septal ring components listed in Table 1 positively regulate Z-ring assembly, notably FtsA, ZipA, and ZapA.

The mechanism by which the Z ring constricts is not known, but several models have been discussed (Romberg and Levin, 2003; Ryan and Shapiro, 2003). The "cinch model" posits that a motor protein functions as a ratchet to slide rigid protofilaments against each other. This model is not considered particularly attractive because there is no strong candidate for the motor protein and the protofilaments appear too unstable. The "bending" model is based on the observation that protofilaments consisting of FtsZ·GTP are straight, but those of FtsZ·GDP are curved (Lu et al., 2000). Thus, GTP hydrolysis might drive conformational changes in FtsZ that shorten the radius of curvature of the Z ring. The "depolymerization" model suggests that progressive loss of FtsZ subunits from the protofilaments drives constriction of the Z ring. The high turnover of FtsZ subunits in the Z ring seems to support the "depolymerization" model as it is easy to envision how rapid remodeling could lead to net loss of FtsZ during constriction (Ryan and Shapiro, 2003). Nevertheless, the rate of remodeling is about the same before and during constriction (Anderson et al., 2004). Moreover, the FtsZ84 (Ts) protein, which has reduced GTPase activity owing to a lesion in the GTP-binding site, supports fairly normal division at the permissive temperature despite the fact that the Z rings in an *ftsZ84*(Ts) mutant turn over about three times more slowly than wild type (Anderson et al., 2004). These observations emphasize the need for further studies of the relationship between turnover and constriction.

Implicit in most discussions of Z-ring constriction is the notion that FtsZ uses GTP hydrolysis to drive cytokinesis by overcoming the outward-directed force of turgor pressure. To the extent that septum assembly involves localized synthesis of new cell wall, however, it is not analogous to pinching a balloon. Localized synthesis could bypass the problem of turgor. Alternatively, synthesis of cell wall could drive membrane constriction with the Z ring serving as a mobile scaffold. Thus, whether the Z ring must generate force to support division is neither settled nor can it be taken for granted (Weiss, 2004).

REGULATION OF Z-RING ASSEMBLY

So far as is known, cell division is regulated primarily at the level of Z-ring assembly. When discussing this regulation, it is helpful to consider two situations separately. One situation is that regulation ensures that the Z ring assembles in the right place and at the right time in normally cycling cells—this sort of regulation is widely conserved among different bacterial species. The other situation is environmental conditions that interdict Z-ring assembly to prevent division under special circumstances—this sort of regulation is less well conserved, because some of these circumstances reflect adaptations restricted to certain bacteria.

The Min System and Nucleoid Occlusion Regulate Z-Ring Assembly during the Cell Cycle

Two independent pathways control Z-ring formation in normally cycling cells (reviewed in Rothfield et al., 2005). One pathway is a poorly characterized phenomenon called nucleoid occlusion and prevents Z rings from assembling prematurely at the center of newborn cells, as this region is occupied (blocked) by the nucleoid. The other pathway is better understood and involves the Min proteins, which prevent Z-ring assembly in the DNA-free regions near the cell poles. In effect, nucleoid occlusion and the Min system work together to ensure that the only permissive site for Z-ring assembly is the DNA-free region that opens up at the midcell as chromosomes begin to segregate to incipient daughter cells (Fig. 1).

In *E. coli*, there are three Min proteins—MinC, MinD, and MinE (de Boer et al., 1989). Deletion of the *min* locus causes about half the cells (depending on the growth conditions) to divide near a pole rather than at the midcell. Polar divisions produce unequal daughter cells; one is somewhat larger than normal and contains two nucleoids, whereas the other is a "minicell" with no nucleoid. The Min system is complex. In brief, MinC is the division inhibitor per se; it binds to FtsZ and prevents Z-ring assembly (Hu et al., 1999). To do so, however, MinC needs to bind to MinD, a membrane-associated ATPase. Interaction of MinC with MinD raises the concentration of MinC at the membrane and improves the interaction of MinC with septal ring assemblies (Johnson et al., 2004). The MinCD inhibitor complex is restricted to the poles by the regulatory protein MinE. Remarkably, MinE restricts MinCD to the poles not by anchoring it there but by causing MinCD to oscillate from pole to pole with a period of about 30 s to 1 min (de Boer et al., 1989; Hu and Lutkenhaus, 1999; Hu et al., 1999; Raskin and de Boer, 1999a, 1999b). The biochemistry underlying oscillation involves reversible polymerization of MinD into helical polymers at the inner face of the cytoplasmic membrane (reviewed in Rothfield et al., 2005). (i) MinD undergoes ATP-dependent assembly into a helical polymer that grows from one pole toward the midcell. This polymer recruits MinC and thereby prevents FtsZ from assembling near that pole. (ii) As the polymer approaches the midcell, it encounters MinE. MinE displaces MinC and stimulates MinD's ATPase activity, causing MinD to depolymerize. (iii) MinD then diffuses through the cytosol, where it rebinds ATP. MinD·ATP assembles into a polymer eminating from opposite pole, where MinE concentrations are lowest. The net effect of MinCD oscillation in *E. coli* is a time-averaged concentration minimum at the midcell.

B. subtilis also has a Min system that suppresses polar divisions, but there are some striking differences as compared with *E. coli*. Namely, MinCD localizes statically to the cell poles, where it binds to the DivIVA protein. (Marston et al., 1998; Marston and Errington, 1999). DivIVA is not homologous to MinE, and the two proteins function differently. MinE of *E. coli* is mobile, is most abundant near the midcell, and triggers disassembly of MinCD. DivIVA of *B. subtilis* associates with the poles and promotes assembly of MinCD. These differences highlight the issue of why the *E. coli* Min system should be so complex. One might

speculate that an oscillator is inherently more robust because it undergoes constant remodeling. This might help to faithfully define division sites after filamentation or other shape changes.

Although it has been appreciated for quite some time that the nucleoid vetoes Z-ring assembly in its immediate vicinity (Woldringh et al., 1991), the molecular basis of this phenomenon has long been mysterious. But recently proteins involved in nucleoid occlusion have been identified in *B. subtilis* and *E. coli*. The *B. subtilis* protein is named Noc (<u>n</u>ucleoid <u>oc</u>clusion), and is a member of the ParB family of DNA-binding proteins (Wu and Errington, 2004). The *E. coli* protein is named SlmA (<u>s</u>ynthetic <u>l</u>ethal with <u>min</u>), and belongs to the <u>s</u>ynthetic <u>l</u>ethal *m*in TetR family of DNA-binding proteins (Bernhardt and de Boer, 2005). Despite their lack of sequence similarity, Noc and SlmA share many functional properties. Both proteins localize to the nucleoid, where they exert a localized inhibition of Z-ring assembly. Both proteins can inhibit Z-ring assembly globally when overproduced. So far, only SlmA has been studied in vitro. Curiously, SlmA promotes bundling of FtsZ protofilaments. This is somewhat paradoxical, because in vivo SlmA seems to destabilize FtsZ polymers. At this point one can infer that SlmA mediates nucleoid occlusion via a direct interaction with FtsZ, but how this interaction prevents Z-ring assembly remains to be determined.

Environmental Signals That Regulate Z-Ring Assembly

Several conditions are known to interdict the normal process of cell division. The regulated step is generally related to Z-ring assembly.

A well-studied and widely conserved instance of environmental regulation of cell division is the SOS response to DNA damage, which causes transient filamentation. In *E. coli*, the underlying mechanism is that DNA damage leads to the cleavage of the LexA repressor, inducing ~30 genes under LexA control (Courcelle et al., 2001). Most of these genes are involved in DNA repair, but *sulA* (also called *sfiA*) encodes a protein that binds to and sequesters FtsZ (Huisman et al., 1984; Mukherjee et al., 1998; Trusca et al., 1998; Cordell et al., 2003). Inhibition of cell division prevents the septum from closing like a guillotine on chromosomes that cannot be segregated because replication forks are stalled at lesions in the DNA. Once the damage has been repaired, LexA reestablishes repression of *sulA* and the Lon protease rapidly degrades whatever SulA protein is present (Mizusawa and Gottesman, 1983). At this point new Z rings can form and division ensues. Recent reports suggest that inactivation of FtsEX or inhibition of FtsI also induces the SOS response (Miller et al., 2004; O'Reilly and Kreuzer, 2004). We discuss these claims when we discuss the respective division proteins later in this chapter.

Another interesting instance of regulating cell division occurs during sporulation of *B. subtilis*. Vegetatively growing *B. subtilis* cells divide in the middle to produce two identical daughters. In contrast, spore formation involves a polar division event that produces a small cell, which ultimately develops into the spore. The switch to polar division is accomplished by moving the Z ring from the midcell to the cell poles. This change occurs via a spiral FtsZ filament that grows from the midcell toward the poles (Ben-Yehuda and Losick, 2002). Reconfiguration of the FtsZ polymer involves increased transcription of *ftsZ* and activation of an FtsZ-associated protein named SpoIIE.

ASSEMBLY OF THE SEPTAL RING: A MODEL FOR PROTEIN-PROTEIN INTERACTION?

As noted above, the first recognized event in cell division is assembly of the FtsZ protein into a ring structure at the inner surface of the cytoplasmic membrane. The FtsZ-binding proteins FtsA, ZipA, and ZapA probably localize to the Z ring as it is assembling and may even associate with FtsZ outside the context of the Z ring (Den Blaauwen et al., 1999; Hale and de Boer, 1999; Rueda et al., 2003; Jensen et al., 2005). The interactions between FtsZ and FtsA, ZipA and ZapA have been convincingly docu-

mented by a variety of methods in several laboratories and have been adequately reviewed elsewhere (Errington et al., 2003; Romberg and Levin, 2003; Goehring and Beckwith, 2005).

Comparatively little is known about how the "late" division proteins are recruited to the septal ring. The earliest approaches to this issue involved studying protein localization in various *E. coli* mutant backgrounds (reviewed in Buddelmeijer and Beckwith, 2002; Errington et al., 2003; Vicente et al., 2006). These studies revealed a set of dependencies that imply the various division proteins localize to the septal ring in a defined order. Remarkably, this order is almost completely linear: FtsZ → [FtsA/ZipA/ZapA] → FtsEX → FtsK → FtsQ → FtsBL → FtsW → FtsI → FtsN → AmiC (Fig. 2). In general, this order of recruitment is interpreted to reflect the assembly pathway for a multiprotein complex, and thus makes predictions about which proteins interact. Moreover, septal targeting domains have been identified in several of these proteins (Fig. 3). These targeting domains presumably mediate protein-protein interactions. It is important to characterize these interactions because they should shed light on how the many proteins that comprise the septal ring work together during the complex process of septum assembly. For example, direct protein-protein interactions could play an important role not only in recruitment of the Fts proteins to the septal ring but also in regulating their activities once they are there.

Recently, several groups have started to look directly at identifying protein-protein interactions among the late proteins. Using immunoprecipitation, Buddelmeijer and Beckwith (2004) showed that FtsQ, FtsL, and FtsB form a complex in *E. coli*. Because this complex was observed in an FtsK depletion strain, where none of the proteins in question can localize to the septal ring, it seems likely that the FtsQLB complex assembles in the membrane independent of the division status of the cell. The FtsQLB complex appears to be widely conserved. The pneumococcal equivalents have been assembled into a trimeric complex in vitro

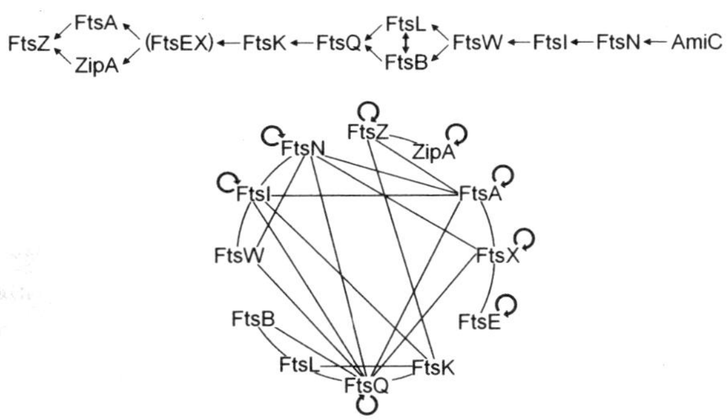

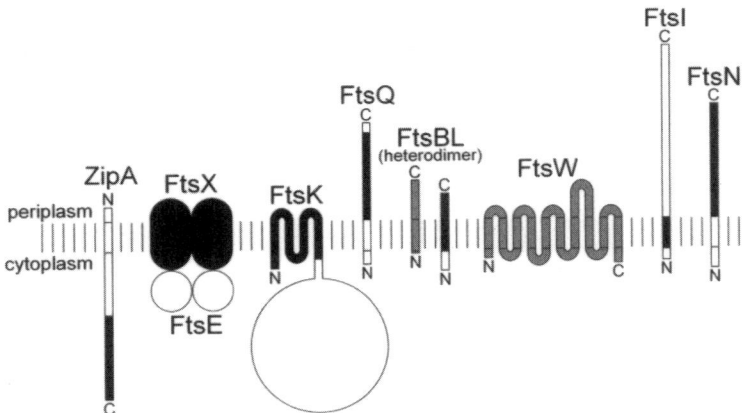

FIGURE 2 Models for assembly of the septal ring. (Top) The order of assembly of proteins into the septal ring as determined by localization dependency. First, FtsZ forms the Z ring. FtsA and ZipA join next, independently of one another. Once both FtsA and ZipA have localized, the remaining proteins join the ring in the order indicated such that localization of each depends upon localization of the proteins to the left of it. ZapA and EnvC are omitted because their dependency relationships have not been established, but since neither is essential, none of the proteins shown are likely to depend on ZapA or EnvC for recruitment. Dependence of FtsK and other late proteins on FtsEX is leaky. (Bottom) Bacterial two-hybrid assays have identified a network of interactions among the division proteins. Lines connect proteins reported to interact in at least one assay, while circular arrows indicate self-interaction (e.g., dimerization). Adapted from *Current Biology* **15**:R514–R526 (Goehring and Beckwith, 2005), copyright 2005, with permission from Elsevier with inclusion of additional data (Arends and Weiss, unpublished).

FIGURE 3 Septal targeting domains in membrane proteins involved in cell division in *E. coli*. For clarity, proteins are not drawn to scale. Regions sufficient for septal localization are shown in black, while those known to be dispensable are white. Grey indicates a protein has not been studied in this regard. In many cases only a limited number of constructs have been characterized, so the targeting domains might be smaller than indicated. Note that targeting information can reside in any domain (cytoplasmic, transmembrane, or periplasmic) and there is no evidence that a targeting motif is shared among different proteins. These findings are consistent with the notion that division proteins localize by a cascade of protein–protein interactions rather than by binding to a common target. Because FtsL must associate with FtsB to localize to the septal ring, the targeting regions identified in FtsL probably mediate complex formation with FtsB rather than septal localization per se. Adapted with permission from the *Journal of Bacteriology* (Wissel et al. [2004]) with inclusion of new data from our lab (Arends and Weiss, unpublished).

(Noirclerc-Savoye et al., 2005). The *B. subtilis* equivalents also appear to form a complex, although whether they do so away from the septal ring has been challenged (Daniel et al., 1998; Daniel and Errington, 2000; Katis et al., 2000; Sievers and Errington, 2000; Robson et al., 2002).

Numerous interactions among Fts proteins have been observed by using bacterial two-hybrid systems based on reconstitution of a phage repressor (Di Lallo et al., 2003), reconstitution of an adenylate cyclase (Karimova et al., 2005), or redirecting proteins to the *E. coli* poles (Corbin et al., 2004). Taken together, these studies imply that there is a network of interactions among the late-division proteins rather than a cascade of pairwise interactions (Fig. 2). The network model is appealing for a variety of reasons, not the least of which is that, to a first approximation, the different studies identify a similar set of interactions. Moreover, the septal ring is a transient structure, and the cooperativity that results from a network of interactions would seem well suited for providing both high affinity and facile disassembly.

Although the bacterial two-hybrid studies have provided an important framework, two important questions still need to be addressed. First, are the interactions direct? This is an issue because most of the fusions were to full-length division proteins that presumably localize to the septal ring, so one cannot easily distinguish simple binary interactions from coassembly into the septal ring (i.e., bridging proteins might be involved). Second, are all the interactions seen in these systems physiological? In some cases the relevant proteins are produced from compatible plasmids with copy numbers of 20 and 200, so overproduction might drive weak, fortuitous interactions that do not occur during cell division.

One particularly perplexing issue is how the late proteins associate with the septal ring. Dependency studies indicate that FtsA and ZipA must both localize to the Z ring before the late proteins can be recruited. The simplest model would be for one or more late proteins to bind

directly to either FtsA or ZipA. Consistent with this, all of the studies find interactions between FtsA and multiple late proteins (Di Lallo et al., 2003; Corbin et al., 2004; Karimova et al., 2005). ZipA has not been carefully tested for interactions, although the discovery of a point mutation in *ftsA* that bypasses the requirement for ZipA in cell division and septal ring assembly argues that ZipA's role in recruitment of downstream proteins is indirect (Geissler et al., 2003). Moreover, whereas FtsA is widely conserved, ZipA appears to be restricted to the γ-proteobacteria, making FtsA the stronger a priori candidate for interaction with "late" proteins. So far so good, but consider the following issues. (i) If late proteins are recruited by binding to FtsA, why is ZipA required for recruitment at all? And why is there a ~17-min gap in slow-growing cells between localization of FtsA and localization of the "late" proteins (Aarsman et al., 2005)? (ii) Because FtsA is associated with the inner face of the cytoplasmic membrane, the most logical interaction targets in FtsQ, FtsI, and FtsN would be the cytoplasmic domains of these proteins. But the cytoplasmic domains of FtsI, FtsN, and FtsQ are not needed for septal localization or cell division (Dai et al., 1996; Addinall et al., 1997; Chen et al., 1999; Wissel et al., 2005). Moreover, according to one study, the most important part of FtsQ, as far as its interaction with FtsA is concerned, appears to be the last 10 residues of the periplasmic (!) domain (Karimova et al., 2005). (iii) A ZapA-FtsQ fusion protein binds directly to FtsZ and allows FtsQ to be "prematurely" targeted to the Z ring in cells that lack FtsA (Goehring et al., 2005). The ZapA-FtsQ fusion recruits FtsI but not FtsN to the septal ring in cells depleted of FtsA. The ability of ZapA-FtsQ to bypass the normal requirement for FtsA in recruitment of FtsI suggests that FtsA's role in this process under normal circumstances is indirect.

In summary, there is general agreement that the Fts proteins localize to the septal ring in a dependent fashion and these proteins exhibit numerous interactions with one another in a variety of experimental systems. Nevertheless, there are discrepancies that will have to be resolved before a definitive interaction model can be drawn. At present it is too early to say which studies have been misinterpreted—it could be the localization data, the structure-function analyses, the two-hybrid systems, or various combinations of any of them.

Finally, note that dependencies do not necessarily correspond in any simple fashion to timing. The timing of Fts protein localization has been studied in *E. coli* cells growing slowly in minimal medium (Den Blaauwen et al., 1999; Aarsman et al., 2005). These studies revealed that the septal ring assembles in two steps. FtsZ, FtsA, and ZipA localized in unison. Then there was a period during which no additional Fts proteins joined the ring; in cells growing with a doubling time of ~85 min, the gap lasted ~17 min. Then the "late" proteins localized essentially in unison. These results raise a fascinating question: why is there a significant time lag between assembly of early and late proteins into the septal ring? One suggestion comes from the observation that FtsZ appears to sponsor synthesis of "preseptal" peptidoglycan in the absence of FtsA and other division proteins (Nanninga, 1991) (see "Septal Peptidoglycan Synthesis," below). Perhaps localization of the "late" proteins depends on synthesis of "preseptal" peptidoglycan (Aarsman et al., 2005). Another suggestion comes from the observation that depleting cells of *S*-adenosylmethionine results in the formation of partial septal ring assemblies containing FtsZ, FtsA, and ZipA but not FtsQ, FtsI, or FtsN (Newman et al., 1998; Wang et al., 2005). Perhaps an Fts protein has to be methylated for the late proteins to localize (Wang et al., 2005).

Two recent large-scale screens have identified "new" proteins that appear to interact with known division proteins (Butland et al., 2005; Stenberg et al., 2005). Further work is needed to establish whether these are bona fide division proteins. Nevertheless, one suspects that the set of division proteins discussed in this chapter is incomplete.

FtsZ-BINDING PROTEINS

FtsA and ZipA both bind to the last ~15 amino acids of FtsZ. These interactions have been characterized by a variety of labs and a variety of methods, and in the case of ZipA there is even a high-resolution structure showing how the tail of FtsZ fits into a shallow cleft on ZipA's C-terminal domain (Mosyak et al., 2000). While it might seem surprising that two proteins would use the same binding site on FtsZ, the fact that FtsZ is so much more abundant than either FtsA or ZipA suggests that there might not be competition in vivo.

Inactivation of either FtsA or ZipA alone prevents division but has only a modest effect on the abundance of Z rings, indicating that, at least in *E. coli*, neither protein is by itself essential for Z-ring assembly or stability (Addinall and Lutkenhaus, 1996; Hale and de Boer, 1999; Liu et al., 1999). However, simultaneous inactivation of FtsA and ZipA leads to the formation of filaments devoid of Z rings (Pichoff and Lutkenhaus, 2002). Thus, FtsA and ZipA are positive regulators of FtsZ assembly, but their functions are somewhat redundant. Additional evidence for ZipA's role in promoting Z-ring formation is that mild overproduction rescues division and Z-ring assembly in an *ftsZ84*(Ts) mutant, and that purified ZipA promotes dramatic bundling of FtsZ protofilaments in vitro (RayChaudhuri, 1999; Hale et al., 2000). To our knowledge, the effect of FtsA on polymer formation by purified FtsZ has not been reported.

Both proteins are capable of anchoring the Z ring to the cytoplasmic membrane, which might be important if FtsZ is to generate force during cytokinesis. ZipA can fulfill this function because it is an integral membrane protein (Hale and de Boer, 1997). FtsA, in contrast, associates peripherally with the membrane by virtue of its last ~15 residues, which form an amphipathic helix (Pichoff and Lutkenhaus, 2005). Curiously, this helix is required not only for association of FtsA with the membrane but also for localization of FtsA to the Z-ring (Yim et al., 2000; Pichoff and Lutkenhaus, 2005).

FtsA and ZipA localize independently to the Z ring, but both proteins are needed for subsequent recruitment of FtsEX, FtsK, and other "late" proteins. The mechanism by which FtsA and ZipA recruit these proteins to the septal ring is under investigation. In the case of FtsA, there are multiple reports that it interacts with several late proteins, especially FtsK, FtsQ, FtsI, and FtsN (Di Lallo et al., 2003; Corbin et al., 2004; Karimova et al., 2005). The ability of ZipA to interact with late proteins does not appear to have been explored, although the fact that ZipA's membrane anchor domain is required for cell division, but not for localization of ZipA to the septal ring, is intriguing in this context (Hale et al., 2000). On the other hand, Margolin and coworkers have argued that ZipA functions indirectly in recruitment of late proteins (Geissler et al., 2003). These researchers isolated a mutant form of FtsA (R286 changed to W) that supports efficient cell division even in a *zipA* null background. This mutant FtsA protein can assume all the essential functions of ZipA, including recruitment of downstream division proteins. Such facile bypass is most plausible if ZipA plays an indirect role in recruitment, by modulating FtsZ or FtsA, for example. Similarly, however, one could argue that FtsA too might function indirectly in recruitment as FtsA is not essential in *B. subtilis* (Beall and Lutkenhaus, 1992; Jensen et al., 2005) and the normal requirement for FtsA in septal ring assembly in *E. coli* can be largely bypassed by artificially tethering FtsQ to the septal ring (Goehring et al., 2005).

The structure of FtsA from *Thermatoga maritima* has been determined by X-ray crystallography (van den Ent and Lowe, 2000), and the similarity to actin originally proposed on the basis of sequence comparisons has been confirmed (Bork et al., 1992). A unique feature of FtsA is a subdomain (designated 1c) not found in actin. This subdomain is dispensable for localization of FtsA to the septal ring but not for recruitment of downstream division proteins and has been proposed to interact with FtsI and FtsN (Corbin et al., 2004; Rico et al., 2004). FtsA from *E. coli, T. maritima, B. subtilis*, and *Strep-*tococcus pneumoniae has been reported to bind ATP, but only the *S. pneumoniae* protein assembles into a polymer in an ATP-dependent fashion (Sanchez et al., 1994; van den Ent and Lowe, 2000; Yim et al., 2000; Feucht et al., 2001; Lara et al., 2005b; Paradis-Bleau et al., 2005). The polymer formed by *S. pneumoniae* FtsA has a very different structure from an actin polymer. Nevertheless, it highlights the possibility that FtsA could have a cytoskeletal function during cytokinesis. Estimates of FtsA abundance range from 50 to 740 in *E. coli*, to ~1,000 in *B. subtilis*, to ~2,000 in *S. pneumoniae* (Wang and Gayda, 1992; Feucht et al., 2001; Rueda et al., 2003; Lara et al., 2005b). These values have implications for the ability of FtsA to form a polymer during division, both because polymerization is concentration dependent and because it would probably take ~1,000 molecules to form a continuous filament around the "waist" of the cell. Given the obvious analogy of FtsA to actin, the role of ATP in FtsA function has received surprisingly little attention. To our knowledge, only one publication has reported on the consequence of mutations in the ATP-binding site (Sanchez et al., 1994). It was concluded that ATP binding might not be essential for FtsA function. This is an astonishing result in view of the strong conservation of the ATP-binding site. It would be interesting to see this line of investigation extended to more mutants.

We now turn our attention to ZapA. ZapA was discovered in *B. subtilis* during a screen to identify proteins that promote assembly of the Z ring (Gueiros-Filho and Losick, 2002). The screen took advantage of the fact that overproduction of MinD in *B. subtilis* is lethal because excess MinD prevents Z rings from assembling anywhere in the cell. The *zapA* gene rescued growth of a strain with excess MinD when overexpressed. Subsequent characterization showed that ZapA localizes to the Z ring, binds directly to FtsZ, and promotes polymerization of FtsZ into protofilaments and bundling of FtsZ protofilaments into higher-order structures (Gueiros-Filho and Losick, 2002; Low et al., 2004). Despite the fact that ZapA is widely conserved, *zapA* null mutants of *B. subtilis* and *E. coli* divide normally (Gueiros-Filho and Losick, 2002; Johnson et al., 2004). Nevertheless, several synthetic phenotypes confirm that ZapA contributes to Z-ring formation in *B. subtilis.* In particular, the absence of ZapA blocked division in cells that have reduced levels of FtsZ or null mutations in *divIVA* or *ezrA*, two other regulators of Z-ring assembly in *B. subtilis* (Gueiros-Filho and Losick, 2002). Why is ZapA not essential? Presumably this is because it does not contribute to recruitment of "late" proteins to the septal ring.

In terms of structure, ZapA is a small, dimeric protein with an N-terminal FtsZ-binding domain and a C-terminal coiled-coil domain (Gueiros-Filho and Losick, 2002; Low et al., 2004). The crystal structure of ZapA from *Pseudomonas aeruginosa* revealed that two dimers associate to form an elongated, antiparallel tetramer (Low et al., 2004). Studies of ZapA in solution suggest that tetramerization might be a crystallization artifact, as it appears to require nonphysiologically high protein concentrations, although the possibility of tetramers forming in the context of the septal ring cannot be excluded. How does ZapA promote Z-ring assembly? ZapA could promote polymerization by binding to adjacent FtsZ monomers within a protofilament and/or promote bundling by cross-linking FtsZ monomers in neighboring protofilaments (Gueiros-Filho and Losick, 2002; Low et al., 2004). The symmetry of ZapA observed in the crystal structure implies that for ZapA to cross-link adjacent protofilaments in the Z ring, these protofilaments would have to adopt an antiparallel arrangement (Low et al., 2004).

AN ABC TRANSPORTER FOR CELL DIVISION?

By homology, FtsEX belongs to the ABC transporter family of proteins, with FtsE being the ATP-binding cassette component and FtsX the membrane component (Gill et al., 1986; Gibbs et al., 1992). Consistent with these predictions, FtsE binds ATP and can be coimmune precipitated with FtsX (de Leeuw et al., 1999).

Unlike most of the division proteins discussed in this chapter, FtsEX is not essential. FtsEX null mutants and depletion strains are mildly filamentous on Luria-Bertani medium containing at least 0.5% NaCl, but form long filaments and ultimately lyse if NaCl is omitted (de Leeuw et al., 1999; Schmidt et al., 2004). This salt remedial phenotype was originally observed with *ftsE*(Ts) mutants. The standard explanation for salt rescue of Ts mutants is that suitable ionic conditions can rescue the folding of a mutant protein, but this explanation does not account for rescue of a null mutant. One early explanation for the salt-remedial phenotype was that FtsEX transports an ion involved in cell division; presumably this ion can enter the cell by a secondary route if the salt concentration is high enough (Gill et al., 1986). We consider this explanation unlikely because there are no charged residues in the TMHs of FtsX (S. J. R. Arends and D. S. Weiss, unpublished data) and there is no obvious unifying theme as to which electrolytes do and do not rescue: NaCl, NaH_2PO_4, Na_2SO_4, KCl, and NH_4Cl all rescue, but K_2SO_4, $MgSO_4$, and $MnCl_2$ do not, nor do the osmolytes sucrose, glycine betaine, or proline (de Leeuw et al., 1999). What then is the basis of the salt-remedial phenotype? FtsEX is largely dispensable for septal ring assembly on media with high salt, but facilitates localization of FtsK and subsequent downstream division proteins in media with low salt (Schmidt et al., 2004). Because protein-protein interactions are often sensitive to ionic conditions, we have suggested that salt and FtsEX play somewhat redundant roles in stabilizing the septal ring (Schmidt et al., 2004).

An important question is whether FtsEX has any role in cell division beyond serving as an assembly factor. It is not obvious why an ABC transporter would be needed for cell division because there is no substrate that is known to be required for division but not for elongation. The sequence of FtsEX is not sufficiently similar to any characterized ABC transporter to suggest what the substrate might be. Sequence comparisons place FtsEX together with importers rather than exporters

(Bouige et al., 2002), but there is no evidence for a periplasmic binding protein and not all ABC systems that group with importers actually function in this capacity. Two examples are the MacAB and LolCDE systems of *E. coli*, both of which are phylogenetically close to FtsEX (Bouige et al., 2002). MacAB is an exporter that confers resistance to macrolides (Kobayashi et al., 2001), while the LolCDE system is not a transporter at all—it is involved in release of lipoproteins from the cytoplasmic membrane (Narita and Tokuda, 2005). These considerations leave us with only ad hoc suggestions such as FtsEX might transport a substrate involved in peptidoglycan synthesis or serve as a lipid flippase. A more radical suggestion is that FtsX serves as a membrane anchor, while FtsE uses ATP hydrolysis to promote constriction of the septal ring.

O'Reilly and Kreuzer (2004) recently reported that transposon insertions in *ftsE* or *ftsX* result in weak constitutive induction of the SOS response. They suggested that FtsEX pumps out a genotoxic substance. We think a viable alternative explanation is that FtsK localizes poorly in the absence of FtsEX (Schmidt et al., 2004). As described below, FtsK promotes chromosome segregation and *ftsK* mutants are partially induced for the SOS response (Schmidt et al., 2004).

COORDINATING SEPTUM ASSEMBLY WITH NUCLEOID SEGREGATION

As already mentioned, nucleoid occlusion is one mechanism for linking cell division to chromosome segregation. Another mechanism involves FtsK, a large protein with three domains: an N-terminal membrane anchor, a flexible linker, and a C-terminal motor domain with ATPase and DNA-binding activities.

The most interesting aspects of FtsK function are related to resolving covalent linkages between sister chromosomes that, if left unresolved, would prevent segregation. Depending on the growth conditions, about 10% of the cells in an *E. coli* population cannot segregate their chromosomes because they are catenated or dimeric (Steiner et al., 1999). Catenanes are resolved by Topo IV. The C-terminal domain of FtsK interacts directly with Topo IV, recruits it to the midcell, and stimulates its decatenase activity (Espeli et al., 2003). Chromosome dimers arise from *recA*-dependent (homologous) recombination between sister chromosomes and are resolved by the XerCD recombinase. XerCD acts at a 28-bp chromosomal site named *dif* near the terminus. FtsK facilitates dimer resolution by two distinct mechanisms. FtsK loads onto DNA in the septal region and translocates that DNA to bring the *dif* sites into alignment for recombination (Capiaux et al., 2002; Corre and Louarn, 2002; Bigot et al., 2004). Upon reaching the *dif* sites, FtsK activates the XerCD recombinase by a direct protein-protein interaction (Aussel et al., 2002; Yates et al., 2003; Massey et al., 2004).

Observations of single molecules of FtsK's motor domain translocating on DNA substrates revealed that FtsK translocates remarkably fast, ~5 kb/s (Pease et al., 2005). Even more remarkable, FtsK translocates directionally on DNA from the *dif* region, as if FtsK "knows" the direction to the nearest *dif* site (Pease et al., 2005). Recently, a sequence motif that directs FtsK to *dif* was identified: GNGNAGGG (or its complement, or both—the data do not distinguish) (Bigot et al., 2005; Levy et al., 2005). When FtsK encounters this motif from the 3′ end of the G-rich strand, it pauses and, ~40% of the time, it reverses direction. The GNG-NAGGG motif effectively points the way to the nearest *dif* site because it is abundant and displays a highly skewed distribution, especially in the *dif* region, where it is always oriented toward *dif* (i.e., the orientation reverses when *dif* is crossed). Although it may be counterintuitive that the motif should cause FtsK to reverse ~40% rather than 100% of the time, mathematical modeling indicates that the lower reversal rate improves the efficiency of the search process (Pease et al., 2005). If the reversal rate were close to 100%, a single misoriented GNG-NAGGG motif (or close match to that concensus) could prevent FtsK from reaching *dif*.

As noted above, only a fraction of the cells in an *E. coli* population need FtsK's assistance to complete chromosome segregation. Nevertheless, *ftsK* is an essential gene. It turns out that the only truly essential part of FtsK is its N-terminal membrane anchor domain (Draper et al., 1998; Wang and Lutkenhaus, 1998). This domain is sufficient for localization to the septal ring, where it is needed for recruitment of several additional essential division proteins (Wang and Lutkenhaus, 1998; Yu et al., 1998; Chen and Beckwith, 2001). Because overproduction of any of several other division proteins (FtsQ, FtsB, FtsA, ZipA, or FtsN) rescues an *ftsK* null mutant, it has been argued that the essential function of this domain is to help assemble or stabilize the septal ring (Geissler and Margolin, 2005).

FtsK belongs to a large family of proteins implicated in DNA translocation. Another well-studied member of this family is SpoIIIE, a protein required for sporulation in *B. subtilis.* Sporulation involves a polar division event to create a small cell that ultimately matures into a spore. SpoIIIE localizes to the polar septum, where it pumps DNA from the mother cell to the nascent forespore (Bath et al., 2000). Unlike FtsK, SpoIIIE does not appear to pump DNA in a sequence-directed fashion. Rather, it is the orientation of SpoIIIE in the polar septum that determines the direction in which DNA will be translocated (Sharp and Pogliano, 2002).

The N-terminal domain of SpoIIIE is involved in membrane fusion (Sharp and Pogliano, 2003; Liu et al., 2006). Whether FtsK participates in membrane fusion during cell division in *E. coli* is not yet known. Although the N-terminal domains of FtsK and SpoIIIE are not strikingly similar in terms of amino acid sequence, they both have four transmembrane helixes (TMHs), a phenylalanine-rich motif near the beginning of TMH 2 and a GGGxxG motif near the beginning of TMH 4 (Errington et al., 2001; Liu et al., 2006). Several *ftsK* point and truncation mutants form deep constrictions, suggesting that these forms of FtsK are specifically defective in the final stages of septum closure; however, an *ftsK* depletion strain forms smooth filaments (Begg et al., 1995; Diez et al., 1997; Wang and Lutkenhaus, 1998; Yu et al., 1998).

SEPTAL PEPTIDOGLYCAN SYNTHESIS

The structure of peptidoglycan (also called murein) and how it is assembled in the periplasm are described in Chapter 11. In brief, peptidoglycan consists of long glycan strands joined by peptide cross-links. So far as is known, the peptidoglycan in the division septum is chemically indistinguishable from that in the cell cylinder (de Jonge et al., 1989; Obermann and Höltje, 1994). The terminal stages of peptidoglycan synthesis involve a lipid-linked disaccharide-pentapeptide precursor called lipid II that is assembled in the cytoplasm. Lipid II is transported across the membrane to the periplasm, where transglycosylases and transpeptidases incorporate the disaccharide-pentapeptide moiety into the existing sacculus. Several details of these incorporation steps are uncertain.

The protein(s) that catalyze translocation of lipid II across the cytoplasmic membrane have yet to be identified, but a host of transglycosylases and transpeptidases are known (see Chapter 11). One of these proteins, FtsI, is clearly required for septal peptidoglycan synthesis but not for elongation (Spratt and Pardee, 1975; Botta and Park, 1981). FtsI, also known as penicillin-binding protein 3 (PBP3), is a monofunctional transpeptidase and thus classified as a "Class B High-Molecular-Weight PBP." PBP3 localizes to the septal ring, where it cross-links septal peptidoglycan and recruits FtsN (Adam et al., 1997; Addinall et al., 1997; Weiss et al., 1997, 1999). The other Class B HMW PBP in *E. coli* is PBP2, which is specifically required for elongation rather than division, although it has been reported to localize to some extent to the division site and may be active there (Den Blaauwen et al., 2003).

It is presently unclear which proteins catalyze transglycosylation of septal peptidoglycan. There are four known candidates in *E. coli*: PBP1a, PBP1b, PBP1c, and MgtA. The first three proteins are bifunctional transglycosylase/transpeptidase enzymes and thus classified as "Class A High-Molecular-Weight PBPs." MgtA, in contrast, is a monofunctional transglycosylase. Because null mutations in any of the four respective genes have little or no discernible effect on cell morphology, it can be concluded that none is uniquely important for cell division. The most interesting reported synthetic phenotype is that simultaneous inactivation of PBP1a and PBP1b is lethal and causes rapid lysis (Yousif et al., 1985; Wientjes and Nanninga, 1991). Although lysis precedes any telltale morphological changes, it is speculated that PBP1a and PBP1b have redundant roles in murein synthesis during both elongation and division: Another hint that PBP1b may be active at the septum is that overproduction of dominant negative forms of PBP1b caused *E. coli* cells to lyse at the division site (Meisel et al., 2003). It would be interesting to know whether any of the known transglycosylases localize to the division site. This is the case in *B. subtilis*, where several Class A HMW PBPs localize to the septal ring (Scheffers et al., 2004). Among these proteins, PBP1 stands out because it shows strong septal localization, and mutants that lack PBP1 grow as filaments in media with low Mg^{2+} (Pedersen et al., 1999).

Copurification experiments in which cell extracts were passed over columns containing immobilized proteins involved in murein metabolism indicated that FtsI is found in a complex with numerous other enzymes involved in synthesis (PBP1b, PBP1c) or degradation (PBP4, PBP7, Slt70, MltB) of peptidoglycan (von Rechenberg et al., 1996; Schiffer and Höltje, 1999; Vollmer et al., 1999). Höltje postulates that FtsI is part of a large murein "holoenzyme" that contains both synthetic and hydrolytic activities (Höltje, 1998). Such a complex would have obvious utility for coordinating synthesis and degradation of peptidoglycan. Nevertheless, of the proteins involved, only FtsI is required for viability or cell division (Denome et al., 1999). Thus, if FtsI ordinarily functions in a complex with these other proteins, it must also be able to get along without them.

The onset of cell division is associated with a switch in sites of peptidoglycan synthesis from diffusely distributed throughout the cell cylinder in elongating cells to predominantly localized synthesis at the midcell during division (Wientjes and Nanninga, 1989). In principle, this switch could be accomplished by channeling substrate to the division site or allosteric activation of the peptidoglycan synthesis machinery in the septal ring. These two possibilities are not mutually exclusive, but so far there is only evidence for the latter. First, FtsI reacts about twice as fast with cephalexin in dividing as compared with nondividing cells (Eberhardt et al., 2003). Cephalexin is an FtsI-selective β-lactam and, more importantly, a substrate analogue. Second, the crystal structure of a bifunctional transglycosylase/transpeptidase from *S. pneumoniae* revealed a completely inaccessible transpeptidase active site, suggesting that interaction with other proteins and/or substrates is needed to open the active site (Macheboeuf et al., 2005). However, the crystal structure of a transpeptidase more closely related to FtsI did not reveal an occluded active site (Pares et al., 1996).

Activation of localized zones of murein synthesis even occurs in the absence of septation, e.g., in *ftsA*, *ftsI*, and *ftsQTs* mutants at the nonpermissive temperature (Taschner et al., 1988; de Pedro et al., 1997). These mutants therefore produce filaments with shallow indentations that can be seen in the electron microscope but are not readily detected by light microscopy. The peptidoglycan in these indentations has been called "preseptal" murein. Synthesis of preseptal murein is not inhibited by β-lactams that inactivate PBP1a, PBP1b, PBP2, or PBP3—essentially all the proteins thought to be required for the final stages of peptidoglycan synthesis! Nanninga has proposed that preseptal murein is synthesized by a penicillin-insensitive peptidoglycan synthesizing (PIPS) activity, but these hypothetical enzymes have yet to be identified (Nanninga, 1991). Alternatively, PIPS might reflect some degree of leakiness in the various ways (Ts mutants, antibiotic treatments) used to inhibit septation. Activation of murein synthesis at potential division sites is not observed in an *ftsZ84*(Ts) mutant (Taschner et al., 1988), implying that the Z ring activates this process. The glycan strands in preseptal murein are on average longer than those found in murein in the cell cylinder (Ishidate et al.,

1998). This could mean that transglycosylation is more processive during the initiation of septation or that lytic transglycosylases are less active in the absence of a Z ring.

Below we discuss several late recruits to the septal ring: FtsI, FtsW, the FtsQLB complex, and FtsN. Of these proteins, only FtsI is unambiguously involved in peptidoglycan synthesis, but multiple lines of circumstantial evidence also implicate FtsW in this process. The case for the involvement of the FtsQLB complex and FtsN is much weaker. In essence, these are all membrane proteins with the bulk of their sequence in the periplasm, and comparison of genomes from various bacteria indicates that they are restricted to organisms that have a peptidoglycan cell wall (but are not found in all such organisms).

FtsI

As mentioned above, FtsI (also called penicillin-binding protein 3, PBP3) is a transpeptidase required for synthesis of septal peptidoglycan. The protein has a simple overall structure: a short N-terminal cytoplasmic tail (~20 amino acids [a.a.]), a single transmembrane helix (~20 a.a.), and a large periplasmic region that includes both a domain of unknown function (~200 a.a.) and the transpeptidase catalytic domain (~340 a.a.) (Bowler and Spratt, 1989). The cytoplasmic domain is not essential—it can be replaced with green fluorescent protein (GFP) without loss of FtsI's ability to support cell division (Wissel et al., 2005). The TMH is the primary targeting domain for septal localization of FtsI (Piette et al., 2004; Wissel and Weiss, 2004; Wissel et al., 2005). Remarkably, Wissel et al. (2005) showed that a 26-a.a. fragment of FtsI that includes the TMH but little else is sufficient to target GFP to the septal ring, although it localizes less efficiently than does the full-length protein. Alanine-scanning mutagenesis revealed that residues important for septal localization lie on one face of the helix (Wissel et al., 2005). This argues for a protein-protein interaction, presumably with FtsW, which has 10 TMHs and appears to be the protein directly responsible for recruiting

FtsI to the septal ring (Mercer and Weiss, 2002). The domain of unknown function has been proposed to interact with other division proteins (Nguyen-Distèche et al., 1998). So far, there is genetic evidence consistent with the possibility that this domain interacts with FtsN (Wissel and Weiss, 2004) and FtsL (Karimova et al., 2005) and also activates the transpeptidase catalytic activity (Marrec-Fairley et al., 2000).

Finally, the catalytic domain is the site of the transpeptidase activity responsible for cross-linking peptidoglycan cell wall. It exhibits homology to other transpeptidases involved in peptidoglycan metabolism (Ghuysen, 1991). In particular, the active site contains a universally conserved serine residue (S307) that forms a covalent bond with the peptide substrate during transpeptidation, a reaction that proceeds via an acyl-enzyme intermediate (Broome-Smith et al., 1985; Houba-Herin et al., 1985; Keck et al., 1985; Nicholas et al., 1985). The catalytic domain also binds β-lactam antibiotics, which mimic a transpeptidase substrate and serve as suicide inhibitors by forming a long-lived covalent adduct with the catalytic serine. Note, however, that the authentic substrates used by FtsI in vivo are not yet known, and transpeptidation has so far been demonstrated only with substrate analogues (Adam et al., 1997). In particular, FtsI does not perform catalysis on lipid II, the canonical precursor for peptidoglycan synthesis (Adam et al., 1997). This finding probably means that FtsI uses a substrate other than lipid II, which might not be surprising considering that FtsI lacks a transglycosylase domain. Alternatively, lack of activity with lipid II could mean that FtsI only performs its normal reactions when in a complex with other division proteins.

The three-dimensional structure of the periplasmic domain from an FtsI homologue, PBP2x from *S. pneumoniae*, has been solved by X-ray crystallography (Pares et al., 1996; Dessen et al., 2001). The catalytic domain has a long groove that presumably accommodates the two peptides to be joined in the cross-linking reaction. The domain of unknown function has a strikingly elongated shape. One end fits into a pocket in the catalytic domain, and two long arms extend into solution, creating a large central cavity that could accommodate another protein in vivo. This arrangement suggests that the domain of unknown function regulates the catalytic domain and that the arms might engage in protein-protein interactions.

Cohen and coworkers observed that selective inactivation of FtsI with β-lactam antibiotics causes weak induction of the SOS response and provides modest protection against lysis (Miller et al., 2004). But another study did not find evidence for induction of genes involved in the SOS response (Arends and Weiss, 2004). The basis for this discrepancy is not clear. *E. coli* only goes through three to five mass doublings in the absence of division before the cells become overtly sick, so timing and growth rate could have a large impact on the outcome of studies of how cells respond to inactivation of FtsI (Ghigo et al., 1999; Weiss et al., 1999; Arends and Weiss, 2004).

FtsW

FtsW belongs to a large family of polytopic membrane proteins found in all bacteria that have a peptidoglycan cell wall (Ikeda et al., 1989; Joris et al., 1990; Margolin, 2000). This family is sometimes called "SEDS" for shape, elongation, division, and sporulation (Henriques et al., 1998). SEDS proteins work in conjunction with a class B high-molecular-weight penicillin-binding protein. *E. coli* has two such pairs, FtsW-FtsI for cell division and RodA-PBP2 for elongation. The genes for these protein pairs are generally cotranscribed and inactivation of either protein in the pair results in the same phenotype—e.g., a division defect (FtsI/FtsW) or an elongation defect (PBP2/RodA).

The topology of FtsW from *E. coli* and *S. pneumoniae* has been studied with computer predictions, gene fusions, and cysteine-accessibility methods (Gerard et al., 2002; Lara and Ayala, 2002). The topology is conserved, there being 10 TMHs and one large periplasmic loop (67 a.a. in *E. coli*) connecting TMH 7 to TMH 8.

The only clearly established function of FtsW is to recruit FtsI to the septal ring (Mercer and Weiss, 2002). FtsI and FtsW appear to interact in bacterial two-hybrid systems (Di Lallo et al., 2003; Karimova et al., 2005). Mutagenesis of the loop connecting TMH 9 to TMH 10 of FtsW impairs recruitment of FtsI (Pastoret et al., 2004). This was interpreted to mean that FtsI probably interacts with the TMH 9–10 loop, although the effect could be indirect.

An important question is whether FtsW has additional functions. There has been persistent speculation that FtsW is involved in translocation of lipid II (Ehlert and Höltje, 1996; Lara and Ayala, 2002). This hypothesis appears to be based on the fact that FtsW has multiple TMHs and works together with the peptidoglycan synthase FtsI. But topology is not a strong argument, and the fact that FtsW partners with FtsI could also be used to argue against the flippase hypothesis, because the direct acceptor of lipid II ought to be a transglycosylase, not a monofunctional transpeptidase such as FtsI. The flippase hypothesis has been tested directly by looking at the effect of FtsW depletion on accumulation of nucleotide-linked precursors for murein synthesis (Lara et al., 2005a). UDP-MurNAc-pentapeptide did not accumulate, suggesting FtsW is not a flippase. One concern with this interpretation is that the cells were still elongating owing to continued function of the FtsW homologue RodA, which might have obscured the result.

The FtsQLB Complex

Although these three proteins differ greatly in size, they have a similar architecture: a small cytoplasmic domain, a single transmembrane helix, and a comparatively large periplasmic domain. As summarized above (see "Assembly of the Septal Ring: a Model for Protein-Protein Interaction?"), multiple lines of evidence indicate that these three proteins form a complex and that this complex interacts with numerous additional division proteins. FtsQ stands out because bacterial two-hybrid analyses indicate that it interacts with at least eight different division proteins, suggesting that FtsQ serves as a central organizer during septal ring assembly (Karimova et al., 2005). Nevertheless, further work is needed to determine which of the reported interactions are direct and which are physiologically significant.

We suspect that the FtsQLB complex may serve only as an assembly factor. The largest domains in FtsL and FtsB are their periplasmic domains, at ~60 and ~80 a.a. These domains consist largely of coiled-coil motif that presumably mediates the association of these two proteins (Guzman et al., 1992; Buddelmeijer et al., 2002). Moreover, the amino acid sequences of FtsL and FtsB homologues from different bacteria are very poorly conserved, to the point that identification of homologues is often based on finding a short open reading frame in the right position (e.g., next to *ftsI* in the case of *ftsL*) that codes for a protein predicted to have the right topology and secondary structure. Enzymatic domains should exhibit stronger conservation than this. The periplasmic domain of FtsQ is ~225 residues in size but exhibits a striking lack of sequence conservation as noted for FtsL and FtsB.

FtsN

FtsN is yet another "late" division protein with a short N-terminal cytoplasmic domain, a single transmembrane helix, and a large periplasmic domain (~265 a.a.) (Dai et al., 1996). Neither the cytoplasmic nor transmembrane domains appear to be essential because replacing them with a cleavable signal sequence from MalE did not prevent FtsN from supporting cell division (Dai et al., 1996). This construct should have resulted in release of FtsN's periplasmic domain into the periplasm as a soluble protein, although depending on the level of overproduction and kinetics of cleavage, some fraction of the fusion protein may have retained the MalE fragment as a membrane anchor. The periplasmic domain is predominantly unstructured, except for the final ~75 amino acids, which fold into a peptidoglycan-binding domain (Ursinus et al., 2004; Yang et al., 2004). The position of FtsN in the recruitment hierarchy suggests it may interact with FtsI and

AmiC, the only two septal ring proteins with clearly established enzymatic activities related to peptidoglycan metabolism (Addinall et al., 1997; Chen and Beckwith, 2001; Bernhardt and de Boer, 2004). Although these findings seem to implicate FtsN somehow in murein metabolism, an FtsN derivative lacking the peptidoglycan-binding domain still supports cell division (Ursinus et al., 2004). Moreover, although analysis of *ftsI* mutants that recruit FtsN poorly suggested a site in FtsI that interacts directly with FtsN (Wissel and Weiss, 2004), the fact that FtsI's presence in the septal ring is not sufficient to recruit FtsN (Goehring et al., 2005) raises the possibility that the FtsI-FtsN interaction is indirect. An early report claimed that inactivation of FtsI's transpeptidase activity with a β-lactam prevented septal localization of FtsN, raising the possibility that FtsN recognizes an FtsI-induced modification of the peptidoglycan. However, we find that FtsN localizes efficiently after inhibition of FtsI (M. C. Wissel and D. S. Weiss, unpublished data).

FtsN was discovered as a multicopy suppressor of an *ftsA*(Ts) mutation (Dai et al., 1993). Curiously, overproduction of FtsN also rescues cell division in *ftsK*(Ts), *ftsQ*(Ts), *ftsI*(Ts), and *ftsEX* depletion strains too (Dai et al., 1993; Draper et al., 1998). The basis of the suppression is not clear. We have also found that FtsN's dependence on upstream proteins for septal localization is somewhat leaky, such that overproduction of FtsN partially rescues localization in several mutant backgrounds—*ftsA*(Ts), *ftsQ*(Ts), and *ftsI*(Ts), but not *ftsZ*(Ts) (M. C. Wissel, K. L. Schmidt, and D. S. Weiss, unpublished data). This finding raises questions about the target recognized by FtsN in the septal ring and about the validity of using GFP-FtsN fusion proteins to infer that a complete septal ring is present in some mutant backgrounds, as the fusion protein could localize even if some "upstream" proteins were absent or present in reduced amounts.

SEPARATION OF DAUGHTER CELLS

Cytokinesis ends with the nascent daughter cells still joined by a shared peptidoglycan wall, which must be selectively hydrolyzed to separate the two cells. As summarized in Chapter 11, *E. coli* has 17 known murein hydrolases. These can be classified based on their enzymatic activities—amidases and endopeptidases cleave peptide cross-links, lytic transglycosylases degrade glycan strands. The physiological roles of the various hydrolases have been difficult to define owing to extensive functional redundancy. Höltje and coworkers constructed a set of mutants lacking various combinations of these enzymes (Heidrich et al., 2001, 2002). A mutant lacking all known lytic transglycosylases formed chains of three to eight cells, with approximately half of the cells being in chains. A mutant lacking all known amidases had a stronger chaining phenotype, with 90 to 100% of the cells in chains of 6 to 24 cells. Although a mutant lacking multiple endopeptidases did not form chains, deleting endopeptidases from strains that lacked lytic transglycosylases or amidases exacerbated the chaining phenotype of those mutants. Taken together, these data indicate that all classes of murein hydrolases contribute to cell separation, with amidases being the most important players. Using completely different approaches, Bernhard and de Boer found that two amidases (AmiA and AmiC) and a probable endopeptidase (EnvC) are important for cell separation and that AmiC and EnvC localize to the septal ring (Bernhardt and de Boer, 2003, 2004).

CONSTRICTION OF THE OUTER MEMBRANE

The outer membrane is considered to invaginate simultaneously with the rest of the cell envelope in *E. coli*. However, a recent study showing that the outer membrane lags behind in *Caulobacter crescentus* (Judd et al., 2005) suggests the sequence of events in *E. coli* ought to be revisited with more modern methods for preservation of the cell's architecture. No outer membrane proteins are known to be specifically required for cell division, but a mechanism for outer membrane constriction in *E. coli* has been proposed based on the fact that two very abundant proteins, Braun's lipoprotein and OmpA, connect the outer membrane to the murein sacculus. This suggests that ongoing incorporation of these proteins during septum assembly would cause the outer membrane to follow the peptidoglycan layer passively. Consistent with this hypothesis, outer membranes invaginate poorly in *lkyD* and *lpo* mutants (Weigand et al., 1976; Fung et al., 1978). These mutants are rather pleiotropic—they shed outer membrane, leak periplasmic proteins, and are permeable to many drugs—making it difficult to be certain that poor invagination is a primary defect.

QUESTIONS FOR THE FUTURE

It is clear that the Z ring orchestrates septal peptidoglycan synthesis and that it does so by recruiting a host of proteins to the division site. But only one of these proteins has a defined role in that process: the transpeptidase FtsI. The functions of the remaining proteins are largely mysterious, except for their role(s) in septal ring assembly. Moreover, it is not clear whether the proteins discussed here constitute a complete set, at least in *E. coli*. A number of proteins known to be involved in division in other bacteria, but lacking apparent homologues in *E. coli*, were not considered in this discussion. Finally, some important activities have yet to be accounted for, including the protein that transports lipid II to the periplasm and (in *E. coli*) the transglycosylase(s) involved in septal murein synthesis. The lack of an outer membrane protein mutant with a clear *fts* phenotype or a specific defect in constriction of the outer membrane is also noteworthy.

It is clear that the initial stages of septal ring assembly involve direct interactions of FtsZ with FtsA, ZipA, and ZapA. It is also clear that interactions among the late proteins play a role in assembly. Less clear, however, is which of the reported late interactions are direct and how these interactions contribute to assembly, regulation, and signaling.

It is clear the Z ring constricts to remain at the leading edge of the developing septum. This could mean FtsZ generates force for cytokinesis, serves as a moving scaffold, or both.

Knowing more about the architecture of the Z ring might clarify these issues, although biophysical methods are likely to be needed to test ideas about force generation.

We have a good framework for bacterial cell division, but many important features are still missing.

ACKNOWLEDGMENTS

We thank Nate Goehring and Waldemar Vollmer for comments that significantly improved the manuscript.

S.J.R.A. was supported by an NIH Training Grant in Biotechnology (T32 GM08365). Work in D.S.W.'s lab has been supported by R01 GM59893 from the NIH.

REFERENCES

Aarsman, M. E., A. Piette, C. Fraipont, T. M. Vinkenvleugel, M. Nguyen-Distèche, and T. den Blaauwen. 2005. Maturation of the *Escherichia coli* divisome occurs in two steps. *Mol. Microbiol.* **55:**1631–1645.

Adam, M., C. Fraipont, N. Rhazi, M. Nguyen-Distèche, B. Lakaye, J. M. Frere, B. Devreese, J. Van Beeumen, Y. van Heijenoort, J. van Heijenoort, and J. M. Ghuysen. 1997. The bimodular G57-V577 polypeptide chain of the class B penicillin-binding protein 3 of *Escherichia coli* catalyzes peptide bond formation from thiolesters and does not catalyze glycan chain polymerization from the lipid II intermediate. *J. Bacteriol.* **179:**6005–6009.

Addinall, S. G., C. Cao, and J. Lutkenhaus. 1997. FtsN, a late recruit to the septum in *Escherichia coli*. *Mol. Microbiol.* **25:**303–309.

Addinall, S. G., and J. Lutkenhaus. 1996. FtsA is localized to the septum in an FtsZ-dependent manner. *J. Bacteriol.* **178:**7167–7172.

Anderson, D. E., F. J. Gueiros-Filho, and H. P. Erickson. 2004. Assembly dynamics of FtsZ rings in *Bacillus subtilis* and *Escherichia coli* and effects of FtsZ-regulating proteins. *J. Bacteriol.* **186:**5775–5781.

Arends, S. J., and D. S. Weiss. 2004. Inhibiting cell division in *Escherichia coli* has little if any effect on gene expression. *J. Bacteriol.* **186:**880–884.

Aussel, L., F. X. Barre, M. Aroyo, A. Stasiak, A. Z. Stasiak, and D. Sherratt. 2002. FtsK is a DNA motor protein that activates chromosome dimer resolution by switching the catalytic state of the XerC and XerD recombinases. *Cell* **108:**195–205.

Bath, J., L. J. Wu, J. Errington, and J. C. Wang. 2000. Role of *Bacillus subtilis* SpoIIIE in DNA transport across the mother cell-prespore division septum. *Science* **290:**995–997.

Beall, B., and J. Lutkenhaus. 1992. Impaired cell division and sporulation of a *Bacillus subtilis* strain with the ftsA gene deleted. *J. Bacteriol.* **174:**2398–2403.

Begg, K. J., S. J. Dewar, and W. D. Donachie. 1995. A new *Escherichia coli* cell division gene, *ftsK*. *J. Bacteriol.* **177:**6211–6222.

Ben-Yehuda, S., and R. Losick. 2002. Asymmetric cell division in *B. subtilis* involves a spiral-like intermediate of the cytokinetic protein FtsZ. *Cell* **109:**257–266.

Bernhardt, T. G., and P. A. de Boer. 2003. The *Escherichia coli* amidase AmiC is a periplasmic septal ring component exported via the twin-arginine transport pathway. *Mol. Microbiol.* **48:**1171–1182.

Bernhardt, T. G., and P. A. de Boer. 2004. Screening for synthetic lethal mutants in *Escherichia coli* and identification of EnvC (YibP) as a periplasmic septal ring factor with murein hydrolase activity. *Mol. Microbiol.* **52:**1255–1269.

Bernhardt, T. G., and P. A. de Boer. 2005. SlmA, a nucleoid-associated, FtsZ binding protein required for blocking septal ring assembly over chromosomes in *E. coli*. *Mol. Cell* **18:**555–564.

Bigot, S., J. Corre, J. M. Louarn, F. Cornet, and F. X. Barre. 2004. FtsK activities in Xer recombination, DNA mobilization and cell division involve overlapping and separate domains of the protein. *Mol. Microbiol.* **54:**876–886.

Bigot, S., O. A. Saleh, C. Lesterlin, C. Pages, M. El Karoui, C. Dennis, M. Grigoriev, J. F. Allemand, F. X. Barre, and F. Cornet. 2005. KOPS: DNA motifs that control *E. coli* chromosome segregation by orienting the FtsK translocase. *EMBO J.* **24:**3770–3780.

Bork, P., C. Sander, and A. Valencia. 1992. An ATPase domain common to prokaryotic cell cycle proteins, sugar kinases, actin, and hsp70 heat shock proteins. *Proc. Natl. Acad. Sci. USA* **89:**7290–7294.

Botta, G. A., and J. T. Park. 1981. Evidence for involvement of penicillin-binding protein 3 in murein synthesis during septation but not during cell elongation. *J. Bacteriol.* **145:**333–340.

Bouige, P., D. Laurent, L. Piloyan, and E. Dassa. 2002. Phylogenetic and functional classification of ATP-binding cassette (ABC) systems. *Curr. Protein Peptide Sci.* **3:**541–559.

Bowler, L. D., and B. G. Spratt. 1989. Membrane topology of penicillin-binding protein 3 of *Escherichia coli*. *Mol. Microbiol.* **3:**1277–1286.

Broome-Smith, J. K., P. J. Hedge, and B. G. Spratt. 1985. Production of thiol-penicillin-binding protein 3 of *Escherichia coli* using a two primer method of site-directed mutagenesis. *EMBO J.* **4:**231–235.

Buddelmeijer, N., and J. Beckwith. 2002. Assembly of cell division proteins at the *E. coli* cell center. *Curr. Opin. Microbiol.* **5:**553–557.

Buddelmeijer, N., and J. Beckwith. 2004. A complex of the *Escherichia coli* cell division proteins FtsL, FtsB and FtsQ forms independently of its localization to the septal region. *Mol. Microbiol.* **52:**1315–1327.

Buddelmeijer, N., N. Judson, D. Boyd, J. J. Mekalanos, and J. Beckwith. 2002. YgbQ, a cell division protein in *Escherichia coli* and *Vibrio cholerae*, localizes in codependent fashion with FtsL to the division site. *Proc. Natl. Acad. Sci. USA* **99:**6316–6321.

Butland, G., J. M. Peregrin-Alvarez, J. Li, W. Yang, X. Yang, V. Canadien, A. Starostine, D. Richards, B. Beattie, N. Krogan, M. Davey, J. Parkinson, J. Greenblatt, and A. Emili. 2005. Interaction network containing conserved and essential protein complexes in *Escherichia coli*. *Nature* **433:**531–537.

Capiaux, H., C. Lesterlin, K. Perals, J. M. Louarn, and F. Cornet. 2002. A dual role for the FtsK protein in *Escherichia coli* chromosome segregation. *EMBO Rep.* **3:**532–536.

Chen, J. C., and J. Beckwith. 2001. FtsQ, FtsL and FtsI require FtsK, but not FtsN, for co-localization with FtsZ during *Escherichia coli* cell division. *Mol. Microbiol.* **42:**395–413.

Chen, J. C., D. S. Weiss, J. M. Ghigo, and J. Beckwith. 1999. Septal localization of FtsQ, an essential cell division protein in *Escherichia coli*. *J. Bacteriol.* **181:**521–530.

Corbin, B. D., B. Geissler, M. Sadasivam, and W. Margolin. 2004. Z-ring-independent interaction between a subdomain of FtsA and late septation proteins as revealed by a polar recruitment assay. *J. Bacteriol.* **186:**7736–7744.

Cordell, S. C., E. J. H. Robinson, and J. Löwe. 2003. Crystal structure of the SOS cell division inhibitor SulA and in complex with FtsZ. *Proc. Natl. Acad. Sci. USA* **100:**7889–7894.

Corre, J., and J. M. Louarn. 2002. Evidence from terminal recombination gradients that FtsK uses replichore polarity to control chromosome terminus positioning at division in *Escherichia coli*. *J. Bacteriol.* **184:**3801–3807.

Courcelle, J., A. Khodursky, B. Peter, P. O. Brown, and P. C. Hanawalt. 2001. Comparative gene expression profiles following UV exposure in wild-type and SOS-deficient *Escherichia coli*. *Genetics* **158:**41–64.

Dai, K., Y. Xu, and J. Lutkenhaus. 1993. Cloning and characterization of *ftsN*, an essential cell division gene in *Escherichia coli* isolated as a multicopy suppressor of *ftsA12*(Ts). *J. Bacteriol.* **175:**3790–3797.

Dai, K., Y. Xu, and J. Lutkenhaus. 1996. Topological characterization of the essential *Escherichia coli* cell division protein FtsN. *J. Bacteriol.* **178:**1328–1334.

Daniel, R. A., and J. Errington. 2000. Intrinsic instability of the essential cell division protein FtsL of *Bacillus subtilis* and a role for DivIB protein in FtsL turnover. *Mol. Microbiol.* **36:**278–289.

Daniel, R. A., E. J. Harry, V. L. Katis, R. G. Wake, and J. Errington. 1998. Characterization of the essential cell division gene *ftsL*(*yIID*) of *Bacillus subtilis* and its role in the assembly of the division apparatus. *Mol. Microbiol.* **29:**593–604.

de Boer, P. A., R. E. Crossley, and L. I. Rothfield. 1989. A division inhibitor and a topological specificity factor coded for by the minicell locus determine proper placement of the division septum in *E. coli*. *Cell* **56:**641–649.

de Jonge, B. L., F. B. Wientjes, I. Jurida, F. Driehuis, J. T. Wouters, and N. Nanninga. 1989. Peptidoglycan synthesis during the cell cycle of *Escherichia coli*: composition and mode of insertion. *J. Bacteriol.* **171:**5783–5794.

de Leeuw, E., B. Graham, G. J. Phillips, C. M. ten Hagen-Jongman, B. Oudega, and J. Luirink. 1999. Molecular characterization of *Escherichia coli* FtsE and FtsX. *Mol. Microbiol.* **31:**983–993.

Den Blaauwen, T., M. E. Aarsman, N. O. Vischer, and N. Nanninga. 2003. Penicillin-binding protein PBP2 of *Escherichia coli* localizes preferentially in the lateral wall and at mid-cell in comparison with the old cell pole. *Mol. Microbiol.* **47:**539–547.

Den Blaauwen, T., N. Buddelmeijer, M. E. Aarsman, C. M. Hameete, and N. Nanninga. 1999. Timing of FtsZ assembly in *Escherichia coli*. *J. Bacteriol.* **181:**5167–5175.

Denome, S. A., P. K. Elf, T. A. Henderson, D. E. Nelson, and K. D. Young. 1999. *Escherichia coli* mutants lacking all possible combinations of eight penicillin binding proteins: viability, characteristics, and implications for peptidoglycan synthesis. *J. Bacteriol.* **181:**3981–3993.

de Pedro, M. A., J. C. Quintela, J. V. Höltje, and H. Schwarz. 1997. Murein segregation in *Escherichia coli*. *J. Bacteriol.* **179:**2823–2834.

Dessen, A., N. Mouz, E. Gordon, J. Hopkins, and O. Dideberg. 2001. Crystal structure of PBP2x from a highly penicillin-resistant *Streptococcus pneumoniae* clinical isolate: a mosaic framework containing 83 mutations. *J. Biol. Chem.* **276:**45106–45112.

Diez, A. A., A. Farewell, U. Nannmark, and T. Nystrom. 1997. A mutation in the *ftsK* gene of *Escherichia coli* affects cell-cell separation, stationary-phase survival, stress adaptation, and expression of the gene encoding the stress protein UspA. *J. Bacteriol.* **179:**5878–5883.

Di Lallo, G., M. Fagioli, D. Barionovi, P. Ghelardini, and L. Paolozzi. 2003. Use of a two-hybrid assay to study the assembly of a complex multicomponent protein machinery: bacterial septosome differentiation. *Microbiology* **149:**3353–3359.

Draper, G. C., N. McLennan, K. Begg, M. Masters, and W. D. Donachie. 1998. Only the N-terminal domain of FtsK functions in cell division. *J. Bacteriol.* **180:**4621–4627.

Eberhardt, C., L. Kuerschner, and D. S. Weiss. 2003. Probing the catalytic activity of a cell division-specific transpeptidase in vivo with beta-lactams. *J. Bacteriol.* **185:**3726–3734.

Ehlert, K., and J. V. Höltje. 1996. Role of precursor translocation in coordination of murein and phospholipid synthesis in *Escherichia coli. J. Bacteriol.* **178:**6766–6771.

Errington, J., J. Bath, and L. J. Wu. 2001. DNA transport in bacteria. *Nat. Rev. Mol. Cell Biol.* **2:**538–545.

Errington, J., R. A. Daniel, and D. J. Scheffers. 2003. Cytokinesis in bacteria. *Microbiol. Mol. Biol. Rev.* **67:**52–65.

Espeli, O., C. Lee, and K. J. Marians. 2003. A physical and functional interaction between *Escherichia coli* FtsK and topoisomerase IV. *J. Biol. Chem.* **278:**44639–44644.

Feucht, A., I. Lucet, M. D. Yudkin, and J. Errington. 2001. Cytological and biochemical characterization of the FtsA cell division protein of *Bacillus subtilis. Mol. Microbiol.* **40:**115–125.

Fung, J., T. J. MacAlister, and L. I. Rothfield. 1978. Role of murein lipoprotein in morphogenesis of the bacterial division septum: phenotypic similarity of *lkyD* and *lpo* mutants. *J. Bacteriol.* **133:**1467–1471.

Geissler, B., D. Elraheb, and W. Margolin. 2003. A gain-of-function mutation in *ftsA* bypasses the requirement for the essential cell division gene *zipA* in *Escherichia coli. Proc. Natl. Acad. Sci. USA* **100:**4197–4202.

Geissler, B., and W. Margolin. 2005. Evidence for functional overlap among multiple bacterial cell division proteins: compensating for the loss of FtsK. *Mol. Microbiol.* **58:**596–612.

Gerard, P., T. Vernet, and A. Zapun. 2002. Membrane topology of the *Streptococcus pneumoniae* FtsW division protein. *J. Bacteriol.* **184:**1925–1931.

Ghigo, J. M., D. S. Weiss, J. C. Chen, J. C. Yarrow, and J. Beckwith. 1999. Localization of FtsL to the *Escherichia coli* septal ring. *Mol. Microbiol.* **31:**725–737.

Ghuysen, J. M. 1991. Serine beta-lactamases and penicillin-binding proteins. *Annu. Rev. Microbiol.* **45:**37–67.

Gibbs, T. W., D. R. Gill, and G. P. Salmond. 1992. Localised mutagenesis of the *fts YEX* operon: conditionally lethal missense substitutions in the FtsE cell division protein of *Escherichia coli* are similar to those found in the cystic fibrosis transmembrane conductance regulator protein (CFTR) of human patients. *Mol. Gen. Genet.* **234:**121–128.

Gill, D. R., G. F. Hatfull, and G. P. Salmond. 1986. A new cell division operon in *Escherichia coli. Mol. Gen. Genet.* **205:**134–145.

Goehring, N. W., and J. Beckwith. 2005. Diverse paths to midcell: assembly of the bacterial cell division machinery. *Curr. Biol.* **15:**R514–R526.

Goehring, N. W., F. Gueiros-Filho, and J. Beckwith. 2005. Premature targeting of a cell division protein to midcell allows dissection of divisome assembly in *Escherichia coli. Genes Dev.* **19:**127–137.

Gueiros-Filho, F. J., and R. Losick. 2002. A widely conserved bacterial cell division protein that promotes assembly of the tubulin-like protein FtsZ. *Genes Dev.* **16:**2544–2556.

Guzman, L. M., J. J. Barondess, and J. Beckwith. 1992. FtsL, an essential cytoplasmic membrane protein involved in cell division in *Escherichia coli. J. Bacteriol.* **174:**7716–7728.

Hale, C. A., and P. A. de Boer. 1997. Direct binding of FtsZ to ZipA, an essential component of the septal ring structure that mediates cell division in *E. coli. Cell* **88:**175–185.

Hale, C. A., and P. A. de Boer. 1999. Recruitment of ZipA to the septal ring of *Escherichia coli* is dependent on FtsZ and independent of FtsA. *J. Bacteriol.* **181:**167–176.

Hale, C. A., A. C. Rhee, and P. A. de Boer. 2000. ZipA-induced bundling of FtsZ polymers mediated by an interaction between C-terminal domains. *J. Bacteriol.* **182:**5153–5166.

Heidrich, C., M. F. Templin, A. Ursinus, M. Merdanovic, J. Berger, H. Schwarz, M. A. de Pedro, and J. V. Höltje. 2001. Involvement of N-acetylmuramyl-L-alanine amidases in cell separation and antibiotic-induced autolysis of *Escherichia coli. Mol. Microbiol.* **41:**167–178.

Heidrich, C., A. Ursinus, J. Berger, H. Schwarz, and J. V. Höltje. 2002. Effects of multiple deletions of murein hydrolases on viability, septum cleavage, and sensitivity to large toxic molecules in *Escherichia coli. J. Bacteriol.* **184:**6093–6099.

Henriques, A. O., P. Glaser, P. J. Piggot, and C. P. Moran, Jr. 1998. Control of cell shape and elongation by the rodA gene in *Bacillus subtilis. Mol. Microbiol.* **28:**235–247.

Höltje, J. V. 1998. Growth of the stress-bearing and shape-maintaining murein sacculus of *Escherichia coli. Microbiol. Mol. Biol. Rev.* **62:**181–203.

Houba-Herin, N., H. Hara, M. Inouye, and Y. Hirota. 1985. Binding of penicillin to thiol-penicillin-binding protein 3 of *Escherichia coli*: identification of its active site. *Mol. Gen. Genet.* **201:**499–504.

Hu, Z., and J. Lutkenhaus. 1999. Topological regulation of cell division in *Escherichia coli* involves rapid pole to pole oscillation of the division inhibitor MinC under the control of MinD and MinE. *Mol. Microbiol.* **34:**82–90.

Hu, Z., A. Mukherjee, S. Pichoff, and J. Lutken-haus. 1999. The MinC component of the division site selection system in *Escherichia coli* interacts with FtsZ to prevent polymerization. *Proc. Natl. Acad. Sci. USA* **96:**14819–14824.

Huisman, O., R. D'Ari, and S. Gottesman. 1984. Cell-division control in *Escherichia coli*: specific induction of the SOS function SfiA protein is sufficient to block septation. *Proc. Natl. Acad. Sci. USA* **81:**4490–4494.

Ikeda, M., T. Sato, M. Wachi, H. K. Jung, F. Ishino, Y. Kobayashi, and M. Matsuhashi. 1989. Structural similarity among *Escherichia coli* FtsW and RodA proteins and *Bacillus subtilis* SpoVE protein, which function in cell division, cell elongation, and spore formation, respectively. *J. Bacteriol.* **171:**6375–6378.

Ishidate, K., A. Ursinus, J. V. Höltje, and L. Roth-field. 1998. Analysis of the length distribution of murein glycan strands in *ftsZ* and *ftsI* mutants of *E. coli. FEMS Microbiol. Lett.* **168:**71–75.

Jensen, S. O., L. S. Thompson, and E. J. Harry. 2005. Cell division in *Bacillus subtilis*: FtsZ and FtsA association is Z-ring independent, and FtsA is required for efficient midcell Z-Ring assembly. *J. Bacteriol.* **187:**6536–6544.

Johnson, J. E., L. L. Lackner, C. A. Hale, and P. A. de Boer. 2004. ZipA is required for targeting of DMinC/DicB, but not DMinC/MinD, complexes to septal ring assemblies in *Escherichia coli. J. Bacteriol.* **186:**2418–2429.

Joris, B., G. Dive, A. Henriques, P. J. Piggot, and J. M. Ghuysen. 1990. The life-cycle proteins RodA of *Escherichia coli* and SpoVE of *Bacillus subtilis* have very similar primary structures. *Mol. Microbiol.* **4:**513–517.

Judd, E. M., L. R. Comolli, J. C. Chen, K. H. Downing, W. E. Moerner, and H. H. McAdams. 2005. Distinct constrictive processes, separated in time and space, divide caulobacter inner and outer membranes. *J. Bacteriol.* **187:**6874–6882.

Karimova, G., N. Dautin, and D. Ladant. 2005. Interaction network among *Escherichia coli* membrane proteins involved in cell division as revealed by bacterial two-hybrid analysis. *J. Bacteriol.* **187:**2233–2243.

Katis, V. L., R. G. Wake, and E. J. Harry. 2000. Septal localization of the membrane-bound division proteins of *Bacillus subtilis* DivIB and DivIC is codependent only at high temperatures and requires FtsZ. *J. Bacteriol.* **182:**3607–3611.

Keck, W., B. Glauner, U. Schwarz, J.K. Broome-Smith, and B.G. Spratt. 1985. Sequences of the active-site peptides of three of the high-Mr penicillin-binding proteins of *Escherichia coli* K-12. *Proc. Natl. Acad. Sci. USA* **82:**1999–2003.

Kobayashi, N., K. Nishino, and A. Yamaguchi. 2001. Novel macrolide-specific ABC-type efflux transporter in *Escherichia coli. J. Bacteriol.* **183:**5639–5644.

Lara, B., and J. A. Ayala. 2002. Topological characterization of the essential *Escherichia coli* cell division protein FtsW. *FEMS Microbiol. Lett.* **216:**23–32.

Lara, B., D. Mengin-Lecreulx, J. A. Ayala, and J. van Heijenoort. 2005a. Peptidoglycan precursor pools associated with MraY and FtsW deficiencies or antibiotic treatments. *FEMS Microbiol. Lett.* **250:**195–200.

Lara, B., A. I. Rico, S. Petruzzelli, A. Santona, J. Dumas, J. Biton, M. Vicente, J. Mingorance, and O. Massidda. 2005b. Cell division in cocci: localization and properties of the *Streptococcus pneumoniae* FtsA protein. *Mol. Microbiol.* **55:**699–711.

Levy, O., J. L. Ptacin, P. J. Pease, J. Gore, M. B. Eisen, C. Bustamante, and N. R. Cozzarelli. 2005. Identification of oligonucleotide sequences that direct the movement of the *Escherichia coli* FtsK translocase. *Proc. Natl. Acad. Sci. USA* **102:**17618–17623.

Liu, N.-J. L., R. J. Dutton, and K. Pogliano. 2006. Evidence that the SpoIIIE DNA translocase participates in membrane fusion during cytokinesis and engulfment. *Mol. Microbiol.* **59:**1097–1113.

Liu, Z., A. Mukherjee, and J. Lutkenhaus. 1999. Recruitment of ZipA to the division site by interaction with FtsZ. *Mol. Microbiol.* **31:**1853–1861.

Low, H. H., M. C. Moncrieffe, and J. Lowe. 2004. The crystal structure of ZapA and its modulation of FtsZ polymerisation. *J. Mol. Biol.* **341:**839–852.

Lu, C., M. Reedy, and H. P. Erickson. 2000. Straight and curved conformations of FtsZ are regulated by GTP hydrolysis. *J. Bacteriol.* **182:**164–170.

Macheboeuf, P., A. M. Di Guilmi, V. Job, T. Vernet, O. Dideberg, and A. Dessen. 2005. Active site restructuring regulates ligand recognition in class A penicillin-binding proteins. *Proc. Natl. Acad. Sci. USA* **102:**577–582.

Margolin, W. 2000. Themes and variations in prokaryotic cell division. *FEMS Microbiol. Rev.* **24:**531–548.

Marrec-Fairley, M., A. Piette, X. Gallet, R. Brasseur, H. Hara, C. Fraipont, J. M. Ghuysen, and M. Nguyen-Distèche. 2000. Differential functionalities of amphiphilic peptide segments of the cell-septation penicillin-binding protein 3 of *Escherichia coli. Mol. Microbiol.* **37:**1019–1031.

Marston, A. L., and J. Errington. 1999. Selection of the midcell division site in *Bacillus subtilis* through MinD-dependent polar localization and activation of MinC. *Mol. Microbiol.* **33:**84–96.

Marston, A. L., H. B. Thomaides, D. H. Edwards, M. E. Sharpe, and J. Errington. 1998. Polar localization of the MinD protein of *Bacillus subtilis* and its role in selection of the mid-cell division site. *Genes Dev.* **12:**3419–3430.

Massey, T. H., L. Aussel, F. X. Barre, and D. J. Sherratt. 2004. Asymmetric activation of Xer site-specific recombination by FtsK. *EMBO Rep.* **5:** 399–404.

Meisel, U., J. V. Höltje, and W. Vollmer. 2003. Overproduction of inactive variants of the murein synthase PBP1B causes lysis in *Escherichia coli*. *J. Bacteriol.* **185:**5342–5348.

Mercer, K. L., and D. S. Weiss. 2002. The *Escherichia coli* cell division protein FtsW is required to recruit its cognate transpeptidase, FtsI (PBP3), to the division site. *J. Bacteriol.* **184:**904–912.

Miller, C., L. E. Thomsen, C. Gaggero, R. Mosseri, H. Ingmer, and S. N. Cohen. 2004. SOS response induction by beta-lactams and bacterial defense against antibiotic lethality. *Science* **305:** 1629–1631.

Mizusawa, S., and S. Gottesman. 1983. Protein degradation in *Escherichia coli*: the *lon* gene controls the stability of *sulA* protein. *Proc. Natl. Acad. Sci. USA* **80:**358–362.

Mosyak, L., Y. Zhang, E. Glasfeld, S. Haney, M. Stahl, J. Seehra, and W. S. Somers. 2000. The bacterial cell-division protein ZipA and its interaction with an FtsZ fragment revealed by X-ray crystallography. *EMBO J.* **19:**3179–3191.

Mukherjee, A., C. Cao, and J. Lutkenhaus. 1998. Inhibition of FtsZ polymerization by SulA, an inhibitor of septation in *Escherichia coli*. *Proc. Natl. Acad. Sci. USA* **95:**2885–2890.

Nanninga, N. 1991. Cell division and peptidoglycan assembly in *Escherichia coli*. *Mol Microbiol.* **5:**791–795.

Narita, S. I., and H. Tokuda. 2005. An ABC transporter mediating the membrane detachment of bacterial lipoproteins depending on their sorting signals. *FEBS Lett.* **580:**1164–1170.

Newman, E. B., L. I. Budman, E. C. Chan, R. C. Greene, R. T. Lin, C. L. Woldringh, and R. D'Ari. 1998. Lack of S-adenosylmethionine results in a cell division defect in *Escherichia coli*. *J. Bacteriol.* **180:**3614–3619.

Nguyen-Distèche, M., C. Fraipont, N. Buddelmeijer, and N. Nanninga. 1998. The structure and function of *Escherichia coli* penicillin-binding protein 3. *Cell. Mol. Life Sci.* **54:**309–316.

Nicholas, R. A., J. L. Strominger, H. Suzuki, and Y. Hirota. 1985. Identification of the active site in penicillin-binding protein 3 of *Escherichia coli*. *J. Bacteriol.* **164:**456–460.

Noirclerc-Savoye, M., A. Le Gouellec, C. Morlot, O. Dideberg, T. Vernet, and A. Zapun. 2005. In vitro reconstitution of a trimeric complex of DivIB, DivIC and FtsL, and their transient co-localization at the division site in *Streptococcus pneumoniae*. *Mol. Microbiol.* **55:**413–424.

Obermann, W., and J. V. Höltje. 1994. Alterations of murein structure and of penicillin-binding proteins in minicells from *Escherichia coli*. *Microbiology* **140:**79–87.

O'Reilly, E. K., and K. N. Kreuzer. 2004. Isolation of SOS constitutive mutants of *Escherichia coli*. *J. Bacteriol.* **186:**7149–7160.

Paradis-Bleau, C., F. Sanschagrin, and R. C. Levesque. 2005. Peptide inhibitors of the essential cell division protein FtsA. *Protein Eng. Des. Sel.* **18:**85–91.

Pares, S., N. Mouz, Y. Petillot, R. Hakenbeck, and O. Dideberg. 1996. X-ray structure of *Streptococcus pneumoniae* PBP2x, a primary penicillin target enzyme. *Nat. Struct. Biol.* **3:**284–289.

Pastoret, S., C. Fraipont, T. den Blaauwen, B. Wolf, M. E. Aarsman, A. Piette, A. Thomas, R. Brasseur, and M. Nguyen-Distèche. 2004. Functional analysis of the cell division protein FtsW of *Escherichia coli*. *J. Bacteriol.* **186:**8370–8379.

Pease, P. J., O. Levy, G. J. Cost, J. Gore, J. L. Ptacin, D. Sherratt, C. Bustamante, and N. R. Cozzarelli. 2005. Sequence-directed DNA translocation by purified FtsK. *Science* **307:**586–590.

Pedersen, L. B., E. R. Angert, and P. Setlow. 1999. Septal localization of penicillin-binding protein 1 in *Bacillus subtilis*. *J. Bacteriol.* **181:**3201–3211.

Pichoff, S., and J. Lutkenhaus. 2002. Unique and overlapping roles for ZipA and FtsA in septal ring assembly in *Escherichia coli*. *EMBO J.* **21:**685–693.

Pichoff, S., and J. Lutkenhaus. 2005. Tethering the Z ring to the membrane through a conserved membrane targeting sequence in FtsA. *Mol. Microbiol.* **55:**1722–1734.

Piette, A., C. Fraipont, T. Den Blaauwen, M. E. Aarsman, S. Pastoret, and M. Nguyen-Distèche. 2004. Structural determinants required to target penicillin-binding protein 3 to the septum of *Escherichia coli*. *J. Bacteriol.* **186:**6110–6117.

Raskin, D. M., and P. A. de Boer. 1999a. MinDE-dependent pole-to-pole oscillation of division inhibitor MinC in *Escherichia coli*. *J. Bacteriol.* **181:** 6419–6424.

Raskin, D. M., and P. A. de Boer. 1999b. Rapid pole-to-pole oscillation of a protein required for directing division to the middle of *Escherichia coli*. *Proc. Natl. Acad. Sci. USA* **96:**4971–4976.

RayChaudhuri, D. 1999. ZipA is a MAP-Tau homolog and is essential for structural integrity of the cytokinetic FtsZ ring during bacterial cell division. *EMBO J.* **18:**2372–2383.

Rico, A. I., M. Garcia-Ovalle, J. Mingorance, and M. Vicente. 2004. Role of two essential domains of *Escherichia coli* FtsA in localization and progression of the division ring. *Mol. Microbiol.* **53:**1359–1371.

Robson, S. A., K. A. Michie, J. P. Mackay, E. Harry, and G. F. King. 2002. The *Bacillus subtilis* cell division proteins FtsL and DivIC are intrinsically unstable and do not interact with one another in the absence of other septasomal components. *Mol. Microbiol.* **44:**663–674.

Romberg, L., and P. A. Levin. 2003. Assembly dynamics of the bacterial cell division protein FTSZ: poised at the edge of stability. *Annu. Rev. Microbiol.* **57**:125–154.

Rothfield, L., A. Taghbalout, and Y. L. Shih. 2005. Spatial control of bacterial division-site placement. *Nat. Rev. Microbiol.* **3**:959–968.

Rueda, S., M. Vicente, and J. Mingorance. 2003. Concentration and assembly of the division ring proteins FtsZ, FtsA, and ZipA during the *Escherichia coli* cell cycle. *J. Bacteriol.* **185**:3344–3351.

Ryan, K. R., and L. Shapiro. 2003. Temporal and spatial regulation in prokaryotic cell cycle progression and development. *Annu. Rev. Biochem.* **72**:367–394.

Sanchez, M., A. Valencia, M. J. Ferrandiz, C. Sander, and M. Vicente. 1994. Correlation between the structure and biochemical activities of FtsA, an essential cell division protein of the actin family. *EMBO J.* **13**:4919–4925.

Scheffers, D. J., L. J. Jones, and J. Errington. 2004. Several distinct localization patterns for penicillin-binding proteins in *Bacillus subtilis. Mol. Microbiol.* **51**:749–764.

Schiffer, G., and J. V. Höltje. 1999. Cloning and characterization of PBP 1C, a third member of the multimodular class A penicillin-binding proteins of *Escherichia coli. J. Biol. Chem.* **274**:32031–32039.

Schmidt, K. L., N. D. Peterson, R. J. Kustusch, M. C. Wissel, B. Graham, G. J. Phillips, and D. S. Weiss. 2004. A predicted ABC transporter, FtsEX, is needed for cell division in *Escherichia coli. J. Bacteriol.* **186**:785–793.

Sharp, M. D., and K. Pogliano. 2002. Role of cell-specific SpoIIIE assembly in polarity of DNA transfer. *Science* **295**:137–139.

Sharp, M. D., and K. Pogliano. 2003. The membrane domain of SpoIIIE is required for membrane fusion during *Bacillus subtilis* sporulation. *J Bacteriol.* **185**:2005–2008.

Sievers, J., and J. Errington. 2000. The *Bacillus subtilis* cell division protein FtsL localizes to sites of septation and interacts with DivIC. *Mol. Microbiol.* **36**:846–855.

Spratt, B. G., and A. B. Pardee. 1975. Penicillin-binding proteins and cell shape in *E. coli. Nature* **254**:516–517.

Steiner, W., G. Liu, W. D. Donachie, and P. Kuempel. 1999. The cytoplasmic domain of FtsK protein is required for resolution of chromosome dimers. *Mol. Microbiol.* **31**:579–583.

Stenberg, F., P. Chovanec, S. L. Maslen, C. V. Robinson, L. L. Ilag, G. von Heijne, and D. O. Daley. 2005. Protein complexes of the *Escherichia coli* cell envelope. *J. Biol. Chem.* **280**:34409–34419.

Stricker, J., P. Maddox, E. D. Salmon, and H. P. Erickson. 2002. Rapid assembly dynamics of the *Escherichia coli* FtsZ-ring demonstrated by fluorescence recovery after photobleaching. *Proc. Natl. Acad. Sci. USA* **99**:3171–3175.

Taschner, P. E., P. G. Huls, E. Pas, and C. L. Woldringh. 1988. Division behavior and shape changes in isogenic *ftsZ, ftsQ, ftsA, pbpB,* and *ftsE* cell division mutants of *Escherichia coli* during temperature shift experiments. *J. Bacteriol.* **170**:1533–1540.

Thanedar, S., and W. Margolin. 2004. FtsZ exhibits rapid movement and oscillation waves in helix-like patterns in *Escherichia coli. Curr. Biol.* **14**:1167–1173.

Trusca, D., S. Scott, C. Thompson, and D. Bramhill. 1998. Bacterial SOS checkpoint protein SulA inhibits polymerization of purified FtsZ cell division protein. *J. Bacteriol.* **180**:3946–3953.

Ursinus, A., F. van den Ent, S. Brechtel, M. de Pedro, J. V. Höltje, J. Lowe, and W. Vollmer. 2004. Murein (peptidoglycan) binding property of the essential cell division protein FtsN from *Escherichia coli. J. Bacteriol.* **186**:6728–6737.

van den Ent, F., and J. Lowe. 2000. Crystal structure of the cell division protein FtsA from *Thermotoga maritima. EMBO J.* **19**:5300–5307.

Varma, A., and K. D. Young. 2004. FtsZ collaborates with penicillin binding proteins to generate bacterial cell shape in *Escherichia coli. J. Bacteriol.* **186:** 6768–6774.

Vicente, M., A. I. Rico, R. Martinez-Arteaga, and J. Mingorance. 2006. Septum enlightenment: assembly of bacterial division proteins. *J. Bacteriol.* **188**:19–27.

Vollmer, W., M. von Rechenberg, and J. V. Höltje. 1999. Demonstration of molecular interactions between the murein polymerase PBP1B, the lytic transglycosylase MltA, and the scaffolding protein MipA of *Escherichia coli. J. Biol. Chem.* **274**:6726–6734.

von Rechenberg, M., A. Ursinus, and J. V. Höltje. 1996. Affinity chromatography as a means to study multienzyme complexes involved in murein synthesis. *Microb. Drug Resist.* **2**:155–157.

Wang, H., and R. C. Gayda. 1992. Quantitative determination of FtsA at different growth rates in *Escherichia coli* using monoclonal antibodies. *Mol. Microbiol.* **6**:2517–2524.

Wang, L., and J. Lutkenhaus. 1998. FtsK is an essential cell division protein that is localized to the septum and induced as part of the SOS response. *Mol. Microbiol.* **29**:731–740.

Wang, S., S. J. Arends, D. S. Weiss, and E. B. Newman. 2005. A deficiency in S-adenosylmethionine synthetase interrupts assembly of the septal ring in *Escherichia coli* K-12. *Mol. Microbiol.* **58**:791–799.

Weigand, R. A., K. D. Vinci, and L. I. Rothfield. 1976. Morphogenesis of the bacterial division septum: a new class of septation-defective mutants. *Proc. Natl. Acad. Sci. USA* **73**:1882–1886.

Weiss, D. S. 2004. Bacterial cell division and the septal ring. *Mol. Microbiol.* **54**:588–597.

Weiss, D. S., J. C. Chen, J. M. Ghigo, D. Boyd, and J. Beckwith. 1999. Localization of FtsI (PBP3) to the septal ring requires its membrane anchor, the Z ring, FtsA, FtsQ, and FtsL. *J. Bacteriol.* **181**:508–520.

Weiss, D. S., K. Pogliano, M. Carson, L. M. Guzman, C. Fraipont, M. Nguyen-Distèche, R. Losick, and J. Beckwith. 1997. Localization of the *Escherichia coli* cell division protein FtsI (PBP3) to the division site and cell pole. *Mol. Microbiol.* **25**:671–681.

Wientjes, F. B., and N. Nanninga. 1989. Rate and topography of peptidoglycan synthesis during cell division in *Escherichia coli*: concept of a leading edge. *J. Bacteriol.* **171**:3412–3419.

Wientjes, F. B., and N. Nanninga. 1991. On the role of the high molecular weight penicillin-binding proteins in the cell cycle of *Escherichia coli*. *Res. Microbiol.* **142**:333–344.

Wissel, M. C., and D. S. Weiss. 2004. Genetic analysis of the cell division protein FtsI (PBP3): amino acid substitutions that impair septal localization of FtsI and recruitment of FtsN. *J. Bacteriol.* **186**:490–502.

Wissel, M. C., J. L. Wendt, C. J. Mitchell, and D. S. Weiss. 2005. The transmembrane helix of the *Escherichia coli* division protein FtsI localizes to the septal ring. *J. Bacteriol.* **187**:320–328.

Woldringh, C. L., E. Mulder, P. G. Huls, and N. Vischer. 1991. Toporegulation of bacterial division according to the nucleoid occlusion model. *Res. Microbiol.* **142**:309–320.

Wu, L. J., and J. Errington. 2004. Coordination of cell division and chromosome segregation by a nucleoid occlusion protein in *Bacillus subtilis*. *Cell* **117**:915–925.

Yang, J. C., F. Van Den Ent, D. Neuhaus, J. Brevier, and J. Löwe. 2004. Solution structure and domain architecture of the divisome protein FtsN. *Mol. Microbiol.* **52**:651–660.

Yates, J., M. Aroyo, D. J. Sherratt, and F. X. Barre. 2003. Species specificity in the activation of Xer recombination at dif by FtsK. *Mol. Microbiol.* **49**:241–249.

Yim, L., G. Vandenbussche, J. Mingorance, S. Rueda, M. Casanova, J. M. Ruysschaert, and M. Vicente. 2000. Role of the carboxy terminus of *Escherichia coli* FtsA in self-interaction and cell division. *J. Bacteriol.* **182**:6366–6373.

Yousif, S. Y., J. K. Broome-Smith, and B. G. Spratt. 1985. Lysis of *Escherichia coli* by beta-lactam antibiotics: deletion analysis of the role of penicillin-binding proteins 1A and 1B. *J. Gen. Microbiol.* **131**:2839–2845.

Yu, X. C., A. H. Tran, Q. Sun, and W. Margolin. 1998. Localization of cell division protein FtsK to the *Escherichia coli* septum and identification of a potential N-terminal targeting domain. *J. Bacteriol.* **180**:1296–1304.

STRUCTURE AND BIOSYNTHESIS OF THE MUREIN (PEPTIDOGLYCAN) SACCULUS

Waldemar Vollmer

▮▮

The essential murein (peptidoglycan) sacculus is located in the periplasm of gram-negative bacteria (Fig. 1A) and protects the cell from rupture by the internal osmotic pressure (turgor). The sacculus is a giant molecule that forms an uninterrupted layer around the cytoplasmic membrane (Weidel and Pelzer, 1964). Hence, sacculi can be isolated and visualized by electron microscopy as thin and empty envelopes with the size and shape of the bacterial cell (Fig. 1B to D). The chemical structure of murein is well known. Murein is a heteropolymer composed of glycan strands that are cross-linked by short peptides, forming a netlike structure. However, presently there is no technique to directly visualize the glycan strands and peptides on an isolated sacculus. The current model of the murein architecture is based on predictions on the conformations of the glycan strands and the peptides and on observed physical parameters like thickness, elasticity, and porosity. Because this fascinating molecule completely surrounds the cytoplasmic membrane, the cell can elongate and divide only if the sacculus is enlarged. The enlargement of the sacculus during growth and division must involve the cleavage of covalent bonds by murein hydrolases and the

insertion of new precursors by murein synthases. Surprisingly, the enlargement of the thin sacculus is accompanied by a massive release of subunits, a process called murein turnover, by more than a dozen different murein hydrolases. Apparently, the synthetic and hydrolytic enzymes facilitating growth and division of the sacculus must be strictly controlled in their activities to ensure the maintenance of cell shape and cell size of newborn cells and to avoid lysis. Furthermore, rod-shaped bacteria like *Escherichia coli* appear to have two alternating phases during the cell cycle. During elongation, murein is enlarged at the cylindrical part of the cell, whereas during cell division, a focused murein synthesis takes place at the septation site to form the murein of the new polar caps of the daughter cells.

This chapter reviews the murein structure and biosynthesis and the possible mechanism(s) of enlargement during growth of gram-negative bacteria. The interested reader may turn to recent books (de Pedro et al., 1993; Ghuysen and Hakenbeck, 1994; Koch, 1995), book chapter (Park, 1996), and reviews (Höltje, 1998; van Heijenoort, 1998; Vollmer and Höltje, 2001, 2004) for a comprehensive overview of the topic. Unless otherwise stated, the presented data were obtained in studies with *E. coli*, which is the most extensively studied

Waldemar Vollmer, Mikrobielle Genetik, Universität Tübingen, 72076 Tübingen, Germany.

The Periplasm
Edited by Michael Ehrmann © 2007 ASM Press, Washington, D.C.

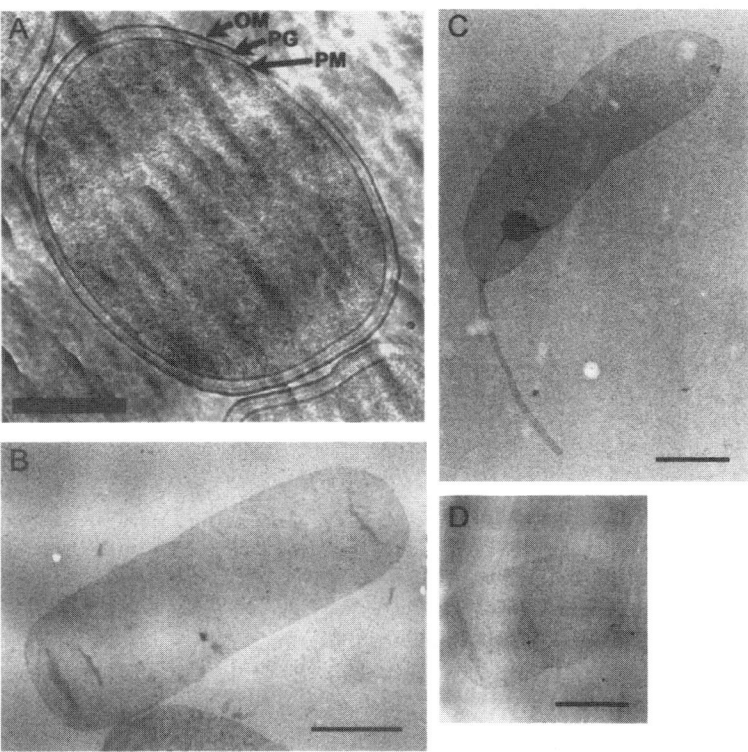

FIGURE 1   (A) A cryo-transmission electron microscopy picture of a frozen-hydrated section of an *E. coli* cell. The murein layer (PG) is embedded in the envelope between the cytoplasmic membrane (PM) and the outer membrane (OM); bar, 200 nm. (Taken from Matias et al. [2003] with permission.) The other transmission electron microscopy pictures show isolated murein sacculi from *E. coli* (B), *Caulobacter crescentus* (C), and *Pseudomonas aeruginosa* (D). B to D, bar, 500 nm.

gram-negative bacterium with respect to murein structure and biosynthesis.

CHEMICAL STRUCTURE OF MUREIN

Murein consists of glycan strands that are cross-linked by short peptides (Schleifer and Kandler, 1972). The linear glycan strands are composed of alternating, β-1,4-linked *N*-acetylglucosamine (GlcNAc) and *N*-acetylmuramic acid (MurNAc) residues, the latter being GlcNAc substituted with D-lactate at C-3 (Fig. 2A). Synthesized from disaccharide units, the glycan strands have an even number of sugar residues. The terminal residues are GlcNAc and 1,6-anhydroMurNAc, that is MurNAc with an intramolecular ether linkage from C-1 to C-6. Depending on the strain and the growth conditions, about 3 to 6% of the murein subunits have a 1,6-anhydroMurNAc glycan strand-terminating residue. Hence, the average degree of oligomerization of the glycan strands is about 25 to 40 disaccharide units (Glauner et al., 1988). Isolated murein glycan strands with up to 30 disaccharide units can be separated by high-pressure liquid chromatography (HPLC) (Harz et al., 1990). The average

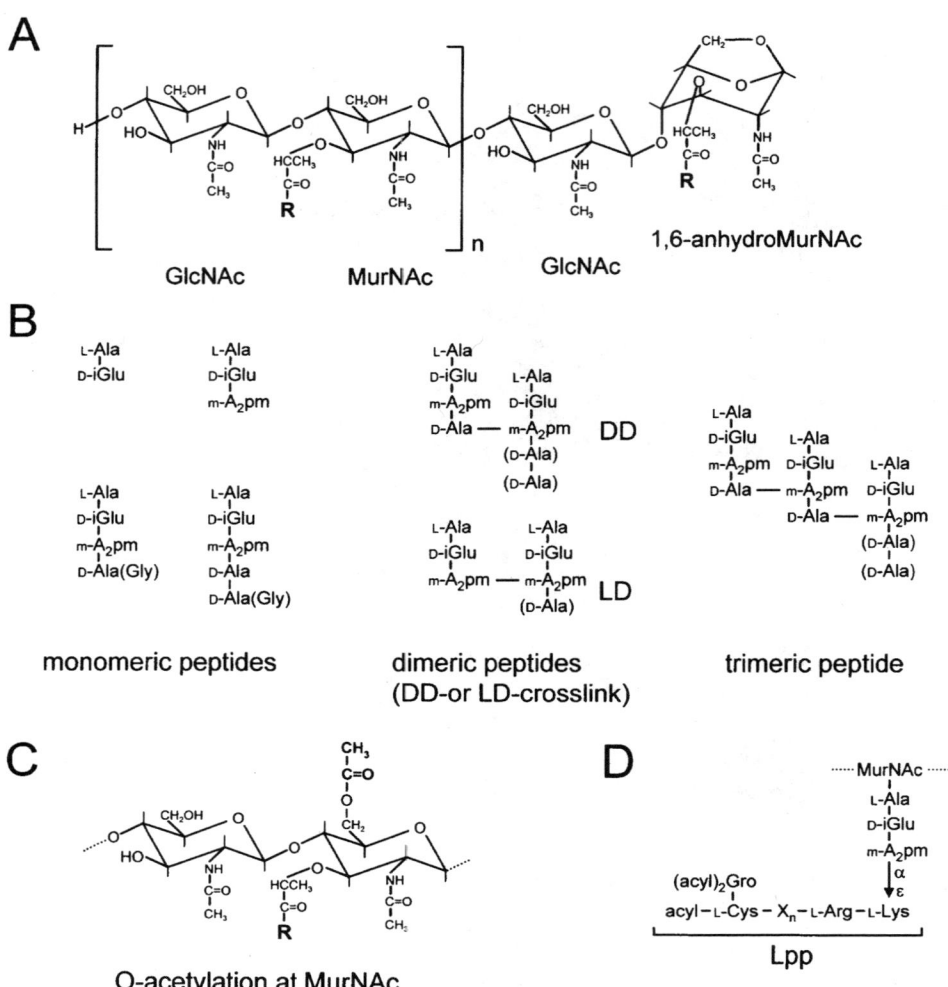

A

GlcNAc   MurNAc   GlcNAc   1,6-anhydroMurNAc

B

monomeric peptides   dimeric peptides (DD- or LD-crosslink)   trimeric peptide

C

O-acetylation at MurNAc

D

Lpp

FIGURE 2 Structure of murein from gram-negative bacteria. (A) The murein glycan strands consist of alternating N-acetylglucosamine (GlcNAc) and N-acetylmuramic acid (MurNAc) residues. A 1,6-anhydroMurNAc residue is present at one chain end. R, peptide. (B) Structures of monomeric, dimeric, and trimeric peptides. There are two types of cross-links (DD and LD). In the murein, the L-Ala residue of the peptide is attached to the lactyl group of the MurNAc residue of the glycan strands. iGlu (iso-glutamate), the γ-carboxyl group of Glu, is linked to the m-A$_2$pm residue. (C) The murein of some bacteria contains an O-acetyl group (in bold) at C-6 of a fraction of the MurNAc residues. (D) Structure of the covalent linkage between Braun's lipoprotein (Lpp) and the tripeptide in the murein.

length of the short glycan strands is 8.9 disaccharide units. The long glycan strands with more than 30 disaccharide units, which represent about 25 to 30% of the total glycan material, have an average length of about 45 disaccharide units.

The peptides contain rare D-amino acids and are attached by an amide linkage to the lactyl group of MurNAc. Their initial sequence is L-Ala-D-iGlu-m-A$_2$pm-D-Ala-D-Ala (pentapeptide), but processing may lead to the di-, tri-, and tetrapeptide (Fig. 2B). In these peptides, the D-Glu residues are linked via their γ-carboxyl group to the L-center of m-A$_2$pm (*meso*-diaminopimelic acid), the di-basic amino acid at position 3 that is required for the cross-linking of adjacent glycan strands. A low fraction of peptides contain Gly instead of D-Ala at position 4 or 5. The chemical structure of the basic murein subunit is shared by nearly all gram-negative bacteria, including a few gram-positive rods (Schleifer and Kandler, 1972). In some species, the free carboxyl groups of D-Glu or m-A$_2$pm (or both) are amidated. Several gram-negative species, e.g., *Neisseria meningitidis* and *Helicobacter pylori*, contain an acetyl group at C-6 of MurNAc (Fig. 2C) (Antignac et al., 2003; Clarke and Dupont, 1992).

The murein lipoprotein (Braun's lipoprotein, Lpp) is anchored by its N-terminal lipid residues to the outer membrane and is covalently linked to the murein, thus connecting the outer membrane with the murein and contributing to the stability of the cell envelope in gram-negative bacteria. Lpp is attached to murein by a peptide bond between the ε-amino group of its C-terminal Lys residue and the L-carboxyl group of the m-A$_2$pm residue in the murein peptide (Fig. 2D) (Braun and Wolff, 1970).

The peptides of adjacent glycan strands may be cross-linked, forming the netlike polymeric murein structure. Most cross-links are between the carboxyl group of D-Ala (position 4) of one peptide and the amino group at the D-center of m-A$_2$pm (position 3) of another peptide (DD-cross-link). A smaller number of cross-links exist between the L-center of m-A$_2$pm of one peptide and the D-center of m-A$_2$pm of the second peptide (LD-cross-link) (Glauner et al., 1988). Depending on the strain and growth conditions, between 40 and 60% of the peptides in the murein are part of cross-links in *E. coli* (Glauner et al., 1988) and in other gram-negative bacteria (Quintela et al., 1995). Next to dimeric peptide structures, there are lower fractions of trimeric and tetrameric structures that contain three or four cross-linked peptides (Fig. 2B).

The studies on the fine structure of murein have revealed that murein is not a homogeneous material, but is composed of more than 50 subunits (disaccharide peptide units) that differ with respect to the length of the peptide chain (di-, tri-, tetra-, pentapeptide), the presence of either D-Ala or Gly at positions 4 or 5, the state of oligomerization (monomer, dimer, trimer, tetramer), the kind of cross-linkage (DD or LD), the presence of 1,6-anhydro-MurNAc residues (glycan strand ends), and the presence of L-Lys-L-Arg (at positions 4 and 5) that remain after proteolytic digestion of Lpp (Glauner, 1988; Glauner et al., 1988). The major subunits in the murein of *E. coli* are the disaccharide tetrapeptide monomer (35.9% of the total material) and the DD-cross-linked bis-disaccharide tetratetrapeptide dimer (27.3% of the total material). An average *E. coli* cell has a total of about 3×10^6 disaccharide peptide subunits in its murein sacculus (Wientjes et al., 1991). Among gram-negative bacteria, there is some variability in the fine structure of the murein, in particular, with respect to the abundance of the LD-cross-linkage and the amount of bound lipoprotein (Quintela et al., 1995).

MACROMOLECULAR PROPERTIES OF MUREIN

Murein is a unique molecule with respect to its physical properties. The sacculi have the same size and shape as the bacterial cells from which they are isolated (Fig. 1B to D). For example, sacculi from *E. coli* cells are rod shaped with a length of 2 to 4 μm and a diameter of 0.5 to 1 μm. Compared with these dimensions, the sacculi are very thin. Cryo-transmission electron microscopy of frozen-hydrated cell sections and atomic force microscopy on isolated sacculi determined a thickness of about 6 nm (Matias et al., 2003; Yao et al., 1999). Small-angle neutron-scattering experiments revealed that 75 to 80% of the surface of isolated sacculi have a thickness of 2.5 nm, and the remaining 20 to 25% of the surface is maximally 7 nm thick (Labischinski et al., 1991).

The murein net is quite elastic and the surface of isolated sacculi can reversibly expand and shrink threefold without rupture (Koch and Woeste, 1992). Isolated sacculi are two- to threefold more deformable in the direction of the long axis of the cell than in the direction perpendicular to the long axis (Yao et al., 1999). In living bacteria, the surface area of the murein was 45% greater than that in the same bacteria upon relaxation of the murein by a sudden destruction of the cytoplasmic membrane (Koch et al., 1987). Apparently, the murein net is expanded to some extent in the living cell because of the cell's turgor pressure. Relaxed murein has pores with a mean radius of 2.06 nm, as determined by the size of fluorescent labeled dextrans that are able to penetrate the murein (Demchick and Koch, 1996). Theoretically, globular proteins of up to a molecular mass of 24 kDa should be able to penetrate relaxed murein. In the living cell, the expanded murein might represent no permeability barrier even for proteins with a molecular mass of up to 50 kDa or more. Many proteins are transported across the periplasm (and, hence, through the murein net) to become inserted into the outer membrane. The outer membrane lipoproteins are transported across the periplasm in a soluble complex with the chaperone LolA (Tokuda and Matsuyama, 2004), which apparently can diffuse through the murein layer. On the other hand, the assembly

of larger complexes through the periplasm, like flagella (Nambu et al., 1999), fimbriae, or complexes secreting proteins or DNA, seems to require the activity of specialized murein hydrolases to locally open meshes in the murein net (Dijkstra and Keck, 1996).

MOLECULAR ARCHITECTURE OF THE MUREIN SACCULUS
Murein is a heterogeneous material and is not crystalline (like chitin), making it impossible to directly determine its structure at high resolution by the presently available techniques. Therefore, the architecture of murein has been modeled on the basis of its chemical composition and physical properties (Barnickel et al., 1979, 1983; Labischinski et al., 1985). A glycan strand of an average length of about 25 to 40 nm is too long for a vertical arrangement in the periplasm. Therefore, most models assume that the glycan strands are arranged in parallel to the cytoplasmic membrane in a planar murein layer. Furthermore, the lactyl-peptide moiety at MurNAc is rather bulky, causing the peptides to protrude in a helical pattern from the glycan strand, with about four disaccharide units being required for one complete turn

FIGURE 3 Architecture of murein. (A) The peptides (arrows) protrude helically from the glycan strands. Light grey bar, GlcNAc; dark grey bar, MurNAc. (B) Model for the architecture of a murein layer. The glycan strands are indicated as bold zigzag lines. Arrows indicate the cross-linked peptides in the murein layer (*xy* plane). The non-cross-linked peptides pointing up and down are shown as dotted lines. A tessera is the smallest unit (pore) formed by two glycan strands and two peptide cross-links. (C) Model of the murein sacculus. The glycan strands (lines) run in the direction perpendicular to the long axis of the cell (*x* direction in B), whereas the peptide cross-links (arrows) are in the direction of the long axis (*y* direction in B). About 70 to 120 glycan strands of average length are required for one circumference, and about 500 to 1,000 glycan strands are arranged in parallel to cover the length of the cell. Most of the surface of the sacculus is made of a single layer. (Reproduced from Höltje [1998] with permission.)

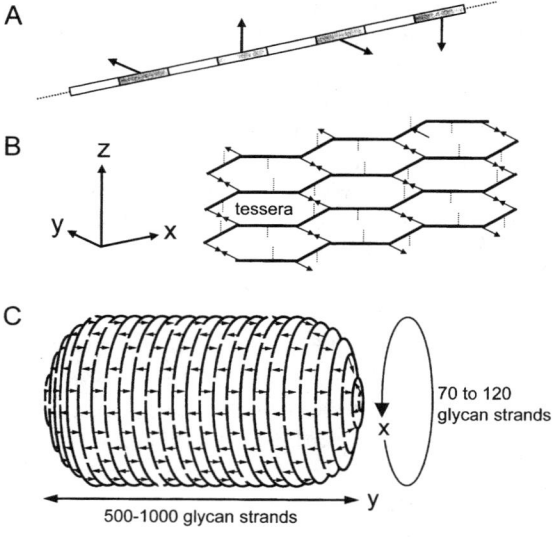

(Fig. 3A). Every second peptide lies in the same plane and can form a cross-link with a peptide protruding from an adjacent glycan strand, by forming a mono-layered netlike structure. Half of the peptides would point up or down and could not be present in cross-links. In the stress-bearing murein, the glycan strands would not be straight but would follow a zigzag line. The smallest pore formed by two glycan strands and two peptide cross-links has been termed "tessera" (Koch, 1995, 1998) (Fig. 3B). The measured thickness of the sacculi (see above) led to the conclusion that 75 to 80% of the surface is single layered, whereas the rest is triple layered.

Molecular modeling suggested that the glycan strands are rather rigid structures, whereas the peptides are flexible. In the favorable conformation, a peptide bends back toward the glycan strand. Under stretched conditions, the peptide adopts a more straight conformation, allowing the murein net to become expanded. The flexibility of the peptides could be the reason for the observed elasticity of murein. The anisotropy in elasticity indicates that the flexible peptides might run predominantly in the direction of the long axis of the cell, whereas the glycan strands run predominantly in the direction perpendicular to the long axis of the cell (Fig. 3C). The available data are not sufficient to determine whether the glycan strands are arranged exactly perpendicular to the long axis of the cell, or if they run around the cell in a helical fashion.

The layered murein cannot be perfect for two reasons. First, compared with the dimensions of the cell, the glycan strands are rather short. About 70 to 120 glycan strands of average length are required to complete one circumference, and 500 to 1,000 glycan strands have to run in parallel to cover the length of the rod-shaped cell. Second, the percentage of cross-linked peptides is slightly lower than the theoretical value of 50%. Apparently, these imperfections (or holes larger than a tessera) can be tolerated without impairing the function of the sacculus as the major stress-bearing structure in the envelope.

THE SACCULUS IS EMBEDDED IN THE CELL ENVELOPE

In *E. coli*, a fraction of about 5 to 9% of the total muropeptide subunits carry a Lpp molecule (depending on strain and growth conditions) (Glauner, 1988), giving rise to about 1.8×10^5 to 3.2×10^5 covalent connections between the murein sacculus and the outer membrane per cell. There could be a similar number per cell of noncovalent interactions between several outer membrane proteins, e.g., OmpA and different porins (OmpC, OmpF), and murein. In addition, macromolecular assemblies such as secretion systems or iron acquisition complexes contribute to the firm anchoring of the outer membrane to the murein. It is possible that adhesion sites ("Bayer junctions") between the inner membrane and the outer membrane or between the inner membrane and the murein exist (Bayer, 1979), but their visualization by electron microscopy depends on the method used for sample preparation (Kellenberger, 1990; Matias et al., 2003). Such adhesion sites might be formed transiently by macromolecular complexes involved in *trans*-envelope transport of substances. In addition, (transient) connections between cytoplasmic membrane-anchored murein synthases or/and newly synthesized murein glycan strands and the murein layer are likely to exist at the sites of growth of the sacculus.

MUREIN SYNTHESIS AND HYDROLYSIS

Murein synthesis starts in the cytoplasm with the formation of UDP-GlcNAc and UDP-MurNAc, followed by the successive ligation of L-Ala, D-Glu, m-A$_2$pm, and D-Ala-D-Ala to UDP-MurNAc, yielding UDP-MurNAc-pentapeptide. All enzymes required for the synthesis of the cytoplasmic precursors are known (van Heijenoort, 1998). For the transport across the cytoplasmic membrane, the precursor is modified by a hydrophobic lipid residue. The transferase MraY catalyzes the formation of undecaprenyl pyrophosphoryl-MurNAc-pentapeptide (lipid I) from UDP-MurNAc-pentapeptide and undecaprenyl phosphate. Next, MurG synthesizes undecaprenyl pyrophosphoryl-(GlcNAc)MurNAc-pentapeptide (lipid II) from lipid I and UDP-GlcNAc. Lipid II is the last monomeric murein precursor, and it is flipped across the cytoplasmic membrane by a yet unknown flippase.

The final steps in murein synthesis take place at the periplasmic side of the cytoplasmic membrane and involve two reactions (Fig. 4). First, the murein glycan strands are oligomerized by transglycosylation, and second, the peptide cross-links are formed by transpeptidation. Because of the absence of ATP and other energy sources in the periplasm, the energy driving both reactions is generated by the release of activating residues: In transglycosylation, the undecaprenyl pyrophosphoryl residue of lipid II is released (Fig. 4A). Undecaprenyl pyrophosphate is recycled by phosphatases (El Ghachi et al., 2005), yielding undecaprenyl phosphate that is required for lipid I synthesis. During transpeptidation, the terminal D-Ala at position 5 of the donor peptide is released with the concomitant formation of a peptidyl-enzyme intermediate complex, followed by the transfer of the peptidyl moiety to the side-chain amino group of m-A$_2$pm of a so-called acceptor peptide (Fig. 4B) (Ghuysen, 1991; Goffin and Ghuysen, 1998). The transpepti-

dases are also capable of breaking the amide bond in penicillin (and other β-lactams) to form a penicilloyl-enzyme intermediate, which, however, is almost completely inert and not transferred to an acceptor molecule. The covalent attachment of penicillin to the active site of the transpeptidases makes them easily detectable as penicillin-binding proteins (PBPs). PBPs belong to the large class of acyl serine transferases and are characterized by the following sequence signature in their PBP domain: NH_2-X_{30-60}-$S*X_2K$(motif 1)-X_{50-90}-(S/Y)X(N/C)(motif 2)-$X_{100-200}$-(K/H)(T/S)G(motif 3)-X_{40-60}-COOH (S*, essential Ser residue, X, variable amino acids). Among the PBPs, there are also enzymes that hydrolyze amide bonds between D-amino acids in the murein (DD-carboxypeptidases and DD-endopeptidases) (Fig. 5A). Serine β-lactamases, which are capable of inactivating penicillin by hydrolyzing the β-lactam bond, are similar in sequence to PBPs and they share the typical sequence signature with PBPs (Joris et al., 1991).

The murein synthases are multimodular proteins that are located in the periplasm and are anchored to the cytoplasmic membrane by a noncleavable signal peptide. They are of relatively low abundances of about 60 to 130 mol-

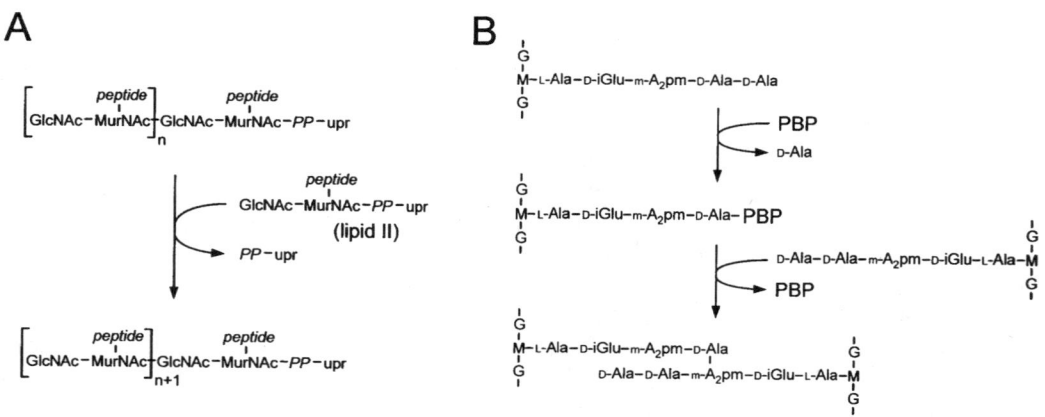

FIGURE 4 Murein synthesis reaction with lipid II as substrate. (A) The glycan strands are oligomerized by transglycosylation. *PP*, pyrophosphate; upr, undecaprenyl. (B) The DD-cross-links are formed by transpeptidation by a penicillin-binding protein (PBP). The intermediate peptidyl-enzyme complex is shown. (left) Donor peptide; (right) acceptor peptide; G, GlcNAc; M, MurNAc.

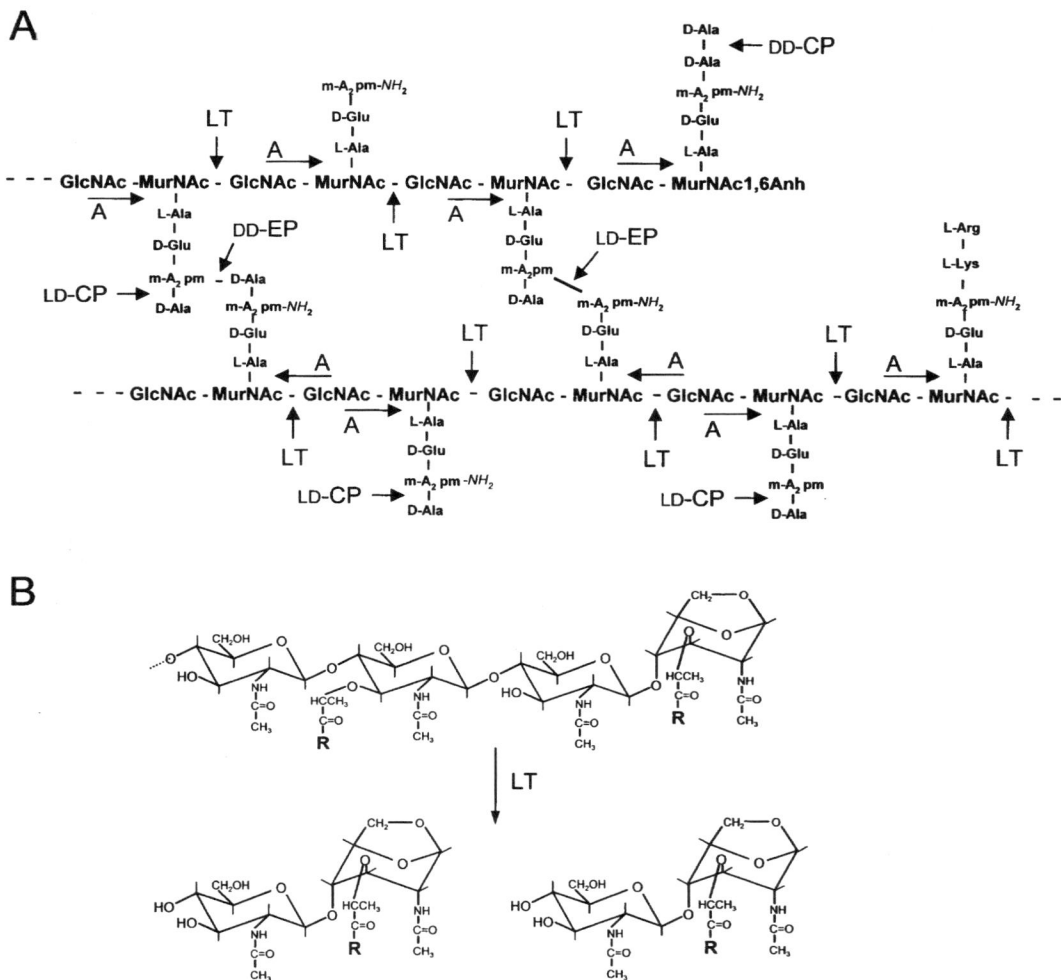

FIGURE 5 Murein hydrolysis. (A) Cleavage sites of different classes of murein hydrolases in high-molecular-weight murein. A, N-acetylmuramyl-L-alanine amidase; LT, lytic transglycosylase; DD-EP, DD-endopeptidase; LD-EP, LD-endopeptidase; DD-CP, DD-carboxypeptidase; LD-CP, LD-carboxypeptidase. (B) Cleavage of a murein glycan strand by an exo-specific lytic transglycosylase (LT) with concomitant formation of a 1,6-anhydro bond at the MurNAc residue.

ecules per cell in *E. coli* (Dougherty et al., 1996). In this bacterium, there are two bifunctional murein synthases that carry (in the periplasm) an N-terminal transglycosylase domain and a C-terminal transpeptidase (PBP) domain, the PBPs1A and 1B (Table 1). Although cells lacking either PBP1A or PBP1B are viable, the absence of both enzymes is lethal for the cell (Matsuhashi et al., 1990). The mono-

functional transpeptidases appear to be specifically involved in different processes: PBP2 is essential for cell elongation and maintenance of the rod shape, whereas the essential cell division protein PBP3 (or FtsI) is active at the site of septation. Both monofunctional PBPs have a noncatalytic periplasmic domain of approximately 200 amino acids that might participate in protein-protein interactions (Terrak et al.,

TABLE 1 The periplasmic murein synthases and hydrolases in *E. coli*

Enzyme[a]	Activity[b]	Gene	Localization/remarks[c]
Murein Synthases			
PBP1A	TG/TP	*ponA* (*mrcA*)	IM anchored
PBP1B(α, β, γ)	TG/TP	*ponB* (*mrcB*)	IM anchored/three isoforms
PBP1C	TG/TP ?	*pbpC*	IM anchored
PBP2	TP	*pbpA* (*mrdA*)	IM anchored/essential for elongation
PBP3	TP	*ftsI* (*pbpB*)	IM anchored/essential for septation
MtgA	TG	*mtgA*	IM anchored
Murein Hydrolases			
Slt70	LT	*sltY*	Periplasm
MltA	LT	*mltA*	OM anchored/lipoprotein
MltB	LT	*mltB*	OM anchored/lipoprotein
MltC	LT	*mltC*	OM anchored/lipoprotein
MltD	LT	*mltD*	OM anchored/lipoprotein
EmtA (MltE)	LT	*emtA*	OM anchored/lipoprotein
AmiA	A	*amiA*	Periplasm
AmiB	A	*amiB*	Periplasm
AmiC	A	*amiC*	Periplasm/septation site
PBP4	DD-EP/CP	*dacB*	Membrane associated
PBP7	DD-EP	*pbpG*	Membrane associated
MepA	DD/LD-EP	*mepA*	Periplasm
PBP5	DD-CP	*dacA*	IM anchored
PBP6	DD-CP	*dacC*	IM anchored
PBP6B	DD-CP	*dacD*	Periplasm or IM anchored
EnvC[d]	Unknown	*envC* (*yibP*)	Periplasm/septal ring
FlgJ[e]	Unknown	*flgJ*	Periplasm/flagellum

[a]For references see Höltje (1998).
[b]TG, transglycosylase; TP, transpeptidase; LT, lytic transglycosylase; A, *N*-acetylmuramyl-L-alanine amidase; DD-EP, DD-endopeptidase; LD-EP, LD-endopeptidase; DD-CP, DD-carboxypeptidase.
[c]IM, inner membrane; OM, outer membrane.
[d]Bernhardt and De Boer (2004).
[e]Nambu et al. (1999).

1999). A bifunctional PBP of minor abundance is PBP1C, which, although having a transpeptidase domain, may function in vivo only as a transglycosylase. *E. coli* has also a monofunctional transglycosylase, MtgA, which is not essential for cell growth and division.

Murein hydrolases are enzymes that cleave covalent bonds in the murein sacculus or in murein fragments (Höltje, 1995). Autolysins are, by virtue of their enzymatic specificities and their periplasmic localization, capable of digesting high-molecular-weight murein to small, soluble fragments (Fig. 5A). Muramidases, e.g., lysozyme, hydrolyze the β1,4 bond between MurNAc and GlcNAc in the murein glycan strands. Unlike lysozyme, lytic transglycosylases are muramidases that cleave the glycosidic linkage under concomitant formation of a 1,6-anhydro bond at the MurNAc residue (Höltje et al., 1975) (Fig. 5B). Of the six known lytic transglycosylases of *E. coli*, only one is soluble (Slt70), whereas five are lipoproteins anchored to the outer membrane (MltA, MltB, MltC, MltD, and EmtA) and face into the periplasm. EmtA is an endo-specific lytic transglycosylase, whereas Slt70, MltA, and MltB are exo-enzymes that remove disaccharide units presumably from the 1,6-anhydroMurNAc glycan strand end. Other autolysins cleave either the bond between the glycan strands and the peptides (the *N*-acetylmuramyl-L-alanine amidases AmiA, AmiB, and AmiC) or hydrolyze peptide bridges (the DD- and/or LD-endopeptidases MepA, PBP4, and PBP7) (Table 1).

Obviously, the activities of autolysins can destroy the integrity of the sacculus, and they must be strictly controlled in the cell to avoid autolysis.

The murein hydrolases are active during growth and division in *E. coli*. In one generation, a cell loses 40 to 50% of its total murein by the activity of hydrolases (Goodell, 1985; Park, 1993). This is a surprisingly high turnover rate, considering the mainly single-layered architecture of murein and the essentiality to maintain constantly the integrity of the sacculus. The chemical structures of the turnover products indicate that all autolytic activities—lytic transglycosylases, *N*-acetylmuramyl-L-alanine amidases, and endopeptidases—participate in murein turnover. In particular, the amidases and lytic transglycosylases are involved in cleavage of the septum during cell division. For example, a mutant lacking three amidases and the soluble lytic transglycosylase Slt70 grows in chains of up to 40 nonseparated cells (Heidrich et al., 2001).

Only a minor fraction of the released murein fragments is lost into the growth medium. Instead, *E. coli* has an efficient uptake and recycling system for murein turnover products. Uptake from the periplasm into the cytoplasm occurs via the unspecific oligopeptide permease Opp and the murein peptide permease Mpp. Once taken up, different cytoplasmic enzymes hydrolyze the murein fragments to compounds that can enter the biosynthetic pathway of murein precursors (Uehara et al., 2005). Some gram-negative bacteria, e.g., *Citrobacter freundii* and *Enterobacter cloacae*, couple in an elegant way murein turnover with signaling for the production of chromosomally encoded β-lactamase, an enzyme that is secreted to degrade exogenous β-lactam antibiotics (Jacobs et al., 1997). The inhibition of murein synthesis by β-lactams causes an uncontrolled activity of autolytic enzymes that is characterized by a sudden increase in murein turnover products that are taken up into the cytoplasm. A transcriptional activator, AmpR, is inactive with bound UDP-MurNAc pentapeptide murein precursor, but becomes activated by binding 1,6-anhydroMurNAc tripeptide turnover product, leading to the expression of AmpC β-lactamase. Mutants without lytic transglycosylases (that produce the 1,6-anhydroMurNAc compounds) or without the permeases required for the uptake of turnover products fail to induce β-lactamase and lyse in the presence of exogenous β-lactam antibiotic.

The *O*-acetylation of MurNAc appears to be a secondary modification, because the murein precursors are not modified in the species with *O*-acetylated murein. The enzyme(s) responsible for *O*-acetylation of murein in gram-negative bacteria as well as the function of this modification for gram-negative bacteria are not known. In gram-positive species such as *Staphylococcus aureus*, the *O*-acetylation of MurNAc residues reduces the activity of exogenous lysozyme, thereby contributing to the bacterial resistance against this ubiquitous murein hydrolase (Bera et al., 2005).

MUREIN SEGREGATION AND MORPHOGENESIS OF *E. COLI*

The topography of the enlargement of the sacculus has been studied by in vivo labeling of murein with radioactive precursor, followed by high-resolution autoradiography of the isolated sacculi by electron microscopy (Woldringh et al., 1987). An easier method with higher resolution is based on the in vivo incorporation of exogenous D-Cys into murein by a yet unknown periplasmic enzyme, followed by biotinylation and immunodetection of the label by electron (or fluorescence) microscopy (de Pedro et al., 1997). Pulse and pulse-chase experiments revealed that a focused murein synthesis takes place at the septation site during division. During cell elongation in the absence of exogenous D-Cys, the label is diluted at the cylindrical part of the sacculus, indicating that new material is incorporated into the existing side wall. At the cell poles there is no incorporation of new material or loss of old murein (turnover), even during long chase periods. These results on the murein segregation pattern are consistent with the observations that during elongation new cross-links are formed in the sacculus, mainly between newly synthesized and old (already existing) material, whereas during cell division there is also cross-linkage between two newly made peptides (Burman and Park, 1984; de Jonge et al., 1989). The segregation pattern of the outer membrane is similar to that of the murein layer, indicating that the enlargement of both layers is coordinated (de Pedro et al., 2003).

Morphogenesis of E. coli seems rather simple with two phases in the cell cycle (Nanninga, 1998; Vollmer and Höltje, 2001). During elongation, the cell is enlarged by growth of the envelope along the cylindrical part with maintenance of a constant diameter. Once the length of the cell has doubled, there is a switch to a focused envelope synthesis at the site of septation for the synthesis of the new polar caps of the daughter cells.

Little is known about how murein enlargement is temporal and topologically controlled. For cell division, more than 12 essential cell division (Fts) proteins assemble at midcell in a ringlike structure (Goehring and Beckwith, 2005; Weiss, 2004). When division progresses, the multiprotein ring (divisome) operates at the leading edge of constriction, where it might control the attachment of new murein in ever smaller concentric circles. One murein synthase, the monofunctional transpeptidase PBP3 (also named FtsI), is essential for division and localizes at the site of division. Therefore, specific inhibition of PBP3 results in filamentous growth. At least one transglycosylase (PBP1A, PBP1B, or MtgA) must participate in murein synthesis during cell division to oligomerize the murein glycan strands. In addition, periplasmic murein hydrolases, especially N-acetylmuramyl-L-alanine amidases, are required to cleave the septum during division. AmiC, which is exported via the Tat pathway, locates to the division site and contributes to the cleavage of the septum (Bernhardt and de Boer, 2003).

During elongation, the monofunctional transpeptidase PBP2 is required, and cells without functional PBP2 lose rod shape and grow spherically. There are other mutants with similar phenotypes and the affected genes have been implicated with the spatial control of the murein synthases during cell elongation. Especially the Mre proteins, which are required to maintain rod shape in E. coli and Bacillus subtilis, have recently attracted attention. MreB was found to be a cytoskeletal protein arranged in a helix underneath the cytoplasmic membrane (Jones et al., 2001). It remains to be determined whether and by which mechanism this protein controls murein synthesis during the elongation phase.

Mutants lacking the DD-carboxypeptidase PBP5 show remarkable morphological defects, including kinks, bends, and even branches (Young, 2003). The effects exacerbate if additional PBPs are deleted, but normal rod shape can always be restored in these multiple mutants by the production of PBP5 from a plasmid-borne dacA gene. The branches arise either from de novo generation of poles at the side wall or from splitting of an existing pole by areas of active murein synthesis (de Pedro et al., 2003). The tips of the deformed regions behave as misplaced, fully functional cell poles (Nilsen et al., 2004). Surprisingly, the combination of dacA deletion with certain ftsZ(ts) mutations results, at restrictive temperature, in E. coli cells with a helical cell shape (Varma and Young, 2004). The molecular mechanism(s) by which PBP5 and FtsZ contribute to maintenance of rod shape and the reason why the other DD-carboxypeptidases (see Table 1) cannot restore rod shape in the absence of PBP5 are not known.

GROWTH MODELS

Several models for the growth of the sacculus during elongation and division have been proposed. Most models have in common that the enlargement of the sacculus is achieved by the insertion of new glycan strands and that both synthesis of new murein and hydrolysis of bonds within the existing sacculus are combined. For the safe enlargement of the stress-bearing murein layer, Koch has proposed that new covalent bonds have to be formed first before hydrolysis of other bonds in the murein net takes place ("make-before-break" strategy) (Koch, 1995).

In the growth model of Park, local hydrolysis of cross-links by murein hydrolases precedes the incorporation of a newly synthesized glycan strand that becomes cross-linked to the neighboring strands in the existing murein layer (Park, 1996) (Fig. 6). Initially, cross-links are formed only between old and new material, but after a while, new glycan strands are inserted next to strands that were incorporated briefly before, resulting in cross-links between two new peptides. This model does not follow the make-before-break strategy and it does not explain the observed murein turnover (the release of fragments from the sacculus during growth). Park proposed that murein turnover is a process of its own for sensing the structure of the sacculus.

According to the "three-for-one" growth model proposed by Höltje, three new glycan strands are attached in a relaxed conformation underneath one glycan strand ("docking" strand) in the sacculus (Fig. 7). Simultaneous removal of the docking strand allows the insertion of the new triplet into the sacculus, by this increasing its surface. During elongation, the middle strand of the new triplet would be preformed by a monofunctional transglycosylase, whereas during cell division, all three glycan strands would be synthesized simultaneously. The three-for-one growth model explains the existence of a minor fraction of trimeric cross-links with a short half-life (which are the attachment sites for the new glycan strand triplet) and the phenomenon of massive murein turnover (the removal of the docking strands), and it is in accordance with the make-before-break strategy. If the glycan strands run

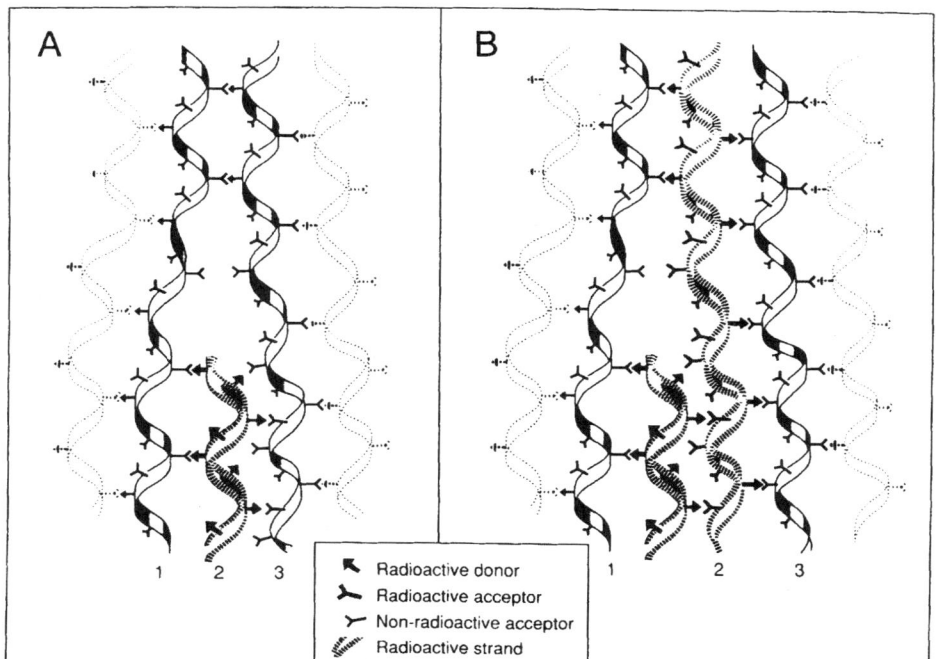

FIGURE 6 Model of the enlargement of the murein layer proposed by Park (1996). (A) Hydrolysis of cross-links precedes the insertion of a new glycan strand, connecting new and old material. (B) After some time, connections between two new strands are also made. (Reproduced from Park [1996] with permission.)

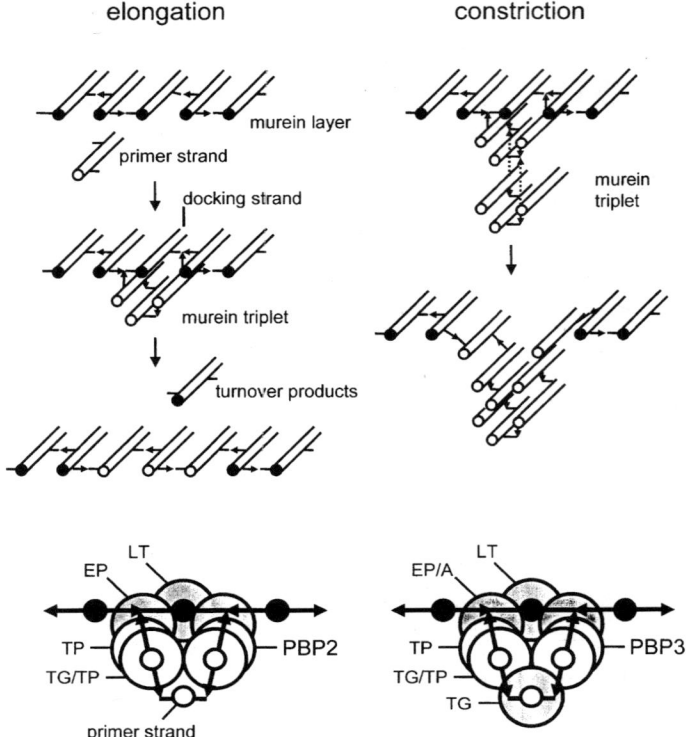

FIGURE 7 Three-for-one growth model proposed by Höltje. A triplet of three new strands is attached to the sacculus and is inserted by the concomitant removal of one old strand (docking strand). During elongation, one strand of the triplet (the primer strand) is preformed (left side). During cell division, all three strands are newly synthesized (right side). Below are shown the hypothetical murein synthesis multienzyme complexes for the enlargement of the sacculus. Complexes active during elongation contain PBP2, and complexes active during cell division contain PBP3. TG, transglycosylase; TP, transpeptidase; EP, endopeptidase; LT, lytic transglycosylase; A, *N*-acetylmuramyl-L-alanine amidase. (Reproduced from Höltje [1998] with permission.)

perpendicular to the long axis of the cell, the diameter of the cell would remain constant if the inserted triplet has the same length as the removed docking strand. The length of the cell would exactly double in one generation, if every second strand at the cylindrical part of the sacculus could serve as a docking strand. A molecular mechanism for the removal of every second strand, which involves DD- and LD-carboxypeptidase activities, to copy the existing sacculus for the transmission of the bacterial size and shape to the next generation has been suggested. Höltje has also proposed that all enzymatic activities are coordinated by the for-

mation of multienzyme complexes consisting of different murein synthases and hydrolases. Different murein synthesis multienzyme complexes would be active either in cell elongation (with PBP2) or in cell division (with PBP3). Incorporation of the hydrolases into complexes with synthases would be the way to control their potentially dangerous activities. Until now, several interactions between murein synthases and hydrolases have been identified in vitro (Romeis and Höltje, 1994; Vollmer et al., 1999) and there is genetic evidence for the existence of such interactions (Meisel et al., 2003).

UNANSWERED QUESTIONS

Despite the large knowledge on the murein structure and on the biosynthetic pathway, many questions still need to be answered. Which enzymes are responsible for the formation of the LD-cross-links, the attachment of Lpp, the incorporation of D-Cys, and the O-acetylation of MurNAc? What is the function of the LD-cross-links? What is the function of the O-acetyl group at MurNAc that is present in several gram-negative species? Are there specific sites on the sacculus where the very long glycan strands (>30 disaccharide units) accumulate, or are they evenly distributed? Where are the sites with higher thickness (three-layered regions?) located on the sacculus? Are these the sites of growth of the sacculus? How does the helically arranged cytoskeletal protein MreB (directly or indirectly) regulate the enlargement of murein to ensure maintenance of rod shape? What is the molecular basis for the requirement of the DD-carboxypeptidase activity of PBP5 to maintain rod shape and to prevent branching of the cells? How are murein synthesis and hydrolysis regulated by cell division proteins during septation? Will it be possible to prove experimentally the mechanism(s) of insertion of precursors into the sacculus? Which interactions between different murein synthases and between murein synthases and hydrolases occur in vivo? Are the murein biosynthetic pathway and the mechanism(s) of growth of the sacculus identical in all gram-negative species? There is still a long way to go until we know how gram-negative bacteria grow.

REFERENCES

Antignac, A., J. C. Rousselle, A. Namane, A. Labigne, M. K. Taha, and I. G. Boneca. 2003. Detailed structural analysis of the peptidoglycan of the human pathogen *Neisseria meningitidis. J. Biol. Chem.* **278**:31521–31528.

Barnickel, G., H. Labischinski, H. Bradaczek, and P. Giesbrecht. 1979. Conformational energy calculation on the peptide part of murein. *Eur. J. Biochem.* **95**:157–165.

Barnikel, G., D. Naumann, H. Bradaczek, H. Labischinski, and P. Giesbrecht. 1983. Computer aided molecular modelling of the three-dimensional structure of bacterial peptidoglycan, p. 61–66. *In* R. Hakenbeck, J.-V. Höltje, and H. Labischinski (ed.), *The Target of Penicillin: The Murein Sacculus of Bacterial Cell Walls. Architecture and Growth.* de Gruyter, New York, N.Y.

Bayer, M. E. 1979. The fusion sites between outer membrane and cytoplasmic membrane of bacteria: their role in membrane assembly and virus infection, p. 167–202. *In* M. Inoue (ed.), *Bacterial Outer Membranes.* John Wiley & Sons, Inc., New York, N.Y.

Bera, A., S. Herbert, A. Jakob, W. Vollmer, and F. Götz. 2005. Why are pathogenic staphylococci so lysozyme resistant? The peptidoglycan O-acetyltransferase OatA is the major determinant for lysozyme resistance of *Staphylococcus aureus. Mol. Microbiol.* **55**:778–787.

Bernhardt, T. G., and P. A. de Boer. 2003. The *Escherichia coli* amidase AmiC is a periplasmic septal ring component exported via the twin-arginine transport pathway. *Mol. Microbiol.* **48**:1171–1182.

Bernhardt, T. G., and P. A. de Boer. 2004. Screening for synthetic lethal mutants in *Escherichia coli* and identification of EnvC (YibP) as a periplasmic septal ring factor with murein hydrolase activity. *Mol. Microbiol.* **52**:1255–1269.

Braun, V., and H. Wolff. 1970. The murein-lipoprotein linkage in the cell wall of *Escherichia coli. Eur. J. Biochem.* **14**:387–391.

Burman, L. G., and J. T. Park. 1984. Molecular model for elongation of the murein sacculus of *Escherichia coli. Proc. Natl. Acad. Sci. USA* **81**:1844–1848.

Clarke, A. J., and C. Dupont. 1992. O-acetylated peptidoglycan: its occurrence, pathobiological significance, and biosynthesis. *Can. J. Microbiol.* **38**:85–91.

de Jonge, B. L., F. B. Wientjes, I. Jurida, F. Driehuis, J. T. Wouters, and N. Nanninga. 1989. Peptidoglycan synthesis during the cell cycle of *Escherichia coli*: composition and mode of insertion. *J. Bacteriol.* **171**:5783–5794.

Demchick, P., and A. L. Koch. 1996. The permeability of the wall fabric of *Escherichia coli* and *Bacillus subtilis. J. Bacteriol.* **178**:768–773.

de Pedro, M. A., J.-V. Höltje, and W. Löffelhardt (ed.). 1993. *Bacterial Growth and Lysis. Metabolism and Structure of the Bacterial Sacculus.* Plenum Press, New York, N.Y.

de Pedro, M. A., J. C. Quintela, J.-V. Höltje, and H. Schwarz. 1997. Murein segregation in *Escherichia coli. J. Bacteriol.* **179**:2823–2834.

de Pedro, M. A., K. D. Young, J.-V. Höltje, and H. Schwarz. 2003. Branching of *Escherichia coli* cells arises from multiple sites of inert peptidoglycan. *J. Bacteriol.* **185**:1147–1152.

Dijkstra, A. J., and W. Keck. 1996. Peptidoglycan as a barrier to transenvelope transport. *J. Bacteriol.* **178**:5555–5562.

Dougherty, T. J., K. Kennedy, R. E. Kessler, and M. J. Pucci. 1996. Direct quantitation of the number of individual penicillin-binding proteins per cell in *Escherichia coli. J. Bacteriol.* **178**:6110–6115.

El Ghachi, M., A. Derbise, A. Bouhss, and D. Mengin-Lecreulx. 2005. Identification of multiple genes encoding membrane proteins with undecaprenyl pyrophosphate phosphatase (UppP) activity in *Escherichia coli. J. Biol. Chem.* **280**:18689–18695.

Ghuysen, J.-M. 1991. Serine beta-lactamases and penicillin-binding proteins. *Annu. Rev. Microbiol.* **45**:37–67.

Ghuysen, J. M., and R. Hakenbeck (ed.). 1994. *Bacterial Cell Wall.* Elsevier Science B.V., Amsterdam, The Netherlands.

Glauner, B. 1988. Separation and quantification of muropeptides with high-performance liquid chromatography. *Anal. Biochem.* **172**:451–464.

Glauner, B., J.-V. Höltje, and U. Schwarz. 1988. The composition of the murein of *Escherichia coli. J. Biol. Chem.* **263**:10088–10095.

Goehring, N. W., and J. Beckwith. 2005. Diverse paths to midcell: assembly of the bacterial cell division machinery. *Curr. Biol.* **15**: R514–R526.

Goffin, C., and J.-M. Ghuysen. 1998. Multimodular penicillin-binding proteins: an enigmatic family of orthologs and paralogs. *Microbiol. Mol. Biol. Rev.* **62**:1079–1093.

Goodell, E. W. 1985. Recycling of murein by *Escherichia coli. J. Bacteriol.* **163**: 305–310.

Harz, H., K. Burgdorf, and J.-V. Höltje. 1990. Isolation and separation of the glycan strands from murein of *Escherichia coli* by reversed phase high-performance liquid chromatography. *Anal. Biochem.* **190**:120–128.

Heidrich, C., M. F. Templin, A. Ursinus, M. Merdanovic, J. Berger, H. Schwarz, M. A. de Pedro, and J.-V. Höltje. 2001. Involvement of N-acetyl-muramyl-L-alanine amidases in cell separation and antibiotic-induced autolysis of *Escherichia coli. Mol. Microbiol.* **41**:167–178.

Höltje, J.-V. 1995. From growth to autolysis: the murein hydrolases in *Escherichia coli. Arch. Microbiol.* **164**:243–254.

Höltje, J.-V. 1998. Growth of the stress-bearing and shape-maintaining murein sacculus of *Escherichia coli. Microbiol. Mol. Biol. Rev.* **62**:181–203.

Höltje, J.-V., D. Mirelman, N. Sharon, and U. Schwarz. 1975. Novel type of murein transglycosylase in *Escherichia coli. J. Bacteriol.* **124**:1067–1076.

Jacobs, C., J. M. Frere, and S. Normark. 1997. Cytosolic intermediates for cell wall biosynthesis and degradation control inducible beta-lactam resistance in gram-negative bacteria. *Cell* **88**:823–832.

Jones, L. J., R. Carballido-Lopez, and J. Errington. 2001. Control of cell shape in bacteria: helical, actin-like filaments in *Bacillus subtilis. Cell* **104**:913–922.

Joris, B., P. Ledent, O. Dideberg, E. Fonze, J. Lamotte-Brasseur, J. A. Kelly, J. M. Ghuysen, and J.-M. Frere. 1991. Comparison of the sequences of class A beta-lactamases and of the secondary structure elements of penicillin-recognizing proteins. *Antimicrob. Agents Chemother.* **35**:2294–2301.

Kellenberger, E. 1990. The 'Bayer bridges' confronted with results from improved electron microscopy methods. *Mol. Microbiol.* **4**:697–705.

Koch, A. L. 1995. *Bacterial Growth and Form.* Chapman & Hall, New York, N.Y.

Koch, A. L. 1998. Orientation of the peptidoglycan chains in the sacculus of *Escherichia coli. Res. Microbiol.* **149**:689–701.

Koch, A. L., S. L. Lane, J. A. Miller, and D. G. Nickens. 1987. Contraction of filaments of *Escherichia coli* after disruption of cell membrane by detergent. *J. Bacteriol.* **169**:1979–1984.

Koch, A. L., and S. Woeste. 1992. Elasticity of the sacculus of *Escherichia coli. J. Bacteriol.* **174**:4811–4819.

Labischinski, H., G. Barnickel, D. Naumann, and P. Keller. 1985. Conformational and topological aspects of the three-dimensional architecture of bacterial peptidoglycan. *Ann. Inst. Pasteur Microbiol.* **136A**:45–50.

Labischinski, H., E. W. Goodell, A. Goodell, and M. L. Hochberg. 1991. Direct proof of a 'more-than-single-layered' peptidoglycan architecture of *Escherichia coli* W7: a neutron small-angle scattering study. *J. Bacteriol.* **173**:751–756.

Matias, V. R., A. Al-Amoudi, J. Dubochet, and T. J. Beveridge. 2003. Cryo-transmission electron microscopy of frozen-hydrated sections of *Escherichia coli* and *Pseudomonas aeruginosa. J. Bacteriol.* **185**:6112–6118.

Matsuhashi, M., M. Wachi, and F. Ishino. 1990. Machinery for cell growth and division: penicillin-binding proteins and other proteins. *Res. Microbiol.* **141**:89–103.

Meisel, U., J.-V. Höltje, and W. Vollmer. 2003. Overproduction of inactive variants of the murein synthase PBP1B causes lysis in *Escherichia coli. J. Bacteriol.* **185**:5342–5348.

Nambu, T., T. Minamino, R. M. Macnab, and K. Kutsukake. 1999. Peptidoglycan-hydrolyzing activity of the FlgJ protein, essential for flagellar rod formation in *Salmonella typhimurium. J. Bacteriol.* **181**:1555–1561.

Nanninga, N. 1998. Morphogenesis of *Escherichia coli. Microbiol. Mol. Biol. Rev.* **62**:110–129.

Nilsen, T., A. S. Gosh, M. B, Goldberg, and K. D. Young. 2004. Branching sites and morphological abnormalities behave as ectopic poles in shape-defective *Escherichia coli. Mol. Microbiol.* **52**:1045–1054.

Park, J. T. 1993. Turnover and recycling in oligopeptide permease-negative strains of *Escherichia coli*: indirect evidence for an alternative permease system and for a monolayered sacculus. *J. Bacteriol.* **175**:7–11.

Park, J. T. 1996. The murein sacculus, p. 48–57. *In* F. C. Neidhardt, R. Curtiss III, J. L. Ingraham, E. C. C. Lin, K. B. Low, B. Magasanik, W. S. Reznikoff, M. Riley, M. Schaechter, and H. E. Umbarger (ed.), Escherichia coli *and* Salmonella: *Cellular and Molecular Biology*, 2nd ed., vol. 1. ASM Press, Washington, D.C.

Quintela, J. C., M. Caparros, and M. A. de Pedro. 1995. Variability of peptidoglycan structural parameters in gram-negative bacteria. *FEMS Microbiol. Lett.* **125:**95–100.

Romeis, T., and J.-V. Höltje. 1994. Specific interaction of penicillin-binding proteins 3 and 7/8 with soluble lytic transglycosylase in *Escherichia coli*. *J. Biol. Chem.* **269:**21603–21607.

Schleifer, K. H., and O. Kandler. 1972. Peptidoglycan types of bacterial cell walls and their taxonomic implications. *Bacteriol. Rev.* **36:**407–477.

Terrak, M., T. K. Ghosh, J. van Heijenoort, J. Van Beeumen, M. Lampilas, J. Aszodi, J. A. Ayala, J.-M. Ghuysen, and M. Nguyen-Disteche. 1999. The catalytic, glycosyl transferase and acyl transferase modules of the cell wall peptidoglycan-polymerizing penicillin-binding protein 1b of *Escherichia coli*. *Mol. Microbiol.* **34:**350–364.

Tokuda, H., and S. Matsuyama. 2004. Sorting of lipoproteins to the outer membrane in *E. coli*. *Biochim. Biophys. Acta* **1693:**5–13.

Uehara, T., K. Suefuji, N. Valbuena, B. Meehan, M. Donegan, and J. T. Park. 2005. Recycling of the anhydro-N-acetylmuramic acid derived from cell wall murein involves a two-step conversion to N-acetylglucosamine-phosphate. *J. Bacteriol.* **187:**3643–3649.

van Heijenoort, J. 1998. Assembly of the monomer unit of bacterial peptidoglycan. *Cell. Mol. Life Sci.* **54:**300–304.

Varma, A., and K. D. Young. 2004. FtsZ collaborates with penicillin-binding proteins to generate bacterial shape in *Escherichia coli*. *J. Bacteriol.* **186:**6788–6774.

Vollmer, W., and J.-V. Höltje. 2001. Morphogenesis of *Escherichia coli*. *Curr. Opin. Microbiol.* **4:**625–633.

Vollmer, W., and J.-V. Höltje. 2004. The architecture of the murein (peptidoglycan) in gram-negative bacteria: vertical scaffold or horizontal layer(s)? *J. Bacteriol.* **186:**5978–5987.

Vollmer, W., M. von Rechenberg, and J.-V. Höltje. 1999. Demonstration of molecular interactions between the murein polymerase PBP1B, the lytic transglycosylase MltA, and the scaffolding protein MipA of *Escherichia coli*. *J. Biol. Chem.* **274:** 6726–6734.

Weidel, W., and H. Pelzer. 1964. Bagshaped macromolecules—a new outlook on bacterial cell walls. *Adv. Enzymol.* **26:**193–232.

Weiss, D. S. 2004. Bacterial cell division and the septal ring. *Mol. Microbiol.* **54:**588–597.

Wientjes, F. B., C. L. Woldringh, and N. Nanninga. 1991. Amount of peptidoglycan in cell walls of gram-negative bacteria. *J. Bacteriol.* **173:**7684–7691.

Woldringh, C., P. Huls, E. Pas, G. H. Brakenhoff, and N. Nanninga. 1987. Topography of peptidoglycan synthesis during elongation and polar cap formation in a cell division mutant of *Escherichia coli* MC43100. *J. Gen. Microbiol.* **133:**575–586.

Yao, X., M. Jericho, D. Pink, and T. J. Beveridge. 1999. Thickness and elasticity of gram-negative murein sacculi measured by atomic force microscopy. *J. Bacteriol.* **181:**6865–6875.

Young, K. D. 2003. Bacterial shape. *Mol. Microbiol.* **49:**571–580.

PERIPLASMIC EVENTS IN THE ASSEMBLY OF BACTERIAL LIPOPOLYSACCHARIDES

Chris Whitfield, Emilisa Frirdich, and Anne N. Reid

12

The surfaces of gram-negative bacteria are typically complex arrays of proteins and glycoconjugates. The amphipathic glycolipid, lipopolysaccharide (LPS), represents a major component of the outer membrane of most gram-negative bacteria and plays an essential role in the barrier properties of the outer membrane (Nikaido, 2003). LPSs also play additional varied roles in the biology of gram-negative pathogens. The prototypical LPS molecule comprises three structurally and functionally distinct domains: lipid A, core oligosaccharide, and O-antigenic polysaccharide (O-polysaccharide) (reviewed in Raetz and Whitfield, 2002). In *Escherichia coli* and *Salmonella enterica*, lipid A has a bis-phosphorylated β-1,6-linked diglucosamine backbone typically carrying six fatty acyl chains and this structure is widely conserved. The core oligosaccharide has a reasonably well-conserved inner region (i.e., lipid A-proximal), typically containing 3-deoxy-D-*manno*-oct-2-ulosonic acid (Kdo) and L-*glycero*-D-*manno*-heptose (heptose) residues, and a more variable outer core domain whose structure differs within and between species. The outer core typically provides the site of attachment for the hypervariable O-polysaccharide. Bacteria from some genera, in particular, mucosal pathogens, produce lipooligosaccharides. This form of LPS lacks O-polysaccharide; instead it contains variable oligosaccharide extensions attached to the core oligosaccharide (Raetz and Whitfield, 2002). Until relatively recently, it was believed that LPS was essential for viability in all those organisms that produce it. The universality of this concept has been proven incorrect by recent research on *Neisseria meningitidis* (Bos and Tommassen, 2005; Steeghs et al., 1998) and *Moraxella catarrhalis* (Peng et al., 2005). These organisms normally produce lipooligosaccharide but can withstand mutations that eliminate its formation.

LPS is synthesized from activated precursors (nucleotide monophospho- and diphospho-sugars) available in the cytoplasm and the assembly process involves a significant number of integral and peripheral membrane proteins. In all cases, the assembly processes span the inner membrane. Located at the periplasmic face of the inner membrane are enzymes that complete the assembly process or carry out post-synthetic modifications of the glycoconjugates. In addition to meeting the challenge of a transenvelope biosynthesis and assembly pathway, the system must accommodate the amphi-

Chris Whitfield, Emilisa Frirdich, and Anne N. Reid, Department of Molecular and Cellular Biology, University of Guelph, Guelph, Ontario N1G 2W1, Canada.

The Periplasm
Edited by Michael Ehrmann © 2007 ASM Press, Washington, D.C.

pathic properties of the product and the radically different chemistries of the structural domains. The composite lipid A-core (known as "rough" LPS) is synthesized independently of the O-polysaccharide and is exported across the inner membrane by the ATP-binding cassette (ABC) transporter, MsbA. The hydrophilic O-polysaccharide is synthesized and exported by a process requiring undecaprenyl phosphate as an essential carrier lipid. This is the same carrier lipid that is involved in the biosynthesis of peptidoglycan (Vollmer and Holtje, 2001) and capsular polysaccharides (Whitfield, 2006). There are three known O-polysaccharide biosynthesis mechanisms that all culminate in an undecaprenyl pyrophosphate-linked polymer at the periplasmic face of the inner membrane. At this point, the lipid A-core and O-polysaccharide biosynthesis pathways converge with a ligation reaction to complete the "smooth" LPS molecule (Color Plate 15). The product is then translocated to the cell surface. Coordinated and efficient assembly pathways for LPS reflect the essential need for LPS in viability of most gram-negative bacteria and the need to sustain a high rate of synthesis for rapid growth.

Here we provide an overview of the assembly pathways for LPS molecules, with particular emphasis on reactions and components occurring in the periplasm. The systems from *E. coli* and *Salmonella* are deliberately emphasized because of the comprehensive available biochemical data and other bacteria are introduced only where they add to the mechanistic knowledge or provide interesting counterpoints.

BIOSYNTHESIS AND EXPORT OF LIPID A-CORE

The base lipid A structure is well established in *E. coli* and *Salmonella* (Color Plate 16) and its widely conserved structure reflects an important role in the barrier properties of the outer membrane (Nikaido, 2003; Raetz and Whitfield, 2002). However, the ability to respond to changing environments, such as replication inside or outside the host, requires a dynamic structure that can adapt to changing environmental conditions. Only in the past decade has it been recognized that at least some modifications of lipid A can occur postsynthetically, in the periplasm and outer membrane, to facilitate an outer membrane remodeling process that is not strictly dependent on de novo synthesis and replacement of existing LPS species. The later-stage timing of such modifying reactions may have the additional advantage of not having to interfere with the critically important constitutive cytoplasmic lipid A biosynthetic pathway.

The biosynthesis of the base lipid A structure occurs at the cytoplasm–inner membrane interface and the process has been largely resolved in *E. coli* and *Salmonella*. Species-specific variations on the generally conserved theme are now being identified. This pathway has been reviewed extensively elsewhere (Raetz and Whitfield, 2002; Trent, 2004). In *E. coli* and *Salmonella*, the lipid A biosynthesis pathway culminates in the formation of hexaacylated bisphosphorylated lipid A bearing two Kdo residues; this molecule is widely referred to as chemotype Re LPS. The sequence of reactions is such that addition of the two Kdo residues from the inner core oligosaccharide is essential for completion of the last two acylation steps in lipid A. The core oligosaccharide is then assembled by sequential glycosyl transfer to the Re LPS acceptor by peripheral membrane proteins that utilize nucleotide diphosphosugar precursors. A given species can produce more than one core type and further diversification is achieved by nonstoichiometric modifications in the inner core (Fig. 1). The timing of the nonstoichiometric addition of these residues is not clearly understood. Biosynthesis of the core oligosaccharide has been reviewed in detail elsewhere (Frirdich and Whitfield, 2005; Raetz and Whitfield, 2002; Whitfield et al., 2003).

Substantial progress has been made recently in understanding the export of lipid A-core, following the identification of the MsbA ABC transporter and establishment of its critical role in the process (Doerrler et al., 2001) (see

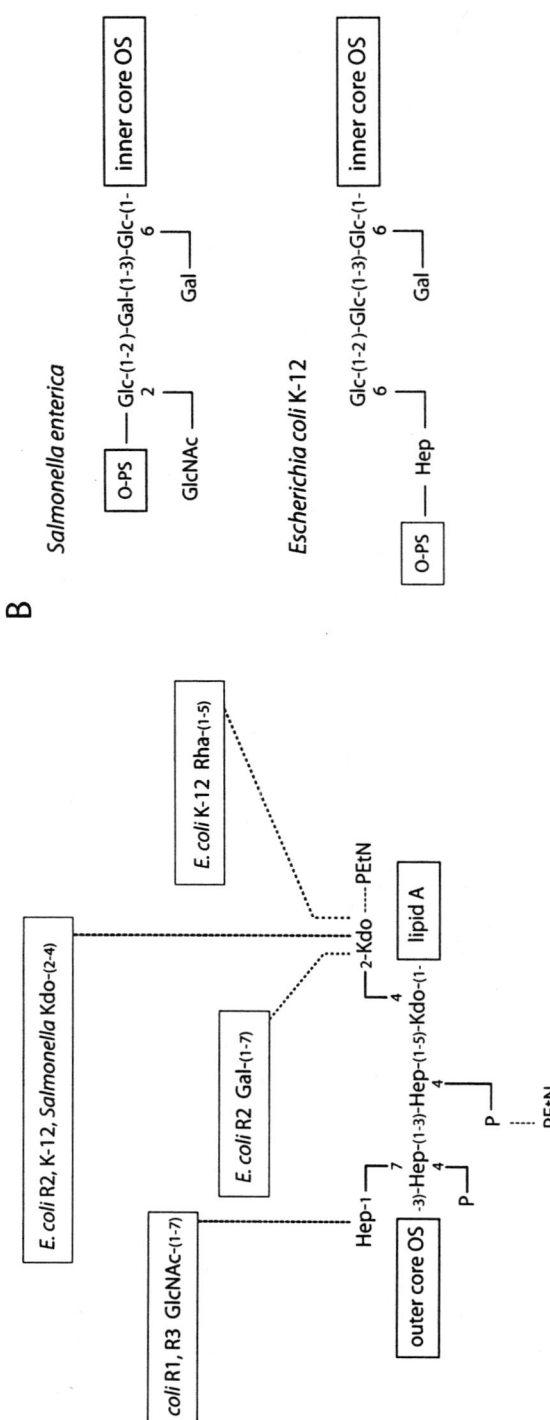

FIGURE 1 Structure of the core oligosaccharide in *S. enterica* and *E. coli*. The inner core is shown in A and the various nonstoichiometric glycose and phosphoethanolamine modifications of the base structure are represented by the dashed lines. Two representative outer core structures are shown in B.

Chapter 17). The first structure of the MsbA homodimer (from *E. coli*) was solved in 2001 (Chang and Roth, 2001) and it has been influential in modeling other ABC transporters. Structures of the vanadate-trapped MsbA (Mg.ADP.Vi) complex from *Salmonella* (Reyes and Chang, 2005) and of the MsbA homolog from *Vibrio cholerae* (Chang, 2003) have followed. The application of site-directed spin labeling is providing additional insight into the details of the transport mechanism and suggests the current crystal structures may provide a limited view of the range of conformers of MsbA evident during the transport cycle (Buchaklian et al., 2004; Buchaklian and Klug, 2005; Dong et al., 2005). In the current working model (Dong et al., 2005; Reyes and Chang, 2005), it is proposed that lipid A core binds to MsbA and the polar parts of the exported molecule enter the lumen of the transporter. ATP binding then drives dimerization of the two nucleotide-binding domains. This traps the export substrate within the lumen and conformational changes in the MsbA dimer result in a flipping of the lipid A-core through 180°. ATP hydrolysis is accompanied by an opening of the formerly closed periplasmic face of the transporter. In this model, the polar sugars are sequestered in the lumen throughout the process, while the acyl chains of lipid A are effectively dragged through the flanking lipid bilayer. Although the description of the essential role of MsbA in lipid A-core export is conclusive, there are conflicting views concerning the possibility that MsbA plays an additional role in the transbilayer movement of glycerophospholipids (for further discussion see Doerrler et al., 2004; Tefsen et al., 2005a).

Regulated Modifications of Lipid A

In *Salmonella*, lipid A structure is modulated in response to environmental cues, some of which reflect the transition to an intracellular lifestyle during infection. Many of the modifications described in vitro have recently been confirmed in bacteria isolated from macrophages (Gibbons et al., 2005). These lipid A alterations are regulated by an elegant system comprising a pair of two-component regulatory systems: the PmrAB and PhoPQ systems. Some PhoP-regulated genes also require the SlyA protein as a coregulator, providing an additional level of control (Navarre et al., 2005). The response regulator PmrA is activated by high concentrations of Fe^{3+} and/or mildly acidic pH conditions that are sensed by the cognate sensor, PmrB (Gunn et al., 1998; Wosten et al., 2000). In addition, PmrA can be activated by low Mg^{2+} and Ca^{2+}, conditions that would be encountered in an intracellular environment (Groisman et al., 1997; Guo et al., 1997). In this situation, PmrA activation requires PhoPQ (Groisman, 2001) and occurs via the intermediate, PmrD (Kato et al., 2003; Kato and Groisman, 2004; Kox et al., 2000). While PhoPQ-regulated genes are part of a larger regulon of genes involved in *Salmonella* virulence (reviewed in Groisman, 2001), PmrA activity is currently confined to LPS modifications. Many of the lipid A modifications seen in *Salmonella* are latent in *E. coli* K-12 but can be induced by growth in media containing metavanadate (Zhou et al., 1999).

The PmrAB proteins take their name from their involvement in resistance to polymyxin B, a cationic lipopeptide antimicrobial compound that interacts with and intercalates between LPS molecules, exploiting the negative charges, to breach the outer membrane (Vaara, 1992). The mode of action of polymyxin is, in a general sense, reflective of a range of cationic antimicrobial peptides produced as an innate defense by plant and animal hosts (Ganz, 2003). Before compromising the inner membrane, cationic peptides bind to the cell surface through hydrophobic and electrostatic interactions and exploitation of the negative charges on LPS, in particular, the lipid A phosphates, plays a major role. Recently, it was shown that PhoPQ also responds to host antimicrobial peptides (Bader et al., 2005).

Resistance to cationic peptides is conferred by modifications that appear to reduce the overall net-negative charge in lipid A. Such modifications mask the phosphate residues, thereby reducing the electrostatic interactions between polymyxin (and cationic peptides) and the outer membrane. These include addition of phosphoethanolamine and 4-amino-4-deoxy-L-arabinose (4-aminoarabinose) (Color Plate 16), although 4-aminoarabinose exerts a more substantial effect in resistance to cationic peptides (reviewed in Raetz and Whitfield, 2002; Trent, 2004). In addition, the PhoPQ system regulates genes involved in modification of the lipid A phosphate groups and changes in acylation (Color Plate 16) (reviewed in Groisman, 2001; Raetz and Whitfield, 2002; Trent, 2004). Addition of palmitate and the formation of 2-hydroxymyristate may increase the number of hydrophobic interactions between the lipid A fatty acyl chains and, as a consequence, decrease the fluidity of the outer membrane. This may potentially influence interactions between cationic peptides and the outer membrane (Guo et al., 1998; Nikaido, 2003). Different lipid A modifications are required for resistance to different antimicrobial peptides (Shi et al., 2004). These changes may be important for survival at different stages of infection, highlighting the need for the complex regulatory pathways. Some modifications have effects that extend beyond altered outer membrane structure. For example, palmitoylation and removal of the β-hydroxymyristate chain at position 3 generate heptaacylated and pentaacylated lipid A molecules, respectively, which show decreased ability to activate cellular LPS receptors (Kawasaki et al., 2004a, 2004b). This suggests that some lipid A modifications may alter host inflammatory responses, promoting immune evasion and increased survival in host tissues.

ADDITION OF 4-AMINOARABINOSE TO LIPID A

The addition of 4-aminoarabinose to lipid A occurs at the periplasmic face of the inner membrane, following export of lipid A-core via MsbA (Doerrler et al., 2004). The genes involved in the synthesis and transfer of 4-aminoarabinose to lipid A are encoded by the *ugd* (*pmrE*) and *pmrHFIJKLM* loci. PmrAB directly regulates the *pmrH* operon (Gunn et al., 1998, 2000) and indirectly affects transcription of *ugd* by a process involving the additional regulatory system, RcsC/YojN/RcsB (Mouslim et al., 2003), again reflecting the sophisticated signal integration in cell-surface remodeling. The gene products functioning in 4-aminoarabinose synthesis or transfer have been renamed with an *arn* designation (reviewed in Raetz and Whitfield, 2002). The 4-aminoarabinose-substitution pathway has received significant attention because of its potential as a target for antibacterial strategies that would render pathogens more susceptible to host cationic peptides.

The activated sugar nucleotide precursor for 4-aminoarabinose modifications is the uridine diphospho derivative, UDP-4-deoxy-4-formamido-L-arabinose and is derived from UDP-glucose in an established cytoplasmic pathway. UDP-glucose is converted to UDP-glucuronic acid by Ugd (PmrE; UDP-glucose dehydrogenase) and then UDP-4-deoxy-4-formamido-L-arabinose is formed by the action of ArnA and ArnB (reviewed in Raetz and Whitfield, 2002). ArnA is a bifunctional enzyme that performs the initial oxidative decarboxylation reaction and final formylation (Breazeale et al., 2005; Gatzeva-Topalova et al., 2004, 2005a, 2005b; Williams et al., 2005). Since sugar nucleotides are confined to the cytoplasm, and the lipid A-modification step is periplasmic, an intermediate molecule is required as the direct donor; this donor has been identified as undecaprenyl phosphoryl-4-aminoarabinose (Trent et al., 2001b). The formamido derivative is transferred from its UDP-activated precursor to undecaprenyl phosphate by ArnC (Breazeale et al., 2005; Trent et al., 2001c), but the absence of formyl groups in the final modified lipid A–core product invokes an essential requirement for a deformylation step at the level of the undecaprenol-linked intermediate. It has been speculated that deformylation is required to maintain the equilibrium of the ArnC reaction in the forward direction or for transfer of the lipid intermediate across the inner membrane (Breazeale et al., 2005). ArnT (PmrK) is the integral inner membrane protein that utilizes undecaprenyl phosphoryl-4-aminoarabinose as the donor to modify lipid A (Trent et al., 2001c). Details of the required acceptor are not fully resolved, but the addition of 4-aminoarabinose is abrogated in a *lpxM* mutant producing pentaacylated lipid A (i.e., lacking a 3′-acyloxyacyl-linked myristate; Color Plate 16) (Tran et al., 2005). The means by which the undecaprenyl phosphoryl-4-aminoarabinose is exported across the inner membrane to the periplasm is currently unknown, although there are other membrane proteins encoded by the *arn* locus that may fulfill this export role. Alternatively, the multiple transmembrane helices in ArnT could potentially indicate a dual role for ArnT as an exporter-transferase, rather than simply reflecting its interaction with a lipid donor and acceptor.

ADDITION OF PHOSPHOETHANOL-AMINE TO LIPID A

Like addition of 4-aminoarabinose, modification of lipid A with phosphoethanolamine depends on functional MsbA, indicative of a periplasmic reaction (Doerrler et al., 2004). The PmrC (EptA) protein is involved in phosphoethanolamine addition to lipid A but, despite its regulation by PmrAB, there are conflicting reports on the extent of the influence of this substitution on resistance to cationic peptides (Lee et al., 2004; Tamayo et al., 2005). PmrC is an inner membrane protein with five to six transmembrane helices at the N terminus and a large periplasmic domain toward the C terminus, containing the putative active site. The C terminus also shares significant similarity to the C terminus of *N. meningitidis* Lpt-3 (LptA) (Lee et al., 2004), a known phosphethanolamine transferase (Mackinnon et al., 2002).

REMOVAL OF THE 1-PHOSPHATE IN LIPID A

An alternative to adding substituents to the lipid A phosphates is the selective removal of the phosphate residue at the 1-position. The 1-dephosphorylation process may provide some advantages during infection, including

decreased immune stimulation and a lowered affinity to antimicrobial peptides. In *Salmonella*, the *ugtL* gene is regulated by PhoPQ and is required for resistance to polymyxin, as well as some (but not all) antimicrobial cationic peptides (Shi et al., 2004). The UgtL protein is an integral inner membrane protein and is involved in the formation of monophosphorylated lipid A, but unfortunately, details of the mechanism of action are unknown.

LpxE is a phosphatase responsible for the removal of the phosphate at position 1 of *Rhizobium leguminosarum* and orthologs of LpxE are present in *Agrobacterium tumefaciens, Mesorhizobium loti, Sinorhizobium meliloti,* as well as in the intracellular bacterial pathogens *Francisella tularensis, Brucella melitensis,* and *Legionella pneumophila* (Karbarz et al., 2003). Topology studies predict that the active site of LpxE faces the periplasm (Karbarz et al., 2003), and this has been verified in the system from *Francisella novicida,* where 1-dephosphorylation occurs after export via MsbA (Wang et al., 2004). *Helicobacter pylori* also modifies its LPS by removal of the 1-phosphate using a membrane-associated 1-phosphatase whose active site faces the periplasm (Tran et al., 2004). Removal of the 1-phosphate group is followed by the addition of phosphoethanolamine. This is in contrast to *E. coli* (Zhou et al., 1999) and *Salmonella* (Zhou et al., 2001), where phosphoethanolamine is added to the 1-phosphate to produce a pyrophosphate-linked moiety.

MODIFICATION OF THE LPS ACYLATION PATTERN

Several enzymes have now been identified that alter the fatty acylation of the lipid A molecule (reviewed in Raetz and Whitfield, 2002; Trent, 2004). These enzymes are localized to the inner membrane, periplasm, and the outer membrane, suggesting these modifications occur at different points in the assembly pathway. Additions directed by inner membrane proteins in *E. coli* include the addition of palmitoleate instead of laurate as a secondary acyl chain on the 2′-β-hydroxymyristate under conditions of cold shock (Carty et al., 1999). Another example is the LpxO-mediated hydroxylation of the 3′-secondary myristate residue in *S. enterica* (Gibbons et al., 2000).

The PagP palmitoyltransferase and the PagL 3-*O*-deacylase are small outer membrane proteins, whose active sites face the outer leaflet of the outer membrane. The changes resulting from their activities decrease lipid A signaling Toll-like receptor 4 (TLR-4) (reviewed in Kawasaki et al., 2004a) and confer resistance to certain cationic antimicrobial peptides. PagP activity increases following membrane damage that leads to translocation of phospholipids to the outer leaflet of the outer membrane (reviewed in Bishop, 2005). PagP homologs with varying substrate or acceptor specificity have been identified in several pathogens including *E. coli*, *Salmonella*, *Erwinia*, *Yersinia*, *Legionella*, and *Bordetella* and some are known to be essential for intracellular infection and virulence. PagL removes a *R*-hydroxymyristate chain from position 3 of lipid A (Trent et al., 2001a) and PagL homologs have now been identified in a variety of gram-negative organisms including *Bordetella bronchiseptica* and *Pseudomonas aeruginosa*, as well as some organisms that are non-pathogenic (Geurtsen et al., 2005). In *P. aeruginosa*, PagL was found to be required for growth in conditions that affect the permeability of the outer membrane, such as in the presence of hydrophobic compounds (Geurtsen et al., 2005). PagL-mediated deacylation does not occur in the presence of 4-aminoarabinose substituted lipid A (Kawasaki et al., 2005). These results further highlight the complex regulation and interdependence of lipid A modifications.

Postsynthetic Modification of the Core Oligosaccharide

Phosphoethanolamine additions are not confined to lipid A; the inner core Kdo and heptose regions of the core oligosaccharide (Fig. 1) are also modified. The genomes of *E. coli* and *Salmonella* contain several orthologs of the *N. meningitidis lpt-3* (*lptA*) gene (Lee et al., 2004; Tamayo et al., 2005), including the lipid A phosphoethanolamine transferase, *pmrC* (*etpA*) described above. *yhjW* (*eptB*) and *yijP* (*cptA*)

have recently been identified as the genes encoding Kdo and heptose-modifying phosphoethanolamine transferases. From sequence similarity shared with PmrC/EptA, EptB is predicted to be an integral inner membrane enzyme with its active site in the periplasm.

EptB is a phosphoethanolamine transferase that modifies the inner core oligosaccharide at KdoII (Fig. 1) (Kanipes et al., 2001; Reynolds et al., 2005). The enzyme is capable of transferring phosphoethanolamine from different phosphoethanolamine species to Re-LPS acceptor, generating stoichiometric amounts of diacylglycerol by-products (Reynolds et al., 2005). The precise role that the phosphoethanolamine substituent on KdoII plays in the biology of the organism is unknown, although Etp shows an unusual Ca^{2+} inducibility (via an unknown regulatory mechanism). An *eptB* mutant in a heptose-deficient strain shows Ca^{2+} hypersensitivity, a phenotype not evident in a strain with a complete core oligosaccharide. It was suggested that phosphoethanolamine addition may be important in reducing the permeability of the *ept* mutant outer membrane to Ca^{2+}, therefore providing tolerance to high levels of Ca^{2+} (Reynolds et al., 2005). A full core oligosaccharide with phosphoethanolamine residues present on heptose I (HepI; Fig. 1) may compensate for the loss of the Ca^{2+}-inducible EptB activity and allow viability of an *eptB* mutant. The *cptA* gene product modifies the inner core in *Salmonella* and is thought to be responsible for addition of phosphoethanolamine to HepI (Fig. 1) (Tamayo et al., 2005), but details of the reaction mechanism and its biological function have yet to be resolved. CptA has only minimal impact on resistance to cationic peptides and virulence in a murine model.

In an unusual modification in *H. pylori*, the KdoII residue of the inner core is removed by a Kdo hydrolase (Stead et al., 2005). This hydrolase is likely also periplasmic, as cleavage of Kdo only occurs on 1-dephosphorylated lipid A species that, as indicated above, are only available after export. The removal of the negatively charged Kdo residue could also potentially influence the immune response and sensitivity to cationic peptides.

BIOSYNTHESIS OF O-ANTIGENS
Despite the diversity of O-polysaccharide structures and three different modes of biosynthesis, relatively limited options are known for initiating synthesis and all of the processes terminate in a conserved ligation reaction. The initiation reaction involves the formation of an undecaprenyl pyrophosphate-linked glycose by transfer of a glycose-1-P residue to undecaprenyl phosphate. Through this process, the sugar-phosphate linkage in the donor molecule is retained in the undecaprenyl pyrophosphate-linked product and serves to drive the subsequent ligation of O-polysaccharide to lipid A-core acceptor. The observation that heterologous O-polysaccharides, representing each of the three biosynthesis pathways, are successfully ligated to the core oligosaccharide of *E. coli* K-12 provides strong support for a common ligation substrate being formed in each system. To the extent this has been characterized, the O-polysaccharide donor is an undecaprenyl pyrophosphate-linked polymer (reviewed in Heinrichs et al., 1998b; Raetz and Whitfield, 2002; Whitfield et al., 1997). The intervening steps between initiation and ligation vary in terms of the mode of O-polysaccharide chain elongation, membrane topology (i.e., reactions at the cytoplasmic or periplasmic face of the inner membrane), and the processes involved in export of undecaprenyl pyrophosphate-linked polymer or its intermediates.

O-Polysaccharide Initiation Reactions
The first initiating enzyme activity to be identified was that of WbaP from *Salmonella* (Osborn et al., 1972). WbaP is a member of a family of polyisoprenyl-phosphate hexose-1-phosphate transferases (Valvano, 2003). WbaP transfers galactose-1-P to undecaprenyl phosphate to generate undecaprenyl pyrophosphoryl-Gal and its homologs also include glucose-1-P transferases. The protein topology for WbaP predicted from sequence data indicates a

membrane-embedded N terminus with four transmembrane helices, a large periplasmic domain, and a substantial cytoplasmic C-terminal domain that is thought to contain the catalytic activity in WbaP (Wang et al., 1996) and its homologs (Pelosi et al., 2005). Another initiating enzyme is WecA (formerly Rfe), a member of the polyisoprenyl-phosphate N-acetyl-hexosamine-1-phosphate transferase family (Price and Momany, 2005; Valvano, 2003). WecA is an integral membrane protein that forms the undecaprenyl pyrophosphoryl-GlcNAc (Kido et al., 1995; Rick et al., 1994). It was initially identified from its involvement in enterobacterial common antigen synthesis, but WecA homologs initiate a variety of O-polysaccharides that contain an acetamido sugar as the initial residue (reviewed in Valvano, 2003). The three pathways for O-polysaccharide biosynthesis that utilize WbaP and WecA homologs diverge immediately following initiation.

Polymerization in a Wzy-Dependent Mechanism
The O-polysaccharides of *Salmonella* serogroups B (Typhimurium) and E1 (Anatum) were among the first investigated in biosynthetic studies. Pioneering studies by M. J. Osborn, H. Nikaido, A. Wright, and P. Robbins and others established the sequence of glycosyl transfer and the direction of chain growth (reviewed in Mäkelä and Stocker, 1984; Raetz and Whitfield, 2002; Schnaitman and Klena, 1993; Whitfield, 1995). In this pathway (Color Plate 17), the steps following WbaP/WecA activity assemble the undecaprenyl pyrophosphate-linked O-repeat unit and the enzymes involved are usually predicted to be peripheral membrane proteins. However, there is still no information concerning the manner in which the various enzymes interact with the membrane and with one another to facilitate their organization into the predicted functional complexes.

Following their assembly, the individual undecaprenyl pyrophosphate-linked O-repeat units are exported to the site of polymerization at the periplasm. This process minimally requires a characteristic Wzx homolog, an integral inner membrane protein providing the putative O-repeat transporter ("flippase"). Experimental data are available to support this assignment. In a model system lacking *wzx*, undecaprenyl pyrophosphate-linked O-repeat units appeared to accumulate at the cytoplasmic face of the inner membrane (Liu et al., 1996). Furthermore, the Wzx homolog from the enterobacterial common antigen biosynthesis system can translocate a synthetic polyprenol-linked intermediate into sealed (everted) inner membrane vesicles (Rick et al., 2003). Wzx proteins specifically recognize the initiating sugar, or the initiating transferase that adds this sugar to undecaprenyl phosphate (Feldman et al., 1999; Marolda et al., 2004). The latter is consistent with data suggesting that specific residues in WbaP and WecA are also involved in the release of undecaprenyl pyrophosphate-linked intermediates into the Wzx-mediated export pathway (Amer and Valvano, 2002; Liu et al., 1996).

Once at the periplasmic face of the cytoplasmic membrane, undecaprenyl pyrophosphate-linked O-repeat units are polymerized by transfer of the nascent chain from its lipid carrier to the nonreducing terminus of the newly exported lipid-linked O-repeat unit (Color Plate 17). The polymer therefore grows by blockwise addition of O-repeat units to the reducing terminus (Bray and Robbins, 1967; Robbins et al., 1967). The Wzy protein is essential for this process and is widely considered to be a polymerase, although its mechanism of action is still unknown. The activity of Wzy is also affected by other factors at the periplasm-outer membrane interface, including TolA and Pal (Gaspar et al., 2000; Vines et al., 2005). A *wzy* mutant produces LPS consisting of a single O-repeat unit ligated to lipid A–core (called "semi-rough" LPS) (Collins and Hackett, 1991). This is a defining feature of the pathway. Wzy proteins are all predicted to be integral membrane proteins with 11 to 13 transmembrane helices and, like the Wzx proteins, they exhibit a conserved predicted membrane topology but little primary sequence similarity (Daniels et al., 1998; Morona et al., 1994). This is not unexpected since Wzy proteins are specific for the cognate O unit and for the precise inter-repeat unit linkage. Resolution of the mechanism of action will require in vitro studies in a reconstituted system, but the low levels of Wzx and Wzy expression (Daniels et al., 1998) limit such approaches.

The Wzz (formerly Cld or Rol) protein is the final component of the assembly system and is required for polymerization control. It dictates the strain-specific modal distribution of O-polysaccharide chain lengths that is reflected in characteristic clusters of bands in sodium dodecyl sulfate-polyacrylamide gel electrophoresis (SDS-PAGE) (reviewed in Morona et al., 2000; Whitfield et al., 1997). In the case of *Salmonella*, isolates may contain two Wzz homologs with differing activities (Murray et al., 2005). Modality in O-polysaccharide chain length is essential for virulence in *E. coli* (Burns and Hull, 1998), *Shigella flexneri* (Morona et al., 2003; Morona and Van Den Bosch, 2003), *Salmonella* (Murray et al., 2003, 2005), and *Yersinia enterocolitica* (Najdenski et al., 2003). The existing models proposed for the underlying process variably implicate Wzz, Wzy, and WaaL. In one model, Wzz acts as a timing "clock," interacting with the Wzy polymerase and modulating its activity between two states that favor either elongation or termination (i.e., transfer to the ligase) (Bastin et al., 1993). In the absence of Wzz, the distribution of chain lengths is independent of size, correlating instead with a constant probability of transfer to lipid A core. Morona et al. have suggested an alternative model in which Wzz acts as a molecular chaperone to assemble a complex consisting of Wzy, the WaaL ligase, and undecaprenyl pyrophosphate-linked O-polysaccharide (Morona et al., 1995). The specific modality would be determined by different kinetics resulting from a Wzz-dependent ratio of Wzy to WaaL. However, in *P. aeruginosa*, modality is established prior to ligation (Daniels et al., 2002), suggesting that if the WaaL protein is involved in the chain length-regulating process, this does not require ligation activity per se.

Wzz homologs show local sequence similarities and share predicted secondary structures (reviewed in Morona et al., 2000; Whitfield et al., 1997). Wzz homologs are located in the cytoplasmic membrane and all have two transmembrane helices flanking a periplasmic loop with a predicted propensity for coiled-coil structure (Morona et al., 2000). The coiled-coil domains may be required for the known oligomerization of Wzz (Daniels and Morona, 1999; Daniels et al., 2002). A protein with similar membrane topology, Wzc, interacts with the related Wzy-dependent polysaccharide biosynthesis system in assembly of E. coli group 1 capsules (Whitfield and Paiment, 2003). Wzc is distinguished from Wzz by the size of the periplasmic region and an additional C-terminal autokinase domain (Doublet et al., 2002; Wugeditsch et al., 2001). Like Wzz, Wzc is required for high-level polymerization. Neither Wzz nor Wzc has any apparent specificity for the repeat-unit structures of their polysaccharide products and must therefore be able to interact with diverse Wzy proteins. Wzc forms a tetrameric structure via extensive interactions in the periplasmic domain (Collins et al., 2005). Despite several efforts to define residues that confer modal specificity on Wzz, no clear picture has yet emerged and residues throughout the primary sequence are important (Daniels and Morona, 1999; Franco et al., 1998; Klee et al., 1997).

O-Antigen Modifications Linked to Wzy-Dependent Pathways

O-Polysaccharides are structurally hypervariable because of variations in the genetic content of the biosynthesis gene locus. As an example, E. coli produces more than 170 serologically distinct O-antigens (Ørskov et al., 1977). However, an additional means of O-antigenic variation seen in some bacteria involves the introduction of postsynthetic modifications to create new O-factors. In many cases, the modifying enzymes are carried by lysogenic or cryptic bacteriophages. Perhaps the best-characterized system is the serotype conversion of S. flexneri O-antigens (reviewed in Allison and Verma, 2000). The base O-polysaccharide structure (serotype Y) is a tetrasaccharide synthesized by a Wzy-dependent pathway. This structure can be modified by phage-encoded O-acetylation and glucosylation, either alone or in combinations and on various residues in the tetrasaccharide O-repeat unit. Comparable processes occur in Salmonella and the classic transducing phage, P22, provides a good example of a serotype-converting phage (reviewed in Vander Byl and Kropinski, 2000).

MODIFICATION BY GLUCOSYLATION

The genetic loci for O-antigen glucosylation comprise three cotranscribed genes and share a common organization. Two genes (now renamed gtrA and gtrB) are conserved and can be exchanged between serotypes, and there is a single variable serotype-specific gene (here designated gtr*). In the working model for the glucosylation process (Color Plate 17B), the direct donor of glucosylation is undecaprenyl phosphoryl-Glc and glucose residues are transferred to the O-polysaccharide by the serotype-specific Gtr* enzyme in the periplasm. GtrB forms the lipid-linked donor by transferring glucose from UDP-glucose to undecaprenyl phosphate at the cytoplasmic face of the inner membrane. GtrA is proposed to be required for export of the donor to the periplasm. The premise for the periplasmic location of the glucosylation reaction is based on studies with Salmonella that first showed undecaprenyl pyrophosphate-linked intermediates provide the acceptor for glucosylation (Takeshita and Mäkelä, 1971). Since the O-repeat unit at the reducing terminus lacked modification, glucosylation must follow polymerization and the resulting acceptors are confined to a periplasmic location in the Wzy-dependent assembly system (see above). However, additional precedent for this type of reaction comes from the comparable 4-aminoarabinose-addition system that modifies lipid A at the periplasmic face of the inner membrane using an undecaprenyl phosphate-linked donor (described above).

The membrane topology of one *S. flexneri* Gtr★ representative, GtrV from serotype 5a, has been investigated in some detail by using a hybrid protein approach (Korres and Verma, 2004). The GtrV protein has nine transmembrane helices. Multiple membrane-spanning helices are a conserved feature of Gtr★, Wzy, and WaaL proteins that all use undecaprenol-linked donors in their reactions. GtrV has a large periplasmic domain toward the N terminus (between helices I and II). The N terminus itself is located in the cytoplasm, whereas the C terminus is periplasmic. GtrV and its homologs (GtrX, GtrI, and GtrII) share overall similarity in topology and all have a conserved reentrant loop after transmembrane helix IV (Korres and Verma, 2004; Lehane et al., 2005). Critical residues conserved in Gtr★ homologs have been identified in periplasmic domains of GtrII, consistent with the proposed location of the active site (Lehane et al., 2005). The glucosyltransferase function of Gtr★ enzymes has yet to be shown directly, but the enzyme is essential for glucosylation in a serotype-specific mechanism.

The GtrB protein comprises two transmembrane helices and a large cytoplasmic N-terminal domain (approximately 230 of its 305 to 309 residues). This topology has been confirmed with hybrid protein fusions (Korres et al., 2005). Membranes from cells expressing GtrB transfer glucose from UDP-glucose to exogenous decaprenol phosphate acceptor, consistent with the functional assignment (Guan et al., 1999). In early studies, the native lipid showed properties expected of undecaprenyl phosphate (Nikaido et al., 1971; Nikaido and Nikaido, 1971). Furthermore, it was predicted to contain a β-linked glucose residue, consistent with the observation that GtrB shares some motifs with inverting glycosyltransferases (Guan et al., 1999). The reaction is therefore equivalent to the formation of undecaprenyl phosphoryl-4-aminoarabinose by ArnT (see above). While the native glucosyl lipid has not been fully characterized, the methods developed for description of the 4-aminoarabinose modification system (Trent et al., 2001b) should be applicable. GtrB homologs share limited similarity with dolichol phosphorylmannose synthase (Guan et al., 1999; Mavris et al., 1997).

The functional data for GtrA are limited. The small (120-residue) GtrA encodes an inner membrane protein that is predicted to contain four transmembrane helices and have N- and C-terminal domains located in the cytoplasm (Korres et al., 2005). There are conflicting results concerning whether GtrA is actually required for glucosylation (reviewed in Allison and Verma, 2000). In some cases, glucosylation can occur in cells expressing Gtr★ alone. Perhaps other exporters for undecaprenyl-linked sugars participate with low efficiency in these situations. In cases where GtrA is essential, the existing data cannot distinguish between GtrA itself being the exporter and a model where Gtr★ itself fulfills this role with, or without, the involvement of GtrA (Guan et al., 1999). Dual glycosyltransferase proteins have also been postulated for the synthase-dependent O-polysaccharide assembly system (see below). Resolution of these details awaits further analyses and synthetic lipid-linked sugars may offer a potential experimental approach.

O-ACETYLATION

O-Acetyltransferases also modify O-polysaccharides. Well-documented examples include OafA, which confers O:5 specificity in *Salmonella* (Slauch et al., 1996); Oac encoded by the *Shigella* seroconverting bacteriophage SF6 (Verma et al., 1991); and the O-acetyltransferase carried by *P. aeruginosa* bacteriophage D3 (Newton et al., 2001). Acetyl-CoA provides the donor for O-acetylation and the acceptor has been proposed to be undecaprenyl pyrophosphate-linked intermediates (discussed in Mäkelä and Stocker, 1984). The use of acetyl-CoA as the donor initially suggested a cytoplasmic location for O-acetylation, but further analysis of O-acetyltransferases has led to the proposal that the active site of these enzymes may be located in the periplasm (reviewed in Clarke et al., 2000). OafA and its homologs are large integral membrane proteins and contain an N-terminal domain that shares some similarity with a human acetyl-CoA transporter. This similarity plays a major role in the suggestion of the periplasmic reaction site, since it provides an enzyme that might be sufficient for both the export of the donor and acetylation of the acceptor. Other O-acetyltransferases are smaller soluble proteins. In each case, the enzyme is accompanied by a member of the AlgI protein family. The latter are integral membrane proteins that may perform a donor export function. Appropriate Wzy-dependent biosynthesis systems are clearly available to begin to test these hypotheses.

ALTERATION OF WZY SPECIFICITY

Another bacteriophage-encoded modification of O-polysaccharide involves modulation of polymerization specificity. The change in anomeric linkage between O-repeat units results in a novel O-factor. Such modifications have been studied in *Salmonella* containing the ε^{15}-seroconverting phage (Losick and Robbins, 1967) and in *P. aeruginosa* phage D3 lysogens where the reaction is accompanied by O-acetylation (Newton et al., 2001). Both systems are proposed to encode inhibitors of the wild-type Wzy polymerases, as well as a novel phage-encoded replacement Wzy enzyme. The D3 system has provided insight into the process and the inhibitor has been identified as Iap, a small (31-residue) hydrophobic inhibitory polypeptide. The novel replacement Wzy protein results in a switch from α- to β-linkages between repeat units. The structural and functional relationships (if any) between the polypeptide Wzy-inhibitors in the two known systems have not been established and more work is required to determine whether these intriguing polypeptides share common structure and/or mechanism.

The ABC Transporter–Dependent Pathway

The ABC transporter-dependent pathway differs fundamentally from Wzy-dependent biosynthesis in that the polymer is completed inside the cytoplasm (Color Plate 18). The O-polysaccharide is extended by processive addition of glycosyl residues to the nonreducing terminus of the undecaprenyl pyrophosphate-linked acceptor. On completion, an ABC transporter exports the polymer across the inner membrane for ligation. There is no involvement of Wzx, Wzy, or Wzz homologs. All ABC transporters consist of two integral membrane protein domains with (typically) six membrane-spanning helices, and two hydrophilic domains containing an ATP-binding motif or Walker box. The organization of these domains varies in different members of the ABC-transporter superfamily (Higgins, 2001). In the O-polysaccharide ABC transporters, Wzm provides the transmembrane domains and although Wzm proteins generally exhibit little primary sequence identity, they have nearly identical hydropathy plots. In contrast, the primary sequences of the ATP-binding Wzt homologs are highly conserved in the N-terminal nucleotide-binding region and may contain a more variable C terminus.

A homolog of WecA serves as the initiating enzyme for all of the currently characterized ABC transporter-dependent O-polysaccharides. By far the best-studied example is the family of polymannan O-polysaccharides of *E. coli* and *Klebsiella pneumoniae*. The WecA product, undecaprenyl pyrophosphoryl-GlcNAc, acts as a primer for chain extension (Kido et al., 1995; Rick et al., 1994). Structural analysis of the O-polysaccharide:core oligosaccharide linkage region indicates that the primer is then extended by the activity of two mannosyltransferases that commit the undecaprenyl pyrophosphoryl-GlcNAc to O-polysaccharide biosynthesis. This generates an acceptor appropriate for addition of the O-repeat unit domain (Vinogradov et al., 2002). The completion of each glycan chain requires a single undecaprenol lipid carrier, rather than the intensive involvement seen in the Wzy-dependent pathway.

Following polymerization, the nascent O-polysaccharide must be exported across the cytoplasmic membrane for ligation, but key questions have surrounded how chain extension is terminated and how modality in O-polysaccharide chain length can be determined in the processive mechanism. Insight into this process came from the finding that the polymannans and several other O-polysaccharides terminate in residues not found in the repeat units (reviewed in Vinogradov et al., 2002). In the case of the *E. coli* polymannans (e.g., serotype O9a; Color Plate 18), the terminal residue is an O-methyl or phosphoryl-O-methyl residue (depending on the serotype) added by a mono- or dual-function WbdD enzyme (Clarke et al., 2004). The WbdD-mediated chain termination reactions establish modal regulation in participation with the ABC transporter because the export process is coupled to chain termination. To achieve this coordination, specific features of the polymer (perhaps the nonreducing terminal residues) are recognized by the ABC transporter, specifically the C-terminal domain of the ATP-binding Wzt protein (Cuthbertson et al., 2005). Details of the export process have not been resolved, but an attractive model can be proposed based on that described above for MsbA, where the hydrophilic glycan is transferred across the inner membrane within the exporter lumen, while the undecaprenol lipid carrier is retained within the bilayer.

The Synthase–Dependent Pathway

Synthases are processive glycosyltransferases that have the capacity to synthesize polymers within one or more linkages. The plasmid-encoded poly-*N*-acetylmannosamine O:54 antigen of *S. enterica* serovar Borreze is currently the only known example of a synthase-dependent O-polysaccharide (Keenleyside et al., 1994; Keenleyside and Whitfield, 1995, 1996), but synthases are involved in biosynthesis of important biological products including cellulose and chitin. Other well-studied examples are the glycosaminoglycans produced by members of the *Pasteurellaceae* and streptococci (reviewed in DeAngelis, 2002), and the type 3 capsules of *Streptococcus pneumoniae* (Cartee et al., 2000, 2001; Forsee et al., 2000). These capsule-synthesizing enzymes provide useful working models for the synthase chain-elongation process, although they apparently do not have a requirement for an undecaprenol-linked acceptor. The direction of chain growth in the existing synthase systems has been subject to debate and, as it turns out, there may be no unified model. Bacterial hyaluronan synthases appear to elongate at the reducing terminus while the vertebrate enzymes add residues to the nonreducing end (Bodevin-Authelet et al., 2005; Tlapak-Simmons et al., 2005). The structure of the O:54 antigen, and the requirement for WecA in its biosynthesis, means that growth can only occur at the nonreducing terminus of an undecaprenol-linked acceptor in a reaction similar to the ABC transporter-dependent pathways.

In the O:54-biosynthesis system, the chain-extending synthase is proposed to be WbbF (formerly RfbB) (Color Plate 19). This enzyme is part of the HasA (hyaluronan synthase) family and all members are integral membrane proteins with shared topology (Heldermon et al., 2001; Keenleyside and Whitfield, 1996). The proteins are believed to catalyze a vectorial polymerization reaction that extends the glycan chain and extrudes it across the inner membrane in a coupled synthesis-export process (DeAngelis, 2002; Keenleyside and Whitfield, 1996). While the involvement of WaaL in the ligation reaction is consistent with the possibility that the exported form of the polymer is undecaprenyl pyrophosphate linked, no information casts light on the exact mechanism of export. It is also unclear how chain length is established. Studies on biosynthesis of the streptococcal type 3 capsule suggest that chain termination reflects an error in the step where the polymer is normally translocated between two active glycosyltransfer sites for the two monomers in the disaccharide repeat unit (Cartee et al., 2000). The balance of sugar nucleotide donors dictates efficient or abortive translocation. Whether this could apply to an enzyme with a single donor (like WbbF) is unknown. It is surprising that despite their importance in biology, the catalytic mechanism(s) of synthases is still controversial (reviewed in Charnock et al., 2001).

LIGATION OF O-POLYSACCHARIDE TO LIPID A-CORE ACCEPTOR

The final step in assembly of smooth LPS involves ligation of O-polysaccharide to the nascent lipid A-core. The ligation reaction occurs at the periplasmic face of the inner membrane (reviewed in Raetz and Whitfield, 2002; Whitfield et al., 1997). MsbA delivers the lipid A-core acceptor to the periplasmic face (see above). The three pathways for O-polysaccharide biosynthesis appear to converge in delivery of the completed undecaprenyl pyrophosphate-linked polymer to this location. The ligation reactions exhibit specificity for the precise acceptor residue in the core oligosaccharide, as well as for other proximal residues (see Heinrichs et al., 1998a; Kaniuk et al., 2004). The ligation reaction mechanism has not been determined and, until recently, WaaL was the only protein known to be required for the process. However, in *Salmonella* specificity for the core oligosaccharide acceptor appears to involve an additional currently unidentified factor, so the process may be more complex than originally thought (Kaniuk et al., 2004).

The primary sequences of WaaL proteins do not offer insight into the reaction mechanism. Collectively they share only low levels of similarity in their primary sequences and this could reflect their specificity for different core oligosaccharide acceptors. However, all WaaL homologs are predicted to be integral membrane proteins with a conserved membrane topology containing eight or more membrane-spanning helices and two large periplasmic domains. The periplasmic domains from different WaaL proteins contain some conserved motifs that are essential for function (Schild et al., 2005), consistent with the periplasmic reaction site and a shared ligase reaction mechanism. Because the ligation donor is an undecaprenyl pyrophosphate-linked product, the reaction mechanism for WaaL may be related to that of the Wzy polymerase and it is notable that they do share some common features (Schild et al., 2005).

TRANSLOCATION OF LPS TO THE CELL SURFACE

Despite the extensive progress that has been made in unraveling the pathways and mechanisms involved in LPS biosynthesis, the processes involved in translocating LPS from the outer face of the inner membrane to the cell surface are still a mystery. Experimental approaches to address this question have been limited because LPS is essential for viability in most bacteria, but the observation that *N. meningitidis* can survive without lipooligosaccharide (Bos and Tommassen, 2005; Steeghs et al., 1998) offers a unique system that can be exploited for translocation experiments.

The Imp protein (Braun and Silhavy, 2002) is an outer membrane (predicted) β-barrel protein that is essential for LPS assembly in the outer membrane (Bos et al., 2004). Initial reports had implicated an Omp85-family member, YaeT, as an LPS translocation component, but these have been controversial (Genevrois et al., 2003; Voulhoux et al., 2003), and in *E. coli*, YaeT appears to have general effects on the folding and assembly of outer membrane proteins (Wu et al., 2005). These conflicting reports reflect the complex interactions between processes involved in the insertion of phospholipids, proteins, and LPS into the outer membrane. It has also been proposed that LPS molecules and some outer membrane proteins interact prior to (or during) translocation of the protein to the outer membrane, with the LPS acting as a "co-chaperone" (Bulieris et al., 2003). This can only occur if the LPS and outer membrane protein translocation processes are temporally and spatially coupled. In an elegant recent study, genetic interactions were demonstrated between *imp* and genes encoding *yaeT* and several outer membrane lipoproteins (Wu et al., 2005). The authors propose a homeostatic mechanism controlling global outer membrane composition and involving the outer membrane lipoprotein, YfgL.

In electron microscopy studies performed several decades ago, it was observed that LPS export to the cell surface occurred at discrete sites that coincide with domains where the inner and outer membranes come into close apposition (Bayer et al., 1982; Mühlradt et al., 1973). The physiological significance of the adhesions sites has been controversial (reviewed in Bayer, 1991; Kellenberger, 1990), but these contact sites are consistent with the presence of scaffold assemblies involved in translocation of macromolecules and there is growing evidence for such structures in many biological systems, including protein secretion, drug efflux, and translocation of capsular polysaccharides. Recent experiments addressing LPS translocation in spheroplasts confirmed the earlier observations (Bayer et al., 1982) that export occurs at domains where inner and outer membranes are attached (Tefsen et al., 2005b). In an intriguing recent observation, Ghosh and Young identified a stable helical disposition of LPS in the outer membrane, confirming and significantly extending prior observations suggesting that the LPS molecules are positionally stable, rather than being subject to extensive lateral mobility (Ghosh and Young, 2005). One interpretation of these observations is that LPS insertion involves movement of the active complexes to follow helical tracks. It has been shown that peptidoglycan is also distributed in a helical arrangement along the side walls of *E. coli* and it will be interesting to see whether the events are coupled and whether the emerging bacterial cytoskeletal components (reviewed in Cabeen and Jacobs-Wagner, 2005) play any role in dictating the sites of LPS insertion. Identification of the various "players" in outer membrane assembly will afford an opportunity to resolve the translocation mechanism, but given that the key proteins may be essential for viability and that extensive interactions may be involved, a combined application of genetic methods, biochemistry, and structural biology will be required.

ACKNOWLEDGMENTS

Work on LPS biosynthesis in the authors' laboratory has been generously supported by the Natural Sciences and Engineering Research Council, the Canadian Institutes of Health Research, and the Canadian Bacterial Diseases Network (NCE Program). C.W. is the grateful recipient of a Canada Research Chair, and E.F. and A.N.R. have been supported by graduate scholarships from the Canadian Institutes of Health Research.

ADDENDUM IN PROOF

LPS biosynthesis and assembly represents a dynamic research topic. While this chapter was being typeset, progress was made in a number of the specific areas covered. While it is impossible to update all aspects, there are two that merit inclusion here.

(i) The processes involved in the translocation of the complete LPS molecule across the periplasm and outer membrane remain a critical open question. Recent work is beginning to shed light on the essential components. Wu et al. (2006) establish a critical role in LPS translocation played by a complex including the outer membrane protein, Imp, and a rare lipoprotein, RlpB. Another report, by Spandero et al., provides preliminary data that implicate additional proteins in LPS assembly (2006). These proteins are encoded by genes adjacent to the CMP-Kdo biosynthesis locus.

(ii) While the biosynthesis and assembly of O antigens are considered to be carried out by an enzyme complex, supporting data are rather limited. In a recent paper, Marolda et al. (2006) provide genetic evidence for interplay and recognition between Wzx, Wzy, and Wzz proteins.

REFERENCES

Allison, G. E., and N. K. Verma. 2000. Serotype-converting bacteriophages and O-antigen modification in *Shigella flexneri*. *Trends Microbiol.* **8:**17–23.

Amer, A. O., and M. A. Valvano. 2002. Conserved aspartic acids are essential for the enzymic activity of the WecA protein initiating the biosynthesis of O-specific lipopolysaccharide and enterobacterial common antigen in *Escherichia coli*. *Microbiology* **148:**571–582.

Bader, M. W., S. Sanowar, M. E. Daley, A. R. Schneider, U. Cho, W. Xu, R. E. Klevit, H. Le Moual, and S. I. Miller. 2005. Recognition of antimicrobial peptides by a bacterial sensor kinase. *Cell* **122:**461–472.

Bastin, D. A., G. Stevenson, P. K. Brown, A. Haase, and P. R. Reeves. 1993. Repeat unit polysaccharides of bacteria: a model for polymerization resembling that of ribosomes and fatty acid synthetase, with a novel mechanism for determining chain length. *Mol. Microbiol.* **7:**725–734.

Bayer, M. E. 1991. Zones of membrane adhesion in the cryofixed envelope of *Escherichia coli*. *J. Struct. Biol.* **107:**268–280.

Bayer, M. H., G. P. Costello, and M. E. Bayer. 1982. Isolation and partial characterization of membrane vesicles carrying markers of the membrane adhesion sites. *J. Bacteriol.* **149:**758–767.

Bishop, R. E. 2005. The lipid A palmitoyltransferase PagP: molecular mechanisms and role in bacterial pathogenesis. *Mol. Microbiol.* **57:**900–912.

Bodevin-Authelet, S., M. Kusche-Gullberg, P. E. Pummill, P. L. DeAngelis, and U. Lindahl. 2005. Biosynthesis of hyaluronan: direction of chain elongation. *J. Biol. Chem.* **280:**8813–8818.

Bos, M. P., B. Tefsen, J. Geurtsen, and J. Tommassen. 2004. Identification of an outer membrane protein required for the transport of lipopolysaccharide to the bacterial cell surface. *Proc. Natl. Acad. Sci. USA* **101:**9417–9422.

Bos, M. P., and J. Tommassen. 2005. Viability of a capsule- and lipopolysaccharide-deficient mutant of *Neisseria meningitidis. Infect. Immun.* **73:**6194–6197.

Braun, M., and T. J. Silhavy. 2002. Imp/OstA is required for cell envelope biogenesis in *Escherichia coli. Mol. Microbiol.* **45:**1289–1302.

Bray, D., and P. W. Robbins. 1967. The direction of chain growth in *Salmonella anatum* O-antigen biosynthesis. *Biochem. Biophys. Res. Commun.* **28:**334–339.

Breazeale, S. D., A. A. Ribeiro, A. L. McClerren, and C. R. Raetz. 2005. A formyltransferase required for polymyxin resistance in *Escherichia coli* and the modification of lipid A with 4-amino-4-deoxy-L-arabinose. Identification and function of UDP-4-deoxy-4-formamido-L-arabinose. *J. Biol. Chem.* **280:**14154–14167.

Buchaklian, A. H., A. L. Funk, and C. S. Klug. 2004. Resting state conformation of the MsbA homodimer as studied by site-directed spin labeling. *Biochemistry* **43:**8600–8606.

Buchaklian, A. H., and C. S. Klug. 2005. Characterization of the Walker A motif of MsbA using site-directed spin labeling electron paramagnetic resonance spectroscopy. *Biochemistry* **44:**5503–5509.

Bulieris, P. V., S. Behrens, O. Holst, and J. H. Kleinschmidt. 2003. Folding and insertion of the outer membrane protein OmpA is assisted by the chaperone Skp and by lipopolysaccharide. *J. Biol. Chem.* **278:**9092–9099.

Burns, S. M., and S. I. Hull. 1998. Comparison of loss of serum resistance by defined lipopolysaccharide mutants and an acapsular mutant of uropathogenic *Escherichia coli* O75:K5. *Infect. Immun.* **66:**4244–4253.

Cabeen, M. T., and C. Jacobs-Wagner. 2005. Bacterial cell shape. *Nat. Rev. Microbiol.* **3:**601–610.

Cartee, R. T., W. T. Forsee, J. W. Jensen, and J. Yother. 2001. Expression of the *Streptococcus pneumoniae* type 3 synthase in *Escherichia coli.* Assembly of type 3 polysaccharide on a lipid primer. *J. Biol. Chem.* **276:**48831–48839.

Cartee, R. T., W. T. Forsee, J. S. Schutzbach, and J. Yother. 2000. Mechanism of type 3 capsular polysaccharide synthesis in *Streptococcus pneumoniae. J. Biol. Chem.* **275:**3907–3914.

Carty, S. M., K. R. Sreekumar, and C. R. Raetz. 1999. Effect of cold shock on lipid A biosynthesis in *Escherichia coli.* Induction at 12°C of an acyltransferase specific for palmitoleoyl-acyl carrier protein. *J. Biol. Chem.* **274:**9677–9685.

Chang, G. 2003. Structure of MsbA from *Vibrio cholera:* a multidrug resistance ABC transporter homolog in a closed conformation. *J. Mol. Biol.* **330:**419–430.

Chang, G., and C. B. Roth. 2001. Structure of MsbA from *E. coli:* a homolog of the multidrug resistance ATP binding cassette (ABC) transporters. *Science* **293:**1793–1800.

Charnock, S. J., B. Henrissat, and G. J. Davies. 2001. Three-dimensional structures of UDP-sugar glycosyltransferases illuminate the biosynthesis of plant polysaccharides. *Plant Physiol.* **125:**527–531.

Clarke, A. J., H. Strating, and N. T. Blackburn. 2000. Pathways for the O-acetylation of bacterial cell wall polysaccharides, p. 187–224. *In* R. J. Doyle (ed.), *Glycomicrobiology.* Kluwer Academic/Plenum Publishers, New York, N.Y.

Clarke, B. R., L. Cuthbertson, and C. Whitfield. 2004. Nonreducing terminal modifications determine the chain length of polymannose O antigens of *Escherichia coli* and couple chain termination to polymer export via an ATP-binding cassette transporter. *J. Biol. Chem.* **279:**35709–35718.

Collins, L. V., and J. Hackett. 1991. Molecular cloning, characterization, and nucleotide sequence of the *rfc* gene, which encodes an O-antigen polymerase of *Salmonella typhimurium. J. Bacteriol.* **173:**2521–2529.

Collins, R. F., K. Beis, B. R. Clarke, R. C. Ford, M. Hulley, J. H. Naismith, and C. Whitfield. 2006. Periplasmic protein-protein contacts in the inner membrane protein, Wzc, form a tetrameric complex required for the assembly of *Escherichia coli* group 1 capsules. *J. Biol. Chem.* **281:**2144–2150.

Cuthbertson, L., J. Powers, and C. Whitfield. 2005. The C-terminal domain of the nucleotide-binding domain protein Wzt determines substrate specificity in the ATP-binding cassette transporter for the lipopolysaccharide O antigens in *Escherichia coli* serotypes O8 and O9a. *J. Biol. Chem.* **280:**30310–30319.

Daniels, C., C. Griffiths, B. Cowles, and J. S. Lam. 2002. *Pseudomonas aeruginosa* O-antigen chain length is determined before ligation to lipid A core. *Environ. Microbiol.* **4:**883–897.

Daniels, C., and R. Morona. 1999. Analysis of *Shigella flexneri* Wzz (Rol) function by mutagenesis and cross-linking: Wzz is able to oligomerize. *Mol. Microbiol.* **34:**181–194.

Daniels, C., C. Vindurampulle, and R. Morona. 1998. Overexpression and topology of the *Shigella flexneri* O-antigen polymerase (Rfc/Wzy). *Mol. Microbiol.* **28:**1211–1222.

DeAngelis, P. L. 2002. Microbial glycosaminoglycan glycosyltransferases. *Glycobiology* **12:**9R–16R.

Doerrler, W. T., H. S. Gibbons, and C. R. Raetz. 2004. MsbA-dependent translocation of lipids across the inner membrane of *Escherichia coli. J. Biol. Chem.* **279:**45102–45109.

Doerrler, W. T., M. C. Reedy, and C. R. Raetz. 2001. An *Escherichia coli* mutant defective in lipid export. *J. Biol. Chem.* **276:**11461–11464.

Dong, J., G. Yang, and H. S. McHaourab. 2005. Structural basis of energy transduction in the transport cycle of MsbA. *Science* **308:**1023–1028.

Doublet, P., C. Grangeasse, B. Obadia, E. Vaganay, and A. J. Cozzone. 2002. Structural organization of the protein-tyrosine autokinase Wzc within *Escherichia coli* cells. *J. Biol. Chem.* **277:** 37339–37348.

Feldman, M. F., C. L. Marolda, M. A. Monteiro, M. B. Perry, A. J. Parodi, and M. A. Valvano. 1999. The activity of a putative polyisoprenol-linked sugar translocase (Wzx) involved in *Escherichia coli* O antigen assembly is independent of the chemical structure of the O repeat. *J. Biol. Chem.* **274:**35129–35138.

Forsee, W. T., R. T. Cartee, and J. Yother. 2000. Biosynthesis of type 3 capsular polysaccharide in *Streptococcus pneumoniae.* Enzymatic chain release by an abortive translocation process. *J. Biol. Chem.* **275:**25972–25978.

Franco, A. V., D. Liu, and P. R. Reeves. 1998. The Wzz (Cld) protein in *Escherichia coli*: amino acid sequence variation determines O-antigen chain length specificity. *J. Bacteriol.* **180:**2670–2675.

Frirdich, E., and C. Whitfield. 2005. Lipopolysaccharide inner core oligosaccharide structure and outer membrane stability in human pathogens belonging to the Enterobacteriaceae. *J. Endotox. Res.* **11:**133–144.

Ganz, T. 2003. Defensins: antimicrobial peptides of innate immunity. *Nat. Rev. Immunol.* **3:**710–720.

Gaspar, J. A., J. A. Thomas, C. L. Marolda, and M. A. Valvano. 2000. Surface expression of O-specific lipopolysaccharide in *Escherichia coli* requires the function of the TolA protein. *Mol. Microbiol.* **38:**262–275.

Gatzeva-Topalova, P. Z., A. P. May, and M. C. Sousa. 2004. Crystal structure of *Escherichia coli* ArnA (PmrI) decarboxylase domain. A key enzyme for lipid A modification with 4-amino-4-deoxy-L-arabinose and polymyxin resistance. *Biochemistry* **43:**13370–13379.

Gatzeva-Topalova, P. Z., A. P. May, and M. C. Sousa. 2005a. Structure and mechanism of ArnA: conformational change implies ordered dehydrogenase mechanism in key enzyme for polymyxin resistance. *Structure* **13:**929–942.

Gatzeva-Topalova, P. Z., A. P. May, and M. C. Sousa. 2005b. Crystal structure and mechanism of

the *Escherichia coli* ArnA (PmrI) transformylase domain. An enzyme for lipid A modification with 4-amino-4-deoxy-L-arabinose and polymyxin resistance. *Biochemistry* **44:**5328–5338.

Genevrois, S., L. Steeghs, P. Roholl, J. J. Letesson, and P. van der Ley. 2003. The Omp85 protein of *Neisseria meningitidis* is required for lipid export to the outer membrane. *EMBO J.* **22:**1780–1789.

Geurtsen, J., L. Steeghs, J. T. Hove, P. van der Ley, and J. Tommassen. 2005. Dissemination of lipid A deacylases (*pagL*) among gram-negative bacteria: identification of active-site histidine and serine residues. *J. Biol. Chem.* **280:**8248–8259.

Ghosh, A. S., and K. D. Young. 2005. Helical disposition of proteins and lipopolysaccharide in the outer membrane of *Escherichia coli. J. Bacteriol.* **187:**1913–1922.

Gibbons, H. S., S. R. Kalb, R. J. Cotter, and C. R. Raetz. 2005. Role of Mg^{2+} and pH in the modification of *Salmonella* lipid A after endocytosis by macrophage tumour cells. *Mol. Microbiol.* **55:** 425–440.

Gibbons, H. S., S. Lin, R. J. Cotter, and C. R. Raetz. 2000. Oxygen requirement for the biosynthesis of the S-2-hydroxymyristate moiety in *Salmonella typhimurium* lipid A. Function of LpxO, a new Fe^{2+}/a-ketoglutarate-dependent dioxygenase homologue. *J. Biol. Chem.* **275:**32940–32949.

Groisman, E. A. 2001. The pleiotropic two-component regulatory system PhoP-PhoQ. *J. Bacteriol.* **183:**1835–1842.

Groisman, E. A., J. Kayser, and F. C. Soncini. 1997. Regulation of polymyxin resistance and adaptation to low-Mg^{2+} environments. *J. Bacteriol.* **179:**7040–7045.

Guan, S., D. A. Bastin, and N. K. Verma. 1999. Functional analysis of the O antigen glucosylation gene cluster of *Shigella flexneri* bacteriophage SfX. *Microbiology* **145:**1263–1273.

Gunn, J. S., K. B. Lim, J. Krueger, K. Kim, L. Guo, M. Hackett, and S. I. Miller. 1998. PmrA-PmrB-regulated genes necessary for 4-aminoarabinose lipid A modification and polymyxin resistance. *Mol. Microbiol.* **27:**1171–1182.

Gunn, J. S., S. S. Ryan, J. C. Van Velkinburgh, R. K. Ernst, and S. I. Miller. 2000. Genetic and functional analysis of a PmrA-PmrB-regulated locus necessary for lipopolysaccharide modification, antimicrobial peptide resistance, and oral virulence of *Salmonella enterica* serovar Typhimurium. *Infect. Immun.* **68:**6139–6146.

Guo, L., K. B. Lim, J. S. Gunn, B. Bainbridge, R. P. Darveau, M. Hackett, and S. I. Miller. 1997. Regulation of lipid A modifications by *Salmonella typhimurium* virulence genes *phoP-phoQ. Science* **276:**250–253.

Guo, L., K. B. Lim, C. M. Poduje, M. Daniel, J. S. Gunn, M. Hackett, and S. I. Miller. 1998. Lipid

A acylation and bacterial resistance against vertebrate antimicrobial peptides. *Cell* **95**:189–198.

Heinrichs, D. E., M. A. Monteiro, M. B. Perry, and C. Whitfield. 1998a. The assembly system for the lipopolysaccharide R2 core-type of *Escherichia coli* is a hybrid of those found in *Escherichia coli* K-12 and *Salmonella enterica*. Structure and function of the R2 WaaK and WaaL homologs. *J. Biol. Chem.* **273**:8849–8859.

Heinrichs, D. E., J. A. Yethon, and C. Whitfield. 1998b. Molecular basis for structural diversity in the core regions of the lipopolysaccharides of *Escherichia coli* and *Salmonella enterica*. *Mol. Microbiol.* **30**:221–232.

Heldermon, C., P. L. DeAngelis, and P. H. Weigel. 2001. Topological organization of the hyaluronan synthase from *Streptococcus pyogenes*. *J. Biol. Chem.* **276**:2037–2046.

Higgins, C. F. 2001. ABC transporters: physiology, structure and mechanism—an overview. *Res. Microbiol.* **152**:205–210.

Kanipes, M. I., S. Lin, R. J. Cotter, and C. R. Raetz. 2001. Ca^{2+}-induced phosphoethanolamine transfer to the outer 3-deoxy-D-manno-octulosonic acid moiety of *Escherichia coli* lipopolysaccharide. A novel membrane enzyme dependent upon phosphatidylethanolamine. *J. Biol. Chem.* **276**:1156–1163.

Kaniuk, N. A., E. Vinogradov, and C. Whitfield. 2004. Investigation of the structural requirements in the lipopolysaccharide core acceptor for ligation of O antigens in the genus *Salmonella*: WaaL "ligase" is not the sole determinant of acceptor specificity. *J. Biol. Chem.* **279**:36470–36480.

Karbarz, M. J., S. R. Kalb, R. J. Cotter, and C. R. Raetz. 2003. Expression cloning and biochemical characterization of a *Rhizobium leguminosarum* lipid A 1-phosphatase. *J. Biol. Chem.* **278**:39269–39279.

Kato, A., and E. A. Groisman. 2004. Connecting two-component regulatory systems by a protein that protects a response regulator from dephosphorylation by its cognate sensor. *Genes Dev* **18**:2302–2313.

Kato, A., T. Latifi, and E. A. Groisman. 2003. Closing the loop: the PmrA/PmrB two-component system negatively controls expression of its posttranscriptional activator PmrD. *Proc. Natl. Acad. Sci. USA* **100**:4706–4711.

Kawasaki, K., R. K. Ernst, and S. I. Miller. 2004a. Deacylation and palmitoylation of lipid A by Salmonellae outer membrane enzymes modulate host signaling through Toll-like receptor 4. *J. Endotox. Res.* **10**:439–444.

Kawasaki, K., R. K. Ernst, and S. I. Miller. 2004b. 3-O-deacylation of lipid A by PagL, a PhoP/PhoQ-regulated deacylase of *Salmonella typhimurium*, modulates signaling through Toll-like receptor 4. *J. Biol. Chem.* **279**:20044–20048.

Kawasaki, K., R. K. Ernst, and S. I. Miller. 2005. Inhibition of *Salmonella enterica* serovar Typhimurium lipopolysaccharide deacylation by aminoarabinose membrane modification. *J. Bacteriol.* **187**:2448–2457.

Keenleyside, W. J., M. B. Perry, L. L. MacLean, C. Poppe, and C. Whitfield. 1994. A plasmid-encoded *rfb*$_{O:54}$ gene cluster is required for biosynthesis of the O:54 antigen in *Salmonella enterica* serovar Borreze. *Mol. Microbiol.* **11**:437–448.

Keenleyside, W. J., and C. Whitfield. 1995. Lateral transfer of *rfb* genes: a mobilizable ColE1-type plasmid carries the *rfb*$_{O:54}$ (O:54 antigen biosynthesis) gene cluster from *Salmonella enterica* serovar Borreze. *J. Bacteriol.* **177**:5247–5253.

Keenleyside, W. J., and C. Whitfield. 1996. A novel pathway for O-polysaccharide biosynthesis in *Salmonella enterica* serovar Borreze. *J. Biol. Chem.* **271**:28581–28592.

Kellenberger, E. 1990. The 'Bayer bridges' confronted with results from improved electron microscopy methods. *Mol. Microbiol.* **4**:697–705.

Kido, N., V. I. Torgov, T. Sugiyama, K. Uchiya, H. Sugihara, T. Komatsu, N. Kato, and K. Jann. 1995. Expression of the O9 polysaccharide of *Escherichia coli*: sequencing of the *E. coli* O9 *rfb* gene cluster, characterization of mannosyl transferases, and evidence for an ATP-binding cassette transport system. *J. Bacteriol.* **177**:2178–2187.

Klee, S. R., B. D. Tzschaschel, K. N. Timmis, and C. A. Guzman. 1997. Influence of different *rol* gene products on the chain length of *Shigella dysenteriae* type 1 lipopolysaccharide O antigen expressed by *Shigella flexneri* carrier strains. *J. Bacteriol.* **179**:2421–2425.

Korres, H., M. Mavris, R. Morona, P. A. Manning, and N. K. Verma. 2005. Topological analysis of GtrA and GtrB proteins encoded by the serotype-converting cassette of *Shigella flexneri*. *Biochem. Biophys. Res. Commun.* **328**:1252–1260.

Korres, H., and N. K. Verma. 2004. Topological analysis of glucosyltransferase GtrV of *Shigella flexneri* by a dual reporter system and identification of a unique reentrant loop. *J. Biol. Chem.* **279**:22469–22476.

Kox, L. F., M. M. Wosten, and E. A. Groisman. 2000. A small protein that mediates the activation of a two-component system by another two-component system. *EMBO J.* **19**:1861–1872.

Lee, H., F. F. Hsu, J. Turk, and E. A. Groisman. 2004. The PmrA-regulated *pmrC* gene mediates phosphoethanolamine modification of lipid A and polymyxin resistance in *Salmonella enterica*. *J. Bacteriol.* **186**:4124–4133.

Lehane, A. M., H. Korres, and N. K. Verma. 2005. Bacteriophage-encoded glucosyltransferase GtrII of *Shigella flexneri*: membrane topology and identification of critical residues. *Biochem. J.* **389**:137–143.

Liu, D., R. Cole, and P. R. Reeves. 1996. An O-antigen processing function for Wzx (RfbX): a promising candidate for O-unit flippase. *J. Bacteriol.* **178:**2102–2107.

Losick, R., and P. W. Robbins. 1967. Mechanism of epsilon-15 conversion studies with a bacterial mutant. *J. Mol. Biol.* **30:**445–455.

Mackinnon, F. G., A. D. Cox, J. S. Plested, C. M. Tang, K. Makepeace, P. A. Coull, J. C. Wright, R. Chalmers, D. W. Hood, J. C. Richards, and E. R. Moxon. 2002. Identification of a gene (*lpt-3*) required for the addition of phosphoethanolamine to the lipopolysaccharide inner core of *Neisseria meningitidis* and its role in mediating susceptibility to bactericidal killing and opsonophagocytosis. *Mol. Microbiol.* **43:** 931–943.

Mäkelä, P. H., and B. A. D. Stocker. 1984. Genetics of lipopolysaccharide, p. 59–137. *In* E. T. Rietschel (ed.), *Handbook of Endotoxin*, vol. I. Elsevier Science Publishers, B.V., Amsterdam, The Netherlands.

Marolda, C. L., L. D. Tatar, C. Alaimo, M. Aebi, and M. A. Valvano. 2006. Interplay of the Wzx translocase and the corresponding polymerase and chain length regulator proteins in the translocation and periplasmic assembly of lipopolysaccharide O antigen. *J. Bacteriol.* **188:**5124–5135.

Marolda, C. L., J. Vicarioli, and M. A. Valvano. 2004. Wzx proteins involved in biosynthesis of O antigen function in association with the first sugar of the O-specific lipopolysaccharide subunit. *Microbiology* **150:**4095–4105.

Mavris, M., P. A. Manning, and R. Morona. 1997. Mechanism of bacteriophage SfII-mediated serotype conversion in *Shigella flexneri*. *Mol. Microbiol.* **26:** 939–950.

Morona, R., C. Daniels, and L. Van Den Bosch. 2003. Genetic modulation of *Shigella flexneri* 2a lipopolysaccharide O antigen modal chain length reveals that it has been optimized for virulence. *Microbiology* **149:**925–939.

Morona, R., M. Mavris, A. Fallarino, and P. A. Manning. 1994. Characterization of the *rfc* region of *Shigella flexneri*. *J. Bacteriol.* **176:**733–747.

Morona, R., and L. Van Den Bosch. 2003. Multicopy *icsA* is able to suppress the virulence defect caused by the *wzz*(SF) mutation in *Shigella flexneri*. *FEMS Microbiol. Lett.* **221:**213–219.

Morona, R., L. Van Den Bosch, and C. Daniels. 2000. Evaluation of Wzz/MPA1/MPA2 proteins based on the presence of coiled-coil regions. *Microbiology* **146:**1–4.

Morona, R., L. Van Den Bosch, and P. A. Manning. 1995. Molecular, genetic, and topological characterization of O-antigen chain length regulation in *Shigella flexneri*. *J. Bacteriol.* **177:**1059–1068.

Mouslim, C., T. Latifi, and E. A. Groisman. 2003. Signal-dependent requirement for the co-activator protein RcsA in transcription of the RcsB-regulated *ugd* gene. *J. Biol. Chem.* **278:**50588–50595.

Mühlradt, P. F., J. Menzel, J. R. Golecki, and V. Speth. 1973. Outer membrane of *Salmonella*. Sites of export of newly synthesized lipopolysaccharide on the bacterial surface. *Eur. J. Biochem.* **35:**471–481.

Murray, G. L., S. R. Attridge, and R. Morona. 2003. Regulation of *Salmonella typhimurium* lipopolysaccharide O antigen chain length is required for virulence; identification of FepE as a second Wzz. *Mol. Microbiol.* **47:**1395–1406.

Murray, G. L., S. R. Attridge, and R. Morona. 2005. Inducible serum resistance in *Salmonella typhimurium* is dependent on *wzz*(*fepE*)-regulated very long O antigen chains. *Microb. Infect.* **7:**1296–1304.

Najdenski, H., E. Golkocheva, A. Vesselinova, J. A. Bengoechea, and M. Skurnik. 2003. Proper expression of the O-antigen of lipopolysaccharide is essential for the virulence of *Yersinia enterocolitica* O:8 in experimental oral infection of rabbits. *FEMS Immunol. Med. Microbiol.* **38:**97–106.

Navarre, W. W., T. A. Halsey, D. Walthers, J. Frye, M. McClelland, J. L. Potter, L. J. Kenney, J. S. Gunn, F. C. Fang, and S. J. Libby. 2005. Co-regulation of *Salmonella enterica* genes required for virulence and resistance to antimicrobial peptides by SlyA and PhoP/PhoQ. *Mol. Microbiol.* **56:**492–508.

Newton, G. J., C. Daniels, L. L. Burrows, A. M. Kropinski, A. J. Clarke, and J. S. Lam. 2001. Three-component-mediated serotype conversion in *Pseudomonas aeruginosa* by bacteriophage D3. *Mol. Microbiol.* **39:**1237–1247.

Nikaido, H. 2003. Molecular basis of bacterial outer membrane permeability revisited. *Microbiol. Mol. Biol. Rev.* **67:**593–656.

Nikaido, H., K. Nikaido, T. Nakae, and P. H. Makela. 1971. Glucosylation of lipopolysaccharide in *Salmonella*: biosynthesis of O antigen factor 12_2. I. Over-all reaction. *J. Biol. Chem.* **246:**3902–3911.

Nikaido, K., and H. Nikaido. 1971. Glucosylation of lipopolysaccharide in *Salmonella*: biosynthesis of O antigen factor 12_2. II. Structure of the lipid intermediate. *J. Biol. Chem.* **246:**3912–3919.

Ørskov, I., F. Ørskov, B. Jann, and K. Jann. 1977. Serology, chemistry, and genetics of O and K antigens of *Escherichia coli*. *Bacteriol. Rev.* **41:**667–710.

Osborn, M. J., M. A. Cynkin, J. M. Gilbert, L. Muller, and M. Singh. 1972. Synthesis of bacterial O-antigens. *Methods Enzymol.* **28:**583–601.

Pelosi, L., M. Boumedienne, N. Saksouk, J. Geiselmann, and R. A. Geremia. 2005. The glucosyl-1-phosphate transferase WchA (Cap8E) primes the capsular polysaccharide repeat unit biosynthesis of *Streptococcus pneumoniae* serotype 8. *Biochem. Biophys. Res. Commun.* **327:**857–865.

Peng, D., W. Hong, B. P. Choudhury, R. W. Carlson, and X. X. Gu. 2005. *Moraxella catarrhalis* bacterium without endotoxin, a potential vaccine candidate. *Infect. Immun.* **73**:7569–7577.

Price, N. P., and F. A. Momany. 2005. Modeling bacterial UDP-HexNAc:polyprenol-P HexNAc-1-P transferases. *Glycobiology* **15**:29R–42R.

Raetz, C. R., and C. Whitfield. 2002. Lipopolysaccharide endotoxins. *Annu. Rev. Biochem.* **71**:635–700.

Reyes, C. L., and G. Chang. 2005. Structure of the ABC transporter MsbA in complex with ADP vanadate and lipopolysaccharide. *Science* **308**:1028–1031.

Reynolds, C. M., S. R. Kalb, R. J. Cotter, and C. R. Raetz. 2005. A phosphoethanolamine transferase specific for the outer 3-deoxy-D-manno-octulosonic acid residue of *Escherichia coli* lipopolysaccharide. Identification of the *eptB* gene and Ca^{2+} hypersensitivity of an *eptB* deletion mutant. *J. Biol. Chem.* **280**:21202–21211.

Rick, P. D., K. Barr, K. Sankaran, J. Kajimura, J. S. Rush, and C. J. Waechter. 2003. Evidence that the *wzxE* gene of *Escherichia coli* K-12 encodes a protein involved in the transbilayer movement of a trisaccharide-lipid intermediate in the assembly of enterobacterial common antigen. *J. Biol. Chem.* **278**:16534–16542.

Rick, P. D., G. L. Hubbard, and K. Barr. 1994. Role of the *rfe* gene in the synthesis of the O8 antigen in *Escherichia coli* K-12. *J. Bacteriol.* **176**:2877–2884.

Robbins, P. W., D. Bray, M. Dankert, and A. Wright. 1967. Direction of chain growth in polysaccharide synthesis. *Science* **158**:1536–1542.

Schild, S., A. K. Lamprecht, and J. Reidl. 2005. Molecular and functional characterization of O antigen transfer in *Vibrio cholerae*. *J. Biol. Chem.* **280**:25936–25947.

Schnaitman, C. A., and J. D. Klena. 1993. Genetics of lipopolysaccharide biosynthesis in enteric bacteria. *Microbiol. Rev.* **57**:655–682.

Shi, Y., M. J. Cromie, F. F. Hsu, J. Turk, and E. A. Groisman. 2004. PhoP-regulated *Salmonella* resistance to the antimicrobial peptides magainin 2 and polymyxin B. *Mol. Microbiol.* **53**:229–241.

Slauch, J. M., A. A. Lee, M. J. Mahan, and J. J. Mekalanos. 1996. Molecular characterization of the *oafA* locus responsible for acetylation of *Salmonella typhimurium* O-antigen: *oafA* is a member of a family of integral membrane trans-acylases. *J. Bacteriol.* **178**:5904–5909.

Sperandeo, P., C. Pozzi, G. Dehó, and A. Polissi. 2006. Non-essential KDO biosynthesis and new essential cell envelope biogenesis genes in the *Escherichia coli yrbG-yhbG* locus. *Res. Microbiol.* **157**:547–558.

Stead, C., A. Tran, D. Ferguson, Jr., S. McGrath, R. Cotter, and S. Trent. 2005. A novel 3-de-oxy-D-manno-octulosonic acid (Kdo) hydrolase that removes the outer Kdo sugar of *Helicobacter pylori* lipopolysaccharide. *J. Bacteriol.* **187**:3374–3383.

Steeghs, L., R. den Hartog, A. den Boer, B. Zomer, P. Roholl, and P. van der Ley. 1998. Meningitis bacterium is viable without endotoxin. *Nature (Lond.)* **392**:449–450.

Takeshita, M., and P. H. Makela. 1971. Glucosylation of lipopolysaccharide in *Salmonella*: biosynthesis of O antigen factor 12_2. 3. The presence of 12_2 determinants in haptenic polysaccharides. *J. Biol. Chem.* **246**:3920–3927.

Tamayo, R., B. Choudhury, A. Septer, M. Merighi, R. Carlson, and J. S. Gunn. 2005. Identification of *cptA*, a PmrA-regulated locus required for phosphoethanolamine modification of the *Salmonella enterica* serovar Typhimurium lipopolysaccharide core. *J. Bacteriol.* **187**:3391–3399.

Tefsen, B., M. P. Bos, F. Beckers, J. Tommassen, and H. de Cock. 2005a. MsbA is not required for phospholipid transport in *Neisseria meningitidis*. *J. Biol. Chem.* **280**:35961–35966.

Tefsen, B., J. Geurtsen, F. Beckers, J. Tommassen, and H. de Cock. 2005b. Lipopolysaccharide transport to the bacterial outer membrane in spheroplasts. *J. Biol. Chem.* **280**:4504–4509.

Tlapak-Simmons, V. L., C. A. Baron, R. Gotschall, D. Haque, W. M. Canfield, and P. H. Weigel. 2005. Hyaluronan biosynthesis by class I streptococcal hyaluronan synthases occurs at the reducing end. *J. Biol. Chem.* **280**:13012–13018.

Tran, A. X., M. J. Karbarz, X. Wang, C. R. Raetz, S. C. McGrath, R. J. Cotter, and M. S. Trent. 2004. Periplasmic cleavage and modification of the 1-phosphate group of *Helicobacter pylori* lipid A. *J. Biol. Chem.* **279**:55780–55791.

Tran, A. X., M. E. Lester, C. M. Stead, C. R. Raetz, D. J. Maskell, S. C. McGrath, R. J. Cotter, and M. S. Trent. 2005. Resistance to the antimicrobial peptide polymyxin requires myristoylation of *Escherichia coli* and *Salmonella typhimurium* lipid A. *J. Biol. Chem.* **280**:28186–28194.

Trent, M. S. 2004. Biosynthesis, transport, and modification of lipid A. *Biochem. Cell Biol.* **82**:71–86.

Trent, M. S., W. Pabich, C. R. Raetz, and S. I. Miller. 2001a. A PhoP/PhoQ-induced lipase (PagL) that catalyzes 3-O-deacylation of lipid A precursors in membranes of *Salmonella typhimurium*. *J. Biol. Chem.* **276**:9083–9092.

Trent, M. S., A. A. Ribeiro, W. T. Doerrler, S. Lin, R. J. Cotter, and C. R. Raetz. 2001b. Accumulation of a polyisoprene-linked amino sugar in polymyxin-resistant *Salmonella typhimurium* and *Escherichia coli*: structural characterization and transfer to lipid A in the periplasm. *J. Biol. Chem.* **276**:43132–43144.

Trent, M. S., A. A. Ribeiro, S. Lin, R. J. Cotter, and C. R. Raetz. 2001c. An inner membrane enzyme in *Salmonella* and *Escherichia coli* that transfers 4-amino-4-deoxy-L-arabinose to lipid A: induction on polymyxin-resistant mutants and role of a novel lipid-linked donor. *J. Biol. Chem.* **276:** 43122–43131.

Vaara, M. 1992. Agents that increase the permeability of the outer membrane. *Microbiol. Rev.* **56:**395–411.

Valvano, M. A. 2003. Export of O-specific lipopolysaccharide. *Front. Biosci.* **8:**s452–s471.

Vander Byl, C., and A. M. Kropinski. 2000. Sequence of the genome of *Salmonella* bacteriophage P22. *J. Bacteriol.* **182:**6472–6481.

Verma, N. K., J. M. Brandt, D. J. Verma, and A. A. Lindberg. 1991. Molecular characterization of the O-acetyl transferase gene of converting bacteriophage SF6 that adds group antigen 6 to *Shigella flexneri*. *Mol. Microbiol.* **5:**71–75.

Vines, E. D., C. L. Marolda, A. Balachandran, and M. A. Valvano. 2005. Defective O-antigen polymerization in *tolA* and *pal* mutants of *Escherichia coli* in response to extracytoplasmic stress. *J. Bacteriol.* **187:**3359–3368.

Vinogradov, E., E. Frirdich, L. L. MacLean, M. B. Perry, B. O. Petersen, J. O. Duus, and C. Whitfield. 2002. Structures of lipopolysaccharides from *Klebsiella pneumoniae*. Elucidation of the structure of the linkage region between core and polysaccharide O chain and identification of the residues at the non-reducing termini of the O chains. *J. Biol. Chem.* **277:**25070–25081.

Vollmer, W., and J. V. Holtje. 2001. Morphogenesis of *Escherichia coli*. *Curr. Opin. Microbiol.* **4:**625–633.

Voulhoux, R., M. P. Bos, J. Geurtsen, M. Mols, and J. Tommassen. 2003. Role of a highly conserved bacterial protein in outer membrane protein assembly. *Science* **299:**262–265.

Wang, L., D. Liu, and P. R. Reeves. 1996. C-terminal half of *Salmonella enterica* WbaP (RfbP) is a galactosyl-1-phosphate transferase domain catalyzing the first step of O-antigen synthesis. *J. Bacteriol.* **178:**2598–2604.

Wang, X., M. J. Karbarz, S. C. McGrath, R. J. Cotter, and C. R. Raetz. 2004. MsbA transporter-dependent lipid A 1-dephosphorylation on the periplasmic surface of the inner membrane: topography of *Francisella novicida* LpxE expressed in *Escherichia coli*. *J. Biol. Chem.* **279:**49470–49478.

Whitfield, C. 1995. Biosynthesis of lipopolysaccharide O-antigens. *Trends Microbiol.* **3:**178–185.

Whitfield, C. 2006. Biosynthesis and assembly of capsular polysaccharides in *Escherichia coli*. *Annu. Rev. Biochem.* **75:**39–68.

Whitfield, C., P. A. Amor, and R. Köplin. 1997. Modulation of surface architecture of gram-negative bacteria by the action of surface polymer:lipid A-core ligase and by determinants of polymer chain length. *Mol. Microbiol.* **23:**629–638.

Whitfield, C., N. Kaniuk, and E. Frirdich. 2003. Molecular insights into the assembly and diversity of the outer core oligosaccharide in lipopolysaccharides from *Escherichia coli* and *Salmonella*. *J. Endotox. Res.* **9:**244–249.

Whitfield, C., and A. Paiment. 2003. Biosynthesis and assembly of group 1 capsular polysaccharides in *Escherichia coli* and related extracellular polysaccharides in other bacteria. *Carbohydr. Res.* **338:**2491–1502.

Williams, G. J., S. D. Breazeale, C. R. Raetz, and J. H. Naismith. 2005. Structure and function of both domains of ArnA, a dual function decarboxylase and a formyltransferase, involved in 4-amino-4-deoxy-L-arabinose biosynthesis. *J. Biol. Chem.* **280:**23000–23008.

Wosten, M. M., L. F. Kox, S. Chamnongpol, F. C. Soncini, and E. A. Groisman. 2000. A signal transduction system that responds to extracellular iron. *Cell* **103:**113–125.

Wu, T., J. Malinverni, N. Ruiz, S. Kim, T. J. Silhavy, and D. Kahne. 2005. Identification of a multicomponent complex required for outer membrane biogenesis in *Escherichia coli*. *Cell* **121:**235–245.

Wu, T., A. C. McCandlish, L. S. Gronenberg, S. S. Chang, T. J. Silhavy, and D. Kahne. 2006. Identification of a protein complex that assembles lipopolysaccharide in the outer membrane of *Escherichia coli*. *Proc. Natl. Acad. Sci. USA* **103:**11754–11759.

Wugeditsch, T., A. Paiment, J. Hocking, J. Drummelsmith, C. Forrester, and C. Whitfield. 2001. Phosphorylation of Wzc, a tyrosine autokinase, is essential for assembly of group 1 capsular polysaccharides in *Escherichia coli*. *J. Biol. Chem.* **276:**2361–2371.

Zhou, Z., S. Lin, R. J. Cotter, and C. R. Raetz. 1999. Lipid A modifications characteristic of *Salmonella typhimurium* are induced by NH_4VO_3 in *Escherichia coli* K12. Detection of 4-amino-4-deoxy-L-arabinose, phosphoethanolamine and palmitate. *J. Biol. Chem.* **274:**18503–18514.

Zhou, Z., A. A. Ribeiro, S. Lin, R. J. Cotter, S. I. Miller, and C. R. Raetz. 2001. Lipid A modifications in polymyxin-resistant *Salmonella typhimurium*: PMRA-dependent 4-amino-4-deoxy-L-arabinose, and phosphoethanolamine incorporation. *J. Biol. Chem.* **276:**43111–43121.

ELECTRON TRANSPORT ACTIVITIES IN THE PERIPLASM

Stuart J. Ferguson

13

The periplasm is the location of numerous enzymes and electron transfer proteins that permit bacteria to use a huge range of electron donors and acceptors in respiratory or photosynthetic processes. It is these periplasmic activities that permit various species of bacteria to use a substantial number of electron donors and acceptors in many varied growth modes. In this chapter we seek to provide an overview of the main periplasmic oxidation and reduction reactions alongside some consideration of how the assembly of these proteins is achieved. Many of these proteins are assembled in the cytoplasm and then exported to the periplasm in a folded state. The export is via the Tat system, which is discussed in more detail in Chapter 2. Understanding of periplasmic oxidation/reduction processes has to be based on knowledge of the physical dimensions of the periplasm and the ways in which periplasmic proteins can interact with the electron transfer system of the cytoplasmic membrane. Thus to set the scene these points are discussed first.

DIMENSIONS AND PHYSICO-CHEMICAL PROPERTIES OF THE PERIPLASM

There has been much debate over the past twenty years about the width of the periplasm. A standard textbook view, based principally on negatively stained electron microscope pictures of *Escherichia coli*, indicated a width of about 70 Å (Ferguson, 1988). The growing realization that the periplasm contained many globular proteins with dimensions in this range, along with recognition that observations on *E. coli*, which is not particularly rich in periplasmic electron transfer processes, ought not to be extrapolated automatically to other organisms, meant that this value came under suspicion (Ferguson, 1990; Graham et al., 1991; Vanwielink and Duine, 1990). The width of the periplasm may of course vary with not only the organism but also the particular growth conditions.

In the case of *E. coli* some recent structural work has given us a ruler with which to size the periplasm. The protein AcrB extends from the cytoplasmic membrane approximately 70 Å into the periplasm. To be functional this multidrug efflux protein needs to dock with an outer membrane protein called TolC. The latter has a long α-helical tube that extends about 100 Å into the periplasm from the outer mem-

Stuart J. Ferguson, Department of Biochemistry, University of Oxford, South Parks Road, Oxford OX1 3QU, United Kingdom.

The Periplasm
Edited by Michael Ehrmann © 2007 ASM Press, Washington, D.C.

brane. The dimensions of these two proteins allow us to conclude that, at least under some conditions, the periplasmic surfaces of the cytoplasmic and outer membranes must be separated by about 170 Å, thus defining the width of the periplasm (Murakami et al., 2002). These two proteins are found in many other bacteria. Where they do occur we can take their presence as an indication for a periplasmic width of 170 Å. In other cases we can only argue by analogy that this is an appropriate estimate for the width of the periplasm.

As this and the other chapters of this book amply demonstrate, a considerable complement of proteins is located in the periplasm. It may well be that the number of such proteins varies significantly depending on the growth conditions, and thus there may be variation in the protein-packing density and/or volume of the periplasm. The crowding of proteins is obviously also encountered in the cytoplasm, where nucleic acids will also compete for space and contribute to the viscosity. In the case of the periplasm it has been suggested that the presence of outer wall saccharides penetrating into the periplasm contributes to a gel-like state within the periplasm (Ferguson, 1988, 1991). Biophysical evidence in support of this property of the periplasm was obtained for *E. coli* (Brass et al., 1986); to what extent this applies to other organisms is not known. Many periplasmic electron transfer reactions are thought to rely on collisions between protein partners; a gel state will restrict such collisions and thus slow electron transfer.

In summary, the environment within the periplasm is very different from the aqueous solutions used in vitro. How this environment affects electron transfer events that are linked to protein-protein interactions is not known. However, it should be borne in mind that high K_m values for periplasmic proteins as electron donors or acceptors and low rates of electron transfer in vitro do not necessarily mean that a reaction cannot occur in the confined environment of the periplasm. The effective concentrations of periplasmic proteins are probably in the millimolar range in some cases.

GENERAL ORGANIZATIONS OF BACTERIAL CYTOPLASMIC MEMBRANE ELECTRON TRANSFER SYSTEMS

Periplasmic oxidation or reduction reactions nearly always have to transfer electrons to or from electron transfer proteins in the cytoplasmic membrane. In general, there are two such "exchange" points for electrons between periplasm and membrane. One is via the quinones. The typical quinones are ubiquinone and menaquinone and each of these can be reduced to the corresponding quinol by electrons originating from familiar respiratory substrates such as NADH and succinate. The dehydrogenases for these substrates are organized such that electrons are transferred from globular domains, in these cases at the cytoplasmic surface of the membrane, to below the phospholipid headgroup region where they can combine with protons and quinones to give quinols. As we shall see, a similar arrangement must allow electrons to flow from periplasmic oxidations to quinones. A key point is that quinones, retained in the hydrocarbon core of the bilayer, can act as electron acceptors from many substrates. Likewise, quinols can be oxidized by different routes, thus contributing to a picture of quinones and quinols providing a junction point.

A very important system for oxidizing quinols is the cytochrome bc_1 complex. The detailed mechanism of this complex is beyond the scope of this chapter (see, e.g., Nicholls and Ferguson, 2002), but a key point in the present context is that electrons leave the complex from the heme group of the cytochrome c_1 component of the complex. This protein has a transmembrane helix but mainly comprises a globular domain that is located outside the bilayer and in the periplasm. Thus electrons on the heme of cytochrome c_1 can be readily transferred to water-soluble proteins in the periplasm. We return to the nature of these proteins below.

The cytochrome bc_1 complex is not only the oxidant for quinols. Structures are known for several proteins, e.g., fumarate and nitrate reductases, which transfer electrons via quinol oxidation sites in transmembrane helices to globular catalytic domains where the substrate reduction occurs. However, oppositely arranged proteins, i.e., with the globular domain in contact with the periplasm are not yet as well described but we return to examples later. Nevertheless, it is clear that there are electron exit routes to the periplasm from quinols that are independent of the cytochrome bc_1 complex. One of these routes is provided by a widespread type of protein usually known as NapC (Roldan et al., 1998). NapC is normally one of the proteins that is coded for within the *nap* operon, which includes genes for the subunits, NapA and NapB, of a periplasmic nitrate reductase. NapC is a *c*-type cytochrome with four covalently attached heme groups. It is currently envisaged that whereas these hemes are in a globular domain that is exposed to the periplasm, these must also be a binding site for the quinol that accepts this reductant from, and delivers quinone to, the hydrophobic core of the bilayer. Strangely, a protein that resembles NapC on the basis of sequence comparisons is often found in association with enzymes other than periplasmic nitrate reductase and thus it appears to be adaptable to the demands of supplying electrons to different periplasmic reductases. Indeed it seems likely that this type of protein can also act to transfer electrons from a periplasmic oxidation reaction to quinone. An example is that of a NapC-like protein that appears to be associated with the hydroxylamine oxidoreductase of ammonia-oxidizing bacteria. On the other hand, in some cases it is clear that electrons can be supplied to periplasmic nitrate reductases independently of NapC. Thus in *Wolinella succinogenes* a periplasmic nitrate reductase is seemingly supplied with electrons by an iron-sulfur protein while an orthologue of NapC has been shown to be required specifically for electron transfer to a nitrite reductase in this organism (Simon et al., 2003). The latter, a penta-heme *c*-type cytochrome, has a very close counterpart in *E. coli* where the electron donor protein is, in contrast, putatively an iron-sulfur protein. Yet another donor to a periplasmic nitrate reductase has been suggested for *Campylobacter jejuni* (Pittman and Kelly, 2005). Hence a pattern seems to be emerging where there is plasticity in the use of electron transfer proteins that can catalyze electron transfer to the periplasm from quinols.

Recently Yanyushin et al. (2005) proposed that some bacteria use a previously unrecognized complex to transfer electrons from quinols to periplasmic, and thus water-soluble, copper proteins. Thus in the organism *Chloroflexus auranticus* a bioinformatics approach has identified a putative complex that might fulfill the role played in other organisms by the bc_1 complex, for which there is no evidence in this organism. This newly identified complex appears to include *c*-type cytochromes, which would face the periplasm, and subunits analogous to molybdenum-containing oxidoreductases. We can conclude that there are several plausible routes for electrons to be transferred from the membrane-bound quinols to the periplasm. As we shall see, some of these components are also involved in transferring electrons from the periplasm to quinones.

The second junction point in electron transfer chains occurs typically on the oxidizing side of the bc_1 complex or any alternatives to this (e.g., Yanyushin et al., 2005). Most gram-negative organisms possess periplasmic soluble *c*-type cytochromes and cupredoxins that can transfer electrons either from the electron transport chain (usually the bc_1 complex) to reductases or to the electron transport chain (usually cytochrome oxidases) from periplasmic dehydrogenases. In some cases water-soluble iron-sulfur proteins known as high-potential iron proteins (HiPiPs) may play this role.

ALTERNATIVE AND REDUNDANT ELECTRON TRANSFER PATHWAYS/ PROTEINS IN THE PERIPLASM

The genome sequences of many gram-negative bacteria show the presence of several periplasmic (as judged by the presence of a targeting sequence) *c*-type cytochromes and cupredoxins. The factors affecting the expression of many of these proteins have usually not been elucidated but functionally some of these proteins seem interchangeable.

A seemingly straightforward example occurs in *Paracoccus pantotrophus*. In vitro the periplasmic cytochrome cd_1 nitrite reductase from this organism is active with either a monoheme cytochrome c_{550} or the cupredoxin pseudoazurin (Pearson et al., 2003). These are also in vivo alternative electron donor proteins for cytochrome cd_1 (Pearson et al., 2003), each assumed to be capable of accepting electrons from the cytochrome bc_1 complex. The molecular basis for this behavior is not immediately obvious as cytochrome c_{550} and pseudoazurin have very different structures. It has been suggested that both cytochrome c_{550} and pseudoazurin have a similar positively charged patch on a hydrophobic surface. For cytochrome c peroxidase from the related bacterium *Paracoccus denitrificans* in vitro studies show that pseudoazurin and cytochrome c_{550} compete with each other for one binding site on the peroxidase (Pauleta et al., 2004). However, in this case there is no evidence yet that cytochrome c_{550} or pseudoazurin can act as alternatives in vivo. It is also not clear that these two electron donor proteins interact with cytochrome cd_1 and cytochrome c peroxidase in the same way. The former has a negatively charged surface patch that has been suggested to interact with the positively charged patch on the donor proteins, a type of interaction termed pseudospecificity (Williams et al., 1995). However, the peroxidase does not have a similar negatively charged patch on its surface. The requirements for electron transfer, at a sufficient rate, from a donor protein to an acceptor may not be very strict. For example, horse heart mitochondrial cytochrome c will often act as a relatively efficient, but obviously nonphysiological, electron transfer partner for bacterial enzymes, as indeed is the case with cytochrome cd_1. In general, we can expect a lack of strict specificity for some of the electron transfer reactions that occur in the periplasm.

What might be the purpose of multiple electron transfer proteins, each seemingly capable of acting between the same pairs of donor and acceptor proteins? An obvious suggestion is that the choice between a copper and a heme protein may depend on the relative availability of copper and iron. While such considerations clearly underpin the selection of plastocyanin or a c-type cytochrome in green algae, there is as yet no comparable evidence for such a rationale in the periplasm of gram-negative bacteria.

Cyclic electron transport in the nonsulfur purple bacteria, in particular *Rhodobacter capsulatus*, provides another example of seeming redundancy among periplasmic electron transfer proteins. For many years before the advent of gene knockout technology it was believed that the periplasmic monoheme cytochrome c_2 was the obligatory electron carrier between the cytochrome bc_1 complex and the photosynthetic reaction center. But as explained later this can be substituted by another cytochrome.

An extreme example of apparent redundancy among electron transfer proteins in the periplasm is provided by the tetraheme c-type cytochrome that is a subunit of a photosynthetic reaction center from *Rhodopseudomonas viridis*. This subunit is believed to accept electrons from cytochrome c_2 and deliver them to the special pair chlorophyll of the reaction center. But many other organisms, including the frequently studied *R. capsulatus* and *R. sphaeroides*, do not have this tetraheme subunit and the cytochrome c_2 is able to donate electrons directly to the special pair. It is ironic that the reaction center from *R. vividis* provided the first crystal structure of a membrane protein, yet, approximately twenty years later, the function of one of its polypeptide components remains obscure. Cytochrome c', which may have a periplasmic role in binding toxic levels of nitric oxide (Cross et al., 2000), and the cytochrome b_{562} of *E. coli* are two other proteins for which structures determined more than 20 years ago have not been complemented by firm identification of periplasmic function.

WHY ARE MANY OXIDATION/ REDUCTION REACTIONS CATALYZED IN THE PERIPLASM?

There are a number of possibilities here. First, periplasmic handling avoids the need to provide transport systems for import of a substrate and export of a product. This leads to a second consideration, which is the exclusion of potentially toxic chemical species from the cytoplasm. A third possible reason is that the energetics of the oxidation/reduction reaction is only compatible with electron delivery to or from the electron transfer chain at the level of the c-type cytochromes. As the latter are periplasmic, it makes sense for redox partners also to be located in the periplasm. Finally, in terms of the chemiosmotic mechanism of energy transduction an oxidation reaction in the periplasm will, if linked to a cytoplasmic site of a reductase reaction, automatically lead to the generation of a proton motive force across the membrane. Part of this generation is often the result of release of protons into the periplasm, but the major contribution comes from the movement of electrons across the membrane. This leads to consideration as to why there are periplasmic reductases that consume protons, seemingly thus contributing to the diminution of the proton motive force. However, this is a very minor effect as the consumption of protons in the periplasm does not involve any change movement across the membrane. Thus, for example, if electrons flow from quinols, via the cytochrome bc_1 complex and periplasmic c-type cytochromes, to a periplasmic reductase, then charge movement across the cytoplasmic membrane will be achieved by the proton motive activity of the cytochrome bc_1 complex. For electron donors that generate cytoplasmic NADH there can also be a contribution from the proton motive activity of an NADH-ubiquinone oxidoreductase (Nicholls and Ferguson, 2002). The consumption in the periplasm of protons will have a negligible effect on the overall magnitude of the steady-state proton motive force, except under conditions of unusually low buffering capacity in the periplasm, and be connected through the outer membrane to the buffering capacity of the extracellular medium. The erroneous belief that consumption of protons by periplasmic reductases is incompatible with the chemiosmotic mechanism is still often encountered.

PERIPLASMIC OXIDATION REACTIONS THAT ARE LINKED TO THE RESPIRATORY CHAIN VIA SOLUBLE c-TYPE CYTOCHROMES ON COPPER PROTEINS, CUPREDOXINS

Feeding electrons into an electron transport chain at the level of the c-type cytochromes or cupredoxins means that generation of a proton motive force will only be possible, as far as we currently understand electron transfer pathways, if the electrons flow to oxygen. This is because the cytochrome oxidases are the only proton motive enzymes that act on the oxidizing side of these periplasmic components. Thus, in general, we might expect only those reductants that have such a positive redox potential that connection to the electron transfer chain is possible only at the level of c-type cytochromes rather than quinone. This point is exemplified by the periplasmic oxidation of ferrous iron as catalyzed, for example, by *Thiobacillus ferrooxidans*. The redox potential of the Fe^{2+}/Fe^{3+} reaction is unlikely to be less than around 700 mV and, even at the acidic pH values at which this organism grows, there will only be a small energy drop to the oxygen/water reaction. Entry of electrons via c-type cytochromes at the oxygen end of the electron transfer system is thus unavoidable in this case (Nicholls and Ferguson, 2002).

It is less obvious why the periplasmic dehydrogenases for methanol and methylamine are connected to the electron transfer chain via c-type cytochromes. Each of these dehydrogenases contains an unusual cofactor, pyrroloquinoline (PQQ) in the case of methanol dehydrogenase and tryptophylquinone (TTQ) in the case of methylamine dehydrogenase. Yet the oxidation/reduction potentials for the methylamine and methanol oxidations to formaldehyde would suggest that the reactions could be linked to the respiratory chain at the quinone/quinol level. The generation of toxic formaldehyde as product suggests a rationale for a periplasmic location for the enzymes. Although the subsequent step of formaldehyde oxidation occurs in the cytoplasm, one may surmise that periplasmic generation followed by diffusion into cytoplasm restricts the formaldehyde concentration in the latter compartment. PQQ-linked oxidations can deliver electrons at the quinone level as exemplified by a glucose dehydrogenase (Goodwin and Anthony, 1998).

A final example of periplasmic oxidation at the level of c-type cytochromes is provided by sulfite oxidation to sulfate. On energetic grounds this too could, in principle, be linked at the quinone level and effectively is so in organisms that are specialized oxidizers of reduced sulfur compounds. However, in less specialized organisms sulfite, along with thiosulfate, oxidation again appears to be associated with periplasmic c-type cytochromes. Although sulfite and thiosulfate oxidation is clearly an important reaction in organisms that thrive by oxidizing a range of reduced sulfur species, a more recent example of periplasmic sulfite oxidation is provided by *Campylobacter jejuni* for which it is only one of many options among its chemoheterotrophic growth options (Myers and Kelly, 2005).

ELECTRON DELIVERY TO PERIPLASMIC REDUCTASES FROM c-TYPE CYTOCHROMES OR CUPREDOXINS

One of the best examples of this category of reactions is the provision of electrons to two reductases of the denitrification pathway, those for nitrite and nitrous oxide. In this case electrons pass from the cytochrome bc_1 complex to either the cupredoxins, azurin, or pseudoazurin or a monoheme c-type cytochrome, all of which are water soluble. These pass electrons to the reductases, the nitrous oxide reductase being an unusual copper enzyme with a, to date, unique arrangement of a sulfur-bridged copper center at the active site, while there are two kinds of mutually exclusive nitrite reductases, one containing two types of heme c and d_1, with the latter being unique to this class of enzyme, and the other a copper enzyme.

Another commonly encountered reaction is the reduction of hydrogen peroxide to water. This reaction is catalyzed by a cytochrome c peroxidase, which, itself a c-type cytochrome, catalyzes electron transfer from c-type cytochromes or cupredoxins to peroxide.

These electron transfer processes again direct our attention to the absence of strict specificity among periplasmic electron transfer proteins. As mentioned above, in the case of the organism *P. denitrificans*, for example, c-type cytochrome and pseudoazurin catalyze periplasmic electron transfer from the cytochrome bc_1 complex to reductases for nitrite and nitrous oxide as well as to cytochrome c peroxidase.

ELECTRON TRANSFER TO PERIPLASMICALLY LOCATED GLOBULAR DOMAINS VIA SOLUBLE c-TYPE CYTOCHROMES, CUPREDOXINS, OR HIGH-POTENTIAL IRON PROTEINS

Two prominent examples here are electron provision to cytochrome oxidases and photosynthetic reaction centers. In each case electrons are transferred within the periplasm from the cytochrome bc_1 complex, or its proposed substitute in some organisms (Yanyushin et al., 2005), to an electron-receiving center, for example, a Cu center in oxidases or the oxidized special pair of chlorophylls in bacterial reaction centers. Depending on the organism, this electron transfer might be catalyzed by a c-type cytochrome, a cupredoxin or a type of water-soluble iron-sulfur protein known as HiPip.

ELECTRON TRANSFER WITHIN THE PERIPLASM BY GLOBULAR DOMAINS OF MEMBRANE-ANCHORED PROTEINS

It was originally believed that in photosynthetic bacteria of the *Rhodobacter* genus electron transfer from the cytochrome bc_1 complex to either the photosynthetic reaction center (light-driven cyclic electron flow) or a terminal oxidase enzyme was catalyzed solely by the water-soluble electron transfer protein cytochrome c_2. However, in *Rhodobacter capsulatus* the loss of this protein through directed mutagenesis did not prevent cyclic electron transfer. It was discovered that the role of the absent soluble cytochrome c was taken over by a membrane-anchored c-type cytochrome known as c_y. Later it was discovered that there is an analogous protein in *R. sphaeroides*, although strangely in one case the membrane-anchored protein can also donate to an oxidase whereas in others it cannot (Daldal et al., 2001, 2003). In the organism *P. denitrificans* there is evidence that electron transfer from the cytochrome bc_1 complex to terminal oxidases can be similarly catalyzed by either water-soluble c-type cytochromes or proteins similar to cytochrome c_y with a membrane tether and a globular domain containing the heme. Many other similar examples are becoming known and such proteins can operate in organisms without a formal periplasm, for example, the archaeon *Sulfolobus acidocaldarius* in which a b-type cytochrome performs pivoting movements with respect to the membrane external surface (Schoepp-Cothenet et al., 2001).

Overall we see a variable pattern in which electron transfer is often achieved in the periplasm through the diffusion of water-soluble proteins whereas in other cases the diffusion is presumably restricted to two dimensions close to the outer surface of the cytoplasmic membrane.

ELECTRON DELIVERY FROM THE PERIPLASM TO MEMBRANE-BOUND QUINONES

Two examples are used to illustrate the general principles. The first is the oxidation of formate by a membrane-bound formate dehydrogenase. The structure of the *E. coli* enzyme is known. There is a globular domain exposed to the periplasm and from which electrons flow into the transmembrane domain, which includes two heme groups. The latter transfer the electrons across the membrane to a binding site for menaquinone which is reduced to menaquinol. In principle, this type of structural organization could be replicated for other oxidation reactions.

A second example is the oxidation of hydroxylamine to nitrite, a reaction that occurs in the ammonia-oxidizing bacteria. The oxidation is catalyzed by a periplasmic water-soluble multi-heme c-type cytochrome, from which the electrons are thought to pass to a tetraheme c-type cytochrome known as cytochrome c_{554}. The next step is electron transfer into the quinols. This is currently deduced to be catalyzed by an orthologue of the NapC protein discussed elsewhere in this chapter.

ELECTRON DELIVERY FROM MEMBRANE-BOUND QUINOLS TO PERIPLASMIC REDUCTASES

As explained earlier, tetraheme c-type cytochromes of the NapC family are widely implicated in transferring electrons to periplasmic reductases. A variation on this type of protein occurs in organisms that have a membrane-bound, but with catalytic site exposed to periplasm, dimethyl sulfoxide (DMSO) or trimethylamine-N-oxide (TMAO) reductase. For example, DorC and TorC are the respective multi-heme c-type cytochromes that are specific donors to these reductases.

An interesting recent example is the chlorate reductase. This has three subunits, a polypeptide with a molybdenum-containing active site, an FeS protein, and a b-type heme (Wolterink et al., 2003); it is closely related to periplasmic enzymes for reduction of selenate and oxidation of ethyl benzene or dimethyl sulfide. Presumably electrons are transferred via an unknown component to or from quinones or quinols to the b-type heme subunit. A recent addition to this family is a periplasmic perchlorate reductase, which is proposed to be linked to the quninones of the electron transport chain by a NapC-type protein (Bender et al., 2005).

The perchlorate reductase is either accompanied by a periplasmic chlorate reductase or is able to reduce chlorate to chlorite itself. There is a periplasmic dismutase that generates chloride and oxygen from the toxic chlorite. This example illustrates the notion that a periplasmic location for electron-transport-linked metabolism is advantageous for handling toxic species; molybdoenzymes including selenate and chlorate reductase are members of this group, which also includes dimethyl sulfoxide reductase (McEwan et al., 2002).

ELECTRON TRANSFER TO AND FROM THE OUTER MEMBRANE

There is general acceptance that the use of some insoluble metal compounds as electron acceptors requires electron transfer to the outer surface of the outer bacterial membrane. The reduction of ferric oxides serves as an example. In the organism *Shewanella frigidimarina* there is evidence that a protein known as CymA, which is actually an orthologue of the NapC discussed earlier, is involved in electron transfer to the periplasm from the membrane-bound quinols. From CymA the electrons are argued to pass via several multi-heme c-type cytochromes, including a decaheme protein that is thought to be associated with the outer membrane (Richardson, 2000).

In *Geobacter sulfurreducens* a monoheme periplasmic c-type cytochrome has been proposed to shuttle electrons to insoluble Fe(III) oxides, but other evidence argues against it (Lloyd et al., 1999). Again, there may be redundancy among electron transfer proteins that can obscure the role of a particular protein. Reduction of Fe(III) and Mn(IV) oxides by *G. sulfurreducens* requires c-type cytochromes associated with the outer membrane; presumably periplasmic c-type cytochromes have to transfer electrons to these carriers (Mehta et al., 2005). Much remains to be learned about the transfer of electrons to and from the outer membrane and insoluble electron donors or acceptors.

SULFATE-REDUCING BACTERIA

In this chapter an attempt has been made to generalize the electron transport system of a bacterium in terms of a quinone and/or a c-type cytochrome entry/exit point to the electron transfer chain. Many organisms use both these systems but some, e.g., *E. coli*, only one, in this case the quinone/quinol junction point.

Sulfate-reducing bacteria present a more uncertain picture as relatively little is known of their cytoplasmic membrane electron transport systems. However, they are important examples of organisms that rely on periplasmic electron transfer reactions. Particularly well studied are the hydrogenase and a tetraheme c-type cytochrome called cytochrome c_3. It is generally accepted that electrons derived by hydrogenase from hydrogen in the periplasm pass to cytochrome c_3 and from there to the enzymes that catalyze reduction of sulfate to sulfide. How the electrons are delivered to these cytoplasmic enzymes is not known and is beyond the scope of this chapter. The cytochrome c_3 structural motifs can also occur elsewhere in periplasmic cytochromes, as discussed by Mowatt and Chapman (2005).

ASSEMBLY OF PERIPLASMIC ELECTRON TRANSFER PROTEINS

At one time it seemed probable that all water-soluble periplasmic proteins were exported from the cytoplasm to the periplasm in an unfolded conformation via the Sec system. Once arrived in the periplasm they would acquire their redox active cofactor, be it a bare metal ion or an organometallic moiety such as heme, and fold. Today, we know that this route is taken by some but by no means all electron transfer proteins. However, a very substantial number of proteins acquire their cofactor and final folded conformation in the cytoplasm from which they are exported in that state via the Tat system to the periplasm. This system handles proteins such as molybdenum-containing enzymes or other proteins with labile cofactors. The Tat system is described in some detail in Chapter 2.

Cytochromes of both the b and c types are exported via the sec system. In the case of the b types that is essentially all that is known, because currently we have no information about the mechanism of heme transfer to the periplasm or of its incorporation into folded cytochromes b. It seems that the specialized d_1 heme is inserted in cytochrome cd_1 nitrite reductase in the periplasm. How the d_1 heme is delivered to the periplasm is another unknown.

Periplasmic c-type cytochromes are assembled by one of two sets of proteins known as systems I and II. Current knowledge about each of these has been reviewed extensively recently (Stevens et al., 2004) and so only a summary is presented here. System I comprises up to eight Ccm (cytochrome c maturation) proteins, but exactly what role is played by each remains mysterious. Possibly some of the components catalyze heme transfer to the periplasm. CcmA and CcmB are often considered as candidates for this role as these two proteins are believed to form an ATP-dependent transporter of the ABC type. However, there is no strong evidence for this role. It is known that heme can become covalently attached to the CcmE protein via a novel bond formed between an original vinyl group of heme and a histidine side chain within the periplasmically located globular domain of this protein that is anchored to the membrane by a transmembrane helix. This heme attachment still occurs if the CcmF, CcmG, and CcmH proteins have been lost through disruption of their genes, thus eliminating any of them as putative heme transport proteins. There is evidence that heme is transferred from CcmE to apocytochromes c both in vivo and in vitro. The chemistry of this transfer is obscure, but it may be relevant that heme can only be attached to c-type cytochromes containing the CXXCH motif and not to a single cysteine variant such as is found in some eukaryotes. A critical point is that the two cysteines of the CXXCH motif do have a propensity to form an intramolecular disulfide bond, something that is very likely accelerated by the DsbA/DsbB system in the periplasm (see Chapter 3). Attachment of heme to CXXCH to give a c-type cytochrome requires that the two cysteines are in the reduced state, i.e., the protein has two thiol groups. The proteins CcmH and CcmG, which is a member of the thioredoxin family, are implicated in the reversing of any disulfide bond formation in the CXXCH motif. This means that CcmG and CcmH must be supplied with reductant. This requirement is met through the action of the

DsbD protein (Chapter 10), which is in turn reduced by cytoplasmic thioredoxin. DsbD has two periplasmic domains that act in sequence to transfer reductant from the transmembrane domain to target proteins. In some bacteria DsbD is replaced by, or supplemented by, a protein known as CcdA, which is essentially similar to the eight transmembrane helices of DsbD. CcdA relies on independent periplasmic partners to play the role fulfilled by the two periplasmic domains of DsbD. A variety of evidence, including in vitro formation of the thioether bonds of a c-type cytochrome, shows that ferrous heme is needed. It is possible that the DsbD/CcmH/CcmG system also plays a role in maintaining this reduced state.

System II is seemingly less complex than System I, with only five components, one of which is usually CcdA, identified so far (Feissner et al., 2005). As with System I, this is believed to be involved in maintenance of the integrity of the thiols of apo cytochromes c, a role in which ResA (CcsX) protein is also involved. The roles of the other two components (ResB/CcsB and ResC/CcsA) are far from clear and as with System I there is no clear evidence as to how heme is translocated to the periplasm. Only in very rare cases are components of System I and System II found in the same organism, but the factors underlying the distribution of these two systems are not known.

An interesting point concerning the formation of c-type cytochromes is that some of these molecules contain multiple hemes per polypeptide chain, in excess of 25 for both System I- and II-dependent substrates. For several reasons, it is believed that the hemes are attached to the unfolded protein and that the three-dimensional structure is acquired subsequently (Stevens et al., 2004).

Some periplasmic copper proteins are also exported by the sec system and thus presumably acquire copper in the periplasm. Prominent here are the cupredoxins, for example, azurin, pseudoazurin, and amicyanin (the latter usually associated with methylamine oxidation), but the nature of the copper insertion process is not known; it may be uncatalyzed and thus spontaneous. Even the copper–containing periplasmic nitrite reductase follows this route as judged by its ready heterologous expression in *E. coli*. More complex copper proteins require more complex biogenesis pathways. Included in this category is the copper-containing nitrous oxide reductase, which has a novel copper cluster at the active site and is exported by the Tat system.

The assembly of purely organic cofactors is best illustrated by methanol and methylamine dehydrogenases in which PPQ and TTQ, respectively, are present. Methanol and several other related alcohol or sugar dehydrogenases contain the noncovalently bound cofactor PQQ (Goodwin and Anthony, 1998; Toyama et al., 2004). This moiety is synthesized in the cytoplasm and transported to combine with the apo protein; the latter is transported separately to the periplasm by the Sec system (Goodwin and Anthony, 1995). On the other hand, methylamine dehydrogenase, which contains the TTQ cofactor (formed by modification and cross-linking of two tryptophan residues), appears to be exported by the Tat system as judged by the N-terminal targeting signal sequence.

AN EXCEPTIONAL ELECTRON TRANSFER PATHWAY TO THE PERIPLASM

Use of certain reduced sulfur compounds as electron donors for respiration depends heavily on periplasmic proteins. For example, oxidation of thiosulfate in *P. pantotrophus* relies on several *c*-type cytochromes as well as a thioredoxin-like protein called SoxW. In addition, there is a requirement for SoxV, which is very similar to CcdA described on the previous page. It is known that SoxV transfers electrons to at least SoxW of possibly several periplasmic partners that are responsible for the maintenance in a reduced state of one or more active site cysteine residues on other proteins involved in the various oxidation steps for reduced sulfur compounds (Appia-Ayme and Berks, 2002; Bardischewsky et al., 2006). One oxidation/reduction step involving thiosulfate

involves the SoxV protein. It is currently thought that SoxV serves to transfer reductant for a reductive step necessary for the full activity of the thiosulfate-oxidizing system.

CONCLUDING REMARKS

It is ironic that Peter Mitchell, the formulator of the chemiosmotic hypothesis to explain electron-transport-linked ATP synthesis, identified and named the periplasm at much the same time (see Ferguson, 1992) as his better known contribution, and yet the role of the periplasm in electron transport processes was overlooked for many years. Thus in 1988 I was able (Ferguson, 1988) to list all the known periplasmic electron transfer proteins in a single short table. No attempt to provide a catalogue has been made in the present article; it would be unmanageably long and redundant, given the ability to search on the Web for general topics such as periplasmic proteins. Thus this chapter has sought to provide some general principles concerning the nature of periplasmic electron transfer pathways and the assembly of the individual proteins. Beyond the topics discussed above, one can also repeat the speculations made previously (Ferguson, 1988, 1992) concerning (i) whether avoidance of providing transport systems is sometimes a rationale for a periplasmic location of an oxidation or reduction reaction and (ii) whether there would be insufficient surface area on the cytoplasmic surface of the membrane to accommodate proteins needed for many of the electron transfer processes. This chapter has only concerned itself with gram-negative organisms and, as earlier (Ferguson, 1988), much of the greater metabolic diversity of gram-negative compared with gram-positive organisms still correlates with periplasmic electron transfer activities. However, it has been argued that there is a periplasm in at least some gram-positive organisms (Merchante et al., 1995), but the presence of truly water-soluble proteins remains rare. Globular proteins, with membrane anchors, do function on the external surface of the membrane in some gram-positive organisms and thus permit processes such as denitrifiction that are nevertheless more commonly associated with gram-negative organisms.

REFERENCES

Appia-Ayone, C., and B. C. Berks. 2002. SoxV, an orthologue of the CcdA disulfide transporter, is involved in thiosulfate oxidation in *Rhodovulum sulfidophilum* and reduces the periplasmic thioredoxin SoxW. *Biochem. Biophys. Res. Commun.* **296:**737–741.

Bardischewsky, F., J. Fischer, B. Holler, and C. Friedrich. 2006. SoxV transfers electrons to the periplasm of *Paracoccus pantotrophus*—an essential reaction for chemotrophic sulphur oxidation. *Microbiology* **152:**465–472.

Bender, K. S., C. Shang, R. Charkraborty, S. M. Belchik, J. D. Coates, and L. A. Achenbach. 2005. Identification, characterization, and classification of genes encoding perchlorate reductase. *J. Bacteriol.* **187:**5090–5096.

Brass, J. M, C. F. Higgins, M. Folley, P. A. Rugman, J. Birmingham, and P. B. Garland. 1986. Lateral diffusion of proteins in the periplasm of *Escherichia coli. J. Bacteriol.* **165:**787–794.

Cross, R., J. Aish, S. J. Paston, R. K. Poole, and J. W. B. Moir. 2000. Cytochrome *c'* from *Rhodobacter capsulatus* confers increased resistance to nitric oxide. *J. Bacteriol.* **182:**1442–1447.

Daldal, F., M. Deshmukh, and R. C. Prince. 2003. Membrane-anchored cytochrome *c* as an electron carrier in photosynthesis and respiration: past, present and future of an unexpected discovery. *Photosynth. Res.* **76:**127–134.

Daldal, F., S. Mandaci, C. Winterstein, H. Myllykallio, K. Duyck, and D. Zannoni. 2001. Mobile cytochrome c_2 and membrane-anchored cytochrome c_y are both efficient electron donors to the cbb_3- and aa_3-type cytochrome c oxidases during respiratory growth of *Rhodobacter sphaeroides. J. Bacteriol.* **183:**2013–2024.

Feissner, R. E., C. S. Beckett, J. A. Loughman, and R. G. Kranz. 2005. Mutations in cytochrome assembly and periplasmic redox pathways in *Bordetella pertussis. J. Bacteriol.* **187:**3941–3949.

Ferguson, S. J. 1988. Periplasmic electron transport reactions, p.151–182. *In* C. Anthony (ed.), *Bacterial Energy Transduction.* Academic Press, San Diego, Calif.

Ferguson, S. J. 1990. Periplasm underestimated. *Trends Biochem. Sci.* **15:**377.

Ferguson, S. J. 1991. The periplasm, p. 311–339. *In* S. Mohan, C. Dow, and J. A. Cole (ed.), *Prokaryotic Structure and Function: a New Perspective.* SGM Symposium 47, Cambridge University Press, Cambridge, United Kingdom.

Goodwin, P. M., and C. Anthony. 1995. The biosynthesis of periplasmic electron-transport proteins in methylotrophic bacteria. Physiology and genetics of PQQ and PQQ-containing enzymes. *Microbiology* **141:**1051–1064.

Goodwin, P. M., and C. Anthony. 1998. The biochemistry, physiology and genetics of PQQ and PQQ-containing enzymes. *Adv. Microbial. Physiol.* **40:**1–80.

Graham, L. L., J. J. Beveridge, and N. Nanninga. 1991. The periplasmic space and the concept of the periplasm. *Trends Biochem. Sci.* **16:**328–329.

Lloyd, J. R., E. L. Blunt-Harris, and D. R. Lovley. 1999. The periplasmic 9.6 kilodalton *c*-type cytochrome of *Geobacter sulfurreducens* is not an electron shuttle to Fe(III). *J. Bacteriol.* **181:**7647–7649.

McEwan, A. G., J. P. Ridge, C. A. McDevitt, and P. Hugenholtz. 2002. The DMSO reductase family of microbial molybdenum enzymes: molecular properties and role in the dissimilatory reduction of toxic elements. *Geomicrobiol. J.* **19:**3–21.

Mehta, T., M. V. Coppi, S. E. Childers, and D. R. Lovley. 2005. Outer membrane c-type cytochromes required for Fe(III) and Mn(IV) oxide reduction in *Geobacter sulfurreducens. Appl. Environ. Microbiol.* **71:**8634–8641.

Merchante, R., H. M. Pooley, and D. Karamata. 1995. A periplasm in *Bacillus subtilis. J. Bacteriol.* **177:**6176–6183.

Mowatt, C. G., and S. K. Chapman. 2005. Multiheme cytochromes—new structures, new chemistry? *Dalton Trans.* **2005:**3381–3389.

Murakami, S., R. Nakashima, E. Yamashita, and A. Yamaguchi. 2002. Crystal structure of bacterial multidrug efflux transporter AcrB. *Nature* **419:** 587–593.

Myers, J. D., and D. J. Kelly. 2005. A sulphite respiration system in the chemoheterotrophic human pathogen *Campylobacter jejuni. Microbiology* **151:**233–242.

Nicholls, D. G., and S. J. Ferguson. 2002. *Bioenergetics 3.* Academic Press, San Diego, Calif.

Pauleta, S. R., A. Cooper, M. Nutley, N. Errington, S. Harding, F. Guerlesquin, C. F. Godhew, I. Moura, J. J. G. Moura, and G. W. Pettigrew. 2004. A copper protein and a cytochrome bind at the same site on bacterial cytochrome *c* peroxidase. *Biochemistry* **43:**14566–14576.

Pearson, I. V., M. D. Page, R. J. M. van Spanning, and S. J. Ferguson. 2003. A mutant of *Paracoccus denitrificans* with disrupted genes coding for cytochrome c_{550} and pseudoazurin establishes these two proteins as the *in vivo* electron donors to cytochrome cd_1 nitrite reductase. *J. Bacteriol.* **185:**6308–6315.

Pittman, M. S., and D. J. Kelly. 2005. Electron transport through nitrate and nitrite reductases in *Campylobacter jejuni. Biochem. Soc. Trans.* **33:**190–192.

Richardson, D. J. 2000. Bacterial respiration: a flexible process for a changing environment. *Microbiology* **146:**551–571.

Roldan, M. D., H. J. Sears, M. P. Cheesman, S. J. Ferguson, A. J. Thomson, B. C. Berks, and

D. J. Richardson. 1998. Spectroscopic characterisation of a novel multiheme c-type cytochrome widely implicated in bacterial electron transport. *J. Biol. Chem.* **273**:28985–28990.

Schoepp-Cothenet, B., M. Schults, F. Baymann, M. Brugna, W. Nitschke, H. Myllykallio, and C. Schmidt. 2001. The membrane-extrinsic domain of cytochrome $b_{558/566}$ from the archaeon *Sulfolobus acidocaldarius* performs pivoting movements with respect to the membrane surface. *FEBS Lett.* **487**:372–376.

Simon, J., M. Sänger, S. C. Schunster, and R. Gross. 2003. Electron transport to periplasmic nitrate reductase (NapA) of *Wolinella succinogenes* is independent of a NapC protein. *Mol. Microbiol.* **49**:69–79.

Stevens, J. M., O. Daltrop, J. W. A. Allen, and S. J. Ferguson. 2004. Cytochrome c biogenesis: chemical and biological enigmas. *Acc. Chem. Res.* **37**:999–1007.

Toyama, H., F. S. Matthews, O. Adachi, and K. Matsushita. 2004. Quinohemoprotein alcohol dehydrogenases: structure, function and physiology. *Arch. Biochem. Biophys.* **428**:10–21.

Vanwielink, J. E., and J. A. Duine. 1990. How big is the periplasmic space? *Trends Biochem. Sci.* **15**:136–137.

Williams, P. A., V. Fulop, Y. C. Leung, C. Chan, J. W. B. Moir, G. Howlett, S. J. Ferguson, S. E. Radford, and J. Hajdu. 1995. Pseudospecific docking surfaces on electron-transfer proteins as illustrated by pseudoazurin, cytochrome c_{550} and cytochrome cd_1 nitrite reductase. *Nat. Struct. Biol.* **2**:975–982.

Wolterink, A. F. W. M., E. Schiltz, P. L. Hagerdoorn, W. R. Hagen, S. W. M. Kengen, and A. J. M. Stams. 2003. Characterization of the chlorate reductase from *Pseudomonas chloritidismutans*. *J. Bacteriol.* **185**:3210–3213.

Yanyushin, M. F., M. C. del Rosario, D. C. Brune, and R. E. Blankenship. 2005. New class of bacterial membrane oxidoreductases. *Biochemistry* **44**:10037–10045.

PERIPLASMIC NITRATE REDUCTION

Jeff A. Cole

14

INTRODUCTION TO BACTERIAL NITRATE REDUCTION

The reduction of nitrate to nitrite is currently regarded as the first step of three major biological processes: one process is assimilatory; the other two fulfill respiratory roles. The assimilation of nitrogen from nitrate into the pool of organic nitrogen compounds requires nitrate transport across the bacterial cytoplasmic membrane followed by a three-stage process that occurs in the cytoplasm: nitrate reduction to nitrite, nitrite reduction to ammonia, and ammonia assimilation, usually by the glutamine synthetase-glutamate synthase pathway (Tempest et al., 1970). The first step of the two other processes, denitrification and the respiratory reduction of nitrate to ammonia, is also the reduction of nitrate to nitrite, but both of these processes can be instigated either in the cytoplasm or outside the membrane in the periplasm. The latter process is the subject of this overview.

DISCOVERY OF PERIPLASMIC NITRATE REDUCTASES

Before 1980, just two types of bacterial nitrate reductase were widely recognized: these were the soluble, cytoplasmic nitrate reductases in-

volved in nitrate assimilation and the membrane-associated respiratory nitrate reductases. Little was known about the assimilatory enzymes (they are frequently referred to as Nas, but are not considered further in this chapter), but already a vast literature had accumulated describing the anaerobically induced nitrate reductase involved in both nitrate reduction to ammonia by *Escherichia coli* and related bacteria, and in denitrification, especially by pseudomonads. All of these enzyme complexes described so far consist of three structural components: the catalytic molybdoprotein, NarG; a nonheme, iron-sulfur protein, NarH, which forms a tight complex with NarG; and a *b*-type cytochrome, NarI, which transfers electrons from the quinol pool to the NarGH complex. These components are invariably encoded in a four or more gene operon, *narGHJI*, in which the fourth component, NarJ, is a pathway-specific chaperone or assembly factor required for the posttranslational assembly of the functional complex (Palmer et al., 1996). These energy-conserving nitrate reductases are synthesized only during anaerobic growth, and they are found predominantly (exclusively?) in facultative anaerobes that encounter high concentrations of nitrate.

Even by 1980 there was unease about reports of anaerobically induced, soluble nitrate

Jeff A. Cole, School of Biosciences, University of Birmingham, Birmingham B15 2TT, United Kingdom.

The Periplasm
Edited by Michael Ehrmann © 2007 ASM Press, Washington, D.C.

reductases, for example, from *Clostridium perfringens* (Seki-Chiba and Ishimoto, 1977) and photosynthetic bacteria (Alef and Klemme, 1985, and references cited therein). However, soluble, periplasmic nitrate reductases were discovered mainly from studies to resolve the unrelated question whether bacteria can denitrify nitrate under aerobic conditions. The seminal paper by Robertson and Kuenen (1984), "Old wine in new bottles," demonstrated aerobic denitrification by the then-named *Thiosphaera pantotropha*—now renamed *Paracoccus pantotrophus* (Rainey et al., 1999). This was followed by equally seminal papers by Ferguson, Richardson, and their colleagues showing that the closely related organism, then known as *Paracoccus denitrificans*, can reduce nitrate during aerobic growth on reduced carbon sources (Sears et al., 1997, 2000), and that the soluble nitrate reductase involved is located in the periplasm (Richardson et al., 1990). Biochemical and genetic analysis of the paracoccus enzyme by the Ferguson group, in which key players included Berks, Richardson, and Moir, revealed a multisubunit complex encoded by a five-gene operon, *napEDABC*, in which the B and C genes encode a diheme and a tetraheme cytochrome *c*, and NapA is the 90-kDa molybdoprotein that is the catalytic subunit (Berks et al., 1995).

In parallel with the Ferguson group, Siddiqui et al. (1993) characterized a similar Nap from *Alcaligenes eutrophus* (later renamed *Ralstonia eutropha*). The arrival of the genomics era within the next few years revealed an ever increasing range of Nap complexes in diverse groups of bacteria, notably the Nap complex of *E. coli* and its close relatives in the γ-proteobacteria (Grove et al., 1996). It rapidly became apparent that, in contrast to the more conserved components of the membrane-associated respiratory nitrate reductase encoded by the *narGHJI* operons, there were at least five levels of diversity among the periplasmic nitrate reductases: their distribution among bacteria of different physiological types; their regulation; their physiological roles; their components; and their genetic context (Potter et al., 2001). Subsequent sections of this chapter that review each of these aspects are followed by a summary of recent developments, including the realization by Richardson and his colleagues that there is a fourth type of nitrate reductase, a membrane-associated group with catalytic sites located in the periplasm, which might explain how the diversity of the Nap enzymes has evolved. Finally, some of the currently unanswered questions are highlighted.

DISTRIBUTION, REGULATION, AND FUNCTIONS OF PERIPLASMIC NITRATE REDUCTASES

In contrast to the well-defined role for nitrate reductase A in energy conservation during anaerobic growth, many diverse functions have been documented for the periplasmic nitrate reductases, some of which are summarized in Table 1. Both *R. eutropha* and *P. pantotrophus* are adaptable organisms that can grow either aerobically or anaerobically. However, any initial suggestion that the periplasmic nitrate reductases might function only in aerobic denitrification in other bacteria was rapidly dispelled. In contrast to the role of Nap in aerobic denitrification in *T. pantotrophus*, Nap has been shown to be essential for denitrification under anaerobic growth conditions in some pseudomonads (Bedzyk et al., 1999).

TABLE 1 Various physiological roles of Nap in different bacteria

Organism	Subunits encoded by operon	Physiological role of Nap
Paracoccus pantotrophus	NapEDCBA	Aerobic denitrification
Escherichia coli	NapFDAGHBC	Nitrate scavenging
Rhodobacter capsulatus	NapKEFDABC	Enhanced survival in dark
Desulfovibrio desulfuricans	NapCMADGH	Survival during sulfate starvation
Pseudomonas strain G179	NapEFDABC	Anaerobic denitrification

How gene expression is regulated frequently provides strong clues to the physiological role of the encoded products; an excellent example of this principle is the regulation of *nap* gene expression during aerobic growth of *P. pantotrophus*. In this organism, the *nap* genes are highly expressed during growth on a reduced carbon compound, for example, butyrate, but are repressed during growth on succinate or malate (Sears et al., 2000). This was the first evidence that Nap fulfills a redox-balancing role in this organism. In contrast, it is the *nar* operon that is induced during anaerobic growth in the presence of a high concentration of nitrate, providing energy for anaerobic growth (Wood et al., 2001). A redox-balancing role for Nap has also been proposed for the photosynthetic bacterium, *Rhodobacter sphaeroides* (Reyes et al., 1996). During growth on a reduced carbon compound, the quinone:quinol redox couple becomes too reduced for the photosynthetic electron transfer chain to function optimally, so Nap is proposed to provide an electron sink to reestablish redox balance. Ellington et al. (2003) demonstrated that, although expression of *nap* cannot support anaerobic growth of chemoheterotrophic cultures of *Rhodobacter capsulatus*, it enhanced survival during incubation in the dark. It was proposed that Nap provides a proton motive force for cell maintenance during dark periods and therefore confers a selective advantage during light-dark transitions.

Periplasmic nitrate reductases were soon discovered in the facultative anaerobe, *E. coli*, and the obligate anaerobe, *Desulfovibrio desulfuricans*, neither of which catalyzes denitrification (Bursakov et al., 1995; Dias et al., 1999; Grove et al., 1996). In these essentially fermentative bacteria, Nap provides the first step in a periplasmic pathway for nitrate reduction to ammonia, in which the partner nitrite reductase is the *c*-type cytochrome, NrfA. One of the fascinating early questions concerning the periplasmic pathway for nitrate reduction to ammonia by *E. coli* was why the expression of genes encoding a nitrate reductase should be repressed by the pathway substrate, nitrate. High concentrations of nitrate are sensed by the two-component regulatory system, NarX-NarL, in which NarX is a membrane-associated environmental sensor that detects and responds to nitrate. Nitrate-bound NarX is an autokinase that transfers its phosphate group to the cytoplasmic transcription factor, NarL. Phosphorylated NarL then represses transcription of both the *nap* and *nrf* operons (*nrf* stands for nitrite reduction by formate; the operon encodes the periplasmic nitrite reductase) but activates transcription of genes for the alternative, cytoplasmic pathway for nitrate reduction to ammonia. The explanation for this regulation was simultaneously resolved in two laboratories by independent but complementary approaches.

Chemostat competition experiments by Potter et al. (1999) demonstrated that, compared with a mutant expressing only the membrane-associated nitrate reductase, a strain expressing the periplasmic nitrate reductase had a selective advantage when only a limited supply of nitrate was available as the terminal electron acceptor for anaerobic growth. Conversely, in nitrate-sufficient media, the strain expressing the membrane-associated nitrate reductase was at a selective advantage. They also showed that the K_m for nitrate of the periplasmic nitrate reductase is significantly lower than that of NarG. The outcome of a chemostat competition experiment is determined by the ratio of V_{max} to K_m for the growth-limiting substrate for the two strains. Table 2 shows that this ratio is higher for a strain expressing only the *nap* operon during nitrate-limited growth, but higher for the strain expressing the *narG* operon when nitrate is abundant.

In parallel experiments, Wang et al. (1999) showed that the alternative environmental sensor for nitrate, NarQ, responds to far lower concentrations of nitrate by phosphorylating the alternative response regulator, NarP. As phosphorylated NarP activates *nap* transcription but phosphorylated NarL represses transcription, low concentrations of nitrate are essential for optimal expression of the periplasmic nitrate reductase, whereas high concentrations are required for synthesis of the alternative nitrate reductase, NarG. The two

TABLE 2 Why an *E. coli* strain expressing a periplasmic nitrate reductase outcompetes a strain expressing only nitrate reductase A during nitrate-limited anaerobic growth[a]

Growth conditions	Enzyme	V_{max}	K_s (μM)	$V_{max}:K_m$ ratio
Excess nitrate	NRA	300	50	6
	NAP	10	10	1
Limiting nitrate	NRA	10	50	0.2
	NAP	50	10	5

[a]Reaction rates are nanomoles of nitrate reduced per minute per milligram of bacterial dry mass.

parallel studies provide a nice example of how studies of transcription regulation often reveal the physiological roles of the encoded products. Furthermore, only *nap* genes rather than *narGHJI* are found in some pathogenic bacteria with small genomes, and Nap accumulates optimally under nitrate-limited conditions similar to those that occur in the bodies of warm-blooded animals, including the essentially anaerobic lower gastrointestinal tract of humans.

E. coli is so far one of the few bacteria in which multiple physiological roles for Nap have been documented experimentally. Brondijk et al. (2004) recently showed that Nap also fulfills a redox-balancing role in *E. coli*.

Only a few periplasmic nitrate reductases have been intensively studied in the laboratory; so much of our current information is deduced from bioinformatic analysis of the genomic databases. These databases in turn are highly biased toward the genomes of pathogenic bacteria, so it is still dangerous to draw general conclusions from the limited data available. With this caveat, the conservation of *nap* gene clusters in pathogens with small genomes, especially when coupled with the absence of the *narGHJI* operon, is strong evidence that conservation of Nap provides a selective advantage for these organisms in the human body. In this sense, Nap is physiologically more important to enteric bacteria than the membrane-associated nitrate reductase A, and hence with good reason is considered to be a pathogenicity determinant.

As far as NarG is concerned, there are only loose links, often at the regulatory level, between the type of nitrate reductase expressed and the type of nitrite reductase to which it is linked. Nitrite produced during nitrate reduc-

tion by NarG can be reduced to nitric oxide by either the cytochrome cd_1 or copper-containing nitrite reductases, or be reduced to ammonia by the cytoplasmic NADH-dependent nitrite reductase, NirB (Berks et al., 1995). Periplasmic nitrate reductases can also couple with different types of nitrite reductase, the main difference being that the periplasmic Nrf replaces the cytoplasmic NirB during anaerobic reduction of nitrite to ammonia. There is, however, an excellent correlation between the components of the various *nap* gene clusters and the type of nitrite reductase present: this relationship is summarized in the next section.

COMPONENTS OF THE *nap* GENE CLUSTERS AND THEIR BIOCHEMICAL FUNCTIONS

By the middle of 2005, twelve different organizations of *nap* gene clusters could be recognized, excluding those from *Campylobacter jejuni*, *Shewanella oneidensis*, and other *Shewanella* species that appear to be scattered rather than closely organized into an operon (Color Plate 20). Table 3 summarizes the sizes and structures of the gene products and the current perception of their functions. Note, however, that extreme variability can be found even within a single genus, the most spectacular example to date being the various *Shewanella* species that have been sequenced (Table 4). At least four different combinations occur in this genus, *napAB, DAB, DABC,* and *DAGHB*; and duplicated *nap* clusters occur in *S. frigidimarina, S. amzonensis, S. putrefaciens,* and *Shewanella* sp. PV4.

Several points immediately emerge from Color Plate 20. First, 12 different types of poly-

TABLE 3 Polypeptides encoded by *nap* gene clusters in various bacteria

Polypeptide	Mass (kDa)	Structure and function	Example reference
NapA	80 or 90–100	Molybdoprotein + 4Fe-4S iron sulfur center; catalytic subunit	Berks et al., 1995; Dias et al., 1999
NapB	16–21	Diheme *c*-type cytochrome; electron donor to NapA in many bacteria	Brige et al., 2002
NapC	22–29	Tetraheme *c*-type cytochrome; quinol dehydrogenase; electron donor to NapB/or ?	Roldan et al., 1998
NapD	9–14	Cytoplasmic, pathway-specific chaperone? Posttranslational modification of NapA? Essential, but function poorly defined	Berks et al.; 1995; Potter and Cole, 1999
NapE	6–7	Unknown function: predicted to be an integral membrane protein	Berks et al., 1995
NapF	16–18	Nonheme iron-sulfur protein: insertion of iron-sulfur center into NapA? Controversial	Olmo-Mira et al., 2004
NapG	18–27	Nonheme iron-sulfur protein: electron transfer from NapH to terminal Nap components, possibly directly to NapA	Simon et al., 2003; Brondijk et al., 2004
NapH	30–34	Nonheme iron-sulfur protein: quinol dehydrogenase in anaerobes	Simon et al., 2003; Brondijk et al., 2004
NapK	6	Unknown function: possibly integral membrane protein in photosynthetic bacteria	Reyes et al., 1998
NapL	34	Possibly secreted; function unknown	Simon et al., 2003
NapM	15	Small tetraheme *c*-type cytochrome similar to those found in *Shewanella* species	Marietou et al., 2005

peptide are encoded by the various *nap* clusters, but only two of them, *napA* and *napD*, have so far always been found together. These encode the catalytic subunit, NapA, and the pathway-specific chaperone, NapD. Very few residues are highly conserved among the various NapD polypeptides, which are essential for

TABLE 4 Variety of combinations of genes in the *nap* clusters of different *Shewanella* strains[a]

Shewanella species or strain	Clusters of *nap* genes present
S. amazonensis	napDABC
S. baltica	napDAB
S. denitrificans	napDABC
S. frigidimarina	napAB
	napDABC
S. oneidensis	napDAGHB
S. putrefaciens CN-32	napDABC
	napDAGHB
Shewanella PV4	napDABC
	napDAGHB

[a]From data compiled by D. Richardson.

the posttranslational assembly of a functional Nap complex (Potter and Cole, 1999). Note, however, the absence of a *napD* gene in the two-gene *nap* operon of *S. frigidimarina*, though *napD* is present in the second *nap* gene cluster in this bacterium (Table 4). Until recently, the diheme cytochrome *c*, NapB, was considered to be an essential component, but there is no *napB* gene in the *D. desulfuricans nap* gene cluster (Marietou et al., 2005). It was only three years ago that Simon and colleagues showed that *Wolinalla succinogenes* lacks the tetraheme cytochrome, NapC, and that neither of the only two tetraheme *c*-type cytochromes encoded elsewhere on the *Wolinella* chromosome is essential for nitrate reduction (Simon et al., 2003). NapK has so far been found only in photosynthetic bacteria, and NapL was identified in *W. succinogenese* (Reyes et al., 1998; Simon et al., 2003).

Also apparent from the figure is that certain components tend to be found in bacteria of a particular physiological type. For example, both of the *c*-type cytochromes, NapB and NapC, are found in nonfermentative bacteria such as *Pseudomonas stutzeri*, *P. pantotrophus*, and *R. eutropha*. Conversely, the nonheme iron-sulfur proteins, NapG and NapH, seem always to be present in obligate anaerobes. Particularly interesting are the seven-gene *nap* operons of facultative anaerobic bacteria such as *E. coli* and the small-genome pathogen, *Haemophilus parainfluenzae*; these bacteria have retained all four components, NapB, C, G, and H. It was recently shown that in *E. coli* electrons are transferred preferentially from menadiol via NapC and NapB to NapA, but there is also a slower rate of electron transfer from ubiquinol via NapH and NapG to NapB-NapA, again via NapC (Brondijk et al., 2002, 2004). While this implies that NapH-NapG forms a ubiquinol dehydrogenase, it is clear that the homologous proteins in *W. succinogenese*, which lacks ubiquinol, function as a menadiol dehydrogenase (Simon et al., 2003). This again emphasizes the danger of trying to draw general conclusions from just a few examples.

NapK has so far been found only in photosynthetic bacteria in which its function is unknown (Reyes et al., 1998). The function of NapE is also still unclear, though some role in posttranslational modification, or a chaperone function, seem plausible. The most enigmatic component is NapF. A role in insertion of a nonheme iron-sulfur center into *R. sphaeroides* NapA has been proposed (Olmo-Mira et al., 2004), but this is difficult to reconcile with the nonessential role of NapF in *E. coli* (Potter and Cole, 1999; Brondijk et al., 2004).

PATHWAYS OF ELECTRON TRANSFER TO THE DIFFERENT PERIPLASMIC NITRATE REDUCTASES

Comprehensive experimental analysis of electron transfer pathways to the periplasmic nitrate reductases has been restricted to only a few species, for example, *W. succinogenes*, *P. pantotrophus*, and *E. coli*, with less extensive studies of a few other bacteria such as *R. sphaeroides*, *R. capsulatus,* and *Shewanella* species. The limited

number of components in the Nap systems of *W. succinogenes* and *P. pantotrophus* limits scope for doubt in the pathways of electron transfer from the quinol pool to NapA. At one extreme, NapA from *W. succinogenes* accepts electrons from menaquinol directly via the iron-sulfur proteins, NapG-NapH, in which NapH is the menaquinol dehydrogenase (Fig. 1a). There is no NapC in this organism (Simon et al., 2003). At the opposite extreme, neither NapG nor NapH is found in the facultatively anaerobic α- and β-proteobacteria such as *P. pantotrophus* and *R. eutropha* (Fig. 1b) (Berks et al., 1995). In these bacteria in which Nap can function in aerobic denitrification, the *c*-type cytochromes, NapC and NapB, rather than iron-sulfur proteins, transfer electrons from the ubiquinol pool to NapA. Although menaquinol is the electron donor in *W. succinogenes* and ubiquinol is the electron donor in *P. pantotrophus*, the situation in *E. coli* illustrates that there is no simple correlation between the type of quinol electron donor and whether iron-sulfur proteins or *c*-type cytochromes transfer electrons to NapA. In *E. coli*, both ubiquinol and menadiol are effective electron donors to the terminal components of the Nap pathway. However, in contrast to the cytoplasmic nitrate reductase, Nar, electrons are transferred more rapidly via the menaquinol pathway, which is mediated by NapC and NapB independently of NapG and NapH, than via ubiquinol, for which NapG and NapH play key roles (Brondijk et al., 2004).

Little structural information is available about the electron transfer proteins that associate with NapA. The exception is the diheme NapB from *Haemophilus influenzae*, for which a structure at 1.25 Å resolution is available (Brige et al., 2002). This reveals the presence of a fold unprecedented in all other classes of *c*-type cytochromes, confirming that NapB belongs to a previously undescribed class of *c*-type cytochrome. Another so-far unique feature is that the two propionate side chains on both heme groups are hydrogen bonded to each other. The authors proposed that there is an efficient, linear electron transfer pathway

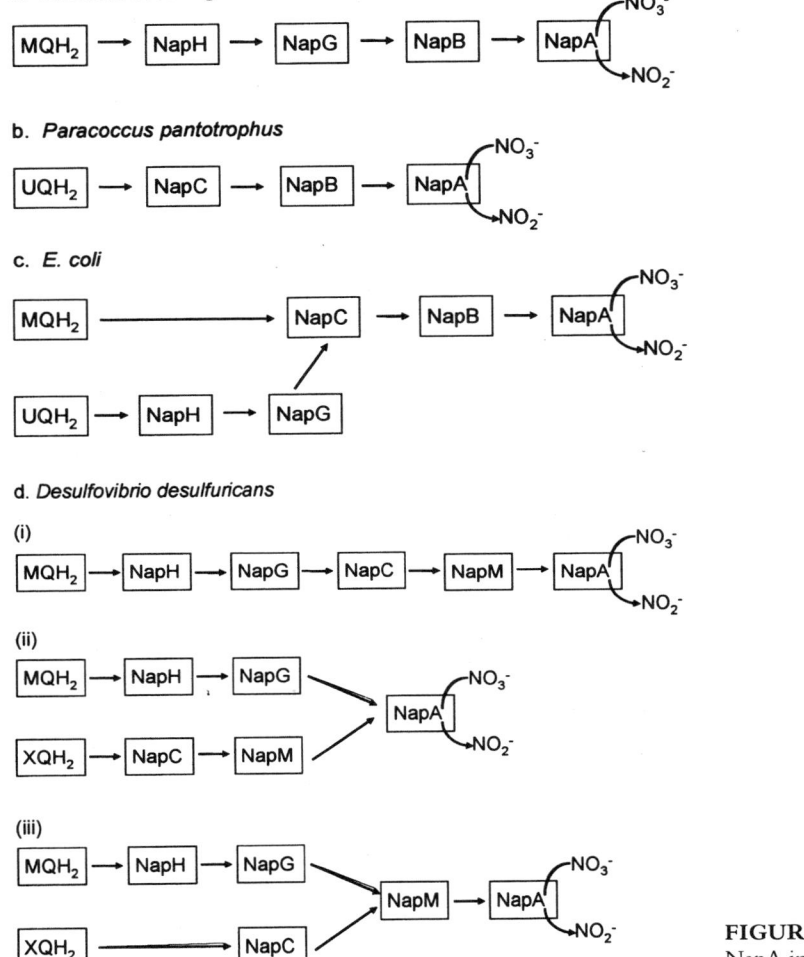

a. *Wolinella succinogenes*

b. *Paracoccus pantotrophus*

c. *E. coli*

d. *Desulfovibrio desulfuricans*

(i)

(ii)

(iii)

FIGURE 1 Electron transfer to NapA in different bacteria.

spanning about 40 Å from NapC via NapB to the catalytic site of NapA

Given that *W. succinogenese* is an obligate anaerobe, pseudomonads are nonfermentative facultative anaerobes and *E. coli* is a more versatile facultative anaerobe that can also survive by fermentation, it will be particularly interesting to dissect the electron transfer pathways to NapA in *D. desulfuricans*. The first published sequence of a *nap* gene cluster in a sulfate-reducing bacterium revealed a six-gene organization, *napC-napM-napA-napD-napG-napH* (Marietou et al., 2005). Notable features of this *nap* gene cluster (which is almost certainly transcribed as a single operon)(A. Marietou and J. Cole, unpublished results) are the lack of a *napB* gene but the inclusion of genes for two other *c*-type cytochromes, *napM* and *napC*, as well and both *napG* and *napH*.

The NapC polypeptide is more similar to the NrfH subgroup of tetraheme cytochromes than to NapC from other bacteria, suggesting that there might be an evolutionary link between the periplasmic nitrite reductases and Nap in obligate anaerobes. NapM is similar to the small tetraheme *c*-type cytochrome from

S. oneidensis that transfers electrons from NapC to NapA. It is also possible that NapM plays a promiscuous role in electron transfer from other electron donors, for example, one of the periplasmic hydrogenases synthesized by *D. desulfuricans* and *D. vulgaris* (Haveman et al., 2004). It is also possible that, as in *S. putrefaciens* MR-1, the small *c*-type cytochrome couples electron transfer to oxidants other than nitrate (Myers and Myers, 1997).

The third gene, *napA*, encodes the catalytic subunits that, with NapA from *Symbiobacterium thermophilum* (Ueda et al., 2004), are the closest known relatives of, and a similar size to, the ferredoxin-dependent assimilatory nitrate reductase, NarB, of *Synechococcus* sp. Superficially the fifth gene obviously encodes NapG with four conserved cysteine clusters, but it lacks a twin-arginine motif for export into the periplasm. Consistent with the lack of a TAT targeting sequence, the start of NapG translation is tightly coupled to the translation stop codon of the preceding gene encoding NapD. The nonheme, iron-sulfur protein, NapG, might be a direct electron donor to *D. desulfuricans* and *S. thermophilum* NapA; if so, it must pass through the TAT system as a mature NapAG complex taking advantage of the NapA TAT signal sequence. In *S. thermophilum* the NapC protein most likely serves to reduce NapG, but in *D. desulfuricans* the presence of a membrane-associated NapH suggests a route of electron transfer from quinol to NapG via NapH similar to that proposed for *W. succinogenes* (Simon et al., 2003). The C-terminal sequence of *D. desulfuricans* NapH also deviates considerably from other NapH sequences, again raising doubts about its substrate specificity. In the absence of direct experimental evidence, at least three possible electron transfer pathways can be proposed for nitrate reduction by sulfate-reducing bacteria.

1. Each of the *nap* genes encodes an essential component of a single linear pathway, as shown in Fig. 1d (i).
2. NapG-NapH and NapC-NapM might provide converging electron transfer pathways to NapA, as shown in Fig. 1d (ii).

3. NapC and NapG-NapH might donate electrons to NapA via a common component, NapM, as shown in Fig. 1d (iii).

During growth on lactate, the most likely primary electron donors for nitrate reduction are lactate (directly or via an NAD^+-dependent lactate dehydrogenase), formate, or hydrogen. Each of these electron donors can conceivably pass electrons directly to the Nap complex (the XH_2 pathways in Fig. 1d) or indirectly via menaquinone.

EVOLUTION OF THE PERIPLASMIC NITRATE REDUCTASES

Early attempts to rationalize the variations in *nap* gene cluster components focused on their different physiological roles and attempted to correlate the presence or absence of specific components with the type of electron donor (ubiquinol or menadiol) available. The midpoint redox potential of the menaquinone-menadiol couple is about 0 mV compared with +70 mV for the ubiquinone-ubiquinol couple. Consequently, many obligate anaerobes synthesize menaquinone, but not ubiquinone, while the opposite is true for aerobes or facultative anaerobes that are unable to grow anaerobically by fermentation. As shown above, however, the *E. coli* Nap system soon provided an exception to this correlation.

It was noted above that periplasmic nitrate reductases are structurally more similar to the cytoplasmic, assimilatory nitrate reductase Nas than to the membrane-bound nitrate reductase A. Jepson et al. (2006) have noted a fourth class of nitrate reductase that, like NarG, is membrane associated, but the catalytic sites are located outside the membrane—not necessarily in the periplasm, as the first examples of this class of nitrate reductase were found in gram-positive bacteria, which of course lack a periplasm. Two examples of this type are NapA from *S. thermophilum*, an uncultured gram-positive thermophile (Ueda et al., 2004), and *Desulfitobacterium hafniense* (http://genome.jgi-psf.org/draft_microbes/desha/desha.home.html), a gram-positive organism important in environmental dehalogenation (Damborsky, 1999).

A smaller, monomeric NapA is also predicted in the facultatively anaerobic δ-proteobacterium, *Anaeromyxobacter dehalogenans*. The *nap* gene cluster in this organism includes a gene encoding a small tetraheme *c*-type cytochrome homologous to NapM. Richardson and Jepson propose a line of evolution of the dimeric NapAB complexes from Nas via a primordial Nap, as shown in Fig. 2 (Jepson et al., 2006). They suggest that the driving force has been the need for NapA to evolve from an enzyme that accepts electrons from low midpoint redox potential iron-sulfur proteins in the reducing environment of the cytoplasm to be able to function in the increasingly oxidizing environment of the periplasm (Damborsky, 1999; Sargent et al., 2002; Jepson et al., 2004, 2006; Marietou et al., 2005). According to this hypothesis, the first evolutionary step would be the acquisition by Nas of an export sequence to generate the smaller version of NapA, as currently found in *S. thermophilum* (Jepson et al., 2006). Next to evolve would have been a Nap complex that could draw electrons from

the membrane-bound electron transfer chain. The absence of a twin-arginine signal peptide at the N terminus of apo-NapG from *D. desulfuricans*, *S. thermophilum*, and *D. hafniense* implies that NapG would associate with NapA in the cytoplasm prior to export into the periplasm directed by the NapA twin-arginine export sequence (Jepson et al., 2006). This is similar to other iron-sulfur proteins, for example, NarH, the electron donor to NarG, which also lacks a targeting sequence (Sargent et al., 2002). To function at higher redox potentials, the low-potential iron-sulfur proteins would then have been replaced by the higher midpoint potential *c*-type cytochromes. Recruitment of genes for the *c*-type cytochromes from the NrfHA complex would result in *nap* gene clusters similar to those currently seen in *A. dehalogenans* and *D. desulfuricans*. For obligate anaerobes such as *W. succinogenes*, electrons would still be delivered by iron-sulfur proteins such as NapGH (Simon et al., 2003; Marietou et al., 2005). Next to evolve would be the dual systems involving both NapGH and NapC/NapB, as

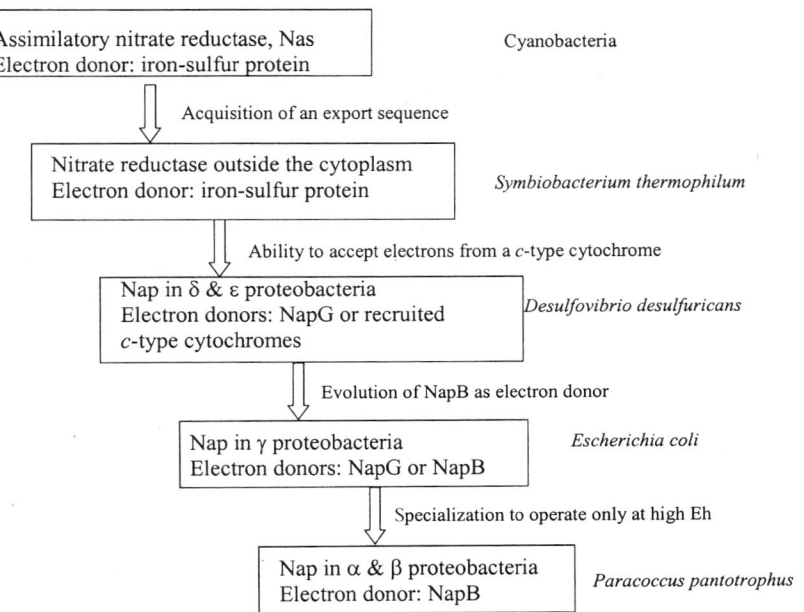

FIGURE 2 Evolution of cytochrome *c*-linked periplasmic nitrate reductases.

found in fermentative facultative anaerobes of the γ-proteobacteria (Fig. 2). Last to evolve were enzymes dedicated to accepting electrons only from the *c*-type cytochromes, NapC and NapB, in the α-proteobacteria (Jepson et al., 2006).

There is some structural information to support this speculation. For example, in contrast to the membrane-bound enzyme in which aspartate provides oxygen ligands to the catalytic molybdenum center, this is replaced by a sulfur cysteine ligand in both Nas and Nap (Arnoux et al., 2003; Bertero et al., 2003; Butler et al., 1999; Dias et al., 1999; Hettmann et al., 2004; Jepson et al., 2004; Jormakka et al., 2004). The limited biochemical data currently available clearly show that the structures of NapA from different bacteria are more variable than that of Nas or NarG. Although the molecular mass of mature NapA from α-, β-, and γ-proteobacteria is about 90 kDa (typically about 800 amino acids in length), NapA from the δ and ε groups is smaller, similar to the 729 amino acids found in Nas from the cyanobacterium, *Synechococcus elongatus*. Furthermore, NapA from α-proteobacteria forms a tight complex with its immediate electron donor, NapB, whereas NapA from *D. desulfuricans* was purified as a single subunit (Bursakov et al., 1995). NapA and NapB in *R. sphaeroides* form a very tight complex with a K_D of 0.5 nM, and from the crystal structure of the complex determined at a resolution of 3.2 Å, it appears that tight complex formation involves two loops at the N- and C-terminal extremities of NapB that adopt an extended conformation and embrace the NapA subunit (Arnoux et al., 2003). At the opposite extreme and also consistent with this proposal is the structure, determined to a resolution of 1.9 Å, of the monomeric NapA from *D. desulfuricans* (Dias et al., 1999). In contrast to the more polar and anionic surfaces of NapA that interact with NapB in the *R. sphaeroides* and *P. pantotrophus* enzymes, the corresponding surfaces of NapA from *D. desulfuricans*, *S. thermophilum*, *D. hafniense*, and *Anaeromyxobacter dehalogenans* are predicted to be hydrophobic and slightly positively charged. As predicted, NapA and NapB are only loosely associated in the *E. coli* enzyme and are readily separated during anion-exchange chromatogaphy (Thomas et al., 1999).

Further support for the ideas of Richardson and Jepson comes from the *nap* gene cluster in *A. dehalogenans*, which includes a gene encoding a small tetraheme *c*-type cytochrome homologous to NapM. Furthermore, the *A. dehalogenans napA* is linked to a *napM* homologue that also clusters with, and might even be in the same operon as, *nrfHA*, genes encoding a tetraheme membrane-anchored protein of the NapC family and a pentaheme cytochrome *c* nitrite reductase, strengthening the evidence for the possible evolutionary origin of *D. desulfuricans* NapM from a primordial NrfHA complex. Finally, note that although the *nap* gene clusters are duplicated in some *Shewanella* species (Table 3), the basic *napDA* unit is linked in *S. amazonensis*, *S. putrefaciens* CN-32, and *Shewanella* PV4 either to *napGHB* or to *napBC*, suggesting that this might represent an intermediate stage in the evolution of the *napEDABC* unit of the α- and β-proteobacteria from the *napFDAGHBC* unit of the γ-proteobacteria.

UNANSWERED QUESTIONS

While it is clear that nitrate reduction in the periplasm is a rapidly expanding area of interest, many basic questions concerning the physiological role of Nap in gram-negative bacteria and the specific functions of many of the components revealed by genomic analysis remain unresolved. Perhaps the most fundamental question is whether there are mechanisms for energy conservation as electrons are transferred from the quinol pool to NapA. Pertinent to this question is why the quinol dehydrogenase, NapGH, in some organisms is coupled directly to NapA, but in other organisms it is linked via NapC. Does this simply reflect progressive evolution of the Nap complex or differences in biochemical function? The only universally present component in *nap* gene clusters apart from *napA* is *napD*. One might reasonably expect the amino acid sequence of NapD to be highly conserved between different bacterial groups, so why is the opposite the case? How can a common biochemical role be fulfilled by such different structures? In view of the vestigial similarity between *Desulfovibrio napMC* and *E. coli nrfAB* (Marietou et al., 2005; Jepson et al., 2006), did Nap evolve from Nrf and, if so, can other similar evolutionary links be recognized? The *D. desulfuricans* NapG lacks a TAT-dependent targeting sequence, so how does it get to the periplasm? If it travels piggyback across the cytoplasmic membrane in a complex with NapA, does this reveal fundamental differences in the mechanism of assembly of different Nap complexes in different bacteria? Sequence-tagged mutagenesis and a growing association of nitrate-reducing sulfate-reducing bacteria with gastric diseases in humans implicate Nap as a pathogenicity determinant, but this is speculation that awaits more direct experimental confirmation. One of the few physiological roles not yet assigned to Nap is nitrate assimilation: does any bacterium rely on Nap and Nrf to generate ammonia for nitrate assimilation? While the list is endless, one final intriguing question is how nitrate reduction in the periplasm of sulfate-reducing bacteria is regulated. Given that the sulfate:sulfite redox couple (or more correctly, the adenosine phosphosulfate:sulfite redox couple) is strongly electronegative compared with the nitrate:nitrite couple, it would be fascinating to determine whether nitrate reduction by *D. desulfuricans* is repressed by sulfate. If so, this would be contrary to one of the central dogmas of anaerobic bacterial metabolism, namely that bacteria have evolved control mechanisms that enable them to use the energetically most favorable electron acceptor first.

REFERENCES

Alef, K., and J. H. Klemme. 1977. Characterization of a soluble NADH–independent nitrate reductase from the photosynthetic bacterium *Rhodopseudomonas capsulata. Z. Naturforsch.* **32:**954–956.

Arnoux, P., M. Sabaty, J. Alric, B. Frangioni, B. Guigliarelli, J. M. Adriano, and D. Pignol. 2003. Structural and redox plasticity in the heterodimeric periplasmic nitrate reductase. *Nat. Struct. Biol.* **10:**928–934.

Bedzyk, L., T. Wang, and R. W. Ye. 1999. The periplasmic nitrate reductase in *Pseudomonas* sp. strain G-179 catalyzes the first step of denitrification. *J. Bacteriol.* **181:**2802–2806.

Berks, B. C., D. J. Richardson, A. Reilly, A. C. Willis, and S. J. Ferguson. 1995. The *napED-ABC* gene cluster encoding the periplasmic nitrate reductase system of *Thiosphaera pantotropha. Biochem. J.* **309:**983–992.

Bertero, M. G., R. A. Rothery, M. Palak, C. Hou, D. Lim, F. Blasco, J. H. Weiner, and N. C. Strynadka. 2003. Insights into the respiratory electron transfer pathway from the structure of nitrate reductase A. *Nat. Struct. Biol.* **10:**681–687.

Brige, A., D. Leys, T. E. Meyer, M. A. Cusanovich, and J. J. Van Beeumen. 2002. The 1.25 Å resolution structure of the diheme NapB subunit of soluble nitrate reductase reveals a novel cytochrome c fold with a stacked heme arrangement. *Biochemistry* **41:**4827–4836.

Brondijk, T. H., D. Fiegen, D. J. Richardson, and J. A. Cole. 2002. Roles of NapF, NapG and NapH, subunits of the *Escherichia coli* periplasmic nitrate reductase, in ubiquinol oxidation. *Mol. Microbiol.* **44:**245–255.

Brondijk, T. H., A. Nilavongse, N. Filenko, D. J. Richardson, and J. A. Cole. 2004. The NapGH components of the periplasmic nitrate reductase of *Escherichia coli* K-12: location, topology, and physiological roles in quinol oxidation and redox balancing. *Biochem. J.* **379:**47–55.

Bursakov, S., M.-Y. Liu, W. J. Payne, J. LeGall, I. Moura, and J. J. G. Moura. 1995. Isolation and preliminary characterisation of a soluble nitrate reductase from the sulphate reducing organism *Desulfovibrio desulfuricans* ATCC 27774. *Anaerobe* **1:**55–60.

Butler, C. S., J. M. Charnock, B. Bennett, H. J. Sears, A. J. Reilly, S. J. Ferguson, C. D. Garner, D. J. Lowe, A. D. Thomson, B. C. Berks, and D. J. Richardson. 1999. Models for molybdenum coordination during the catalytic cycle of periplasmic nitrate reductase from *Paracoccus denitrificans* derived from EPR and EXAFS spectroscopy. *Biochemistry* **38:**9000–9012.

Damborsky, J. 1999. Tetrachloroethene-dehalogenating bacteria. *Folia Microbiol. (Praha)* **44:**247–262.

Dias, J. M., M. E. Than, A. Humm, R. Huber, G. P. Bourenkov, H. D. Bartunik, S. Bursakov, J. Calvete, J. Caldeira, C. Carneiro, J. J. Moura, I. Moura, and M. J. Romao. 1999. Crystal structure of the first dissimilatory nitrate reductase at 1.9 A solved by MAD methods. *Structure* **7:**65–79.

Ellington, M. J., D. J. Richardson, and S. J. Ferguson. 2003. *Rhodobacter capsulatus* gains a competitive advantage from respiratory nitrate reduction during light-dark transitions. *Microbiology* **149:**941–948.

Grove, J., S. Tanapongpipat, G. Thomas, L. Griffiths, H. Crooke, and J. Cole. 1996. *Escherichia coli* K-12 genes essential for the synthesis of c-type cytochromes and a third nitrate reductase located in the periplasm. *Mol. Microbiol.* **19**:467–481.

Haveman, S. A., E. A. Greene, C. P. Stillwell, J. K. Voordouw, and G. Voordouw. 2004. Physiological and gene expression analysis of inhibition of *Desulfovibrio vulgaris* Hildenborough by nitrite. *J. Bacteriol.* **186**:7944–7950.

Hettmann, T., R. A. Siddiqui, C. Frey, T. Santos-Silva, M. J. Romao, and D. Diekmann. 2004. Mutagenesis study on amino acids around the molybdenum centre of the periplasmic nitrate reductase from *Ralstonia eutropha. Biochem. Biophys. Res. Commun.* **320**:1211–1219.

Jepson, B. N., L. J. Anderson, L. M. Rubio, C. J. Taylor, C. S. Butler, E. Flores, A. Herrero, J. N. Butt, and D. J. Richardson. 2004. Tuning a nitrate reductase for function. The first spectropotentiometric characterization of a bacterial assimilatory nitrate reductase reveals novel redox properties. *J. Biol. Chem.* **31**:32212–32218.

Jepson, B. J. N., A. Marietou, S. Mohan, J. A. Cole, C. S. Butler, and D. J. Richardson. 2006. Evolution of the soluble nitrate reductase: defining the monomeric periplasmic nitrate reductase subgroup. *Biochem. Soc. Trans.* **34**:122–126.

Jormakka, M., D. Richardson, B. Byrne, and S. Iwata. 2004. Architecture of NarGH reveals a structural classification of Mo-bisMGD enzymes. *Structure* **12**:95–104.

Marietou, A., D. Richardson, J. Cole, and S. Mohan. 2005. Nitrate reduction by *Desulfovibrio desulfuricans*: a periplasmic nitrate reductase that lacks NapB, but includes a unique tetraheme *c*-type cytochrome, NapM. *FEMS Microbiol Lett.* **248**:217–225.

Myers, C. R., and J. M. Myers. 1997. Cloning and sequence of *cymA*, a gene encoding a tetraheme cytochrome *c* required for reduction of iron(III), fumarate, and nitrate by *Shewanella putrefaciens* MKR-1. *J. Bacteriol.* **179**:1143–1152.

Olmo-Mira, M. F., M. Gavira, D. J. Richardson, F. Castillo, C. Moreno-Vivian, and M. D. Roldan. 2004. NapF is a cytoplasmic iron-sulfur protein required for Fe-S cluster assembly in the periplasmic nitrate reductase. *J. Biol. Chem.* **26**:49727–49735.

Palmer, T., C. L. Santini, C. Iobbi-Nivol., D. J. Eaves, D. H. Boxer, and G. Giordano. 1996. Involvement of the *narJ* and *mob* gene products in distinct steps in the biosynthesis of the molybdoenzyme nitrate reductase in *Escherichia coli. Mol. Microbiol.* **20**:875–884.

Potter, L., H. Angove, D. Richardson, and J. Cole. 2001. Nitrate reduction in the periplasm of Gram-negative bacteria. *Adv. Microb. Physiol.* **45**:51–112.

Potter, L., P. Millington, G. Thomas, and J. Cole. 1999. Competition between *Escherichia coli* strains expressing either a periplasmic or a membrane-bound nitrate reductase: does Nap confer a selective advantage during nitrate-limited growth? *Biochem. J.* **344**:77–84.

Potter, L. C., and J. A. Cole. 1999. Essential roles for the products of the *napABCD* genes, but not *napFGH*, in periplasmic nitrate reduction by *Escherichia coli* K-12. *Biochem. J.* **344**:69–76.

Rainey, F. A., D. P. Kelly, E. Stackebrandt, J. Burhardt, A. Hiraishi, Y. Katayama, and A. P. Wood. 1999. A re-evaluation of the taxonomy of *Paracoccus denitrificans* and a proposal for the combination *Paracoccus pantotrophus* comb. nov. *Int. J. Syst. Bacteriol.* **49**:645-651.

Reyes, F., M. Gavira, F. Castillo, and C. Moreno-Vivian. 1998. Periplasmic nitrate reducing system of the phototrophic bacterium *Rhodobacter sphaeroides* DSM 158: transcriptional and mutational analysis of the *napKEFDABC* gene cluster. *Biochem. J.* **331**:897–904.

Reyes, F., M. D. Roldan, W. Klipp, F. Castillo, and C. Moreno-Vivian. 1996. Isolation of periplasmic nitrate reductase genes from *Rhodobacter sphaeroides* DSM 158: structural and functional differences among prokaryotic nitrate reductases. *Mol. Microbiol.* **19**:1307–1318.

Richardson, D. J., A. G. McEwan, M. D. Page, J. B. Jackson, and S. J. Ferguson. 1990. The identification of cytochromes involved in the transfer of electrons to the periplasmic NO_3^- reductase of *Rhodobacter capsulatus* and resolution of a soluble NO_3^--reductase-cytochrome-c_{552} redox complex. *Eur. J. Biochem.* **194**:263–270.

Robertson, L. A., and J. G. Kuenen. 1984. Aerobic denitrification—old wine in new bottles? *Antonie van Leeuwenhoek* **50**:525–544.

Roldan, M. D., H. J. Sears, S. J. Ferguson, M. R. Cheeseman, A. J. Thomson, B. C. Berks, and D. J. Richardson. 1998. Spectroscopic characterisation of a novel tetra-heme *c*-type cytochrome widely implicated in bacterial electron transport. *J. Biol. Chem.* **273**:28785–28790.

Sargent, F., B. C. Berks, and T. Palmer. 2002. Assembly of membrane-bound respiratory complexes by the Tat protein-transport system. *Arch. Microbiol.* **178**:77–84.

Sears, H. J., G. Sawers, B. C. Berks, S. J. Ferguson, and D. J. Richardson. 2000. Control of periplasmic nitrate reductase gene expression (*napEDABC*) from *Paracoccus pantotrophus* in response to oxygen and carbon substrates. *Microbiology* **146**:2977–2985.

Sears, H. J., S. Spiro, and D. J. Richardson. 1997. Effect of aeration and carbon substrate on expresion of the periplasmic and membrane-bound nitrate reductases of *Paracoccus denitrificans. Microbiology* **143**:3765–3774.

Seki-Chiba, S., and M. Ishimoto. 1977. Studies on nitrate reductase of *Clostridium perfringens*. Purification, some properties, and effect of tungstate on its formation. *J. Biochem. (Tokyo)* **82:**1663–1671.

Siddiqui R. A., U. Warnecke-Eberz, A. Hengsberger, B. Schneider, S. Kostka, and B. Friedrich. 1993. Structure and function of a periplasmic nitrate reductase in *Alcaligenes eutrophus* H16. *J Bacteriol.* **175:**5867–5876.

Simon, J., M. Sänger, S. C. Schuster, and R. Gross. 2003. Electron transport to periplasmic nitrate reductase (NapA) of *Wolinella succinogenese* is independent of a NapC protein. *Mol. Microbiol.* **49:**69–79.

Tempest, D. W., J. L. Meers, and C. M. Brown. 1970. Synthesis of glutamate in *Aerobacter aerogenes* by a hitherto unknown route. *Biochem. J.* **117:**405–407.

Thomas, G., L. Potter, and J. Cole. 1999. The periplasmic nitrate reductase of *Escherichia coli*: a heterodimeric molybdoprotein with a double-arginine signal sequence and an unusual leader peptide cleavage site. *FEMS Microbiol. Lett.* **174:**167–171.

Ueda, K., A. Yamashita, J. Ishikawa, T. O. Watsuji, H. Ikeda, M. Hattori, and T. Beppu. 2004. Genome sequence of *Symbiobacterium thermophilum*, an uncultivatable bacterium that depends on microbial commensalisms. *Nucleic Acids Res.* **32:**4937–4944.

Wang, H., C. P. Tseng, and R. P. Gunsalus. 1999. The *napF* and *narG* nitrate reductase operons in *Escherichia coli* are differentially expressed in response to submicromolar concentrations of nitrate but not nitrite. *J. Bacteriol.* **181:**5303–5308.

Wood, N. J., T. Alizadeh, S. Bennett, J. Pearce, S. J. Ferguson, and J. W. Moir. 2001. Maximal expression of membrane-bound nitrate reductase in *Paracoccus* is induced by nitrate via a third FNR-like regulator named NarR. *J. Bacteriol.* **183:**3606–3613.

THE BIOSYNTHESIS OF
THE MOLYBDENUM COFACTOR
AND ITS INCORPORATION
INTO MOLYBDOENZYMES

Silke Leimkühler

15

Molybdenum cofactor (Moco) biosynthesis is an ancient, ubiquitous, and highly conserved pathway leading to the biochemical activation of molybdenum. Moco is the essential component of a group of redox enzymes, which are diverse in terms of their phylogenetic distribution and their architectures, both at the overall level and in their catalytic geometry. A wide variety of transformations are catalyzed by these enzymes at carbon, sulfur, and nitrogen atoms, which include the transfer of an oxo group or two electrons to or from the substrate. More than 40 molybdoenzymes have been identified in bacteria, archaea, plants, and animals to date. Some of the better-known Moco-containing enzymes include sulfite oxidase, xanthine dehydrogenase, and aldehyde oxidase in humans; assimilatory nitrate reductase in plants; and dissimilatory nitrate reductase, dimethyl sulfoxide (DMSO) reductase, and formate dehydrogenase in bacteria and archaea. In molybdoenzymes Mo is coordinated to a dithiolene group on the 6-alkyl side chain of a pterin called molybdopterin (MPT). The biosynthesis of Moco can be divided into three general steps present in all organisms: (i) formation of the so-called precursor Z, (ii) formation of MPT from precursor Z, and (iii) insertion of molybdenum to form Moco. A fourth step is found in bacteria by additional modification of Moco with the attachment of GMP, AMP, IMP, or CMP to the phosphate group of MPT. Mutations in the human Moco biosynthetic genes lead to death in early childhood. The affected patients show severe neurological abnormalities such as attenuated growth of the brain, seizures, and dislocated ocular lenses. So far no effective therapy is available to cure the disease. This review focuses on the biosynthesis of Moco in bacteria and its incorporation into specific target proteins.

MOLYBDENUM IN BIOLOGICAL SYSTEMS

Molybdenum is the only second-row transition metal that is required by most living organisms, and the few species that do not require molybdenum use tungsten, which lies immediately below molybdenum in the periodic table (Pope et al., 1980; Stiefel, 1993, 2002). Molybdenum is required for cofactors of redox enzymes that catalyze reactions at carbon, sulfur, and nitrogen atoms, including nitrogenase and nitrate reductase, the key enzymes in the nitrogen cycle. There are two distinct types of molybdoenzymes: Molybdenum nitrogenase has a unique

Silke Leimkühler, Universität Potsdam, Institut für Biochemie und Biologie, AG Biochemie (Proteinanalytik), Karl-Liebknecht Strasse 24-25, Haus 25, 14476 Potsdam, Germany.

The Periplasm
Edited by Michael Ehrmann © 2007 ASM Press, Washington, D.C.

molybdenum–iron–sulfur cluster, the $[Fe_4S_3]$–(bridging-S)$_3$-$[MoFe_3S_3]$ center called FeMoco. It catalyzes the reduction of atmospheric dinitrogen to ammonia. All other molybdoenzymes are oxidoreductases that transfer an oxo group or two electrons to or from the substrate. They bind Moco in which molybdenum is coordinated to a dithiolene group on the 6-alkyl side chain of MPT (Fig. 1) (Rajagopalan, 1991, 1996, 1997; Rajagopalan et al., 1993).

In contrast to the multinuclear FeMoco in nitrogenase, the active site of mainly all other characterized Moco-containing enzymes (using either tungsten or molybdenum) is mono-nuclear, with a single equivalent of the metal (Hille, 1996).

Molybdenum is chemically similar to tungsten and vanadium, both of which can replace molybdenum in the cofactors of a few enzymes, such as tungstopterin enzymes of hyperthermophiles (Johnson et al., 1996; Kletzin and Adams, 1996) and vanadium nitrogenase (Eady, 1995). Molybdenum and tungsten are present in the environment as the soluble oxyanions molybdate, MoO_4^{2-}, and tungstate, WO_4^{2-}. At low concentrations vanadium is present in neutral solutions as a mixture of $H_2VO_4^-$ and HVO_4^{2-} (Willsky, 1990). Molybdenum is

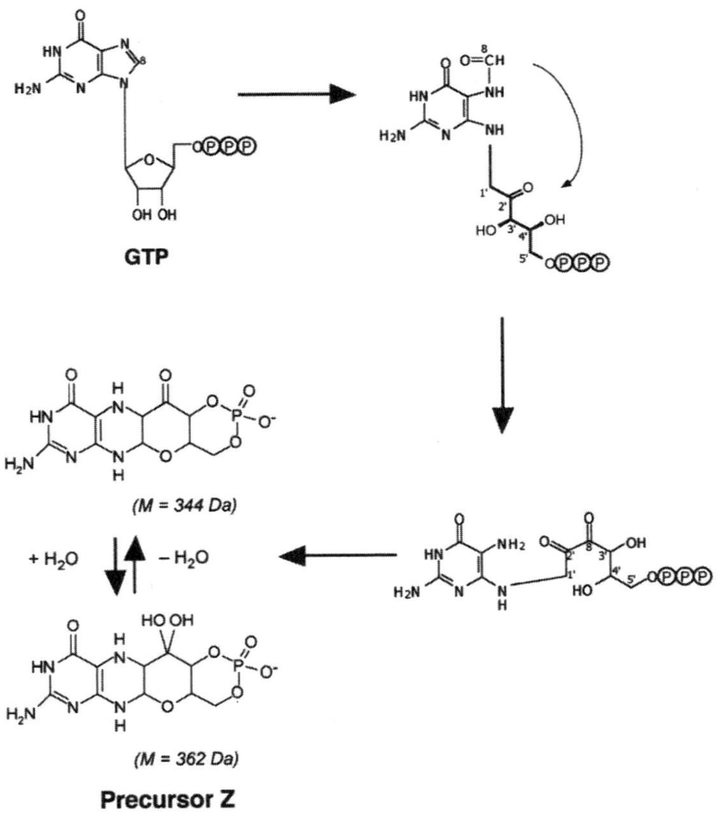

FIGURE 1 Synthesis of precursor Z from GTP. All carbon atoms of the GTP are found within precursor Z. The C8 atom transferred as formyl is inserted between the C2′ and C3′ atoms of the ribose. Precursor Z is shown in the tetrahydropyrano form and in a hydrated product with a geminal diol at the C1′ position.

the most abundant transition metal in seawater (110 nM); thus, molybdenum has been widely incorporated into biological systems. However, the distribution of molybdenum on land is uneven. Its concentration in soil water is usually less than 50 nM and may be very low, in particular, in acid soils with high iron concentrations (Pau and Lawson, 2002). Both molybdenum and tungsten have a chemical versatility that is useful to biological systems (Pilato and Stiefel, 1999): they are redox-active under physiological conditions (ranging between oxidation states VI and IV); because the V valence state is also accessible, they can act as transducers between obligatory two-electron and one-electron oxidation-reduction systems; and they can catalyze reactions such as the hydroxylation of carbon centers under more moderate conditions than are required by other systems (Hille, 1996).

MOLYBDATE TRANSPORT

Mutations that affect Moco biosynthesis were identified by exploiting the ability of nitrate reductase to convert chlorate to the toxic product chlorite. Chlorate-resistant mutants lack nitrate reductase activity and therefore survive when grown in the presence of chlorate (Stewart, 1988).

The first high-affinity molybdate transport mutant, *Escherichia coli modC* (formerly called *chlD*) (Shanmugam et al., 1992), was identified because molybdoenzyme activity could be restored by addition of molybdenum at a 1,000-fold higher concentration (>0.1 mM) than that sufficient for wild-type cells (Shanmugam et al., 1992). The *modC* mutant phenotype is pleiotropic for all molybdoenzymes. *E. coli* ModC is the ATP-binding protein of a high-affinity molybdenum uptake system belonging to the ATP-binding cassette (ABC) superfamily of transporters that is responsible for the transport of a wide variety of molecules in prokaryotes and eukaryotes (Davidson and Chen, 2004; Higgins, 1992). In their most common form in bacteria ABC transporters consist of two integral membrane proteins (ModB for molybdate transport in *E. coli*) and two hy-

drophilic peripheral membrane proteins (the ATP-binding protein ModC for molybdate transport in *E. coli*). In gram-negative bacteria, a periplasmic binding protein captures the substrate and delivers it to the transporter complex in the inner membrane (ModA for the molybdate transport system in *E. coli*). Both molybdate and tungstate bind to the *E. coli* ModA protein in the periplasm as a tetrahedral complex, which is held by seven hydrogen bonds formed between the oxygen of the bound anion and the protein groups from two domains (Hu et al., 1997). For molybdate-dependent regulation, the *modABC* operon expression is controlled by the ModE protein, which binds to the operator region of the *modABC* operon in its molybdate-bound form (McNicholas et al., 1997). ModE also enhances the transcription of molybdenum-dependent enzymes like DMSO reductase (McNicholas et al., 1998) or nitrate reductase A (Self et al., 1999) and also of the molybdenum cofactor biosynthesis operon *moaABCDE* (Anderson et al., 2000). ModE is a homodimer and each monomer can be subdivided into four structural domains: the N-terminal DNA-binding domain, a linker domain, and two molybdate-binding (mop) domains (Schuttelkopf et al., 2003). ModE binds two molybdate molecules per dimer and molybdate binding subsequently results in extensive conformational changes of the domain, thus enabling DNA binding (Schuttelkopf et al., 2003).

For the synthesis of molybdoenzymes, the cell needs to activate the molybdate after transport into the cell to an appropriate form and incorporate it into either FeMoco or into MPT, the organic part of Moco. In Moco, the metal is coordinated by a unique dithiolene group of the cofactor, which coordinates molybdenum and tungsten in the same manner.

THE BIOSYNTHESIS OF MOCO IN *E. COLI*

Much has been learned about Moco biosynthesis from studies of Moco mutants in *E. coli* where five loci have been implicated in the pleiotrophy of the molybdoenzymes: *moa*, *mob*, *mod*, *moe*, and *mog*, comprising 14 genes (Rajagopalan, 1996; Shanmugam et al., 1992). With the exception of *mod* encoding the genes for the high-affinity molybdate transport system, all of these are involved in the biosynthesis of Moco. The biosynthesis of Moco can be divided into four steps in *E. coli* (Color Plate 21): (i) formation of precursor Z, (ii) formation of MPT from precursor Z, (iii) insertion of molybdenum to form Moco, and (iv) additional modification of Moco with the attachment of GMP, forming the MPT-guanine dinucleotide cofactor (MGD). Each of these steps is described in detail below.

The Formation of Precursor Z

The biosynthesis of Moco starts from 5′-GTP (Color Plate 21 and Fig. 1), which results in the formation of precursor Z, the first stable intermediate of Moco biosynthesis (Hanzelmann and Schindelin, 2004; Johnson et al., 1989; Wuebbens and Rajagopalan, 1993, 1995). Precursor Z is an oxygen-sensitive 6-alkyl pterin with a cyclic phosphate group, originating from GTP. In contrast to other pteridine biosynthetic pathways, formation of precursor Z is cyclohydrolase I and II independent. It was shown that the C8 atom of 5′-GTP is not released as formate but is retained and incorporated in a rearrangement reaction and inserted between the 2′- and 3′-ribose carbon atoms (Wuebbens and Rajagopalan, 1993), thus forming the four-carbon atoms of the pyrano ring that is typical for MPT (Fig. 1).

In *E. coli* the gene products MoaA and MoaC were identified to be essential for the biosynthesis of precursor Z. MoaA belongs to the *S*-adenosylmethionine (SAM)-dependent radical enzyme superfamily, members of which catalyze the formation of protein and/or substrate radicals by reductive cleavage of SAM by a [4Fe-4S] cluster (Sofia et al., 2001). MoaA in fact assembles two oxygen-sensitive [4Fe-4S] clusters, one typical for SAM-dependent radical enzymes at the N terminus and an additional C-terminal cluster unique to MoaA proteins, as shown in the human homologue MOCS1A (Hanzelmann et al., 2004). The

crystal structure of *Staphylococcus aureus* MoaA showed an incomplete $(\alpha\beta)_6$ TIM barrel type formed by the N-terminal part of the protein (Hanzelmann and Schindelin, 2004). The lateral opening of the incomplete TIM barrel is covered by residues of the C-terminal part of the protein including a second $[4Fe-4S]^{2+}$ cluster. Both FeS clusters are present at a distance of about 17 Å and create a large active site pocket constructed from predominantly basic residues. The non-cysteinyl-ligated unique Fe site of the C-terminal [4Fe-4S] cluster is believed to be involved in the binding and activation of 5′-GTP (Hanzelmann and Schindelin, 2004). The structure of MoaC revealed that it is present as a hexamer composed of two dimers with a putative active site located at the dimer interface (Wuebbens et al., 2000). Since some radical SAM-dependent enzymes require another protein onto which the radical is transferred, it has been speculated that MoaC might act in a similar function. However, nothing is known about the exact role of MoaC for precursor Z synthesis so far.

Recently, it was shown that precursor Z most likely exists as the tetrahydropyranopterin form and possesses a geminal diol at the C1′ position (Fig. 1), both characteristics being important for maintaining the stereochemistry of the pterin C6 position and providing the directed reactivity for the insertion of the two sulfurs for MPT formation (Santamaria-Araujo et al., 2004).

The Formation of MPT from Precursor Z

For the formation of MPT from precursor Z, two sulfur atoms are incorporated to the C1′ and C2′ positions of precursor Z, a reaction catalyzed by MPT synthase (Pitterle et al., 1993). The purification of MPT synthase identified a heterotetrameric enzyme, consisting of two small (~8,750 Da) and two large subunits (~16,850 Da), encoded by *moaD* and *moaE*, respectively (Pitterle and Rajagopalan, 1989, 1993). It was shown that the small subunit of MPT synthase carries the sulfur in form of a thiocarboxylate group at its C-terminal glycine

(Pitterle and Rajagopalan, 1993). The high-resolution crystal structure of MPT synthase revealed that the two MoaE subunits form a central dimer, and the MoaD subunits are located at opposite ends of each dimer (Rudolph et al., 2001) (Color Plate 22). A pocket with highly conserved amino acids in MoaE can be seen in proximity to the C terminus of MoaD, most likely building the precursor Z-binding site. From the structure the question arose whether the addition of the dithiolene sulfurs occurs independently at both active sites or whether a hemisulfurated intermediate is transferred from one active site to the other for the addition of the second sulfur atom. Wuebbens and Rajagopalan (2003) were able to identify the hemisulfurated intermediate that is tightly associated with MPT synthase, which makes it more likely that each precursor Z molecule remains bound at a single active site until conversion of MPT is completed and that an exchange of carboxylated and thiocarboxylated MoaD occurs while the intermediate is bound at the same active site (Color Plate 22).

For the synthase to act catalytically, it is necessary to regenerate its transferable sulfur. Since an *E. coli moeB* mutant contained the nonsulfurated form of the synthase, it appeared that the MoeB protein is involved in the activation of the synthase by sulfur transfer, thus it was named MPT synthase sulfurase. It was shown that MoeB and ATP are necessary for the reactivation of MPT synthase (Pitterle and Rajagopalan, 1993). MoaD shares structural similarities to ubiquitin (Rudolph et al., 2001) and MoeB sequence similarities to the ubiquitin-activating enzyme E1 (McGrath et al., 1991; Rajagopalan, 1997). However, biochemical data and the crystal structure of the MoeB-MoaD complex revealed that the interaction of MoeB with MoaD resembles only the first step of the ubiquitin-targeted degradation of proteins (Leimkühler et al., 2001; Rudolph et al., 2001). It was shown that MoeB solely activates the C terminus of MoaD by formation of an acyladenylate and no thioester intermediate was identified between MoaD and MoeB (Leimkühler et al., 2001). Subsequently, the ac-

tivated MoaD acyl adenylate is converted to a thiocarboxylate by action of a persulfide-containing protein. In *E. coli*, the sulfur donor is L-cysteine, but the specific protein involved in sulfur transfer to MPT synthase has not been identified (Leimkühler and Rajagopalan, 2001a). In *E. coli* any of the three L-cysteine desulfurase proteins, namely IscS, CSD, or CsdB, was able to donate the sulfur for the thiocarboxylation of MoaD in a defined in vitro system (Leimkühler and Rajagopalan, 2001a).

In humans, the homologous protein to *E. coli* MoeB, named MOCS3, contains an additional domain with homologies to rhodaneses at its C terminus. Since it was shown that the rhodanese-like domain acts as direct sulfur donor to human MPT synthase, transferring thiosulfate-sulfur via a protein-bound persulfide intermediate (Matthies et al., 2004), it is believed that in *E. coli* a separate rhodanese-like protein might perform the same reaction.

In summary (Color Plate 22), the partitioning of MoaD between MoaE and MoeB is governed by the carboxylate versus thiocarboxylate status of the C-terminal glycine of MoaD. For the reactivation of MPT synthase, MoaD-carboxylate dissociates from its complex with MoaE to form a stable adenylate complex with MoeB. In the adenylate complex, MoaD is susceptible to sulfuration, which proceeds by a sulfur transfer pathway from L-cysteine by the action of a persulfide-forming sulfurtransferase. However, it is yet unknown how the sulfurtransferase interacts with the MoeB-MoaD complex to perform the sulfur transfer reaction. After the formation of the MoaD-thiocarboxylate group, MoaD dissociates from MoeB and reassociates with MoaE to form an active MPT synthase, which is able convert precursor Z to MPT (Rudolph et al., 2003).

Insertion of Molybdenum into MPT To Form Moco

After synthesis of the dithiolene moiety in MPT, the chemical backbone is built for binding and coordination of the molybdenum atom. The mechanism of molybdenum insertion remains one of the most enigmatic aspects of Moco biosynthesis and nothing is known about whether molybdate must undergo intracellular processing before insertion into Moco.

In the step of molybdenum insertion in *E. coli*, the gene products of *moeA* and *mogA* are involved. *E. coli mogA* mutants express a molybdate-repairable phenotype with molybdate concentrations of 0.1 to 1 mM (Grunden and Shanmugam, 1997). MogA was the first protein of Moco biosynthesis for which the crystal structure was solved (Liu et al., 2000); however, it is still unclear which form of molybdenum is inserted. The crystal structure of MoeA showed a two-domain structure, with one domain being structurally related to MogA (Schrag et al., 2001; Xiang et al., 2001). It was suggested that *E. coli* MoeA and MogA proteins are both essential for the incorporation of molybdenum in the cofactor, thus indicating a multistep reaction of molybdenum chelation. The likely function of MogA is to prepare nascent MPT produced by MPT synthase in an ATP-dependent manner for MoeA-mediated Mo-ligation (Nichols and Rajagopalan, 2002, 2005).

In *Arabidopsis thaliana*, the Cnx1 protein contains a C-terminal domain with homologies to *E. coli* MogA and an N-terminal domain with homologies to *E. coli* MoeA. The high-resolution MPT-bound structure of the Cnx1 G domain has recently been determined (Kuper et al., 2004). Besides defining the MPT-binding site a novel and unexpected adenylated form of MPT (MPT-AMP) was identified, and it was shown that the Cnx1 G-domain catalyzed the adenylation of MPT in a reaction dependent on Mg^{2+} and ATP (Llamas et al., 2004). Since the Cnx1 E domain is required for the subsequent Mg-dependent molybdenum-insertion reaction, it was proposed that the G and the E domains act sequentially during Moco biosynthesis. MPT-AMP formed by the G domain is directly transferred and processed by the E domain (Llamas et al., 2004). In the crystal structure of Cnx1-G copper bound to the MPT dithiolene sulfurs was identified, which is proposed to protect the dithiolene group before molybdenum insertion (Kuper et al., 2004).

However, it was shown that active Moco can be synthesized in vitro by ligation of molybdate to MPT de novo from precursor Z and MPT without the requirement of any additional factors (Leimkühler and Rajagopalan, 2001b). Under the molybdate concentrations used in the assay (>5 mM), the involvement of MoeA and MogA was not required for the generation of the active form of Moco (Leimkühler and Rajagopalan, 2001b). This shows that the ATP-dependent activation of MPT and molybdenum is only required under physiological concentrations and that in vitro copper is not needed to protect the dithiolene group of MPT.

Additional Modification of Moco with the Attachment of GMP

Moco formed in *E. coli* is further modified by the covalent addition of GMP to the C4' atom of MPT via a pyrophosphate bond, a reaction catalyzed by the MobA protein (Lake et al., 2000; Palmer et al., 1996). In other prokaryotes, variants of the cofactor containing CMP, AMP, or IMP linked to MPT were identified (Börner et al., 1991; Johnson et al., 1990; Rajagopalan and Johnson, 1992). In contrast, eukaryotes solely incorporate the MPT form of the cofactor into their molybdoenzymes. It was shown that in most *E. coli* molybdoenzymes the molybdenum atom is coordinated by the dithiolene groups of two MGD molecules, forming the bis-MGD cofactor (Hilton and Rajagopalan, 1996).

The *mob* locus, encoding *mobA* and *mobB*, has been described to be involved in the attachment of GTP for the formation of the MGD cofactor (Johnson et al., 1991; Palmer et al., 1994). The *mobB* gene is not essential for cofactor biosynthesis, since *mobA* alone can fully complement the deletion of both *mob* genes (Palmer et al., 1996). MobA catalyzes the conversion of MPT and GTP to MGD (Palmer et al., 1998), while MobB is a GTP-binding protein with weak intrinsic GTPase activity (Eaves et al., 1997). In vitro, MobA, GTP, $MgCl_2$, and Mo-MPT are sufficient for the formation and insertion of bis-MGD into *Rhodobacter sphaeroides* DMSO reductase (Temple and Rajagopalan, 2000). The crystal structure of MobA, which is a monomer, showed a nucleotide-binding Rossman fold formed by the N-terminal half of the protein (Lake et al., 2000; Stevenson et al., 2000). It was suggested that the addition of the dinucleotide to the cofactor occurs after the insertion of molybdenum into MPT (Nichols and Rajagopalan, 2005; Temple and Rajagopalan, 2000).

Surprisingly, a molybdoenzyme named YedY was structurally and biochemically characterized from *E. coli*, which was shown to bind the MPT form of Moco without the addition of a nucleotide. The protein consists of a soluble catalytic subunit termed YedY with homologies to sulfite oxidases, which is likely anchored to the periplasmic site of the membrane by a heme-containing *trans*-membrane subunit termed YedZ. YedY represents the only molybdoenzyme isolated from *E. coli* so far characterized by the presence of the MPT form of Moco (Brokx et al., 2005; Loschi et al., 2004).

The Insertion of bis-MGD into *E. coli* Molybdoenzymes

The crystal structures of different molybdoenzymes showed that the cofactor is deeply buried in the protein, which suggested that the insertion of the cofactor is coupled to the final steps of protein folding and assembly of the enzyme (Boyington et al., 1997; Chan et al., 1995; Enroth et al., 2000; Kisker et al., 1997a; Ramão et al., 1995; Schindelin et al., 1996). Thus, the question arose how the cofactor is stored in the cell and shuttled to its user enzymes.

Specific chaperones for the insertion of the bis-MGD cofactor into *E. coli* trimethylamine *N*-oxide (TMAO) reductase and nitrate reductase A have been described. In *E. coli*, TorA, a member of the DMSO reductase family, is the main respiratory enzyme for the TMAO reduction when the cells are grown anaerobically in the presence of TMAO (Ilbert et al., 2003; Tranier et al., 2003). TorA is located in the periplasm and receives electrons from TorC, a pentahemic c-type cytochrome (Gon et al., 2001). TorA and TorC are encoded by the *tor-CAD* operon, which is induced in the presence of TMAO (Mejean et al., 1994). TorA crosses the inner membrane by the Tat machinery in a folded state (see below), meaning that bis-MGD is inserted into the apoprotein in the cytoplasm before translocation (Santini et al., 1998). TorD was shown to be a specific chaperone for TorA maturation that interacts with apo-TorA and generates a form that is able to receive bis-MGD (Genest et al., 2005). A second role was also attributed to TorD during the translocation process of TorA by the Tat translocon to prevent export of immature TorA (Jack et al., 2004; Oresnik et al., 2001). During this proofreading mechanism, TorD binds the signal peptide of TorA and exhibits a quality control activity to ensure that only matured TorA is translocated. TorD from *Shewanella massilia* was crystallized and its three-dimensional structure revealed an all-helical architecture of the TorD dimer showing no similarity with other known protein structures (Ilbert et al., 2004). No binding of bis-MGD to TorD was reported, which makes it likely that TorD binds to apoTorA and modifies its conformation to a competent state, facilitating its maturation. It was shown that TorD possesses a high specificity toward its partner since DmsD, required for the activity of DMSO reductase in *E. coli*, could not replace TorD during TorA maturation, like TorD is unable to act on DmsA (Ilbert et al., 2003). A possible coevolution of molybdoenzymes of the DMSO-reductase family and of their dedicated chaperones has been suggested (Ilbert et al., 2004).

The formation of active membrane-bound nitrate reductase A in *E. coli*, a member of the DMSO reductase family, requires the presence of three subunits, NarG, NarH, and NarI, as well as a fourth protein, NarJ, which is not part of the active nitrate reductase (Blasco et al., 1998). NarG is the catalytic domain-binding bis-MGD, NarH is an [FeS] cluster containing an electron transfer unit, and NarI is a heme-containing membrane anchor subunit. In *narJ*-strains, both NarG and NarH subunits are associated in an unstable and inactive NarGH complex. The NarJ protein specifically recognizes the catalytic NarG subunit and in the absence of NarJ, no bis-MGD is present in the NarGH complex. For NarJ an association with the nitrate reductase apoprotein was reported, acting as a specific chaperone on NarG involved in bis-MGD insertion (Blasco et al., 1998). Upon insertion of the molybdenum cofactor into the aponitrate reductase, NarJ is then dissociated from the activated enzyme. It has been suggested that NarJ plays an essential role in allowing interactions between NarG and the Moco biosynthesis proteins MogA, MoeA, MobA, and MobB (Vergnes et al., 2004). It was shown that MogA, MoeA, MobA, and MobB form a complex and were able to deliver the mature Moco onto aponitrate reductase A in a NarJ-assisted process (Vergnes et al., 2004).

Identification of a Protein Required for Moco Insertion into *Rhodobacter capsulatus* Xanthine Dehydrogenase

Xanthine dehydrogenase (XDH) is a complex metalloflavoprotein that catalyzes the hydroxylation of hypoxanthine and xanthine, the last two steps in the formation of urate, using a water molecule as ultimate source of oxygen incorporated into the product. DNA sequence analysis of genes encoding XDH in *R. capsulatus* revealed two open reading frames, designated *xdhA* and *xdhB* (Leimkühler et al., 1998). It was shown that XdhA binds the two [2Fe-2S] clusters and the FAD cofactor, whereas XdhB binds the MPT form of Moco. Immediately downstream of *xdhB* a third gene was identified, designated *xdhC*, which is cotranscribed with *xdhAB*. Interposon mutagenesis revealed that the *xdhC* gene product is required for XDH activity, but XdhC is not a subunit of active XDH (Leimkühler et al., 1998). By using recombinant *R. capsulatus* XDH purified to homogeneity, the crystal structure of the enzyme was solved, showing that the cofactor is deeply buried in the protein, which suggested that insertion of Moco occurs by help of a chaperone-like protein coupled to the final steps of maturation (Truglio et al., 2002).

Analysis of the Moco content of inactive XDH purified from an *R. capsulatus xdhC* mu-

tant strain showed that in the absence of XdhC, Moco is not inserted into XDH, but XDH still consisted of a $(\alpha\beta)_2$ heterotetramer with a full complement of FeS clusters and FAD (Leimkühler and Klipp, 1999). An interaction of XdhC with apo-XDH was not identified; however, analyses revealed a different conformation of MPT-free XDH in *xdhC*- and Moco$^-$-deficient *R. capsulatus* mutant strains (Leimkühler and Klipp, 1999). It was assumed that during Moco biosynthesis, the MPT-free apo-XDH stays in a suitable "open" conformation, which enables XdhB to bind mature Moco. Thus, the role of XdhC was proposed to act as an XDH-specific Moco carrier protein, a Moco insertase, or a chaperone involved in proper folding during or after the insertion of Moco into XDH. This shows that not only enzymes of the DMSO reductase family binding the bis-MGD form of Moco require a chaperone-assisted maturation of the protein. In conclusion, it seems likely that each prokaryotic molybdoenzyme has its own system-specific chaperone that plays a special role in Moco insertion and target protein folding.

MOCO CARRIER PROTEINS IN EUKARYOTES

The presence of system-specific chaperones involved in Moco insertion for molybdoenzymes seems to be restricted to prokaryotes, since so far no chaperone has been identified for eukaryotic molybdoenzymes. However, oxygen seems to be a major factor of free Moco inactivation, which makes it likely that during its biosynthesis Moco stays protein bound until its insertion into the specific target proteins.

In *Chlamydomonas reinhardtii* a Moco carrier protein (MocoCP) has been identified (Witte et al., 1998). The homotetrameric MocoCP was able to bind and transfer Moco directly to aponitrate reductase and could protect Moco against inactivation under aerobic conditions. Also an interaction of MocoCP and nitrate reductase was shown (Ataya et al., 2003). A protein with similar characteristics was also identified in *Vicia faba* (Aguilar et al., 1992). However, so far, no indication of the existence of such a protein exists in prokaryotes.

OVERVIEW OF THE MONONUCLEAR MOLYBDENUM ENZYMES

To date, more than 40 enzymes containing Moco have been identified in bacteria, archaea, fungi, plants, and animals, presenting a group of structurally and biochemically related proteins catalyzing different redox reactions (Hille, 1996).

The mononuclear molybdenum enzymes are categorized on the basis of the structures of their molybdenum centers, dividing them into three families, each with a distinct active site structure and type of reaction catalyzed (Fig. 2): the xanthine oxidase (XO) family, the sulfite oxidase family, and the DMSO reductase family (Hille, 1996, 2002; Kniemeyer and Heider, 2001). The XO family is characterized by an $LMo^{VI}OS(OH)$ core in the oxidized state, with one equivalent of the pterin cofactor (designated L) coordinated to the metal. These enzymes (including XDH) typically catalyze the hydroxylation of carbon centers. Enzymes of the sulfite oxidase family coordinate a sin-gle equivalent of the pterin cofactor with an $LMo^{VI}O_2(S\text{-}Cys)$ core in its oxidized state (the cysteine ligand is provided by the polypeptide). Members of this family (including sulfite oxidase and nitrate reductase) catalyze the transfer of an oxygen atom either to or from the substrate. The DMSO-reductase family is diverse in both structure and function, but all members have two equivalents of the pterin cofactor bound to the metal. The molybdenum coordination sphere is usually completed by a single M=O group with a sixth ligand in the $L_2M^{VI}O(X)$ core. The sixth ligand, X, can be a serine, a cysteine, a selenocysteine, or a hydroxide and/or water molecule. The reactions catalyzed by members of this family frequently involve oxygen-atom transfer, but dehydrogenation reactions also occur.

There has been considerable debate about the possible catalytic role of the pterin cofactor in molybdoenzymes. In XDH/XO and sulfite oxidase the cofactor is fully buried in the

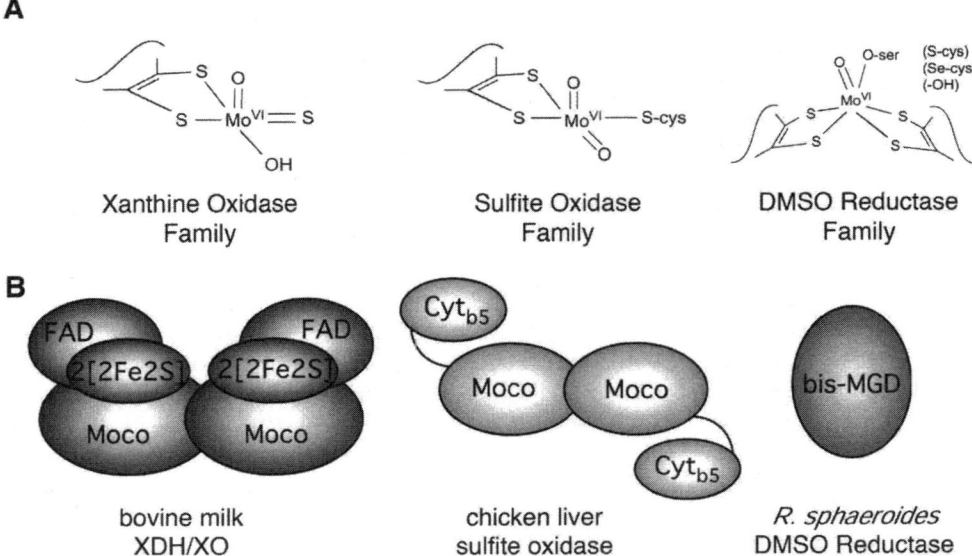

FIGURE 2 Active site structures of mononuclear molybdenum enzymes. (A) The three molybdenum-containing enzyme families are divided into the xanthine oxidase, sulfite oxidase, and DMSO-reductase families according to their active site structures. (B) Representative models of subunit and cofactor composition of molybdenum-containing enzymes, from which the X-ray structures have been solved: bovine milk XDH/XO (Enroth et al., 2000), chicken liver sulfite oxidase (Kisker et al., 1997b), and *R. sphaeroides* DMSO reductase (Schindelin et al., 1996).

enzyme and not exposed to the solvent-access channel through which substrate enters the active site, precluding its direct participation in catalysis (Enroth et al., 2000; Kisker et al., 1997a; Truglio et al., 2002). Instead, it is likely that the cofactor modulates the reactivity and/or the reduction potential of the center, in addition to having a role in electron transfer either into or out of the metal center. Such a role in electron transfer is clear for the molybdenum-containing hydroxylases (e.g., XDH), because the distal amino group of the pterin cofactor hydrogen bonds to a cysteine residue of the nearer iron-sulfur center (Romão et al., 1995).

The crystal structures of several molybdenum- and tungsten-containing enzymes have been reported, with the structure of the tungsten-containing aldehyde-ferredoxin oxidoreductase from *Pyrococcus furiosus* being the first (Chan et al., 1995). The structures of representative enzymes of each class of molybdoenzymes are shown in Fig. 2. These enzymes have a complex overall architecture, and generally contain multiple redox-active centers with the sole exception of monomeric DMSO reductase, binding solely bis-MGD (Schindelin et al., 1996; Schneider et al., 1996).

Mainly all molybdoenzymes identified in *E. coli* belong to the DMSO-reductase family of molybdenum enzymes binding bis-MGD. The sole exception so far is YedY, which belongs to the sulfite oxidase family and binds the MPT form of the cofactor (Loschi et al., 2004). Based on sequence homologies, the DMSO-reductase family has been divided into three types (McDevitt et al., 2002). Furthermore, enzymes of these three types can be distinguished by their first domain. Type I enzymes, exemplified by the soluble, periplasmic nitrate reductase (NapA) and the formate dehydrogenase (FdhG), contain a [4Fe-4S] iron-sulfur cluster at the N terminus of the first domain, whereas the type II enzymes, such as the membrane-bound DMSO reductase (DmsA), still conserve a cysteine-rich motif but are thought not to bind an [Fe-S] cluster. Finally, this conserved motif is absent in the first domain of type III

proteins, comprising monomeric enzymes such as TorA, which is located in the periplasm.

Localization of Molybdoenzymes in *E. coli*

Another striking feature of most *E. coli* molybdoenzymes is that they are either membrane associated or located in the periplasm. It was shown that in *E. coli* the proteins for the biosynthesis of Moco are located in the cytoplasm and that the insertion of Moco occurs before the translocation of molybdoenzymes either to the membrane or the periplasmic space. The recently discovered Tat (twin-arginine translocation) pathway is dedicated to the transport of folded proteins (see Chapter 2). Proteins are targeted to the Tat pathway by N-terminal signal peptides harboring consecutive, essentially invariant, arginine residues within an SR-RxFLK consensus motif (Berks, 1996). This signal sequence has been identified in *E. coli* in the DMSO reductase and a homologue (DmsA and DmsB, YnfE/F and YnfG), formate dehydrogenases (FdnG and FdnH, FdoG and FdoH), periplasmic nitrate reductase and its electron donor protein (NapA and NapG), TMAO reductase and a homologue (TorA and TorZ), and the sulfite oxidase homologue (YedY) (Berks et al., 2005).

Most molybdoenzymes in *E. coli* are part of respiratory systems, and sometimes more than one respiratory system is produced for a given substrate (Fig. 3). Reduction of nitrate can be carried out by at least three respiratory systems (Cole, 1996): at high concentrations of nitrate, only the membranous NarGHI system is synthesized (Stewart, 1988), whereas at very low concentrations the Nap system is produced (Brondijk et al., 2004). The operon encoding a third system, *narZYWV* (not shown), is expressed during the early stationary phase independent of the presence of nitrate (Chang et al., 1999). TMAO is reduced to the volatile compound trimethylamine (TMA) by at least two respiratory systems, the TorCAD and DmsABC systems (Bilous and Weiner, 1988; Mejean et al., 1994). Formate dehydrogenase catalyzes the oxidation of formate to CO_2 and

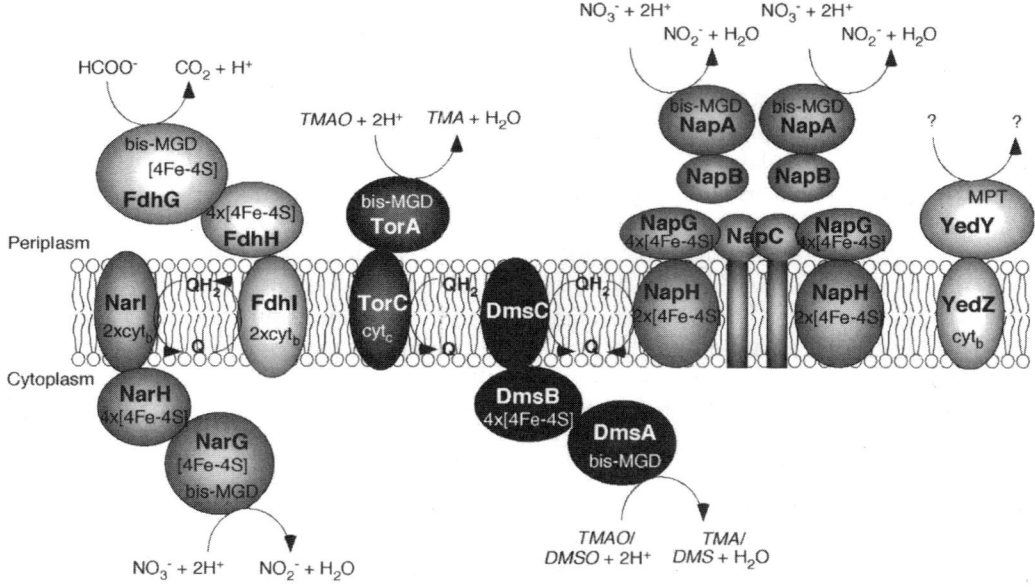

FIGURE 3 Proposed localization of *E. coli* molybdoenzymes that are part of respiratory systems. FdhGHI is a component of the nitrate respiratory pathway, in which formate oxidation is coupled to nitrate reduction (NarGHI) via lipid-soluble quinone/hydroquinone (Q/QH_2). TMAO is reduced to TMA by at least two respiratory systems, TorCAD and DmsABC (also reducing DMSO to dimethyl sulfide [DMS]). NapABCGH is produced under nitrate-limiting conditions and are believed to be involved in redox balancing. The physiological substrates for YedY, a member of the sulfite-oxidase family binding the MPT form of Moco, are not known to date.

H^+. *E. coli* contains two structurally related but differentially expressed respiratory formate dehydrogenases: formate dehydrogenase-O (not shown) and formate dehydrogenase-N (Jormakka et al., 2003). FdhGHI is a component of the nitrate respiratory pathway, where, under anaerobic conditions, formate oxidation is coupled to nitrate reduction (NarG) via lipid-soluble quinone (Jormakka et al., 2003). FdoGHI and NarZYV are the corresponding isoenzymes that are additionally present under aerobic conditions to ensure rapid adaptation during a shift from aerobiosis to anaerobiosis (Abaibou et al., 1995). A characteristic of formate dehydrogenases is an intrinsic selenocysteine residue that acts as ligand to molybdenum in the bis-MGD cofactor (Fig. 2).

CONCLUSIONS AND PERSPECTIVES

The biosynthesis of Moco is conserved in all organisms (with the exception of *Saccharomyces cerevisiae* that apparently possesses no molybdoenzymes) and genes for Moco biosynthesis are found in bacteria, archaea, fungi, plants, and animals. While the proteins for Moco biosynthesis are usually located in the cytoplasm, transport of molybdate requires membrane-bound and periplasmic components in bacteria. It is still an unsolved question whether molybdate, after transport into the cytoplasm, is transformed to an activated species for incorporation into formed MPT. Additionally, the direct sulfur donor for the thiocarboxylate group on MoaD, which subsequently generates the dithiolenes in MPT for Mo coordination, has not been identified yet. Protection of synthesized Moco by attachment to carrier molecules seems likely, but specific carrier proteins and the chaperone-assisted mechanism of insertion into the apoproteins remain to be elucidated. Thus, even more than ten years after the report of the first crystal structure of a

Moco-containing enzyme, which confirmed the proposed structure of Moco derived by chemical methods by the pioneering work of Rajagopalan and coworkers that started more than 30 years ago, a lot of open questions remain. While in bacteria a wide range of variations of Moco exist, the dinucleotide form of Moco has not been identified in eukaryotes so far. Bacteria usually contain a large variety of different molybdoenzymes catalyzing specific, usually nonessential redox reactions. In humans, possessing solely three molybdoenzymes, a defect in Moco biosynthesis is lethal because of the loss of sulfite oxidase activity.

Note also that proteins of the bacterial Moco biosynthesis pathway served as evolutionary ancestor for specific pathways in eukaryotes: recent structural studies of the bacterial MoaD-MoeB complex have demonstrated that the activation step during the ubiquitin-dependent degradation of proteins was derived from the simpler and universally distributed MoaD and MoeB proteins of Moco biosynthesis. The MoaD protein is believed to be the evolutionary ancestor of ubiquitin, and MoeB is believed to be the ancestor of E1-like proteins (Schindelin, 2005).

Thus the elucidation of Moco biosynthesis, the analysis of assembly of molybdoenzymes, insertion of Moco into the apoproteins, and translocation of the mature proteins to different compartments are fascinating topics of ongoing research.

ACKNOWLEDGMENTS

I thank Karsten Krepinsky for critical reading of the manuscript and helpful suggestions.

The research of S.L.'s laboratory was supported by grants from the Deutsche Forschungsgemeinschaft and the Fonds der Chemischen Industrie.

REFERENCES

Abaibou, H., J. Pommier, S. Benoit, G. Giordano, and M. A. Mandrand-Berthelot. 1995. Expression and characterization of the *Escherichia coli fdo* locus and a possible physiological role for aerobic formate dehydrogenase. *J. Bacteriol.* **177:** 7141–7149.

Aguilar, M., K. Kalakoutskii, J. Cardenas, and E. Fernandez. 1992. Direct transfer of molybdopterin cofactor to aponitrate reductase from a carrier protein in *Chlamydomonas reinhardtii*. *FEBS Lett.* **307:**162–163.

Anderson, L. A., E. McNairn, T. Lubke, R. N. Pau, and D. H. Boxer. 2000. ModE-dependent molybdate regulation of the molybdenum cofactor operon moa in *Escherichia coli*. *J. Bacteriol.* **182:**7035–7043.

Ataya, F. S., C. P. Witte, A. Galvan, M. I. Igeno, and E. Fernandez. 2003. Mcp1 encodes the molybdenum cofactor carrier protein in *Chlamydomonas reinhardtii* and participates in protection, binding, and storage functions of the cofactor. *J. Biol. Chem.* **278:**10885–10890.

Berks, B. C. 1996. A common export pathway for proteins binding complex redox cofactors? *Mol. Microbiol.* **22:**393–404.

Berks, B. C., T. Palmer, and F. Sargent. 2005. Protein targeting by the bacterial twin-arginine translocation (Tat) pathway. *Curr. Opin. Microbiol.* **8:**174–181.

Bilous, P. T., and J. H. Weiner. 1988. Molecular cloning and expression of the *Escherichia coli* dimethyl sulfoxide reductase operon. *J. Bacteriol.* **170:** 1511–1518.

Blasco, F., J. P. Dos Santos, A. Magalon, C. Frixon, B. Guigliarelli, C. L. Santini, and G. Giordano. 1998. NarJ is a specific chaperone required for molybdenum cofactor assembly in nitrate reductase A of *Escherichia coli*. *Mol. Microbiol.* **28:**435–447.

Börner, G., M. Karrasch, and R. K. Thauer. 1991. Molybdopterin adenine dinucleotide and molybdopterin hypoxanthine dinucleotide in formylmethanofuran dehydrogenase from *Methanobacterium thermoautotrophicum* (Marburg). *FEBS Lett.* **290:**31–34.

Boyington, J. C., V. N. Gladyshev, S. V. Khangulov, T. C. Stadtman, and P. D. Sun. 1997. Crystal structure of formate dehydrogenase H: catalysis involving Mo, molybdopterin, selenocysteine, and an Fe$_4$S$_4$ cluster. *Science* **275:**1305–1308.

Brokx, S. J., R. A. Rothery, G. Zhang, D. P. Ng, and J. H. Weiner. 2005. Characterization of an *Escherichia coli* sulfite oxidase homologue reveals the role of a conserved active site cysteine in assembly and function. *Biochemistry* **44:**10339–10348.

Brondijk, T. H., A. Nilavongse, N. Filenko, D. J. Richardson, and J. A. Cole. 2004. NapGH components of the periplasmic nitrate reductase of *Escherichia coli* K-12: location, topology and physiological roles in quinol oxidation and redox balancing. *Biochem. J.* **379:**47–55.

Chan, M. K., S. Mukund, A. Kletzin, M. W. W. Adams, and D. C. Rees. 1995 Structure of a hyperthermophilic tungstopterin enzyme, aldehyde ferredoxin oxidoreductase. *Science* **267:**1463–1469.

Chang, L., L. I. Wei, J. P. Audia, R. A. Morton, and H. E. Schellhorn. 1999. Expression of the *Escherichia coli* NRZ nitrate reductase is highly growth phase dependent and is controlled by RpoS, the alternative vegetative sigma factor. *Mol. Microbiol.* **34:**756–766.

Cole, J. 1996. Nitrate reduction to ammonia by enteric bacteria: redundancy, or a strategy for survival during oxygen starvation? *FEMS Microbiol. Lett.* **136:**1–11.

Davidson, A. L., and J. Chen. 2004. ATP-binding cassette transporters in bacteria. *Annu. Rev. Biochem.* **73:**241–68.

Eady, R. R. 1995. Vanadium nitrogenase of *Azotobacter.*, p. 363–405. *In* H. Sigel and A. Sigel (ed.), *Metal Ions in Biological Systems. Vanadium and Its Role in Life.* Marcel Dekker, New York, N.Y.

Eaves, D. J., T. Palmer, and D. H. Boxer. 1997. The product of the molybdenum cofactor gene *mobB* of *Escherichia coli* is a GTP-binding protein. *Eur. J. Biochem.* **246:**690–697.

Enroth, C., B. T. Eger, K. Okamoto, T. Nishino, and E. F. Pai. 2000. Crystal structures of bovine milk xanthine dehydrogenase and xanthine oxidase: structure-based mechanism of conversion. *Proc. Natl. Acad. Sci. USA* **97:**10723–10728.

Genest, O., M., Ilbert, V. Mejean, and C. Iobbi-Nivol. 2005. TorD, an essential chaperone for TorA molybdoenzyme maturation at high temperature. *J. Biol. Chem.* **280:**15644–15648.

Gon, S., M. T. Giudici-Orticoni, V. Mejean, and C. Iobbi-Nivol. 2001. Electron transfer and binding of the c-type cytochrome TorC to the trimethylamine N-oxide reductase in *Escherichia coli. J. Biol. Chem.* **276:**11545–11551.

Grunden, A. M., and K. T. Shanmugam. 1997. Molybdate transport and regulation in bacteria. *Arch. Microbiol.* **168:**345–354.

Hanzelmann, P., H. L. Hernandez, C. Menzel, R. Garcia-Serres, B. H. Huynh, M. K. Johnson, R. R. Mendel, and H. Schindelin. 2004. Characterization of MOCS1A, an oxygen-sensitive iron-sulfur protein involved in human molybdenum cofactor biosynthesis. *J. Biol. Chem.* **279:**34721–34732.

Hanzelmann, P., and H. Schindelin. 2004. Crystal structure of the S-adenosylmethionine-dependent enzyme MoaA and its implications for molybdenum cofactor deficiency in humans. *Proc. Natl. Acad. Sci. USA* **101:**12870–12875.

Higgins, C. F. 1992. ABC transporters: from microorganisms to man. *Annu. Rev. Cell. Biol.* **8:** 67–113.

Hille, R. 1996. The mononuclear molybdenum enzymes. *Chem. Rev.* **96:**2757–2816.

Hille, R. 2002. Molybdenum and tungsten in biology. *Trends Biochem. Sci.* **27:**360–367.

Hilton, J. C., and K. V. Rajagopalan. 1996. Identification of the molybdenum cofactor of dimethyl sulfoxide reductase from *Rhodobacter sphaeroides* f. sp. *denitrificans* as bis(molybdopterin guanine dinucleotide)molybdenum. *Arch. Biochem. Biophys.* **325:**139–143.

Hu, Y., S. Rech, R. P. Gunsalus, and D. C. Rees. 1997. Crystal structure of the molybdate binding protein ModA. *Nat. Struct. Biol.* **4:**703–707.

Ilbert, M., V. Mejean, M. T. Giudici-Orticoni, J. P. Samama, and C. Iobbi-Nivol. 2003. Involvement of a mate chaperone (TorD) in the maturation pathway of molybdoenzyme TorA. *J. Biol. Chem.* **278:**2878–2879.

Ilbert, M., V. Mejean, and C. Iobbi-Nivol. 2004. Functional and structural analysis of members of the TorD family, a large chaperone family dedicated to molybdoproteins. *Microbiology* **150:**935–943.

Jack, R. L., G. Buchanan, A. Dubini, K. Hatzixanthis, T. Palmer, and F. Sargent. 2004. Coordinating assembly and export of complex bacterial proteins. *EMBO J.* **23:**3962–3972.

Johnson, J. L., L. W. Indermaur, and K. V. Rajagopalan. 1991. Molybdenum cofactor biosynthesis in *Escherichia coli*. Requirement of the *chlB* gene product for the formation of molybdopterin guanine dinucleotide. *J. Biol. Chem.* **266:**12140–12145.

Johnson, J. L., K. V. Rajagopalan, and O. Meyer. 1990. Isolation and characterization of a second molybdopterin dinucleotide: molybdopterin cytosine dinucleotide. *Arch. Biochem. Biophys.* **283:**542–545.

Johnson, J. L., M. M. Wuebbens, and K. V. Rajagopalan. 1989. The structure of a molybdopterin precursor. Characterization of a stable, oxidized derivative. *J. Biol. Chem.* **264:**13440–13447.

Johnson, M. K., D. C. Rees, and M. W. Adams. 1996. Tungstoenzymes. *Chem. Rev.* **96:**2817–2840.

Jormakka, M., B. Byrne, and S. Iwata. 2003. Formate dehydrogenase—a versatile enzyme in changing environments. *Curr. Opin. Struct. Biol.* **13:**418–423.

Kisker, C., H. Schindelin, A. Pacheco, W. A. Wehbi, R. M. Garrett, K. V. Rajagopalan, J. H. Enemark, and D. C. Rees. 1997a. Molecular basis of sulfite oxidase deficiency from the structure of sulfite oxidase. *Cell* **91:**973–983.

Kisker, C., H. Schindelin, and D. C. Rees. 1997b. Molybdenum-cofactor-containing enzymes: structure and mechanism. *Ann. Rev. Biochem.* **66:**233–267.

Kletzin, A., and M. W. Adams. 1996. Tungsten in biological systems. *FEMS Microbiol. Rev.* **18:**5–63.

Kniemeyer, O., and J. Heider. 2001. Ethylbenzene dehydrogenase, a novel hydrocarbon-oxidizing molybdenum/iron-sulfur/heme enzyme. *J. Biol. Chem.* **276:**21381–21386.

Kuper, J., A. Llamas, H. J. Hecht, R. R. Mendel, and G. Schwarz. 2004. Structure of the molybdopterin-bound Cnx1G domain links molybdenum and copper metabolism. *Nature* **430:**803–806.

Lake, M. W., C. A. Temple, K. V. Rajagopalan, and H. Schindelin. 2000. The crystal structure of the *Escherichia coli* MobA protein provides insight into molybdopterin guanine dinucleotide biosynthesis. *J. Biol. Chem.* **275:**40211–40217.

Leimkühler, S., M. Kern, P. S. Solomon, A. G. McEwan, G. Schwarz, R. R. Mendel, and W. Klipp. 1998. Xanthine dehydrogenase from the phototrophic purple bacterium *Rhodobacter capsulatus* is more similar to its eukaryotic counterparts than to prokaryotic molybdenum enzymes. *Mol. Microbiol.* **27:**853–869.

Leimkühler, S., and W. Klipp. 1999. Role of XDHC in molybdenum cofactor insertion into xanthine dehydrogenase of *Rhodobacter capsulatus*. *J. Bacteriol.* **181:**2745–2751.

Leimkühler, S., and K. V. Rajagopalan. 2001a. An *Escherichia coli* NifS-like sulfurtransferase is required for the transfer of cysteine sulfur in the *in vitro* synthesis of molybdopterin from precursor Z. *J. Biol. Chem.* **276:**22024–22031.

Leimkühler, S., and K. V. Rajagopalan. 2001b. *In vitro* incorporation of nascent molybdenum cofactor into human sulfite oxidase. *J. Biol. Chem.* **276:**1837–1844.

Leimkühler, S., M. M. Wuebbens, and K. V. Rajagopalan. 2001. Characterization of *Escherichia coli* MoeB and its involvement in the activation of MPT synthase for the biosynthesis of the molybdenum cofactor. *J. Biol. Chem.* **276:**34695–34701.

Liu, M. T., M. M. Wuebbens, K. V. Rajagopalan, and H. Schindelin. 2000. Crystal structure of the gephyrin-related molybdenum cofactor biosynthesis protein MogA from *Escherichia coli*. *J. Biol. Chem.* **275:**1814–1822.

Llamas, A., R. R. Mendel, and G. Schwarz. 2004. Synthesis of adenylated molybdopterin: an essential step for molybdenum insertion. *J. Biol. Chem.* **279:**55241–55246.

Loschi, L., S. J. Brokx, T. L. Hills, G. Zhang, M. G. Bertero, A. L. Lovering, J. H. Weiner, and N. C. Strynadka. 2004. Structural and biochemical identification of a novel bacterial oxidoreductase. *J. Biol. Chem.* **279:**50391–50400.

Matthies, A., K. V. Rajagopalan, R. R. Mendel, and S. Leimkühler. 2004. Evidence for the physiological role of a rhodanese-like protein for the biosynthesis of the molybdenum cofactor in humans. *Proc. Natl. Acad. Sci. USA* **101:**5946–5951.

McDevitt, C. A., P. Hugenholtz, G. R. Hanson, and A. G. McEwan. 2002. Molecular analysis of dimethyl sulphide dehydrogenase from *Rhodovulum sulfidophilum*: its place in the dimethyl sulphoxide reductase family of microbial molybdopterin-containing enzymes. *Mol. Microbiol.* **44:**1575–1587.

McGrath, J. P., S. Jentsch, and A. Varshavsky. 1991. *UBA1:* an essential yeast gene encoding ubiquitin-activating enzyme. *EMBO J.* **10:**227–236.

McNicholas, P. M., M. M. Mazzotta, S. A. Rech, and R. P. Gunsalus. 1998. Functional dissection of the molybdate-responsive transcription regulator, ModE, from *Escherichia coli*. *J. Bacteriol.* **180:**4638–4643.

McNicholas, P. M., S. A. Rech, and R. P. Gunsalus. 1997. Characterization of the ModE DNA-binding sites in the control regions of *modABCD* and *moaABCDE* of *Escherichia coli*. *Mol. Microbiol.* **23:**515–524.

Mejean, V., C. Iobbi-Nivol, M. Lepelletier, G. Giordano, M. Chippaux, and M. C. Pascal. 1994. TMAO anaerobic respiration in *Escherichia coli*: involvement of the *tor* operon. *Mol. Microbiol.* **11:**1169–1179.

Nichols, J., and K. V. Rajagopalan. 2002. *Escherichia coli* MoeA and MogA. Function in metal incorporation step of molybdenum cofactor biosynthesis. *J. Biol. Chem.* **277:**24995–25000.

Nichols, J. D., and K. V. Rajagopalan. 2005. In vitro molybdenum ligation to molybdopterin using purified components. *J. Biol. Chem.* **280:**7817–7822.

Oresnik, I. J., C. L. Ladner, and R. J. Turner. 2001. Identification of a twin-arginine leader-binding protein. *Mol. Microbiol.* **40:**323–331.

Palmer, T., I. P. Goodfellow, R. E. Sockett, A. G. McEwan, and D. H. Boxer. 1998. Characterisation of the *mob* locus from *Rhodobacter sphaeroides* required for molybdenum cofactor biosynthesis. *Biochim. Biophys. Acta* **1395:**135-140.

Palmer, T., C.-L. Santini, C. Iobbi-Nivol, D. J. Eaves, D. H. Boxer, and G. Giordano. 1996. Involvement of the *narJ* and *mob* gene products in the biosynthesis of the molybdoenzyme nitrate reductase in *Escherichia coli*. *Mol. Microbiol.* **20:**875–884.

Palmer, T., A. Vasishta, P. W. Whitty, and D. H. Boxer. 1994. Isolation of protein FA, a product of the *mob* locus required for molybdenum cofactor biosynthesis in *Escherichia coli*. *Eur. J. Biochem.* **222:**687–692.

Pau, R. N., and D. M. Lawson. 2002. Transport, homeostasis, regulation, and binding of molybdate and tungstate to proteins. *Met. Ions Biol. Syst.* **39:**31–74.

Pilato, R. S., and E. I. Stiefel. 1999. Molybdenum and tungsten enzymes, p. 81–152. *In* J. Reedijk and E. Bouwman. (ed.), *Bioinorganic Catalysis*, 2nd ed. Dekker, New York, N.Y.

Pitterle, D. M., J. L. Johnson, and K. V. Rajagopalan. 1993. *In vitro* synthesis of molybdopterin from precursor Z using purified converting factor. Role of protein-bound sulfur in formation of the dithiolene. *J. Biol. Chem.* **268:**13506–13509.

Pitterle, D. M., and K. V. Rajagopalan. 1989. Two proteins encoded at the *chlA* locus constitute the converting factor of *Escherichia coli chlA1. J. Bacteriol.* **171:**3373–3378.

Pitterle, D. M., and K. V. Rajagopalan. 1993. The biosynthesis of molybdopterin in *Escherichia coli.* Purification and characterization of the converting factor. *J. Biol. Chem.* **268:**13499–13505.

Pope, M. T., E. R. Still, and J. P. Williams. 1980. A comparison between the chemistry and biochemistry of molybdenum and related elements, p. 3–40. *In* M. P. Coughlan (ed.), *Molybdenum and Molybdenum-Containing Enzymes.* Pergamon Press, Oxford, United Kingdom.

Rajagopalan, K. V. 1991. Novel aspects of the biochemistry of the molybdenum cofactor, p. 215–290. *In* A. Meister (ed.), *Advances in Enzymology and Related Areas of Molecular Biology,* vol. 64. John Wiley and Sons, New York, N.Y.

Rajagopalan, K. V. 1996. Biosynthesis of the molybdenum cofactor, p. 674–679. *In* F. C. Neidhardt, R. Curtiss III, J. L. Ingraham, E. C. C. Lin, K. B. Low, B. Magasanik, W. S. Reznikoff, M. Riley, M. Schaechter, and H. E. Umbarger (ed.), Escherichia coli *and* Salmonella: *Cellular and Molecular Biology,* 2nd ed., vol. 1. ASM Press, Washington, D.C.

Rajagopalan, K. V. 1997. Biosynthesis and processing of the molybdenum cofactors. *Biochem. Soc. Trans.* **25:**757–761.

Rajagopalan, K. V., and J. L. Johnson. 1992. The pterin molybdenum cofactors. *J. Biol. Chem.* **267:** 10199–10202.

Rajagopalan, K. V., J. L, Johnson, M. M. Wuebbens, D. M. Pitterle, J. C. Hilton, T. R. Zurick, and R. M. Garrett. 1993. Chemistry and biology of the molybdenum cofactors. *Adv. Exp. Med. Biol.* **338:**355–362.

Ramão, M. J., M. Archer, I. Moura, J. J. G. Moura, J. LeGall, R. Engh, M. Schneider, P. Hof, and R. Huber. 1995. Crystal structure of the xanthine oxidase-related aldehyde oxido-reductase from *D. gigas. Science* **270:**1170–1176.

Rudolph, M. J., M. M. Wuebbens, K. V. Rajagopalan, and H. Schindelin. 2001. Crystal structure of molybdopterin synthase and its evolutionary relationship to ubiquitin activation. *Nat. Struct. Biol.* **8:**42–46.

Rudolph, M. J., M. M. Wuebbens, O. Turque, K. V. Rajagopalan, and H. Schindelin. 2003. Structural studies of molybdopterin synthase provide insights into its catalytic mechanism. *J. Biol. Chem.* **278:**14514–14522.

Santamaria-Araujo, J. A., B. Fischer, T. Otte, M. Nimtz, R. R. Mendel, V. Wray, and G. Schwarz. 2004. The tetrahydropyranopterin structure of the sulfur- and metal-free molybdenum cofactor precursor. *J. Biol. Chem.* **279:**15994–15999.

Santini, C. L., B. Ize, A. Chanal, M. Muller, G. Giordano, and L. F. Wu. 1998. A novel sec-independent periplasmic protein translocation pathway in *Escherichia coli. EMBO J.* **17:**101–112.

Schindelin, H. 2005. Evolutionary origin of the activation step during ubiquitin-dependent protein degradation, p. 21–43. *In* R. J. Mayer, A. Ciechanover, and M. Rechsteiner (ed.), *Protein Degradation: Ubiquitin and the Chemistry of Life,* vol. I. Wiley-VCH, Weinheim, Germany.

Schindelin, H., C. Kisker, J. Hilton, K. V. Rajagopalan, and D. C. Rees. 1996. Crystal structure of DMSO reductase: redox-linked changes in molybdopterin coordination. *Science* **272:**1615–1621.

Schneider, F., J. Löwe, R. Huber, H. Schindelin, C. Kisker, and J. Knäblein. 1996. Crystal structure of dimethyl sulfoxide reductase from *Rhodobacter capsulatus* at 1.88 Å resolution. *J. Mol. Biol.* **263:**53–69.

Schrag, J. D., W. Huang, J. Sivaraman, C. Smith, J. Plamondon, R. Larocque, A. Matte, and M. Cygler. 2001. The crystal structure of *Escherichia coli* MoeA, a protein from the molybdopterin synthesis pathway. *J. Mol. Biol.* **310:**419–431.

Schuttelkopf, A. W., D. H. Boxer, and W. N. Hunter. 2003. Crystal structure of activated ModE reveals conformational changes involving both oxyanion and DNA-binding domains. *J. Mol. Biol.* **326:**761–767.

Self, W. T., A. M. Grunden, A. Hasona, and K. T. Shanmugam. 1999. Transcriptional regulation of molybdoenzyme synthesis in *Escherichia coli* in response to molybdenum: ModE-molybdate, a repressor of the *modABCD* (molybdate transport) operon is a secondary transcriptional activator for the *hyc* and *nar* operons. *Microbiology* **145**(Pt 1):41–55.

Shanmugam, K. T., V. Stewart, R. P. Gunsalus, D. H. Boxer, J. A. Cole, M. Chippaux, J. A. DeMoss, G. Giordano, E. C. C. Lin, and K. V. Rajagopalan. 1992. Proposed nomenclature for the genes involved in molybdenum metabolism in *Escherichia coli* and *Salmonella typhimurium. Mol. Microbiol.* **6:**3452–3454.

Sofia, H. J., G. Chen, B. G. Hetzler, J. F. Reyes-Spindola, and N. E. Miller. 2001. Radical SAM, a novel protein superfamily linking unresolved steps in familiar biosynthetic pathways with radical mechanisms: functional characterization using new analysis and information visualization methods. *Nucleic Acids Res.* **29:**1097–1106.

Stevenson, C. E., F. Sargent, G. Buchanan, T. Palmer, and D. M. Lawson. 2000. Crystal structure of the molybdenum cofactor biosynthesis protein MobA from *Escherichia coli* at near-atomic resolution. *Struct. Fold Des.* **8:**1115–1125.

Stewart, V. 1988. Nitrate respiration in relation to facultative metabolism in enterobacteria. *Microbiol. Rev.* **52:**190–232.

Stiefel, E. I. 1993. Molybdenum enzymes, cofactors and chemistry, p. 1–18. *In* E. I. Stiefel, D. Coucouvanis, and W. E. Newton (ed.), *Molybdenum Enzymes, Cofactors and Model Systems*. American Chemical Society, Washington, D.C.

Stiefel, E. I. 2002. The biogeochemistry of molybdenum and tungsten. *Met. Ions Biol. Syst.* **39:**1–30.

Temple, C. A., and K. V. Rajagopalan. 2000. Mechanism of assembly of the bis(molybdopterin guanine dinucleotide)molybdenum cofactor in *Rhodobacter sphaeroides* dimethyl sulfoxide reductase. *J. Biol. Chem.* **275:**40202–40210.

Tranier, S., C. Iobbi-Nivol, C. Birck, M. Ilbert, I. Mortier-Barriere, V. Mejean, and J. P. Samama. 2003. A novel protein fold and extreme domain swapping in the dimeric TorD chaperone from *Shewanella massilia*. *Structure (Camb)* **11:**165–174.

Truglio, J. J., K. Theis, S. Leimkuhler, R. Rappa, K. V. Rajagopalan, and C. Kisker. 2002. Crystal structures of the active and alloxanthine-inhibited forms of xanthine dehydrogenase from *Rhodobacter capsulatus*. *Structure (Camb)* **10:**115–125.

Vergnes, A., K. Gouffi-Belhabich, F. Blasco, G. Giordano, and A. Magalon. 2004. Involvement of the molybdenum cofactor biosynthetic machinery in the maturation of the *Escherichia coli* nitrate reductase A. *J. Biol. Chem.* **279:**41398–41403.

Willsky, G. R. 1990. Vanadium in the bioshere, p. 1–24. *In* N. D. Chasteen (ed.), *Vanadium in Biological Systems: Physiology and Biology*. Kluwer Academic Publishers, Dordrecht, The Netherlands.

Witte, C. P., M. I. Igeno, G. Schwarz, and E. Fernandez. 1998. The *Chlamydomonas reinhardtii* MoCo carrier protein is multimeric and stabilizes molybdopterin cofactor in a molybdate charged form. *FEBS Lett.* **431:**205–209.

Wuebbens, M. M., and K. V. Rajagopalan. 1993. Structural characterization of a molybdopterin precursor. *J. Biol. Chem.* **268:**13493–13498.

Wuebbens, M. M., and K. V. Rajagopalan. 1995. Investigation of the early steps of molybdopterin biosynthesis in *Escherichia coli* through the use of *in vivo* labeling studies. *J. Biol. Chem.* **270:**1082–1087.

Wuebbens, M. M., and K. V. Rajagopalan. 2003. Mechanistic and mutational studies of *Escherichia coli* molybdopterin synthase clarify the final step of molybdopterin biosynthesis. *J. Biol. Chem.* **278:**14523–14532.

Wuebbens, M. M., M. T. Liu, K. Rajagopalan, and H. Schindelin. 2000. Insights into molybdenum cofactor deficiency provided by the crystal structure of the molybdenum cofactor biosynthesis protein MoaC. *Struct. Fold Des.* **8:**709–718.

Xiang, S., J. Nichols, K. V. Rajagopalan, and H. Schindelin. 2001. The crystal structure of *Escherichia coli* MoeA and its relationship to the multifunctional protein gephyrin. *Structure* **9:**299–310.

TRANSFER OF ENERGY AND INFORMATION ACROSS THE PERIPLASM IN IRON TRANSPORT AND REGULATION

Volkmar Braun and Susanne Mahren

16

Two periplasmic functions are discussed: (i) the transfer of energy from the cytoplasmic membrane across the periplasm to the outer membrane, and (ii) the transfer of information from the outer membrane across the periplasm through the cytoplasmic membrane into the cytoplasm.

When outer membrane proteins become energized, their structure changes so that they can function as transporters for substrates that are present at very low concentrations or that are too large to diffuse through the pores formed by the porins. These proteins must also become energized to serve as receptors for certain phages and colicins, which results in phage infection and colicin killing. Although the sensitivity of *Escherichia coli* to phages and colicins provided the first hints of the requirement for the TonB protein in the cytoplasmic membrane and its mode of action, receptor functions are not discussed in this chapter. Rather, energization is discussed in the context of the transporter activities.

A more recently discovered process is the transfer of information from the cell surface into the cytoplasm, involving transcription reg-

ulation of certain genes by extracytoplasmic function (ECF) sigma factors. The first and the most detailed studies have been on the ferric citrate transport system.

A compilation of TonB-dependent transporter activities and colicin receptor activities can be found in *Iron Transport in Bacteria* (edited by J. H. Crosa, A. R. Mey, and S. M. Payne, 2004, ASM Press) and the special issue of the journal *Biochimie*, "Bacterial Derived Antimicrobial Toxins" [84(nos. 5–6): 2002], respectively. Recent reviews are cited in the text.

ENERGY-COUPLED IRON TRANSPORT ACROSS THE OUTER MEMBRANE OF GRAM-NEGATIVE BACTERIA

Conformational Changes in Outer Membrane Iron Transporters

To enter gram-negative bacteria, substrates must cross three compartments: the outer membrane, the periplasm, and the cytoplasmic membrane. Most substrates diffuse through outer membrane porins into the periplasm and then are actively transported, i.e., with energy coupling, across the cytoplasmic membrane. In contrast, iron in the form of Fe^{3+} or Fe^{3+} incorporated into siderophores, transferrin, lactoferrin, heme, or heme proteins is actively transported across the outer membrane by proteins

Volkmar Braun and Susanne Mahren, Mikrobiologie/Membranphysiologie, Auf der Morgenstelle 28, D-72076 Tübingen, Germany.

that recognize their substrates with high specificity. Fe^{3+}, Fe^{3+} siderophores, and heme are transported, whereas heme must be released from heme proteins and Fe^{3+} from transferrin and lactoferrin before transport. In the latter cases, the outer membrane proteins, designated receptors to indicate their binding capacity or transporters to point to their transport function, have three functions: mobilization of heme and iron from the proteins, binding, and subsequent transport. In some bacteria an additional protein, called a hemophore, releases heme from hemoglobin and delivers it to the transporters (Wandersman and Delepelaire, 2004). The K_d values for Fe^{3+} siderophores binding to the transporters are in the nanomolar range (Cao et al., 2003; Scott et al., 2001). The K_d value for heme is estimated to be in the micromolar range, but that of the heme-hemophore complex bound to the outer membrane transporters is a few nanomolar (Letoffe et al., 1999).

To date, the crystal structures of six energy-coupled outer membrane transporters have been determined: FhuA (Ferguson et al., 1998; Locher et al., 1998), FepA (Buchanan et al., 1999), and FecA (Ferguson et al., 2002; Yue et al., 2003) of *E. coli* K-12 transport ferrichrome, ferric enterobactin, and ferric citrate, respectively; FpvA (Cobessi et al., 2005a) and FptA (Cobessi et al., 2005b) of *Pseudomonas aeruginosa* transport ferric pyoverdine and ferric pyochelin, respectively; and BtuB transports vitamin B_{12} (Chimento et al., 2003, 2005). They all form a β-barrel composed of 22 transmembrane β-strands and a globular domain composed of a mixed four-stranded β-sheet, called the cork, plug, or hatch. The cork inserts from the periplasm into the β-barrel and completely occludes the pore formed by the β-barrel.

FhuA was cocrystallized with a lipopolysaccharide molecule found in the outer leaflet of the outer membrane (Ferguson et al., 1998), from which the exact position of FhuA in the outer membrane can be derived. The location of FhuA is representative for all outer membrane transporters, which share the typical structure of transmembrane β-strands and gir-

dles of aromatic amino acids along the outer and inner borders of the lipid bilayer. In the crystal structures, the substrates are located in a pocket of the transporters well above the outer membrane bilayer. The pockets are lined with amino acids that reflect the polar or apolar nature of the substrates. For example, the highly charged diferric dicitrate is bound to FecA by one aromatic residue and ten polar residues, four of which are arginine residues located in surface loops, the β-barrel, and the apices of the cork. In contrast, uncharged ferrichrome is bound by only one arginine residue, but by seven aromatic amino acids, one glycine residue, and one glutamine residue.

Upon binding of the substrates, the transporters undergo major changes in the loops exposed to the cell surface and in the N-terminal region exposed to the periplasm, and small movements of the cork (1 to 2 Å). FhuA and FecA are the only iron transporters whose crystal structures were determined in the unloaded form and in the form loaded with the ferric siderophores. Whether FepA in the crystal structure was loaded or unloaded is not clear. The FpvA crystal structure contains iron-free pyoverdine, and the FptA crystal structure contains ferric pyochelin. Binding of ferrichrome to FhuA and diferric dicitrate to FecA unwinds a short helix (residues 24 to 29 in FhuA). This switch helix is fixed by hydrophobic interactions to a periplasmic pocket (Color Plate 23). In FhuA, the unwinding is accompanied by a large movement of 17 Å out of the pocket, resulting in an extended conformation. Substrate-loaded FecA seems to move similarly since this region does not provide a defined electron density map, which suggests an extended flexible conformation. A disordered conformation has also been observed in FpvA loaded with ferric pyochelin (Cobessi et al., 2005a) and was inferred from EPR data of substrate-loaded BtuB. These results suggest that this form is preferentially assumed for transport of the substrates.

The crystal structure of FecA was determined in the unloaded form and loaded with diferric dicitrate and with dicitrate (Ferguson et al., 1998; Yue et al., 2003). Dicitrate binds similarly but not identically with ferric dicitrate, it does not cause strong movements of the N-terminal region, and it does not induce the strong movements of loop 7 (11 Å) and loop 8 (15 Å) observed upon binding of diferric dicitrate (Yue et al., 2003). The loop movement closes the entrance of diferric dicitrate to its binding site and prevents its escape to the medium, which guarantees vectorial transport into the periplasm when the cork moves relative to the β-barrel and opens a pore. Despite a large surplus (10^3-fold) of citrate over iron in the transport assays, which is required to obtain the diferric dicitrate form, competition of citrate and ferric citrate for binding to FecA does not substantially reduce the transport rate. It is unclear whether binding of dicitrate to FecA and binding of iron-free pyoverdine to its transporters have physiological meaning.

Energy Transfer from the Cytoplasmic Membrane across the Periplasm into the Outer Membrane

Despite substantial structural changes in the outer membrane transporters, substrate binding does not open the pore in the β-barrel. Release of the substrates from their strong binding and opening of the pores requires energy. Since there is no energy source in the outer membrane, the energy must come from the cytoplasmic membrane. Early studies have demonstrated that TonB is involved in iron transport (Wang and Newton, 1971) and that the proton motive force serves as energy source for TonB-dependent processes (Bradbeer, 1993; Hancock and Braun, 1976). Binding of substrates to the outer membrane transporters occurs independently of TonB, but subsequent release requires TonB. TonB is only active in cooperation with two other proteins, ExbB and ExbD. Earlier statements that ExbB and ExbD are nonessential auxiliary proteins came from the observation that *exbB exbD* mutants showed residual TonB activities. These activities were, however, provided by the homologous TolQ TolR proteins, which can substitute for ExbB and ExbD to a certain degree (Braun

and Herrmann, 1993). TonB is a transmembrane protein with the N terminus in the cytoplasm, followed by a transmembrane segment and a large portion in the periplasm. ExbD is arranged similarly as TonB, whereas ExbB crosses the cytoplasmic membrane three times and has the N terminus in the periplasm (Color Plate 23).

TonB interacts with the outer membrane transporters as shown indirectly by mutations in a short six-residue segment at the N terminus of outer membrane transporters, called the TonB box, which are suppressed by two different mutations in residue 160 of TonB. The two regions thereby identified by genetic suppression analysis were then shown to interact physically by spontaneous in vivo cross-linking between cysteine residues introduced into region 160 of TonB and the TonB box of BtuB (Cadieux and Kadner, 1999) and FecA (Ogierman and Braun, 2003).

The sequence of the TonB box is not well conserved. For example, the TonB box of FecA (DALTV) can be functionally replaced with the TonB box of FhuA (DTITV) and FepA (DTIVV) and retains its ability to induce transcription of the transport genes and to transport ferric citrate (Ogierman and Braun, 2003). However, FecA carrying the FepA TonB box transports iron at a lower rate than wild-type FecA. Replacement of TonB box residues of FecA with cysteine residues is tolerated, except for one replacement (D80C), which supports the theory that mainly the secondary structure of the TonB box rather than the sequence is important for function (Cadieux and Kadner, 1999).

Region 160 of TonB cross-linked via disulfide bridges to the TonB box of FecA shows no enhancement in binding in the presence of the substrate, ferric citrate. This is unexpected because conversion of the FecA N terminus from an ordered to a disordered conformation upon binding of diferric dicitrate, as shown by the FecA crystal structures, is considered to be important for binding of FecA to TonB. Various experimental approaches have revealed enhanced interaction of TonB with FepA, FhuA, and BtuB in the presence of the respective substrate. However, substantial interaction is always observed in the absence of substrate. Nevertheless, ligand-loaded transporters preferentially bind to TonB (summarized and discussed in Postle and Kadner, 2003).

Constructing a model of the interaction of TonB with outer membrane transporters and the mechanism of energy transfer is hampered by data that cannot be reconciled easily. For example, the TonB:ExbB:ExbD ratio in the cell was determined to be 1:7:2 (Postle and Kadner, 2003), but it is not evident whether this is the stoichiometry of the complex. Whether TonB forms monomers, dimers, or multimers is not clear. Dimers of full-length TonB were found in vivo by using a genetic approach whereby dimer formation of a hybrid protein between TonB and the DNA-binding region of ToxR is required for activation of cholera toxin gene transcription (Sauter et al., 2003). The C-terminal periplasmic fragment (residues 164 to 239) dimerizes, whereas the entire periplasmic fragment 33 to 239 does not. Crystal structures of TonB periplasmic fragments form monomers or dimers depending on the length of the fragments (Koedding et al., 2005); TonB125–239 and TonB145–239 are monomeric, whereas TonB155–239 and TonB164–239 are dimeric. NMR analysis of TonB103–239 reveals a monomeric fragment (Peacock et al., 2004, and references therein). The in vitro data on the periplasmic fragments agree with the in vivo data, which suggests that the dimer of full-length TonB observed in vivo reflects the real structure. However, this does not necessarily imply that the dimeric crystal structure of the shorter fragments reflects the true dimeric structure. In fact, the disulfide cross-linkage of cysteine residues that replace aromatic amino acids in a functionally important aromatic cluster does not agree with predicted cross-linkages derived from the dimeric crystal structures (Ghosh and Postle, 2005). Longer TonB fragments interact with FhuA more efficiently than shorter fragments in vitro and inhibit FhuA activity in vivo, whereas shorter fragments do not. Surface plasmone resonance revealed first the formation of a 1:1 complex of TonB and FhuA and then the formation of a 2:1 complex (Khursigara et al., 2005a). Deletion of the proline-rich region in TonB results in the formation of only the 1:1 complex, yet this fragment is still active (Khursigara et al., 2005b).

It is currently impossible to decide whether the various results reflect different TonB states during the reaction cycle—energization, deenergization, interaction with ExbB and ExbD within the cytoplasmic membrane and the periplasm, and binding to and release from transporters—or whether some results are experimental artifacts caused in vitro by the use of truncated forms that lack the N-terminal membrane anchor, which is essential for TonB activity. Current thinking proposes that TonB reacts to the proton motive force of the cytoplasmic membrane with the help of the ExbB and ExbD proteins. TonB assumes an energized conformation, stores potential energy, interacts with outer membrane transporters and receptors, and allosterically changes their conformation. Evidence for structural changes in TonB in response to the proton motive force or depending on the presence of ExbB/ExbD was obtained by various experimental approaches (Postle and Kadner, 2003). Most recently, spontaneous in vivo disulfide cross-linking between residues in a cluster of aromatic residues in the C-proximal end of TonB depended on the presence of ExbB/ExbD and a functional TonB (Ghosh and Postle, 2005).

Experimental data support a model in which TonB shuttles between the outer membrane and the cytoplasmic membrane. After sucrose density gradient fractionation, about the same amounts of TonB are found in the outer membrane and in the cytoplasmic membrane. In an experiment in which a cysteine residue introduced into the cytoplasmic N terminus of TonB was labeled with a fluorescent dye that does not diffuse through the cytoplasmic membrane (Larsen et al., 2003), labeling was higher in an *exbB tolQ* double mutant. This agrees with the finding that in the absence of ExbB/ExbD and TolQ/TolR, TonB is only in the outer membrane. Either ExbB/ExbD, and in their absence TolQ/TolR, fix TonB to the cytoplasmic membrane or energization is required for insertion into the cytoplasmic membrane. Since dissipation of the proton motive force does not change the distribution of TonB between the two membranes, it is more likely that TonB is attached to ExbB/ExbD (Letain and Postle, 1997). However, if shuttling is involved in energy transduction, one would expect a strong proton motive force-dependent change in TonB distribution between the two membranes.

Less important, but typical for the lack of understanding, is the finding that TonB can be cross-linked to the outer membrane proteins OmpA and Lpp, which suggests specific sites of interactions with TonB. However, *ompA* or *lpp* deletion mutants do not influence TonB activity. Taking these and other nondiscussed data together, it seems premature to propose a detailed structural model of the mechanism of TonB function, even though such working models are important for designing proper experiments.

Regardless of the mechanism of TonB function, interaction with energized TonB must induce conformational changes in the outer membrane proteins that alter the stereochemistry of the amino acid residues at the substrate binding sites such that the substrates are released. At the same time, the cork moves relative to the β-barrel, which energetically is made possible by approximately 75 water molecules located in the large interface between the cork and the β-barrel. Most of the water molecules are hydrogen bonded to the cork and the β-barrel, and approximately one-third of the molecules bridge the cork to the β-barrel (Chimento et al., 2005; Faraldo-Gomez et al., 2003).

To date, there is no experimental evidence that the cork moves out of the β-barrel. Fixation of the FhuA cork to the β-barrel by a disulfide bridge between residues located close to the periplasmic entrance of the β-barrel (residues 27 and 533) abolishes ferrichrome transport (Endriß et al., 2003). However, fixation by disulfide bridges between residues lo-

cated close to a predicted channel-forming region (residues 109 and 356, or residues 112 and 383) does not reduce the transport rate (Eisenhauer et al., 2005). The disulfide bridge between residues 27 and 533 tethers the cork to the barrel so that the strong ferrichrome-induced movement of the N terminus, including residue 22, cannot take place. This probably inactivates FhuA transport. On the other hand, FhuA transport activity is largely retained when residues 23 to 30 or 24 to 31 are deleted (Endriß et al., 2003). This is all the more surprising because this region involves the switch helix (residues 24 to 29), which one would expect to be important for conferring structural changes to FhuA when energized TonB binds to the FhuA TonB box (residues 7 to 11). FhuA tolerates partial truncation of this potentially energy-transducing region.

Eighty-four of 110 completely sequenced bacterial genomes encode TonB-dependent transporters (Koebnik, 2005). These transporters are particularly abundant in *Bacteroides* species (90 to 115 proteins), *Caulobacter crescentus* (65 proteins), xanthomonads (34 to 66 proteins), and pseudomonads (25 to 35 proteins). They transport more than just iron and vitamin B_{12}, the substrates hitherto identified. *C. crescentus* actively transports maltodextrins across the outer membrane (Neugebauer et al., 2005), and *Bacteroides fragilis* and *Bacteroides thetaiotaomicron* are predicted to actively transport sugars across the outer membrane (Xu et al., 2003).

SIGNALING ACROSS THE PERIPLASM

Signaling from the cell surface to the cytoplasm was first discovered in the ferric citrate transport system (summarized in Braun, 1997; Braun and Mahren, 2005; Braun et al., 2003). Ferric citrate induces transcription of the *fecABCDE* transport genes without entering the cytoplasm. Binding to the FecA outer membrane protein is sufficient for transcription initiation. This stands in contrast to the two-component regulatory systems in which the substrates must enter at least the periplasm to induce transcription. No phosphorylation/

dephosphorylation, as observed in two-component systems, is involved in transcription regulation of the *fec* transport genes.

A six-component system regulates *fec* transport gene transcription: FecA in the outer membrane; TonB, ExbB, and ExbD in the periplasm and the cytoplasmic membrane; FecR spanning the cytoplasmic membrane; and FecI in the cytoplasm (Color Plate 23). Diferric dicitrate, not dicitrate, serves as inducer. This agrees with the strong movement of loops 7 and 8 upon binding of diferric dicitrate (Ferguson et al., 2002) but not upon binding of dicitrate (Yue et al., 2003). Loops 7 and 8 are essential for induction and transport; their deletion abolishes both activities (Sauter et al., 2003).

Induction and transport by FecA are very similar processes as shown by the difficulty in obtaining mutants that have lost only one function. Screening for induction-positive but transport-negative mutants resulted in only one mutant, designated *fecA4* (L157P, N529D, R611C), which induces TonB-independent *fec* transport gene transcription, but does not transport ferric citrate. Two other mutants, I593F and W122C, constitutively transcribe *fec* transport genes (Härle et al., 1995; U. Stroeher, unpublished results). The properties of the mutants show that transport is not required for transcription initiation and that signaling and transport are distinct FecA functions.

Signal Transduction in the Periplasm

The information that diferric dicitrate is bound to FecA must be conveyed to the cytoplasm to induce transcription of the *fec* transport genes. A signal is transmitted across the outer membrane by FecA, which interacts with FecR in the periplasm (Color Plates 23 and 24). FecR transfers the signal into the cytoplasm.

In contrast to other TonB-dependent outer membrane transporters, FecA contains N-terminal to the TonB box a 79-residue peptide, designated the signaling domain, which is located in the periplasm. The signaling domain is essential for signaling but dispensable for transport (Kim et al., 1997). It interacts with the C-proximal region of FecR, which is essential for signaling. Interaction was shown by retention of FecA by an N-terminally His-tagged FecR bound to Ni-agarose and binding of the FecA N terminus to the FecR C terminus in a bacterial two-hybrid system. $FecA_{1-79}$, which consists of the first 79 amino acids of the mature protein, interacts with $FecR_{101-317}$, which consists of the entire periplasmic portion of FecR (Enz et al., 2000). $FecR_{237-317}$ is sufficient for binding to $FecA_{1-79}$. Mutant FecR (D138E, V197A) induces *fec* gene transcription in the absence of ferric citrate and binds more strongly to FecA. Preferentially C-proximal mutations in FecR abolish binding to FecA and induction of *fec* transport gene transcription. The mutations FecR(L245E), FecR (L269G), and FecR(F284L) are suppressed by mutations in the signaling domain of FecA, FecA(G39R), and FecA(D43E) (Enz et al., 2003; U. W. Stroeher and V. Braun, unpublished data). The degree of FecA-FecR binding correlates with the degree of induction. Selection of FecA mutations that no longer bind to FecR in the two-hybrid system yields mutations located on one side of the $FecA_{1-79}$ crystal structure (Color Plate 25). These mutations are in the same region as the suppressor mutations: this occurrence strongly supports the conclusion that the region in which the mutations accumulate forms the FecA interface to FecR. Taken together, these data clearly demonstrate interaction of the periplasmic FecA and FecR domains for induction of *fec* transport gene transcription.

REGULATION OF *fecIR* AND *fecABCDE* TRANSCRIPTION IN THE CYTOPLASM

The signal transmitted by FecR across the cytoplasmic membrane activates the FecI sigma factor (Color Plate 26), which belongs to the type 4 or ECF sigma factors of the sigma 70 family. Interaction of the cytoplasmic domain of FecR ($FecR_{1-85}$) with FecA initiates transcription of the *fecABCDE* transport genes. $FecR_{1-85}$ binds to FecI and causes constitutive transcription of *fec* transport genes in the absence of diferric dicitrate, FecA, TonB, ExbB, and ExbD. Single amino acid replacements render FecR$_{1-85}$ inactive and unable to bind to FecI, e.g., three tryptophan residues highly conserved in antisigma factors of the FecR type replaced by arginine residues by random mutagenesis. FecR$_{1-85}$ may simulate the conformation of FecR after reaction to the signal received from FecA occupied with diferric dicitrate. In this conformation, FecR binds to FecI such that FecI assumes the active conformation that recruits the RNA polymerase and binds to the promoter upstream of the *fecA* gene and initiates transcription of the *fecABCDE* genes. Alternatively, binding of FecR$_{1-85}$ to FecI might prevent FecI from inactivation by precipitation or degradation by proteases. Mutants in domain 4 of FecI are impaired in binding to FecR$_{1-85}$. FecI binds to the β′-subunit of the RNA polymerase, and mutants in domain 2.2 of FecI show a reduced β′ binding (Mahren and Braun, 2003).

The FecI-RNA polymerase complex binds to the promoter upstream of *fecA,* as demonstrated by DNA mobility band-shift experiments and competition experiments between wild-type and mutant promoter regions (Angerer et al., 1995; Ochs et al., 1996). Single-nucleotide mutations that reduce binding of FecI-RNA polymerase and *fec* transcription initiation are located around +11 of the *fecA* transcript (Angerer and Braun, 1998; Angerer et al., 1995). Northern hybridization experiments reveal a main transcript derived from *fecA* and much lower yields of transcripts encompassing the downstream *fecBCDE* genes (Enz et al., 1995). The *fecABCDE* genes form an operon with FecA as the major product, which is logical because basal levels of FecA must be synthesized even under noninducing conditions since FecA is required to start induction.

In addition to the positive regulation of *fec* transport gene transcription by diferric dicitrate, negative regulation by the Fur protein, which acts as repressor when loaded with Fe^{2+}, also occurs. Footprinting scans reveal binding of Fe^{2+}–Fur between bp +1 and −38 of the *fecA* coding strand (Angerer and Braun, 1998). In addition, Fe^{2+}–Fur binds to the promoter of *fecI* between bp +6 and −23 (Angerer and Braun, 1998). Fe^{2+}–Fur repression is the only regulation of *fecIR* transcription that does not respond to diferric dicitrate in the medium. Cells first recognize growth-limiting iron concentrations in the medium. They synthesize FecIR, but transcription of the ferric citrate transport genes is not initiated. In the presence of ferric citrate in the medium and low iron supply, transcription is initiated by the basal level of FecA in the outer membrane and then by signaling through FecIR synthesized under the de-repressed conditions. If enough iron is in the cells, Fe^{2+}–Fur represses *fecIR* and *fecABCDE* transcription.

OCCURRENCE OF THE FERRIC CITRATE TRANSPORT SYSTEM

Genes homologous to the *E. coli* K-12 *fec* genes have been found in sequenced bacterial genomes, but many lack some genes and therefore do not constitute a complete *fec* transport and regulatory system. Only a few of the bacterial species grow on ferric citrate as sole iron source, transport ferric citrate, or are regulated by ferric citrate.

Apart from *E. coli* K-12, ferric citrate transport systems were demonstrated in *E. coli* B and *Shigella flexneri* (Luck et al., 2001) and *Klebsiella pneumoniae* and *Photorhabdus luminescens* (Mahren et al., 2005). The genome of *Erwinia caratovora* susp. *atroseptica* SCR11043 encodes a complete *fec* system. The genomes of *E. coli* 0157:H7 strain EDL933, *E. coli* CFT073, and *Salmonella enterica* do not contain the *fec* operon. Of three examined *K. pneumoniae* strains, only one contains an inducible FecA protein. In an *Enterobacter aerogenes* strain, no *fec* system is found; however, functional *fecIR* regulatory genes have been detected (Mahren et al., 2005).

Apparently, those strains that contain a *fec* system acquired it by horizontal gene transfer. Only single strains of a genus contain a complete *fec* system, and incomplete *fec* operons and inactive mutated *fec* genes occur. The *fec* operons are flanked by IS elements involved in gene transfer. For example, the *fec* locus of *E. coli* K-12 is flanked upstream by an IS*1* element and downstream by an IS*911* element that is disrupted by an IS*30* element and a truncated IS*2* insertion. The *fec* sequences are 99.9 to 100% identical among strains that share 78 to 87% identity among other genes.

OCCURRENCE OF Fec-TYPE TRANSCRIPTION REGULATION

Although the *fec* system occurs rarely, its presence in *E. coli* K-12 fortunately allowed a novel regulatory mechanism to be unraveled. To date, ECF sigma factors are the most frequently occurring sigma factors (Helmann, 2002). The Fec-type of regulation is a subtype of the ECF regulation. Dependence of sigma factor activity on a cytoplasmic membrane protein of the FecR type occurs in the heme uptake system of *Bordetella avium*, in which RhuR (equivalent to FecR) is required for RhuI (equivalent to FecI) activity (Kirby et al., 2004). Heme binding to the outer membrane protein BhuR initiates transcription, and an N-terminal fragment of RhuR leads to heme-independent *bhuR* transcription. In *Ralstonia solanacearum*, the outer membrane protein PrhA, the cytoplasmic membrane protein PrhR, and the sigma factor PrhI are at the beginning of a regulatory cascade that controls the plant hypersensitivity response (Brito et al., 2002). Mutations in any of these genes delay disease formation in plants, and a mutant carrying an N-terminal PrhR fragment is fully pathogenic. These properties suggest dependence of PrhI activity on PrhR.

In contrast, a sigma/antisigma mechanism is indicated in the regulation of heme uptake by *Serratia marcescens*, in which HasS (equivalent to FecR) reduces HasI-mediated transcription of the *hasR* gene, which encodes the outer membrane transport protein (Rossi et al., 2003). In *Bordetella bronchiseptica*, two genes, *bupI* and *bupR*, regulate expression of the BfrZ outer membrane protein, which was identified in a Fur repressor assay. *bupI* overexpressed from a multicopy plasmid triggers *bfrZ* transcription, which is reduced in the presence of *bupR*

(Pradel and Locht, 2001). In *P. aeruginosa*, the PvdS sigma factor is active in the absence of the FpvR cytoplasmic transmembrane protein (Redly and Poole, 2005), and in *Pseudomonas putida*, the PupI sigma factor is active in the absence of the PupR transmembrane protein (Koster et al., 1994). In both strains, the cytoplasmic transmembrane proteins reduce sigma factor activity. These findings agree with an anti-sigma-factor activity of these FecR homologues.

A characteristic feature of outer membrane transporters involved in signaling for transcription initiation is an N-terminal extension at the TonB box, as identified in FecA (Kim et al., 1997). The FpvA ferric pyoverdine transporter of *P. aeruginosa* contains such an extension. Its deletion abolishes transcription initiation but retains transport (Shen et al., 2002). Isolated FpvA mutants are either transport or signaling deficient, which shows that the two functions are distinct and independent from each other (James et al., 2005). The FpvR signal transducer across the cytoplasmic membrane is unique in functioning as an antisigma factor for two sigma factors. It contains 67 cytoplasmic amino acid residues that interact with PvdS and FpvI (Redly and Poole, 2005).

Gene arrangements, e.g., *fecAIR*, are predicted to occur particularly frequently in *B. thetaiotaomicron* (23 cases), *B. fragilis* (16 cases), *Nitrosomonas europaea* (15 cases), *Pseudomonas fluorescens* (18 cases), *P. putida* (11 cases) (Martinez-Bueno et al., 2002), and *P. aeruginosa* (8 cases). Many of the FecIRA homologs are not involved in iron transport, the systems hitherto identified, e.g., the hypersensitive response mentioned earlier. Sequence comparisons predict that the homologs of *Bacteroides* strains are part of sugar signaling and transport systems (Xu et al., 2003). A comprehensive computer analysis of the 108 completely sequenced bacterial genomes known to date yields 115 *fecIRA* homologs in 26 genomes with a gene arrangement as in *E. coli* K-12 (Koebnik, 2005). Additional loci with different gene arrangements also occur, e.g., *fecIR* homologs in tandem with the same or opposite polarity, and genes for the outer membrane proteins, called transducers, adjacent upstream of the *fecI* homologs with the same or opposite polarity, or located elsewhere on the genome. It is predicted that the FecIRA homologs function similarly to the FecIRA proteins in the sense that the inducer binds to an outer membrane protein that functions as a transducer of a signal. The FecR homolog receives the signal from the FecA homolog and transduces the signal across the cytoplasmic membrane. In the cytoplasm, the FecR homolog interacts with the FecI homolog, which recruits the RNA polymerase and specifically directs it to only one or two promoters.

The periplasm—once considered not more than a space between the outer membrane and the cytoplasmic membrane—is now known to be a compartment through which energy and information flow. Understanding of the underlying mechanisms will require much sophisticated work since the three compartments cannot be taken apart without loss of essential aspects of periplasmic functions.

ACKNOWLEDGMENTS

We thank Karen A. Brune for critically reading the manuscript and Michael Braun for predicting the FecI and FecR structures.

This work was supported by the Deutsche Forschungsgemeinschaft and the Fonds der Chemischen Industrie.

REFERENCES

Angerer, A., and V. Braun. 1998. Iron regulates transcription of the *Escherichia coli* ferric citrate transport genes directly and through the transcription initiation proteins. *Arch. Microbiol.* **169**:483–490.

Angerer, A., S. Enz, M. Ochs, and V. Braun. 1995. Transcriptional regulation of ferric citrate transport in *Escherichia coli* K-12. FecI belongs to a new subfamily of σ^{70}-type factors that respond to extracytoplasmic stimuli. *Mol. Microbiol.* **18**:163–174.

Braun, V. 1997. Surface signaling: novel transcription initiation mechanism string from the cell surface. *Arch. Microbiol.* **167**:325–331.

Braun, V., and C. Herrmann. 1993. Evolutionary relationship of uptake systems for biopolymers in *Escherichia coli*: cross-complementation between the TonB-ExbB-ExbD and the TolA-TolQ-TolR proteins. *Mol. Microbiol.* **8**:261–268.

Braun, V., and S. Mahren. 2005. Transmembrane transcriptional control (surface signalling) of the *Escherichia coli* Fec type. *FEMS Microbiol. Rev.* **29**:673–684.

Braun, V., S. Mahren, and M. Ogierman. 2003. Regulation of the FecI-type ECF sigma factor by transmembrane signalling. *Curr. Opin. Microbiol.* **6**:173–180.

Bradbeer, C. 1993. The proton motive force drives the outer membrane transport of cobalamin in *Escherichia coli*. *J. Bacteriol.* **175**:3146–3150.

Brito B., D. Aldon, P. Barberis, C. Boucher, and S. Genin. 2002. A signal transfer system through three compartments transduces the plant cell contact–dependent signal controlling *Ralstonia solanacearum hrp* genes. *Mol. Plant. Microb. Interact.* **15**:109–119

Buchanan, S. K., B. S. Smith, L. Venkatramani, D. Xia, L. Esser, M. Palnitkar, R. Chakraborty, D. van der Helm, and J. Deisenhofer. 1999. Crystal structure of the outer membrane active transporter FepA from *Escherichia coli*. *Nature Struct. Biol.* **6**:56–63.

Cadieux, N., and R. J. Kadner. 1999. Site-directed disulfide bonding reveals an interaction site between energy-coupling protein TonB and BtuB, the outer membrane cobalamin transporter. *Proc. Natl. Acad. Sci. USA* **96**:10673–10678.

Campbell, E. A., J. L. Tupy, T. M. Gruber, S. Wang, M. M. Shatrp, C. A. Gross, and S. A. Darst. 2003. Crystal structure of *Escherichia coli* σ^E with the cytoplasmic domain of its anti-σ RseA. *Mol. Cell* **11**:1067–1078.

Cao, Z., P. Warfel, S. M. C. Newton, and P. E. Klebba. 2003. Spectroscopic observations of ferric enterobactin transport. *J. Biol. Chem.* **278**:1022–1028.

Chimento, D. P., R. J. Kadner, and M. C. Wiener. 2005. Comparative structural analysis of TonB-dependent outer membrane transporters: implications for the transport cycle. *Proteins* **59**:240–251.

Chimento, D. P., A. K. Mohanty, R. J. Kadner, and M. C. Wiener. 2003. Substrate-induced transmembrane signaling in the cobalamin transporter BtuB. *Nature Struct. Biol.* **10**:394–401.

Cobessi, D., H. Celia, N. Folschweiller, I. Schalk, M. Abdallah, and F. Pattus. 2005a. The crystal structure of the pyoverdine outer membrane receptor FpvA from *Pseudomonas aeruginosa* at 3.6 Å resolution. *J. Mol. Biol.* **347**:121–134.

Cobessi, D., H. Celia, and F. Pattus. 2005b. Crystal structure at high resolution of ferric-pyochelin and its membrane receptor FptA from *Pseudomonas aeruginosa*. *J. Mol. Biol.* **347**:121–134.

Crosa, J. H., A. R. Mey, and S. M. Payne (ed.). 2004. *Iron Transport in Bacteria.* ASM Press, Washington, D.C.

Eisenhauer, H. A., S. Shames, P. D. Pawelek, and J. W. Coulton. 2005. Siderophore transport through *Escherichia coli* outer membrane receptor FhuA with disulfide-tethered cork and barrel domains. *J. Biol. Chem.* **280**:30574–30580.

Endriβ, F., M. Braun, H. Killmann, and V. Braun. 2003. Mutant analysis of the *Escherichia coli* FhuA protein reveals sites of FhuA activity. *J. Bacteriol.* **185**:4683–4692.

Enz, S., H. Brand, C. Orellana, S. Mahren, and V. Braun. 2003. Sites of interaction between the FecA and FecR signal transduction proteins of ferric citrate transport in *Escherichia coli* K-12. *J. Bacteriol.* **185**:3745–3752.

Enz, S., V. Braun, and J. H. Crosa. 1995. Transcription of the region encoding the ferric dicitrate-transport system in *Escherichia coli*: similarity between promoters for *fecA* and for extracytoplasmic function sigma factors. *Gene* **163**:13–18.

Enz, S., S. Mahren, U. H. Stroeher, and V. Braun. 2000. Surface signaling in ferric citrate transport gene induction: interaction of the FecA, FecR, and FecI regulatory proteins. *J. Bacteriol.* **182**:637–646.

Faraldo-Gomez, J. D., G. R. Smith, and M. S. P. Sansom. 2003. Molecular dynamics simulations of the bacterial outer membrane protein FhuA: a comparative study of the ferrichrome-free and bound states. *Biophys. J.* **85**:1406–1420.

Ferguson, A. D., E. Hofmann, J. W. Coulton, K. Diederichs, and W. Welte. 1998. Siderophore-mediated iron transport: crystal structure of FhuA with bound lipopolysaccharide. *Science* **282**:2215–2220.

Ferguson, A. D., R. Chakraborty, B. S. Smith, L. Esser, D. van der Helm, and J. Deisenhofer. 2002. Structural basis of gating by the outer membrane transporter FecA. *Science* **295**:1715–1719.

Garcia-Herrero, A., and H. J. Vogel. 2005. NMR solution structure of the periplasmic signaling domain of the TonB-dependent outer membrane transporter FecA. *Mol. Microbiol.* **58**:1226–1237.

Ghosh, J., and K. Postle. 2005. Disulphide trapping of an in vivo energy-dependent conformation of *Escherichia coli* TonB protein. *Mol. Microbiol.* **55**:276–288.

Härle, C., K. Insook, A. Angerer, and V. Braun. 1995. Signal transfer through three compartments: transcription initiation of the *Escherichia coli* ferric citrate transport system from the cell surface. *EMBO J.* **14**:1430–1438.

Hancock, R. E. W., and V. Braun. 1976. Nature of the energy requirement for the irreversible adsorption of bacteriophages T1 and φ 80 to *Escherichia coli*. *J. Bacteriol.* **125**:409–415.

Helmann, J. D. 2002. The extracytoplasmic function (ECF) sigma factors. *Adv. Microb. Physiol.* **46**:47–110.

James, E. H., P. A. Beare, L. W. Martin, and I. L. Lamont. 2005. Mutational anlysisis of a bifunctional ferrisiderophore receptor and signal-transducing protein from *Pseudomonas aeruginosa*. *J. Bacteriol.* **187**:4514–4520.

Khursigara, C. M., G. De Grescenco, P. D. Pawelek, and J. W. Coulton. 2005a. Kinetic analyses reveal multiple steps in forming TonB-FhuA complexes from *Escherichia coli*. *Biochemistry* **44**:3441–3453.

Khursigara, C. M., G. De Grescenco, P. D. Pawelek, and J. W. Coulton. 2005b. Deletion of proline-rich region of TonB disrupts formation of a 2:1 complex with FhuA, an outer membrane receptor of *Escherichia coli*. *Protein Sci.* **14**:1266–1273.

Kim, I., A. Stiefel, S. Plantör, A. Angerer, and V. Braun. 1997. Transcription induction of the ferric citrate transport genes via the N-terminus of the FecA outer membrane protein, the Ton system and the electrochemical potential of the cytoplasmic membrane. *Mol. Microbiol.* **23**: 333–344.

Kirby A. E., N. D. King, and T. D. Connell. 2004. RhuR, an extracytoplasmic function sigma factor activator, is essential for heme-dependent expression of the outer membrane heme and hemoprotein receptor of *Bordetella avium*. *Infect. Immun.* **72**: 896–907.

Koebnik, R. 2005. TonB-dependent trans-envelope signalling: the exception or the rule? *Trends Microbiol.* **13**:343–347.

Koedding, J., P. Howard, L. Kaufmann, P. Polzer, A. Lustig, and W. Welte. 2004. Dimerization of TonB is not essential for its binding the outer membrane siderophore receptor FhuA of *Escherichia coli*. *J. Biol. Chem.* **279**:9978–9986.

Koster, M., W. van Klompenburg, W. Bitter, J. Leong, and P. Weisbeek. 1994. Role for the outer membrane ferric siderophore receptor PupB in signal transduction across the bacterial cell envelope. *EMBO J.* **13**:2805–2813.

Larsen, R. A., T. E. Letain, and K. Postle. 2003. In vivo evidence of TonB shuttling between the cytoplasmic and outer membrane in *Escherichia coli*. *Mol. Microbiol.* **49**:211–218.

Letain T. E., and K. Postle. 1997. TonB protein appears to transducer energy by shuttling between the cytoplasmic membrane and the outer membrane in *Escherichia coli*. *Mol. Microbiol.* **24**:271–283.

Letoffe S., K. Wecker, M. Delepelaire, P. Delepelaire, and C. Wandersman. 1999. Interactions of HasA, a bacterial hemophore, with hemoglobin and with its outer membrane receptor HasR. *Mol. Microbiol.* **33**:564–555.

Locher, K. P., B. Rees, R. Koebnik, A. Mitschler, L. Moulinier, J. P. Rosenbusch, and D. Moras. 1998. Transmembrane signaling across the ligand-gated FhuA receptor: crystal structures of free and ferrichrome-bound states reveal allosteric changes. *Cell* **95**:771–778.

Luck, S., S. A. Turner, K. Rajakumar, H. Sakellaris, and B. Adler. 2001. Ferric dicitrate transport system (Fec) of *Shigella flexneri* 2a YSH6000 is encoded on a novel pathogenicity island carrying multiple antibiotic resistance genes. *Infect. Immun.* **69:**6012–6021.

Mahren, S., and V. Braun. 2003. The FecI extra-cytoplasmic–function sigma factor of *Escherichia coli* interacts with the β′ subunit of RNA polymerase. *J. Bacteriol.* **185:**1796–1802.

Mahren, S., H. Schnell, and V. Braun. 2005. Occurrence and regulation of the ferric citrate transport system in *Escherichia coli* B, *Klebsiella pneumoniae*, *Enterobacter aerogenes* and *Photorhabdus luminescens*. *Arch. Microbiol.* **184:**175–186.

Martinez-Bueno, M. A., R. Tobes, M. Rey, and J. l. Ramos. 2002. Detection of multiple extra-cytoplasmic function (ECF) sigma factors in the genome of *Pseudomonas putida* KT2440 and their counterparts in *Pseudomonas aeruginosa*. PAO1. *Environ. Microbiol.* **4:**842–855.

Neugebauer, H., C. Herrmann, W. Kammer, G. Schwarz, A. Nordheim, and V. Braun. 2005. ExbBD-dependent transport of maltodextrins through the novel MalA protein across the outer membrane of *Caulobacter crescentus*. *J. Bacteriol.* **187:**8300–8311.

Ochs, M., A. Angerer, S. Enz, and V. Braun. 1996. Surface signaling in transcriptional regulation of the ferric citrate transport system of *Escherichia coli*: mutational analysis of the alternative sigma factor FecI supports its essential role in *fec* transport gene transcription. *Mol. Gen. Genet.* **250:**455–456.

Ogierman, M., and V. Braun. 2003. Interactions between the outer membrane ferric citrate transporter FecA and TonB: Studies of the FecA TonB box. *J. Bacteriol.* **185:**1870–1885.

Peacock, P. S., A. M. Weljie, S. P. Howard, F. D. Price, and H. J. Vogel. 2004. The solution structure of the C-terminal domain of TonB and interaction studies with TonB box peptides. *J. Mol. Biol.* **345:**1185–1197.

Postle, K., and R. J. Kadner. 2003. Touch and go: tying TonB to transport. *Mol. Microbiol.* **49:**869–882.

Pradel, E., and C. Locht. 2001. Expression of the putative siderophore receptor gene *bfrZ* is con-trolled by the extracytoplasmic-function sigma factor BupI in *Bordetella bronchiseptica*. *J. Bacteriol.* **183:**2910–2917.

Redly, G. A., and K. Poole. 2005. FpvIR control of *fpvA* pyoverdine receptor gene expression in *Pseudomonas aeruginosa*: demonstration of an interaction between FpvI and FpvR and identification of mutations in each compromising this interaction. *J. Bacteriol.* **187:**5648–5657.

Rossi, M. S., A. Paquelin, J. M. Ghigo, and C. Wandersman. 2003. Haemophore-mediated signal transduction across the bacterial cell envelope in *Serratia marcescens*: the inducer and the transported substrate are different molecules. *Mol. Microbiol.* **48:**1467–1480.

Sauter, A., S. P. Howard, and V. Braun. 2003. In vivo evidence for TonB dimerization. *J. Bacteriol.* **185:**5747–5754.

Scott, D. C., Z. Cao, Z. Qi, M. Bauler, J. D. Igo, S. M. C. Salete, and P. E. Klebba. 2001. Exchangeability of N-termini in the ligand-gated porins of *Escherichia coli*. *J. Biol. Chem.* **276:**13025–13033.

Shen, J., A. Meldrum, and K. Poole. 2002. FpvA receptor involvement in pyoverdine biosynthesis in *Pseudomonas aeruginosa*. *J. Bacteriol.* **184:**3268–3275.

Wandersman, C., and P. Delepelaire. 2004. Bacterial iron sources: from siderophores to hemophores. *Annu. Rev. Microbiol.* **58:**611–647.

Wang, C. C., and A. Newton. 1971. An additional step in the transport of iron defined by the *tonB* locus of *Escherichia coli*. *J. Biol. Chem.* **246:**2147–2151.

Yue, W. W., S. Grizot, and S. K. Buchanan. 2003. Structural evidence for iron-free citrate and ferric citrate binding to the TonB-dependent outer membrane transporter FecA. *J. Mol. Biol.* **332:**353–368.

Xu, J., M. K. Bjursell, J. Himrod, S. Deng, L. K. Carmichael, H. C. Chiang, L. V. Hooper, and J. I. Gordon. 2003. A genomic view of the human *Bacteroides thetaiotaomicron* symbiosis. *Science* **299:**2074–2076.

PERIPLASMIC ABC TRANSPORTERS

Elie Dassa

17

The envelope of gram-negative bacteria is a complex architecture consisting of three layers: the cytoplasmic or inner membrane, the rigid peptidoglycan network, and the outer membrane (Bos and Tommassen, 2004). The peptidoglycan is immersed into the periplasmic space, an aqueous compartment. Membrane transport proteins regulate the exchange of metabolites between the external medium, the periplasm, and the cytosol. Primary active transporters, such as ATP-binding cassette (ABC) transporters, directly utilize the free energy released upon the hydrolysis of ATP to pump substrates against a concentration gradient. Secondary active transporters, including lactose permease (Abramson et al., 2003) and the multidrug efflux pump AcrB (Touze et al., 2004), couple electrochemical proton gradients to substrate transport across the cytoplasmic membrane. Ion channels (Kung and Blount, 2004), such as potassium and chloride channels, function as highly selective pores that open and close in response to specific stimuli, allowing ions to rapidly flow down a gradient.

The ABC superfamily of proteins is one of the largest families of paralogous sequences.

About 5% of the entire *Escherichia coli* genome encode components of ABC systems (Linton and Higgins, 1998). Systems in this superfamily have the common property to share a highly conserved ATP-hydrolyzing domain or protein, the ABC, characterized unequivocally by four short sequence motifs: the so-called Walker motifs A and B, which presumably constitute the nucleotide-binding fold; the signature motif, which is distinctive of ABC ATPases and located upstream of the Walker B motif; and a downstream fourth motif, sometimes called the Switch region, which carries a highly conserved histidine residue (Schneider and Hunke, 1998). ABC systems are widespread among living organisms and they have been detected in all genera of the three kingdoms of life with a remarkable conservation in the primary sequence and the organization of their constitutive domains or subunits (Holland and Blight, 1999, Saurin et al., 1999). The fundamental biological role of the ABC is to couple the energy of ATP hydrolysis to an impressively large variety of essential biological phenomena comprising not only transmembrane transport but also several non-transport-related processes such as translation elongation (Chakraburtty, 2001), translation regulation (Vazquez de Aldana et al., 1995), ribosome biogenesis (Dong et al., 2004), and DNA repair (Hopfner and

Elie Dassa, Unité des Membranes Bactériennes CNRS URA2172, Département de Microbiologie Fondamentale et Médicale, Site Fernbach, Institut Pasteur 25, 75724 Paris Cedex 15, France.

The Periplasm
Edited by Michael Ehrmann © 2007 ASM Press, Washington, D.C.

Tainer, 2003). Overexpression of certain ABC transporters has been described in cancer cell lines and tumors that display the multidrug resistance phenotype (Gottesman et al., 2002). Mutations in 17 genes encoding ABC proteins have been involved in the manifestation of about 22 human disorders such as cystic fibrosis, Tangier's disease, retinitis pigmentosa, Startgardt disease, hyperinsulinemic hypoglycemia, immunodeficiencies, etc. (Dean and Annilo, 2005). Although ABC systems deserved much attention because they are involved in severe human inherited diseases, it should be kept in mind that they have been discovered and characterized in detail in prokaryotes since the 1970s. The most extensively analyzed systems have been the high-affinity histidine and maltose ABC uptake systems of *Salmonella enterica* serovar Typhimurium and *Escherichia coli*.

ORGANIZATION AND FUNCTIONS OF ABC SYSTEMS

ABC systems could be functionally divided into three main categories: importers, exporters, and systems involved in nontransport processes such as regulation of gene expression or DNA repair. Beside the large diversity of substrates handled and the difference in the polarity of transport, ABC transporters share a common organization consisting of two hydrophobic membrane-spanning or integral membrane (IM) domains and two hydrophilic cytoplasmic domains carrying the ABC, peripherally associated with IM on the cytosolic side of the membrane (Color Plates 27 and 28).

In the vast majority of ABC exporters (or type I secretion systems), the IM and ABC domains are fused in a single polypeptide chain (Color Plate 28) such as the *E. coli* hemolysin exporter (Blight and Holland, 1990) or the mammalian TAP1 protein (Beck et al., 1992). These systems are involved in the secretion of various molecules such as peptides, lipids, hydrophobic drugs, polysaccharides, and proteins in both prokaryotes and eukaryotes. However, several systems involved in antibiotic or drug resistance (Kaur and Russell, 1998), bacteriocin immunity (Rince et al., 1997), nodulation

(Vazquez et al., 1993), and membrane polysaccharide biogenesis (Pavelka et al., 1994) have been reported to carry IM and ABC domains on different polypeptides. These systems have been proposed to mediate the export of antibiotics, bacteriocins, and polysaccharides.

In importers, the four domains are in general determined by independent polypeptide chains (Color Plate 27). ABC importers constitute the family of substrate-binding, protein-dependent transporters that mediate the uptake of small nutrients and they are found exclusively in prokaryotes (Boos and Lucht, 1996).

ABC proteins involved in nontransport cellular processes like translation of mRNA and DNA repair do not have identified IM domains but contain two ABC domains fused together (Dassa and Bouige, 2001; Kerr, 2004).

ABC systems have a great impact on bacterial physiology and their dysfunction may have strong deleterious effects. Virtually any bacterial ABC system could be important for viability, virulence, or pathogenesis. High-affinity iron-uptake ABC systems were recognized a long time ago as important effectors of virulence in several pathogens (Henderson and Payne, 1994). ABC exporters provide protection against noxious substances by pumping the molecules out of the bacterial cytosol and they participate in virulence by mediating the secretion of toxins and bacteriocins. ABC systems contribute to the building of the bacterial envelope by mediating the export of lipids, membrane proteins, and polysaccharides.

In addition to their implication in transport, ABC systems are involved in the regulation of several physiological processes. A direct regulatory role should be clearly distinguished from the fact that impairment of transport may have indirect consequences on bacterial physiology. For example, conjugative transfer of the plasmid pCF10 by *Enterococcus faecalis* donor cells occurs in response to a peptide sex pheromone, cCF10, secreted by recipients. Inactivation of the chromosomal *opp* operon abolished response at physiological concentrations of pheromone (Leonard et al., 1996). One example of a possible direct effect is the *trkE* gene, located within the *sapABCDF* operon, encoding a putative peptide transporter, which is known to be involved in the modulation of activity of the TrkH potassium channel (Dosch et al., 1991; Parra-Lopez et al., 1994; Stumpe and Bakker, 1997). This effect could be similar to that exerted by the mammalian SUR1 ABC transport on the regulation of inwardly rectifying potassium channel subunits (KIR6.x) (Aguilar-Bryan et al., 1998; Ashcroft, 2000). However, the best documented example of regulation mediated by an ABC transporter is given by the role of protein MalK in the regulation of the transcription of maltose-regulated genes and in the regulation of maltose transport by inducer exclusion. Transcription of the maltose regulon of *E. coli* K-12 is regulated by the positive activator, MalT (Boos and Bohm, 2000). In the presence of ATP and of the inducer maltotriose, MalT binds upstream of maltose-dependent promoters and activates transcription. Recent data suggest that MalK and maltotriose compete for MalT binding (Joly et al., 2004). MalT/MalK interaction might involve two distinct contact sites on each partner. These sites would be located in domains DT1 and DT3 of MaIT, and in the nucleotide-binding domain and the regulatory domain of MaIK. Such a two-point interaction model would explain how the regulatory activity of MalK might be coupled to transport (Richet et al., 2005).

AN OVERVIEW OF PROKARYOTE ABC BINDING PROTEIN-DEPENDENT UPTAKE SYSTEMS

ABC import systems are also called binding protein-dependent (BPD) transport systems. In addition to the basic core structure of ABC transporters, they require for proper function an extracytoplasmic substrate-binding protein (BP). This component is located in the periplasmic space of gram-negative bacteria. BPs are released from bacteria by a cold osmotic shock procedure and the corresponding transport systems are transiently inactivated due to the loss of the protein. In all cases studied, the released proteins were shown to bind substrates with a high affinity. If such proteins are reintroduced into the periplasm of cells lacking these components, either because they were osmotically shocked or because of a mutation, transport could be restored (Brass et al., 1981). In gram-positive bacteria, which do not have a periplasmic space, the substrate-binding protein is an extracellular lipoprotein, bound to the external face of the cytoplasmic membrane by an N-terminal acyl glyceryl cysteine (Gilson et al., 1988). This anchor is responsible for maintaining the substrate-binding protein in the close vicinity of the membrane components. BPD transporters are extremely diverse in their substrate specificities. Most transport systems are specific for a single substrate or for a family of structurally related substrates. On the other hand, the oligopeptide transport system of *Lactococcus lactis* has the capacity to mediate the uptake of peptides from 4 up to at least 18 residues (Lanfermeijer et al., 1999). The nature of the substrates handled by ABC transporters is extraordinarily wide, including mono- and oligosaccharides, organic and inorganic ions, amino acids and short peptides, iron-siderophores, metals, polyamine cations, opines, and vitamins. These systems are very efficient since they are able to concentrate nutrients up to 104-fold even when the concentration of the nutrient in the external medium is below the micromolar range. As a consequence, BPD transporters are scavenging systems, able to extract trace elements from the environment.

Substrates Cross the Outer Membrane through Three Types of Proteins

To be efficiently transported into the cytoplasm, a nutrient should cross the different layers of the bacterial envelope. In gram-negative bacteria, substrates should first pass through the outer membrane and they can use three different pathways: generalized porins, specialized porins, and high-affinity outer membrane receptors. Most small substrates, with a molecular mass less than 650 Da, cross the outer membrane through the nonspecific or generalized porins like OmpF or OmpC porins of *Entero-*

bacteriaceae. The importance of such porins in transport processes is highlighted by the fact that mutants lacking these proteins are pleiotropically affected in the utilization of several substrates (Bavoil et al., 1977). When the size of the substrate exceeds the size handled by generalized porins, a specific or specialized porin is used. The best example known so far is maltoporin, the *lamB* gene product that is essential for the transport of maltodextrins of more than three glucose residues (Szmelcman and Hofnung, 1975). In contrast with general porins, the genes coding for such specialized porins are often genetically linked to the regions encoding the rest of the transporter and their expression is tightly coregulated. The systems for uptake of iron-siderophores, heme compounds, or vitamin B_{12} are faced with the complication that the substrates are in exceedingly low amounts in the environment (Wandersman and Delepelaire, 2004). The molecular mass of the transported molecules is about 700 to 1,000 Da, over the size limit of porins. For these reasons, Fe^{3+}-siderophore compounds are first bound by high-affinity outer membrane receptors, which are also channels that translocate the substrates into the periplasmic space (Ferguson and Deisenhofer, 2004). Substrates are released from the high-affinity receptor by the virtue of an energy expense.

A very peculiar system, described in *Sphigomonas* sp., was found to be expressed in alginate-induced cells. Alginate is a high-molecular-weight polysaccharide (25,000 Da), which is thought to enter the cell in an intact form since alginate-degrading enzymes are located exclusively in the cytoplasm (Hashimoto et al., 1999). A specific outer membrane organelle, called "pit," is thought to mediate alginate uptake through the outer membrane. Alginate-induced outer membrane proteins are similar to TonB-dependent receptors, opening the possibility that some members of this family might be involved in polysaccharide uptake (Hashimoto et al., 2005). Indeed, it was shown that an outer membrane protein SusC, similar to TonB-dependent receptors, was maltose inducible and essential for maltose and starch uptake in *Bacteroides thetaiotaomicron* (Reeves et al., 1996). More recently, an energy-dependent TonB-independent, ExbBD-dependent transport of maltose was demonstrated in *Caulobacter crescentus* (Neugebauer et al., 2005).

Substrates Are Captured by Substrate-Binding Proteins with High Affinity

With the exception of iron-siderophore transport systems discussed above, the substrate recognition site with the highest specificity is on the substrate-binding proteins in other BDP transporters. For instance, two different substrate-binding proteins are working with the same set of cytoplasmic components, as shown for the histidine and the lysine-arginine-ornithine (LAO)-BPs, which use the same HisMPQ cytoplasmic membrane transporter (Higgins and Ames, 1981). Only few substrate-binding lipoproteins have been characterized biochemically and structurally from gram-positive organisms (Levdikov et al., 2005; Schafer et al., 2004) and *Archaea* (Schiefner et al., 2004). Most of the knowledge on BPs comes from the study of proteins from gram-negative bacteria. The amount of the binding proteins over the cytoplasmic membrane components is usually high. For instance, the periplasmic concentration of maltose-binding protein could be as high as 1 mM. One of the roles of BPs would be to maintain a high concentration of bound substrate at the close vicinity of the outer face of the cytoplasmic membrane (Boos and Shuman, 1998). A recent report suggests that BPs are monomeric and bind substrates with affinities ranging from 0.1 to 1 μM. Studies on the kinetics of binding revealed also that there was one substrate-binding site per molecule of protein. The three-dimensional structure of several BPs was determined, most notably in the laboratory of F. A. Quiocho (Quiocho and Ledvina, 1996). From all these structural studies, it appeared that all BPs adopt a similar folding pattern, made of two globular domains or lobes, called the N and the C lobe since they contain the N- and the C-terminal ends of the protein. Each lobe is composed of a so-called α-β-fold including plated β-sheets surrounded by α-helices and connected by loops. The two lobes are connected in several ways and the ligand-binding site is located at the interface between the two domains. Two different structural subclasses were recognized originally, which differ in the topology by which the polypeptide is distributed and in the number and order of β-strands in each domain (Fukami-Kobayashi et al., 1999). Class I is composed of simple sugar- (arabinose, ribose) binding proteins and branched-chain amino acid-binding proteins. Class II contains the maltose-maltodextrins-, phosphate- and sulfate-binding proteins. Recently, a novel class III was recognized and it groups the divalent cation- (Mn^{2+}, Zn^{2+}) and iron-siderophore-binding proteins. In this class, the two lobes consist of a central five-stranded β-sheet surrounded by α-helices and the domains are connected by a single α-helix spanning the length of the protein (Borths et al., 2002). With the possible exception of class III proteins (Karpel et al., 1991), BPs adopt two different conformations: a ligand-free open form and a ligand-bound closed form, which interconvert through a relatively large bending motion around the hinge (Sharff et al., 1992) (Color Plate 29). This conformational change is important for inducing the signaling cascade that leads ultimately to substrate translocation through the cytoplasmic membrane (reviewed in Davidson, 2002). Some BPs are also chemoreceptors involved in chemotaxis by signaling the presence of substrates to chemotactic transducers. This is the case among others for the maltose-, the galactose-, and the ribose-binding proteins. Maltose BP interacts with Tar (Kossman et al., 1988) and galactose and ribose BPs interact with Trg (Yaghmai and Hazelbauer, 1993).

A Cytoplasmic Membrane ABC Transporter Conveys the Substrate into the Cytoplasm

One or two hydrophobic inner membrane proteins (IM) are components of BPD transport systems. They are proposed to constitute a channel through which substrates cross the cytoplasmic membrane. The two proteins form a heterodimeric complex as it was demonstrated in the maltose and histidine transport systems (Davidson and Nikaido, 1991; Kerppola et al., 1991). For systems with only one gene encoding such proteins as in the vitamin B_{12} transport system of *E. coli*, a homodimer of BtuC constitutes the transmembrane region of the transporter (Color Plate 28) (Locher et al., 2002). These proteins are integral membrane proteins, and computer-assisted predictions of secondary structure and topology indicate that they span the cytoplasmic membrane. In well-characterized systems, they were shown to be accessible to proteolytic digestion from the cytoplasmic and the periplasmic sides of the membrane. In the vast majority of cases, they contain six transmembrane segments, predicted to be folded in an α-helical conformation and joined by loops of variable size. The N and C termini of the proteins are, in general, pointing toward the cytoplasm. There are few exceptions to this general scheme, such as the BtuC protein of the vitamin B_{12} transport system that is made of 10 transmembrane segments (Locher et al., 2002) or the ProW protein from the glycine-betaine transport system where the N terminus is in the periplasmic space (Haardt and Bremer, 1996). Several groups have made considerable efforts to experimentally establish the topological disposition of such proteins. Gene fusion approaches, using periplasmic enzymes that are active only in the periplasm (alkaline phosphatase or β-lactamase) as reporters, were instrumental in these studies (Hennessey and Broome-Smith, 1993; Traxler et al., 1993). In general, the results of such experimental analyses were consistent with the prediction of transmembrane segments and topology as inferred from the primary structure (von Heijne, 1994). However, there is no easy way to accurately identify the residues that are at the boundaries of the membrane. Almost all IMs of BPD transporters display a short conserved motif, the EAA motif located at about 100 residues from the C terminus. The motif is hydrophilic and it was found to reside in a cytoplasmic loop located between the penultimate and the ante-penultimate transmembrane segment in all proteins with a known topology (Saurin et al., 1994). The conservation of this motif argues for an important functional role and it was suggested that it could constitute a site of interaction with the conserved cytoplasmic ATPase (Dassa and Muir, 1993; Kerppola and Ames, 1992).

The ATPase subunit is the most conserved component of ABC systems. In oligopeptide and branched-chain amino acid transporters, the two ABC subunits are usually encoded by two different genes (Detmers et al., 2001); ribose or arabinose (monosaccharide) trans-porters have a single ATPase made of two du-plicated and fused subunits (Schneider, 2001). All other transporters have a single gene en-coding the ABC protein and experimental ev-idence demonstrates that two such subunits are present in the complete transporter. High-resolution structural models became available for HisP and MalK, the ATPase of the histidine and maltose ABC transporters, respectively (Diederichs et al., 2000; Hung et al., 1998), and for a dozen of other ABC ATPases of ATPase domains. The overall folding design of these proteins is similar, and it consists of two re-gions: a RecA-like domain, which contains the nucleotide-binding Walker A and Walker B motif, and a helical domain specific to ABC proteins, which contains the ABC family signa-ture motif, LSGGQ. The residues involved in the interaction of the ATPase with the conserved EAA region of transmembrane pro-teins are located in the helical region (Hunke et al., 2000; Mourez et al., 1997). In the crystal structure of the BtuCD complex, the EAA loop of BtuC forms the contact region with the ATP-binding cassette (Color Plate 28) (Locher et al., 2002).

Inventory and Classification of ABC Systems and Reconstruction of the Evolution of the Family

To understand the complexity and the diversity of ABC systems, computer-assisted methods have been applied. These methods were instru-mental in the primary definition of the super-family based on sequence comparisons. How-ever, in most cases, ABC systems of a given organism (Braibant et al., 2000; Linton and Higgins, 1998; Quentin et al., 1999) or ABC systems with clear functional similarity (Fath and Kolter, 1993; Hughes, 1994; Kuan et al., 1995) were compared. The presence of the highly conserved ATPase domain allowed un-dertaking more global comparisons (Paulsen et al., 1998; Saurin et al., 1999). The latter pub-lication, which constituted the first global study specifically devoted to the ABC superfamily, was recently updated with the analysis of about 600 ATPase proteins or domains (Dassa and Bouige, 2001). The sequences segregate in 29 clusters or families.

The major finding is that the ABC proteins or domains fall into three main subdivisions or classes: class 1 comprises systems with fused ABC and IM domains, class 2 comprises sys-tems with two duplicated fused ABC domains and no transmembrane domains, and class 3 contains systems with IM and ABC domains carried by independent polypeptide chains (Dassa and Bouige, 2001). This clustering matches fairly well with the three functional divisions of ABC systems, which are exporters, systems involved in nontransport processes, and importers, respectively. Therefore, the diver-gence between importers, exporters, and other systems probably occurred once in the history of ABC systems. However, in addition to bind-ing protein-dependent importers, class 3 con-tains several transporters whose function is unknown or that could not be conclusively re-lated to import. Systems of the DRA family (involved in drug and antibiotic resistance) and the CLS family (involved in the biogenesis of capsular polysaccharides, lipopolysaccharides, and teichoic acids) have been suggested to par-ticipate in the export of such molecules. The fact that these transporters are clustered with the BPD systems may suggest either that they are not directly involved in the export of their substrates or, alternately, that the transport polarity of some families may change during evolution.

Each class of ABC transporters is composed of systems identified in the three kingdoms of life: archaea, bacteria, and eukarya. The separation between eukaryotic and prokaryotic systems occurred at the very end of the branches of the tree. This suggests that ABC systems began to specialize very early, probably before the separation of the three kingdoms of living organisms, and that functional constraints on the ABC domain were responsible for the conservation of sequences.

A good correlation exists between the sequences and the overall function of ABC systems. This is probably because ABC domains segregate mostly according to sequence differences in the so-called helical domain that lies between the Walker motifs A and B (Schneider and Hunke, 1998). In the maltose transporter, we have demonstrated that this region is critical for the interaction between ABC and IM (Hunke et al., 2000; Mourez et al., 1997). More generally, the relationship between the sequence and the function would reflect constraints imposed by the interaction of ABC ATPase with their partners.

From this analysis, a hypothetical scenario on the evolution of ABC systems could be proposed (Dassa and Bouige, 2001; Saurin et al., 1999). The ancestor "progenote" cell already had all classes of ABC systems. Prokaryotes inherited all ABC classes. Eukaryotes probably acquired IM-ABC and ABC-IM (class 1) and ABC2 (class 2) systems from the symbiotic bacteria that are the putative ancestors of organelles. Note that most eukaryote IM-ABC systems were specifically targeted to organelle membranes that probably descend from a prokaryote ancestor. For instance, the mammalian TAP proteins, involved in the presentation of antigenic peptides to the class I major histocompatibility complex, are inserted into the endoplasmic reticulum and the ALD proteins, putatively involved in the export of very-long-chain fatty acids from the cytosol into peroxisomes, are targeted to the peroxisomal membrane. From genes encoding IM-ABC or ABC-IM systems, eukaryotes developed specific systems by several independent duplication-fusion events, as

for instance those that led to the constitution of the proteins of the PDR (fungal pleiotropic drug resistance family) (ABC-IM)2 and the P-glycoprotein-like proteins (IM-ABC)2.

FAMILIES OF ABC IMPORTERS

The MET Family Specific for Metallic Cations

This family is composed of systems involved in the uptake of various metallic cations, such as iron, manganese, and zinc (Claverys, 2001). Putative systems belonging to the MET family were found in the genomes of bacteria and archaea and in the cyanelle genome of the photoautotrophic protist *Cyanophora paradoxa*. The ATPases of these systems are strongly related to those of iron-siderophore uptake systems (ISVH family, see below), suggesting that they arose from a common ancestor (Saurin et al., 1999). Weaker but significant similarities could be detected between IM of the MET and ISVH families. The substrate-binding proteins belong to a family of proteins called LraI (for lipoprotein antigen receptor I), which have been involved in the adhesion of bacteria to host cell-surface structures (Fenno et al., 1995) and to other bacteria (Kolenbrander et al., 1994). In pathogenic gram-positive bacteria, MET family importers participate in virulence (Dintilhac et al., 1997; Janulczyk et al., 2003; Marra et al., 2002), suggesting that these permeases are of particular importance during infection of host tissues, in which concentrations of metal ions are generally low.

The ISVH Family Specific for Iron-Siderophores, Vitamin B_{12}, and Hemin

The substrates handled by the ISVH family systems are quite different. Their common characteristic is to chelate iron (ferrichrome, enterobactin, achromobactin, anguibactin, citrate, exochelin, hemin, and vibriobactin) or cobalt (vitamin B_{12}) (Wandersman and Delepelaire, 2004). All these systems are associated with TonB-dependent high-affinity outer membrane receptors in gram-negative bacteria (Faraldo-Gomez and Sansom, 2003). In addition to their role in iron uptake, some TonB-dependent outer membrane receptors, such as the *E. coli* FecA receptor for ferric citrate and the *Serratia marscecens* HasR receptor for heme, participate to a new kind of regulatory response mediated by a sigma/anti-sigma system. Such receptors are characterized by the presence of an N-terminal peptidic extension, which was shown to interact with the anti-sigma periplasmic domain, thereby triggering the transcription of the cognate transport system (Biville et al., 2004; Enz et al., 2003). Once released from the outer membrane receptor, the substrate is translocated through the inner membrane thanks to an ABC BPD importer (Braun and Braun, 2002).

The OSP Family Specific for Di- and Oligosaccharides and Polyols

The OSP family includes transport systems for maltooligosaccharides, cyclodextrins, trehalose/maltose, cellobiose/cellotriose, arabinose oligomers, and lactose. Members of this family also transport several polyols: mannitol, arabitol, sorbitol (glucitol), and glycerol 3-phosphate (Schneider, 2001). Polyols such as trehalose and glucosyl glycerol are known as osmoprotectants in cyanobacteria (Mikkat and Hagemann, 2000). Some systems can mediate the uptake of several oligosaccharides, such as the raffinose/melibiose/isomaltotriose system of *Streptococcus mutans* (Russell et al., 1992). Systems of this family have a highly conserved organization comprising a BP, two IMs, and one ABC. In *Streptomyces reticuli*, it was demonstrated that a single ABC MsiK is involved in the energizing of two different transporters specific for maltose and cellobiose (Schlösser et al., 1997). This property might be general for gram-positive OSP transporters since several completely sequenced genomes display a large excess of IMs and BPs over ABCs (Quentin et al., 1999). The best-characterized system of this family, the *E. coli* maltose/maltodextrin transporter energized by MalK, constitutes a model for understanding the molecular functional mechanism of ABC transporters (Davidson and Chen, 2004).

The MOI Family Specific for Mineral and Organic Ions

The MOI family includes transport systems for inorganic anions such as thiosulfate and sulfate (Kertesz, 2001), molybdate (Self et al., 2001), and organic anions such as polyamines (Igarashi et al., 2001) and thiamine (Webb et al., 1998). Members of this family also transport ferric iron (Adhikari et al., 1996; Sanders et al., 1994). However, iron might be transported as a salt since crystals of the iron-binding protein of *Haemophilus influenzae* show that iron is coordinated by water and phosphate (Bruns et al., 1997). The ABC component of importers specific for phosphate cluster apart from the MOI family. However, the IMs are clustered with the IMs of the MOI family. The MOI family is the largest family of BPD systems. Opines like mannopines and chrysopine are transported by MOI family systems similar to the polyamine transporters. Most systems of the MOI family have two IMs, but ferric iron transporters have the two IM domains fused into a single polypeptide chain (IM2), while molybdate and thiamine transporters have only one IM.

The OTCN Family Involved in the Uptake of Osmoprotectants, Taurine (Alkyl Sulfonates), Alkyl Phosphonates, Phosphites, Hypophosphites, Cyanate, and Nitrate

This family is composed of systems involved in the transport of apparently unrelated solutes. ABC and IM are grouped in single clusters, respectively. Analysis of BP sequences led to the identification of two nonoverlapping clusters.

The first cluster groups systems involved in the transport of osmoprotectants, consisting of small modified peptides that contain an N,N,N'-trimethyl ammonium group like glycine-betaine, choline, and carnitine (Hosie and Poole, 2001). The properties of the transporters specific for osmoprotectants were recently reviewed (Kempf and Bremer, 1998). The most characterized system is the osmoregulatory ProU system of *E. coli*, determining a glycine-betaine transporter, which comprises genes encoding ProV (ABC), ProW (IM), and ProX (BP). They display an organization typical of BPD transporters. The BusA betaine transporter of *L. lactis* constitutes a remarkable exception to this organization scheme, where an extracytoplasmic domain corresponding to the BP is fused to the C terminus of the IM (Obis et al., 1999).

The second cluster is composed of systems involved, respectively, in the uptake of nitrate, cyanate, *N*-alkylsulfonates, alkylphosphonates, phosphates, and hypophosphites.

The OPN Family Specific for Di- and Oligopeptides, Oligosaccharides, and Nickel

Oligopeptides constitute an important source of nutrients and several systems are also involved in cell-cell communication (Detmers et al., 2001). Members of the OPN family have been found in all prokaryotic genera and are characterized by the fact that the two ABC subunits are encoded by different genes. Most of the transporters of this family are specific for di- or oligopeptides, recognizing a large panel of different peptides. In *E. coli*, it was recently demonstrated that a putative oligopeptide transporter encoded by the *yliABCD* operon is required for the uptake of glutathione (Suzuki et al., 2005).

OPN family systems have been reported to catalyze the import of nickel in a limited number of bacteria. Nickel is an essential cofactor for several enzymatic reactions. The Nik system of *E. coli* provides the Ni^{2+} ion for the anaerobic biosynthesis of hydrogenases and is similar in its composition and in the primary sequence of its components to the oligopeptide ABC transporters (Navarro et al., 1993). Nik importers appear to be more restricted in their distribution than oligopeptide transporters since homologues could be identified only in about 15 genomes.

In fact the substrate specificity of this conserved family of transporters is remarkably broad. Oligopeptide-like transporters have been implicated in the uptake of a class of opines such as agrocinopines, agropinic, and mannopinic acids (Hayman et al., 1993). Recently, the transport of disaccharides such as maltose and cellobiose in the hyperthermophilic archaeon *Sulfolobus sulfataricus* was dem-

onstrated to depend on a subgroup of OPN family transporters (Elferink et al., 2001). Clearly, the annotation of OPN family systems should be reevaluated in genome databases.

The PAO Family Specific for Polar Amino Acids and Opines

The PAO family includes transport systems for amino acids that have polar or charged side chains: lysine, histidine, ornithine, arginine, glutamine, glutamate, cystine, and diaminopimelic acid (Hosie and Poole, 2001). The best characterized system of this family is the *S. enterica* serovar Typhimurium histidine transporter, which is energized by HisP, the first ABC protein whose crystal structure was reported with a resolution of 1.5 Å (Hung et al., 1998). Opines such as octopine (N^2-(1-carboxyethyl)-L-arginine) and nopaline (N^2-(1,3-dicarboxypropyl)-L-arginine) are transported in agrobacteria by PAO family transporters (Kim and Farrand, 1997; Lyi et al., 1999). Typical systems have in general two IMs with the exception of the cystine and the glutamine-specific systems, which have only one IM. The *L. lactis* GlnPQ system has a nonconventional organization. GlnP is a multidomain protein with two BP domains fused to the IM domain (BP2IM), and GlnQ is the ABC protein, collectively leading to a transporter with four putative extracellular substrate recognition sites. Only the N-proximal BP domain in GlnP appeared to be critical for glutamine and glutamate uptake in vivo and in vitro (Schuurman-Wolters and Poolman, 2005).

The BPs specific to glutamine are homologous to the extracellular portion of eukaryote ionotropic glutamate receptors. Recent studies indicated that glutamate receptors share the fundamental mechanism of amino acid recognition with the bacterial PAO family BPs (Lampinen et al., 1998).

The HAA Family Specific for Hydrophobic Branched-Chain Amino Acids and Amides

The HAA family is composed of systems specific for the transport of the hydrophobic amino acids leucine, isoleucine, and valine

(Hosie and Poole, 2001). A transport system involved in the uptake of urea and short-chain aliphatic amines in *Methylophilus methylotrophus* belongs to this family (Mills et al., 1998). This system is homologous to the *Synechocystis* and *Anabaena* systems for the uptake of neutral amino acids Ala, Val, Phe, Ile, and Leu (Montesinos et al., 1997). It is therefore possible that the urea transporter of *M. methylotrophus* could also transport such amino acids. Systems of the HAA family have a characteristic organization made up of one or several BPs, two IMs, and two ABCs.

The eukaryote γ-aminobutyric acid type B (GABA(B)) receptors and the related metabotropic glutamate receptor-like family of G-protein-coupled receptors have their extracellular domains homologous to the bacterial leucine-binding protein. Furthermore, the effect of point mutations can be explained by the Venus flytrap model, which proposes that the initial step in the activation of the receptor by the agonist results from the closure of the two lobes of the binding domain (Galvez et al., 1999).

The MOS Family Specific for Monosaccharides

The MOS family systems are involved in the uptake of monosaccharides (pentoses and hexoses) like arabinose, D-allose, galactose, ribose, and xylose. The typical organization of these systems consists of one BP, one IM, and one ABC. The ABC subunit is made up of homologous halves, suggesting that a primordial gene duplication and subsequent fusion event occurred in the generation of the ancestral MOS system (Schneider, 2001). In the *Bacillus subtilis*, *Treponema pallidum*, *Borrelia burgdorferi*, *Archeoglobus fulgidus*, and *Aeropyrum pernix* sequenced genomes, several putative MOS family transporters were identified. However, the putative operons encoding these systems were apparently devoid of a typical substrate-binding protein. Rather, they were associated with secreted proteins homologous to a family of lipoproteins of unknown function, the so-called basic membrane proteins C (BMPC) (Gorbacheva et al., 2000), which constitute potent immuno-

gens in pathogenic bacteria. Psi-Blast analyses show that these lipoproteins display significant similarity to MOS family substrate-binding proteins (E. Dassa, unpublished results). We therefore speculate that at least some BMPCs might be involved in the uptake of an as-yet unidentified monosaccharide.

The DLM Family Specific for D- and L-Methionine and Methionine Derivatives

This family, previously called ABCY, was entered recently into the tribe of ABC BPD transporters and some members were demonstrated to be involved in the high-affinity uptake of D-methionine in *E. coli* (Gal et al., 2002; Merlin et al., 2002) and of D-/L-methionine and methionine sulfoxide in *B. subtilis* (Hullo et al., 2004). Systems of the DLM family have the same overall organization as BPD transporters with the difference that the extracytoplasmic protein is a lipoprotein, even in gram-negative bacteria. IMs display strong similarity to those of the OTCN family and have the characteristic EAA motif found exclusively in a cytoplasmic loop of IMs of the BPD import systems. ATPases of the family cluster near the ATPases of the PAO family. This unusual feature of the components of ABCY systems might indicate that they originate from the association of components from different families of BPD transporters. BPs display a slight similarity to BPs of the OTCN family and belong to a family of surface lipoproteins that includes the NlpA lipoprotein of *E. coli*. The crystal structure of the *Treponema pallidum* TP32 lipoprotein revealed the presence of a molecule of L-methionine in the substrate-binding cleft (Deka et al., 2004). In most genomes several copies of DLM family systems were found, suggesting the possibility that they recognize substrates different from methionine. In *Salmonella enteritidis*, the ABCY family SfbABC system was located in a pathogenicity islet of 4 kb that is inducible by iron limitation and by acidic pH and it was found that inactivation of the *sfbA* gene encoding LPP resulted in a mutant that is avirulent and induces protective immunity in BALB/c mice (Pattery et al., 1999). The membrane-associated lipoprotein-9 GmpC from *Staphylococcus aureus* was shown to bind the dipeptide GlyMet via specific side-chain interactions, suggesting a physiological role in the utilization of methionine-containing dipeptides (Williams et al., 2004).

LESSONS FROM GENOME COMPARISONS

The complete nucleotide sequence of several genomes is presently available and efforts have been developed to build the complete inventories of ABC systems in yeast (Decottignies and Goffeau, 1997), *E. coli* (Dassa et al., 1999; Linton and Higgins, 1998), *B. subtilis* (Quentin et al., 1999), *Mycobacterium tuberculosis* (Braibant et al., 2000), *Caenorhabditis elegans* (Sheps et al., 2004), and *Oryza sativa* (Garcia et al., 2004). Global comparisons of the ABC protein content of several genomes have also been made (Paulsen et al., 1998, 2000; Tomii and Kanehisa, 1998). In the course of the constitution of ABCISSE, our database of ABC systems, we have also analyzed the composition of about 200 completely sequenced genomes (Color Plate 30).

Genes encoding the ATPase of ABC systems constitute about 2% of total genes in most *Bacteria* and *Archaea*. In contrast, this ratio is reduced to about 4‰ in eukaryotes. Bacteria with a small genome (0.5 to 1.5 Mb) have about 15 ABC systems. These are mostly intracellular parasites growing inside host cells. The presence of homologous host genes or the availability of a metabolite can lead to gene inessentiality and to the subsequent gene disruption or deletion. It is therefore possible that the intersection of the ABC systems that are common in these species constitutes the minimal requirement of ABC systems for life. As a general rule, the number of ABC systems increases linearly with the size of the genome. This notion is in agreement with the observation that the number of transporters of all categories (ion gradient–driven, PTS, ABC, facilitators) is approximately proportional to genome size (Paulsen et al., 1998). There are, however,

some exceptions. The genome of *Thermotoga maritima* shows a very high content of ABC systems as compared with species of similar genome size. This is due to the extensive amplification of operons encoding ABC systems putatively involved in the uptake of oligosaccharides (11 systems) and oligopeptides (12 systems), leading to functional specialization. Two putative maltose transporters were shown to display different substrate specificities and affinities (the first for maltose, maltotriose, and mannotetraose and the second for maltose, maltotriose, and trehalose) and to be expressed under different growth conditions (Nanavati et al., 2005). The genomes of soil bacteria such as *Mesorhizobium loti* or *Agrobacterium tumefaciens* contain more than 200 ABC ATPases, including a majority of putative ABC importers, whose abundance probably reflects the ability of these bacteria to cope with fluctuating environmental conditions. On the other hand, the genome of *M. tuberculosis* (4.4 Mb) has only 38 systems. This number is significantly lower than that found in *E. coli* (4.6 Mb, 78 systems) or in *B. subtilis* (4.2 Mb, 84 systems). Since it was found that the total number of transporters was fairly constant among prokaryotes (Paulsen et al., 2000), this means that bacteria with low ABC systems contents compensate for this deficiency by a higher number of transporters from other functional categories. Eukaryotes display a smaller number of ABC systems with respect to genome size as compared with prokaryotes and this is particularly evident in the case of *Saccharomyces cerevisiae*, a free-living microorganism that shares with bacteria almost the same ecological niches. Indeed, high-affinity binding, protein-dependent importers are lacking in eukaryotes.

A closer look at the distribution of ABC systems among the sequenced genomes reveals several interesting features. Class 1 ABC systems (exporters with fused ABC and IM domains) are underrepresented in the genomes of *Bacteria* and are almost absent from the genomes of *Archaea*. By contrast, they represent the major fraction of ABC systems in eukaryotes. Class 2 ABC systems (ABC2 organization, no IM domains) are found in all genomes, even in the smallest ones. This observation establishes the physiological importance of this class of ABC systems, which contains proteins experimentally or putatively involved in gene expression regulation. The number of class 2 systems by genome ranges from 1 to 8 when the genome sizes vary from 0.58 to 132.5 Mb. Class 3 systems (mostly importers) are quasi-exclusively found in prokaryote genomes with one exception: the ABCA subfamily of eukaryote systems (Broccardo et al., 1999), which is composed, among others, of the human ABC1 (ABCA1) and ABCR (ABCA4) proteins involved in cholesterol and N-N-retinylidene-phosphatidylethanolamine transport, respectively (Orso et al., 2000; Weng et al., 1999). Incomplete class 3 systems are also found in the genomes of eukaryotes and are probable remnants of BPD transporters present on the genome of the ancestor of organelles.

In the past decade, a tremendous amount of genetic information was gathered from genome-sequencing projects. Our phylogenetic classification faces the complication of the extreme plasticity of functions displayed by ABC transporters. An effort is required to experimentally analyze in detail the functions of ABC transporters. In addition, sequence analyses should be refined to sort out new families of transport systems that may correspond to yet uncharacterized functions.

REFERENCES

Abramson, J., I. Smirnova, V. Kasho, G. Verner, H. R. Kaback, and S. Iwata. 2003. Structure and mechanism of the lactose permease of *Escherichia coli. Science* **301**:610–615.

Adhikari, P., S. A. Berish, A. J. Nowalk, K. L. Veraldi, S. A. Morse, and T. A. Mietzner. 1996. The *fbpABC* locus of *Neisseria gonorrhoeae* functions in the periplasm-to-cytosol transport of iron. *J. Bacteriol.* **178**:2145–2149.

Aguilar-Bryan, L., J. P. T. Clement, G. Gonzalez, K. Kunjilwar, A. Babenko, and J. Bryan. 1998. Toward understanding the assembly and structure of KATP channels. *Physiol. Rev.* **78**:227–245.

Ashcroft, S. J. H. 2000. The beta-cell K-ATP channel. *J. Membr. Biol.* **176**:187–206.

Bavoil, P., H. Nikaido, and K. von Meyenburg. 1977. Pleiotropic transport mutants of *Escherichia coli* lack porin, a major outer membrane protein. *Mol. Gen. Genet.* **158**:23–33.

Beck, S., A. Kelly, E. Radley, F. Khurshid, R. P. Alderton, and J. Trowsdale. 1992. DNA sequence analysis of 66 kb of the human MHC class II region encoding a cluster of genes for antigen processing. *J. Mol. Biol.* **228**:433–441.

Biville, F., H. Cwerman, S. Letoffe, M. S. Rossi, V. Drouet, J. M. Ghigo, and C. Wandersman. 2004. Haemophore-mediated signalling in *Serratia marcescens*: a new mode of regulation for an extra cytoplasmic function (ECF) sigma factor involved in haem acquisition. *Mol. Microbiol.* **53**:1267–1277.

Blight, M. A., and I. B. Holland. 1990. Structure and function of haemolysin B, P-glycoprotein and other members of a novel family of membrane translocators. *Mol. Microbiol.* **4**:873–880.

Boos, W., and A. Bohm. 2000. Learning new tricks from an old dog—MalT of the *Escherichia coli* maltose system is part of a complex regulatory network. *Trends Genet.* **16**:404–409.

Boos, W., and J. M. Lucht. 1996. Periplasmic binding protein-dependent ABC transporters, p. 1175–1209. *In* F. C. Neidhardt, R. Curtiss III, J. L. Ingraham, E. C. C. Lin, K. B. Low, B. Magasanik, W. S. Reznikoff, M. Riley, M. Schaechter, and H. E. Umbarger (ed.), Escherichia coli *and* Salmonella: Cellular and Molecular Biology, 2nd ed., vol. 1. ASM Press, Washington, D.C.

Boos, W., and H. Shuman. 1998. Maltose/maltodextrin system of *Escherichia coli*: transport, metabolism, and regulation. *Microbiol. Mol. Biol. Rev.* **62**:204–229.

Borths, E. L., K. P. Locher, A. T. Lee, and D. C. Rees. 2002. The structure of *Escherichia coli* BtuF and binding to its cognate ATP binding cassette transporter. *Proc. Natl. Acad. Sci. USA* **99**:16642–16647.

Bos, M. P., and J. Tommassen. 2004. Biogenesis of the gram-negative bacterial outer membrane. *Curr. Opin. Microbiol.* **7**:610–616.

Braibant, M., P. Gilot, and J. Content. 2000. The ATP binding cassette (ABC) transport systems of *Mycobacterium tuberculosis. FEMS Microbiol. Rev.* **24**:449–467.

Brass, J. M., W. Boos, and R. Hengge. 1981. Reconstitution of maltose transport in *malB* mutants of *Escherichia coli* through calcium-induced disruptions of the outer membrane. *J. Bacteriol.* **146**:10–17.

Braun, V., and M. Braun. 2002. Active transport of iron and siderophore antibiotics. *Curr. Opin. Microbiol.* **5**:194–201.

Broccardo, C., M. Luciani, and G. Chimini. 1999. The ABCA subclass of mammalian transporters. *Biochim. Biophys. Acta* **1461**:395–404.

Bruns, C. M., A. J. Nowalk, A. S. Arvai, M. A. McTigue, K. G. Vaughan, T. A. Mietzner, and D. E. McRee. 1997. Structure of *Haemophilus influenzae* Fe(+3)-binding protein reveals convergent evolution within a superfamily. *Nat. Struct. Biol.* **4:**919–924.

Chakraburtty, K. 2001. Translational regulation by ABC systems. *Res. Microbiol.* **152:**391–399.

Claverys, J. P. 2001. A new family of high-affinity ABC manganese and zinc permeases. *Res. Microbiol.* **152:**231–243.

Dassa, E., and P. Bouige. 2001. The ABC of ABCs: a phylogenetic and functional classification of ABC systems in living organisms. *Res. Microbiol.* **152:**211-229.

Dassa, E., M. Hofnung, I. T. Paulsen, and M. H. Saier, Jr. 1999. The *Escherichia coli* ABC transporters: an update. *Mol. Microbiol.* **32:**887–889.

Dassa, E., and S. Muir. 1993. Membrane topology of MalG, an inner membrane protein from the maltose transport system of *Escherichia coli. Mol. Microbiol.* **7:**29–38.

Davidson, A. L. 2002. Mechanism of coupling of transport to hydrolysis in bacterial ATP-binding cassette transporters. *J. Bacteriol.* **184:**1225–1233.

Davidson, A. L., and J. Chen. 2004. ATP-binding cassette transporters in bacteria. *Annu. Rev. Biochem.* **73:**241–268.

Davidson, A. L., and H. Nikaido. 1991. Purification and characterization of the membrane-associated components of the maltose transport system from *Escherichia coli. J. Biol. Chem.* **266:**8946–8951.

Dean, M., and T. Annilo. 2005. Evolution of the ATP-binding cassette (ABC) transporter superfamily in vertebrates. *Annu. Rev. Genomics Hum. Genet.* **6:**123–142.

Decottignies, A., and A. Goffeau. 1997. Complete inventory of the yeast ABC proteins. *Nat. Genet.* **15:**137–145.

Deka, R. K., L. Neil, K. E. Hagman, M. Machius, D. R. Tomchick, C. A. Brautigam, and M. V. Norgard. 2004. Structural evidence that the 32-kilodalton lipoprotein (Tp32) of *Treponema pallidum* is an L-methionine-binding protein. *J. Biol. Chem.* **279:**55644–55650.

Detmers, F. J. M., F. C. Lanfermeijer, and B. Poolman. 2001. Peptides and ATP binding cassette peptide transporters. *Res. Microbiol.* **152:**245–258.

Diederichs, K., J. Diez, G. Greller, C. Muller, J. Breed, C. Schnell, C. Vonrhein, W. Boos, and W. Welte. 2000. Crystal structure of MalK, the ATPase subunit of the trehalose/maltose ABC transporter of the archaeon *Thermococcus litoralis. EMBO J.* **19:**5951–5961.

Dintilhac, A., G. Alloing, C. Granadel, and J. P. Claverys. 1997. Competence and virulence of *Streptococcus pneumoniae*—*adc* and *psaA* mutants exhibit a requirement for Zn and Mn resulting from inactivation of putative ABC metal permeases. *Mol. Microbiol.* **25:**727–739.

Dong, J. S., R. Lai, K. Nielsen, C. A. Fekete, H. F. Qiu, and A. G. Hinnebusch. 2004. The essential ATP-binding cassette protein RLI1 functions in translation by promoting preinitiation complex assembly. *J. Biol. Chem.* **279:**42157–42168.

Dosch, D. C., G. L. Helmer, S. H. Sutton, F. F. Salvacion, and W. Epstein. 1991. Genetic analysis of potassium transport loci in *Escherichia col*: evidence for three constitutive systems mediating uptake of potassium. *J. Bacteriol.* **173:**687–696.

Elferink, M. G. L., S. V. Albers, W. N. Konings, and A. J. M. Driessen. 2001. Sugar transport in *Sulfolobus solfataricus* is mediated by two families of binding protein-dependent ABC transporters. *Mol. Microbiol.* **39:**1494–1503.

Enz, S., H. Brand, C. Orellana, S. Mahren, and V. Braun. 2003. Sites of interaction between the FecA and FecR signal transduction proteins of ferric citrate transport in Escherichia coli K-12. *J. Bacteriol.* **185:**3745–3752.

Faraldo-Gomez, J. D., and M. S. Sansom. 2003. Acquisition of siderophores in gram-negative bacteria. *Nat. Rev. Mol. Cell. Biol.* **4:**105–116.

Fath, M. J., and R. Kolter. 1993. ABC transporters −bacterial exporters. *Microbiol. Rev.* **57:**995–1017.

Fenno, J. C., A. Shaikh, G. Spatafora, and P. Fives-Taylor. 1995. The *fimA* locus of *Streptococcus parasanguis* encodes an ATP-binding membrane transport system. *Mol. Microbiol.* **15:**849–863.

Ferguson, A. D., and J. Deisenhofer. 2004. Metal import through microbial membranes. *Cell* **116:**15–24.

Fukami-Kobayashi, K., Y. Tateno, and K. Nishikawa. 1999. Domain dislocation: a change of core structure in periplasmic binding proteins in their evolutionary history. *J. Mol. Biol.* **286:**279–290.

Gal, J., A. Szvetnik, R. Schnell, and M. Kalman. 2002. The metD D-methionine transporter locus of *Escherichia coli* is an ABC transporter gene cluster. *J. Bacteriol.* **184:**4930–4932.

Galvez, T., M. L. Parmentier, C. Joly, B. Malitschek, K. Kaupmann, R. Kuhn, H. Bittiger, W. Froestl, B. Bettler, and J. P. Pin. 1999. Mutagenesis and modeling of the GABA(B) receptor extracellular domain support a Venus flytrap mechanism for ligand binding. *J. Biol. Chem.* **274:**13362–13369.

Garcia, O., P. Bouige, C. Forestier, and E. Dassa. 2004. Inventory and comparative analysis of rice and *Arabidopsis* ATP-binding cassette (ABC) systems. *J. Mol. Biol.* **343:**249–265.

Gilson, E., G. Alloing, T. Schmidt, J. P. Claverys, R. Dudler, and M. Hofnung. 1988. Evidence for high affinity binding protein-dependent transport systems in Gram-positive bacteria and in *Mycoplasma. EMBO J.* **7:**3971–3974.

Gorbacheva, V. Y., H. P. Godfrey, and F. C. Cabello. 2000. Analysis of the *bmp* gene family in *Borrelia burgdorferi* sensu lato. *J. Bacteriol.* **182:**2037–2042.

Gottesman, M. M., T. Fojo, and S. E. Bates. 2002. Multidrug resistance in cancer: role of ATP-dependent transporters. *Nat. Rev. Cancer* **2:**48–58.

Haardt, M., and E. Bremer. 1996. Use of phoA and lacZ fusions to study the membrane topology of ProW, a component of the osmoregulated ProU transport system of Escherichia coli. *J. Bacteriol.* **178:**5370–5381.

Hashimoto, W., J. S. He, Y. Wada, H. Nankai, B. Mikami, and K. Murata. 2005. Proteomics-based identification of outer-membrane proteins responsible for import of macromolecules in *Sphingomonas* sp A1: alginate-binding flagellin on the cell surface. *Biochemistry* **44:**13783–13794.

Hashimoto, W., K. Momma, H. Miki, Y. Mishima, E. Kobayashi, O. Miyake, S. Kawai, H. Nankai, B. Mikami, and K. Murata. 1999. Enzymatic and genetic bases on assimilation, depolymerization, and transport of heteropolysaccharides in bacteria. *J. Biosci. Bioeng.* **87:**123–136.

Hayman, G. T., S. B. Von Bodman, H. Kim, P. Jiang, and S. K. Farrand. 1993. Genetic analysis of the agrocinopine catabolic region of *Agrobacterium tumefaciens* Ti plasmid pTiC58, which encodes genes required for opine and agrocin 84 transport. *J. Bacteriol.* **175:**5575–5584.

Henderson, D. P., and S. M. Payne. 1994. *Vibrio cholerae* iron transport system: roles of heme and siderophore iron transport in virulence and identification of a gene associated with multiple iron transport systems. *Infect. Immun.* **62:**5120–5125.

Hennessey, E. S., and J. K. Broome-Smith. 1993. Gene fusion techniques for determining membrane protein topology. *Curr. Opin. Struct. Biol.* **3:**524–531.

Higgins, C. F., and G. F. L. Ames. 1981. Two periplasmic transport proteins which interact with a common membrane receptor show extensive homology: complete nucleotide sequences. *Proc. Natl. Acad. Sci. USA* **78:**6038–6042.

Holland, I. B., and M. A. Blight. 1999. ABC-ATPases, adaptable energy generators fuelling transmembrane movement of a variety of molecules organisms from bacteria to humans. *J. Mol. Biol.* **293:**381–399.

Hopfner, K. P., and J. A. Tainer. 2003. Rad50/SMC proteins and ABC transporters: unifying concepts from high-resolution structures. *Curr. Opin. Struct. Biol.* **13:**249–255.

Hosie, A. H. F., and P. S. Poole. 2001. Bacterial ABC transporters of amino acids. *Res. Microbiol.* **152:**259–270.

Hughes, A. L. 1994. Evolution of the ATP-binding-cassette transmembrane transporters of vertebrates. *Mol. Biol. Evol.* **11:**899–910.

Hullo, M. F., S. Auger, E. Dassa, A. Danchin, and I. Martin-Verstraete. 2004. The metNPQ operon of *Bacillus subtilis* encodes an ABC permease transporting methionine sulfoxide, D- and L-methionine. *Res. Microbiol.* **155:**80–86.

Hung, L. W., I. X. Y. Wang, K. Nikaido, P. Q. Liu, G. F. L. Ames, and S. H. Kim. 1998. Crystal structure of the ATP-binding subunit of an ABC transporter. *Nature* **396:**703–707.

Hunke, S., M. Mourez, M. Jehanno, E. Dassa, and E. Schneider. 2000. ATP modulates subunit-subunit interactions in an ATP-binding cassette transporter (MalFGK2) determined by site-directed chemical cross-linking. *J. Biol. Chem.* **275:**15526–15534.

Igarashi, K., K. Ito, and K. Kashiwagi. 2001. Polyamine uptake systems in *Escherichia coli. Res. Microbiol.* **152:**271–278.

Janulczyk, R., S. Ricci, and L. Bjorck. 2003. MtsABC is important for manganese and iron transport, oxidative stress resistance, and virulence of *Streptococcus pyogenes. Infect. Immun.* **71:**2656–2664.

Joly, N., A. Bohm, W. Boos, and E. Richet. 2004. MalK, the ATP-binding cassette component of the *Escherichia coli* maltodextrin transporter, inhibits the transcriptional activator MalT by antagonizing inducer binding. *J. Biol. Chem.* **279:**33123–33130.

Karpel, R., T. Alon, G. Glaser, S. Schuldiner, and E. Padan. 1991. Expression of a sodium proton antiporter (NhaA) in Escherichia coli is induced by Na^+ and Li^+ ions. *J. Biol. Chem.* **266:**21753–21759.

Kaur, P., and J. Russell. 1998. Biochemical coupling between the DrrA and DrrB proteins of the doxorubicin efflux pump of *Streptomyces peucetius. J. Biol. Chem.* **273:**17933–17939.

Kempf, B., and E. Bremer. 1998. Uptake and synthesis of compatible solutes as microbial stress responses to high-osmolality environments. *Arch. Microbiol.* **170:**319–330.

Kerppola, R. E., and G. F. L. Ames. 1992. Topology of the hydrophobic membrane-bound components of the histidine periplasmic permease. Comparisons with other members of the family. *J. Biol. Chem.* **267:**2329–2336.

Kerppola, R. E., V. K. Shyamala, P. Klebba, and G. F. L. Ames. 1991. The membrane-bound proteins of periplasmic permeases form a complex. Identification of the histidine permease HisQMP complex. *J. Biol. Chem.* **266:**9857–9865.

Kerr, I. D. 2004. Sequence analysis of twin ATP binding cassette proteins involved in translational control, antibiotic resistance, and ribonuclease L inhibition. *Biochem. Biophys. Res. Commun.* **315:**166–173.

Kertesz, M. A. 2001. Bacterial transporters for sulfate and organosulfur compounds. *Res. Microbiol.* **152:**279–290.

Kim, H., and S. K. Farrand. 1997. Characterization of the *acc* operon from the nopaline-type Ti plasmid pTic58, which encodes utilization of agrocinopines A and B and susceptibility to agrocin 84. *J. Bacteriol.* **179:**7559–7572.

Kolenbrander, P. E., R. N. Andersen, and N. Ganeshkumar. 1994. Nucleotide sequence of the *Streptococcus gordonii* PK488 coaggregation adhesin gene, *scaA*, and ATP-binding cassette. *Infect. Immun.* **62:**4469–4480.

Kossman, M., C. Wolff, and M. Manson. 1988. Maltose chemoreceptor of *Escherichia coli*: interaction of maltose-binding protein and the Tar signal transducer. *J. Bacteriol.* **170:**4516–4521.

Kuan, G., E. Dassa, W. Saurin, M. Hofnung, and M. H. Saier. 1995. Phylogenetic analyses of the ATP-binding constituents of bacterial extracytoplasmic receptor-dependent ABC-type nutrient uptake permeases. *Res. Microbiol.* **146:**271–278.

Kung, C., and P. Blount. 2004. Channels in microbes: so many holes to fill. *Mol. Microbiol.* **53:**373–380.

Lampinen, M., O. Pentikainen, M. S. Johnson, and K. Keinanen. 1998. AMPA receptors and bacterial periplasmic amino acid-binding proteins share the ionic mechanism of ligand recognition. *EMBO J.* **17:**4704–4711.

Lanfermeijer, F. C., A. Picon, W. N. Konings, and B. Poolman. 1999. Kinetics and consequences of binding of nona- and dodecapeptides to the oligopeptide binding protein (OppA) of Lactococcus lactis. *Biochemistry* **38:**14440–14450.

Leonard, B. A. B., A. Podbielski, P. J. Hedberg, and G. M. Dunny. 1996. *Enterococcus faecalis* pheromone binding protein, PrgZ, recruits a chromosomal oligopeptide permease system to import sex pheromone cCF10 for induction of conjugation. *Proc. Natl. Acad. Sci. USA* **93:**260–264.

Levdikov, V. M., E. V. Blagova, J. A. Brannigan, L. Wright, A. A. Vagin, and A. J. Wilkinson. 2005. The structure of the oligopeptide-binding protein, AppA, from *Bacillus subtilis* in complex with a nonapeptide. *J. Mol. Biol.* **345:**879–892.

Linton, K. J., and C. F. Higgins. 1998. The *Escherichia coli* ATP-binding cassette (ABC) proteins. *Mol. Microbiol.* **28:**5–13.

Locher, K. P., A. T. Lee, and D. C. Rees. 2002. The *E. coli* BtuCD structure: a framework for ABC transporter architecture and mechanism. *Science* **296:**1091–1098.

Lyi, S. M., S. Jafri, and S. C. Winans. 1999. Mannopinic acid and agropinic acid catabolism region of the octopine-type Ti plasmid pTi15955. *Mol. Microbiol.* **31:**339–347.

Marra, A., S. Lawson, J. S. Asundi, D. Brigham, and A. E. Hromockyj. 2002. In vivo characterization of the *psa* genes from *Streptococcus pneumoniae* in multiple models of infection. *Microbiology* **148:**1483–1491.

Merlin, C., G. Gardiner, S. Durand, and M. Masters. 2002. The *Escherichia coli metD* locus encodes an ABC transporter which includes Abc (MetN),

YaeE (MetI), and YaeC (MetQ). *J. Bacteriol.* **184:**5513–5517.

Mikkat, S., and M. Hagemann. 2000. Molecular analysis of the *ggtBCD* gene cluster of *Synechocystis sp* strain PCC6803 encoding subunits of an ABC transporter for osmoprotective compounds. *Arch. Microbiol.* **174:**273–282.

Mills, J., N. R. Wyborn, J. A. Greenwood, S. G. Williams, and C. W. Jones. 1998. Characterisation of a binding-protein-dependent, active transport system for short-chain amides and urea in the methylotrophic bacterium *Methylophilus methylotrophus*. *Eur. J. Biochem.* **251:**45–53.

Montesinos, M. L., A. Herrero, and E. Flores. 1997. Amino acid transport in taxonomically diverse cyanobacteria and identification of two genes encoding elements of a neutral amino acid permease putatively involved in recapture of leaked hydrophobic amino acids. *J. Bacteriol.* **179:**853–862.

Mourez, M., M. Hofnung, and E. Dassa. 1997. Subunit interactions in ABC transporters: a conserved sequence in hydrophobic membrane proteins of periplasmic permeases defines an important site of interaction with the ATPase subunits. *EMBO J.* **16:**2066–2077.

Nanavati, D. M., T. N. Nguyen, and K. M. Noll. 2005. Substrate specificities and expression patterns reflect the evolutionary divergence of maltose ABC transporters in *Thermotoga maritima*. *J. Bacteriol.* **187:**2002–2009.

Navarro, C., L. F. Wu, and M. A. Mandrand-Berthelot. 1993. The *nik* operon of *Escherichia coli* encodes a periplasmic binding protein-dependent transport system for nickel. *Mol. Microbiol.* **9:**1181–1191.

Neugebauer, H., C. Herrmann, W. Kammer, G. Schwarz, A. Nordheim, and V. Braun. 2005. ExbBD-dependent transport of maltodextrins through the novel MalA protein across the outer membrane of Caulobacter crescentus. *J. Bacteriol.* **187:**8300–8311.

Obis, D., A. Guillot, J. C. Gripon, P. Renault, A. Bolotin, and M. Y. Mistou. 1999. Genetic and biochemical characterization of a high-affinity betaine uptake system (BusA) in *Lactococcus lactis* reveals a new functional organization within bacterial ABC transporters. *J. Bacteriol.* **181:**6238–6246.

Orso, E., C. Broccardo, W. E. Kaminski, A. Bottcher, G. Liebisch, W. Drobnik, A. Gotz, O. Chambenoit, W. Diederich, T. Langmann, T. Spruss, M. F. Luciani, G. Rothe, K. J. Lackner, G. Chimini, and G. Schmitz. 2000. Transport of lipids from Golgi to plasma membrane is defective in Tangier disease patients and ABC1-deficient mice. *Nat. Genet.* **24:**192–196.

Parra-Lopez, C., R. Lin, A. Aspedon, and E. A. Groisman. 1994. A *Salmonella* protein that is required for resistance to antimicrobial peptides and transport of potassium. *EMBO J.* **13:**3964–3972.

Pattery, T., J. P. Hernalsteens, and H. De Greve. 1999. Identification and molecular characterization of a novel *Salmonella enteritidis* pathogenicity islet encoding an ABC transporter. *Mol. Microbiol.* **33:**791–805.

Paulsen, I. T., L. Nguyen, M. K. Sliwinski, R. Rabus, and M. H. Saier. 2000. Microbial genome analyses: comparative transport capabilities in eighteen prokaryotes. *J. Mol. Biol.* **301:**75–100.

Paulsen, I. T., M. K. Sliwinski, and M. H. Saier, Jr. 1998. Microbial genome analyses: global comparisons of transport capabilities based on phylogenies, bioenergetics and substrate specificities. *J. Mol. Biol.* **277:**573–592.

Pavelka, M. J., S. F. Hayes, and R. P. Silver. 1994. Characterization of KpsT, the ATP-binding component of the ABC transporter involved with the export of capsular polysialic acid in *Escherichia coli* K1. *J. Biol. Chem.* **269:**20149–20158.

Quentin, Y., G. Fichant, and F. Denizot. 1999. Inventory, assembly and analysis of *Bacillus subtilis* ABC transport systems. *J. Mol. Biol.* **287:**467–484.

Quiocho, F. A., and P. S. Ledvina. 1996. Atomic structure and specificity of bacterial periplasmic receptors for active transport and chemotaxis: variation of common themes. *Mol. Microbiol.* **20:**17–25.

Reeves, A. R., J. N. D'Elia, J. Frias, and A. A. Salyers. 1996. A *Bacteroides thetaiotaomicron* outer membrane protein that is essential for utilization of maltooligosaccharides and starch. *J. Bacteriol.* **178:**823–830.

Reyes, C. L., and G. Chang. 2005. Structure of the ABC transporter MsbA in complex with ADP-vanadate and lipopolysaccharide. *Science* **308:**1028–1031.

Richet, E., N. Joly, and O. Danot. 2005. Two domains of MalT, the activator of the *Escherichia coli* maltose regulon, bear determinants essential for anti-activation by MalK. *J. Mol. Biol.* **347:**1–10.

Rince, A., A. Dufour, P. Uguen, J. P. Lepennec, and D. Haras. 1997. Characterization of the lacticin 481 operon: the *Lactococcus lactis* genes lctF, lctE, and lctG encode a putative ABC transporter involved in bacteriocin immunity. *Appl. Environ. Microbiol.* **63:**4252–4260.

Russell, R. R. B., J. Aduse-Opoku, I. C. Sutcliffe, L. Tao, and J. J. Ferretti. 1992. A binding protein-dependent transport system in *Streptococcus mutans* responsible for multiple sugar metabolism. *J. Biol. Chem.* **267:**4631–4637.

Sanders, J. D., L. D. Cope, and E. J. Hansen. 1994. Identification of a locus involved in the utilization of iron by *Haemophilus influenzae*. *Infect. Immun.* **62:**4515–4525.

Saurin, W., M. Hofnung, and E. Dassa. 1999. Getting in or out: early segregation between importers and exporters in the evolution of ATP-binding cassette (ABC) transporters. *J. Mol. Evol.* **48:**22–41.

Saurin, W., W. Köster, and E. Dassa. 1994. Bacterial binding protein-dependent permeases: characterization of distinctive signatures for functionally related integral cytoplasmic membrane proteins. *Mol. Microbiol.* **12:**993–1004.

Schafer, K., U. Magnusson, F. Scheffel, A. Schiefner, M. O. J. Sandgren, K. Diederichs, W. Welte, A. Hulsmann, E. Schneider, and S. L. Mowbray. 2004. X-ray structures of the maltose-maltodextrin-binding protein of the thermoacidophilic bacterium *Alicyclobacillus acidocaldarius* provide insight into acid stability of proteins. *J. Mol. Biol.* **335:**261–274.

Schiefner, A., G. Holtmann, K. Diederichs, W. Welte, and E. Bremer. 2004. Structural basis for the binding of compatible solutes by ProX from the hyperthermophilic archaeon *Archaeoglobus fulgidus*. *J. Biol. Chem.* **279:**48270–48281.

Schlösser, A., T. Kampers, and H. Schrempf. 1997. The *Streptomyces* ATP-binding component MsiK assists in cellobiose and maltose transport. *J. Bacteriol.* **179:**2092–2095.

Schneider, E. 2001. ABC transporters catalyzing carbohydrate uptake. *Res. Microbiol.* **152:**303–310.

Schneider, E., and S. Hunke. 1998. ATP-binding-cassette (ABC) transport systems: functional and structural aspects of the ATP-hydrolyzing subunits/domains. *FEMS Microbiol. Rev.* **22:**1–20.

Schuurman-Wolters, G. K., and B. Poolman. 2005. Substrate specificity and ionic regulation of GlnPQ from *Lactococcus lactis*. An ATP-binding cassette transporter with four extracytoplasmic substrate-binding domains. *J. Biol. Chem.* **280:**23785–23790.

Self, W. T., A. M. Grunden, A. Hasona, and K. T. Shanmugam. 2001. Molybdate transport. *Res. Microbiol.* **152:**311–321.

Sharff, A. J., L. E. Rodseth, J. C. Spurlino, and F. A. Quiocho. 1992. Crystallographic evidence of a large ligand-induced hinge-twist motion between the 2 domains of the maltodextrin binding protein involved in active transport and chemotaxis. *Biochemistry* **31:**10657–10663.

Sheps, J. A., S. Ralph, Z. Zhao, D. L. Baillie, and V. Ling. 2004. The ABC transporter gene family of *Caenorhabditis elegans* has implications for the evolutionary dynamics of multidrug resistance in eukaryotes. *Genome Biol* **5:**R15.

Stumpe, S., and E. P. Bakker. 1997. Requirement of a large K+-uptake capacity and of extracytoplasmic protease activity for protamine resistance of *Escherichia coli*. *Arch. Microbiol.* **167:**126–136.

Suzuki, H., T. Koyanagi, S. Izuka, A. Onishi, and H. Kumagai. 2005. The *yliA*, *-B*, *-C*, and *-D* genes of *Escherichia coli* K-12 encode a novel glutathione importer with an ATP-binding cassette. *J. Bacteriol.* **187:**5861–5867.

Szmelcman, S., and M. Hofnung. 1975. Maltose transport in *Escherichia coli* K-12: involvement of the bacteriophage λ receptor. *J. Bacteriol.* **124:**112–118.

Tomii, K., and M. Kanehisa. 1998. A comparative analysis of ABC transporters in complete microbial genomes. *Genome Res.* **8:**1048–1059.

Touze, T., J. Eswaran, E. Bokma, E. Koronakis, C. Hughes, and V. Koronakis. 2004. Interactions underlying assembly of the *Escherichia coli* AcrAB-TolC multidrug efflux system. *Mol. Microbiol.* **53:** 697–706.

Traxler, B., D. Boyd, and J. Beckwith. 1993. The topological analysis of integral cytoplasmic membrane proteins. *J. Membr. Biol.* **132:**1–11.

Vazquez de Aldana, C. R., M. J. Marton, and A. G. Hinnebusch. 1995. GCN20, a novel ATP binding cassette protein, and GCN1 reside in a complex that mediates activation of the eIF-2 alpha kinase GCN2 in amino acid-starved cells. *EMBO J.* **14:**3184–3199.

Vazquez, M., O. Santana, and C. Quinto. 1993. The NodI and NodJ proteins from *Rhizobium* and *Bradyrhizobium* strains are similar to capsular polysaccharide secretion proteins from Gram-negative bacteria. *Mol. Microbiol.* **8:**369–377.

von Heijne, G. 1994. Membrane proteins: from sequence to structure. *Annu. Rev. Biophys. Biomol. Struct.* **23:**167–192.

Wandersman, C., and P. Delepelaire. 2004. Bacterial iron sources: from siderophores to hemophores. *Annu. Rev. Microbiol.* **58:**611–647.

Webb, E., K. Claas, and D. Downs. 1998. *thiBPQ* encodes an ABC transporter required for transport of thiamine and thiamine pyrophosphate in *Salmonella typhimurium. J. Biol. Chem.* **273:**8946–8950.

Weng, J., N. L. Mata, S. M. Azarian, R. T. Tzekov, D. G. Birch, and G. H. Travis. 1999. Insights into the function of Rim protein in photoreceptors and etiology of Stargardt's disease from the phenotype in ABCR knockout mice. *Cell* **98:**13–23.

Williams, W. A., R. G. Zhang, M. Zhou, G. Joachimiak, P. Gornicki, D. Missiakas, and A. Joachimiak. 2004. The membrane-associated lipoprotein-9 GmpC from *Staphylococcus aureus* binds the dipeptide GlyMet via side chain interactions. *Biochemistry* **43:**16193–16202.

Yaghmai, R., and G. L. Hazelbauer. 1993. Strategies for differential sensory responses mediated through the same transmembrane receptor. *EMBO J.* **12:**1897–1905.

ANTIMICROBIAL AND STRESS RESISTANCE

Keith Poole

18

Microbes often exist in hostile environments (in nature and, for pathogenic organisms, in hosts) where they are exposed to deleterious compounds (antimicrobials in natural and clinical settings, solvents and heavy metals at sites of pollution, detergents in the mammalian gut) and stresses (e.g., oxidative, acid) with the potential to compromise cell growth and survival. Owing to the presence of porin channels in the outer membrane (OM) that surrounds the typical gram-negative bacterium, many of these toxins/stressors readily access the periplasm, where they can adversely impact periplasmic functions and from which they can ultimately access and target essential processes operating in the cytosol. As such, bacteria have evolved protective mechanisms that operate in the periplasm and these include, predominantly, efflux (of antimicrobials, solvents, and metals), detoxification (of metals and reactive oxygen species [ROS]), and sequestration (of metals). The periplasm also serves as a useful compartment for sensing environmental toxins and/or stresses and initiating signal transduction pathways for the recruitment of appropriate resistance mechanisms and adaptive responses. Both protective and signaling functions play vital roles in bacterial cell survival, in nature and in plant and animal hosts, in the latter instance contributing significantly to bacterial virulence. This chapter highlights the current literature vis-à-vis periplasmic mechanisms of antimicrobial and stress resistance and periplasmic constituents of signaling pathways that influence expression of resistance (periplasmic and other) determinants.

EFFLUX

Increasingly, efflux-mediated exclusion is recognized as a major determinant of bacterial resistance to exogenous toxic compounds (i.e., antibiotics, biocides, heavy metals, detergents, and organic solvents) (Poole, 2005). Five families of antimicrobial efflux systems have been described in gram-negative bacteria, including the major facilitator (MF) superfamily, the ATP-binding cassette (ABC) family, the resistance-nodulation-division (RND) family, the small multidrug resistance (SMR) family, and the multidrug and toxic compound extrusion (MATE) family (Putman et al., 2000). Unlike most other families, pumps of the RND family are cell envelope-spanning and comprise three components, including a cytoplasmic membrane (CM)-associated drug/toxin-proton antiporter (the RND component), a periplasmic membrane fusion protein (the MFP

Keith Poole, Department of Microbiology and Immunology, Queen's University, Kingston, Ontario, Canada K7L 3N6.

The Periplasm
Edited by Michael Ehrmann © 2007 ASM Press, Washington, D.C.

component), and an OM-spanning channel (the OM factor [OMF]) (Fig. 1). Such a tripartite, envelope-spanning arrangement is, however, seen on occasion with other pump families (e.g., the ABC-type macrolide-specific MacAB-TolC system in *Escherichia coli* [Poole, 2004a] and the MF-type EmrAB-TolC [Poole, 2004a] [Fig. 1] and VceCAB [Borges-Walmsley et al., 2005] multidrug efflux systems of *E. coli* and *Vibrio cholerae*, respectively). Envelope-spanning pumps function to deliver cytosolic and/or periplasmic toxins to the cell exterior, using the OM barrier to restrict their reentry. RND-type pumps are broadly distributed among gram-negative bacteria where they are significant determinants of solvent tolerance and resistance to detergents (e.g., bile), heavy metals, and antimicrobials, including antibiotics and biocides (Table 1).

Multidrug

Many RND-type pumps are multidrug exporters that are capable of extruding a wide range of clinically relevant antibiotics (e.g., the well-characterized MexAB-OprM and AcrAB-TolC pumps of *Pseudomonas aeruginosa* and *E. coli*, respectively) (Table 1) and, so, contribute to antibiotic resistance in clinical strains of an increasingly large number of gram-negative

bacterial species (Poole, 2004a, 2005). Although implicated in both intrinsic and acquired multidrug resistance in lab and clinical isolates of a variety of gram-negative bacteria (Poole, 2004a), antibiotic-exporting RND pumps are most often cited as significant determinants of resistance to the fluoroquinolone class of antimicrobials (Poole, 2005). Still, the RND family MexXY/OprM efflux system of *P. aeruginosa* is highlighted as an important determinant of intrinsic and acquired aminoglycoside resistance in lab and clinical, especially cystic fibrosis, isolates (Sobel et al., 2003), being one example of a very limited number of gram-negative RND pumps able to accommodate this class of antimicrobial (Poole, 2005). The MexAB-OprM system of this organism has also been cited as a determinant of resistance to β-lactams in clinical isolates (Poole, 2005).

Despite their significance as determinants of antibiotic resistance, such multidrug exporters also, in many instances, export biocides, solvents, detergents, and in a limited number of examples, plant-derived antimicrobials (phytoalexins and isoflavanoids), metabolic inhibitors, organometallic compounds (tributyltin), homoserine lactones involved in quorum sensing, and possibly virulence factors (Poole, 2005) (Fig. 1, Table 1). The latter, in particular, speaks

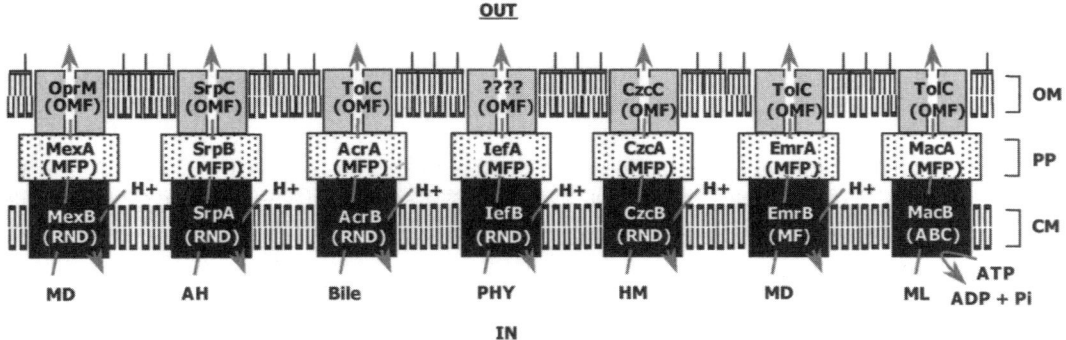

FIGURE 1 Schematic representation of periplasm-spanning gram-negative efflux systems that accommodate agents with antimicrobial activity. Pumps are highlighted either as representative of the different families of such efflux systems known to operate in gram-negative bacteria or to emphasize the substrates accommodated by these pumps. The figure is not intended to suggest that substrates are exported from the cytoplasm. MD, multiple drugs; AH, aromatic hydrocarbons (i.e., solvents); PHY, phytolectins; HM, heavy metals; ML, macrolides; OM, outer membrane; PP, periplasm; CM, cytoplasmic membrane; ????, the OMF component has yet to be identified.

TABLE 1 Substrate profiles and distribution of RND family pumps in gram-negative bacteria

Substrate	Representative pump(s)	Organisms	Reference(s)
Antimicrobials (including antibiotics and/ or biocides)	MexAB-OprM, AcrAB-TolC	*Pseudomonas aeruginosa, Escherichia coli,* most if not all gram-negative organisms	Poole, 2004a, 2005
Heavy metals (Cd, Zn, and sometimes Co)	CzcCBA	*Wautersia metallidurans, P. aeruginosa, Pseudomonas fluorescens, Alcaligenes* sp., *Xanthomonas campestris,*[a] *Legionella pneumophila, Pseudomonas putida,*[a] *Ralstonia solanacearum*[a]	Hassan et al., 1999; Mergeay et al., 2003; Nies, 2003; Rossbach et al., 2000
Ni	CnrCBA	*W. metallidurans*	Mergeay et al., 2003; Nies, 2003
Cu/Ag	CusC(F)BA	*E. coli, Helicobacter pylori, W. metallidurans,*[a] *P. putida*[a]	Canovas et al., 2003; Rensing and Gross, 2003; Waidner et al., 2002
Ag	SilCBA[b]	*Salmonella enterica* serovar Typhimurium, *E. coli* (K-12 and H7:O157), *Serratia marcescens, Klebsiella pneumoniae*	Silver, 2003
Solvents (a.k.a. aromatic hydrocarbons)	SrpABC, MexAB-OprM, AcrAB-TolC	*P. putida, P. aeruginosa, E. coli,* many gram-negative bacteria	Fernandes et al., 2003; Poole, 2005; Ramos et al., 2002
Bile	AcrAB-TolC[c]	*E. coli, S. enterica* serovar Typhimurium, *Campylobacter jejuni, Neisseria gonorrhoeae, Vibrio cholerae*	Poole, 2005
Tributyltin	TbtABM	*Pseudomonas stutzeri*	Jude et al., 2004
Isoflavonoids/ phytoalexins[d]	IefAB/AcrAB	*Agrobacterium tumefaciens, Erwinia amylovora*	Poole, 2005

[a]Genes encoding homologue(s) of the indicated system(s) have been identified in the organism's genome sequence although a role in resistance has not yet been verified.
[b]A hybridization study documented the presence of *silCBA* homologous DNA in a variety of unnamed enteric bacteria of clinical origin (Silver, 2003).
[c]Also contributes to resistance to SDS.
[d]Plant-produced antimicrobials used in defense against infection.

to RND pumps as other than drug-exporting/resistance determinants. Indeed, an increasing number of studies point to these pumps having other than antimicrobial export as their intended function (Poole, 2004a, 2005). The observation, for example, that many RND systems of the *Enterobacteriaceae* provide for resistance to bile and are, in some instances, induced in response to this agent, provides support for these systems being determinants of protection in vivo against this gut "detergent" (Poole, 2005). RND pumps in *Agrobacterium tumefaciens* and *Erwinia amylovora* known to accommodate antibiotics are, in fact, inducible by and provide protection against plant-derived

antimicrobial compounds—again, antibiotics are not the intended substrates. That RND pump overproduction in *P. aeruginosa* can compromise bacterial "fitness" (Sanchez et al., 2002) and that pump-overproducing mutants of *P. aeruginosa* can be selected in animal models of infection in the absence of antimicrobial selection (Join-Lambert et al., 2001) also point to in vivo functions for these pumps independent of antimicrobial efflux and resistance. Still, the identity of putative nonantimicrobial substrates and the nature of the intended function for these efflux systems remain, in most instances, unknown. And while a few RND-type MDR efflux systems are antibiotic inducible (e.g., MexXY, by the ribosome-targeting agents macrolides, aminoglycosides, tetracyclines, and chloramphenicol [Jeannot et al., 2005], and MexCD-OprJ by the antiseptics benzalkonium chloride and chlorhexidine [Morita et al., 2003]), it appears that this is in response to effects imparted on the cell by these antimicrobials and is not a response to the antimicrobials themselves (i.e., the antimicrobials are not the intended efflux substrates).

While many RND-type multidrug efflux systems accommodate and provide resistance to solvents (Ramos et al., 2002; Fernandes et al., 2003) and, indeed, solvent-antibiotic coresistance is often used as marker of RND multidrug pump expression in gram-negative bacteria (Randall et al., 2001), RND pumps specific for and inducible by solvents are also known (e.g., the highly homologous SrpABC, TtgDEF, and SpeABC efflux systems described in many *Pseudomonas putida* isolates) (Fernandes et al., 2003; Poole, 2004c). Whether these are, in fact, intended as solvent-specific exporters or their expression counters some damage/defect imparted by solvent accumulation (in membranes) in bacterial cells is unclear.

A number of tripartite RND family pumps accommodate biocides (antiseptics, disinfectants) including quaternary ammonium compounds (QACs; e.g., benzalkonium chloride), chlorhexidine, and most especially triclosan (Poole, 2005). Indeed, RND-type efflux systems in *P. aeruginosa* are the primary determinants of this organism's resistance to triclosan (Chuanchuen et al., 2003). Tributyltin, an antifouling agent found in, e.g., marine paints, is exported by an RND family multidrug transporter of *Pseudomonas stutzeri* that also accommodates antibiotics (Jude et al., 2004).

Heavy Metals
There are several examples of efflux-mediated resistance to heavy metals in gram-negative bacteria involving tripartite, cell envelope-spanning RND-type efflux systems (reviewed in Nies, 2003) (Table 1). Perhaps the best-studied organism with respect to heavy metal resistance/efflux is *Wautersia metallidurans* (formerly known as

Ralstonia metallidurans, *Ralstonia eutropa*, and *Alcaligenes eutrophus*), an organism known to colonize industrial sediments, soils, or wastes with high heavy metal content and to possess two large plasmids bearing a myriad of genes, particularly efflux genes, for metal resistance (reviewed in Mergeay et al., 2003). This organism expresses multiple RND-type pumps responsible for resistance to Cd, Zn, Co, Ni, Cu, and Ag, with individual pumps able to pump out multiple heavy metals (e.g., the well-characterized CzcCBA pump that provides resistance to Cd, Zn, and Co) (Table 1). A chromosome-encoded homologue of this system has been described in *P. aeruginosa* where it also contributes to Zn, Cd, and Co resistance (Perron et al., 2004).

Inducible resistance to Cu has been reported in *E. coli*, mediated by the CusCBA RND-type efflux system encoded by the chromosomal $cusC(F)BA$ locus that has homologues in a limited number of other organisms, including *W. metallidurans* (Rensing and Grass, 2003; Mergeay et al., 2003) (Table 1). This locus is also inducible by Ag^+ (weakly) and is implicated in resistance to this metal (Silver and colleagues have dubbed the locus *agr* [Ag resistance] to reflect its similarity to the plasmid-borne Ag^+-resistance locus, *silCBA*, found in clinical isolates of *Salmonella* and *Klebsiella* and other enteric organisms [Silver, 2003]). Indeed, the *cus* designation (Cu and silver) provided by Nies (2003) reflects its dual role in Cu and Ag resistance. Intriguingly, the characteristically tripartite RND-MFP-OMF pump, CusCBA, actually functions with a fourth component, CusF, a periplasmic Cu-binding protein. Cus CBA can promote Cu resistance in the absence of CusF, though the latter is required for full activity of this efflux system (Rensing and Grass; 2003; Nies, 2003). A related periplasmic Ag (and Cu-?)-binding protein, SilE, is encoded by the *sil* locus in some *silCBA*-carrying enterics (Silver, 2003), although it is presently unclear if it contributes to SilCBA-mediated Ag efflux and resistance. Finally, an unrelated periplasmic heavy metal (Zn, Cu, Co, Ni)-binding protein, CzcE, is also encoded by the *czc* locus in *W. metallidurans*, although it appears not to function in CzcCBA-mediated heavy metal efflux, its loss in mutant strains having no impact on resistance (Grosse et al., 2004).

RND Pump Structure

Crystal structures of individual RND (*E. coli* AcrB [Murakami and Yamaguchi, 2003]), MFP (*P. aeruginosa* MexA [Higgins et al., 2004; Akama et al., 2004a]), and OMF (*E. coli* TolC [Koronakis et al., 2004] and *P. aeruginosa* OprM [Akama et al., 2004b]) components have now been reported, and models for the assembled structures of tripartite RND-MFP-OMF efflux systems that span the periplasm have been proposed (Eswaran et al., 2004; Murakami and Yamaguchi, 2003) (Color Plate 31). The channel-forming TolC and OprM proteins assemble as trimers that span the OM as a β-barrel and extend into the periplasm as an α-helical barrel, with the proximal, periplasmic ends of the proteins in close apposition to the cognate RND components (Color Plate 31), though possibly not in direct, physical association (Touze et al., 2004). Measuring ca. 140 Å in length, these channels are open at the distal (extracellular) end and taper almost to a close at the proximal (periplasmic) end, presumably necessitating conformational changes in the proteins to accommodate substrate molecules during the export process (Koronakis et al., 2004). The AcrB RND component is also trimeric and adopts a "jellyfish-like" shape with a 50-Å-thick transmembrane region and a 70-Å headpiece that protrudes into the periplasm (Murakami and Yamaguchi, 2003). This headpiece has a funnel-like opening at the top (where it likely approaches near to TolC) connected to a central cavity at the bottom that opens to the periplasm via three vestibules (at each monomer-monomer interface) (e.g., Color Plate 32A) that likely play a role in substrate recognition by and/or access to the AcrAB-TolC pump (Murakami and Yamaguchi, 2003). MexB is highly homologous to AcrB and modeling of the MexB protein indicates that it adopts the same trimeric jellyfish structure (Middlemiss and Poole, 2004). The N-termi-nal two-thirds of MexA has been crystallized, corresponding to the region(s) implicated in OprM binding and multimerization (a bacterial two-hybrid study confirmed the ability of MexA to self-associate [Nehme et al., 2004] as does the available crystal data [Higgins et al., 2004; Akama et al., 2004a]). MexA exists in an elongated state (89 Å in length) characterized by three linear domains: an α-barrel, an lipoyl domain, and a 47-Å α-helical hairpin, the latter of which is implicated in MexA subunit-subunit interaction (Nehme et al., 2004). The C-terminal domain of the molecule that is absent in the available crystal structure has previously been implicated in binding to the cognate RND component (i.e., MexB) (Nehme and Poole, 2005), in agreement with data confirming the involvement of the C-terminal region of the AcrA MFP of *E. coli* in binding its cognate RND component, AcrB (Elkins and Nikaido, 2003). Although the number of MexA/MFP monomers in a functional RND-MFP-OMF pump is as yet only speculative, 9 MexA subunits appear to be sufficient to form a ring/collar that circumscribes the proximal and distal ends, respectively, of OMF and RND components (Higgins et al., 2004), thereby stabilizing the complex and forming a conduit that spans the periplasm from the CM to the OM (Color Plate 31) (Higgins et al., 2004; Akama et al., 2004a). Still, the crystal structure of the AcrB RND pump component reveals a cleft in the periplasmic region of each monomer of the functional trimer that has been proposed as a site of interaction with the MFP component (i.e., AcrA) (Murakami and Yamaguchi, 2003). This suggests that 3 MFPs occur in the tripartite complex, with an RND:MFP:OMF stoichiometry, then, of 1:1:1 (Murakami and Yamaguchi, 2003). A mutation in the cleft region of the three-dimensional model of the MexB RND component that has been shown to compromise MexB function (Middlemiss and Poole, 2004) was suppressed by a number of mutations in the C-terminal end of the cognate MFP component, MexA (Nehme and Poole, 2005). This is consistent both with the

C-terminal end of the MexA playing a role in interaction with its RND component and with the cleft region of MexB, indeed, being a site for interaction with MexA.

Evidence for Periplasmic Capture of Substrate Molecules by RND-Type Pumps

A variety of mutant and gene-splicing studies have confirmed that the periplasmic domains of the CM-spanning RND components of tripartite RND-MFP-OMF pumps are responsible for substrate recognition (Mao et al., 2002; Tikhonova et al., 2002; Elkins and Nikaido, 2002), suggesting that pump substrates are recognized (and captured) from the periplasm. In agreement with this, the aforementioned crystal structure of AcrB revealed the existence of vestibules at the monomer interfaces of this trimeric RND component, at the plane of the periplasmic face of the CM, and contiguous with the periplasm (Color Plate 32A) (Murakami and Yamaguchi, 2003). These vestibules are clearly implicated as portals of substrate entry to the central cavity of this protein, which appears be the site of substrate binding (Yu et al., 2003) and, indeed, mutations in the vestibule region influence substrate specificity (Middlemiss and Poole, 2004). That antimicrobials known not to cross the CM (e.g., the β-lactam carbenicillin) are nonetheless accommodated by RND-type pumps (e.g., MexAB-OprM) (Poole, 2004a) is also consistent with periplasmic capture of substrates by these pumps. A study showing that coexpression of a single-component, CM-associated antibiotic exporter able only to accommodate cytosolic drug and a tripartite RND-type multidrug exporter produced a multiplicative enhancement of resistance can best be explained as the latter accommodating substrate in the periplasm but not the cytosol (Palmer, 2003). Further evidence in support of periplasmic substrate capture comes from the demonstration that certain RND-MFP-OMF efflux systems operate with periplasmic substrate-binding proteins (e.g., the periplasmic CusF Cu-binding protein that functions with the CusCBA Cu exporter [see above; Color Plate 32B]). A study of the AcrD component of the AcrAD-TolC aminoglycoside exporter of *E. coli* reconstituted into multilamellar proteoliposomes also confirmed the ability of this RND component to capture periplasmic substrate (Aires and Nikaido, 2005). Finally, the observation that the Cu sensitivity of *E. coli* CopA ATPase mutants is not ameliorated by the presence of CusCFBA argues that this RND-type efflux system is unable to replace CopA in exporting Cu^+ from the cytosol, its efflux activity apparently restricted to periplasmic Cu^+ (Rensing and Grass, 2003). That *cus* mutants, too, are only adversely impacted vis-à-vis Cu resistance when the periplasmic CueO is also absent is additional support for CusCBA functioning only in export of periplasmic Cu, CueO being present and active in the periplasm only. Current models suggest, then, that under aerobic conditions CueO functions to detoxify periplasmic Cu^+ following its export there from the cytosol by CopA, while under anaerobic conditions (or other circumstances where CueO might be compromised) CusCFBA functions to accommodate periplasmic Cu^+ delivered there by CopA (Rensing and Grass, 2003) (Color Plate 32B). It may be, in fact, that all RND-MFP-OMF efflux systems act exclusively on periplasmic substrates, originating either outside the cell (i.e., captured upon entry before crossing into the cytosol) or in the cytosol where they are delivered to the periplasm by other exporters or via diffusion across the CM. Still, AcrD in proteoliposomes appeared to be able to capture cytosolic substrates (Aires and Nikaido, 2005) although it is uncertain whether this is reflective of the in vivo operation of this RND component or whether it truly captured cytosolic substrate versus substrate that had, for example, penetrated the membrane (i.e., reached the periplasmic leaflet of the CM bilayer where it might then be accessed by the vestibules). CzcCBA metal efflux system appears, also, to accommodate cytoplasmic Zn^{2+} and Cd^{2+} (Legatzki et al., 2003).

METAL TRAFFICKING, SEQUESTRATION, AND DETOXIFICATION

While many heavy metals are, at low levels, required for the activity of bacterial enzymes/proteins, their adverse effects (at higher levels) on a variety of bacterial macromolecules and the ability of some (e.g., Cu, Fe) to promote the generation of cell-damaging ROS requires that their levels in bacteria be strictly controlled (homeostasis). To this end, metal fluxes into/out of the cell are regulated, with a variety of periplasmic components playing a vital role in this process. Determinants of metal resistance involving trafficking, sequestration, and/or detoxification have been reported in a number of bacteria, and generally involve resistance to mercury or copper (Table 2).

Mercury

Inducible resistance to mercury, encoded by the *mer* operon present on a variety of mobile genetic elements and ultimately mediated by the cytosolic MerA reductase that reduces Hg^{2+} to nontoxic and volatile Hg^0, depends on many transport proteins to deliver Hg^{2+} to the reductase (Fig. 2A) (Barkay et al., 2003). One of these, MerP, is a periplasmic Hg^{2+}-binding protein that captures periplasmic Hg^{2+} and delivers it to the CM-associated Hg^{2+} transporter, MerT, and thence to MerA. So-called broad-spectrum *mer* operons (provide resistance to both Hg^{2+} and organomercurials like phenymercury [PM]) contain an additional gene, *merG*, that encodes a periplasmic protein essential for PM (but not Hg^{2+}) resistance. Although its exact function is unknown, mutants lacking *merG* accumulate increased levels of PM, suggesting that its activity somehow limits uptake of or cell permeably to organomercurials (Kiyono and Pan-Hou, 1999).

Copper

The plasmid-borne *pcoABCDE* locus of *E. coli* (a homologue of the plasmid-borne *copABCD* locus of *Pseudomonas syringae* that lacks, however, a *pcoE* homologue) encodes a Cu-inducible and generally periplasmic system for Cu resistance/homeostasis (Fig. 2B) (reviewed in Rensing and Grass, 2003). PcoA (CopA) is a multicopper oxidase whose contribution to Cu resistance has, until recently, been unclear and proposed to include Cu sequestration (PcoA binds 11 Cu atoms/protein) and/or oxidation of catechol siderophores that then sequester periplasmic Cu (Rensing and Grass, 2003). The recent identification, however, of CueO (also a multicopper oxidase whose function in *E. coli* can be substituted by PcoA, suggesting a similar activity) as a cuprous oxidase ($Cu^+ \rightarrow Cu^{2+}$) (Singh et al., 2004) strongly supports PcoA functioning in the detoxification of Cu^+ via oxidation to Cu^{2+}, a much less toxic form of the ion. PcoB (CopB) is an OM-associated protein that together with PcoA is essential for Pco-mediated Cu resistance, with the remaining components only needed for full resistance. The role of PcoB in resistance is unclear, although it may be involved in directing OM-associated Cu to PcoA and/or Cu export across the OM. PcoC is a periplasmic Cu-binding protein that may deliver Cu to PcoA but also appears to direct this metal to the CM-associated transporter, PcoD, which is proposed to deliver it to cytosolic apo-PcoA, reminiscent of MerT delivery of Hg^{2+} to MerA. Apo-PcoA appears to be exported to the periplasm via the twin-arginine translocation (Tat) pathway that is specific for prefolded, often metal cofactored proteins (see Chapter 2 for more information on this export pathway). This suggests that the protein is able to bind cytosolic Cu and deliver it to the periplasm (following its export from the cytosol) where it can then be oxidized to Cu^{2+}. Still, effective Cu sequestration may be part of its detoxification function, inasmuch as CopABCD-expressing *P. syringae* exposed to high levels of Cu turns blue and accumulates Cu in the periplasm to levels that can approach 12% of the dry weight of the cell (Puig et al., 2002). This phenotype has been seen in other *Pseudomonas* spp. where periplasmic Cu sequestration has also been implicated as a resistance mechanism (Gilotra and Srivastava, 1997). *pco/cop* homologues have been described in other organisms (Table 2) and a chromoso-

TABLE 2 Non-efflux-mediated periplasmic resistance mechanisms

Resistance	Constituent(s)	Activity	Organism	Reference(s)
Cu	CopABCD[a,b]	Metal trafficking and sequestration or detoxification	*Pseudomonas syringae, Xanthomonas campestris, Wautersia metallidurans, Ralstonia solanacearum,[c] Pseudomonas aeruginosa,[d] Pseudomonas putida*	Adaikkalam and Swarup, 2005; Canovas et al., 2003; Cooksey, 1994; Mergeay et al., 2003
	PcoABCD[d]/E	Metal trafficking and sequestration or detoxification	*Escherichia coli, Desulfovibrio* sp.	Karnachuk et al., 2003; Rensing and Grass, 2003
	?[e]	Sequestration[e]	*P. putida*	Saxena et al., 2002
	?[f]	Sequestration[f]	*Pseudomonas picketti*	Gilotra and Srivastava, 1997
	CueO	Multicopper oxidase; detoxification	*E. coli*	Rensing and Grass, 2003
Oxidative stress	SodC, Sod	Cu/Zn cofactored superoxide dismutase; detoxification	*Salmonella enterica* serovar Typhimurium,[g] *Salmonella* spp.,[g] *E. coli, Caulobacter crescentus, Vibrio cholerae, Neisseria meningitidis, Hemophilus* spp., *Actinobacillus pleuropneumoniae, Legionella pneumophila, Pasteurella haemolytica,[h] Brucella* sp.,[h] *Photobacterium leiognathi,[h] Rhodobacter sphaeroides*	Desideri and Falconi, 2003; Kroll et al., 1995
	Sod	Fe cofactored superoxide dismutase; detoxification	*Desulfovibrio vulgaris, Photobacterium damselae* subsp. *piscicida,[i,j] Aquaspirillum magnetotacticum[i]*	Barnes et al., 1999a; Fournier et al., 2003; Short and Blakemore, 1989
	?[k]	Mn cofactored superoxide dismutase; detoxification	*Aeromonas hydrophila, Aeromonas salmonicida,[k] A. magnetotacticum[k]*	Barnes et al., 1999b; Leclere et al., 2004; Short and Blakemore, 1989
	KatA, KatB,[l] KatE,[n] CatF[p]	Catalase; detoxification	*Helicobacter pylori, Vibrio fischeri, Brucella abortus,[m] P. syringae,[o] P. aeruginosa,[q] Brucella melitensis, Edwardsiella tarda,[q] Rhizobium meliloti[r]*	Brown et al., 1995; Gee et al., 2004; Harris and Hazel, 2003; Herouart et al., 1996; Klotz and Hutcheson, 1992; Sha et al., 1994; Srinivasa et al., 2003; Visick and Ruby, 1998
	KatA, KatP,[s] KatG[t]	Bifunctional catalase-peroxidase; detoxification	*L. pneumophila, E. coli* O157:H7,[s] *C. crescentus, E. coli*	Amemura-Maekawa et al., 1999; Brunder et al., 1996; Heimberger and Eisenstark, 1988; Schnell and Steinman, 1995
	Ccp	Cytochrome *c* peroxidase; detoxification	*Paracoccus denitrificans, Pseudomonas stutzeri, Campylobacter jejuni,[u] Neisseria gonorrhoeae,[v] Bacteroides fragilis*	Goodhew et al., 1990; Henrixson and DiRita, 2004; Herren et al., 2003; Seib et al., 2004

(Continued)

TABLE 2 *Continued*

Resistance	Constituent(s)	Activity	Organism	Reference(s)
	Tpx[u,x]	Thiol peroxidase; detoxification	*E. coli*	Cha et al., 1996
	Gst	Glutathione-*S*-transferase; detoxification[y]	*Proteus mirabilis*,[q] *Ochrobactrum anthropi*[q]	Allocati et al., 2003; Tamburro et al., 2004
	MsrA	Peptide methionine sulfoxide reductase[z]; detoxification	*O. anthropi*[q]	Tamburro et al., 2004
	Hyd	[Fe] dehydrogenase	*D. vulgaris*	Fournier et al., 2004
SDS[aa]	MdoAB	Membrane-derived oligosaccharides[bb]	*E. coli*	Rajagopal et al., 2003
Lysozyme	Ivy/YkfE	Periplasmic lysozyme inhibitor	*E. coli*	Deckers et al., 2004; Masschalck and Michiels, 2003
Organic solvents	Cti	*cis-trans*-isomerase[cc]	*P. putida*	Junker and Ramos, 1999; Ramos-Gonzalez et al., 2001

[a]CopABCD and PcoABCD are highly homologous, Cu-inducible systems, although the former lacks a PcoE counterpart.

[b]CopABCD is plasmid-encoded in all but *P. putida* and *P. aeruginosa*.

[c]A Cu-inducible *copABCD* homologue has been identified on a megaplasmid in this organism together with additional *cop* genes (*copK* and *copC*), which have been shown in proteomic studies to be Cu inducible and periplasmic, with CopK also shown to bind Cu (Mergeay et al., 2003). A role for these in Cu resistance remains to be determined.

[d]A chromosomal locus encoding a CopABCD homologue has been identified in this organism, although a role in Cu resistance has not yet been verified.

[e]The Cu resistance gene(s) have not been identified, although preliminary studies suggest that efflux and periplasmic Cu sequestration are responsible for resistance in this copper mine isolate.

[f]The Cu resistance determinant(s) have not been identified, although preliminary studies indicate that it is plasmid encoded and that the isolate accumulates Cu in the periplasm, consistent with a sequestration method of resistance.

[g]The *sodC* gene common to all salmonellae is dubbed *sodC2* to distinguish it from an additional prophage-linked *sodC* gene, *sodC1*, found in some highly virulent strains. A third, also prophage-associated *sodC* gene, *sodC3*, has been described in *S. enterica* serovar Typhimurium LT2 (Battistoni, 2003).

[h]A CuZn superoxide dismutase activity has been detected and the corresponding *sod* gene cloned although the localization of the activity was not assessed.

[i]A periplasmic Fe superoxide dismutase activity has been detected, but the corresponding *sod* gene has not been identified or cloned.

[j]Formerly *Pasteurella piscicida*.

[k]A periplasmic Mn superoxide dismutase activity has been detected, but the corresponding *sod* gene has not been identified or cloned.

[l]Of *Edwardsiella tarda*.

[m]A gene encoding a periplasmic catalase has been recovered, but no gene designation was provided.

[n]Of *Brucella melitensis*.

[o]Multiple periplasmic catalase activities have been detected and a catalase gene cloned, with the nucleotide sequence showing evidence of a signal peptide in the predicted protein.

[p]Of *P. syringae*.

[q]Activity detected in both cytoplasm and periplasm.

[r]Now known as *Sinorhizobium meliloti*.

[s]The *E. coli* O157:H7 KatP catalase is plasmid encoded.

[t]The KatG catalases of *C. crescentus* and *E. coli* are present in both the cytoplasm and the periplasm.

[u]A putative periplasmic cytochrome *c* perixodase gene has been identified, although localization and activity of the protein have not been assessed.

[v]Present in the periplasm as a lipoprotein anchored to the cytoplasmic membrane.

[w]Homologues identified in *Hemophilus influenzae* and *V. cholerae*, although activity and/or localization of the putative Tpx proteins was not assessed.

[x]Functions generally as a lipid hydroperoxide peroxidase. *tpx* mutants were more sensitive to oxidative damage and showed severely reduced viability at stationary phase (Cha et al., 2004).

[y]Probably provides protection against oxidative stress through role in biodegradation of aromatic xenobiotics that would otherwise elicit the generation of damaging reactive oxygen species. *gst* mutants of *P. mirabilis* were also more sensitive to H_2O_2.

[z]Induced by aromatic xenobiotics that generate reactive oxygen species. MsrA catalyzes the reduction of free and protein-bound methionone-sulfoxide residues to methionine, restoring their biological activity (i.e., likely counters damage caused by ROS).

[aa]SDS resistance is a common property of the *Enterobacteriaceae*. In light of studies in *E. coli* documenting the ability of SDS to cross the OM and enter the periplasm (Nickerson and Aspedon, 1992), mechanism(s) of protecting periplasmic components from this detergent are clearly necessary.

[bb]Membrane-derived oligosaccharides (MDOs) are osmoregulated periplasmic glucans. Several lines of evidence correlated lack of or reduced MDO synthesis and SDS susceptibility in this study, although the mechanism(s) by which MDOs promoted SDS resistance are unknown.

[cc]Catalyzes the *cis-trans*-isomerization of esterified fatty acids in membrane phospholipids in response to solvent exposure. This has the effect of countering the increase membrane fluidity that results from solvent interaction with membranes.

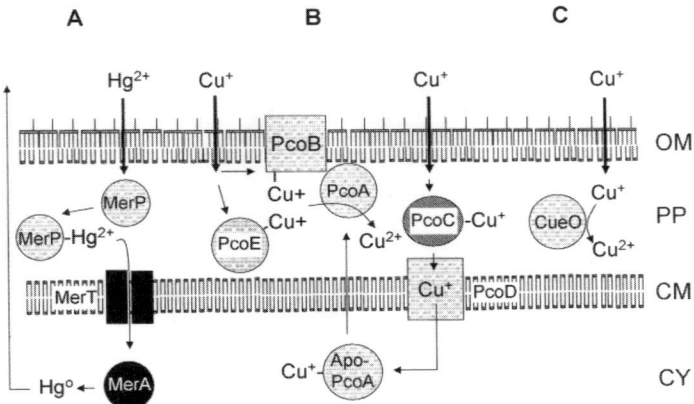

FIGURE 2 Periplasmic metal trafficking and detoxification. (A) MerP binds Hg^{2+} in the periplasm and delivers it to CM transporter MerT, which delivers it to the MerA mercury reductase in the cytoplasm. Reduction to Hg^0 renders Hg nontoxic and volatile. (B) PcoABCDE functions in the detoxification of periplasmic Cu^+, absolutely dependent on PcoA, a multicopper oxidase that likely oxidizes toxic Cu^+ to less toxic Cu^{2+}. PcoB is an OM protein of unknown function that may be involved in the delivery of Cu to PcoA or export of Cu across the OM. PcoE is a periplasmic Cu^+-binding protein that likely shuttles Cu^+ to PcoA. PcoCD functions in the binding of Cu^+ in the periplasm and its subsequent transport across the CM where it may be delivered to apo-PcoA in the cytoplasm, prior to its export to the periplasm. (C) CueO is a periplasmic multicopper oxidase with Cu^+ oxidase activity that functions to oxidize Cu^+ to Cu^{2+}. OM, outer membrane; PP, periplasm; CM, cytoplasmic membrane; CY, cytosol.

mal *copABCD* locus involved in Cu resistance has recently been reported in *P. putida* (Adaikkalam and Swarup, 2005).

The aforementioned multicopper/cuprous oxidase, CueO, is coregulated with a CM-associated P-type ATPase, CopA (unrelated to CopA of the *P. syringae* CopABCD Cu resistance determinant), that functions in the export of Cu^+ from the cytosol. These two proteins constitute the Cue system that functions in the periplasmic delivery (CopA) and detoxification (by oxidation to Cu^{2+}; CueO) (Fig. 2C) of cytosolic Cu^+ under aerobic conditions (CueO activity is strictly oxygen dependent) (Rensing and Grass, 2003). Under anaerobic conditions, the CusCFBA efflux system (see above) effectively replaces CueO to export (versus detoxify) periplasmic Cu^+, inasmuch as *cusC(F)BA* mutants show increased susceptibility to Cu under anaerobic but not aerobic conditions (Rensing and Grass, 2003).

RESISTANCE TO OXIDATIVE STRESS

Bacteria growing aerobically are unavoidably exposed to ROS such as superoxide (O_2^-) and peroxide (H_2O_2) that form by accident and are released into the cytosol (and periplasm?) when molecular oxygen adventitiously oxidizes redox enzymes (Imlay, 2003). ROS compromise bacterial growth as a result of direct/indirect damage to proteins, lipids, and nucleic acids. Historically, cells have protected themselves from ROS via the production of cytoplasmic Fe- or Mn-cofactored superoxide dismutases (Fe-SOD, Mn-SOD) ($2O_2^- + 2H^+ \rightarrow H_2O_2 + O_2$), catalases ($2H_2O_2 \rightarrow H_2O + O_2$), and peroxidases ($AH_2 + H_2O_2 \rightarrow A + 2H_2O$) that function to eliminate ROS (Imlay, 2003). ROS can also be generated in the periplasm and perhaps enter this compartment from the cell exterior, and not surprisingly, SOD, catalase, and peroxidase activities have been detected in the periplasm of many gram-negative bacteria (Table 2).

Superoxide Dismutase

Unlike cytoplasmic SOD, the bulk of periplasmic SODs are copper-zinc cofactored (CuZn-SOD) (reviewed in Desideri and Falconi, 2003), reminiscent of eukaryotic systems, although a few use Fe or Mn as cofactors (Table 2). The CuZn (a.k.a. SodC) and other periplasmic SODs do not protect cells from O_2^- generated inside cells, though they are important determinants of resistance to exogenously generated ROS (Korshunov and Imlay, 2002; Battistoni, 2003; San Mateo et al., 1998; Leclere et al., 2004). With few exceptions (e.g., in *Legionella pneumophila* where loss of SodC impacts viability in stationary phase [St John and Steinman, 1996]), periplasmic CuZn SODs are not required for viability under normal lab growth conditions, an observation consistent with minimal or no periplasmic ROS being generated endogenously under these conditions (Battistoni, 2003). While it has been accepted in general that periplasmic SodCs function to protect sensitive periplasmic targets from the action of O_2^-, assuming this anion is not capable of permeating lipid membranes (i.e., the cytoplasmic membrane), studies in *E. coli* have clearly shown that exogenously generated O_2^- can, in fact, enter the cytosol and damage cytosolic targets (Korshunov and Imlay, 2002). Moreover, SodC was shown to protect *E. coli*

cells from the cytosolic damage caused by exogenously supplied O_2^-, indicating that this enzyme may also function generally in gram-negative bacteria in blocking external O_2^- entry into the cytosol (Korshunov and Imlay, 2002).

Clearly, a major potential source of external ROS (for animal pathogens, particularly facultative intracellular ones) comes from host immune cells (e.g., phagocytes) whose antibacterial activity involves the generation of a so-called "respiratory burst." As such, SodC has been implicated as a virulence determinant, functioning to protect bacteria from oxidative challenge by phagocytic cells (Battistoni, 2003). Consistent with this, *sodC* mutants of several bacterial species (e.g., *Salmonella enterica* serovar Typhimurium [Krishnakumar et al., 2004], *Salmonella enterica* serovar Choleraesuis [Ammendola et al., 2005; Sansone et al., 2002], *Brucella abortus* [Gee et al., 2005], and *Haemophilus ducreyi* [San Mateo et al., 1999]) have been shown to be attenuated for virulence in animal infection models and/or to be compromised for intracellular survival within macrophages (reviewed in Battistoni, 2003).

Catalase

A number of periplasmic catalases have been identified in gram-negative bacteria, including

TABLE 3 Classification and properties of β-lactamases of gram-negative bacteria

Ambler classification	Type of enzyme	Preferred substrates
A	Restricted-spectrum β-lactamase	Penicillins, cephalosporins
	Extended-spectrum β-lactamase	Penicillins, narrow-spectrum and extended-spectrum cephalosporins, monobactams
	Carbapenemase	Penicillins, cephalosporins, carbapenems, monobactams; sometimes extended-spectrum β-lactams
B	Carbapenemase	Most β-lactams, including carbapenems and extended-spectrum β-lactams and fourth-generation cephalosporins
C	Expanded-spectrum cephalosporinase	Penicillins, narrow- and extended-spectrum cephalosporins, cephamycins, monobactams
D	Narrow-spectrum penicillinase	Penicillins, cloxacillin
	Extended-spectrum β-lactamase	Penicillins, cloxacillin, extended-spectrum β-lactams, sometimes monobactams or fourth-generation cephalosporins
	Carbapenemase	Penicillins, oxacillin, carbapenems

bifunctional catalase-peroxidases that have been described in *L. pneumophila, E. coli, Caulobacter crescentus,* and *E. coli* O157:H7 (where the gene is plasmid encoded) (Table 2). While these enzymes generally promote resistance to H_2O_2-mediated cell killing in vitro (Gee et al., 2004; Srinivasa Rao et al., 2003; Barnes et al., 1999b; Herouart et al., 1996), few studies have addressed a contribution to protection from killing in vivo. Still, periplasmic catalase-deficient mutants of *Edwardsiella tarda* (Srinivasa Rao et al., 2003) and *Aeromonas salmonicida* subsp. *salmonicida* (Barnes et al., 1999b) were shown to be attenuated for virulence in animal models of infection, and a periplasmic catalase-deficient mutant of *L. pneumophila* showed reduced growth in macrophages, indicative of impaired pathogenesis (Bandyopadhyay et al., 2003).

Peroxidase

Cytochrome *c* peroxidase (Ccp) catalyzes the reduction of H_2O_2 to H_2O using the periplasmic *c*-type cytochrome as the electron donor. This enzyme has been described in many gram-negative bacteria, although a role in protection from H_2O_2 killing has only been confirmed in a few instances (e.g., *Neisseria gonorrhoeae* [Seib et al., 2004]). Resistance to organic peroxides but not H_2O_2 was, however, provided by the Cpp of *Bacteroides fragilis* (Herren et al., 2003). A putative *ccp* gene has been identified in *Campylobacter jejuni* and although a role in H_2O_2 (or other peroxide) protection was not addressed, mutant studies revealed its importance for in vivo survival (Hendrixson and DiRita, 2004). An oxygen stress-inducible "thiol" peroxidase (Tpx) has been identified in the periplasm of *E. coli* and shown to enhance survival of the organism under conditions of oxygen stress (Cha et al., 1996, 2004).

Others

Recently, an oxidative stress-inducible periplasmic [Fe] dehydrogenase of the anaerobic sulfur-reducing bacterium *Desulfovibrio vulgaris* Hildenburough has been implicated in resistance to oxidative stress by a possible involvement in O_2 removal in the periplasm via its reduction to H_2O (Fournier et al., 2004). Periplasmic glutathione-*S*-transferase (Gst) and peptide methionine sulfoxide reductase (MsrA) have also been implicated in ROS resistance in *Ochrobactrum anthropi* and *Proteus mirabilis*, probably indirectly via degradation of xenobiotics known to generate ROS (Gst) (Allocati et al., 2003; Tamburro et al., 2004) or by reversing oxidative damage to proteins (MsrA) (Tamburro et al., 2004).

β-LACTAMASE

β-Lactamases are hydrolytic enzymes that disrupt the amide bond of the characteristic 4-membered β-lactam ring of the β-lactam class of antibiotics (i.e., penicillins, cephalosporins, carbapenems, monobactams, and penems), thereby rendering the antimicrobial inactive (Fisher et al., 2005; Poole, 2004b). Broadly distributed among gram-negative bacteria, these periplasmic enzymes are major determinants of bacterial β-lactam resistance, their cleavage of these agents serving to protect the periplasmic β-lactam targets (i.e., the so-called penicillin-binding proteins [PBPs] that function as enzymes of cell wall biosynthesis) from their inhibitory activity (Poole, 2004b). Four molecular classes of β-lactamases are known, dubbed A to D, and include both metal-dependent (Zn^{2+}-requiring; class B) and metal-independent (active site serine; classes A, C, and D) enzymes (Fisher et al., 2005) (Table 3). Gram-negative β-lactamases can be narrow spectrum like the original plasmid-encoded class A TEM and SHV and class D OXA enzymes that predominate in *Enterobacteriaceae* or broad spectrum like the chromosomal and, increasingly, plasmid-encoded class C AmpC enzymes found in many *Enterobacteriaceae* and nonfermenting gram-negative bacteria (e.g., *P. aeruginosa, Acinetobacter baumannii*), the widely distributed and generally plasmid-encoded extended-spectrum β-lactamases (ESBL) of classes A and D, and plasmid-encoded class A, B, and D carbapenemases (Table 3) (Poole, 2004b; Jacoby and Munoz-Price, 2005). The class B metalloenzymes are of particular concern owing to the ability of some of these to hydrolyze and thus provide resistance to virtually all classes of β-lactams.

OTHER MECHANISMS

Periplasmic membrane-derived oligosaccharides (MDOs) normally associated with osmotolerance in *E. coli* and other gram-negative bacteria have also been implicated in detergent (i.e., sodium dodecyl sulfate, SDS) resistance in *E. coli*, possibly via an electrostatic repulsion mechanism, inasmuch as these anionic MDOs only promote resistance to anionic detergents like SDS (Rajagopal et al., 2003). Lysozyme is an enzyme that cleaves peptidoglycan, rendering bacterial cells osmotically sensitive and prone to lysis. A periplasmic lysozyme inhibitor has been reported in *E. coli* where it contributes to lysozyme resistance under conditions where the OM barrier that normally excludes this enzyme is compromised, although the details of inhibition are unknown (Deckers et al., 2004). Finally, a component of solvent (i.e., toluene) resistance in *P. putida* has been shown to be a periplasmic *cis-trans*-isomerase (Cti) that functions in the *cis-trans*-isomerization of esterified fatty acids in response to solvent exposure (Junker and Ramos, 1999). Apparently, solvents cause an increase in membrane fluidity that is countered by increased levels of *trans*-fatty acids in membranes.

PERIPLASMIC CONTROL OF RESISTANCE DETERMINANTS

Bacteria sense their immediate environments and trigger appropriate responses through the up- or down-regulation of specific genes and/or their products. Because the periplasm of gram-negative bacteria is contiguous with the external milieu, owing to the presence of channel-forming porins in the OM, the periplasm often reflects the immediate environment of the cell and, as such, extracellular conditions can accurately be detected by sensor components present in the periplasm. Two major families of bacterial regulatory systems that detect and respond to external (or at least extracytosolic) signals are the two-component regulatory and extracytoplasmic sigma factor (ECF)-based systems. Two-component regulatory systems are composed of a CM-spanning sensor kinase with domains in the periplasm and cytosol, a topology that permits transmission of signals from the periplasm to the cytosol, and a response regulator partner whose phosphorylation by its cognate kinase serves to impact its DNA-binding activity and, thus, control expression of target genes (reviewed in Bijlsma and Groisman, 2003). Sensing of environmental signals by the periplasmic domain of the sensor kinase triggers autophosphorylation of the cytosolic domain and subsequent phosphotransfer to the response regulator that controls target gene expression (e.g., Fig. 3A). ECF sigma factors function to direct RNA polymerase to target genes, and their activities are controlled by CM-spanning anti-ECF sigma factors, which are themselves responsive to environmental signals, or by signal-responsive anti-anti-sigma factors that indirectly control ECF sigma factor activity via control of the cognate anti-sigma factor (Helmann, 2002) (e.g., Fig. 3B).

A variety of two-component systems are implicated in the control of bacterial multidrug efflux systems, in general, of the RND family (Table 4), although the nature of the signal(s) that serve to recruit these efflux systems is unknown. The identification of these signals may, however, facilitate the elucidation of the natural functions of these systems, their contribution to drug resistance likely being an unintended consequence of their intended activities in the cell (Poole, 2005). The well-known PhoPQ two-component locus found in several bacteria (Groisman, 2001) plays a positive but complex and as yet not fully clarified role in promoting resistance to polycationic antimicrobials including polymyxins, cationic antimicrobial peptides associated with innate immunity, and, in some instances (e.g., *P. aeruginosa*), aminoglycosides. Apparently, this involves, to some extent at least, modification of lipopolysaccharide (i.e., neutralization of anionic charges in the LPS that are the sites of polycation binding) (Groisman, 2001; Shi et al.,

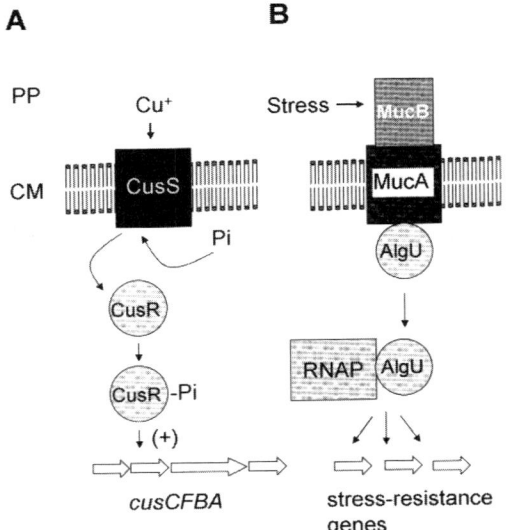

FIGURE 3 Regulation of resistance/stress determinants through signal recognition in the periplasm (PP) and transduction across the cytoplasmic membrane (CM). (A) CusSR as a representative two-component regulatory system that mediates the periplasmic Cu^+ inducibility of the CusCBA/F efflux system. The CM-spanning histidine kinase, CusS, senses Cu^+ in the periplasm and autophosphorylates a cytoplasmic domain before phosphorylating the response regulator CusR, which in turn activates *cusCFBA* gene expression. (B) AlgU-MucAB as a representative ECF sigma factor (AlgU) controlled by a CM-spanning anti-sigma factor (MucA). Periplasmic MucB interacts with and is believed to control the activity of MucA in response to certain periplasmic stress signals. AlgU sequestration (physically and functionally) by MucA is thus alleviated in response to periplasmic stress, permitting AlgU recruitment of RNA polymerase apoenzyme and subsequent transcription of target (i.e., stress response) genes. It is not clear, however, that this requires physical dissociation of AlgU from MucA.

2004). In some organisms, too, PhoPQ plays a role in tolerance to acid and oxidative stresses (Table 4). A second two-component system controlled by PhoPQ, PmrAB, also contributes to resistance to polymyxins in a variety of organisms (Table 4), again via their control of genes involved in LPS modification (Moskowitz et al., 2004; McPhee et al., 2003; Gunn et al., 2000). Many of the determinants of heavy metal resistance, including those in-

volved in efflux and metal trafficking/detoxification, are metal inducible and this is mediated by two-component regulatory systems responsive to the metal substrates of these resistance determinants (Table 4) (e.g., *cusCBA* regulated by CusSR in response to Cu) (Fig. 3A). Two-component systems controlling resistance to acid and oxidative stress and to serum are also known (Table 4). The RseA-RpoE anti-sigma factor/ECF sigma factor pair implicated in the periplasmic stress response in *E. coli* and other enteric organisms (Duguay and Silhavy, 2004) (see Chapter 6 for more information on RpoE) has been shown to influence resistance to heavy metals, antimicrobial peptides, and oxidative stress (Table 4), and a homologue of this system, AlgU-MucA, is implicated in tolerance to oxidative stress in *P. aeruginosa* (Table 4 and Fig. 3B).

CONCLUDING REMARKS

In many ways the periplasm serves as a buffer for the cytosol where numerous essential processes take place, protecting these from deleterious environmental toxic agents and stresses that might otherwise access this vital compartment. Clearly, efflux via trans-periplasm RND-type pumps is a major determinant of resistance to a great many environmental toxins, but is also a major problem vis-à-vis resistance to antimicrobials in the clinic. While crystal structures of some of these are now available, there is still limited information regarding assembly of these tripartite systems (and so the structural organization of the functional unit), and the details of substrate recognition and capture in the periplasm are generally lacking. Whether such pumps capture cytosolic as well as periplasmic substrates has yet to be fully resolved, as does the intended or natural function of the multidrug efflux members of this family of pumps. Despite years of study, the specific mechanism(s) of PcoABCDE/CopABCD-mediated resistance to Cu and the exact roles of all constituents remain to be elucidated. Finally, the plethora of periplasmic enzymes for protection against ROS emphasizes the significance of environmental (versus cell-generated)

TABLE 4 Two-component regulatory systems involved in resistance to antimicrobials, heavy metals, and environmental stresses[a]

Resistance or tolerance	Regulatory genes	Target genes	Organisms(s)	Reference(s)
Multidrug	smeRS	smeABC	*Stenotrophomonas maltophilia*	Li et al., 2002
	baeSR	mdtABC, acrD	*E. coli*	Nishino et al., 2005
	evgAS	yhiUV, emrKY	*E. coli*	Nishino et al., 2003
	adeRS	adeABC	*Acinetobacter baumannii*	Marchand et al., 2004
	cpxRS	acrD	*E. coli*	Hirakawa et al., 2003
Polycationic antimicrobials	pmrAB	Genes involved in LPS modification[b]	*S. enterica* serovar Typhimurium, *P. aeruginosa, Yersinia pestis, E. coli, Erwinia carotovora*	Gunn et al., 2000; Hyytiainen et al., 2003; McPhee et al., 2003; Moskowitz et al., 2004; Winfield and Groisman, 2004; Winfield et al., 2005
	phoPQ	Genes involved in LPS modification[b]	*S. enterica* serovar Typhimurium, *P. aeruginosa, Y. pestis, Yersinia pseudotuberculosis, Erwinia chrysanthemi, Shigella flexneri, E. coli, N. meningitidis*	Groisman, 2001; Johnson et al., 2001; Llama-Palacios et al., 2003; Marfarlane et al., 2000; Marceau et al., 2004; Winfield et al., 2005
Antimicrobial peptides	rpoE	?[c]	*S. enterica* serovar Typhimurium	Crouch et al., 2005
Heavy metals	czcRS	czcCBA	*W. metallidurans, P. aeruginosa*	Mergeay et al., 2003; Perron et al., 2004
	cusRS	cusC(F)BA	*E. coli*	Rensing and Grass, 2003
	pcoRS	pcoABCD	*E. coli*	Rensing and Grass, 2003
	copRS	copABCD	*P. syringae, X. campestris,[d] W. metallidurans[e]*	Cooksey, 1994; Mergeay et al., 2003
	silRS	silCBA	*S. enterica* serovar Typhimurium, *E. coli* (K-12 and O157:H7)	Silver, 2003
	cnrYX[f]	cnrCBA	*W. metallidurans*	Mergeay et al., 2003
	rseA−rpoE[g]	ompC[h]	*E. coli*	Egler et al., 2005
Acid	actSR	?[c]	*Sinorhizobium medicae[i]*	Fenner et al., 2004
	phoBR	asr	*E. coli*	Suziedeliene et al., 1999
	phoPQ	Multiple genes	*E. coli, S. enterica* serovar Typhimurium, *E. chrysanthemi, Y. pestis*	Llama-Palacios et al., 2003; Oyston et al., 2000; Zwir et al., 2005
Oxidative stress	phoPQ	?[c]	*Y. pestis*	Oyston et al., 2000
	rpoE−RseA[f,g]	?[c]	*E. coli, S. enterica* serovar Typhimurium, *Vibrio angustum*	Hild et al., 2000; McDougland et al., 2002; Testerman et al., 2002
	algU−mucA[f,j]	?[c]	*P. aeruginosa, P. syringae*	Keith and Bender, 1999; Schurr et al., 1996
	gacS/gacA	?[c]	*P. fluorescens, Pseudomonas chlororaphis*	Heeb et al., 2005; Kang et al., 2004
Serum	arcAB	?[c]	*Haemophilus influenzae*	Souza-Hart et al., 2003

[a]Unless otherwise indicated, the highlighted regulatory protein pairs include a CM-spanning sensor kinase with a sensing domain in the periplasm and a cytoplasmic response regulator.

[b]Includes genes responsible for addition of 4-aminoarabinose to LPS.

[c]Unknown.

[d]Also known as *Xanthomonas axonopodis*.

[e]The *cop* genes of *W. metallidurans* are Cu inducible, although a role in Cu resistance has not yet been demonstrated.

[f]Regulatory partners are an ECF sigma factor and its CM-spanning cognate anti-sigma factor.

[g]Part of a *rpoE⁻ rseABC* operon involved in a periplasmic stress response. Binding of the periplasmic RseB to the CM-spanning anti-sigma factor RseA enhances its sequestration of the RpoE ECF sigma factor. Sensing of periplasmic "stress" factors by RseB titrates it off RseA, possibly assisting in the subsequent release of RpoE, which then activates expression of the RpoE "modulon."

[h]A *rpoE* mutant was more susceptible to Zn, Cd, and Cu and much of this was related, somehow, to a reduction in OmpC levels—*ompC* expression declined in a *rpoE* knockout and an *ompC* deletion was as susceptible to Cu and Cd as a *rpoE* or *rpoE/ompC* mutant (Egler et al., 2005).

[i]Formerly *Rhizobium meliloti*.

[j]*algU* is part of an operon, *algU-mucABCD*. The periplasmic MucB interacts with and likely modulates the activity of the CM-spanning anti-sigma factor MucA, which in turn controls the activity of the ECF sigma factor AlgU (a.k.a. AlgT/σ22).

oxidative stress in the life of gram-negative bacteria, and while these are promoted as virulence determinants (protect pathogens from host-produced ROS during infection), it is far from clear that this is their only function.

REFERENCES

Adaikkalam, V., and S. Swarup. 2005. Characterization of *copABCD* operon from a copper-sensitive *Pseudomonas putida* strain. *Can. J. Microbiol.* **51:** 209–216.

Aires, J. R., and H. Nikaido. 2005. Aminoglycosides are captured from both periplasm and cytoplasm by the AcrD multidrug efflux transporter of *Escherichia coli. J. Bacteriol.* **187:**1923–1929.

Akama, H., M. Kanemaki, M. Yoshimura, T. Tsukihara, T. Kashiwagi, H. Yoneyama, S. I. Narita, A. Nakagawa, and T. Nakae. 2004b. Crystal structure of the drug-discharge outer membrane protein, OprM, of *Pseudomonas aeruginosa*: dual modes of membrane anchoring and occluded cavity end. *J. Biol. Chem.* **17:**52816–52819.

Akama, H., T. Matsuura, S. Kashiwagi, H. Yoneyama, S. Narita, T. Tsukihara, A. Nakagawa, and T. Nakae. 2004a. Crystal structure of the membrane fusion protein, MexA, of the multidrug transporter in *Pseudomonas aeruginosa. J. Biol. Chem.* **279:**25939–25942.

Allocati, N., B. Favaloro, M. Masulli, M. F. Alexeyev, and C. Di Ilio. 2003. *Proteus mirabilis* glutathione S-transferase B1-1 is involved in protective mechanisms against oxidative and chemical stresses. *Biochem. J.* **373:**305–311.

Amemura-Maekawa, J., S. Mishima-Abe, F. Kura, T. Takahashi, and H. Watanabe. 1999. Identification of a novel periplasmic catalase-peroxidase KatA of *Legionella pneumophila. FEMS Microbiol. Lett.* **176:**339–344.

Ammendola, S., M. Ajello, P. Pasquali, J. S. Kroll, P. R. Langford, G. Rotilio, P. Valenti, and A. Battistoni. 2005. Differential contribution of *sodC1* and *sodC2* to intracellular survival and pathogenicity of *Salmonella enterica* serovar Choleraesuis. *Microb. Infect.* **7:**698–707.

Bandyopadhyay, P., B. Byrne, Y. Chan, M. S. Swanson, and H. M. Steinman. 2003. *Legionella pneumophila* catalase-peroxidases are required for proper trafficking and growth in primary macrophages. *Infect. Immun.* **71:**4526–4535.

Barkay, T., S. M. Miller, and A. O. Summers. 2003. Bacterial mercury resistance from atoms to ecosystems. *FEMS Microbiol. Rev.* **27:**355–384.

Barnes, A. C., M. C. Balebona, M. T. Horne, and A. E. Ellis. 1999a. Superoxide dismutase and catalase in *Photobacterium damselae* subsp. *piscicida* and their roles in resistance to reactive oxygen species. *Microbiology* **145:**483–494.

Barnes, A. C., T. J. Bowden, M. T. Horne, and A. E. Ellis.1999b. Peroxide-inducible catalase in *Aeromonas salmonicida* subsp. *salmonicida* protects against exogenous hydrogen peroxide and killing by activated rainbow trout, *Oncorhynchus mykiss* L., macrophages. *Microb. Pathog.* **26:**149–158.

Battistoni, A. 2003. Role of prokaryotic Cu,Zn superoxide dismutase in pathogenesis. *Biochem. Soc. Trans.* **31:**1326–1329.

Bijlsma, J. J., and E. A. Groisman. 2003. Making informed decisions: regulatory interactions between two-component systems. *Trends Microbiol.* **11:**359–366.

Borges-Walmsley, M. I., D. Du, K. S. McKeegan, G. J. Sharples, and A. R. Walmsley. 2005. VceR regulates the *vceCAB* drug efflux pump operon of *Vibrio cholerae* by alternating between mutually exclusive conformations that bind either drugs or promoter DNA. *J. Mol. Biol.* **349:**387–400.

Brown, S. M., M. L. Howell, M. L. Vasil, A. J. Anderson, and D. J. Hassett. 1995. Cloning and characterization of the *katB* gene of *Pseudomonas aeruginosa* encoding a hydrogen peroxide-inducible catalase: purification of KatB, cellular localization, and demonstration that it is essential for optimal resistance to hydrogen peroxide. *J. Bacteriol.* **177:**6536–6544.

Brunder, W., H. Schmidt, and H. Karch. 1996. KatP, a novel catalase-peroxidase encoded by the large plasmid of enterohaemorrhagic *Escherichia coli* O157:H7. *Microbiology* **142:**3305–3315.

Canovas, D., I. Cases, and V. de Lorenzo. 2003. Heavy metal tolerance and metal homeostasis in *Pseudomonas putida* as revealed by complete genome analysis. *Environ. Microbiol.* **5:**1242–1256.

Cha, M. K., H. K. Kim, and I. H. Kim. 1996. Mutation and mutagenesis of thiol peroxidase of *Escherichia coli* and a new type of thiol peroxidase family. *J. Bacteriol.* **178:**5610–5614.

Cha, M. K., W. C. Kim, C. J. Lim, K. Kim, and I. H. Kim. 2004. *Escherichia coli* periplasmic thiol peroxidase acts as lipid hydroperoxide peroxidase and the principal antioxidative function during anaerobic growth. *J. Biol. Chem.* **279:**8769–8778.

Chuanchuen, R., R. R. Karkhoff-Schweizer, and H. P. Schweizer. 2003. High-level triclosan resistance in *Pseudomonas aeruginosa* is solely a result of efflux. *Am. J. Infect. Control* **31:**124–127.

Cooksey, D. A. 1994. Molecular mechanisms of copper resistance and accumulation in bacteria. *FEMS Microbiol. Rev.* **14:**381–386.

Crouch, M. L., L. A. Becker, I. S. Bang, H. Tanabe, A. J. Ouellette, and F. C. Fang. 2005. The alternative sigma factor sigma is required for resistance of *Salmonella enterica* serovar Typhimurium to anti-microbial peptides. *Mol. Microbiol.* **56:**789–799.

Deckers, D., B. Masschalck, A. Aertsen, L. Callewaert, C. G. Van Tiggelen, M. Atanassova, and C. W. Michiels. 2004. Periplasmic lysozyme inhibitor contributes to lysozyme resistance in *Escherichia coli. Cell. Mol. Life Sci.* **61:**1229–1237.

Desideri, A., and M. Falconi. 2003. Prokaryotic Cu,Zn superoxide dismutases. *Biochem. Soc. Trans.* **31:**1322–1325.

Duguay, A. R., and T. J. Silhavy. 2004. Quality control in the bacterial periplasm. *Biochim. Biophys. Acta* **1694:**121–134.

Egler, M., C. Grosse, G. Grass, and D. H. Nies. 2005. Role of the extracytoplasmic function protein family sigma factor RpoE in metal resistance of *Escherichia coli. J. Bacteriol.* **187:**2297–2307.

Elkins, C. A., and H. Nikaido. 2002. Substrate specificity of the RND-type multidrug efflux pumps AcrB and AcrD of *Escherichia coli* is determined predominately by two large periplasmic loops. *J. Bacteriol.* **184:**6490–6498.

Elkins, C. A., and H. Nikaido, 2003. Chimeric analysis of AcrA function reveals the importance of its C-terminal domain in its interaction with the AcrB multidrug efflux pump. *J. Bacteriol.* **185:**5349–5356.

Eswaran, J., E. Koronakis, M. K. Higgins, C. Hughes, and V. Koronakis. 2004. Three's company: component structures bring a closer view of tripartite drug efflux pumps. *Curr. Opin. Struct. Biol* **14:**741–747.

Fenner, B. J., R. P. Tiwari, W. G. Reeve, M. J. Dilworth, and A. R. Glenn. 2004. *Sinorhizobium medicae* genes whose regulation involves the ActS and/or ActR signal transduction proteins. *FEMS Microbiol. Lett.* **236:**21–31.

Fernandes, P., B. S. Ferreira, and J. M. Cabral. 2003. Solvent tolerance in bacteria: role of efflux pumps and cross-resistance with antibiotics. *Int. J. Antimicrob. Agents* **22:**211–216.

Fisher, J. F., S. O. Meroueh, and S. Mobashery. 2005. Bacterial resistance to β-lactam antibiotics: compelling opportunism, compelling opportunity. *Chem. Rev.* **105:**395–424.

Fournier, M., Z. Dermoun, M. C. Durand, and A. Dolla. 2004. A new function of the *Desulfovibrio vulgaris* Hildenborough [Fe] hydrogenase in the protection against oxidative stress. *J. Biol. Chem.* **279:**1787–1793.

Fournier, M., Y. Zhang, J. D. Wildschut, A. Dolla, J. K. Voordouw, D. C. Schriemer, and G. Voordouw. 2003. Function of oxygen resistance proteins in the anaerobic, sulfate-reducing bacterium *Desulfovibrio vulgaris* Hildenborough. *J. Bacteriol.* **185:**71–79.

Gee, J., M. E. Kovach, V. K. Grippe, S. Hagius, J. V. Walker, P. H. Elzer, and R. M. Roop. 2004. Role of catalase in the virulence of *Brucella melitensis* in pregnant goats. *Vet. Microbiol.* **102:**111–115.

Gee, J. M., M. W. Valderas, M. E. Kovach, V. K. Grippe, G. T. Robertson, W. L. Ng, J. M. Richardson, M. E. Winkler, and R. M. Roop. 2005. The *Brucella abortus* Cu,Zn superoxide dismutase is required for optimal resistance to oxidative killing by murine macrophages and wild-type virulence in experimentally infected mice. *Infect. Immun.* **73:**2873–2880.

Gilotra, U., and S. Srivastava. 1997. Plasmid-encoded sequestration of copper by *Pseudomonas pickettii* strain US321. *Curr. Microbiol* **34:**378–381.

Goodhew, C. F., I. B. Wilson, D. J. Hunter, and G. W. Pettigrew. 1990. The cellular location and specificity of bacterial cytochrome c peroxidases. *Biochem. J.* **271:**707–712.

Groisman, E. A. 2001. The pleiotropic two-component regulatory system PhoP-PhoQ. *J. Bacteriol.* **183:**1835–1842.

Grosse, C., A. Anton, T. Hoffmann, S. Franke, G. Schleuder, and D. H. Nies. 2004. Identification of a regulatory pathway that controls the heavy-metal resistance system Czc via promoter *czcN* p in *Ralstonia metallidurans. Arch. Microbiol.* **182:**109–118.

Gunn, J. S., S. S. Ryan, J. C. Van Velkinburgh, R. K. Ernst, and S. I. Miller. 2000. Genetic and functional analysis of a PmrA-PmrB-regulated locus necessary for lipopolysaccharide modification, antimicrobial peptide resistance, and oral virulence of *Salmonella enterica* serovar Typhimurium. *Infect. Immun.* **68:**6139–6146.

Harris, A. G., and S. L. Hazell. 2003. Localisation of *Helicobacter pylori* catalase in both the periplasm and cytoplasm, and its dependence on the twin-arginine target protein, KapA, for activity. *FEMS Microbiol. Lett.* **229:**283–289.

Hassan, M. T., D. van der Lelie, D. Springael, U. Romling, N. Ahmed, and M. Mergeay. 1999. Identification of a gene cluster, *czr*, involved in cadmium and zinc resistance in *Pseudomonas aeruginosa. Gene* **238:**417–425.

Heeb, S., C. Valverde, C. Gigot-Bonnefoy, and D. Haas. 2005. Role of the stress sigma factor RpoS in GacA/RsmA-controlled secondary metabolism and resistance to oxidative stress in *Pseudomonas fluorescens* CHA0. *FEMS Microbiol. Lett.* **243:**251–258.

Heimberger, A., and A. Eisenstark. 1988. Compartmentalization of catalases in *Escherichia coli. Biochem. Biophys. Res. Commun.* **154:**392–397.

Helmann, J. D. 2002. The extracytoplasmic function (ECF) sigma factors. *Adv. Microb. Physiol* **46:**47–110.

Hendrixson, D. R., and V. J. DiRita. 2004. Identification of *Campylobacter jejuni* genes involved in commensal colonization of the chick gastrointestinal tract. *Mol. Microbiol.* **52:**471–484.

Herouart, D., S. Sigaud, S. Moreau, P. Frendo, D. Touati, and A. Puppo. 1996. Cloning and characterization of the *katA* gene of *Rhizobium* *meliloti* encoding a hydrogen peroxide-inducible catalase. *J. Bacteriol.* **178:**6802–6809.

Herren, C. D., E. R. Rocha, and C. J. Smith. 2003. Genetic analysis of an important oxidative stress locus in the anaerobe *Bacteroides fragilis. Gene* **316:** 167–175.

Higgins, M. K., E. Bokma, E. Koronakis, C. Hughes, and V. Koronakis. 2004. Structure of the periplasmic component of a bacterial drug efflux pump. *Proc. Natl. Acad. Sci USA* **101:**9994–9999.

Hild, E., K. Takayama, R. M. Olsson, and S. Kjelleberg. 2000. Evidence for a role of *rpoE* in stressed and unstressed cells of marine *Vibrio angustum* strain S14. *J. Bacteriol.* **182:**6964–6974.

Hirakawa, H., K. Nishino, T. Hirata, and A. Yamaguchi. 2003. Comprehensive studies of drug resistance mediated by overexpression of response regulators of two-component signal transduction systems in *Escherichia coli. J. Bacteriol.* **185:**1851–1856.

Hyytiainen, H., S. Sjoblom, T. Palomaki, A. Tuikkala, and P. E. Tapio. 2003. The PmrA-PmrB two-component system responding to acidic pH and iron controls virulence in the plant pathogen *Erwinia carotovora* ssp. *carotovora. Mol. Microbiol.* **50:**795–807.

Imlay, J. A. 2003. Pathways of oxidative damage. *Annu. Rev. Microbiol.* **57:**395–418.

Jacoby, G. A., and L. S. Munoz-Price. 2005. The new β-lactamases. *N. Engl. J. Med.* **352:**380–391.

Jeannot, K., M. L. Sobel, F. El Garch, K. Poole, and P. Plesiat. 2005. Induction of the MexXY efflux pump in *Pseudomonas aeruginosa* is dependent on drug-ribosome interaction. *J. Bacteriol.* **187:** 5341–5346.

Johnson, C. R., J. Newcombe, S. Thorne, H. A. Borde, L. J. Eales-Reynolds, A. R. Gorringe, S. G. Funnell, and J. J. McFadden. 2001. Generation and characterization of a PhoP homologue mutant of *Neisseria meningitidis. Mol. Microbiol.* **39:** 1345–1355.

Join-Lambert, O. F., M. Michea-Hamzehpour, T. Köhler, F. Chau, F. Faurisson, S. Dautrey, C. Vissuzaine, C. Carbon, and J. C. Pechère. 2001. Differential selection of multidrug efflux mutants by trovafloxacin and ciprofloxacin in an experimental model of *Pseudomonas aeruginosa* acute pneumonia in rats. *Antimicrob. Agents Chemother.* **45:** 571–576.

Jude, F., C. Arpin, C. Brachet-Castang, M. Capdepuy, P. Caumette, and C. Quentin. 2004. TbtABM, a multidrug efflux pump associated with tributyltin resistance in *Pseudomonas stutzeri. FEMS Microbiol. Lett.* **232:**7–14.

Junker, F., and J. L. Ramos. 1999. Involvement of the cis/trans isomerase Cti in solvent resistance of *Pseudomonas putida* DOT-T1E. *J. Bacteriol.* **181:** 5693–5700.

Kang, B. R., B. H. Cho, A. J. Anderson, and Y. C. Kim. 2004. The global regulator GacS of a biocontrol bacterium *Pseudomonas chlororaphis* O6 regulates transcription from the rpoS gene encoding a stationary-phase sigma factor and affects survival in oxidative stress. *Gene* **325:**137–143.

Karnachuk, O. V., S. Y. Kurochkina, D. Nicomrat, Y. A. Frank, D. A. Ivasenko, E. A. Phyllipenko, and O. H. Tuovinen. 2003. Copper resistance in *Desulfovibrio* strain R2. *Antonie Van Leeuwenhoek* **83:**99–106.

Keith, L. M., and C. L. Bender. 1999. AlgT (σ^{22}) controls alginate production and tolerance to environmental stress in *Pseudomonas syringae. J. Bacteriol.* **181:**7176–7184.

Kiyono, M., and H. Pan-Hou. 1999. The *merG* gene product is involved in phenylmercury resistance in *Pseudomonas* strain K-62. *J. Bacteriol.* **181:**726–730.

Klotz, M. G., and S. W. Hutcheson. 1992. Multiple periplasmic catalases in phytopathogenic strains of *Pseudomonas syringae. Appl. Environ. Microbiol.* **58:** 2468–2473.

Koronakis, V., J. Eswaran, and C. Hughes. 2004. Structure and function of TolC: the bacterial exit duct for proteins and drugs. *Annu. Rev. Biochem.* **73:**467–489.

Korshunov, S. S., and J. A. Imlay. 2002. A potential role for periplasmic superoxide dismutase in blocking the penetration of external superoxide into the cytosol of Gram-negative bacteria. *Mol. Microbiol.* **43:**95–106.

Krishnakumar, R., M. Craig, J. A. Imlay, and J. M. Slauch. 2004. Differences in enzymatic properties allow SodCI but not SodCII to contribute to virulence in *Salmonella enterica* serovar Typhimurium strain 14028. *J. Bacteriol.* **186:**5230–5238.

Kroll, J. S., P. R. Langford, K. E. Wilks, and A. D. Keil. 1995. Bacterial [Cu,Zn]-superoxide dismutase: phylogenetically distinct from the eukaryotic enzyme, and not so rare after all! *Microbiology* **141:**2271–2279.

Leclere, V., M. Bechet, and R. Blondeau. 2004. Functional significance of a periplasmic Mn-superoxide dismutase from *Aeromonas hydrophila. J. Appl. Microbiol.* **96:**828–833.

Legatzki, A., G. Grass, A. Anton, C. Rensing, and D. H. Nies. 2003. Interplay of the Czc system and two P-type ATPases in conferring metal resistance to *Ralstonia metallidurans. J. Bacteriol.* **185:**4354–4361.

Li, X.-Z., L. Zhang, and K. Poole. 2002. SmeC, an outer membrane multidrug efflux protein of *Stenotrophomonas maltophilia. Antimicrob. Agents Chemother.* **46:**333–343.

Llama-Palacios, A., E. Lopez-Solanilla, C. Poza-Carrion, F. Garcia-Olmedo, and P. Rodriguez-Palenzuela. 2003. The *Erwinia chrysanthemi phoP-phoQ* operon plays an important role in growth at low pH, virulence and bacterial survival in plant tissue. *Mol. Microbiol.* **49:**347–357.

Macfarlane, E. L., A. Kwasnicka, and R. E. Hancock. 2000. Role of *Pseudomonas aeruginosa* PhoP-phoQ in resistance to antimicrobial cationic peptides and aminoglycosides. *Microbiology* **146:**2543–2554.

Mao, W., M. S. Warren, D. S. Black, T. Satou, T. Murata, T. Nishino, N. Gotoh, and O. Lomovskaya. 2002. On the mechanism of substrate specificity by resistance nodulation division (RND)-type multidrug resistance pumps: the large periplasmic loops of MexD from *Pseudomonas aeruginosa* are involved in substrate recognition. *Mol. Microbiol.* **46:**889–901.

Marceau, M., F. Sebbane, F. Ewann, F. Collyn, B. Lindner, M. A. Campos, J. A. Bengoechea, and M. Simonet. 2004. The *pmrF* polymyxin-resistance operon of *Yersinia pseudotuberculosis* is upregulated by the PhoP-PhoQ two-component system but not by PmrA-PmrB, and is not required for virulence. *Microbiology* **150:**3947–3957.

Marchand, I., L. Damier-Piolle, P. Courvalin, and T. Lambert. 2004. Expression of the RND-type efflux pump AdeABC in *Acinetobacter baumannii* is regulated by the AdeRS two-component system. *Antimicrob. Agents Chemother.* **48:**3298–3304.

Masschalck, B., and C. W. Michiels. 2003. Antimicrobial properties of lysozyme in relation to foodborne vegetative bacteria. *Crit. Rev. Microbiol.* **29:**191–214.

McDougald, D., L. Gong, S. Srinivasan, E. Hild, L. Thompson, K. Takayama, S. A. Rice, and S. Kjelleberg. 2002. Defences against oxidative stress during starvation in bacteria. *Antonie Van Leeuwenhoek* **81:**3–13.

McPhee, J. B., S. Lewenza, and R. E. Hancock. 2003. Cationic antimicrobial peptides activate a two-component regulatory system, PmrA-PmrB, that regulates resistance to polymyxin B and cationic antimicrobial peptides in *Pseudomonas aeruginosa. Mol. Microbiol.* **50:**205–217.

Mergeay, M., S. Monchy, T. Vallaeys, V. Auquier, A. Benotmane, P. Bertin, S. Taghavi, J. Dunn, L. D. van der, and R. Wattiez. 2003. *Ralstonia metallidurans*, a bacterium specifically adapted to toxic metals: towards a catalogue of metal-responsive genes. *FEMS Microbiol. Rev.* **27:**385–410.

Middlemiss, J. K., and K. Poole. 2004. Differential impact of MexB mutations on substrate selectivity of the MexAB-OprM multidrug efflux pump of *Pseudomonas aeruginosa. J. Bacteriol.* **186:**1258–1269.

Morita, Y., T. Murata, T. Mima, S. Shiota, T. Kuroda, T. Mizushima, N. Gotoh, T. Nishino, and T. Tsuchiya. 2003. Induction of *mexCD-oprJ* operon for a multidrug efflux pump by disinfectants in wild-type *Pseudomonas aeruginosa* PAO1. *J. Antimicrob. Chemother.* **5:**991–994.

Moskowitz, S. M., R. K. Ernst, and S. I. Miller. 2004. PmrAB, a two-component regulatory system of *Pseudomonas aeruginosa* that modulates resistance to cationic antimicrobial peptides and addition of aminoarabinose to lipid A. *J. Bacteriol.* **186:**575–579.

Murakami, S., and A. Yamaguchi. 2003. Multidrug-exporting secondary transporters. *Curr. Opin. Struct. Biol.* **13:**443–452.

Nehme, D., X. Z. Li, R. Elliot, and K. Poole. 2004. Assembly of the MexAB-OprM multidrug efflux system of *Pseudomonas aeruginosa*: identification and characterization of mutations in *mexA* compromising MexA multimerization and interaction with MexB. *J. Bacteriol.* **186:**2973–2983.

Nehme, D., and K. Poole. 2005. Interaction of the MexA and MexB components of the MexAB-OprM multidrug efflux system of *Pseudomonas aeruginosa*: identification of MexA extragenic suppressors of a T578I mutation in MexB. *Antimicrob. Agents Chemother.* **49:**4375–4378.

Nickerson, K. W., and A. Aspedon. 1992. Detergent-shock response in enteric bacteria. *Mol. Microbiol.* **6:**957–961.

Nies, D. H. 2003. Efflux-mediated heavy metal resistance in prokaryotes. *FEMS Microbiol. Rev.* **27:**313–339.

Nishino, K., T. Honda, and A. Yamaguchi. 2005. Genome-wide analyses of *Escherichia coli* gene expression responsive to the BaeSR two-component regulatory system. *J. Bacteriol.* **187:**1763–1772.

Nishino, K., Y. Inazumi, and A. Yamaguchi. 2003. Global analysis of genes regulated by EvgA of the two-component regulatory system in *Escherichia coli. J. Bacteriol.* **185:**2667–2672.

Oyston, P. C., N. Dorrell, K. Williams, S. R. Li, M. Green, R. W. Titball, and B. W. Wren. 2000. The response regulator PhoP is important for survival under conditions of macrophage-induced stress and virulence in *Yersinia pestis. Infect. Immun.* **68:**3419–3425.

Palmer, M. 2003. Efflux of cytoplasmically acting antibiotics from gram-negative bacteria: periplasmic substrate capture by multicomponent efflux pumps inferred from their cooperative action with single-component transporters. *J. Bacteriol.* **185:**5287–5289.

Perron, K., O. Caille, C. Rossier, C. Van Delden, J. L. Dumas, and T. Kohler. 2004. CZCR-CZCS: a two component system involved in heavy metal and carbapenem resistance in *Pseudomonas aeruginosa. J. Biol. Chem.* **279:**8761–8768.

Poole, K. 2004a. Efflux-mediated multiresistance in Gram-negative bacteria. *Clin. Microbiol. Infect.* **10:**12–26.

Poole, K. 2004b. Resistance to β-lactam antibiotics. *Cell. Mol. Life Sci.* **61:**2200–2223.

Poole, K. 2004c. Efflux pumps, p. 635–674. *In* J.-L. Ramos (ed.), *Pseudomonas*, vol. I. Genomics, life style and molecular architecture. Kluwer Academic/Plenum Publishers, New York, N.Y.

Poole, K. 2005. Efflux-mediated antimicrobial resistance. *J. Antimicrob. Chemother.* **56:**20–51.

Puig, S., E. M. Rees, and D. J. Thiele. 2002. The ABCDs of periplasmic copper trafficking. *Structure (Camb.)* **10:**1292–1295.

Putman, M., H. W. van Veen, and W. N. Konings. 2000. Molecular properties of bacterial multidrug transporters. *Microbiol. Mol. Biol. Rev.* **64:**672–693.

Rajagopal, S., N. Eis, M. Bhattacharya, and K. W. Nickerson. 2003. Membrane-derived oligosaccharides (MDOs) are essential for sodium dodecyl sulfate resistance in *Escherichia coli*. *FEMS Microbiol. Lett.* **223:**25–31.

Ramos, J. L., E. Duque, M. T. Gallegos, P. Godoy, M. I. Ramos-Gonzalez, A. Rojas, W. Teran, and A. Segura. 2002. Mechanisms of solvent tolerance in gram-negative bacteria. *Annu. Rev. Microbiol.* **56:**743–768.

Ramos-Gonzalez, M. I., P. Godoy, M. Alaminos, A. Ben Bassat, and J. L. Ramos. 2001. Physiological characterization of *Pseudomonas putida* DOT-T1E tolerance to *p*-hydroxybenzoate. *Appl. Environ. Microbiol.* **67:**4338–4341.

Randall, L. P., S. W. Cooles, A. R. Sayers, and M. J. Woodward. 2001. Association between cyclohexane resistance in *Salmonella* of different serovars and increased resistance to multiple antibiotics, disinfectants and dyes. *J. Med. Microbiol.* **50:**919–924.

Rensing, C., and G. Grass. 2003. *Escherichia coli* mechanisms of copper homeostasis in a changing environment. *FEMS Microbiol. Rev.* **27:**197–213.

Rossbach, S., M. L. Kukuk, T. L. Wilson, S. F. Feng, M. M. Pearson, and M. A. Fisher. 2000. Cadmium-regulated gene fusions in *Pseudomonas fluorescens*. *Environ. Microbiol.* **2:**373–382.

St John, G., and H. M. Steinman. 1996. Periplasmic copper-zinc superoxide dismutase of *Legionella pneumophila*: role in stationary-phase survival. *J. Bacteriol.* **178:**1578–1584.

Sanchez, P., J. F. Linares, B. Ruiz-Diez, E. Campanario, A. Navas, F. Baquero, and J. L. Martinez. 2002. Fitness of *in vitro* selected *Pseudomonas aeruginosa nalB* and *nfxB* multidrug resistant mutants. *J. Antimicrob. Chemother.* **50:**657–664.

San Mateo, L. R., M. M. Hobbs, and T. H. Kawula. 1998. Periplasmic copper-zinc superoxide dismutase protects *Haemophilus ducreyi* from exogenous superoxide. *Mol. Microbiol.* **27:**391–404.

San Mateo, L. R., K. L. Toffer, P. E. Orndorff, and T. H. Kawula. 1999. Neutropenia restores virulence to an attenuated Cu,Zn superoxide dismutase-deficient *Haemophilus ducreyi* strain in the swine model of chancroid. *Infect. Immun.* **67:**5345–5351.

Sanchez, P., J. F. Linares, B. Ruiz-Diez, E. Campanario, A. Navas, F. Baquero, and J. L. Martinez. 2002. Fitness of *in vitro* selected *Pseudomonas aeruginosa nalB* and *nfxB* multidrug resistant mutants. *J. Antimicrob. Chemother.* **50:**657–664.

Sansone, A., P. R. Watson, T. S. Wallis, P. R. Langford, and J. S. Kroll. 2002. The role of two periplasmic copper- and zinc-cofactored superoxide dismutases in the virulence of *Salmonella choleraesuis*. *Microbiology* **148:**719–726.

Saxena, D., N. Joshi, and S. Srivastava. 2002. Mechanism of copper resistance in a copper mine isolate *Pseudomonas putida* strain S4. *Curr. Microbiol.* **45:**410–414.

Schnell, S., and H. M. Steinman. 1995. Function and stationary-phase induction of periplasmic copper-zinc superoxide dismutase and catalase/peroxidase in *Caulobacter crescentus*. *J. Bacteriol.* **177:**5924–5929.

Schurr, M. J., H. Yu, J. M. Martinez-Salazar, J. C. Boucher, and V. Deretic. 1996. Control of AlgU, a member of the sigma E-like family of stress sigma factors, by the negative regulators MucA and MucB and *Pseudomonas aeruginosa* conversion to mucoidy in cystic fibrosis. *J. Bacteriol.* **178:**4997–5004.

Seib, K. L., H. J. Tseng, A. G. McEwan, M. A. Apicella, and M. P. Jennings. 2004. Defenses against oxidative stress in *Neisseria gonorrhoeae* and *Neisseria meningitidis*: distinctive systems for different lifestyles. *J. Infect. Dis.* **190:**136–147.

Sha, Z., T. J. Stabel, and J. E. Mayfield. 1994. *Brucella abortus* catalase is a periplasmic protein lacking a standard signal sequence. *J. Bacteriol.* **176:**7375–7377.

Shi, Y., M. J. Cromie, F. F. Hsu, J. Turk, and E. A. Groisman. 2004. PhoP-regulated *Salmonella* resistance to the antimicrobial peptides magainin 2 and polymyxin B. *Mol. Microbiol.* **53:**229–241.

Short, K. A., and R. P. Blakemore. 1989. Periplasmic superoxide dismutases in *Aquaspirillum magnetotacticum*. *Arch. Microbiol.* **152:**342–346.

Silver, S. 2003. Bacterial silver resistance: molecular biology and uses and misuses of silver compounds. *FEMS Microbiol. Rev.* **27:**341–353.

Singh, S. K., G. Grass, C. Rensing, and W. R. Montfort. 2004. Cuprous oxidase activity of CueO from *Escherichia coli*. *J. Bacteriol.* **186:**7815–7817.

Sobel, M. L., G. A. McKay, and K. Poole. 2003. Contribution of the MexXY multidrug transporter to aminoglycoside resistance in *Pseudomonas aeruginosa* clinical isolates. *Antimicrob. Agents Chemother.* **47:**3202–3207.

Souza-Hart, J. A., W. Blackstock, V. Di Modugno, I. B. Holland, and M. Kok. 2003. Two-component systems in *Haemophilus influenzae*: a regulatory role for ArcA in serum resistance. *Infect. Immun.* **71:**163–172.

Srinivasa Rao, P. S., Y. Yamada, and K. Y. Leung. 2003. A major catalase (KatB) that is required for resistance to H2O2 and phagocyte-mediated killing in *Edwardsiella tarda*. *Microbiology* **149:**2635–2644.

Suziedeliene, E., K. Suziedelis, V. Garbenciute, and S. Normark. 1999. The acid-inducible *asr* gene in *Escherichia coli:* transcriptional control by the *phoBR* operon. *J. Bacteriol.* **181:**2084–2093.

Tamburro, A., I. Robuffo, H. J. Heipieper, N. Allocati, D. Rotilio, C. Di Ilio, and B. Favaloro. 2004. Expression of glutathione S-transferase and peptide methionine sulphoxide reductase in *Ochrobactrum anthropi* is correlated to the production of reactive oxygen species caused by aromatic substrates. *FEMS Microbiol. Lett.* **241:**151–156.

Testerman, T. L., A. Vazquez-Torres, Y. Xu, J. Jones-Carson, S. J. Libby, and F. C. Fang. 2002. The alternative sigma factor σ^E controls antioxidant defences required for *Salmonella* virulence and stationary-phase survival. *Mol. Microbiol.* **43:**771–782.

Tikhonova, E. B., Q. Wang, and H. I. Zgurskaya. 2002. Chimeric analysis of the multicomponent multidrug efflux transporters from Gram-negative bacteria. *J. Bacteriol.* **184:**6499–6507.

Touze, T., J. Eswaran, E. Bokma, E. Koronakis, C. Hughes, and V. Koronakis. 2004. Interactions underlying assembly of the *Escherichia coli* AcrAB-TolC multidrug efflux system. *Mol. Microbiol.* **53:**697–706.

Visick, K. L., and E. G. Ruby. 1998. The periplasmic, group III catalase of *Vibrio fischeri* is required for normal symbiotic competence and is induced both by oxidative stress and by approach to stationary phase. *J. Bacteriol.* **180:**2087–2092.

Waidner, B., K. Melchers, I. Ivanov, H. Loferer, K. W. Bensch, M. Kist, and S. Bereswill. 2002. Identification by RNA profiling and mutational analysis of the novel copper resistance determinants CrdA (HP1326), CrdB (HP1327), and CzcB (HP 1328) in *Helicobacter pylori*. *J. Bacteriol.* **184:**6700–6708.

Winfield, M. D., and E. A. Groisman. 2004. Phenotypic differences between *Salmonella* and *Escherichia coli* resulting from the disparate regulation of homologous genes. *Proc. Natl. Acad. Sci. USA* **101:**17162–17167.

Winfield, M. D., T. Latifi, and E. A. Groisman. 2005. Transcriptional regulation of the 4-amino-4-deoxy-L-arabinose biosynthetic genes in *Yersinia pestis*. *J. Biol. Chem.* **280:**14765–14772.

Yu, E. W., G. McDermott, H. I. Zgurskaya, H. Nikaido, and D. E. Koshland, Jr. 2003. Structural basis of multiple drug-binding capacity of the AcrB multidrug efflux pump. *Science* **300:**976–980.

Zwir, I., D. Shin, A. Kato, K. Nishino, T. Latifi, F. Solomon, J. M. Hare, H. Huang, and E. A. Groisman. 2005. Dissecting the PhoP regulatory network of *Escherichia coli* and *Salmonella enterica*. *Proc. Natl. Acad. Sci USA* **102:**2862–2867.

OSMOREGULATION IN THE PERIPLASM

Jean-Pierre Bohin and Jean-Marie Lacroix

19

Many bacteria can survive brutal changes in environmental osmotic pressure. Bacteria respond to these changes both passively, essentially due to the water permeability of cellular membranes, and actively by transporting and/or synthesizing new molecules whose accumulation may help the cell to restore sufficient content of water in its various compartments to allow resumption of cell growth (see Wood, 1999, for a recent review). While the first passive phase may be complete in less that 1 min, the duration of adaptive responses may be 10 to 60 min, or even more, before the stressed cells reach a new steady state determined by medium conditions. During these events the periplasmic space is subject to important changes in water content and volume (Oliver, 1996; Koch, 1998; Cayley et al., 2000). However, the syntheses of a very limited number of specifically periplasmic components have been reported to be subject to osmoregulation. In *Escherichia coli*, periplasmic proteins, as for example members of the *mal* and *pho* regulons, are controlled by complex regulatory networks (Case et al., 1986) that globally adapt envelope components (including outer and inner membrane constituents) to various medium factors, including osmotic pressure. A periplasmic trehalase is induced in *E. coli* under high-osmolarity growth conditions (Boos et al., 1987). This protein is directly connected to osmotic adaptation in the cytoplasm with the synthesis of the compatible solute trehalose (Csonka, 1989; see below). OsmY is a small periplasmic protein of unknown function; its two BON domains could interact with phospholipids present on both faces of the periplasm (Yeats and Bateman, 2003). Its gene *osmY* is induced by both osmotic and growth phase signals and *osmY* inactivation results in slightly increased sensitivity to hyperosmotic conditions (Yim et al., 1994). Recently, the HtrA (DegP) protease of *E. coli*, a heat-shock–inducible periplasmic protein, was shown to be osmoregulated by repression of its gene at low osmolarity (Forns et al., 2005). Actually, the only periplasmic components whose synthesis was found to be induced by low osmotic pressure in various bacterial species are osmoregulated periplasmic glucans (OPGs) even though OPG osmoregulation suffers noticeable exceptions.

OPG structures found in different bacterial species share several common characteristics: (i) D-glucose is the only constituent sugar; (ii) glucose units are linked, at least partially, by β-glycosidic bonds; (iii) there are a limited

Jean-Pierre Bohin and Jean-Marie Lacroix, Unité de Glyco-biologie Structurale et Fonctionnelle, UMR CNRS-USTL 8576, IFR118, Université des Sciences et Technologies de Lille, 59655 Villeneuve d'Ascq Cedex, France.

The Periplasm
Edited by Michael Ehrmann © 2007 ASM Press, Washington, D.C.

number of glucose units (5 to 24); (iv) in most cases, but with a few exceptions, OPG concentration in the periplasm increases in response to a decrease of environmental osmolarity. In many, but not all, species the OPG backbones can be decorated to various extents with a variety of substituents. These substituents appear to belong to two classes: (i) residues originating from membrane phospholipids like phosphoglycerol, phosphoethanolamine, or phosphocholine; and (ii) residues originating from intermediary metabolism like acetyl, succinyl, and methylmalonyl. For the latter, acyl-coenzymes A are the likely donor molecules, but this has not been demonstrated (Bohin, 2000).

OPGs were detected during the study of phospholipid turnover in *E. coli* by E. P. Kennedy's group (Van Golde et al., 1973). In this bacterium, rapid phosphatidylglycerol turnover is associated with the transfer of *sn*-1-phosphoglycerol to a new class of oligosaccharides, named as a consequence "membrane-derived oligosaccharides" (MDOs) (reviewed in Kennedy, 1996). The homology between the linear β-glucans found in the periplasm of *E. coli* and the cyclic β-glucans found in the periplasm of *Agrobacterium tumefaciens* was established in 1986 with the demonstration that the syntheses of both kinds of molecules are osmoregulated (Miller et al., 1986). We now know that these compounds are found in many Proteobacteria and most of them do not contain any substituent derived from membrane lipids. The term MDOs is confusing because these compounds are part of the envelope but may not be derived from the membrane. For these reasons, the term OPGs was preferred and the gene and protein nomenclatures were changed accordingly (Lequette et al., 2004).

OPG STRUCTURES

OPG structures were described in several species of the alpha, beta, and gamma subdivisions of the Proteobacteria. No information is available for other gram-negative bacteria. Beyond their common features OPGs show an unexpected structural diversity. Four families

(I to IV) have been defined on the basis of backbone organization (Bohin, 2000).

OPGs of family I were found in *E. coli* (Kennedy, 1996, and references therein), *Pseudomonas syringae* (Talaga et al. 1994), *Erwinia chrysanthemi* (Cogez et al., 2001), *Salmonella enterica* serovar Typhimurium (F. Norel, V. Robbe-Saule, A. Bohin, and J.-P. Bohin, unpublished data), and *Pseudomonas aeruginosa* (Y. Lequette, E. Rollet, A. Delangle, E. P. Greenberg, and J.-P. Bohin, unpublished data). The structures found in these bacteria are very similar. In *E. coli*, OPGs are heterogeneous in size and appear to range from 5 to 12 glucose residues, with the principal species containing 8 or 9 glucose residues. The sugar backbone consisting of β-1,2-linked glucose units is highly branched by the irregular addition of single glucose units attached by β-1,6 linkages (Fig. 1). A heterogeneous substitution by phosphoglycerol, succinyl, and phosphoethanolamine residues is superimposed on this heterogeneity of backbone structures. However, the resulting population of OPGs can be resolved in discrete groups with similar charge/mass ratios by anion-exchange chromatography (Fig. 1) (Kennedy, 1996; Lacroix et al., 1999; Lequette et al., 2004). A different pattern of substitution was found in *E. chrysanthemi* where low levels of succinyl and acetyl residues were detected (Cogez et al., 2001). No substitution was found in *P. syringae* whose OPGs are essentially neutral (Talaga et al., 1994).

OPGs of family II were found in various members of the family *Rhizobiaceae* (Breedveld and Miller, 1994, and references therein). *Agrobacterium*, *Rhizobium*, *Sinorhizobium*, *Brucella*, and *Mesorhizobium* species synthesize periplasmic glucans with similar structures (Briones et al., 1997; Choma and Komaniecka, 2003). In these genera, OPGs are composed of a cyclic β-1,2-glucan backbone containing 17 to 25 glucose residues. Much larger molecules (up to 40 glucose units) were detected within cultures of a particular strain of *Sinorhizobium meliloti* (Breedveld and Miller, 1994). The predominant substituent on OPGs of *S. meliloti* and *A. tumefaciens* strains is phosphoglycerol, but substitu-

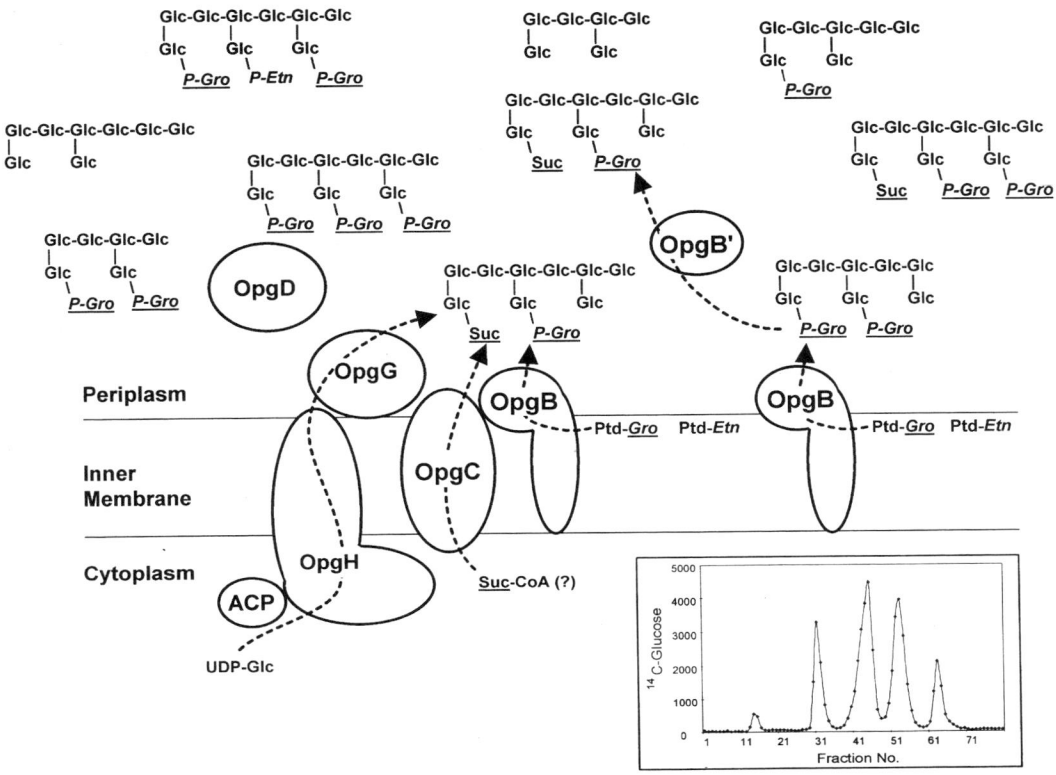

FIGURE 1 Working model of the OPG biosynthetic complex of *E. coli*. The variety of backbone structures and patterns of substitution are schematically represented (see text for details). In the inset is a typical DEAE-Sephacel anion-exchange column chromatography profile of [U-¹⁴C]glucose-labeled OPGs according to Lacroix et al. (1999). Each peak consists of various backbone structures with similar charge/mass ratios (A. Bohin and J.-P. Bohin, unpublished observation).

tions by succinyl and methylmalonyl residues were also found in other related species or strains. The degree of substitution may vary greatly among different strains and the stage of growth also influences the degree of substitution. Consequently, OPGs of family II may be neutral or moderately to highly anionic molecules.

OPGs of family III were first described in *Bradyrhizobium japonicum* as β-1,6- and β-1,3-cyclic glucans containing 10 to 13 glucose units per ring (Rolin et al., 1992; Breedveld and Miller, 1994). Very similar OPGs were found in *Azorhizobium caulinodans* (Komaniecka and Choma, 2003) and in *Azospirillum brasilense* as judged from NMR analysis (Altabe et al., 1998).

In this latter case, however, high-performance anion-exchange chromatography allowed the separation of three distinct structures: glucan I, made of 12 glucose units linked by three β-1,3, eight β-1,6, and one β-1,4 linkage; glucan II, derived from glucan I by the addition of a glucose linked by an α-1,3 linkage; and glucan III, derived from glucan II by the addition of a 2-O-methyl group onto the α-linked glucose unit. Thus, the OPGs of family III differ from those of family II not only by the nature of the glycosidic linkage but also by a strict control of the ring size. Although OPGs of family III are predominantly uncharged, OPGs of *B. japonicum* can also become substituted with phosphocholine (Rolin et al., 1992).

OPGs of family IV have very similar structural features. They were found in *Ralstonia solanacearum* (Talaga et al., 1996), *Xanthomonas campestris* (Talaga et al., 1996; York, 1995), and *Rhodobacter sphaeroides* (Talaga et al., 2002). These OPGs are cyclic and they have a unique degree of polymerization (DP = 13, 16, and 18, respectively). One linkage is α-1,6 whereas all the other glucose residues are linked by β-1,2 linkages. The presence of this α-1,6 linkage induces structural constraints in this kind of molecule, which contrast with the very flexible structures of the cyclic all β-1,2-OPGs of family II (Lippens et al., 1998). While OPGs found in *X. campestris* and *R. solanacearum* are not substituted, those found in *R. sphaeroides* can be substituted by one to seven succinyl ester residues and by one or two acetyl residues.

Thus, OPG diversity is very high not only among different bacterial species, due to different genomic capacity (see below), but also in a particular bacterial strain, because control of the backbone structure is often not stringent and because backbone substitution (if any) is heterogeneous.

GENOMIC OVERVIEW OF OPG BIOSYNTHESIS

Genetic analyses of OPG synthesis were done in a limited number of bacterial species and OPG genes were defined with unrelated names: *mdo* in *E. coli* (Bohin and Kennedy, 1984a) and *chv* in *A. tumefaciens* (for chromosomal virulence; Puvanesarajah et al., 1985). Later, mutants were also obtained in other genera: *R. sphaeroides* (Cogez et al., 2002) and *X. campestris* (Minsavage et al., 2004) where the genes were named *opg*. Complementation experiments were used to get homologous genes in related organisms of the Rhizobiales group, essentially *S. meliloti* (Dylan et al., 1986) and *B. japonicum* (Bhagwat et al., 1993) where the genes were named *ndv*, for nodule development. This strategy also allowed the isolation of a locus similar to *chvB* and named *cviB* in *Azospirillum brasilense* (Raina et al., 1995). From sequence data, it clearly appeared that if the *chv* and *ndv* genes of *A. tumefaciens* and *S. meliloti* were highly similar, they

did not share any significant similarity with the *opg* genes. Moreover, the *ndv* genes found in *B. japonicum* share little similarity with some *opg* genes but no similarity with the *chv* genes. Thus, three distinct sets of genes have been described and similar genes can be recognized in various completely sequenced genomes of Proteobacteria (Table 1). For clarity, only genes playing a role in the backbone synthesis are mentioned. The first set of genes includes *opgG* (abbreviated G), *opgH* (H), *opgD* (D), and *opgI* (I). G, H, and D were described initially in *E. coli* and I in *R. sphaeroides*. Phylogenetic analyses were previously published for H (Cogez et al., 2002) and G and D (Lequette et al., 2004). The second set of genes includes *ndvB* (BB), *ndvC* (CB), and *ndvD* (DB) initially described in *B. japonicum*; the third set includes *chvA* (AA) and *chvB* (BA) initially described in *A. tumefaciens*.

Several conclusions can be drawn from the data shown in Table 1. Several bacterial species do not possess any of the OPG genes already described, as exemplified by *Yersinia pestis*. These bacteria probably do not synthesize OPGs, but this needs to be established. In absence of OPGs, one may wonder if other periplasmic components can substitute for OPGs. The H gene that encodes a protein belonging to the glycosyltransferase family 2 (http://afmb.cnrs-mrs.fr/CAZY/GT_2.html) is present in many bacteria of the alpha, beta, and gamma subdivisions. The G and D genes that encode accessory proteins (see below) are diversely distributed with all the possible combinations; only G, G and D, or only D. In some cases, two copies of H, G, or D are detected. The I gene that encodes another accessory protein is only present in *Rhodobacter*. All three BB and CB and DB genes are found only in related alpha-proteobacteria in addition to H and G. But BB and/or CG, which encode proteins belonging to glycosyl hydrolases family 17 (http://afmb.cnrs-mrs.fr/CAZY/GH_17.html), can also be found in *Pseudomonas*. The BA gene encodes a protein belonging to the glycoside hydrolase of family 94 (http://afmb.cnrs-mrs.fr/CAZY/GH_94.html) and the AA gene encodes an

TABLE 1 Occurrences of OPG genes in various representative genomes[a]

Species	Subdivision	G	H	D	I	BB	CB	DB	BA	AA
Escherichia coli	gamma	1	1	1						
Salmonella enterica	gamma	1	1	1						
Nitrosomonas europaea	beta	1	1							
Erwinia chrysanthemi	gamma	1	1							
Yersinia pseudotuberculosis	gamma	1	1							
Yersinia pestis	gamma									
Azotobacter vinelandii[b]	gamma	2	2	1			1			
Pseudomonas syringae	gamma	1	1	1		1				
Pseudomonas aeruginosa	gamma	1	1			1[c]	1[c]		1[d]	
Wigglesworthia glossinidia	gamma		1							
Ralstonia solanacearum	beta	1	1	2						
Rhodospirillum rubrum[b]	alpha		1	1						
Rhodobacter sphaeroides	alpha	1	1	1	1				1	
Rhodopseudomonas palustris	alpha	1	1			1	1	1		1
Bradyrhizobium japonicum	alpha	1	1			1	1	1		1
Vibrio cholerae	gamma	1	1							
Shewanella oneidensis	gamma	2	1	1						
Xanthomonas campestris	gamma		1	1					1	
Xylella fastidiosa	gamma		1	1						
Agrobacterium tumefaciens	alpha								1	1
Sinorhizobium meliloti	alpha								1	1
Brucella melitensis	alpha								1	1
Azospirillum brasilense[b]	alpha								1	
Bartonella henselae	alpha								1	1
Burkholderia fungorum[b]	beta								1	1

[a]The BLASTp program was used to search in the nonredundant database sequences highly similar to OpgG (G), OpgH (H), and OpgD (D) from *E. coli*; OpgI (I) from *R. sphaeroides*; NdvB (BB), NdvC (CB), and NdvD (DB) from *B. japonicum*; and ChvB (BA) and ChvA (AA) from *A. tumefaciens*. Lines of the table are ordered from the top to the bottom according to the BLAST score with OpgH when present, then to the BLAST score with ChvB.

[b]Bacterial species whose genomic sequencing is not yet complete.

[c]The same protein exhibits similar levels of homology with BB and CB.

[d]A mobile genomic island.

ABC transporter. These genes are found together among the *Rhizobiaceae* and *Burkholderia*. The AA gene is found in association with the BB, CG, and DB genes for the transport of cyclic glucans. The BA gene is found in various other bacteria. However, as long as the functions of such genes are not demonstrated, their assignations to OPG synthesis would be hazardous because they can encode biosynthetic as well as degradative enzymes. For example, an *ndvB* defective mutant of *R. sphaeroides* synthesizes normal OPGs (Cogez et al., 2002).

There is no simple correlation between the natures of the OPG genes present and the structures of OPG produced by a particular bacterial species (see below) and sequence-based predictions are not (yet) possible.

MECHANISMS OF OPG BIOSYNTHESIS

In *E. coli*, two specific proteins are necessary for the glucan backbone biosynthesis. They are encoded by the *opgGH* operon (Lacroix et al., 1991). In vitro, the presence of OpgH in inner membrane vesicles is necessary to obtain the production of linear β-1,2-polyglucose chains from the precursor UDP-glucose. For unknown reasons, the acyl carrier protein, which normally functions in fatty acid synthesis, is also necessary to this activity (Thérisod et al., 1986). OpgH (97 kDa) has eight transmembrane segments and three large cytoplasmic regions (Debarbieux et al., 1997). The central cytoplasmic region shows structural features of a glucosyltransferase of family 2 where several aspartic acid residues are necessary to OPG synthesis (Lequette, 2002). It was postulated that the eight transmembrane segments in OpgH could form a channel for OPG translocation to the periplasm during synthesis (Debarbieux et al., 1997). OpgG (56 kDa) is a periplasmic protein whose function has not been established. In the in vitro test of OpgH activity, OpgG was absent and the product showed a higher degree of polymerization and no branch (Thérisod et al., 1986; Kennedy, 1996). However, OpgG-defective mutants are unable to form either mature OPG molecules or any precursor forms (Lacroix et al., 1991). The 2.5-Å crystal structure of OpgG was recently reported (Hanoulle et al., 2004). The protein is composed of two distinct β-sandwich domains. The N-terminal domain shares some similarities with carbohydrate-active enzyme and the C-terminal domain could be implicated in interactions with other molecules. These data support the hypothesis that OpgG interacts with OpgH for the translocation of nascent molecules and catalyzes the addition of branches.

In *E. coli*, a third protein, OpgD, was recently described (Lequette et al., 2004). It is the product of a paralog of *opgG*. When opgD was inactivated, normal amounts of OPGs were recovered but the backbone structure was altered, with a higher degree of polymerization. Thus, OpgD seems to control (or interfere with) the OPG biosynthetic machinery.

The functions of genes similar to the *E. coli opgG and opgH* genes have been demonstrated in *P. syringae* (Loubens et al., 1993) and *E. chrysanthemi* (Page et al., 2001), which produce linear OPGs of family I, but also in *R. sphaeroides* (Cogez et al., 2002) and (very likely) in *X. campestris* (Minsavage et al., 2004), which produce cyclic OPGs of family IV. However, *opgD* can be absent from certain genomes, and in some cases, *opgG* is absent but *opgD* is present (see above). Therefore, one can envisage that interactions between OpgH, on the one hand, and OpgG/OpgD, on the other hand, have evolved differently in different species.

Very similar OPG genes were found in *A. tumefaciens* (*chvB* and *chvA*), *S. meliloti* (*ndvB*

and *ndvA*), and *Brucella abortus* (*cgs* and *cgt*) (Puvanesarajah et al., 1985; Dylan et al., 1986; Breedveld and Miller, 1994, Iñon de Iannino et al., 1998; Roset et al., 2004). These genes are strictly homologous and the genes from one species can complement OPG biosynthetic defects in other species. In *A. tumefaciens* and *S. meliloti* the two genes are adjacent to each other, but they are separated by 857 bp in *B. abortus*. The biosynthetic protein (BA or Cgs) is a very large cytoplasmic membrane protein (316 to 319 kDa). The *B. abortus* protein was recently shown to be composed of six transmembrane segments determining four large cytoplasmic domains and three very small periplasmic regions (Ciocchini et al., 2004). The biosynthetic enzymatic activity can be assayed efficiently in vitro. Unpurified membrane preparations are able to catalyze the formation of cyclic β-1,2-glucans from UDP-glucose and a high-molecular-weight membrane protein is labeled when radioactive UDP-glucose is present (Altabe et al., 1994; Breedveld and Miller, 1994). The second protein (AA or Cgt) is a 66- or 67-kDa inner membrane protein that shares amino acid sequence similarity with several ATP-binding cassette transporters. When the protein is absent, cyclic glucans are formed but they are not substituted. Since substitution occurs in the periplasm (see below), it can be concluded that the AA function is the translocation of the cyclic molecules toward the periplasm.

Using the hypothesis that structurally different OPGs could functionally compensate for OPG defect in a *S. meliloti ndvB* mutant, Bhagwat et al. (1993) identified a gene they named *ndvB* from *B. japonicum*. Later, two other genes were characterized, *ndvC* (Bhagwat et al., 1996) and *ndvD* (Chen et al., 2002), which form with *ndvB* a locus of three monocistronic genes. NdvB (102 kDa), NdvC (62 kDa), and NdvD (26 kDa) are predicted to be membrane bound, with NdvB and NdvC having several transmembrane segments and NdvD only one. When *ndvB* was inactivated, no OPGs were synthesized in vivo and mutant membrane preparations failed to produce glucans in vitro. In contrast, when *ndvC* was inactivated, normal amounts of

OPGs were produced, but their structures contained almost only β,1-3 linkages. Inactivation of *ndvD* abolished OPG synthesis in vivo but did not affect the glucan synthesis by membrane preparations in vitro. Thus, if NdvB and NdvC are most probably two membrane-bound biosynthetic enzymes, the function of NdvD remains elusive.

OPG SUBSTITUTION ENZYMES

In *E. coli*, the OPGs produced by the biosynthetic enzymes are heterogeneous, and this variety of backbones can further be modified by 1, 2, or 3 residues of phosphoglycerol, and/or succinyl, and/or phosphoethanolamine. Thus, a majority of the OPG molecules have a high anionic character (up to 5 negative charges), while a minority of them are neutral (Fig. 1). To date, only two genes have been recognized as participating in OPG substitution, *opgB* (Jackson et al., 1984) and *opgC* (Lacroix et al., 1999). Phosphoglycerol transfer has been shown to occur in the periplasmic compartment (Bohin and Kennedy, 1984b). Two phosphoglycerol transferase (PGT) activities have been measured in vitro: the membrane-bound PGT I and the periplasmic PGT II. Soluble OPGs behave as phosphoglycerol acceptor for the latter but not for the former enzyme. Thus, Jackson and Kennedy (1983) have proposed a two-step model to account for phosphoglycerol substitution: first, PGT I transfers residues from the membrane phosphatidylglycerol to nascent OPG molecules; second, PGT II swaps residues from one OPG molecule to another. OpgB (85 kDa), corresponding to PGT I, consists of a large periplasmic domain anchored in the inner membrane by three putative transmembrane segments. This protein was found highly sensitive to proteolytic cleavage and PGT II is actually a soluble form of OpgB (Y. Lequette, E. Lanfroy, A. Bohin, V. Cogez, J.-M. Lacroix, and J.-P. Bohin, unpublished data). Whether this OpgB′ form is simply an artifact or is of functional importance remains obscure. However, if one considers the probable viscosity of the periplasm (see below), OpgB′ should greatly facilitate the transfer of phosphoglyc-

erol residues from the surface of the inner membrane toward OPG molecules located in the external part of the periplasm (Fig. 1). OpgC (44 kDa) is a polytopic protein with ten putative transmembrane segments. OpgC was proposed to catalyze the transfer of succinyl residues from the cytoplasmic side of the membrane to the nascent glucan backbones on the periplasmic side of the membrane (Lacroix et al., 1999). Nothing is known about the transfer of phosphoethanolamine residues.

Phosphoglycerol and succinyl transfer was also shown to occur in the periplasmic compartment of *S. meliloti* (Wang et al., 1999). Phosphoglycerol transfer depends on *cgmB*. CgmB (71 kDa) is a soluble protein that shows only very limited similarity with OpgB. Moreover, this enzyme can transfer phosphoglycerol residues to purified molecules of OPG. Thus, the two proteins appear very different in the way they modify OPGs.

In *R. sphaeroides*, *opgC* is cotranscribed with the backbone biosynthetic genes (Cogez et al., 2002). OpgC$_{Rsp}$ (44 kDa) exhibits stretches of hydrophobic amino acids over entire length and eleven transmembrane segments are predicted. Thus, OpgC$_{Rsp}$ and OpgC$_{Eco}$ share several common characters, but they do not show any significant sequence similarities.

In conclusion, despite the fact that they carry out very similar functions, OPG substitution enzymes described to date are not phylogenetically related, but they appear to result from convergent evolution.

REGULATION OF OPG BIOSYNTHESIS

OPG synthesis has been shown to be osmotically regulated in a wide range of Proteobacteria (Bohin, 2000) except for *B. abortus* (Briones et al., 1997), *A. caulinodans* (Komaniecka and Choma, 2003), *S. meliloti* strain GR4, and strains of *Rhizobium leguminosarum*. The observations in *R. leguminosarum* were complicated because these strains excrete large amounts of OPG when grown in media of high osmolarity, suggesting outer membrane modifications in such conditions (Breedveld and Miller, 1994).

In general, the amount of OPG increases as the osmolarity of the medium decreases. For example, in *E. coli*, the amount of periplasmic glucans represents up to 5% of the dry weight of the cell in a medium of low osmolarity (50 to 100 mosM) while it decreases until 0.5% in media of high osmolarity (600 to 700 mosM).

Osmotic regulation takes place at an early stage of the OPG backbone biosynthesis at the transcriptional level and/or at the posttranslational level depending on the species considered. In vitro, the membrane-bound activities of extracts isolated from *S. meliloti*, *R. leguminosarum*, *A. tumefaciens*, and *E. coli* grown in media of low or high osmolarity were very similar and were very sensitive to elevation of ionic strength. In *E. coli*, the rate of OPG synthesis measured in a *zwf*, *pgi* mutant strain that is unable to synthesize UDP-glucose unless glucose is added to the medium falls abruptly after a sudden increase of osmolarity (Lacroix, 1989). Under the same conditions, phosphoglycerol substitution is severely reduced but is not osmoregulated in *E. coli* (Bohin and Kennedy, 1984b). Increase of the medium osmolarity is correlated with an increase of the ionic strength in the cytoplasm and the glucosyltransferase activity of OpgH is strongly inhibited by high ionic strength in vitro (Kennedy, 1996). A similar behavior was reported for glucan polymerase activities in the family *Rhizobiaceae* (Breedveld and Miller, 1994). OPGs are necessary for hypoosmotic adaptation of most rhizobiaceaes since mutants devoid of OPGs are impaired for growth in media of low osmolarity. These data are consistent with the model that OPG synthesis is only regulated at the enzymatic level. Nevertheless, in *E. coli*, growth of mutants devoid of OPG remains unaffected in media of low osmolarity and when the osmolarity of the medium decreases suddenly, almost one generation time is necessary to adjust the level of OPG. This observation is inconsistent with a regulation that is limited to enzymatic modulation. Thus, transcription of the *opgGH* operon of *E. coli* was found to be osmoregulated (Lacroix et al., 1991) and membrane-bound activity of *A. brasilense* was reduced in

extracts of cells grown in a medium of high osmolarity (Altabe et al., 1994).

Independently of this osmotic regulation, a feedback regulation exerted by the end product was observed. After addition of glucose to the *E. coli zwf*, *pgi* strain, OPGs are accumulated rapidly until they reach the concentration observed in the wild-type strain. A similar control is exerted at low and high external osmolarity, that is, at different OPG periplasmic concentrations. Since the membrane-bound glucosyltransferase OpgH is the ultimate target of this control, one can envisage that OpgH senses directly or indirectly a particular subfraction of the OPGs with similar properties whatever the external conditions. Alternatively, OpgH, as other membrane-bound proteins, could sense mechanical deformation or alterations in the physicochemistry of the membrane bilayer (Poolman et al., 2002) due to OPG accumulation. In vitro, membrane-associated glucosyltransferase activity of *R. leguminosarum* was severely reduced after addition of 15 mM OPG. Thus, in these strains, secreted OPGs escape from feedback control and explain the large accumulation of OPG in the medium (Breedveld et al., 1992).

However, factors other than osmolarity affect the regulation of OPG biosynthesis in several bacterial species. OPG synthesis was found upregulated by a diffusible signal factor in *X. campestris* pv. campestris (Vojnov et al., 2001). The *opgH* gene was downregulated in *Xanthomonas axonopodis* pv. citri grown in planta (Astua-Monge et al., 2005), but *opgH* was found induced in *Xylella fastidiosa* during biofilm formation (de Souza et al., 2004), and *opgG* is one of the 153 genes induced during growth of *Ralstonia solanacearum* in tomato (Brown and Allen, 2004). Finally, the *opgGH* operon of *E. coli* was found to be activated by σ^E, the sigma factor that responds to misfolded proteins in the cell envelope (Dartigalongue et al., 2001).

Little is known about genes and regulation of genes other than those implicated in backbone synthesis perhaps because the presence and the OPG level depend on the expression and/or the activity of biosynthetic enzymes implicated in the backbone synthesis. In *S. meliloti*, the *cgmB* gene, encoding the phosphoglyceroltransferase, was shown to be osmoregulated (Wang et al., 1999), while the *opgB* gene, encoding its functional homolog in *E. coli*, is not osmoregulated (Bohin and Kennedy, 1984b).

The *opgD* gene of *E. coli* is regulated by the growth phase because OpgD was preferentially observed in stationary phase (Link et al., 1997). This gene contains a C in position -13 of its promoter (one important characteristic of σ^S dependence) and is partially transcribed by the σ^S-containing RNA polymerase (Lequette, 2002). σ^S is activated when growth slows down and particularly in stationary-growth phase (Hengge-Aronis, 2002). However, *opgD* was not observed among the forty-one genes whose expression was reduced in a σ^S-defective mutant (Lacour and Landini, 2004). The *opgD* homolog of *X. axonopodis* pv. citri was shown to be overexpressed in planta (Astua-Monge et al., 2005).

Regulation of OPG synthesis is therefore quite complex not only because its osmoregulation is not conserved, but also because of the existence of additional regulatory factors.

CONNECTION WITH TREHALOSE SYNTHESIS

Trehalose is a nonreducing disaccharide of 1-1-linked glucose units. It is the most widespread disaccharide in nature, occurring in bacteria, fungi, insects, and plants (Kolbe et al., 2005). It serves as a carbon source and/or can be accumulated as a protectant against high osmolarity, heat, and desiccation (Strom and Kaasen, 1993). In *E. coli*, externally supplied trehalose is only used as a carbon source and not as an osmoprotectant because of the presence of a highly active periplasmic trehalase (the product of *treA*) that is induced at high osmolarity and a PTS system for trehalose that is synthesized at low osmolarity and repressed at high osmolarity (Boos et al. 1987). Thus, osmoprotective trehalose must be endogenously synthesized for accumulation in response to media of high osmolarity. Trehalose is synthesized from UDP-glucose (like OPGs) and glucose 6-phosphate and accumulated up to 400 mM concentration (Strom and Kaasen, 1993). Because accumulation of OPGs and trehalose occurs in media of low and high osmolarity, respectively, no competition for UDP-glucose exists for the synthesis of both kinds of molecules. Because OPGs are not degraded within cells, they never serve as a source of glucose for trehalose synthesis. In contrast, one could imagine that trehalose serves as a glucose donor for OPG biosynthesis after hydrolysis by the cytoplasmic trehalase, the *treF* product, in media of low osmolarity despite the low affinity of TreF for trehalose (K_m, 1.9 mM) (Horlacher et al., 1996). In *E. coli*, little is known about the fate of trehalose once cells are shifted to media of low osmolarity. Nevertheless, release of osmoprotectants such as betaine and trehalose is known to occur through mechanosensitive channels in response to drastic osmotic downshock (Levina et al., 1999), rendering the hydrolysis of the trehalose by TreF hypothetical under extreme conditions. In *B. japonicum*, in vivo NMR studies on cells in stationary-growth phase clearly reveal that no connection exists between OPGs and trehalose (Pfeffer et al., 1994). After a hypoosmotic shock, OPGs are synthesized in stationary-growth phase from glycogen synthesized in the exponential-growth phase from glucose. When no glucose is present in the medium, no synthesis of OPGs occurs. Nevertheless, trehalose was synthesized in media of high osmolarity but was released in the medium during the hypoosmotic shock. Thus, the disaccharide cannot serve as a precursor for OPG biosynthesis. Because osmoprotectants are often released into the medium after an osmotic downshock and because OPGs cannot be degraded, one can consider that, in a majority of bacterial species, trehalose and OPGs cannot serve as precursors for each other.

OPGs AND PATHOGENICITY

Mutants defective in OPG synthesis were primarily obtained during the screening or the selection of attenuated or avirulent mutants of plant or animal pathogens. This is strong evidence of their potential role during pathogenesis. Mutants of *A. tumefaciens* (*chvA* and *chvB*; Puvanesarajah et al., 1985), *P. syringae* pv. syringae (*hrpM*; Mills and Mukhopadhyay, 1990), and *X. campestris* pv. vesicatoria (*opgH*; Minsavage et al., 2004) are completely or severely impaired in their virulence toward their host plants. The *P. syringae* mutant, which is severely impaired in its ability to grow in the plant host, also failed to elicit a nonhost hypersensitive response when inoculated on tobacco. Similarly, mutants of *S. meliloti* (Dylan et al., 1986) form defective nodules where tissue differentiation occurs with no bacterial invasion. As mentioned earlier, OPGs can be recovered in the external medium under certain conditions. However, addition of purified OPGs to the inoculum cannot restore attachment and virulence or symbiotic ability of family *Rhizobiaceae* *opg* mutants.

Random mutations affecting OPG biosynthesis were also recovered during investigation of animal pathogens. Thus, *opgH* was found among chromosomal loci required by *Yersinia enterocolitica* at an early stage of the infection process (Young and Miller, 1997). One of several *P. aeruginosa* PA14 mutants severely impaired in their virulence toward *Caenorhabditis elegans* possesses transposon insertion in *opgH* (Mahajan-Miklos et al., 1999). This mutant has also a dramatic effect in the mouse model, causing no mortality, and is severely affected in its ability to grow in *Arabidopsis* leaves.

Recently, a mutant of *P. aeruginosa* was identified that, while still capable of forming biofilms with normal architecture, does not develop high-level biofilm-specific resistance to antibiotics (Mah et al., 2003). The mutation was identified in the gene *ndvB* (BB type; see Table 1) and this mutant is impaired in the synthesis of a glucan, the structure of which remains to be established. Is this glucan corresponding to OPGs? While the authors favor this hypothesis and discuss the possibility that its cyclic nature allows sequestration of antibiotics, objections can be raised. OPGs of the closely related species *P. syringae* belong to family I and both species have highly similar *opgGH*

genes. Moreover, genes similar to *opgGH* and to *ndvB* coexist in *R. sphaeroides* (BA type; see Table 1), and only *opgGH* are implicated in the synthesis of a cyclic OPG. Thus, further structural investigations are required for final clarification. Nevertheless, one can imagine that *P. aeruginosa* can produce two kinds of OPGs, both simultaneously or one or the other in response to different environmental signals.

Transposon insertion in a gene homologous to *opgB* was found in a highly attenuated mutant of *S. enterica* serovar Typhimurium (Valentine et al., 1998). Unfortunately, it remains uncertain if the observed phenotype was the consequence of this mutation. When cassettes were inserted into *opgB* or *opgGH*, and the resulting OPG phenotypes being checked, no severe attenuation of *Salmonella* virulence in mice was observed (F. Norel, V. Robbe-Saule, A. Bohin, and J.-P. Bohin, unpublished observation).

The implication of OPGs in pathogenicity was further confirmed by directed inactivation of *opg* genes in various species. Nodule development was severely impaired in *ndvB* mutants of *B. japonicum*, which are devoid of OPGs, and to a lesser degree in *ndvC* mutants, which produce OPGs with altered structures (Dunlap et al., 1996). Similarly, Opg⁻ mutants of *E. chrysanthemi* are nonvirulent on potato tubers or chicory leaves, and this lack of virulence was correlated with their inability to grow in the plant host (Page et al., 2001). Moreover, since during coinoculation of mutant and wild-type strains OPGs from the latter are almost certainly released in the environment, and since those free OPGs cannot complement the mutant defect, we can conclude that OPGs must be located in the periplasm to play their role in virulence. This is consistent with the observations that virulence of *B. abortus* depends on transport of OPGs to the periplasm (Roset et al., 2004) and virulence of *A. tumefaciens* cannot be restored by coinoculation of mutant bacteria with purified OPGs (Swart et al., 1993).

Arellano-Reynoso et al. (2005) have recently reported that Opg⁻ mutants of *B. abortus*, an intracellular pathogen, are rapidly cleared by macrophages. Moreover, preincubation of the mutant with purified OPGs partially restored the ability to evade phagosome-lysosome fusion.

In conclusion, when the role of OPGs in plant-bacteria interactions was tested, they always appeared to be an essential component of the bacterial cell envelope required for virulence or symbiosis. For animal pathogens, the situation appears more complex probably because these bacteria have evolved more sophisticated mechanisms to invade or destroy host cells.

PLEIOTROPIC PHENOTYPES OF OPG-DEFECTIVE MUTANTS

OPG-defective mutants characterized in various bacterial species have in common a highly pleiotropic phenotype, which is indicative of a global alteration of their envelope properties. Whatever the bacterial species considered, Opg⁻ colonies generally exhibit a mucoid phenotype and a defect in motility and chemotaxis. This defect was shown to be the consequence of a reduced number of flagella in *E. coli* (Fiedler and Rotering, 1988). In a similar way, Opg⁻ mutants were found more sensitive to hydrophobic antibiotics in *S. meliloti* (Dylan et al., 1990a), biliary salts in *E. chrysanthemi* (Page et al., 2001), sodium dodecyl sulfate in *E. coli* (Rajagopal et al., 2003), or resistant to endogenously produced lysis protein of phage MS2 in *E. coli* (Höltje et al., 1988). Additional phenotypic changes, depending on medium composition, were found in proteins of the envelope in *E. coli* (Fiedler and Rotering, 1988) and A. *tumefaciens* (Cangelosi et al., 1990). Moreover, Opg⁻ mutants of *E. chrysanthemi* produce and secrete lower amounts of protease, cellulase, and pectate lyases (Page et al., 2001).

The doubling times during growth in hypoosmotic media are twice as high in *opg* null mutants of *A. tumefaciens* (Cangelosi et al., 1990) and *S. meliloti* (Dylan et al., 1990a) when compared with the wild type, and increasing the medium osmolarity compensates for that defect. In contrast, growth of Opg⁻ mutants in

E. coli (Bohin and Kennedy, 1984b), *E. chrysanthemi* (Page et al., 2001), or *B. japonicum* (Chen et al., 2002) is only slightly affected.

Several groups have isolated suppressor mutations that restore a wild-type phenotype of *opg* mutants. *E. coli* chemotactic pseudorevertants were mutated in the *ompB* locus, mainly involved in osmoregulation of the outer membrane porins (Fiedler and Rotering, 1988). However, this locus is not involved in regulation of OPG synthesis. Similarly, pseudorevertants of *S. meliloti ndv* mutants were selected for restoration of osmotolerance, motility, or symbiosis. Pseudorevertants for vegetative properties regained only a slight symbiotic ability while symbiotic pseudorevertants were unrestored for vegetative properties and were still highly impaired in the first steps of the symbiotic interaction (Dylan et al., 1990b).

Thus, OPGs may serve in osmoadaptation in some cases, but they also appear to play other roles that are not directly related to osmoadaptation.

OPGs AND STRESS SIGNALING

OPGs appear to be important intrinsic components of the gram-negative bacterial envelope, which can be essential in extreme conditions found in nature, especially when bacteria must interact with a eukaryotic host. What is then the fundamental function of these compounds?

Stock et al. (1977), in their seminal study of the enterobacterial periplasm, have shown this compartment comprises 20 to 40% of the total cell volume and that a Donnan equilibrium exists between the periplasm and the extracellular medium, allowing isoosmolality between periplasm and cytoplasm. Thus, E. P. Kennedy proposed that OPGs function as periplasmic osmoprotectants on the basis of their anionic character, where OPGs would participate in the Donnan equilibrium through the outer membrane, and he found, according to this hypothesis, that OPG synthesis is osmoregulated in *E. coli* (Kennedy, 1982).

However, conflicting data and interpretations have been published concerning the volume and concentration of periplasm (Pfeffer et al., 1994; Oliver, 1996; Koch, 1998; Wood, 1999; Cayley et al., 2000). It is necessary to make a distinction between transient states following rapid upshift or downshift of external osmolarity and steady-state adaptation to a particular medium where the bacterial cell can grow. Only the first situations can be considered as stress conditions, essentially because fluxes of water are extremely rapid, as those of small molecules in the case of abrupt downshift in osmolarity. Cayley et al. (2000) calculated the OPG concentrations in steady-state cells of *E. coli* grown at different external osmolarities. The obtained values, from 50 mM at 70 mosM to 3 mM at 800 mosM, are in the same range as the estimation of 15 mM reported for cells of the family *Rhizobiaceae* (Breedveld and Miller, 1994). The amount of OPGs is sufficient to quantitatively explain the observed Donnan potential maintained across the outer membrane (Sen et al., 1988). However, even in the case of *E. coli*, the anionic character of OPGs is only effective when ionic strength is high (Koch, 1988), and OPGs of other bacteria are uncharged molecules. Moreover, *E. coli* mutants defective in OPG substitution (*opgB*, *opgC*) as mutants defective in OPG backbone synthesis (*opgH* or *opgD*) do not show a particular osmosensitive phenotype (Bohin and Kennedy, 1984a). More intriguing is the fact that during osmotic stresses OPG concentration is only slowly adjusted to that found in steady state: after abrupt downshift it takes *E. coli* cells at least half a generation time to accumulate OPGs up to the tenfold-higher steady-state level (Lacroix, 1989). After osmotic upshift the OPG content decreases to a tenfold-lower level as a consequence of dilution during successive cell divisions, because degradation of OPGs was never observed (J.-P. Bohin, unpublished observations; Ruby and McCabe, 1988). Thus, in some species in which osmoregulation of OPG synthesis occurs this event might function rather in the detection of diluted media than in osmoadaptation per se. Similarly, OmpC and OmpF are osmoregulated by the EnvZ-OmpR two-component

system, but no change in osmotolerance has been associated with defects in the porins or their regulation.

Alternatively, OPGs may have a structural role in the envelope organization by interacting with other structural components like phospholipids and/or peptidoglycan (Banta et al., 1998). The periplasm is generally described as a highly viscous gel-like compartment where peptidoglycan, protein junctions between inner and outer membranes, periseptal and polar annuli would drastically limit molecular movements. However, free movement of the jellyfish protein within the periplasm has been reported (Santini et al., 2001). Similarly, the use of high-resolution NMR spectroscopy under magic angle spinning revealed that the major fraction of OPGs detected in vivo in *R. solanacearum* undergoes significant rotational diffusion (Wieruszeski et al., 2001). Therefore, OPGs are probably not stably associated with macromolecules.

OPGs could also be seen as molecular agents necessary to maintain a minimal periplasmic space at low medium osmolarity. However, *E. coli opg* mutant cells exhibited a larger periplasmic space than wild-type cells when subjected to gentle plasmolysis (Höltje et al., 1988). Thus, they might be involved in the interactions between the outer and inner membranes. OPGs induce the porin channels to close (Delcour et al., 1992), while changing the Donnan potential has no effect on the permeability of porin channels (Sen et al., 1988).

Another hypothesis is that OPGs might represent a macromolecular crowding determinant for protein folding, association, and function (Mogensen and Otzen, 2005). From this point of view, it is noteworthy that the *opgGH* transcription is controlled, at least in part, by σ^E whose synthesis is induced following accumulation of unfolded proteins in the periplasm. An *opgH* mutant exhibits partially increased expression of *osmY* (Bohringer et al., 1995). Similarly, in a *dsbA* mutant that does not form disulfide bonds in cell envelope proteins, a stress is perceived that results in a reduced amount of OpgG and a higher amount of OsmY, two proteins which do not contain disulfide bonds (Hiniker and Bardwell, 2004). In this respect, *dsbA* cells behave as if they perceived a high external osmolarity.

OPGs may also have a function as signals sensed by specific proteins present in the periplasmic compartment or bound to one or the other membrane. As discussed previously, when bacteria are shifted from a diluted medium to a concentrated environment, OPGs are slowly diluted. During the first steps of infection, bacteria that would have been exposed previously to media of low osmolarities would possess, in their periplasmic space, OPG concentrations higher than characteristic for their new environment. If OPG concentration could be monitored by sensor proteins, it could be used as a kind of internal standard providing information on the number of cell divisions, and, indirectly, of bacteria at very low population densities. This could control the expression of a set of genes according to the increase of cell population and it would be, in some way, the counterpart of the quorum-sensing systems involved in the control of gene expression at high cell density.

Finally, OPGs were proposed as a possible signal for two *E. coli* inner membrane sensor proteins: EnvZ (Fiedler and Rotering, 1988) and RcsC (Ebel et al., 1997). Several phenotypic changes observed in OPG-defective cells (for example, increased synthesis of exopolysaccharides and decreased number of flagella) could result from activation of the Rcs sensor/regulator system (Majdalani and Gottesman, 2005). The Rcs regulon appears essential to control the timing of infection processes and formation of biofilms. According to the proposed model, RcsC (a histidine kinase) and RcsD sense a signal and cooperate in a phosphorelay to the response regulator RcsB. In an *opgH* mutant this system is constitutively activated, and RcsC or/and RcsD could sense directly the OPGs present in the periplasm (Ebel et al., 1997). In addition, RcsF, a small lipoprotein inserted in the outer membrane, plays a critical role in signal transduction from the cell

surface to RcsC (Majdalani et al., 2005). Thus, OPGs may globally affect periplasm structural properties or directly modulate the RcsF-RcsC protein interplay.

REFERENCES

Altabe, S. G, N. Iñon de Iannino, D. de Mendoza, and R. A. Ugalde. 1994. New osmoregulated β-(1-3),β(1-6) glucosyltransferase(s) in *Azospirillum brasilense. J. Bacteriol.* **176:**4890–4898.

Altabe, S. G., P. Talaga, J.-M. Wieruszeski, G. Lippens, R. Ugalde, and J.-P. Bohin. 1998. Periplasmic glucans of *Azospirillum brasilense*, p. 390. *In* C. Elmerich, A. Kondorosi, and W. E. Newton (ed.), *Biological Nitrogen Fixation for the 21st Century.* Kluwer Academic Publishers, Dordrecht, The Netherlands.

Arellano-Reynoso, B., N. Lapaque, S. Salcedo, G. Briones, A. E. Ciocchini, R. Ugalde, E. Moreno, I. Moriyon, and J.-P. Gorvel. 2005. Cyclic beta-1,2-glucan is a *Brucella* virulence factor required for intracellular survival. *Nat. Immunol.* **6:**618–625.

Astua-Monge, G., J. Freitas-Astua, G. Bacocina, J. Roncoletta, S. A. Carvalho, and A. Machado. 2005. Expression profiling of virulence and pathogenicity genes of *Xanthomonas axonopodis* pv. citri. *J. Bacteriol.* **187:**1201–1205.

Banta, L. M., J. Bohne, S. D. Lovejoy, and K. Dostal. 1998. Stability of the *Agrobacterium tumefaciens* VirB10 protein is modulated by growth temperature and periplasmic osmoadaptation. *J. Bacteriol.* **180:**6597–6606.

Bhagwat, A. A., K. C. Gross, R. E. Tully, and D. L. Keister. 1996. Beta-glucan synthesis in *Bradyrhizobium japonicum*: characterization of a new locus (*ndvC*) influencing β-(1→6) linkages. *J. Bacteriol.* **178:**4635–4642.

Bhagwat, A. A., R. E. Tully, and D. L. Keister. 1993. Identification and cloning of a cyclic β-(1→3),(1→6)-D-glucan synthesis locus from *Bradyrhizobium japonicum. FEMS Microbiol. Lett.* **114:**139–144.

Bohin, J.-P. 2000. Osmoregulated periplasmic glucans in Proteobacteria—a minireview. *FEMS Microbiol. Lett.* **186:**11–19.

Bohin, J.-P., and E. P. Kennedy. 1984a. Mapping of a locus (*mdoA*) that affects the biosynthesis of membrane-derived oligosaccharides in *Escherichia coli. J. Bacteriol.* **157:**956–957.

Bohin, J.-P., and E. P. Kennedy. 1984b. Regulation of the synthesis of membrane-derived oligosaccharides in *Escherichia coli.* Assay of phosphoglycerol transferase I in vivo. *J. Biol. Chem.* **259:**8388–8393.

Bohringer, J., D. Fischer, G. Mosler, and R. Hengge-Aronis. 1995. UDP-glucose is a potential intracellular signal molecule in the control of expression of sigma S and sigma S-dependent genes in *Escherichia coli. J. Bacteriol.* **177:**413–422.

Boos, W., U. Ehmann, E. Bremer, A. Middendorf, and P. Postma. 1987. Trehalase of *Escherichia coli.* Mapping and cloning of its structural gene and identification of the enzyme as a periplasmic protein induced under high osmolarity growth conditions. *J. Biol. Chem.* **262:**13212–13218.

Breedveld, M. W., and K. J. Miller. 1994. Cyclic β-glucans of the family Rhizobiaceae. *Microbiol. Rev.* **58:**145–161.

Breedveld, M. W., L. P. T. M. Zevenhuizen, and A. J. B. Zehnder. 1992. Synthesis of β-(1,2)-glucans by *Rhizobium leguminosarum* biovar trifolii TA-1: factors influencing excretion. *J. Bacteriol.* **174:**6336–6342.

Briones, G., N. Iñon de Iannino, M. Steinberg, and R. A. Ugalde. 1997. Periplasmic cyclic 1,2-β-glucan in *Brucella* spp. is not osmoregulated. *Microbiology* **143:**1115–1124.

Brown, D. G., and C. Allen. 2004. *Ralstonia solanacearum* genes induced during growth in tomato: an inside view of bacterial wilt. *Mol. Microbiol.* **53:**1641–1660.

Cangelosi, G. A., G. Martinetti, and E. W. Nester. 1990. Osmosensitivity phenotypes of *Agrobacterium tumefaciens* mutants that lack periplasmic β-1,2-glucan. *J. Bacteriol.* **172:**2172–2174.

Case, C. C., B. Bukau, S. Granett, M. R. Villarejo, and W. Boos. 1986. Contrasting mechanisms of *envZ* control of *mal* and *pho* regulon genes in *Escherichia coli. J. Bacteriol.* **166:**706–712.

Cayley, D. S., H. J. Guttman, and M. T. Record, Jr. 2000. Biophysical characterization of changes in amounts and activity of *Escherichia coli* cell and compartment water and turgor pressure in response to osmotic stress. *Biophys. J.* **78:**1748–1764.

Chen, R., A. A. Bhagwat, R. Yaklich, and D. L. Keister. 2002. Characterization of *ndvD*, the third gene involved in the synthesis of cyclic β-(1→3), (1→6)-D-glucans in *Bradyrhizobium japonicum. Can. J. Microbiol.* **48:**1008–1016.

Choma, A., and I. Komaniecka. 2003. Characterization of *Mesorhizobium huakuii* cyclic beta-glucan. *Acta Biochim. Pol.* **50:**1273–1281.

Ciocchini, A. E., M. S. Roset, N. Iñon de Iannino, and R. A. Ugalde. 2004. Membrane topology analysis of cyclic glucan synthase, a virulence determinant of *Brucella abortus. J. Bacteriol.* **186:**7205–7213.

Cogez, V., E. Gak, A. Puskas, S. Kaplan, and J.-P. Bohin. 2002. The *opgGIH* and *opgC* genes of *Rhodobacter sphaeroides* form an operon that controls backbone synthesis and succinylation of osmoregulated periplasmic glucans. *Eur. J. Biochem.* **269:**2473–2484.

Cogez, V., P. Talaga, J. Lemoine, and J.-P. Bohin. 2001. Osmoregulated periplasmic glucans of *Erwinia chrysanthemi. J. Bacteriol.* **183:**3127–3133.

Csonka, L. N. 1989. Physiological and genetic responses of bacteria to osmotic stress. *Microbiol. Rev.* **53:**121–147.

Dartigalongue, C., D. Missiakas, and S. Raina. 2001. Characterization of the *Escherichia coli* σ^E regulon. *J. Biol. Chem.* **276:**20866–20875.

de Souza, A. A., M. A. Takita, H. D. Coletta-Filho, C. Caldana, G. M. Yanai, N. H. Muto, R. C. de Oliveira, L. R. Nunes, and M. A. Machado. 2004. Gene expression profile of the plant pathogen *Xylella fastidiosa* during biofilm formation in vitro. *FEMS Microbiol. Lett.* **237:**341–353.

Debarbieux, L., A. Bohin, and J.-P. Bohin. 1997. Topological analysis of the membrane-bound glucosyltransferase, MdoH, required for osmoregulated periplasmic glucan synthesis in *Escherichia coli. J. Bacteriol.* **179:**6692–6698.

Delcour, A. H., J. Adler, C. Kung, and B. Martinac. 1992. Membrane-derived oligosaccharides (MDO's) promote closing of an *E. coli* porin channel. *FEBS Lett.* **304:**216–220.

Dunlap, J., E. Minami, A. A. Bhagwat, D. L. Keister, and G. Stacey. 1996. Nodule development induced by mutants of *Bradyrhizobium japonicum* defective in cyclic β-glucan synthesis. *Mol. Plant Microbe Interact.* **9:**546–555.

Dylan, T., D. R. Helinski, and G. S. Ditta. 1990a. Hypoosmotic adaptation in *Rhizobium meliloti* requires β-(1→2)-glucan. *J. Bacteriol.* **172:**1400–1408.

Dylan, T., L. Ielpi, S. Stanfield, L. Kashyap, C. Douglas, M. Yanofsky, E. Nester, D. R. Helinski, and G. Ditta. 1986. *Rhizobium meliloti* genes required for nodule development are related to chromosomal virulence genes in *Agrobacterium tumefaciens. Proc. Natl. Acad. Sci. USA* **83:**4403–4407.

Dylan, T., P. Nagpal, D. R. Helinski, and G. S. Ditta. 1990b. Symbiotic pseudorevertants of *Rhizobium meliloti ndv* mutants. *J. Bacteriol.* **172:**1409–1417.

Ebel, W., G. J. Vaughn, H. K. Peters III, and J. E. Trempy. 1997. Inactivation of *mdoH* leads to increased expression of colanic acid capsular polysaccharide in *Escherichia coli. J. Bacteriol.* **179:**6858–6861.

Fiedler, W., and H. Rotering. 1988. Properties of *Escherichia coli* mutants lacking membrane-derived oligosaccharides. *J. Biol. Chem.* **263:**14684–14689.

Forns, N., A. Juarez, C. Madrid. 2005. Osmoregulation of the HtrA (DegP) protease of *Escherichia coli:* an Hha-H-NS complex represses HtrA expression at low osmolarity. *FEMS Microbiol. Lett.* **251:**75–80.

Hanoulle, X., E. Rollet, B. Clantin, I. Landrieu, C. Ödberg-Ferragut, G. Lippens, J.-P. Bohin,

and V. Villeret. 2004. Structural analysis of *Escherichia coli* OpgG, a protein required for the biosynthesis of osmoregulated periplasmic glucans. *J. Mol. Biol.* **342:**195–205.

Hengge-Aronis, R. 2002. Stationary phase gene regulation: what makes an *Escherichia coli* promoter σ^S-selective? *Curr. Opin. Microbiol.* **5:**591–595.

Hiniker, A., and J. C. Bardwell. 2004. In vivo substrate specificity of periplasmic disulfide oxidoreductases. *J. Biol. Chem.* **279:**12967–12973.

Höltje, J. V., W. Fiedler, H. Rotering, B. Walderich, and J. van Duin. 1988. Lysis induction of *Escherichia coli* by the cloned lysis protein of the phage MS2 depends on the presence of osmoregulatory membrane-derived oligosaccharides. *J. Biol. Chem.* **263:**3539–3541.

Horlacher, R., K. Uhland, W. Klein, M. Ehrmann, and W. Boos. 1996. Characterization of a cytoplasmic trehalase of *Escherichia coli*. *J. Bacteriol.* **178:**6250–6257.

Iñon de Iannino, N., G. Briones, M. Tolmasky, and R. A. Ugalde. 1998. Molecular cloning and characterization of *cgs*, the *Brucella abortus* cyclic β(1-2) glucan synthetase gene: genetic complementation of *Rhizobium meliloti ndvB* and *Agrobacterium tumefaciens chvB* mutants. *J. Bacteriol.* **180:**4392–4400.

Jackson, B. J., and E. P. Kennedy. 1983. The biosynthesis of membrane-derived oligosaccharides. A membrane-bound phosphoglycerol transferase. *J. Biol. Chem.* **258:**2394–2398.

Jackson, B. J., J.-P. Bohin, and E. P. Kennedy. 1984. Biosynthesis of membrane-derived oligosaccharides: characterization of *mdoB* mutants defective in phosphoglycerol transferase I activity. *J. Bacteriol.* **160:**976–981.

Kennedy, E. P. 1982. Osmotic regulation and the biosynthesis of membrane-derived oligosaccharides in *Escherichia coli*. *Proc. Natl. Acad. Sci. USA* **79:**1092–1095.

Kennedy, E. P. 1996. Membrane derived oligosaccharides (periplasmic beta-D-glucans) of *Escherichia coli*, p. 1064–1074. In F. C. Neidhardt, R. Curtiss III, J. L. Ingraham, E. C. C. Lin, K. B. Low, B. Magasanik, W. S. Reznikoff, M. Riley, M. Schaechter, and H. E. Umbarger (ed.), Escherichia coli *and* Salmonella. *Cellular and Molecular Biology*, 2nd ed. ASM Press, Washington, D.C.

Koch, A. L. 1998. The biophysics of the gram-negative periplasmic space. *Crit. Rev. Microbiol.* **24:**23–59.

Kolbe, A., A. Tiessen, H. Schluepmann, M. Paul, S. Ulrich, and P. Geigenberger. 2005. Trehalose 6-phosphate regulates starch synthesis via posttranslational redox activation of ADP-glucose pyrophosphorylase. *Proc. Natl. Acad. Sci. USA* **102:**11118–11123.

Komaniecka, I., and A. Choma. 2003. Isolation and characterization of periplasmic cyclic beta-glucans of *Azorhizobium caulinodans*. *FEMS Microbiol. Lett.* **227:**263–269.

Lacour, S., and P. Landini. 2004. σ^S-Dependent gene expression at the onset of stationary phase in *Escherichia coli*: function of σ^S-dependent genes and identification of their promoter sequences. *J. Bacteriol.* **186:**7186–7195.

Lacroix, J.-M. 1989. Etude génétique et physiologique de la régulation osmotique de la biosynthèse du MDO chez *Escherichia coli*. Ph.D. thesis. Université de Paris-Sud, Centre d'Orsay, France.

Lacroix, J.-M., E. Lanfroy, V. Cogez, Y. Lequette, A. Bohin, and J.-P. Bohin. 1999. The *mdoC* gene of *Escherichia coli* encodes a membrane protein that is required for succinylation of osmoregulated periplasmic glucans. *J. Bacteriol.* **181:**3626–3631.

Lacroix, J.-M., I. Loubens, M. Tempête, B. Menichi, and J.-P. Bohin. 1991. The *mdoA* locus of *Escherichia coli* consists of an operon under osmotic control. *Mol. Microbiol.* **5:**1745–1753.

Lequette, Y., C. Ödberg-Ferragut, J.-P. Bohin, and J.-M. Lacroix. 2004. Identification of *mdoD*, an *mdoG* paralog which encodes a twin-arginine-dependent periplasmic protein that controls osmoregulated periplasmic glucan backbone structures. *J. Bacteriol.* **186:**3695–3702.

Lequette, Y. 2002. Biosynthèse des glucanes périplasmiques osmorégulés chez *Escherichia coli*: analyse fonctionnelle des protéines MdoG et MdoH et caractérisation de deux nouvelles activités. Ph.D. thesis. Université des Sciences et Technologie de Lille, France.

Levina, N., S. Totemeyer, N. R. Stokes, P. Louis, M. A. Jones, and I. R. Booth. 1999. Protection of *Escherichia coli* cells against extreme turgor by activation of MscS and MscL mechanosensitive channels: identification of genes required for MscS activity. *EMBO J.* **18:**1730–1737.

Link, A. J., K. Robison, and G. M. Church. 1997. Comparing the predicted and observed properties of proteins encoded in the genome of *Escherichia coli* K-12. *Electrophoresis* **18:**1259–1313.

Lippens, G., J.-M. Wieruszeski, D. Horvath, P. Talaga, and J.-P. Bohin. 1998. Slow dynamics of the cyclic osmoregulated periplasmic glucan of *Ralstonia solanacearum* as revealed by heteronuclear relaxation studies. *J. Am. Chem. Soc.* **120:**170–177.

Loubens, I., L. Debarbieux, A. Bohin, J.-M. Lacroix, and J.-P. Bohin. 1993. Homology between a genetic locus (*mdoA*) involved in the osmoregulated biosynthesis of periplasmic glucans in *Escherichia coli* and a genetic locus (*hrpM*) controlling pathogenicity of *Pseudomonas syringae*. *Mol. Microbiol.* **10:**329–340.

Mah, T. F., B. Pitts, B. Pellock, G. C. Walker, P. S. Stewart, and G. A. O'Toole. 2003. A genetic basis for *Pseudomonas aeruginosa* biofilm antibiotic resistance. *Nature* **426:**306–310.

Mahajan-Miklos, S., M.-W. Tan, L. G. Rahme, and F. M. Ausubel. 1999. Molecular mechanisms of bacterial virulence elucidated using a *Pseudomonas aeruginosa-Caenorhabditis elegans* pathogenesis model. *Cell* **96:**47–56.

Majdalani, N., and S. Gottesman. 2005. The Rcs phosphorelay: a complex signal transduction system. *Annu. Rev. Microbiol.* **59:**379–405.

Majdalani, N., M. Heck, V. Stout, and S. Gottesman. 2005. Role of RcsF in signaling to the Rcs phosphorelay pathway in *Escherichia coli*. *J. Bacteriol.* **187:**6770–6778.

Miller, K. J., E. P. Kennedy, and V. N. Reinhold. 1986. Osmotic adaptation in Gram-negative bacteria: possible role for periplasmic oligosaccharides. *Science* **231:**48–51.

Mills, D., and P. Mukhopadhyay. 1990. Organization of the *hrpM* locus of *Pseudomonas syringae* pv. syringae and its potential function in pathogenesis, p. 47–57. *In* S. Silver, A. M. Chakrabarty, B. Iglewski, and S. Kaplan (ed.), Pseudomonas: *Biotransformation, Pathogenesis, and Evolving Biotechnology*. ASM Press, Washington, D.C.

Minsavage, G. V., M. B. Mudgett, R. E. Stall, and J. B. Jones. 2004. Importance of *opgH*$_{Xcv}$ of *Xanthomonas campestris* pv. vesicatoria in host-parasite interactions. *Mol. Plant Microbe Interact.* **17:**152–161.

Mogensen, J. E., and D. E. Otzen. 2005. Interactions between folding factors and bacterial outer membrane proteins. *Mol. Microbiol.* **57:**326–346.

Oliver, D. B. 1996. Periplasm, p. 88–103. *In* F. C. Neidhardt, R. Curtiss III, J. L. Ingraham, E. C. C. Lin, K. B. Low, B. Maganasik, W. S. Reznikoff, M. Riley, M. Schaechter, and H. E. Umbarger (ed.), *Escherichia coli and Salmonella. Cellular and Molecular Biology*, 2nd ed. ASM Press, Washington, D.C.

Page, F., S. Altabe, N. Hugouvieux-Cotte-Pattat, J.-M. Lacroix, J. Robert-Baudouy, and J.-P. Bohin. 2001. Osmoregulated periplasmic glucan synthesis is required for *Erwinia chrysanthemi* pathogenicity. *J. Bacteriol.* **183:**3134–3141.

Pfeffer, P. E., G. Becard, D. B. Rolin, J. Uknalis, P. Cooke, and S. Tu. 1994. In vivo nuclear magnetic resonance study of the osmoregulation of phosphocholine-substituted β-1,3;1,6 cyclic glucan and its associated carbon metabolism in *Bradyrhizobium japonicum* USDA 110. *Appl. Environ. Microbiol.* **60:**2137–2146.

Poolman, B., P. Blount, J. H. A. Folgering, R. H. E. Friesen, P. C. Moe, and T. van der Heide. 2002. How do membrane proteins sense water stress? *Mol. Microbiol.* **44:**889–902.

Puvanesarajah, V., F. M. Schell, G. Stacey, C. J. Douglas, and E. W. Nester. 1985. Role for 2-linked-β-D-glucan in the virulence of *Agrobacterium tumefaciens*. *J. Bacteriol.* **164:**102–106.

Raina, S., R. Raina, T. V. Venkatesh, and H. K. Das. 1995. Isolation and characterization of a locus from *Azospirillum brasilense* Sp7 that complements the tumorigenic defect of *Agrobacterium tumefaciens chvB* mutant. *Mol. Plant. Microbe Interact.* **8:**322–326.

Rajagopal, S., N. Eis, M. Bhattacharya, and K. W. Nickerson. 2003. Membrane-derived oligosaccharides (MDOs) are essential for sodium dodecyl sulfate resistance in *Escherichia coli*. *FEMS Microbiol. Lett.* **223:**25–31.

Rolin, D. B., P. E. Pfeffer, S. F. Osman, B. S. Szwergold, F. Kappler, and A. J. Benesi. 1992. Structural studies of a phosphocholine substituted β-(1,3);(1,6) macrocyclic glucan from *Bradyrhizobium japonicum* USDA 110. *Biochim. Biophys. Acta* **1116:**215–225.

Roset, M. S., A. E. Ciocchini, R. A. Ugalde, and N. Iñon de Iannino. 2004. Molecular cloning and characterization of *cgt*, the *Brucella abortus* cyclic β-1,2-glucan transporter gene, and its role in virulence. *Infect. Immun.* **72:**2263–2271.

Ruby, E. G., and J. B. McCabe. 1988. Metabolism of periplasmic membrane-derived oligosaccharides by the predatory bacterium *Bdellovibrio bacteriovorus* 109J. *J. Bacteriol.* **170:**646–652.

Santini, C.-L., A. Bernadac, M. Zhang, A. Chanal, B. Ize, C. Blanco, and L.-F. Wu. 2001. Translocation of jellyfish green fluorescent protein via the Tat system of *Escherichia coli* and change of its periplasmic localization in response to osmotic upshock. *J. Biol. Chem.* **276:**8159–8164.

Sen, K., J. Hellman, and H. Nikaido. 1988. Porin channels in intact cells of *Escherichia coli* are not affected by Donnan potentials across the outer membrane. *J. Biol. Chem.* **263:**1182–1187.

Stock, J. B., B. Rauch, and S. Roseman. 1977. Periplasmic space in *Salmonella typhimurium* and *Escherichia coli*. *J. Biol. Chem.* **252:**7850–7861.

Strom, A. R., and I. Kaasen. 1993. Trehalose metabolism in *Escherichia coli*: stress protection and stress regulation gene expression. *Mol. Microbiol.* **8:**205–210.

Swart, S., G. Smit, B. J. J. Lugtenberg, and J. W. Kijne. 1993. Restoration of attachment, virulence and nodulation of *Agrobacterium tumefaciens chvB* mutants by rhicadhesin. *Mol. Microbiol.* **10:**597–605.

Talaga, P., B. Fournet, and J.-P. Bohin. 1994. Periplasmic glucans of *Pseudomonas syringae* pv. syringae. *J. Bacteriol.* **176:**6538–6544.

Talaga, P., B. Stahl, J.-M. Wieruszeski, F. Hillenkamp, S. Tsuyumu, G. Lippens, and J.-P. Bohin. 1996. Cell-associated glucans of *Burkhol-*

deria solanacearum and *Xanthomonas campestris* pv. citri: a new family of periplasmic glucans. *J. Bacteriol.* **178:**2263–2271.

Talaga, P., V. Cogez, J.-M. Wieruszeski, B. Stahl, J. Lemoine, G. Lippens, and J.-P. Bohin. 2002. Osmoregulated periplasmic glucans of the freeliving photosynthetic bacterium *Rhodobacter sphaeroides.* *Eur. J. Biochem.* **269:**2464–2472.

Thérisod, H., A. C. Weissborn, and E. P. Kennedy. 1986. An essential function for acyl carrier protein in the biosynthesis of membrane-derived oligosaccharides of *Escherichia coli. Proc. Natl. Acad. Sci. USA* **83:**7236–7240.

Valentine, P. J., B. P. Devore, and F. Heffron. 1998. Identification of three highly attenuated *Salmonella typhimurium* mutants that are more immunogenic and protective in mice than a prototypical *aroA* mutant. *Infect. Immun.* **66:**3378–3383.

Van Golde, L. M. G., H. Schulman, and E. P. Kennedy. 1973. Metabolism of membrane phospholipids and its relation to a novel class of oligosaccharides in *Escherichia coli. Proc. Natl. Acad. Sci. USA* **70:**1368–1372.

Vojnov, A. A., H. Slater, M. A. Newman, M. J. Daniels, and J. M. Dow. 2001. Regulation of the synthesis of cyclic glucan in *Xanthomonas campestris* by a diffusible signal molecule. *Arch. Microbiol.* **176:**415–420.

Wang, P., C. Ingram-Smith, J. A. Hadley, and K. J. Miller. 1999. Cloning, sequencing, and characterization of the *cgmB* gene of *Sinorhizobium meliloti* involved in cyclic β-glucan biosynthesis. *J. Bacteriol.* **181:**4576–4583.

Wieruszeski, J.-M., A. Bohin, J.-P. Bohin, and G. Lippens. 2001. In vivo detection of the cyclic osmoregulated periplasmic glucan of Ralstonia solanacearum by high-resolution magic angle spinning NMR. *J. Magn. Reson.* **151:**1–6.

Wood, J. M. 1999. Osmosensing by bacteria: signals and membrane-based sensors. *Microbiol. Mol. Biol. Rev.* **63:**230–262.

Yeats, C., and A. Bateman. 2003. The BON domain: a putative membrane-binding domain. *Trends Biochem. Sci.* **28:**352–355.

Yim, H. H., R. L. Brems, and M. Villarejo. 1994. Molecular characterization of the promoter of *osmY,* an *rpoS*-dependent gene. *J. Bacteriol.* **176:**100–107.

York, W. S. 1995. A conformational model for cyclic β-(1→2)-linked glucans based on NMR analysis of the β-glucans produced by *Xanthomonas campestris. Carbohydr. Res.* **278:**205–225.

Young, G. M., and V. L. Miller. 1997. Identification of novel chromosomal loci affecting *Yersinia enterocolitica* pathogenesis. *Mol. Microbiol.* **25:**319–328.

PRACTICAL
IMPLICATIONS

IV

PRACTICAL APPLICATIONS FOR PERIPLASMIC PROTEIN ACCUMULATION

John C. Joly and Michael W. Laird

20

The use of the periplasm for recombinant protein production/biotechnological purposes offers advantages and disadvantages compared with the cytoplasm. By this chapter in the book, the reader is no doubt aware of the proteins and enzymatic activities that can assist newly translocated proteins in attaining their proper structure. While the periplasm may lack a conventional nucleotide-hydrolyzing enzyme that promotes protein folding such as an Hsp60 or 70 family member, there are proteins identified that do aid in disulfide bond formation and isomerization and also proline isomerization (Wulfing and Pluckthun, 1994; Frech and Schmid, 1995; Missiakas and Raina, 1997; Kadokura et al., 2003). Additionally other periplasmic resident proteins may have a productive function identified from in vitro studies showing prevention of aggregation (Richarme and Caldas, 1997).

The fact that the periplasm is an oxidizing environment compared with the cytoplasm makes the periplasm useful for secreting proteins that require disulfide bond formation to obtain their functional, correct structure (Derman and Beckwith, 1991; Rietsch and Beck-

with, 1998). Mammalian secretory proteins with multiple cysteines require an oxidizing environment for the formation of disulfide bonds (Hwang et al., 1992). In eukaryotic cells this process occurs in the endoplasmic reticulum which allows for compartmentalization of the oxidative folding process. The formation of disulfide bonds can result in a tightly folded structure that is now resistant to proteases. When secretory proteins containing disulfide bonds in the final structure are produced without signal peptides, they remain in the cytoplasm and can aggregate and form inclusion bodies (Ejima et al., 1999). This is an undesirable outcome if the protein is difficult to solubilize and refold into its active conformation, especially if the protein contains multiple nonlinear disulfide bonds. Alternatively, it can be desirable if the protein may be readily solubilized and refolded (Chang and Swartz, 1993; Clark, 2001). The packing of the protein into the inclusion body can protect it from proteolysis, allowing significant accumulation of the protein (Joly et al., 1998). Production hosts have been engineered to yield altered redox properties in the cytoplasm rendering the environment less reducing and more oxidizing (Derman et al., 1993; Prinz et al., 1997). These strains can produce proteins containing disulfide bonds by permitting disulfide bond for-

John C. Joly, Early Stage Cell Culture Department, Genentech, Inc., South San Francisco, CA 94080. *Michael W. Laird*, Late Stage Cell Culture Department, Genentech, Inc., South San Francisco, CA 94080.

The Periplasm
Edited by Michael Ehrmann © 2007 ASM Press, Washington, D.C.

mation in the more oxidizing cytoplasmic environment (Bessette et al., 1999; Levy et al., 2001). While this route of producing correctly folded recombinant proteins has been made feasible, it is still unproven whether high levels of protein production can be reached. By high levels, we mean hundreds of milligrams per liter to grams per liter (Zhu et al., 1996; Joly et al., 1998; Tong et al., 2001; Cooksey et al., 2005; Laird et al., 2005). In general, this is the benchmark set in assembling an industrial production process.

The biotechnology industry has exploited the unique properties of the periplasm to produce industrially relevant products (Chang et al., 1987; Joly et al., 1998; Laird et al., 2005). Recombinant protein production is usually undertaken for one of two reasons. First, small milligram quantities of protein are needed for in vitro purposes, such as enzymatic studies, for structural or biochemical investigations, or for diagnostic purposes. Second, large gram to kilogram quantities are needed for industrial applications such as enzymes for consumer goods (Gerritse et al., 1998) or for therapeutic uses (Chang et al., 1987; Joly et al., 1998; Laird et al., 2004a, 2005). Each application has its own unique concerns. For instance, the production of an in vitro diagnostic protein would probably not be concerned with an authentic amino terminus while the production of human therapeutics usually requires this. The tactic of secreting a protein to the periplasm can produce a mature authentic amino terminus duplicating the naturally occurring protein (Chang et al., 1987). The main reasons for using the periplasm as a site of accumulation for recombinant protein production is that it offers the production of a native mature amino terminus and disulfide bonds are readily formed and isomerized to produce properly folded, biologically active structures. In addition, simply targeting the candidate protein to the periplasm may help shield it from unwarranted proteolysis (Talmadge and Gilbert, 1982; Joly et al., 1998).

Certain products are easier to produce as properly folded and active structures than others. Proteins that contain simple disulfide patterns are good candidates for periplasmic accumulation. While the periplasm has the systems in place to form and isomerize disulfides, the occurrence of endogenous host proteins containing more than two disulfide bonds is limited. This implies that there is not a tremendous need for the cell to accommodate multiple disulfide bond-containing proteins. Strains can be engineered to improve the capacity of the host to cope with increased production of disulfide-containing proteins (Joly et al., 1998; Qiu et al., 1998; Bessette et al., 1999, 2001). Hormones, peptides, and antibody fragments can be produced in gram per liter yields that are soluble and properly folded (Chang et al., 1987; Carter et al., 1992; Humphreys, 2003; Laird et al., 2005). The use of pro sequences is sometimes needed for efficient production because of folding interactions between pro and mature sequences (Baker et al., 1993). Recombinant proteins that lack pro sequences appear to be better candidates for periplasmic production.

We would be negligent in not pointing out some very impressive work over the past few years of examples of complex protein production. Complex proteins can have more than two disulfides and the pattern of disulfide pairing is not based on consecutive cysteine pairing. Some of these proteins contain dimeric interchain disulfides with challenging disulfide patterns (Dracheva et al., 1995). Full-length humanized antibodies are one example in recent years that has been successfully produced in the periplasm (Simmons et al., 2002). With five disulfides per heavy chain, two in the variable region and one in each of the CH1, CH2, and CH3 domains, and two disulfides per light chain and with two interchain disulfides between each heavy and light chain, there are sixteen correctly paired disulfides in an intact human immunoglobulin G1 (IgG1) antibody (see Chapter 21). These complex examples may also be produced in gram per liter quantities, but sometimes the yield is only a few milligrams per liter. Unfortunately, there is no exact science to predicting which candidate will be more successfully produced based on the primary sequence alone. In general, smaller polypeptides are usually more readily produced than large ones, but again there are exceptions to this statement (Squires et al., 2004). Most proteins can be successfully translocated across the cytoplasmic membrane to the periplasm with the aid of several considerations noted below.

CRITICAL CYTOPLASMIC EVENTS THAT AFFECT PERIPLASMIC PROTEIN ACCUMULATION

All protein biosynthesis begins in the cytoplasm. For efficient periplasmic protein accumulation to occur, events in the cytoplasm related to translation initiation and protein folding of the mature domain need to be coordinated. The presence of a signal peptide directs the preprotein for export into the periplasm (von Heijne, 1990). The signal peptide targets the preprotein for interactions with factors that keep the preprotein from reaching an export-incompatible structure (Weiss et al., 1988; Watanabe and Blobel, 1989a). These signal peptide binding proteins also guide the preprotein toward the export machineries. Two of the most well-known examples of proteins interacting with preproteins are SecB and Ffh (Watanabe and Blobel, 1989b; Kumamoto, 1991; Luirink et al., 1992; Phillips and Silhavy, 1992). SecB funnels proteins to SecA and the general secretory pathway (Hartl et al., 1990), while Ffh partners with an RNA component, Ffs, to target proteins to the FtsY pathway (Ribes et al., 1990; Miller et al., 1994). SecB appears to act posttranslationally, while the Ffh–Ffs complex appears to act cotranslationally. SecB has been shown to be involved with several preproteins destined for the periplasm (Kumamoto and Francetic, 1993), while the Ffh complex appears to interact with membrane proteins more than periplasmic targeted proteins (Seluanov and Bibi, 1997; Ulbrandt et al., 1997). Nevertheless, the initial studies of Ffh interactions were conducted with the well-known periplasmic protein β-lactamase.

A striking example showing the delicate interplay of cytoplasmic events and their effects on periplasmic export and yield was shown by Simmons and Yansura (1996). In this paper the authors demonstrated that nucleotide changes in the translation initiation region of the gene profoundly influenced periplasmic protein accumulation. The changes were in the Shine Dalgarno sequence and the first six codons of the signal sequence. None of the nucleotide changes altered the protein sequence so that the same protein signal sequence was used each time. By altering the translational strength through the use of silent codon changes without changing the actual primary protein sequence, the amount of protein accumulating in the periplasm was modulated. If the translational strength was too high, precursor protein buildup in the cytoplasm was observed. There appeared to be an optimum level of translation allowing for efficient periplasmic accumulation without precursor buildup. Some candidate proteins exhibited broader ranges than others in the translational strength allowing efficient export. One explanation for those results is there is a bottleneck in the cytoplasm or in the export machinery prior to signal peptide cleavage (see Fig. 1 for model). High levels of the preprotein may overwhelm, or sequester, a cytoplasmic factor responsible for keeping the protein in an export-competent conformation. Insufficient levels of the factors needed for such activities would result in the precursor folding into an export-incompatible form or even forming inclusion bodies. There are many documented examples of cytoplasmic proteins having positive effects on periplasmic protein accumulation of endogenous or recombinant proteins (Phillips and Silhavy, 1990; Wild et al., 1992; Perez-Perez et al., 1995; Berges et al., 1996; Wild et al., 1996). A possible molecular explanation is that the cytoplasmic resident protein is preventing the preprotein from negative pathways, such as premature folding and inclusion body formation, and keeping the protein in an unfolded state and allowing efficient export. A common strategy to seek improvements in periplasmic protein production is changing signal sequences (Rathore et al., 1996; Pritchard et al., 1997; Mori et al., 1998; Humphreys et al., 2000; Laird et al., 2004b). The alternative signal sequences may interact

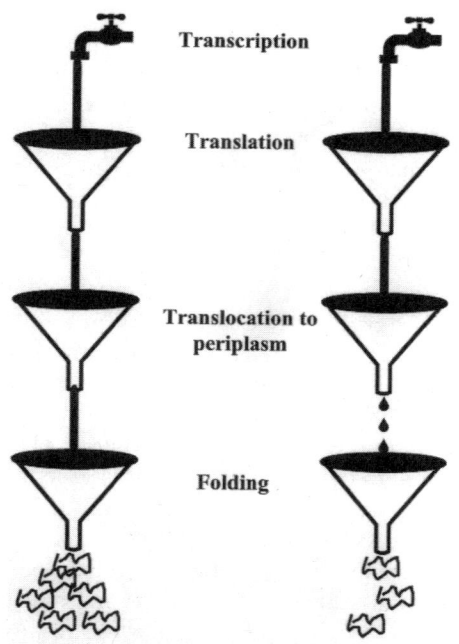

FIGURE 1 Conceptual model of protein secretion in *E. coli*. The spigot represents the transcriptional induction of the recombinant gene and the funnels represent the major processes that are involved in periplasmic protein accumulation. Each major process has several subprocesses that can be limiting, resulting in decreased yield of the intended protein. In this example there is a limitation in the cytoplasmic membrane secretion step, which could manifest as an increase in precursor protein accumulation.

with different cytoplasmic proteins or interact differently with the mature protein. The alternative signals could also target the protein for post- or cotranslational translocation. Another possibility is that by changing the signal sequence, the experimentalist has changed the translational strength of the mRNA affecting the amount of protein produced. Alternatively, by changing the nucleotide sequence, one may have changed mRNA secondary structures that affect translation speed or mRNA stability (Le Calvez et al., 1996). Both of those factors contribute to how strongly the translation initiation region performs.

When conditions are optimized for periplasmic accumulation, a huge percentage of the total cell protein can be exported to the periplasm. Percentages in the 20 to 40% range can be attained when there are seemingly no bottlenecks (Joly et al., 1998; Jeong and Lee, 2001; Cooksey et al., 2004; Laird et al., 2005). These examples are not dependent on overexpression of the general secretory pathway, demonstrating that the endogenous levels of the general secretory machinery can handle enormous throughput. Attempts to improve periplasmic accumulation by overexpressing cytoplasmic membrane components have generally not led to productive outcomes. There needs to be coordinated overexpression of the components, otherwise degradation of the components in excess occurs (Taura et al., 1993). Even when coordinating the elevated expression of the membrane components, it is not always evident that the increased yield is due to the overexpressed machinery (Perez-Perez et al., 1994).

Another export system to the periplasm has been identified in bacteria that operates separately from the general secretory pathway. The twin–arginine translocation system (TAT system) has unique signal sequences but, more interestingly, exports proteins in their final folded forms (Berks et al., 2003). The TAT genes encoding for the secretion components have been identified and TAT-dependent endogenous proteins have also been studied. The proteins may contain cofactors for their activity and these cofactors assemble with the protein in the cytoplasm prior to export. There are a limited number of examples of such TAT-dependent proteins and many are synthesized only under specific environmental conditions (Berks et al., 2000, 2003). The normal throughput of the TAT system appears to be much less than that of the general secretory pathway. Overexpression of the TAT export machinery has been achieved and shown to improve the yields of recombinant proteins targeted for the periplasm with TAT signal sequences (Barrett et al., 2003). The use of the TAT system allows one to use the folding environment of the cytoplasm instead of the periplasm to fold the recombinant protein followed by export of the folded protein (DeLisa et al., 2002, 2003). While feasibility has been demonstrated for recombinant periplasmic protein production using the TAT system, it is still not yet clear if an economically viable process can be developed. Levels of exported recombinant protein appear to be limited at this time. Further improvements in the export machinery throughput could make this option an attractive one because of the ability to translocate folded proteins. Coupling TAT-mediated export with strains that can form disulfide bonds in the cytoplasm could take advantage of cytoplasmic factors for protein folding followed by export of the folded protein.

THE FOLDING ENVIRONMENT IN THE PERIPLASM

The periplasm of bacteria is best thought of as a mesh or gel environment rather than a soluble simple compartment (Hobot et al., 1984). Small molecular solutes can freely equilibrate between the exterior solution and the periplasm although positively charged solutes can be preferentially concentrated in the periplasm. It seems that the negative charges from the lipid components like phospholipids and lipid A attract or require positive ions from the medium (Nikaido and Vaara, 1985). After export from the cytoplasmic membrane machinery, periplasmic proteins need to fold into their proper conformations and form disulfide bonds to lock the conformation into place and form protease-resistant structures. There are three possible fates for newly secreted proteins upon release into the periplasmic space (see Fig. 2). First, proteins can misfold and aggregate forming insoluble inclusion bodies. Second, proteins can be proteolyzed by endogenous periplasmic and membrane proteases where the protease domain of membrane proteases is located in the periplasm. Lastly, proteins can successfully fold into the correct conformation and exist as soluble, biologically active proteins. This last fate is usually preferred by those trying to exploit the periplasm as a site of recombinant protein production. With soluble active protein, solubilizing and refolding proteins from an insoluble form are not required. Instead, it becomes

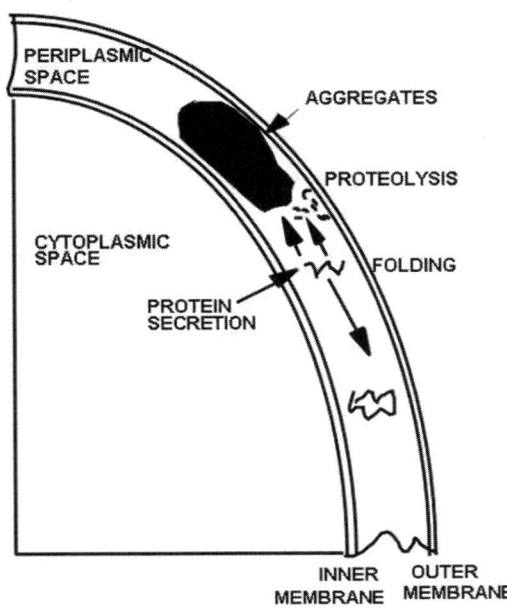

FIGURE 2 Possible fates of proteins after secretion into the periplasm. Proteins have three potential fates in the periplasm: (i) aggregation and inclusion body formation, (ii) proteolysis, and (iii) folding to their proper conformation.

the responsibility of the purification process to selectively bind and elute the target—usually with the aid of several separation steps—to isolate a homogeneous purified protein. The first fate discussed above is not fatal to successful recombinant protein production. Packaging the protein into inclusion bodies can lead to very high production and the protein is protected from the aforementioned proteolysis. If the target protein can be solubilized and folded with an appreciable yield, productive processes can be put together to make economically viable processes (Chang and Swartz, 1993; Winter et al., 2002; Xue et al., 2005). One important consideration in refolding proteins is the solubility of the protein during the refolding steps. Some proteins are poorly soluble during the refolding process, necessitating large volumes for batch processes and therefore large quantities of chemicals. The protein concentration problem during the refolding process can be defeated by slowly feeding the denatured form into the refolding tank, since it is the concentration of the intermediates that is critical, while the correctly folded form is usually quite soluble and may not interfere with refolding intermediates (Rudolph and Fischer, 1990; Buchner et al., 1992). Therefore, slow feeding of the unfolded form of the protein into the refolding solution can minimize the number of poorly soluble folding intermediates at any one time.

OVEREXPRESSION AND DELETION OF ACTIVITIES TO ACHIEVE HIGHER YIELDS OF RECOMBINANT PROTEIN PRODUCTION

The aim of scientists and engineers producing recombinant proteins is to maximize the yield of protein per batch of culture volume. The more protein recovered from each batch decreases the total number of batches necessary and therefore decreases the need to construct and buy more production tanks or facilities to produce the candidate protein. The ability to defer investment in more facilities-related items when several processes are competing for time in the same manufacturing plant is a powerful economic argument for boosting protein yields. Additionally, quality control costs are usually fixed per batch, such that decreasing the total number of batches required per year decreases the total quality control costs associated with the product. Therefore, increasing the yield of protein is important for those trying to produce recombinant proteins for industrial purposes. To maximize the yield of protein per batch of culture volume one can try to overexpress activities that are beneficial to protein folding or delete activities that negate protein accumulation. The enzymes identified as being involved in disulfide bond production and isomerization have been overexpressed by several groups to improve the yield of protein (Humphreys et al., 1996; Joly et al., 1998; Qiu et al., 1998; Bessette et al., 1999; Zhan et al., 1999; Jeong and Lee, 2000; Bessette et al., 2001). It appears that the normal level of these enzymes is sufficient for growth under usual laboratory conditions, but it can be insufficient for the production of many mammalian proteins (Joly et al., 1998;

Qiu et al., 1998). In *E. coli* there are not many examples of periplasmic proteins containing more than two disulfide bonds. A few notable exceptions are AppA, Agp, periplasmic RNase I, and presumably MepA (Hiniker and Bardwell, 2004; Berkmen et al., 2005). This is in contrast to secreted proteins from mammals that can contain great numbers of disulfide bonds and very complex patterns of these disulfides (Qui et al., 1998). *E. coli* proteins such as PhoA and DsbC have simple linear patterns of cysteine pairing such that, upon secretion to the periplasm, the first cysteine pairs with the second and the third pairs with the fourth. If disulfide bond formation is rapid, then there may be little opportunity for endogenous proteins to mispair, obviating the need for large amounts of enzymes needed to isomerize mispaired cysteines. This is not the case when eukaryotic proteins such as tissue plasminogen activators and immunoglobulin proteins are secreted to the periplasm (Qiu et al., 1998; Simmons et al., 2002). Boosting the levels of enzymes is necessary for efficient production of such complex proteins and can make production processes once thought only to be possible in mammalian production systems a reality (Bessette et al., 2001).

Outside of the *dsb* genes, there are not many examples of successful overexpression of proteins positively influencing recombinant protein production. A couple of notable exceptions influencing antibody fragment production are Skp (Thome and Muller, 1991) and FkpA (Horne and Young, 1995). These proteins were identified in genetic screens selecting for more efficient phage display of antibody fragments (Bothmann and Pluckthun, 1998, 2000). While these proteins appear to aid in the production of antibody fragments (Bothmann and Pluckthun, 1998; Hayhurst and Harris, 1999; Bothmann and Pluckthun, 2000), only FkpA has been shown to enhance the expression of a nonantibody fragment as a fusion molecule produced in the cytosol (Scholz et al., 2005). Perhaps the β-sheet structures of antibody proteins are what these periplasmic proteins recognize, similar to the structure of *E. coli* outer membrane proteins (Chen and Henning, 1996; De Cock et al., 1999; Walton and Sousa, 2004) and help correctly fold them into their final conformation. There are several other peptidyl-prolyl isomerases for periplasmic proteins such as SurA (Lazar and Kolter, 1996), PpiA (Liu and Walsh, 1990), and PpiD (Dartigalongue and Raina, 1998), yet we are not aware of the successful demonstration of the involvement of these proteins in recombinant periplasmic protein production. The lack of an identified chaperone system in the periplasm has been hindering the advancement of recombinant periplasmic protein accumulation. Some proteins have been shown to prevent citrate synthase aggregation, but it is unclear how powerful these are at preventing negative interactions in protein production. The DegP protease has been shown to contain a chaperone activity in addition to being a protease (Spiess et al., 1999). Attempts to overproduce a protease inactive version and still take advantage of the chaperone activity failed to improve antibody fragment production (J. Gunson and J. C. Joly, unpublished observations) but can rescue outer membrane protein-folding lethality mutations by sequestering the protein (Misra et al., 2000).

The other strategy highlighted earlier is the removal of deleterious activities for recombinant protein production. The common choices for deleting genes to improve production are proteases. Several periplasmic proteases have been successfully deleted without impairing the general health and viability of the hosts for protein production (Strauch and Beckwith, 1988; Baneyx and Georgiou, 1991). Combinations of protease knockouts can be constructed that retain vigorous growth and production properties (Fuh et al., 1990; Meerman and Georgiou, 1994; Joly et al., 1998). Removing these proteases is one way to deal with the *cpx* and σ^E responses that sense periplasmic and outer membrane stress (Danese et al., 1995; Connolly et al., 1997; Danese and Silhavy, 1997; Pogliano et al., 1997). These response pathways upregulate proteases and folding aids to combat misfolded proteins (Raivio and Silhavy, 2001; Duguay and Silhavy, 2004). The gene most influenced by these pathways is the *degP* gene (Danese et al., 1995). Deletion of the *degP* gene does lead to impaired growth at elevated temperatures but when combined with an *ompT* gene deletion, the strain grows well at elevated temperatures (N. McFarland and M. W. Laird, unpublished observations). Since many protein production processes are run at 30°C instead of 37°C, this is often not an issue. One notable exception is human growth hormone that is produced into a partially soluble, biologically active form in the periplasm at 37°C (Chang et al., 1987). Other proteases that can be eliminated include Protease III (Baneyx and Georgiou, 1991) and Prc (Silber and Sauer, 1994), the products of the *ptr* and *prc* genes, respectively. The knockout of the *ptr* gene is important for yields of insulin and insulin–like growth factor 1 (Baneyx and Georgiou 1991; Joly et al., 1998). The Prc protease normally cleaves the FtsI protein (Hara et al., 1989), but it has also been shown to cleave kappa light chains produced during antibody fragment synthesis (Chen et al., 2004). The deletion of the *prc* gene improves yields of intact antibody fragments. Unfortunately, the simple deletion of *prc* leads to several other undesirable phenotypes, including salt sensitivity and impaired growth at 37°C (Hara et al., 1991). To overcome the *prc* deletion phenotype, a compensatory suppressor mutation was identified in a gene called *spr* (suppressor of *prc*). This suppressor relieves most of the undesirable properties and still lacks the proteolytic function (Hara et al., 1996). Antibody fragments can be produced at several grams per liter in strains that contain both the *degP* and *prc* deletions combined with the *spr* suppressor (Chen et al., 2004).

The removal of the *ompT* gene is often important for recovering both periplasmically and cytoplasmically produced recombinant proteins without proteolysis (Baneyx and Georgiou, 1990, 1991; Laird et al., 2004a). The OmpT protein is an outer membrane protease with the active domain on the exterior of the cell (Sugimura and Nishihara, 1988). This nuisance protease activity is difficult to inhibit and generally degrades proteins during recovery and purification (Grodberg and Dunn, 1987). It is common to use hosts lacking *ompT* to avoid such an issue (Wadensten et al., 1991; Meerman and Georgiou, 1994; Laird et al., 2004a). The common *E. coli* B laboratory strain, BL21, used for protein production lacks the *ompT* gene (Studier et al., 1990).

GENETIC SELECTIONS AND SCREENS FOR IMPROVED RECOMBINANT PROTEIN YIELDS

Most of the secretion and folding catalysts in *E. coli* were identified by using mutant or fusion proteins of *E. coli* origin. There have been less than a handful of published attempts to utilize heterologous proteins in genetic selections and schemes to identify components that influence recombinant protein production. The first example was developed from the basis of previous work that showed the production of biologically active, human tPA in the periplasm of *E. coli* required proper disulfide bond isomerization (Qiu et al., 1998). Armed with this knowledge, a novel selection scheme using a variant tPA molecule was developed to identify *E. coli* mutants that exhibited enhanced disulfide bond isomerization activity (Zhan et al., 2004). The scheme identified mutant strains with increased DsbC expression levels. Genetic mapping showed that the lesions affected RNase E processing activity and ultimately increased the stability of the labile *dsbC* mRNA (Zhan et al., 2004).

Another approach to identify components influencing recombinant protein production was employed by Bass et al. (1996). This genetic selection scheme utilized a protease-deficient mutant with a conditional lethal phenotype (*prc*; Hara et al., 1991) in an attempt to identify additional proteases that could suppress the lethality. Using a random *E. coli* plasmid library, multicopy suppressors of the *prc*-conditional lethal phenotype were obtained. Characterization of these suppressors revealed two novel protease genes, *hhoA* (*degQ*) and *hhoB* (*degS*), which encode proteins homologous to the periplasmic serine protease, DegP (Bass et al., 1996). Additional studies confirmed the rela-

tionship between DegQ/S and DegP (Waller and Sauer, 1996) and provided further evidence as to their substrate specificities and in vivo roles (Kolmar et al., 1996). This knowledge of additional proteases in the periplasm may help tailor future recombinant production hosts.

As mentioned earlier, the Pluckthun laboratory has successfully used phage display of antibody fragments to select for improved antibody folding in the periplasm. Two proteins that were identified, FkpA and Skp, could improve not only phage-displayed antibody fragments but soluble, folded periplasmic forms of the antibody fragments (Bothmann and Pluckthun, 1998, 2000). Finally, an unsuccessful approach to identify "foldases" that influence heterologous protein production was employed using human placental alkaline phosphatase (hPAP). Previous work had shown that when hPAP was produced in the *E. coli* periplasm, most of the protein was found in inclusion bodies. Some biological activity was detected, however, suggesting that there was a small amount of soluble hPAP protein produced (Beck and Burtscher, 1994). Coexpression of PpiA or DsbA did not yield an increase in soluble hPAP (Beck and Burtscher, 1994). For this alternative approach, hPAP was produced with an *E. coli*-derived signal peptide behind a tightly regulated, inducible promoter in an *E. coli* host devoid of alkaline phosphatase activity. The hPAP produced in this system exhibited similar results to those described by Beck and Burtscher (1994) in regard to solubility and activity (M.W. Laird and J.C. Joly, unpublished observations). For the selection scheme, multiple approaches were applied, including the isolation of spontaneous mutants, chemical and transposon mutagenesis, and screening for multicopy suppressors using an *E. coli* plasmid library. Cells expressing hPAP were cultured on agar plates containing a small concentration of inducer and the chromogenic indicator, 5-bromo-4-chloro-3-indolylphosphate (XP). Under these initial conditions, cells exhibited "white" colonies (enzymatically inactive) with the idea of the scheme to identify "blue" colonies (enzymatically active) upon implementing the various pronged approaches. Hundreds of blue colonies were obtained from the different approaches. However, upon characterization, all of the blue colonies were the result of an increase in hPAP transcription, not improved protein folding.

USEFUL TOOLS, LEAKY HOSTS, AND PURIFICATION CONSIDERATIONS

A strategic advantage for producing biotherapeutic proteins in microbial systems is that one can create deletions and overproduction plasmids because the entire genomic sequence of organisms such as *E. coli* is known and powerful tools exist for the precise manipulation of genes. One can remove whole genes, parts of genes, or specific residues fairly easily provided that activity is not essential for growth on commonly used laboratory media (Metcalf et al., 1994; Link et al., 1997; Posfai et al., 1999). One can specifically delete a gene and not disrupt an entire operon with carefully designed strategies. In addition, these targeted genetic lesions can be generated in an unmarked fashion and therefore not compromise antibiotic resistances generally necessary for plasmid construction and maintenance. The ability to precisely tailor production hosts is an important tool for those trying to improve yields in recombinant processes.

The ability to precisely modify the bacterial genome to enhance recombinant protein production is also useful in protecting the heterologous protein production factory from enemy invasion. In other words, by eliminating genes encoding bacteriophage receptors and outer membrane protein channels, the recombinant host's natural predators can be neutralized. One prominent outer membrane protein, TonA, transports Fe^{3+} into the cell (Carmel et al., 1990) yet also serves as the receptor for the bacteriophages T1 (Hancock and Braun, 1976), T5 (Weidel and Kellenberger, 1955), UC-1 (Lundrigan et al., 1983), and φ80 (Hancock and Braun, 1976). In addition, the TonA channel permits the passage of colicin M (Braun et al., 1980) and the antibiotic albomycin (Braun and Herrmann, 1993). Deletion of *tonA*

does not lead to any deleterious growth defects in semidefined (Joly et al., 1998; Simmons et al., 2002; Cooksey et al., 2005; Laird et al., 2005) or defined minimal media fermentations (Laird et al., 2004a, 2004b, 2005). Again, with the tools available for precise gene deletion and allele substitution, other outer membrane proteins with specific functions that also serve as bacteriophage receptors could be modified in the production host genome to further minimize these threats. The idea is to retain the prescribed protein function and lose the ability to support bacteriophage infection (OmpC, Xiong et al., 1996; OmpF, Traurig and Misra, 1999). The elimination of bacteriophage receptors could potentially save time and money by eliminating lost production runs due to bacteriophage contamination and subsequent decontamination of the production facilities.

Another strategy employed to produce recombinant proteins targeted to the periplasm is the use of leaky hosts. These hosts are impaired in their outer membrane such that elevated concentrations of proteins are found in the media (Hancock, 1984; Levengood–Freyermuth et al., 1993; Bernadac et al., 1998; Suzuki et al., 1999; Rinas and Hoffmann, 2004). The idea is to increase the volume of the periplasm and therefore decrease the protein concentration of poorly soluble intermediates. The leaky condition is caused either by mutations in outer membrane proteins and proteins that connect with outer membrane proteins (Webster 1991; Young and Silver, 1991) or in the lipid constitution of the outer membrane (Irvin et al., 1975; Parker et al., 1992; Schnaitman and Klena, 1993; Kloser et al., 1998). While recombinant proteins can be recovered from the external medium, the use of such hosts is limited since outer membrane impairment generally results in compromised growth and protein production. These hosts are not as robust as their wild-type counterparts in fermentors.

The use of leaky hosts to recover the recombinant protein from the medium is an attempt to simplify the recovery of the protein from the cells. A highly desired goal of microbial production of proteins is the complete export of the protein outside the cell. While this can be accomplished with certain fungi, it remains a difficult problem in prokaryotic gram-negative hosts. The common techniques for purifying proteins from the periplasm at large scale involves rupturing the cytoplasmic membrane as well as the outer membrane and releasing the entire contents of the cells. Since the cytoplasm contains approximately 80% of the total cellular protein composition (Pugsley, 1993), the release of its contents makes it more difficult for the purification process. There are dedicated export systems in other prokaryotes but these often involve the use of special sequences to tag the proteins destined for the export machinery (Binet et al., 1997; Blight and Holland, 1994; Pugsley et al., 1997). The sequences may not be removed during the export process, necessitating further processing steps to produce a native sequence. Feasibility can be demonstrated with some of these dedicated export systems, but we are not aware of any industrial processes utilizing this route of production (Blight and Holland, 1994).

Another route to recovery of protein after a simple centrifugation step is the use of phage lysozymes to disrupt the peptidoglycan and cause the periplasmic contents to be released into the medium (Dabora and Cooney, 1991; Leung and Swartz, 2001). Phage lysozymes need access to the peptidoglycan and do so through holin proteins that insert into the cytoplasmic membrane (Young and Blasi, 1995). The use of the lysozymes and holin proteins has to be tightly regulated and turned on only after the desired protein production process has stopped. It is a difficult balance to achieve but has been demonstrated (J. C. Joly, unpublished observations). The problem is the efficiency of the protein recovered and reproducibly controlling the lysis process, especially at large scale. Premature triggering of the activities results in loss of the culture and a wasted production run. Not only is this a loss of time, but obviously money too. Recovery of the protein from the external medium or after a disruption of only the outer membrane and peptidoglycan is still a goal that has not been put

into practice on a large scale. Common methods for isolating periplasmic fractions at small scale such as spheroplasting (Malamy and Horecker, 1964; Neu and Heppel, 1964) or osmotic shock treatment (Neu and Heppel, 1965) are not practically accomplished once cultures reach volumes at and above 10 liters.

CHEMICAL STRATEGIES FOR IMPROVING PERIPLASMIC PROTEIN FOLDING

The solute composition of the periplasm can be manipulated by adjusting the medium composition. The outer membrane of gram-negative bacteria is porous to solutes 600 Da and less (Nakae and Nikaido, 1975). This fact can be exploited when overexpressing recombinant proteins and secreting them to the periplasm. Over the years numerous studies have tried to improve protein folding by adding medium components such as sucrose, glycerol, betaine, proline, arginine, etc. These components can have productive effects in vivo much as they have been shown to improve protein folding in vitro. The positive effects on protein folding in vitro were obviously the impetus to be applied to the in vivo situation. The components to add fall into two categories. The first category consists of osmoprotectants such as sucrose, glycerol, betaine, and sorbitol (Blackwell and Horgan, 1991; Barth et al., 2000). There are several explanations why these solutes may help protein folding in the periplasm. They may induce protective stress responses due to high osmolality or could interact directly with a folding intermediate. The second category includes small thiol/disulfide reagents to help promote disulfide bond formation and isomerization. Oxidized and reduced glutathione is often added to improve the yield of periplasmic proteins since these naturally occurring molecules are thought to be the in vivo currency for thiol/disulfide reagents. While this is the case in the endoplasmic reticulum of eukaryotic cells, gram-negative bacteria contain micromolar concentrations of glutathione, but this is usually found in the cytoplasm rather than the periplasm (Fahey et al., 1978). Some glutathione may leak into the external medium upon cellular division in gram-negative bacteria (Owens and Hartmann, 1986). A proposed glutathione transporter from the cytoplasm to the periplasm has recently been documented (Pittman et al., 2005). Finally, the combination of overexpression of periplasmic components participating in protein-folding pathways and the inclusion of chemical additives has been reported and shown to have positive benefits (Wunderlich and Glockshuber, 1993; Sandee et al., 2005). Another combined approach successfully used chemical additives with the secretion of a normally cytoplasmic chaperone (Schaffner et al., 2001). Multiple approaches to the problem of periplasmic protein folding may be needed to optimize the yield of soluble and biologically active protein exported to the periplasm.

ACKNOWLEDGMENTS

The authors thank numerous colleagues past and present at Genentech working on the problem of microbial production of recombinant proteins over the years. Many discussions and interactions have helped shape our thinking about protein production.

REFERENCES

Baker, D., A. K. Schiau, and D. A. Agard. 1993. The role of pro regions in protein folding. *Curr. Opin. Cell. Biol.* **5:**966–970.

Baneyx, F., and G. Georgiou. 1990. *In vivo* degradation of secreted fusion proteins by the *Escherichia coli* outer membrane protease OmpT. *J. Bacteriol.* **172:**491–494.

Baneyx, F., and G. Georgiou. 1991. Construction and characterization of *Escherichia coli* strains deficient in multiple secreted proteases: protease III degrades high-molecular weight substrates *in vivo*. *J. Bacteriol.* **173:**2696–2703.

Barrett, C. M. L., N. Ray, J. D. Thomas, C. Robinson, and A. Bolhuis. 2003. Quantitative export of a reporter protein, GFP, by the twin-arginine translocation pathway in *Escherichia coli*. *Biochem. Biophys. Res. Commun.* **304:**279–284.

Barth, S., M. Huhn, B. Matthey, A. Klimka, E. A. Galinski, and A. Engert. 2000. Compatible-solute-supported periplasmic expression of functional recombinant proteins under stress conditions. *Appl. Environ. Microbiol.* **66:**1572–1579.

Bass, S., Q. Gu, and A. Christen. 1996. Multicopy suppressors of prc mutant Escherichia coli include two HtrA (DegP) protease homologs (HhoAB), DksA, and a truncated R1pA. *J. Bacteriol.* **178:**1154–1161.

Beck, R., and H. Burtscher. 1994. Expression of human alkaline phosphatase in *Escherichia coli*. *Protein Exp. Purif.* **5:**192–197.

Berges, H., E. Joseph-Liauzun, and O. Fayet. 1996. Combined effects of the signal sequence and the major chaperone proteins on the export of human cytokines in *Escherichia coli*. *Appl. Environ. Microbiol.* **62:**55–60.

Berkmen, M., D. Boyd, and J. Beckwith. 2005. The nonconsecutive disulfide bond of *Escherichia coli* phytase (AppA) renders it dependent on the protein-disulfide isomerase, DsbC. *J. Biol. Chem.* **280:**11387–11394.

Berks, B. C., T. Palmer, and F. Sargent. 2000. The Tat protein export pathway. *Mol. Microbiol.* **35:**260–274.

Berks, B. C., T. Palmer, and F. Sargent. 2003. The Tat protein translocation pathway and its role in microbial physiology. *Adv. Microb. Physiol.* **47:**187–254.

Bernadac, A., M. Gavioli, J. C. Lazzaroni, S. Raina, and R. Lloubes. 1998. *Escherichia coli* tol-pal mutants form outer membrane vesicles. *J. Bacteriol.* **180:**4872–4878.

Bessette, P. H., F. Aslund, J. Beckwith, and G. Georgiou. 1999. Efficient folding of proteins with multiple disulfide bonds in the *Escherichia coli* cytoplasm. *Proc. Natl. Acad. Sci. USA* **96:**13703–13708.

Bessette, P. H., J. Qiu, J. C. A. Bardwell, J. R. Swartz, and G. Georgiou. 2001. Effect of sequences of the active-site dipeptides of DsbA and DsbC on *in vivo* folding of multidisulfide proteins in *Escherichia coli*. *J. Bacteriol.* **183:**980–988.

Binet, R., S. Letoffe, J. M. Ghigo, P. Delepelaire, and C. Wandersman. 1997. Protein secretion by gram-negative bacterial ABC exporters—a review. *Gene* **192:**7–11.

Blackwell, J. R., and R. Horgan. 1991. A novel strategy for the production of a highly expressed recombinant protein in an active form. *FEBS Lett.* **295:**10–12.

Blight, M. A., and I. B. Holland. 1994. Heterologous protein secretion and the versatile Escherichia coli haemolysin translocator. *Trends Biotechnol.* **12:**450–455.

Bothmann, H., and A. Pluckthun. 1998. Selection for a periplasmic factor improving phage display and functional periplasmic expression. *Nat. Biotechnol.* **16:**376–380.

Bothmann, H., and A. Pluckthun. 2000. The periplasmic *Escherichia coli* peptidylprolyl *cis,trans*-isomerase FkpA. *J. Biol. Chem.* **275:**17100–17105.

Braun, V., J. Frenz, K. Hantke, and K. Schaller. 1980. Penetration of colicin M into cells of *Escherichia coli*. *J. Bacteriol.* **142:**162–168.

Braun, V., and C. Herrmann. 1993. Evolutionary relationship of uptake systems for biopolymers in *Escherichia coli*: cross-complementation between the TonB-ExbB-ExbD and the TolA-TolQ-TolR proteins. *Mol. Microbiol.* **8**:261–268.

Buchner, J., I. Pastan, and U. Brinkmann. 1992. A method for increasing the yield of properly folded recombinant fusion proteins: single-chain immunotoxins from renaturation of bacterial inclusion bodies. *Anal. Biochem.* **205**:263–270.

Carmel, G., D. Hellstern, D. Henning, and J. W. Coulton. 1990. Insertion mutagenesis of the gene encoding the ferrichrome-iron receptor of *Escherichia coli* K-12. *J. Bacteriol.* **172**:1861–1869.

Carter, P., R. F. Kelley, M. L. Rodrigues, B. Snedecor, M. Covarrubias, M. D. Velligan, W. L. Wong, A. M. Rowland, C. E. Kotts, M. E. Carver, M. Yang, J. H. Bourell, H. M. Shepard, and D. Henner. 1992. High level *E. coli* expression and production of a bivalent humanized antibody fragment. *Bio/Technology* **10**:163–167.

Chang, C. N., M. Rey, B. Bochner, H. Heyneker, and G. Gray. 1987. High-level secretion of human growth hormone by *Escherichia coli*. *Gene* **55**:189–196.

Chang, J. Y., and J. R. Swartz. 1993. Single-step solubilization and folding of IGF-I aggregates from *Escherichia coli*, p. 178–188. *In* J. L. Cleland (ed.), *Protein Folding: In Vivo and In Vitro*. American Chemical Society, Washington, D.C.

Chen, C., B. Snedecor, J. C. Nishihara, J. C. Joly, N. McFarland, D. C. Andersen, J. E. Battersby, and K. M. Champion. 2004. High-level accumulation of a recombinant antibody fragment in the periplasm of *Escherichia coli* requires a triple-mutant (*degP prc spr*) host strain. *Biotechnol. Bioeng.* **85**:463–474.

Chen, R., and U. Henning. 1996. A periplasmic protein (Skp) of *Escherichia coli* selectively binds a class of outer membrane proteins. *Mol. Microbiol.* **19**:1287–1294.

Clark, E. D. B. 2001. Protein refolding for industrial processes. *Curr. Opin. Biotechnol.* **12**:202–207.

Connolly, L., A. De Las Penas, B. M. Alba, and C. A. Gross. 1997. The response to extracytoplasmic stress in *Escherichia coli* is controlled by partially overlapping pathways. *Genes Dev.* **11**:2012–2021.

Cooksey, B. A., G. C. Sampey, J. L. Pierre, X. Zhang, J. D. Karwoski, G. H. Choi, and M. W. Laird. 2004. Production of biologically active *Bacillus anthracis* edema factor in *Escherichia coli*. *Biotechnol. Prog.* **20**:1651–1659.

Dabora, R. L., and C. L. Cooney. 1990. Intracellular lytic enzyme systems and their use for disruption of *Escherichia coli*. *Adv. Biochem. Eng. Biotechnol.* **43**:11–30.

Danese, P. N., and T. J. Silhavy. 1997. The σ^E and the Cpx signal transduction systems control the synthesis of periplasmic protein-folding enzymes in *Escherichia coli*. *Genes Dev.* **11**:1183–1193.

Danese, P. N., W. B. Snyder, C. L. Cosma, L. J. B. Davis, and T. J. Silhavy. 1995. The *cpx* two-component signal transduction pathway of *Escherichia coli* regulates transcription of the gene specifying the stress-inducible periplasmic protease, DegP. *Genes Dev.* **9**:387–398.

Dartigalongue, C., and S. Raina. 1998. A new heat-shock gene, *ppiD*, encodes a peptidyl-prolyl isomerase required for folding of outer membrane proteins in *Escherichia coli*. *EMBO J.* **17**:3968–3980.

De Cock, H., U. Schafer, M. Potgeter, R. Demel, M. Muller, and J. Tommassen. 1999. Affinity of the periplasmic chaperone Skp of *Escherichia coli* for phospholipids, lipopolysaccharides and non-native outer membrane proteins. Role of Skp in the biogenesis of outer membrane protein. *Eur. J. Biochem.* **259**:96–103.

DeLisa, M. P., P. Samuelson, T. Palmer, and G. Georgiou. Genetic analysis of the twin arginine translocator secretion pathway in bacteria. *J. Biol. Chem.* **277**:29825–29381.

DeLisa, M. P., D. Tullman, and G. Georgiou. 2003. Folding quality control in the export of proteins by the bacterial twin-arginine translocation pathway. *Proc. Natl. Acad. Sci. USA* **100**:6115–6120.

Derman, A. I., and J. Beckwith. 1991. *Escherichia coli* alkaline phosphatase fails to acquire disulfide bonds when retained in the cytoplasm. *J. Bacteriol.* **173**:7719–7722.

Derman, A. I., W. A. Prinz, D. Belin, and J. Beckwith. 1993. Mutations that allow disulfide bond formation in the cytoplasm of *Escherichia coli*. *Science* **262**:1744–1747.

Dracheva, S., R. E. Palermo, G. D. Powers, and D. S. Waugh. 1995. Expression of soluble human interleukin-2 receptor alpha-chain in *Escherichia coli*. *Protein Expr. Purif.* **6**:737–747.

Duguay, A. R., and T. J. Silhavy. 2004. Quality control in the bacterial periplasm. *Biochim. Biophys. Acta* **1694**:121–134.

Ejima, D., M. Watanabe, Y. Sato, M. Date, N. Yamada, and Y. Takahara. 1999. High yield refolding and purification process for recombinant human interleukin-6 expressed in *Escherichia coli*. *Biotechnol. Bioeng.* **62**:301–310.

Fahey, R. C., W. C. Brown, W. B. Adams, and M. B. Worsham. 1978. Occurrence of glutathione in bacteria. *J. Bacteriol.* **133**:1126–1129.

Frech, C., and F. X. Schmid. 1995. DsbA-mediated disulfide bond formation and catalyzed prolyl isomerization in oxidative protein folding. *J. Biol. Chem.* **270**:5367–5374.

Fuh, G., M. G. Mulkerrin, S. Bass, N. McFarland, M. Brochier, J. H. Bourell, D. R. Light, and J. A. Wells. 1990. The human growth hormone receptor. *J. Biol. Chem.* **265**:3111–3115.

Gerritse, G., R. Ure, F. Bizoullier, and W. J. Quax. 1998. The phenotype enhancement method identifies the Xcp outer membrane secretion machinery from *Pseudomonas aclaligenes* as a bottleneck for lipase production. *J. Biotechnol.* **64:**23–38.

Grodberg, J., and J. J. Dunn. 1988. *ompT* encodes the *Escherichia coli* outer membrane protease that cleaves T7 RNA polymerase during purification. *J. Bacteriol.* **170:**1245–1253.

Hancock, R. E. W. 1984. Alterations in outer membrane permeability. *Annu. Rev. Microbiol.* **38:**237–264.

Hancock, R. W., and V. Braun. 1976. Nature of the energy requirement for the irreversible adsorption of bacteriophages T1 and phi80 to *Escherichia coli*. *J. Bacteriol.* **125:**409–415.

Hara, H., N. Abe, M. Nakakouji, Y. Nishimura, and K. Horiuchi. 1996. Overproduction of penicillin-binding protein 7 suppresses thermosensitive growth defect at low osmolarity due to an *spr* mutation of *Escherichia coli*. *Microb. Drug Resist.* **2:**63–72.

Hara, H., Y. Nishimura, J. I. Kato, H. Suzuki, H. Nagasawa, A. Suzuki, and Y. Hirota. 1989. Genetic analyses of processing involving C-terminal cleavage in penicillin-binding protein 3 of *Escherichia coli*. *J. Bacteriol.* **171:**5882–5889.

Hara, H., Y. Yamamoto, A. Higashitani, H. Suzuki, and Y. Nishimura. 1991. Cloning, mapping, and characterization of the *Escherichia coli prc* gene, which is involved in C-terminal processing of penicillin-binding protein 3. *J. Bacteriol.* **173:**4799–4813.

Hartl, F. U., S. Lecker, E. Scheibel, J. P. Hendrick, and W. Wickner. 1990. The binding of SecB to SecA to SecY/E mediates preprotein targeting to the *E. coli* membrane. *Cell* **63:**269–279.

Hayhurst, A., and W. J. Harris. 1999. *Escherichia coli skp* chaperone coexpression improves solubility and phage display of single-chain antibody fragments. *Protein Exp. Purif.* **15:**336–343.

Hiniker, A., and J. C. A. Bardwell. 2004. *In vivo* substrate specificity of periplasmic disulfide oxidoreductases. *J. Biol. Chem.* **279:**1296–1297.

Hobot, J. A., E. Carlemalm, W. Villiger, and E. Kellenberger. 1984. Periplasmic gel: new concept resulting from the reinvestigation of bacterial cell envelope ultrastructure by new methods. *J. Bacteriol.* **160:**143–152.

Horne, S. M., and K. D. Young. 1995. *Escherichia coli* and other species of the Enterobacteriaceae encode a protein similar to the family of Mip-like FK506-binding proteins. *Arch. Microbiol.* **163:**357–365.

Humphreys, D. P. 2003. Production of antibodies and antibody fragments in *Escherichia coli* and a comparison of their functions, uses and modification. *Curr. Opin. Drug Discov. Devel.* **6:**188–196.

Humphreys, D. P., M. Sehdev, A. P. Chapman, R. Ganesh, B. J. Smith, L. M. King, D. J.

Glover, D. G. Reeks, and P. E. Stephens. 2000. High-level periplasmic expression in *Escherichia coli* using a eukaryotic signal peptide: importance of codon usage at the 5′ end of the coding sequence. *Protein Exp. Purif.* **20:**252–264.

Humphreys, D. P., N. Weir, A. Lawson, A. Mountain, and P. A. Lund. 1996. Co-expression of human protein disulphide isomerase (PDI) can increase the yield of an antibody Fab′ fragment expressed in *Escherichia coli*. *FEBS Lett.* **380:**194–197.

Hwang, C., A. J. Sinskey, and H. F. Lodish. 1992. Oxidized redox state of glutathione. *Science* **257:**1496–1502.

Irvin, R. T., A. K Chatterjee, K. E. Sanderson, and J. W. Costerton. 1975. Comparison of the cell envelope structure of a lipopolysaccharide-defective (heptose-deficient) strain and a smooth strain of *Salmonella typhimurium*. *J. Bacteriol.* **124:**930–941.

Jeong, K. J., and S. Y. Lee. 2000. Secretory production of human leptin in *Escherichia coli*. *Biotechnol. Bioeng.* **67:**398–407.

Jeong, K. J., and S. Y. Lee. 2001. Secretory production of human granulocyte colony-stimulating factor in *Escherichia coli*. *Protein Exp. Purif.* **23:**311–318.

Joly, J. C, W. S. Leung, and J. R. Swartz. 1998. Overexpression of *Escherichia coli* oxidoreductases increases recombinant insulin-like growth factor-1 accumulation. *Proc. Natl. Acad. Sci. USA* **95:**2773–2777.

Kadokura, H., F. Katzen, and J. Beckwith. 2003. protein disulfide bond formation in prokaryotes. *Annu. Rev. Biochem.* **72:**111–135.

Kloser, A., M. Laird, M. Deng, and R. Misra. 1998. Modulations in lipid A and phospholipid biosynthesis pathways influence outer membrane protein assembly in *Escherichia coli* K-12. *Mol. Microbiol.* **27:**1003–1008.

Kolmar, H., P. R. H. Waller, and R. T. Sauer. 1996. The DegP and DegQ periplasmic endoproteases of *Escherichia coli*: specificity for cleavage sites and substrate conformation. *J. Bacteriol.* **178:**5925–5929.

Kumamoto, C. A. 1991. Molecular chaperones and protein translocation across the *Escherichia coli* inner membrane. *Mol. Microbiol.* **5:**19–22.

Kumamoto, C. A., and O. Francetic. 1993. Highly selective binding of nascent polypeptides by an *Escherichia coli* chaperone protein *in vivo*. *J. Bacteriol.* **175:**2184–2188.

Laird, M. W., K. Cope, R. Atkinson, M. Donahoe, K. Johnson, and M. Melick. 2004a. Keratinocyte growth factor-2 production in an *ompT*-deficient *Escherichia coli* K-12 mutant. *Biotechnol. Prog.* **20:**44–50.

Laird, M. W., G. C. Sampey, K. Johnson, D. Zukauskas, J. Pierre, J. S. Hong, B. A. Cooksey, Y. Li, O. Galperina, J. D. Karwoski, and R. N. Burke. 2005. Optimization of BLyS production and purification from *Escherichia coli*. *Protein Exp. Purif.* **39:**237–246.

Laird, M. W., D. Zukauskas, K. Johnson, G. C. Sampey, H. Olsen, J. D. Karwoski, B. A. Cooksey, G. H. Choi, J. Askins, A. Tsai, J. Pierre, and W. Gwinn. 2004b. Production and purification of *Bacillus anthracis* protective antigen from *Escherichia coli*. *Protein Exp. Purif.* **38**:145–152.

Lazar, S. W., and R. Kolter. 1996. SurA assists the folding of *Escherichia coli* outer membrane proteins. *J. Bacteriol.* **178**:1770–1773.

Le Calvez, H., J. M. Green, and D. Baty. 1996. Increased efficiency of alkaline phosphatase production levels in *Escherichia coli* using a degenerate Pe1B signal sequence. *Gene* **170**:51–55.

Levengood-Freyermuth, S. K., E. M. Click, and R. E. Webster. 1993. Role of the carboxyl-terminal domain of TolA in protein import and integrity of the outer membrane. *J. Bacteriol.* **175**:222–228.

Leung, W. S., and J. R. Swartz. 2001. Process for bacterial production of polypeptides. U.S. Patent 6,258,560 B1.

Levy, R., R. Weiss, G. Chen, B. L. Iverson, and G. Georgiou. 2001. Production of correctly folded Fab antibody fragment in the cytoplasm of *Escherichia coli trxB gor* mutants via the coexpression of molecular chaperones. *Protein Exp. Purif.* **23**:338–347.

Link, A. J., D. Phillips, and G. M. Church. 1997. Methods for generating precise deletions and insertions in the genome of wild-type *Escherichia coli*: application to open reading frame characterization. *J. Bacteriol.* **179**:6228–6237.

Liu, J., and C. T. Walsh. 1990. Peptidyl-prolyl cis-trans-isomerase from *Escherichia coli*: a periplasmic homolog of cyclophilin that is not inhibited by cyclosporin A. *Proc. Natl. Acad. Sci. USA* **87**:4028–4032.

Luirink, J., S. High, H. Wood, A. Giner, D. Tollervey, and B. Dobberstein. 1992. Signal-sequence recognition by an *Escherichia coli* ribonucleoprotein complex. *Nature* **359**:741–743.

Lundrigan, M. D., J. H. Lancaster, and C. F. Earhart. 1983. UC-1, a new bacteriophage that uses the *tonA* polypeptide as its receptor. *J. Virol.* **45**:700–707.

Malamy, M. H., and B. L. Horecker. 1964. Release of alkaline phosphatase from cells of *Escherichia coli* upon lysozyme spheroplast formation. *Biochemistry* **3**:1889–1893.

Meerman, H. J., and G. Georgiou. 1994. Construction and characterization of a set of *E. coli* strains deficient in all known loci affecting the proteolytic stability of secreted recombinant proteins. *Bio/Technology* **12**:1107–1110.

Metcalf, W. W., W. Jiang, and B. L. Wanner. 1994. Use of the rep technique for allele replacement to construct new *Escherichia coli* hosts for maintenance of R6Kγ origin plasmids at different copy numbers. *Gene* **138**:1–7.

Miller, J. D., H. D. Bernstein, and P. Walter. 1994. Interaction of *E. coli* Ffh/4.5S ribonucleoprotein and FtsY mimics that of mammalian signal recognition particle and its receptor. *Nature* **367**:657–659.

Misra, R., M. Castillokeller, and M. Deng. 2000. Overexpression of protease-deficient DegP$_{S210A}$ rescues the lethal phenotype of *Escherichia coli* OmpF assembly mutants in a *degP* background. *J. Bacteriol.* **182**:4882–4888.

Missiakas, D., and S. Raina. 1997. Protein folding in the bacterial periplasm. *J. Bacteriol.* **179**: 2465-2471.

Mori, T., K. R. Gustafson, L. K. Pannell, R. H. Shoemaker, L. Wu, J. B. McMahon, and M. R. Boyd. 1998. Recombinant production of cyanovirin-N, a potent human immunodeficiency virus-inactivating protein derived from a cultured cyanobacterium. *Protein Exp. Purif.* **12**:151–158.

Nakae, T., and H. Nikaido. 1975. Outer membrane as a diffusion barrier in *Salmonella typhimurium*. Penetration of oligo- and polysaccharides into isolated outer membrane vesicles and cells with degraded peptidoglycan layer. *J. Biol. Chem.* **250**: 7359–7365.

Neu, H. C., and L. A. Heppel. 1964. The release of ribonuclease into the medium when *Escherichia coli* cells are converted to spheroplasts. *J. Biol. Chem.* **239**:3893–3900.

Neu, H. C., and L. A. Heppel. 1965. The release of enzymes from *Escherichia coli* by osmotic shock and during the formation of spheroplasts. *J. Biol. Chem.* **240**:3685–3692.

Nikaido, H., and M. Vaara. 1985. Molecular basis of bacterial outer membrane permeability. *Microbiol. Rev.* **49**:1–32.

Owens, R. A., and P. E. Hartman. 1986. Export of glutathione by some widely used *Salmonella typhimurium* and *Escherichia coli* strains. *J. Bacteriol.* **168**:109–114.

Parker, C. T., A. W. Kloser, C. A. Schnaitman, M. A. Stein, S. Gottesman, and B. W. Gibson. 1992. Role of the *rfaG* and *rfaP* genes in determining the lipopolysaccharide core structure and cell surface properties of *Escherichia coli* K-12. *J. Bacteriol.* **174**:2525–2538.

Perez-Perez, J., G. Marquez, J. L. Barbero, and J. Gutierrez. 1994. Increasing the efficiency of protein export in *Escherichia coli*. *Bio/Technology* **12**:178–180.

Perez-Perez, J., C. Martinez-Caja, J. L. Barbero, and J. Gutierrez. 1995. DnaK/DnaJ supplementation improves the periplasmic production of human granulocyte-colony stimulating factor in *Escherichia coli*. *Biochem. Biophys. Res. Commun.* **210**:524–529.

Phillips, G. J., and T. J. Silhavy. 1990. Heat-shock proteins DnaK and GroEL facilitate export of LacZ hybrid proteins in *E. coli*. *Nature* **344**:882–884.

Phillips, G. J., and T. J. Silhavy. 1992. The *E. coli ffh* gene is necessary for viability and efficient protein export. *Nature* **359:**744–746.

Pittman, M. S., H. C. Robinson, and R. K. Poole. 2005. A bacterial glutathione transporter (*Escherichia coli* CydDC) exports reductant to the periplasm. *J. Biol. Chem.* **280:**32254–32261.

Pogliano, J., A. S. Lynch, D. Belin, E. C. C. Lin, and J. Beckwith. 1997. Regulation of *Escherichia coli* cell envelope proteins involved in protein folding and degradation by the Cpx two-component system. *Genes Dev.* **11:**1169–1182.

Posfai, G., V. Kolisnychenko, Z. Bereczki, and F. R. Blattner. 1999. Markerless gene replacement in *Escherichia coli* stimulated by a double-strand break in the chromosome. *Nucleic Acids Res.* **27:** 4409–4415.

Prinz, W. A., F. Aslund, A. Holmgren, and J. Beckwith. 1997. The role of the thioredoxn and glutaredoxin pathways in reducing protein disulfide bonds in the *Escherichia coli* cytoplasm. *J. Biol. Chem.* **272:** 15661–15667.

Pritchard, M. P., R. Ossetian, D. N. Li, C. J. Henderson, B. Burchell, C. R. Wolf, and T. Friedberg. 1997. A general strategy for the expression of recombinant human cytochrome P450s in *Escherichia coli* using bacterial signal peptides: expression of CYP3A4, CYP2A6, and CYP2E1. *Arch. Biochem. Biophys.* **345:**342–354.

Pugsley, A. P. 1993. The complete general secretory pathway in gram-negative bacteria. *Microbiol. Rev.* **57:**50–108.

Pugsley, A. P., O. Francetic, K. Hardie, O. M. Possot, N. Sauvonnet, and A. Seydel. 1997. Pullulanase: model protein substrate for the general secretory pathway of gram-negative bacteria. *Folia Microbiol.* **42:**184–192.

Qiu, J., J. R. Swartz, and G. Georgiou. 1998. Expression of active human tissue-type plasminogen activator in Escherichia coli. *Appl. Env. Microbiol.* **64:**4891–4896.

Raivio, T. L., and T. J. Silhavy. 2001. Periplasmic stress and ECF sigma factors. *Annu. Rev. Microbiol.* **55:**591–624.

Rathore, D., S. K. Nayak, and J. K. Batra. 1996. Expression of ribonucleolytic toxin restrictocin in *Escherichia coli*: purification and characterization. *FEBS Lett.* **392:**259–262.

Ribes, V., K. Romisch, A. Giner, B. Dobberstein, and D. Tollervey. *E. coli* 4.5S RNA is part of a ribonucleoprotein particle that has properties related to signal recognition particle. *Cell* **63:**591–600.

Richarme, G., and T. D. Caldas. 1997. Chaperone properties of the bacterial periplasmic substrate-binding proteins. *J. Biol. Chem.* **272:**15607–15612.

Rietsch, A., and J. Beckwith. 1998. The genetics of disulfide bond metabolism. *Annu. Rev. Genet.* **32:** 163–184.

Rinas, U., and F. Hoffmann. 2004. Selective leakage of host-cell proteins during high-cell density cultivation of recombinant and non-recombinant *Escherichia coli*. *Biotechnol. Prog.* **20:**679–687.

Rudolph, R., and S. Fischer. 1990. Process for obtaining renatured proteins. U.S. Patent 4,933,434.

Sandee, D., S. Tungpradabkul, Y. Kurokawa, K. Fukui, and M. Takagi. 2005. Combination of Dsb coexpression and an addition of sorbitol markedly enhanced soluble expression of single-chain Fv in *Escherichia coli*. *Biotechnol. Bioeng.* **91:**418–424.

Schaffner, J., J. Winter, R. Rudolph, and E. Schwarz. 2001. Cosecretion of chaperones and low-molecular-size medium additives increases the yield of recombinant disulfide-bridged proteins. *Appl. Env. Microbiol.* **67:**3994–4000.

Schnaitman, C. A., and J. D. Klena. 1993. Genetics of lipopolysaccharide biosynthesis in enteric bacteria. *Microbiol. Rev.* **57:**655–682.

Scholz, C., P. Schaarschmidt, A. M. Engel, H. Andreas, U. Schmitt, E. Faatz, J. Balbach, and F. X. Schmid. 2005. Functional solubilization of aggregation-prone HIV envelope proteins by covalent fusion with chaperone modules. *J. Mol. Biol.* **345:**1229–1241.

Seluanov, A., and E. Bibi. 1997. FtsY, the prokaryotic signal recognition particle receptor homologue, is essential for biogenesis of membrane proteins. *J. Biol. Chem.* **272:**2053–2055.

Silber, K. R., and R. T. Sauer. 1994. Deletion of the *prc* (*tsp*) gene provides evidence for additional tail-specific proteolytic activity in *Escherichia coli* K-12. *Mol. Gen. Genet.* **242:**237–240.

Simmons, L. C., D. Reilly, L. Klimkowski, T. S. Raju, G. Meng, P. Sims, K. Hong, R. L. Shields, L. A. Damico, P. Rancatore, and D. G. Yansura. 2002. Expression of full-length immunoglobulins in *Escherichia coli*: rapid and efficient production of aglycosylated antibodies. *J. Immunol. Methods* **263:**133–147.

Simmons, L. C., and D. G. Yansura. 1996. Translational level is a critical factor for the secretion of heterologous proteins in *Escherichia coli*. *Nat. Biotechnol.* **14:**629–634.

Spiess, C., A. Beil, and M. Ehrmann. 1999. A temperature-dependent switch from chaperone to protease in a widely conserved heat shock protein. *Cell* **97:**339–347.

Squires, C., D. Retallack, L. Chew, T. Ramseier, J. C. Schneider, and H. Talbot. 2004. Heterologous protein production in *P. fluorescens*. *Bioproc. Int.* **2:**54–59.

Strauch, K., and J. Beckwith. 1988. An *Escherichia coli* mutation preventing degradation of abnormal periplasmic proteins. *Proc. Natl. Acad. Sci. USA* **85:** 1576–1580.

Studier, F. W., A. H. Rosenberg, J. J. Dunn, and J. W. Dubendorff. 1990. Use of T7 RNA polymerase to direct expression of cloned genes. *Methods Enzymol.* **185:**60–69.

Sugimura, K., and T. Nishihara. 1988. Purification, characterization, and primary structure of *Escherichia coli* protease VII with specificity for paired basic residues: identity of protease VII and OmpT. *J. Bacteriol.* **170:**5625–5632.

Suzuki, H., W. Hashimoto, and H. Kumagai. 1999. Glutathione metabolism in *Escherichia coli*. *J. Mol. Catal. B: Enzymat.* **6:**175–184.

Talmadge, K., and W. Gilbert. 1982. Cellular location affects protein stability in *Escherichia coli*. *Proc. Natl. Acad. Sci. USA* **79:**1830–1833.

Taura, T., T. Baba, Y. Akiyama, and K. Ito. 1993. Determinants of the quantity of the stable SecY complex in the *Escherichia coli* cell. *J. Bacteriol.* **175:**7771–7775.

Thome, B. M., and M. Muller. 1991. Skp is a periplasmic *Escherichia coli* protein requiring SecA and SecY for export. *Mol. Microbiol.* **5:**2815–2817.

Tong, W.-Y., S.-J. Yao, Z.-Q. Zhu, and J. Yu. 2001. An improved procedure for production of epidermal growth factor from recombinant *E. coli*. *Appl. Microbiol. Biotechnol.* **57:**674–679.

Traurig, M., and R. Misra. 1999. Identification of bacteriophage K20 binding regions of OmpF and lipopolysaccharide in *Escherichia coli* K-12. *FEMS Microbiol. Lett.* **181:**101–108.

Ulbrandt, N. D., J. A. Newitt, and H. D. Bernstein. 1997. The *E. coli* signal recognition particle is required for the insertion of a subset of inner membrane proteins. *Cell* **88:**187–196.

Von Heijne, G. 1990. The signal peptide. *J. Membr. Biol.* **115:**195–201.

Wadensten, H., A. Ekebacke, B. Hammarberg, E. Holmgren, C. Kalderen, M. Tally, T. Moks, M. Uhlen, S. Josephson, and M. Hartmanis. 1991. Purification and characterization of recombinant human insulin-like growth factor II (IGF-II) expressed as a secreted fusion protein in *Escherichia coli*. *Biotechnol. Appl. Biochem.* **13:**412–421

Waller, P. R. H., and R. T. Sauer. 1996. Characterization of *degQ* and *degS*, *Escherichia coli* genes encoding homologs of the DegP protease. *J. Bacteriol.* **178:**1146–1153.

Walton, T. A., and M. C. Sousa. 2004. Crystal structure of Skp, a prefoldin-like chaperone that protects soluble and membrane proteins from aggregation. *Mol. Cell* **15:**367–374.

Watanabe, M., and G. Blobel. 1989a. SecB functions as a cytosolic signal recognition factor for protein export in *E. coli*. *Cell* **58:**695–705.

Watanabe, M., and G. Blobel. 1989b. Binding of a soluble factor of *Escherichia coli* to preproteins does not require ATP and appears to be the first step in protein export. *Proc. Natl. Acad. Sci. USA* **86:**2248–2252.

Webster, R. E. 1991. The *tol* gene products and the import of macromolecules into *Escherichia coli*. *Mol. Microbiol.* **5:**1005–1011.

Weidel, W., and E. Kellenberger. 1955. The *E. coli* B-receptor for the phage T5. II. Electron microscopic studies. *Biochim. Biophys. Acta* **17:**1–9.

Weiss, J. B., P. H. Ray, and P. J. Bassford, Jr. 1988. Purified SecB protein of Escherichia coli retards folding and promotes membrane translocation of the maltose-binding protein *in vitro. Proc. Natl. Acad. Sci. USA* **85:**8978–8982.

Wild, J., E. Altman, T. Yura, and C. A. Gross. 1992. DnaK and DnaJ heat shock proteins participate in protein export in *Escherichia coli*. *Genes Dev.* **6:** 1165–1172.

Wild, J., P. Rossmeissl, W. A. Walter, and C. A. Gross. 1996. Involvement of the DnaK-DnaJ-GrpE chaperone team in protein secretion in *Escherichia coli. J. Bacteriol.* **178:**3608–3613.

Winter, J., H. Lilie, and R. Rudolph. 2002. Renaturation of human proinsulin—a study on refolding and conversion to insulin. *Anal. Biochem.* **310:**148–155.

Wulfing, C., and A. Pluckthun. 1994. Protein folding in the periplasm of *Escherichia coli*. *Mol. Microbiol.* **12:**685–692.

Wunderlich, M., and R. Glockshuber. 1993. *In vivo* control of redox potential during protein folding catalyzed by bacterial protein disulfide-isomerase (DsbA). *J. Biol. Chem.* **268:**24547–24550.

Xiong, X., J. N. Deeter, and R. Misra. 1996. Assembly-defective OmpC mutants of *Escherichia coli* K-12. *J. Bacteriol.* **178:**1213–1215.

Xue, X., Z. Wang, Z. Yan, J. Shi, W. Han, and Y. Zhang. 2005. Production and purification of recombinant human BLyS mutant from inclusion bodies. *Protein Exp. Purif.* **42:**194–199.

Young, K., and L. L. Silver. 1993. Leakage of periplasmic enzymes from *envA1* strains of *Escherichia coli. J. Bacteriol.* **173:**3609–3614.

Young, R., and U. Blasi. 1995. Holins: form and function in bacteriophage lysis. *FEMS Microbiol. Rev.* **17:**191–205.

Zhan, X., J. Gao, C. Jain, M. J. Cieslewicz, J. R. Swartz, and G. Georgiou. 2004. Genetic analysis of disulfide bond isomerization in *Escherichia coli*: expression of DsbC is modulated by RNase E-dependent mRNA processing. *J. Bacteriol.* **186:**654–660.

Zhan, X., M. Schwaller, H. F. Gilbert, and G. Georgiou. 1999. Facilitating the formation of disulfide bonds in the *Escherichia coli* periplasm via coexpression of yeast protein disulfide isomerase. *Biotechnol. Prog.* **15:**1033–1038.

Zhu, Z., G. Zapata, R. Shalaby, B. Snedecor, H. Chen, and P. Carter. 1996. High level secretion of a humanized bispecific diabody from *E. coli*. *Bio/Technology* **14:**192–196.

PERIPLASMIC EXPRESSION OF ANTIBODY FRAGMENTS

David P. Humphreys

21

GENERAL OVERVIEW

Antibodies and their fragments have been found to be of enormous academic, therapeutic, and commercial value. Their modular protein structure has made them amenable to functional diversification by protein engineering. Their success has driven the investigation and comparison of numerous expression technologies and host organisms as diverse as *Escherichia coli*, yeasts, mammalian cells, and transgenic plants and animals. These expression technologies have been developed to increase the simplicity, speed, yield, and scale of production and reduce process running costs while maintaining consistent protein quality. The question to be addressed in this chapter is why one might choose to express antibody fragments in the periplasm of *E. coli* rather than using the multitude of other available organisms? Experience has shown that *E. coli* has much to offer over other expression hosts. *E. coli* offers a combination of the potential for large-scale, low-cost, and high-speed protein production built on decades of familiarity with the genetics, molecular biology, physiology, and regulatory issues of this relatively simple organism. These advantages are especially pronounced

when the protein to be expressed is an antibody fragment since *E. coli* is equally capable of being used for small laboratory-scale expression studies with a minimal investment of time, knowledge, and capital through to large-scale industrial productions (Humphreys and Glover, 2001). This chapter focuses on the biology and practical techniques that can help in the expression of antibody fragments in *E. coli*.

E. coli is unable to perform N-linked or O-linked glycosylation. In addition, *E. coli* is not especially well adapted for the expression of very large, disulfide bond–rich and multimeric proteins. Together these features mean that *E. coli* cannot be used to make full-length antibodies with a naturally glycosylated Fc fragment. As we will see, however, expression limitations posed by *E. coli* have been overcome by engineering of the host or of the antibody product to augment or avoid certain of these biological deficiencies.

CHOICE OF THE PERIPLASM AS THE SUBCELLULAR TARGET FOR ANTIBODY FRAGMENT PRODUCTION

There are three locations for the expression of soluble proteins in *E. coli*: the cytoplasm, the periplasm, and secretion into the extracellular media. In addition, proteins can also be ex-

David P. Humphreys, UCB-Celltech, Slough SL1 4EN, United Kingdom.

The Periplasm
Edited by Michael Ehrmann © 2007 ASM Press, Washington, D.C.

pressed in an insoluble form in the cytoplasm or periplasm. The periplasm is often preferred for expression of soluble antibody fragments and some of the factors for this choice are discussed below.

The Protein Contains Essential Disulfide Bonds

Since the periplasm is the naturally oxidizing subcellular location, peptides should be directed to or through the periplasm if they are to acquire these bonds. *E. coli* strains that are mutated in the cytoplasmic proteins thioredoxin (*trxB*) and glutaredoxin (*gor*) are, however, also capable of disulfide formation in the cytoplasm (Prinz et al., 1997). These strains suffer a serious reduction in growth rate unless the double *trxB/gor* mutation is accompanied by a compensating mutation in the *ahpC* gene that encodes for a peroxiredoxin (Bessette et al., 1999). Even so these strains are found to be incompatible with use of tetracycline as a selection agent (Jurado et al., 2002).

Retention of a Correct N Terminus

Expression in the cytoplasm results in the variable retention of the N-terminal initiator methionine (Hexham et al., 2001). This may be a problem for antibody fragments destined for a commercial or therapeutic use or where it might alter antigen affinity or immunogenicity. Since peptides are directed to the periplasm by an accurately cleaved N-terminal extension "signal peptide," the problem of N-terminal heterogeneity can be avoided. Further information on signal peptide function and choice is given later in this chapter.

Exposure of the Protein to Extracellular Chemical Agents

The periplasm is accessible to media components ≤500 Da in size, hence the protein can be exposed to chemical agents such as salts, pH, redox molecules, and mild chaotropes. These can be useful in aiding the folding, modification, or experimental investigation of the protein of interest. Additional information on

the use of media additions is given later in this chapter.

Simplification of Protein Purification

Proteins can become highly diluted when secreted into the culture media. This results in high-volume capture or filtration/concentration steps. Also, some proteins can become physically damaged by the shear forces generated by some microbial fermentation processes (Harrison et al., 1998; Mukherjee et al., 2004; Chou et al., 2005). Hence, the periplasm acts as a convenient means of encapsulation of the protein of interest. Simple harvest of the cells by centrifugation or filtration can result in a significant volumetric concentration. Following harvest, the periplasmic proteins can be released with varying degrees of selectivity by different chemical or mechanical extraction methods. The periplasm is the site of expression for only ~1/10 of the soluble proteins of *E. coli* and does not contain any DNA or RNA that can interfere with purification. Therefore combination of periplasmic expression with carefully considered harvest and extraction regimes can result in a very useful concentration and partial purification/enrichment of the periplasmic protein of interest.

EXPRESSION OF ANTIBODY FRAGMENTS IN THE PERIPLASM

Uses of Antibody Fragments

The principal characteristic that makes antibodies useful is the amazing specificity and affinity of their interaction with antigen. An antigen can be a linear or three-dimensional peptide motif, small nonpeptidic hapten, or complex chemical structure. This property makes antibodies useful for laboratory research and industrial and clinical applications. For example, in the laboratory, antibodies can be used in immunoblotting, enzyme-linked immunosorbent assay (ELISA), affinity purification, and reverse cloning methods. In the industrial setting antibodies can be the critical detection reagent in a wide range of highly sensitive detection and measurement kits, in machines, and for use in bioremediation efforts. In the clinic, antibodies can be used ex vivo in test kits for the detection of disease-related antigens and in blood cell immunotherapies. In vivo antibodies have found uses as imaging agents, for therapeutic targeting, neutralization and signaling, and as catalytic antibodies (see Glennie and Johnson, 2000; Gavilondo and Larrick, 2000, for reviews).

For many of these applications the presence of glycosylation in the Fc constant regions are either unnecessary or are actually unhelpful. Therefore, in these applications aglycosylated full-length antibodies or one of the various smaller formats of antibody fragment may be used. In all of these cases, expression in the periplasm of *E. coli* must be considered as one method of primary interest.

Antibody Structure

A representation of immunoglobulin G1 (IgG1) as a prototypical antibody including the well-established nomenclatures for the various domains and subdomains is shown in Fig. 1.

Antibodies are composed of several globular domains that are rich in β-sheet and have a conserved disulfide bond—called the Ig fold (Padlan, 1994; Halaby et al., 1999). This core domain is inherently stable in its folded form, a quality that makes it suitable for expression in the periplasm of *E. coli* since it is essentially protease insensitive when fully folded and can withstand some of the physicochemical insults exerted during fermentation, primary recovery, and purification. IgG1 antibodies are composed of 12 highly related Ig folds that are connected by short flexible interdomain "linkers" known as elbow regions or longer highly flexible/functional linkers known as hinges.

This modular design has been utilized by protein engineers whose two main aims have been to design minimal antigen-binding functions (to reduce expression problems), or antibody fragments with altered functionality such as bispecificity. Most antibodies are composed of two light chains and two heavy chains. Each

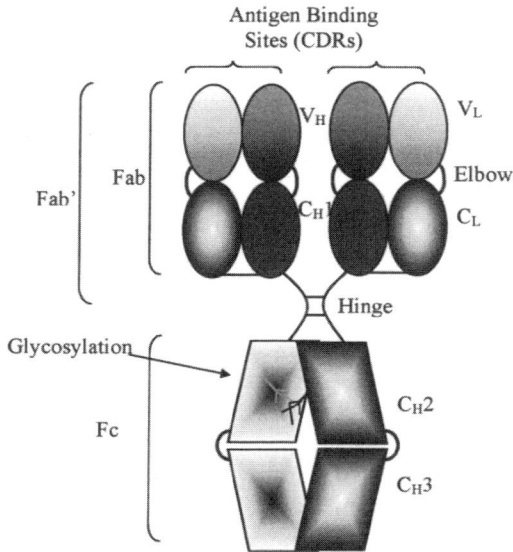

FIGURE 1 Schematic representation of full-length IgG1 (human γ1 isotype). Native IgG consists of two heavy-chain and two light-chain polypeptides. Each heavy chain has four domains: one variable (V_H) and three constant (C_H1-hinge-C_H2-C_H3). Each light chain has two domains: one variable (V_L) and one constant (C_L). The two heavy chains are covalently linked by disulfide bonds between hinges and each light chain is attached to a heavy chain by a disulfide bond. In native human IgG carbohydrate is attached to each heavy chain at Asn297 of the C_H2 domains.

chain has a "variable" N-terminal region (fragment variable, Fv) composed of two Ig folds, vL and vH, respectively. The variable regions each have three hypervariable loops called CDRs (complementarity-determining regions) at one end of the protein that together form a common six-loop exposed antigen-binding surface. The Fv region is followed by a different number of constant regions, 1 in the case of light chain and 3 in the case of heavy chain. The two most N-terminal constant regions (cKappa or lambda for the light chain and CH1 for the heavy chain) appear to provide some physical support for the Fv to ensure that the antigen-binding Fv region is kept in the correct tertiary structure. These four N-terminal domains can be released from full-length

IgG by digestion with the protease papain. The fragment produced is called the Fab fragment (Fragment, antigen binding). Pepsin cuts slightly to the C terminus of papain and so releases a dimeric Fab′ fragment, or F(ab′)$_2$. Both enzymes also release the C-terminal four Ig fold constant domain or Fc fragment (Fragment, crystallizable). The Fc fragment is N-glycosylated in the CH2 domain and is found to confer the bulk of effector functions (along with the hinge and the sugar group) such as serum half-life, complement activation, and recruitment of immune cells. In addition to these antibodies composed of four polypeptides, there are certain species such as the camelids (camels, llama) and sharks that have additional classes of antibody composed of just a dimer of heavy chains. In these "heavy chain" antibodies all of the antigen-binding activity is provided by just 3 CDRs of the V$_{HH}$. As we shall see, smaller antigen-binding fragments are attractive to the antibody expression scientist.

Types of Antibody Fragments

The altered functionalities achievable with smaller fragments (e.g., monomeric binding, lack of effector functions, rapid tissue penetration, rapid systemic clearance) have been found to be increasingly useful for a variety of experimental, industrial, and clinical purposes. Examples of these and other more highly engineered structures are shown in Table 1. The smaller sized fragments have found good fit with *E. coli* expression because, in general, the organism has not evolved to secrete large, multidomain, multimeric proteins into its periplasm. Evolution has already honed the structure of antibodies in terms of the key basic parameters that are desirable in any highly expressed protein: thermal stability, solubility, protease resistance, and ability to acquire the native state efficiently. In heterologously expressed Fab′ fragments these characteristics can be faithfully reproduced. In other fragments, the protein engineer tries to make best use of the intrinsic strength of the antibody structure. However, the desire to create useful highly engineered

structures and altered functionalities has necessarily resulted in some trade-off from these good basic physicochemical characteristics. The results of this trade-off can be, for example, increased exposure to proteolysis, increased aggregation during production in vivo and storage in vitro, and reduced physical stability. The aim of the expression scientist is to minimize the impact of these trade-offs to maximize the productive capability of the expression host.

Fab′ fragments were among the first to have been produced in *E. coli*. These molecules are encoded by the two separate genes of the light and heavy chain. The exact point chosen for termination of the heavy chain can result in the production of truly monomeric Fab fragments that contain none of the hinge cysteines (Better et al., 1988) or if the stop codon is encoded after one or more of the hinge cysteines, then the Fab′ remains capable of forming dimeric (more avid) F(ab′)$_2$ via interchain disulfide bonds (Better et al., 1993). Where the dimeric F(ab′)$_2$ species is the desired product, then further hinge engineering is found to be able to increase the amount formed both in vivo in the periplasm and in vitro upon reduction/oxidation (Rodriguez et al., 1993; Humphreys et al., 1997, 1998). The attraction of the Fab or Fab′ (in contrast to some of the more heavily engineered antibody fragments) is that it is substantially native in sequence and is physically robust. This means that it has less opportunity to offer immunogenic sequences if used therapeutically and suffers from very little of the expression trade-off already mentioned. Fab′ can easily be produced at levels required for laboratory experimentation without optimization and at up to 1 to 2 g/liter in optimized fermentations (Carter et al., 1992). Fab′ is thermally and chemically stable up to ≥60°C and ≥2 M GdHCl (Röthlisberger et al., 2005), which makes it possible to pass them through tough bioprocessing and modification regimes. The presence of one or more hinge cysteines enables chemical modification with an agent such as polyethylene glycol (PEG) (Chapman et al., 1999), toxin (Melton 1996; Garnett, 2001), radionuclide chelating group (King et al., 1994),

TABLE 1 Antibody fragments produced in *E. coli*[a]

Name	Structure	Strengths	Weaknesses	Reference(s)
Single domain Ab, sdAb		Smallest antigen-binding unit, of the camelid vHH or human vH origin. Long CDR3 of camelids can interfere with enzyme active sites. High levels of production, physical stability.	Only 3 CDRs available for Ag binding, purification by tag Protein A or L.	Riechmann and Muyldermans, 1999; Lauwereys et al., 1998; Ewert et al., 2002
Fv and dsFv		Small single domains expressed well in *E. coli*. Can be stabilized by interdomain disulfide bonds and can form the basis of peptide-linked dimeric proteins.	Fv are relatively unstable and can suffer from domain exchange unless the domain interface is engineered.	Skerra and Plückthun, 1988; Schmiedl et al., 2000b
scFv		High yield. Small size/good tumor penetration.	In vitro aggregation (storage and affinity measurement issues). Potential immunogenicity of linkers, peptide junctions, exposed surfaces. Purification with affinity tags (immunogenicity). Low thermal stability.	Bird et al., 1988; Huston et al., 1988; Nieba et al., 1997; Arndt et al., 1998
Diabody		Small size and avidity beneficial for tumor penetration. Potential for bispecificity that can be further elaborated with peptide linking. High yields demonstrated.	Purification issues as for scFv. Purification problems increased if bispecific protein is required.	Holliger et al., 1997; Zhu et al., 1996, 1997; Völkel et al., 2001
Triabody		Small size and avidity beneficial for tumor penetration. Potential for multispecificity.	High yields not demonstrated. Purification issues as for scFv with increased problems if multispecific protein is required.	Atwell et al., 1999; reviewed by Todorovska et al., 2001
scFv-zipper		High yields possible. Small size and avidity beneficial for tumor penetration. Potential for heterodimerization.	Potential immunogenicity of linkers, peptide junctions, exposed surfaces, and coil. Purification with affinity tags (immunogenicity).	Pack and Plückthun, 1992; Rodrigues et al., 1992; Kostelny et al., 1992; Richter et al., 2001
Minibody (scFv linked to C_H3)		Intermediate size and avidity confer good balance between serum half-life and avidity for tumor penetration.	Potential immunogenicity of linkers, peptide junctions, exposed surfaces.	Hu et al., 1992; Lo et al., 1992
ScAb (scFv linked to C_L)		More stable than scFv	Potential immunogenicity of linkers, peptide junctions, exposed surfaces. Purification with affinity tags. Low thermal stability.	Grant et al., 1995; Dooley et al., 1998

(Continued)

TABLE 1 *Continued*

Name	Structure	Strengths	Weaknesses	Reference(s)
Linear F(ab')$_2$		Production of an avid F(ab')$_2$ in a single polypeptide/process.	Low yield. Purification of full-length F(ab')$_2$ difficult. Steric obstruction of second antigen-binding site.	Zapata et al., 1995
Bispecific miniantibodies (scFv linked to both C$_L$ and C$_H$1)		Potential for producing an avid or bispecific F(ab')$_2$ in a single polypeptide/process.	Potential immunogenicity of linkers, peptide junctions, exposed surfaces. Purification with affinity tags (immunogenicity). Potential for C$_L$-driven homodimers. Low yield.	Müller et al., 1998
Fab'		High yielding. Readily purified. Demonstrated in vitro modifications through hinge cysteine. Basis for dimeric and bispecific F(ab')$_2$.	Monomeric binding. Short serum half-life of unmodified Fab'.	Carter et al., 1992; Forsberg et al., 1997; Humphreys et al., 1998; Chapman et al., 1999; Glennie et al., 1987
Kappabody		Bispecific binding in a simple robust expressible backbone	CDR engineering may not always be successful, formation of inappropriate homodimers, occlusion of some antigens from double binding.	Ill et al., 1997
Fab-scFv		Bispecificity and potential for trispecificity	Potential immunogenicity of linkers, peptide junctions, and exposed surfaces. Purification with affinity tags (immunogenicity). Low yield. Steric problems for 2nd antigen binding demonstrated.	Lu et al.; 2002

Key for Table 1:
Disulfide bond ▬

Peptide linker ╱

Antigen A ◯

Antigen B ◇

[a]Adapted with permission from The Thomson Corporation and David P. Humphreys: Production of antibodies and antibody fragments in *Escherichia coli* and a comparison of their functions, uses and modification. *Curr. Opin. Drug Disc. Dev.* (2003) **6**(2):188–196. Copyright 2003, The Thomson Corporation.

or homo- and heterodimerization (Glennie et al., 1987).

Somewhat reduced in size from Fab' is the Fv domain. Fv domains can be expressed as two separate noncovalently associated v-regions. However, due to the relatively low-affinity interaction between vL and vH the association between the two polypeptides can be transitory at low protein concentrations. One approach to solve this problem has been to engineer a cysteine into each v-region such that they can form a stabilizing disulfide bond (Glockshuber et al., 1990). This approach causes a reduction in yield and does not ensure that only vL:vH heterodimers are formed. A more practical solution to this problem that has found widespread use involves linking the two v-regions with a short 15- to 20-amino-acid (aa) linker (Bird et al, 1998; Huston et al., 1988). The peptide is most typically 3 or 4 repeats of (GGGGS)n in an aim to combine sufficient length, flexibility, and hydrophilicity. Other linker sequences have been tested to provide improved expression and DNA-cloning restriction sites (Alfthan et al., 1995; Freund et al., 1993). The linker approach also makes the expression cassette a single gene, which simplifies optimization of expression. scFv are generally expressed at reasonable levels because of their small size and so have found favor with many academic researchers. There has been considerable effort to try and select or engineer Fv frameworks that have higher intrinsic expression and stability properties (reviewed by Wörn and Plückthun, 2001). The findings of such studies can also have a beneficial impact on the expression of other types of antibody fragments.

Shortening of the linker to 5 to 12 aa causes a steric restriction of the relative positions of the v-regions such that they are forced into a dimeric structure that is commonly called a diabody. If just one v-region pair is used, then the diabody formed exhibits homodimeric binding, whereas if two different v-region pairs are used, then the result is a bispecific diabody (Holliger et al., 1997). If the linker is shortened further to 0 to 2 aa, then the steric restriction is such that a trimeric structure known as a triabody is produced (Atwell et al., 1999). Hence with scFv, simple changes in linker length can result in a spectrum of size, avidity, and multispecificity, making them very versatile and well-used antibody fragments. scFv-based antibody fragments make excellent reagents and also have found special favor in

the therapeutic arena as imaging agents since their small size and potential for avidity mean that they can achieve a good target tissue/blood ratio very rapidly. This in an important safety feature when powerful radionuclides are used, and it also means that dosing and imaging can be performed within a single hospital visit.

There are multiple further examples of antibody constructs (shown in Table 1) that have been engineered for altered serum half-life, avidity, and bi- or multispecificity. Almost all of them are based on the key structural and functional components found in Fab' and scFv fragments: use of peptide linkers to connect or sterically restrain Ig domains, use of interchain disulfides to improve protein robustness, and use of homo- and hetero-dimerization domains such as cKappa:cKappa or cKappa:CH1. For all of the individual merit that they each contain, a review of the literature shows that increasing the complexity or "nonnaturalness" of the proteins has tended to lead to decreasing yields in *E. coli*, presumably due to increased proteolysis, aggregation, and translocation issues. Many of these proteins are effectively only of academic interest since they are produced in low quantities (micrograms to a few milligrams per liter) and, due to their complexity, must be purified by using affinity tag or antigen-based regimes that can only reasonably be used on a small scale. They are not discussed in any greater depth in this chapter.

One expression advance of note is that of the production of full-length aglycosylated IgG (Simmons et al., 2002a). Aglycosylated IgG lacks, for the greater part, effector functions except that binding to FcRn is retained. Thus these proteins have the long serum half-life characteristic of antibodies that can make them useful for in vivo therapeutic applications. One early report had shown that the Fc fragment was extensively proteolyzed if retained within the periplasm (Kitai et al., 1988), but here one v-region pair has been expressed as IgG1, IgG2, and IgG4b isotypes while five different v-region pairs have been successfully expressed on the IgG1 scaffold.

Combination of fine balancing of light-chain and heavy-chain expression, use of proteolytically deficient strains, and high-cell-density fermentation has enabled purified yields of IgG of up to 150 mg/liter. The light-chain and heavy-chain genes were under the independent control of two *phoA* promoters. Fine control over translation initiation strength for both polypeptides was exerted by codon differences in the signal peptide-coding region. Balance of LC and HC has also been shown to be important in the optimization of Fab' expression (Humphreys et al., 2002). Furthermore, yields of up to 881 mg/liter of an IgG1 have been achieved with the additional coexpression of DsbA and C, while coexpression of FkpA was also found to have some positive effect (Simmons et al., 2002b). Mutation of the hinge cysteines to serine also enabled 2- to 4-fold increases in periplasmic yield of an IgG1 (Reilly and Yansura, 2004).

PRACTICAL CONSIDERATIONS FOR MAXIMIZING PERIPLASMIC EXPRESSION

Plasmid Factors
An important factor for enabling useful levels of protein expression is good selection or design of the expression plasmid. The key areas considered here are: (i) plasmid copy number/origin of replication, (ii) constitutive versus inducible promoter, (iii) resistance/selection marker, and (iv) design of the coding region.

Copy Number/Origin of Replication
The optimal assembly of plasmid components will be affected by the characteristics of the protein of interest and the type of expression regime to be used. These have a direct impact on the choice of origin of replication. For example, small-scale (5 to 1,000 ml) and rapid (induction time of a few hours, overnight) expression methods may be best suited by a high-copy-number plasmid (100 to 500 copies per cell, e.g., pUC-based plasmids). By contrast, expression regimes that are longer in duration or where the intention is to express an antibody fragment that may be somewhat toxic to the

cell may favor lower copy numbers (20 to 50 copies per cell, e.g., pACYC) (Chang and Cohen, 1978). In extreme conditions a copy number of 1 can be achieved with single integration events into the F' plasmid or chromosome. At the other extreme, runaway replication plasmids can be used to give copy numbers of ~1,000 (Nordström and Uhlin, 1992). Very-high-copy number plasmids can be especially useful when combined with a "leaky" promoter such as *trp* (Hallewell and Emtage, 1980) that does not require the addition of exogenous chemical inducer. The RK2 replicon has also been used to give controllable low, medium, and high plasmid copy numbers to enable optimization of copy number (Sletta et al., 2004).

Promoter Choice
Choice of promoter is perhaps the most obvious plasmid variable to be considered. An excellent and detailed review of promoters can be found elsewhere (Makrides, 1996) so here we limit ourselves to discussion of the general concepts surrounding promoter choice and focus on some key examples.

One can divide promoters into those that are constitutively expressed and those that are inducible. Inducible promoters are perhaps more preferable since biomass can be accumulated prior to induction of protein expression, thereby increasing volumetric yield. Inducible promoters can be further divided into those induced by consumable agents (e.g., lactose, arabinose), by nonmetabolizable agents (e.g., IPTG, IAA, benzoic acid, Hg^{2+}), or by culture conditions (change in temperature, pH, osmotic stress, depletion of media component such as tryptophan or phosphate).

In practice, one finds that each promoter has both pros and cons. For example, metabolizable components generally have the benefits of being natural, safe, and cheap compounds, but since they are consumed during the induction process their levels may need to be carefully monitored or controlled. Nonmetabolizable agents may not suffer in this regard, but the molecules themselves can be mildly toxic to bacteria. Induction by change of culture conditions can also be problematical. For example, where depletion of a media component is being used, it may be difficult to ensure depletion at the optimal time during a fermentation process, or indeed on a small scale to know when or if depletion/induction has actually occurred. The control of physical parameters such as temperature, pH, or salt concentration enables greater temporal control, but such approaches are generally deemed to be difficult to implement satisfactorily at large scale and may also induce undesirable or variable stress responses. For example, the rapid heating or cooling of large-volume fermenters may not be physically possible from an engineering perspective and rapid and pronounced changes in critical process parameters such as pH or ionic strength could clearly have profound effects on the general health and productive capacity of the cells.

On balance, the combined use of the *tac* promoter with either lactose or IPTG as inducers is probably the most widely used. The *lac* promoter uses the wild-type -35 and -10 sequence from the lactose operon while the stronger *tac* promoter uses the -35 sequence of *trp* and the -10 sequence of *lac*. There are two different *tac* sequences that differ slightly in sequence and strength of induction: *tac*I and *tac*II being 11 times and 7 times stronger, respectively, than *lac* and $\sim$3 times stronger than *trp* (de Boer et al., 1983). The *phoA* promoter along with depletion of phosphate has also been used to good effect for the production of antibody fragments (Carter et al., 1992; Simmons et al., 2002a). Depletion of PO_4 induces a whole set of proteins in the *pho* regulon (Torriani, 1990) including alkaline phosphatase (PhoA) and results in a much reduced cell division rate due to its metabolic importance. Reduction of free PO_4 to 0.1 to 600 μM induces the phoA promoter. Amino acid substitutions in the periplasmic scavenger of free PO_4 (PhoS/PstS) are able to desensitize this system such that induction occurs at higher concentrations of free PO_4, something that may be desirable for practical reasons in fermentation processes (Bass and Swartz, 1994).

Sometimes the key feature defining the choice of promoter is the need for very tightly regulated control prior to induction. Examples might include production of semilethal proteins or in vivo turnover/depletion studies. Promoters that are very tightly controlled include *tetA* (tetracycline; Skerra 1994), *araBAD* (arabinose; Greenfield et al., 1978; Guzman et al., 1995; Morgan-Kiss et al., 2002), and the T7/pLys/BL21 (DE3) system (IPTG; Studier, 1991).

For very strong and rapid induction the powerful T7 promoter is useful. In the usual system the gene to be expressed is under the direct control of a T7 promoter. The gene for the T7 RNA polymerase is then typically put under the control of the lacUV5 promoter placed on the chromosome of a recipient strain. IPTG is used to induce the production of the T7 RNA polymerase and hence indirectly the gene of interest. Hence, this is an amplified induction system. In one adaptation, the T7 RNA polymerase is under the control of the heat-inducible λP_L and λP_R tandem promoter (Chao et al., 2002a), and in another, the T7 RNA polymerase is under the control of the *ara*BAD promoter (Chao et al., 2002b). Another temperature-inducible plasmid responds to a decrease in temperature from $\sim$37°C to 28°C by causing an increase in plasmid copy number and hence an increase in gene dosage and the potential for protein expression (Trepod and Mott, 2002).

Resistance/Selection Marker

Resistance marker may not appear to be a critical plasmid parameter. However, if plasmid selection is needed during biomass accumulation and protein expression, then use of chloramphenicol and ampicillin/carbenicillin resistance genes may be unwise. β-Lactamase and chloramphenicol acetyltransferase are produced at very high levels and both are found to accumulate in the media (probably due to leakiness of the outer membrane and/or cell lysis) to levels that can deplete media antibiotics. Bacteriostatic antibiotics such as ampicillin are further penalized by the inability to kill cells that have lost the plasmid or resistance marker. If antibiotic is depleted, then these plasmid-free cells can outgrow plasmid-bearing cells. Outcompetition can be especially pronounced if the expression regime places a large metabolic or fitness burden on the cells. For these reasons the bactericidal antibiotics kanamycin and tetracycline may be preferential: the product of the kanamycin resistance gene (neomycin phosphotransferase II [NPT II]) is cytoplasmic but does not appear to leach as much as CAT while the efflux pump protein product of the tetracycline gene is an inner membrane protein with 12 *trans*-membrane-spanning regions.

Alternatives to the use of antibiotics are available. Auxotrophic genetic complementation can be employed to ensure that cells retain a plasmid. One system mimics that used for retention of the F′ episome. Strains such as JM109 and TG2 have a chromosomal deletion in the *proAB* genes involved in the proline synthesis pathway, which is complemented by a copy of *proAB* on the F′. Hence, colonies only form on amino acid-free solid media if this chromosomal lesion is complemented by the presence of an F′. This idea has been similarly employed in the context of plasmid retention (Fiedler and Skerra, 2001). In principle, essential genes or those that confer a loss of fitness on functional loss can be used in the same manner. Other strategies include using operator-repressor titration for plasmid maintenance (Durany et al., 2005).

Choice of Signal Peptide/Signal Peptide Coding Region

Signal peptides are the mechanism for directing the protein of interest into the periplasm. These can be of the shorter ~21-aa type that direct proteins through the *sec* pathway (Mori and Ito, 2001) or the sometimes longer (20 to 100 aa) type that direct folded proteins through the *Tat* pathway (Robinson and Bolhuis, 2001).

The important features of signal peptides of the *sec* type are well understood:

1. Generally 19 to 21 aa in length
2. A short ~5- to 6-aa positively charged N-terminal region due to the presence of the $-NH_2$ terminus and lysine or arginine residues

3. A longer central hydrophobic region of ~8 to 12 aa. The C-terminal end of the hydrophobic stretch is often punctuated by a turn-promoting residue such as glycine or proline.
4. A C-terminal 5- to 6-aa cleavage site with a high degree of sequence nonrandomness where the cleavage occurs after the most C-terminal residue of the tripeptide small-X-small such as A-X-A or A-X-G (Nielsen et al., 1997).

The first mature amino acid after the signal peptide can detrimentally affect the efficiency of cleavage of the signal peptide and hence yield antibody fragment (von Heijne et al., 1990; Laforet and Kendall, 1991), but since antibodies are secreted proteins they tend to have reasonably efficient mature N-terminal residues. Small or acidic side-chain amino acids are preferred at this position, while large hydrophobic and basic side chains are problematical (Nielsen et al., 1997). Furthermore, the presence of certain amino acids, in particular, arginine and lysine, in the first 20 to 30 aa of the mature domain can also reduce the efficiency of secretion to the periplasm (Li et al., 1988; Andersson and von Heijne, 1991; Kajava et al., 2000). Similarly, the presence of any double-arginine motif that resembles that of the Tat secretion motif (S/T-RRXFLK) might also cause reductions of secretion efficiency (Cristobal et al., 1999). If the N terminus of the mature protein particularly disfavors signal peptide cleavage, then reengineering of the mature N terminus or coexpression of signal peptidase I (Lep) may be helpful (van Dijl et al., 1991).

Many sequences can be used as useful signal peptides and they are generally highly expressed proteins of the periplasm or outer membrane of *E. coli* or the closely related organisms *Erwinia* and *Salmonella* (Sjöström et al., 1987, for review; Skerra and Plückthun, 1991; Klein et al., 1992; Kurokawa et al., 2001). Examples of commonly used signal peptides are shown in Table 2. Studies have demonstrated the practical limits of the various subdomains of signal peptides, such as number of N-terminal charges (Puziss et al., 1992), length, and

TABLE 2 Protein sequences of useful signal peptides

Protein	Subcellular location	Signal peptide sequence
β-Lactamase, bla	Periplasmic	MSIQHFRVALIPFFAAFCLPVFA
Alkaline phosphatase, AP	Periplasmic	MKQSTIALALLPLLFTPVTKA
Phosphate-binding protein, PhoS	Periplasmic	MKVMRTTVATVVAATLSMSAFSVFA
Maltose-binding protein, MalE	Periplasmic	MKIKTGARILALSALTTMMFSASALA
λ-Receptor, LamB	Outer membrane	MMITLRKLPLAVAVAAGVMSAQAMA
Pectate lyase B, PelB	Periplasmic/secreted	MKYLLPTAAAGLLLLAAQPAMA
Outer membrane protein A, OmpA	Outer membrane	MKKTAIAIAVALAGFATVAQA
Heat-stable toxin II, stII	Excreted	MKKNIAFLLASMFVFSIATNAYA
M13 major coat protein	Inner membrane	MKKSLVLKASVAVATLVPMLSFA

hydrophobicity profile of the central region (Chou and Kendall, 1990; Hikita and Mizushima, 1992). Indeed, due to the high degree of sequence and functional conservation, several examples of signal peptide use between species as diverse as plant, insect, mammal, and virus have been demonstrated (Humphreys et al., 2000; Tan et al., 2002). Although it may be unnecessary to change to a bacterial sequence, they are usually found to be the most efficient. Of equal importance is the nucleotide composition of the signal peptide. This is discussed in more detail below.

Optimization of Translation

A good translation unit is essential if one is to be able to get a high level of antibody fragment expression. The consensus sequence for ribosome binding (Shine-Dalgarno site) and approximate spacing to (5 to 13 nt), and preference of initiator codon (ATG > GTG ≫ TTG) are well established (Shine and Dalgarno, 1974; Ringquist et al., 1992; Gold, 1988, for review). Furthermore, codon usage differs between species (Wada et al., 1991) and since many genes encoding antibody fragments will be of non-*E. coli* origin (e.g., mouse, rat, rabbit, or synthetic), one might consider the process of codon optimization.

Of primary importance is the effect of codon choice at the 5′ end of the coding region—which in the case of periplasmic proteins is the signal peptide coding region. The effect of codon usage here is twofold: ribosomal stalling events caused by "rare" codons toward the 5′ have a strong impact upon the translation rate of the whole mRNA (Goldman et al., 1995). Also, hairpin loops encoded within the 5′-untranslated region (UTR):signal peptide coding region can occlude the Shine-Dalgarno site (Wood et al., 1984). It is for this reason that 5′-UTRs and 5′ signal peptide codons tend to be rich in A and T nucleotides. The importance and difficulty in prediction of good signal peptide coding regions in the context of a particular 5′-UTR have been demonstrated several times by various random mutagenesis/selection regimes (Stemmer et al., 1993; Le Calvez et al., 1996; Simmons et al., 1996; Humphreys et al., 2000). One also needs to avoid problems caused by multiple repeats of "rare" codons elsewhere in the coding sequence. These have also been demonstrated in experimental systems to be capable of reducing protein expression (see Kane, 1995, for review).

Finally, it is worth highlighting that there are also differences in the prevalence of (TAA > TGA > TAG) and the efficiency of the three termination codons. These observations can in part be rationalized because there are two release factors in *E. coli*, each of which recognizes two stop codons: RF-1 recognizes TAA and TAG while RF-2 recognizes TAA and TGA. TGA is the least efficient stop codon and can lead to small amounts of readthrough and frameshifts (MacBeath and Kast, 1998; Wenthzel et al., 1998). Experiments have also shown that the nucleotides following the stop codon are nonrandom and can exert an additional effect on the efficiency of termination. TAA(T) appears to be the best combination (Poole et al., 1995; Björnsson et al., 1996).

Expression of Multimeric Proteins

Expression of multimeric proteins such as Fab' fragments, IgG, and those fragments based on the cKappa:CH1 or leucine zipper heterodimerization domains adds an extra level of complexity to the expression problem. The peptides involved in the final 4° structure need to be presented for association in the periplasm in approximately equal functional quantities. The controlled relative expression of two polypeptides can be achieved in the following ways:

1. Dual plasmid experiments where compatible origins of replication and resistance markers are used and one protein of interest is encoded on each plasmid. For example, the colE1 and pACYC origins can be used with the ampicillin and chloramphenicol resistance markers (Humphreys et al., 1995, 1996).

2. Dual promoter plasmids where both polypeptides are encoded on the same plasmid but are under the control of two separate promoters. In one example both polypeptides were both under the control of the *tac* promoter (Humphreys et al., 2002) while in another example polypeptides were under the control of the *phoA* promoter (Simmons and Yansura, 1996).

3. Dicistronic plasmids where the translation units for both polypeptides are under the control of a single promoter and are encoded on the same mRNA. The dicistronic approach has been widely used for Fab' expression (Carter et al., 1992; Better et al., 1993; see Corisdeo and Wang, 2004, for a comparison with dual promoters).

Dicistronic and dual promoter approaches are perhaps more practical than the use of dual plasmids since the individual copy number of compatible plasmids present simultaneously can vary in cultures. Unlike dual promoter approaches, dicistronic approaches do not allow independent temporal or strength control over the induction of transcription of each polypeptide. The expression of both polypeptides is necessarily simultaneous, but control over the relative expression of each polypeptide can be achieved by engineering the signal peptide coding regions, intergenic sequence, or other coding elements.

Augmentation of the Secretion Apparatus and Periplasmic Environment

Nascent polypeptides are exposed to an oxidizing, media-influenced, and protein-rich environment as soon as they emerge from the translocon pore. In this environment disulfide bonds have to be formed or isomerized and peptide-proline bonds may need to be isomerized. Regions of secondary structure need to form and arrange correctly into the correct tertiary structure before they interact with irrelevant hydrophobic sequences or are exposed to proteases. Many of the gene products involved in the catalysis, aiding or interfering in these processes, have been identified. Hence it is possible to augment or ablate these helpful or unhelpful processes by overexpression or interference with these genes. Key proteins involved in the translocation, folding, and proteolysis of polypeptides and the influence of external media components are discussed below.

SECRETION APPARATUS

The inner membrane translocon of *E. coli* is composed of the integral membrane proteins SecY, E, and G. SecY and E form the pore protein while SecG is an ancillary accelerator (Tziatzios et al., 2004). SecD, SecF, and YajC can also act as accelerator proteins (Duong and Wickner, 1997). SecY has been overexpressed alone and with SecE and shown to increase the yield of periplasmic IL-6 (Pérez-Pérez et al., 1994). Coexpression of SecD and F has also been tested (Pérez-Pérez et al., 1996). Having passed through the translocon pore, the signal peptide is removed by signal peptidase I (Lep).

Coexpression of Lep has been shown to help increase the expression of fusions between secreted proteins of *Bacillus* and β-lactamase where signal peptide cleavage was thought to be rate limiting (van Dijl et al., 1991).

FOLDING FACTORS

Numerous periplasmic proteins have been shown to be intimately involved in the folding and prevention of aggregation of periplasmic proteins. These include Dsb proteins that form and isomerize disulfide bonds, PPIases that catalyze the slow *trans* → *cis* isomerization of peptidyl-prolyl bonds, general chaperones such as FkpA (Richarme and Caldas, 1997), and specific chaperones such as Skp and PapD (Jones et al., 1997). A practical review of plasmid resources available for coexpression studies has been compiled (Baneyx and Palumbo, 2003).

There are at least 7 members of the Dsb family (disulfide bond forming) in *E. coli* that form a web of oxidoreductase and isomerase functions in the periplasm (see Bardwell, 1994, for review). In brief, DsbA primarily acts as an oxidant by causing disulfide bond formation and is recycled by contact with the inner membrane protein DsbB. DsbC and DsbG can also act as oxidants in vitro, but they primarily function as disulfide isomerases in vivo analogous to protein disulfide isomerase (PDI) of the endoplasmic reticulum (Sone et al., 1997). DsbC is recycled by contact with DsbD. All of these have been coexpressed with periplasmic heterologous proteins with variable results (Knappik et al., 1993; Humphreys et al., 1995, 1996; Ostermeier et al., 1996; Joly et al., 1998; Zhang et al., 2002).

Most *E. coli* periplasmic proteins do not contain disulfide bonds, and those that do tend to have only 1 or 2 disulfides. Hence the Dsb proteins of *E. coli* are perhaps understandably less well evolved to cope with complex disulfide structures than the related proteins of the endoplasmic reticulum. It seems plausible that these pathways are best augmented when there is an inherent propensity to form incorrect disulfide bonds. Examples of problem proteins

include expression of Fab′ of the γ4 isotype or complex proteins such as tissue plasminogen activator (tPA) that has up to 17 disulfide bonds. Expression of human γ4 isotype Fab′ in the periplasm was found to be lower yielding than an almost identical Fab′ of the γ1 isotype. These two isotypes have a different relative position of the intradomain and hinge cysteines. The reduction in yield was rescued by coexpression of human PDI, suggesting that disulfide bond isomerization was an issue (Humphreys et al., 1996). Periplasmic yields of IgG1 were improved both by coexpression of DsbA+C or mutation of the two hinge cysteines to serine (Simmons et al., 2002b; Reilly and Yansura, 2004). Removal of free cysteines from scFv has also been shown to improve periplasmic yields (Kipriyanov et al., 1997; Schmiedl et al., 2000a). Perhaps the most impressive augmentation of the periplasmic disulfide bond environment by coexpression was when the two central amino acids of the active site motif (CXXC) that is common to these oxidase/isomerase proteins was optimized in both DsbA and DsbC to increase the production of tPA (Bessette et al., 2001). Data also suggest that impressive improvements can be seen when DsbA and C are coexpressed with their respective recycling proteins DsbB and D (Kurokawa et al., 2001; Sandee et al., 2005). Altogether these observations suggest that the ability to form aberrant intra- and interchain disulfides can have a deleterious effect on yield that may be rescued by mutagenesis of the antibody fragment or selected coexpression of Dsb proteins.

Skp (OmpH) is a small trimeric protein whose function appears to be the chaperoning and transport of OMPs across the periplasmic space to the outer membrane (Chen and Henning, 1996; Schäfer et al., 1999). It has been widely used in coexpression studies where the effect on the yield of the protein of interest has been mixed (Bothmann and Plückthun, 1998; Hayhurst and Harris, 1999; Mavrangelos et al., 2001). Coexpression of Skp does, however, commonly result in an increase in the general health of the cells as evidenced by the attainment of higher OD_{600}. This feature might itself be of some practical use when trying to express antibody fragments in flask cultures. Skp is thought to act in a pathway parallel to that of SurA (Rizzitello et al., 2001).

The presence of proline within polypeptides can pose an expression problem since all Xaa-Pro peptide bonds are found in the *trans* form after leaving the ribosome, but certain protein structures require this bond in the *cis* form. Seven percent of all prolyl-peptide bonds in folded proteins are *cis* (see Schmid, 2002, for review). Antibodies are known to contain *cis* bonds, and since rotation around this bond is sterically restricted, it can become the rate-limiting step for the folding of antibodies and their fragments (Skerra and Plückthun, 1991; Jäger and Plückthun, 1997; Thies et al., 1999). Enzymes that catalyze this rearrangement are called PPIases (peptidyl-prolyl *cis-trans*-isomerases). The periplasm has four proteins with PPIase activity: three are soluble proteins (FkpA, SurA, and PpiA) and one (PpiD) is an inner membrane protein.

PpiD (RotD) is anchored to the inner membrane and contains a periplasmic domain with PPIase activity (Dartigalongue and Raina, 1998). PpiD and SurA (RotC) both appear to have particular roles in the biogenesis of outer membrane proteins and so perhaps coexpression of either of these proteins is unlikely to have a major effect on the expression of periplasmic antibody fragments (Lazar and Kolter, 1996; Bitto and McKay, 2003; Hennecke et al., 2005; reviewed by Mogensen and Otzen, 2005).

Little is known about the in vivo function of PpiA (RotA) although its gene appears to be dispensable without causing perturbation of the periplasmic environment (Kleerebezem et al., 1995). Coexpression of PpiA with scFv was found to have no effect on yield (Knappik et al., 1993). On the other hand, FkpA (FKBP12 family member) and SurA are both known to have dual PPIase and chaperone functions (Missiakas et al., 1996; Ramm and Plückthun, 2000; Behrens et al., 2001). FkpA is dimeric in solution and has a C-terminal PPIase domain

(Arié et al., 2001; Saul et al., 2004). Coexpression of FkpA with antibody fragments has been found to be beneficial in some instances (Simmons et al., 2002b; Zhang et al., 2003) even when the protein expressed lacks *cis*-peptidyl-prolyl bonds (Bothmann and Plückthun, 2000). This suggested that the major benefit from FkpA with regard to antibody fragment expression came from its general chaperone function rather than its PPIase activity.

A further set of three genes (*yheS*, *wecB*, and *wecE*) has been identified as being able to increase the periplasmic yield of scFv (Belin et al., 2004). The function of YheS is unknown while both WecB and WecE are known to be involved in glycolipid synthetic pathways. The benefit of some of these mutations was found to be additive to the benefits of coexpression of FkpA or Skp. This led to the suggestion that the absence of WecB/E caused a partial induction of a periplasmic stress response pathway such as σ^e (Dartigalongue et al., 2001), CpxA/R (Pogliano et al., 1997), or BaeR/S (Raffa and Raivio, 2002), and hence indirectly caused expression of useful proteins such as FkpA, Skp, and DsbA/C.

Proteolysis

Proteins that are unable to fold correctly in the periplasm may either aggregate and form inclusion bodies or become turned over by proteases. In these situations, use of strains that are deficient in certain proteases may help to increase the yield of both soluble or insoluble protein. Certain key proteases have been disrupted specifically to help increase the expression of antibody fragments: the outer membrane OmpT/Protease VII (*ompT*), the periplasmic proteases; Protease III (*ptr3*), DegP (*htrA*), Tail-specific protease/Prc (*tsp*), and even the cytoplasmic Lon protease (*lon*). DegP has dual protease and chaperone functions that can be functionally separated by temperature or a single active site S210A mutation (Spiess et al., 1999). In practice, the deletion of proteases from the genome causes a loss of fitness to the cell that can limit their practical use. This underlines the positive function that proteases perform within the periplasm (Meerman and

Georgiou, 1994; Park et al., 1999). In one Fab′-related example, a spontaneous suppressor mutation (*spr*) was required to overcome such a problem after deletion of the Tsp protease (Chen et al., 2004). There are several widely available protease-deficient strains that may be of practical use for those experiencing severe expression problems. For example, wild-type B strains are deficient for the Lon protease and BL-21, which is derived from a B strain, is additionally deficient in OmpT (Studier et al., 1990; Meerman and Georgiou, 1994; Casali, 2003).

Translational Fusions

When engineering an antibody fragment for expression in the periplasm, one must also consider methods for its detection, purification, multimerization, and the addition of other functionalities. The engineering of single cysteines enables the conjugation of other reagents or proteins while N- or C-terminal tags or fusion partners can also aid their expression, solubilization, immunodetection, and purification. A wide variety of tags are available that range in size and conferred functionality (Lichty et al., 2005). Some fragments are easily purified with standard ion-exchange or ProteinA/G/L chromatography. More heavily engineered or complex fragments, e.g., heterospecific triabodies, can only reasonably be detected or purified with the use of translational fusions or tags. A list of useful fusion partners is shown in Table 3. If the protein of interest is to be used in an in vivo or crystallographic function, then it is sometimes also preferable to remove the translational fusion. Examples of potential cleavage methods are shown in Table 4. Residual cleavage-site amino acids remain an issue with some of these approaches.

Media Effects

The outer membrane is porous to solutes ≤500 Da. This means that media components can have a direct physical effect on the folding of the polypeptide. Solutes such as sucrose and raffinose reduced the aggregation of periplasmic β-lactamase (Bowden and Georgiou, 1988), glycine betaine and sorbitol have been shown to be helpful in increasing the yield of soluble periplasmic immunotoxins (Barth et al., 2000), and scFv (Sandee et al., 2005), while L-arginine, ethylurea, and acetamide have aided the production of plasminogen activator (Schäffner et al., 2001). Reducing agents such as glutathione and glutaredoxin can be used to alter the redox balance of the periplasm and hence alter the spectrum of disulfide formation and oxidation (Humphreys et al., 1995). Hence, special media additions might help to reduce a problem with periplasmic expression of an antibody fragment.

Extraction of Proteins from the Periplasm

Having achieved desirable levels of expression in the periplasm, it is perhaps worth describing how to recover the protein of interest from the periplasm. A notional advantage of periplasmic expression over extracellular expression is the ability to concentrate the product simply by harvesting the cells by centrifugation or filtration and avoiding vortex-type protein damage. An additional advantage over expression in the cytoplasm is the ability to avoid release of large quantities of DNA, RNA, and cytoplasmic proteins that may complicate the purification process. The aim of the extraction process is therefore to release the maximum amount of heterologous protein while minimizing the release of other host proteins and biomolecules.

In practice two groups of methods have been employed to try and disrupt or destabilize the outer membrane and cell wall while leaving the inner membrane largely intact:

1. Mechanical disruption such as by sonication, passage through a pressure system such as a French press or Dounce device, freeze-thaw fracture, or osmotic shock. With the exception of osmotic shock, mechanical methods used alone can lack sufficient specificity for the outer membrane/cell wall to effect controlled release of periplasmic proteins. They are often more effective when combined with chemical methods.

TABLE 3 Examples of translational fusion partners for periplasmic proteins

Name	Uses[a]	Comments	Reference(s)
His tag	ID, AP, SSP, LSP	Inexpensive and rechargeable Ni^{2+} or Cu^{2+} columns are ideal for purifications and antibodies are available for immunodetection.	Hochuli et al., 1988
Strep tag	ID, AP	9-aa StrepI tag AYRHPQFGG and 10 aa StrepII tag SNWSHPQFEK. Affinity for stretpavidin 2.7×10^{-4}.	Schmidt and Skerra, 1993, 1994; Voss and Skerra, 1997
Albumin-binding motifs	ID, SSP	22- to 46-aa C-terminal motifs can be used to bind to albumin immobilized on ELISA plates or columns.	König and Skerra, 1998; Dennis et al., 2002
Alkaline phosphatase	ID, MM	47-kDa periplasmic protein dimerizes into its active form and so can be used for direct substrate detection.	Wels et al., 1992; Ducancel et al., 1993
c-myc tag	ID, AP	Ab 9E10 binds to EQKLISEEDL	Munro and Pelham, 1986; Evan et al., 1985
Calmodulin-binding peptide	ID, SSP	Binds to immobilized calmodulin in the presence of Ca^{2+}, eluted with EGTA. 27-aa minimal binding sequence.	Stofko-Hahn et al., 1992
Cellulose-binding domain	SSP	CBP is a ~45-kDa bacterial exoenzyme that binds to cellulose, eluted with H_2O.	Ong et al., 1989
Chitin-binding domain	SSP	5-kDa binding domain from *Bacillus circulans* binds to immobilized chitin beads.	Chong et al., 1997; Hayhurst, 2000; Blank et al., 2002
DsbA	SM	21-kDa periplasmic protein used as a solubilization motif.	Collins-Racie et al., 1995; Zhang et al., 1998
FLAG tag	ID, AP	DYLDDDDK of 'FLAG' or DYKDHDGDYKDHDIDYKDDDDK for '3× FLAG.' Encodes enterokinase cleavage site. Ab M1 sees N-terminal FLAG in a Ca^{2+}-dependent manner, Ab M2 sees N- and C-terminal FLAG and is Ca^{2+} independent.	Brizzard et al., 1994; Knappik and Plückthun, 1994
g3 peptide tag	ID, IP, AP	ATDYGAAIDGF bound by Mab 10C31 at 5×10^{-7} M	Beckmann et al., 1998
GFP	ID	~30-kDa GFP does not seem to be well expressed or attain its fully fluorescent form in the periplasm but is useful for real-time ID.	Casey et al., 2000; Yi et al., 2004; Griep et al., 1999
HA tag	ID, IP, AP	YPYDVPDYA is the target for ID antibodies.	
Leucine zippers and coiled coils	MM	Helical motifs for di- and tetramerization from Fos/Jun and GCN4	Kostelny et al., 1992; Pack and Plückthun, 1992; Pack et al., 1995; Arndt et al., 2001
Maltose-binding protein	ID, IP, SSP, SM	MBP is an ~40-kDa periplasmic protein. Binds to immobilized starch, eluted with maltose.	di Guan et al., 1988; Hayhurst, 2000
Peptide 38 tag	ID, IP, AP	LPSDR motif recognized by antibody 2E11 with 4.9×10^{-6} affinity.	Böldicke et al., 2000
Polycysteine	SSP, LSP	Tetra-Cys. potential for disulfide interference or capping.	Persson et al., 1988

(Continued)

TABLE 3 *Continued*

Name	Uses[a]	Comments	Reference(s)
Polyionic peptides	MM	Poly-Lys and Poly-Glu for dimerization	Richter et al., 2001
Polyphenyl-alanine	SSP, LSP	11 × F. Hydrophobicity expression issue.	Persson et al., 1988
Protein A-binding domain	SSP	Z domain of staphylococcal protein A is encoded by ~58 aa in two antiparallel helices. Binds between CH2 and CH3 of Fc with 2×10^{-8} M affinity.	Nilsson et al., 1987
Thermally responsive polypeptides	SSP, LSP	(VPGXG)n motif has a sharp precipitation point above ~32°C that allows high-volume enrichment/purification.	Meyer and Chilkoti, 1999; Fong et al., 2002
V5 tag	ID, IP, AP	GKPIPNPLLGLDST is the target for ID antibodies.	

[a]ID, Immunodetection; IP, immunoprecipitation; AP, affinity purification; SSP, small-scale purification; LSP, larger-scale purification; SM, solubilization motif; MM, multimerization motif.

2. Chemical disruption with agents that specifically target the outer membrane and cell wall. Key classes of such molecules include detergents, buffers such as Tris and peptides that interact with and disrupt the lipid leaflet, metal ion–chelating agents that destabilize the outer membrane and cell wall, enzymes such as lysozyme that destroy key structural bonds within the cell wall, pH, and monovalent cations (Harrison et al., 1991; see Vaara, 1992, for review). The most commonly used agents are Tris, EDTA, and lysozyme.

TABLE 4 Examples of fusion partners and cleavage regimes for periplasmic expression of antibody fragments

Name	Cleavage site	Comments	Reference(s)
2A protease	KGDIKSY/G	Requires DTT, leaves 1 C-terminal postcleavage aa.	Walker et al., 1994
3C protease	E(T or V)LFQ/GP	Requires DTT, leaves 2 C-terminal postcleavage aa.	Walker et al., 1994
Cu^{2+} ions	Cleaves at DK/TH or DK/SH	Slow cleavage improved by warm alkaline conditions. DKTH is a natural upper hinge antibody sequence.	Humphreys et al., 2000
Cyanogen bromide	Cleaves at Met	Safety issues, derivatization of His, poor specificity.	Haught et al., 1998
Enterokinase	DDDDK/	Encoded within the FLAG sequence, does not leave C-terminal postcleavage aa.	Collins-Racie et al., 1995
Factor Xa protease	I(E or D)GR/X	Commonly used enzyme leaves 1 C-terminal postcleavage aa.	Nagai and Thogersen, 1984
Inteins	X/C(X)nQ/NA	Self-cleavable under mild reaction conditions, activated by thiol agents.	Chong et al., 1998; Sydor et al., 2002
o-Iodobenzoic acid	Cleaves at Trp	Very poor specificity	Hara and Yamakow, 1996
Subtilisin (modified)	PGAAHY/X	Mutation H64A of subtilisin narrowed the site specificity	Carter and Wells, 1987
Tev protease	EXXYXQS/	Tobacco etch virus protein does not leave C-terminal postcleavage aa.	Dougherty et al., 1988; Mondigler and Ehrmann, 1996
Thrombin	LVPR/GS	Leaves 2 C-terminal postcleavage aa.	McKenzie et al., 1991

A brief overview of some useful methods is shown in Table 5. Periplasmic extracts resulting from these procedures may not be suitable for using directly in the first stage of the purification or "capture" step. For example, the presence of EDTA may strip Ni^{2+} or Cu^{2+} off columns used for purification of His-tagged proteins, the pH or ionic content of the extractate may disable binding of the protein to ion-exchange matrices, and the presence of trace amounts of cell wall debris and nucleic acid may block columns or filter discs. Therefore certain pretreatments such as dilution, dialysis, buffer exchange, pH adjustment, centrifugation, filtration, or treatment with DNase I might be necessary before continuing with a purification step. With thermally stable fragments such as Fab' incubation of the extraction at elevated temperatures as high as ~65°C can preclear a substantial percentage of host proteins, thereby achieving a useful enrichment of heterologous protein (Weir and Bailey, 1997).

Tris/EDTA

This method is probably the best to use with large volumes and to achieve a partial enrichment of heterologous over host proteins. Resuspend cells in buffer (100 mM Tris pH 7.4, 10 mM EDTA) and incubate with rapid agitation overnight or for at least 4 h. Agitation can be performed at temperatures from 25 to 65°C to alter the profile of proteins recovered in the periplasmic fraction. Collect the periplasmic supernatant fraction after centrifugation at 13,000 rpm for 10 min in microfuge tubes or up to an hour for larger volumes.

SUCROSE/EDTA

This method is a good combination of mild conditions and minimal manipulation time that makes it amenable for medium and large volumes. Wash the pellet from 1 volume of culture with 1 volume of ice-cold 10 mM Tris·Cl (pH 8.0) and then resuspend in 0.1 to 1 volume of ice-cold 30 mM Tris (pH 8.0), 20% sucrose, 10 mM EDTA. Add 25 μl of lysozyme at 2 mg/ml. Incubate on ice for 30 to 60 min with gentle agitation. Centrifuge at 13,000 rpm/4°C in a microfuge for 3 min and take the supernatant off as the periplasmic fraction. A cytoplasmic fraction can then be prepared by resuspending the pellet in 500 μl of 10 mM Tris·Cl (pH 8.0), 100 mM NaCl, 1 mM $MgCl_2$ with fresh DNase I added to 1 μg/ml. The resuspension is freeze/thawed three times before centrifugation at 13,000 rpm/4°C in a microfuge for 3 min, and the supernatant is taken off as a cytoplasmic fraction.

COLD OSMOTIC SHOCK

This is the best method for getting very clean separation of periplasmic, cytoplasmic, and insoluble (membrane) fractions. It is relatively labor intense and can only be used reliably on

TABLE 5 Comparison of periplasmic extraction methods

Method	Volume	Speed	Handling time	Physical insult	Notes	Reference
Tris/EDTA	~0.5 ml–000's l	Slow (h)	Few minutes	Mild (overnight agitation)	Can be combined with high temperature	Weir and Bailey, 1997
Sucrose/EDTA	~1–100 ml	Medium (30–60 min)	Few minutes	Gentle (stand/agitate on ice)	Requires ≥13,000g centrifugation	Oliver and Beckwith, 1982
Cold osmotic shock	0.5–50 ml	Fast (sample number dependent)	5–15 min	Bursts outer membrane, speed of resuspension limits volume	Speed of pellet resuspension limits volume	Neu and Heppel, 1965
Homogenization	~100 ml–000's l	Slow (volume related)	Volume related	Pressure dependent	Requires capital equipment	Harrison, 1991

small volumes. Pellet 1 ml of culture in a microfuge tube for 1 min and then resuspend it in 150 μl of ice-cold spheroplast buffer (100 mM Tris·Cl [pH 8.0], 500 mM sucrose, 0.5 mM EDTA). Store on ice for 5 min before taking a 50 μl "whole cell" sample. Pellet with 1 min of centrifugation, discard the supernatant, removing the last few drops of liquid with a pipette before resuspending with vigorous pipetting or vortex mixing in 100 μl of ice-cold deionized water for 30 s. Add 5 μl of ice-cold 20 mM $MgCl_2$, mix, then pellet by centrifugation for 2 min. Take off the supernatant as "periplasmic fraction" before resuspending the pellet in 150 μl of of ice-cold spheroplast buffer and adding 15 μl of fresh lysozyme from a 2 mg/ml stock. Incubate on ice for 5 min and then centrifuge for 6 min. Carefully remove and discard the supernatant before resuspending the pellet in 600 μl of 10 mM Tris·Cl (pH 8.0). Freeze-thaw the resuspension three times, add 20 μl of 1 M $MgCl_2$ and 6 μl of DNase I (fresh stock at 1 mg/ml). Mix to degrade DNA and then centrifuge for 25 min. The supernatant is removed as the "cytoplasmic fraction" and the pellet is resuspended in 150 μl of 10 mM Tris·Cl (pH 8.0) as the "membrane-insoluble" fraction.

SUMMARY

The endoplasmic reticulum of eukaryotes is the natural site for the folding and assembly of antibodies. The disulfide formation, peptidyl-prolyl *cis-trans*-isomerization, and acquisition of secondary, tertiary, and quaternary structure that occurs in the ER can also occur in the periplasm of *E. coli*. *E. coli* is an attractive expression host for many reasons. It is a relatively well understood, safe organism that grows rapidly in a wide variety of simple and inexpensive media. It is equally suitable for small-scale laboratory expression experiments and large-scale industrial production systems. These factors have been the logic and driving force behind use of the periplasm of *E. coli* for the expression of antibody fragments.

This chapter has described various designs of antibody fragments along with the engineered adaptations and improvements to both the fragment and the host. The aims of these engineered changes have been to maximize heterologous protein yield and quality and to simplify the expression regime. Expression technologies continuously improve and compete with each other. However, we should expect to see the continued use of *E. coli* for the production of antibody fragments because of its intrinsic practical and economic values: large-scale, high-productivity, short-duration fermentations using simple growth and harvest conditions. Further study of the fundamental biology of protein translocation, folding, modification, stability, and degradation is expected to continue to contribute to future practical improvements in the design, use, and expression level of antibodies and antibody fragments in *E. coli*.

REFERENCES

Alfthan, K., K. Takkinen, D. Sizmann, H. Söderlund, and T. T. Teeri. 1995. Properties of a single-chain antibody containing different linker peptides. *Protein Eng.* **8:**725–731.

Andersson, H., and G. Von Heijne. 1991. A 30-residue-long "export initiation domain" adjacent to the signal sequence is critical for protein translocation across the inner membrane of *Escherichia coli*. *Proc. Natl. Acad. Sci. USA* **88:**9751–9754.

Arié, J. P., N. Sassoon, and J. M. Betton. 2001. Chaperone function of FkpA, a heat shock prolyl isomerase, in the periplasm of *Escherichia coli*. *Mol. Microbiol.* **39:**199–210.

Arndt, K. M., K. M. Müller, and A. Plückthun. 1998. Factors influencing the dimer to monomer transition of an antibody single-chain Fv fragment. *Biochemistry* **37:**12918–12926.

Arndt, K. M., K. Müller, and A. Plückthun. 2001. Helix-stabilized Fv (hsFv) antibody fragments: substituting the constant domains of a Fab fragment for a heterodimeric coiled-coil domain. *J. Mol. Biol.* **312:**221–228.

Atwell, J. L., K. A. Breheney, L. J. Lawrence, A. J. McCoy, A. A. Kortt, and P. J. Hudson. 1999. scFv multimers of the anti-neuraminidase antibody NC10: length of the linker between Vh and Vl domains dictates precisely the transition between diabodies and triabodies. *Protein Eng.* **12:**597–604.

Baneyx, F., and J. L. Palumbo. 2003. Improving heterologous protein folding *via* molecular chaperone and foldase co-expression. *Methods Mol. Biol.* **205:**171–197.

Bardwell, J. C. A. 1994. Building bridges: disulphide bond formation in the cell. *Mol. Microbiol.* **14:**199–205.

Barth, S., M. Huhn, B. Matthey, A. Klimka, E. A. Galinski, and A. Engbert. 2000. Compatible-solute-supported periplasmic expression of functional recombinant proteins under stress conditions. *Appl. Environ. Microbiol.* **66:**1572–1579.

Bass, S., and J. R. Swartz. 1994. Method of controlling polypeptide production in bacterial cells. U.S. patent 5,304,472.

Beckmann, C., B. Haase, K. N. Timmis, and M. Tesar. 1998. Multifunctional g3p-peptide tag for current phage display systems. *J. Immunol. Methods* **212:**131–138.

Behrens, S., R. Maier, H. De Cock, F. X. Schmid, and C. A. Gross. 2001. The SurA periplasmic PPIase lacking its parvulin domains functions *in vivo* and has chaperone activity. *EMBO J.* **20:**285–294.

Belin, P., J. Dassa, P. Drevet, E. Lajeunesse, A. Savatier, J. C. Boulain, and A. Menez. 2004. Toxicity based selection of *Escherichia coli* mutants for functional recombinant protein production: application to an antibody fragment. *Protein Eng. Des. Sel.* **17:**491–500.

Bessette, P. H., F. Aslund, J. Beckwith, and G. Georgiou. 1999. Efficient folding of proteins with multiple disulfide bonds in the *Escherichia coli* cytoplasm. *Proc. Natl. Acad. Sci. USA* **96:**13703–13708.

Bessette, P. H., J. Qiu, J. C. A. Bardwell, J. R. Swartz, and G. Georgiou. 2001. Effects of sequences of the active-site dipeptides of DsbA and DsbC on *in vivo* folding of multidisulfide proteins in *Escherichia coli. J. Bacteriol.* **183:**980–988.

Better, M., S. L. Bernhard, S. P. Lei, D. M. Fishwild, J. A. Lane, S. F. Carroll, and A. H. Horwitz. 1993. Potent anti-CD5 ricin A chain immunoconjugates from bacterially produced Fab′ and F(ab′)₂. *Proc. Natl. Acad. Sci. USA* **90:**457–461.

Better, M., C. P. Chang, R. R. Robinson, and H. Horwitz. 1988. *Escherichia coli* secretion of an active chimeric antibody fragment. *Science* **240:**1041–1043.

Bird, R. E., K. D. Hardman, J. W. Jacobson, S. Johnson, B. M. Kaufman, S. M. Lee, T. Lee, S. H. Pope, G. S. Riordan, and M. Whitlow. 1988. Single-chain antibody-binding sites. *Science* **242:**423–426.

Bitto, E., and D. B. McKay. 2003. The periplasmic molecular chaperone protein SurA binds a peptide motif that is characteristic on integral outer membrane proteins. *J. Biol. Chem.* **278:**49316–49322.

Björnsson, A., S. Mottagui-Tabar, and L. A. Isaksson. 1996. Structure of the C-terminal end of the nascent peptide influences translation termination. *EMBO J.* **15:**1696–1704.

Blank, K., P. Lindner, B. Diefenbach, and A. Plückthun. 2002. Self-immobilizing recombinant antibody fragments for immunoaffinity chromatography: generic, parallel, and scalable protein purification. *Protein Expr. Purific.* **24:**313–322.

Böldicke, T., F. Struck, F. Schaper, W. Tegge, H. Solbeck, B. Villbrandt, P. Lankenau, and M. Bocher. 2000. A new peptide-affinity tag for the detection and affinity purification of recombinant proteins with a monoclonal antibody. *J. Immunol. Methods* **240:**165–183.

Bothmann, H., and A. Plückthun. 1998. Selection for a periplasmic factor improving phage display and functional periplasmic expression. *Nat. Biotechnol.* **16:**376–380.

Bothmann, H., and A. Plückthun. 2000. The periplasmic *Escherichia coli* peptidyl-prolyl cis/trans-isomerase FkpA. *J. Biol. Chem.* **275:**17100–17105.

Bowden, G. A., and G. Georgiou. 1988. The effect of sugars on β-lactamase aggregation in *Escherichia coli. Biotechnol. Prog.* **4:**97–101.

Brizzard, B. L., R. G. Chubet, and D. L. Vizard. 1994. Immunoaffinity purification of FLAG epitope-tagged bacterial alkaline phosphatase using a novel monoclonal antibody and peptide elution. *Biotechniques* **16:**730–735.

Carter, P., R. F. Kelley, M. L. Rodrigues, B. Snedecor, M. Covarrubias, M. D. Velligan, W. L. T. Wong, A. M. Rowland, C. E. Kotts, M. E. Carver, M. Yang, J. H. Bourell, H. M. Shepard, and D. Henner. 1992. High level *Escherichia coli* expression and production of a bivalent humanized antibody fragment. *Biotechnology* **10:**163–167.

Carter, P., and J. A. Wells. 1987. Engineering enzyme specificity by "substrate-assisted catalysis." *Science* **237:**394–399.

Casali, N. 2003. *Escherichia coli* host strains. *Methods Mol. Biol.* **235:**27–48.

Casey, J. L., A. M. Coley, L. M. Tilley, and M. Foley. 2000. Green fluorescent antibodies: novel *in vitro* tools. *Protein Eng.* **13:**445–452.

Chang, A. C. Y., and S. N. Cohen. 1978. Construction and characterization of amplifiable multicopy DNA cloning vehicles derived from the p15A cryptic miniplasmid. *J. Bacteriol.* **134:**1141–1156.

Chao, Y. P., C. J. Chiang, and W. B. Hung. 2002b. Stringent regulation and high level expression of heterologous genes in *Escherichia coli* using T7 system controllable by the *araBAD* promoter. *Biotechnol. Prog.* **18:**394–400.

Chao, Y. P., W. S. Law, P. T. Chen, and W. B. Hung. 2002a. High production of heterologous proteins in *Escherichia coli* using the thermo-regulated T7 expression system. *Appl. Microb. Biotechnol.* **58:**446–453.

Chapman, A. P., P. Antoniw, M. Spitali, S. West, S. Stephens, and D. J. King. 1999. Therapeutic antibody fragments with prolonged *in vivo* half-lives. *Nat. Biotechnol.* **17:**780–783.

Chen, C., B. Snedecor, J. C. Nishihara, J. C. Joly, N. McFarland, D. C. Andersen, J. E. Battersby, and K. M. Champion. 2004. High-level accumulation of a recombinant antibody fragment in the periplasm of *Escherichia coli* requires a triple-mutant (*degP prc spr*) host strain. *Biotechnol. Bioeng.* **85**:463–474.

Chen, R., and U. Henning. 1996. A periplasmic protein (Skp) of *Escherichia coli* selectively binds a class of outer membrane proteins. *Mol. Microbiol.* **19**:1287–1294.

Chong, S., F. B. Mersha, D. G. Comb, M. E. Scott, D. Landry, C. A. Vence, F. B. Perler, J. Benner, R. B. Kucera, C. A. Hirvonen, J. J. Pelletier, H. Paulus, and M. Q. Xu. 1997. Single-column purification of free recombinant proteins using a self-cleavable affinity tag dereived from a protein splicing element. *Gene* **192**:271–281.

Chong, S., G. E. Montello, A. Zhang, E. J. Cantor, W. Liao, M. Q. Xu, and J. Benner. 1998. Utilizing the C-terminal cleavage activity of a protein splicing element to purify recombinant proteins in a single chromatographic step. *Nucleic Acids Res.* **26**:5109–5115.

Chou, D. K., R. Krishnamurthy, T. W. Randolph, J. F. Carpenter, and M. C. Manning. 2005. Effect of Tween[20] and Tween[80] on the stability of albutropin during agitation. *J. Pharm. Sci.* **94**:1368–1381.

Chou, M. M., and D. A. Kendall. 1990. Polymeric sequences reveal a functional interrelationship between hydrophobicity and length of signal peptides. *J. Biol. Chem.* **265**:2873–2880.

Collins-Racie, L. A., J. M. McColgan, K. L. Grant, E. A. Diblasio-Smith, J. M. McCoy, and E. R. Lavallie. 1995. Production of recombinant bovine enterokinase catalytic subunit in *Escherichia coli* using the novel secretory fusion partner DsbA. *Biotechnology* **13**:982–987.

Corisdeo, S., and B. Wang. 2004. Functional expression and display of an antibody Fab fragment in *Escherichia coli*: study of vector designs and culture conditions. *Protein Expr. Purific.* **34**:270–279.

Cristobal, S., J. W. De Gier, H. Nielsen, and G. Von Heijne. 1999. Competition between Sec and TAT dependent protein translocation in *Escherichia coli*. *EMBO J.* **18**:2982–2990.

Dartigalongue, C., D. Missiakas, and S. Raina. 2001. Characterization of the *Escherichia coli* σ^E regulon. *J. Biol. Chem.* **276**:20866–20875.

Dartigalongue, C., and S. Raina. 1998. A new heat-shock gene, *ppiD*, encodes a peptidyl-prolyl isomerase required for folding of outer membrane proteins in *Escherichia coli*. *EMBO J.* **17**:3968–3980.

de Boer, H. A., L. J. Comstock, and M. Vasser. 1983. The *tac* promoter: a functional hybrid derived from the *trp* and *lac* promoters. *Proc. Natl. Acad. Sci. USA* **80**:21–25.

Dennis, M. S., M. Zhang, Y. G. Meng, M. Kadkhodayan, D. Kirchhofer, D. Combs, and L. A. Damico. 2002. Albumin binding as a general strategy for improving the pharmacokinetics of proteins. *J. Biol. Chem.* **277**:35035–35043.

di Guan, C., P. Li, P. D. Riggs, and H. Inouye. 1988. Vectors that facilitate the expression and purification of foreign peptides in *Escherichia coli* by fusion to maltose-binding protein. *Gene* **67**:21–30.

Dooley, H., S. D. Grant, W. J. Harris, and A. J. Porter. 1998. Stabilization of antibody fragments in adverse environments. *Biotechnol. Appl. Biochem.* **28**:77–83.

Dougherty, W. G., J. C. Carrington, S. M. Cary, and T. D. Parks. 1988. Biochemical and mutational analysis of a plant virus polyprotein cleavage site. *EMBO J.* **7**:1281–1287.

Ducancel, F., D. Gillet, A. Carrier, E. Lajeunesse, A. Menez, and J. C. Boulain. 1993. Recombinant colorimetric antibodies: construction and characterization of a bifunctional F(ab)$_2$/alkaline phosphate conjugate produced in *Escherichia coli*. *Biotechnology* **11**:601–605.

Duong, F., and W. Wickner. 1997. Distinct catalytic roles of the SecYE, SecG and SecDFyajC subunits of the preprotein translocase holoenzyme. *EMBO J.* **16**:2756–2768.

Durany, O., P. Bassett, A. M. E. Weiss, R. M. Cranenburgh, P. Ferrer, J. Lopéz-Santin, C. De Mas, and J. A. J. Hanak. 2005. Production of fuculose-1-phosphate aldolase using operator-repressor titration for plasmid maintenance in high cell density *Escherichia coli* fermentations. *Biotechnol. Bioeng.* **91**:460–467.

Evan, G. I., G. K. Lewis, G. Ramsay, and J. M. Bishop. 1985. Isolations of monoclonal antibodies specific for human-*myc* proto-oncogene product. *Mol. Cell. Biol.* **5**:3610–3616.

Ewert, S., C. Cambillau, K. Conrath, and A. Plückthun. 2002. Biophysical properties of camelid Vhh domains compared to those of human Vh3 domains. *Biochemistry* **41**:3628–3636.

Fiedler, M., and A. Skerra. 2001. *proAB* complementation of an auxotrophic E. *coli* strain improves plasmid stability and expression yield during fermenter production of a recombinant antibody fragment. *Gene* **274**:111–118.

Fong, R. B., Z. Ding, A. S. Hoffman, and P. S. Stayton. 2002. Affinity separation using an Fv antibody fragment–"smart" polymer conjugate. *Biotechnol. Bioeng.* **79**:271–276.

Forsberg, G., M. Forsgren, M. Jaki, M. Norin, C. Sterky, A. Enhorning, K. Larsson, M. Ericsson, and P. Björk. 1997. Identification of framework residues in a secreted recombinant antibody fragment that control production level and localization in *Escherichia coli*. *J. Biol. Chem.* **272**:12430–12436.

Freund, C., A. Ross, B. Guth, A. Plückthun, and T. A. Holak. 1993. Characterization of the linker peptide of the single-chain Fv fragment of an antibody by NMR spectroscopy. *FEBS Lett.* **320:**97–100.

Garnett, M. C. 2001. Targeted drug conjugates: principles and progress. *Adv. Drug Delivery Rev.* **53:** 171–216.

Gavilondo, J. V., and J. W. Larrick. 2000. Antibody engineering at the millennium. *Biotechniques* **29:** 128–145.

Glennie, M. J, and P. W. M. Johnson. 2000. Clinical trials of antibody therapy. *Immunology Today* **21:**403-410.

Glennie, M. J., H. M. McBride, A. T. Worth, and G. T. Stevenson. 1987. Preparation and performance of bispecific F(ab′γ)$_2$ antibody containing thioether-linked Fab′γ fragments. *J. Immunol.* **139:** 2367–2375.

Glockshuber, R., M. Malia, I. Pfitzinger, and A. Plückthun. 1990. A comparison of strategies to stabilize immunoglobulin Fv fragments. *Biochemistry* **29:**1362–1367.

Gold, L. 1988. Posttranslational regulatory mechanisms in *Escherichia coli. Annu. Rev. Biochem.* **57:** 199–233.

Goldman, E., A. H. Rosenberg, G. Zubay, and F. W. Studier. 1995. Consecutive low-usage leucine codons block translation only when near the 5′ end of a message in *Escherichia coli. J. Mol. Biol.* **245:**467–473.

Grant, S. D., P. M. Cupit, D. Learmonth, F. R. Byrne, B. M. Graham, A. J. R. Porter, and W. J. Harris. 1995. Expression of monovalent and bivalent antibody fragments in *Escherichia coli. J. Hematother.* **4:**383–388.

Greenfield, L., T. Boone, and G. Wilcox. 1978. DNA sequence of the *araBAD* promoter in *Escherichia coli* B/r. *Proc. Natl. Acad. Sci. USA* **75:**4724–4728.

Griep, R. A., C. Van Twisk, J. M. Van Der Wolf, and A. Schots. 1999. Fluobodies: green fluorescent single-chain Fv fusion proteins. *J. Immunol. Methods* **230:**121–130.

Guzman, L. M., D. Belin, M. J. Carson, and J. Beckwith. 1995. Tight regulation, modulation, and high-level expression by vectors containing the arabinose PBAD promoter. *J. Bacteriol.* **177:**4121–4130.

Halaby, D. M., A. Poupon, and J. P. Mornon. 1999. The immunoglobulin fold family: sequence analysis and 3D structure comparisons. *Protein Eng.* **12:**563–571.

Hallewell, R. A., and S. Emtage. 1980. Plasmid vectors containing the tryptophan operon promoter suitable for efficient regulated expression of foreign genes. *Gene* **9:**27–47.

Hara, S., and M. Yamakawa. 1996. Production in *Escherichia coli* of moricin, a novel type antibacterial peptide from the silkworm, *Bombyx mori. Biochem. Biophys. Res. Commun.* **220:**664–669.

Harrison, J. S., A. Gill, and M. Hoare. 1998. Stability of a single-chain Fv antibody fragment when exposed to a high shear environment combined with air-liquid interfaces. *Biotechnol. Bioeng.* **59:** 517–519.

Harrison, S. T. L., J. S. Dennis, and H. A. Cahse. 1991. Combined chemical and mechanical processes for the disruption of bacteria. *Bioseparation* **2:**95–105.

Haught, C., G. D. Davis, R. Subramanian, K. W. Jackson, and R. G. Harrison. 1998. Recombinant production and purification of novel antisense antimicrobial peptide in *Escherichia coli. Biotechnol. Bioeng.* **57:**55–61.

Hayhurst, A. 2000. Improved expression characteristics of single-chain Fv fragments when fused downstream of the *Escherichia coli* maltose-binding protein or upstream of a single immunoglobulin-constant domain. *Protein Expr. Purific.* **18:**1–10.

Hayhurst, A., and W. J. Harris. 1999. *Escherichia coli* Skp chaperone coexpression improves solubility and phage display of single-chain antibody fragments. *Protein Expr. Purific.* **15:**336–343.

Hennecke, G., J. Nolte, R. Volkmer-Engert, J. Schneider-Mergener, and S. Behrens. 2005. The periplasmic chaperone SurA exploits two features characteristic of integral outer membrane proteins for selective substrate recognition. *J. Biol. Chem.* **280:**23540–23548.

Hexham, J. M., V. King, D. Dudas, P. Graff, M. Mahnke, Y. K. Wang, J. F. Goetschy, D. Plattner, M. Zurini, F. Bitsch, P. Lake, and M. E. Digan. 2001. Optimization of the anti-(human CD3) immunotoxin DT389-scFv(UCHT1) N-terminal sequence to yield a homogeneous protein. *Biotechnol. Appl. Biochem.* **34:**183–187.

Hikita, C., and S. Mizushima. 1992. Effects of total hydrophobicity and length of the hydrophobic domain of a signal peptide on the *in vitro* translocation efficiency. *J. Biol. Chem.* **267:**4882–4888.

Hochuli, E., W. Bannwarth, H. Dobeli, R. Gentz, and D. Stuber. 1988. Genetic approach to facilitate purification of recombinant proteins with a novel metal chelate adsorbent. *Biotechnology* **6:** 1321–1325.

Holliger, P., M. Wing, J. D. Pound, H. Bohlen, and G. Winter. 1997. Retargeting serum immunoglobulins with bispecific diabodies. *Nat. Biotechnol.* **15:**632–636.

Hu, S. Z., L. Shively, A. Raubitschek, M. Sherman, L. E. Williams, J. Y. C. Wong, J. E. Shively, and A. M. Wu. 1996. Minibody: a novel engineered anti-carcinoembryonic antigen antibody fragment (single-chain Fv-CH3) which exhibits rapid, high-level targeting of xenografts. *Cancer Res.* **56:**3055–3061.

Humphreys, D. P. 2003. Production of antibodies and antibody fragments in *Escherichia coli* and a comparison of their functions, uses and modification. *Curr. Opin. Drug Discov. Devel.* **6:**188–196.

Humphreys, D. P., B. Carrington, L. C. Bowering, R. Ganesh, M. Sehdev, B. J. Smith, L. M. King, D. G. Reeks, A. Lawson, and A. G. Popplewell. 2002. A plasmid system for optimization of Fab' production in *Escherichia coli:* importance of balance of heavy chain and light chain synthesis. *Protein Expr. Purific.* **26:**309–320.

Humphreys, D. P., A. P. Chapman, D. G. Reeks, V. Lang, and P. E. Stephens. 1997. Formation of dimeric Fabs in *Escherichia coli:* effect of hinge size and isotype, presence of interchain disulphide bond, Fab' expression levels, tail piece sequences and growth conditions. *J. Immunol. Methods* **209:**193–202.

Humphreys, D. P., and D. J. Glover. 2001. Therapeutic antibody production technologies: molecules, applications, expression and purification. *Curr. Opin. Drug Discov. Devel.* **4:**172–185.

Humphreys, D. P., L. M. King, S. M. West, A. P. Chapman, M. Sehdev, M. W. Redden, D. J. Glover, B. J. Smith, and P. E. Stephens. 2000. Improved efficiency of site-specific copper(II) ion-catalysed protein cleavage effected by mutagenesis of cleavage site. *Protein Eng.* **3:**201–206.

Humphreys, D. P., O. M. Vetterlein, A. P. Chapman, D. J. King, P. Antoniw, A. J. Suitters, D. G. Reeks, T. A. H. Parton, L. M. King, B. J. Smith, V. Lang, and P. E. Stephens. 1998. F(ab')₂ molecules made from *E. coli* produced Fab' with hinge sequences conferring increased serum permanence times in an animal model. *J. Immunol. Methods* **217:**1–10.

Humphreys, D. P., N. Weir, A. Mountain, and P. A. Lund. 1995. Human protein disulfide isomerase functionally complements a *dsbA* mutation and enhances the yield of pectate lyase C in *Escherichia coli. J. Biol. Chem.* **270:**28210–28215.

Humphreys, D. P., N. Weir, A. Lawson, A. Mountain, and P. A. Lund. 1996. Co-expression of human protein disulphide isomerase (PDI) can increase the yield of an antibody Fab' fragment expressed in *Escherichia coli. FEBS Lett.* **380:**194–197.

Huston, J. S., D. Levinson, M. Mudgett-Hunter, M. S. Tai, J. Novotný, M. N. Margolies, R. J. Ridge, R. E. Bruccoleri, E. Haber, R. Crea, and H. Oppermann. 1988. Protein engineering of antibody binding sites: recovery of specific activity in an anti-digoxin single-chain Fv analogue produced in *Escherichia coli. Proc. Natl. Acad. Sci. USA* **85:**5879–5883.

Ill, C. R., J. N. Gonzales, E. K. Houtz, J. R. Ludwig, E. D. Melcher, J. E. Hale, R. Pourmand, V. M. Keivens, L. Myers, K. Beidler, P. Stuart, S. Cheng, and R. Radhakrishnan. 1997. Design and construction of a hybrid immunoglobulin domain with properties of both heavy and light chain variable regions. *Protein Eng.* **10:**949–957.

Jäger, M., and A. Plückthun. 1997. The rate-limiting steps for the folding of an antibody scFv fragment. *FEBS.* **418:**106–110.

Joly, J. C., W. S. Leung, and J. R. Swartz. 1998. Overexpression of *Escherichia coli* oxidoreductases increases recombinant insulin–like growth factor-I accumulation. *Proc. Natl. Acad. Sci. USA* **95:**2773–2777.

Jones, C. H., P. N. Danese, J. S. Pinkner, T. J. Silhavy, and S. J. Hultgren. 1997. The chaperone-assisted membrane release and folding pathway is sensed by two signal transduction systems. *EMBO J.* **16:**6394–6406.

Jurado, P., D. Ritz, J. Beckwith, V. De Lorenzo, and L. A. Fernandez. 2002. Production of functional single-chain Fv antibodies in the cytoplasm of *Escherichia coli. J. Mol. Biol.* **320:**1–10.

Kajava, A. V., S. N. Zolov, A. E. Kalinin, and M. A. Nesmeyanova. 2000. The net charge of the first 18 residues of the mature sequence affects protein translocation across the cytoplasmic membrane of gram-negative bacteria. *J. Bacteriol.* **182:**2163–2169.

Kane, J. F. 1995. Effects of rare codon clusters on high-level expression of heterologous proteins in *Escherichia coli. Curr. Biol.* **6:**494–500.

King, D. J., A. Turner, A. P. H. Farnsworth, J. R. Adair, R. J. Owens, B. Pedley, D. Baldock, K. A. Proudfoot, A. D. G. Lawson, N. R. A. Beeley, K. Millar, T. A. Millican, B. A. Boyce, P. Antoniw, A. Mountain, R. H. J. Begent, D. Shochat, and G. T. Yarranton. 1994. Improved tumour targeting with chemically cross-linked recombinant antibody fragments. *Cancer Res.* **54:**6176–6185.

Kipriyanov, S. M., G. Moldenhauer, A. C. R. Martin, O. A. Kupruyanova, and M. Little. 1997. Two amino acid mutations in an anti-human CD3 single chain Fv antibody fragment that affect the yield of bacterial secretion but not the affinity. *Protein Eng.* **10:**445–453.

Kitai, K., T. Kudo, S. Nakamura, T. Masegi, Y. Ichikawa, and K. Horikoshi. 1988. Extracellular production of human immunoglobulin G Fc region (hIg-Fc) by *Escherichia coli. Appl. Microbiol. Biotechnol.* **28:**52–56.

Kleerebezem, M., M. Heutink, and J. Tommassen. 1995. Characterization of an *Escherichia coli* rotA mutant, affected in periplasmic peptidyl-prolyl cis/trans isomerase. *Mol. Microbiol.* **18:**313–320.

Klein, B. K., J. O. Polazzi, C. S. Devine, S. H. Rangwala, and P. O. Olins. 1992. Effects of signal peptide changes on the secretion of bovine somatotropin (bST) from *Escherichia coli. Protein Eng.* **5:**511–517.

Knappik, A., C. Krebber, and A. Plückthun. 1993. The effect of folding catalysts on the *in vivo* folding process of different antibody fragments expressed in *Escherichia coli. Biotechnology* **11:**77–83.

Knappik, A., and A. Plückthun. 1994. An improved affinity tag based on the FLAG peptide for the detection and purification of recombinant antobody fragments. *Biotechniques* **17:**754–761.

König, T., and A. Skerra. 1998. Use of an albumin-binding domain for the selective immobilisation of recombinant capture antibody fragments on ELISA plates. *J. Immunol. Methods* **218:**73–83.

Kostelny, S. A., M. S. Cole, and J. Y. Tso. 1992. Formation of a bispecific antibody by the use of leucine zippers. *J. Immunol.* **148:**1547–1553.

Kurokawa, Y., H. Yanagi, and T. Yura. 2001. Overproduction of bacterial protein disulfide isomerase (DsbC) and its modulator (DsbD) markedly enhances periplasmic production of human nerve growth factor in *Escherichia coli. J. Biol. Chem.* **276:**14393–14399.

Laforet, G. A., and D. A. Kendall. 1991. Functional limits of conformation, hydrophobicity, and steric constraints in prokaryotic signal peptide cleavage regions. *J. Biol. Chem.* **266:**1326–1334.

Lauwereys, M., M. A. Ghahroudi, A. Desmyter, J. Kinne, W. Holzer, E. De Genst, L. Wyns, and S. Muyldermans. 1998. Potent enzyme inhibitors derived from dromedary heavy-chain antibodies. *EMBO J.* **17:**3512–3520.

Lazar, S. W., and R. Kolter. 1996. SurA assists the folding of *Escherichia coli* outer membrane proteins. *J. Bacteriol.* **178:**1770–1773.

Le Calvez, H., J. Green, and D. Baty. 1996. Increased efficiency of alkaline phosphatase production levels in *Escherichia coli* using a degenerate PelB signal sequence. *Gene* **170:**51–55.

Li, P., J. Beckwith, and H. Inouye. 1988. Alteration of the amino terminus of the mature sequence of a periplasmic protein can severely affect protein export in *Escherichia coli. Proc. Natl. Acad. Sci. USA* **85:**7685–7689.

Lichty, J. J., J. L. Malecki, H. D. Agnew, D. J. Michelson-Horowitz, and S. Tan. 2005. Comparison of affinity tags for protein purification. *Protein Expr. Purific.* **41:**98–105.

Lo, K. M., A. Roy, S. F. Foley, J. T. Coll, and S. D. Gillies. 1992. Expression and secretion of an assembled tetrameric CH2-deleted antibody in *E. coli. Hum. Antibodies Hybridomas* **3:**123–128.

Lu, D., X. Jimenez, H. Zhang, P. Bohlen, L. Witte, and Z. Zhu. 2002. Fab-scFv fusion protein: an efficient approach to production of bispecific antibody fragments. *J. Immunol. Methods* **267:**213–226.

MacBeath, G., and P. Kast. 1998. UGA readthrough artifacts—when popular gene expression systems need a pATCH. *Biotechniques* **24:**789–794.

Makrides, S. C. 1996. Strategies for achieving high-level expression of genes in *Escherichia coli. Microbiol. Rev.* **60:**512–538.

Mavrangelos, C., M. Thiel, P. J. Adamson, D. J. Millard, S. Nobbs, H. Zola, and I. C. Nicholson. 2001. Increased yield and activity of soluble single-chain antibody fragments by combining high-level expression and the Skp chaperonin. *Protein Expr. Purific.* **23:**289–295.

McKenzie, K. R., E. Adams, W. J. Britton, R. J. Garsia, and A. Basten. 1991. Sequence and immunogenicity of the 70-kDa heat shock protein of *Mycobacterium leprae. J. Immunol.* **147:**312–319.

Meerman, H. J., and G. Georgiou. 1994. Construction and characterization of a set of *E. coli* strain deficient in all known loci affecting the proteolytic stability of secreted recombinant proteins. *Biotechnology* **12:**1107–1110.

Melton, R. G. 1996. Preparation and purification of antibody-enzyme conjugates for therapeutic applications. *Adv. Drug Deliv. Rev.* **22:**289–301.

Meyer, D. E., and A. Chilkoti. 1999. Purification of recombinant proteins by fusion with thermally-responsive polypeptides. *Nat. Biotechnol.* **17:**1112–1115.

Missiakas, D., J. M. Betton, and S. Raina. 1996. New components of protein folding in extracytoplasmic compartments of *Escherichia coli* SurA, FkpA and Skp/OmpH. *Mol. Microbiol.* **21:**871–884.

Mogensen, J. E., and D. E. Otzen. 2005. Interactions between folding factors and bacteral outer membrane proteins. *Mol. Microbiol.* **57:**326–346.

Mondigler, M., and M. Ehrmann. 1996. Site-specific proteolysis of the *Escherichia coli* SecA protein *in vivo. J. Bacteriol.* **178:**2986–2988.

Morgan-Kiss, R. M., C. Wadler, and J. E. Cronan, Jr. 2002. Long-term and homogeneous regulation of the *Escherichia coli* araBAD promoter by use of a lactose transporter of relaxed specificity. *Proc. Natl. Acad. Sci. USA* **99:**7373–7377.

Mori, H., and K. Ito. 2001. The Sec protein-translocation pathway. *Trends Microbiol.* **9:**494–500.

Mukherjee, K. J., D. C. D. Rowe, N. A. Watkins, and D. K. Summers. 2004. Studies of single-chain antibody expression in quiescent *Escherichia coli. Appl. Environ. Microbiol.* **70:**3005–3012.

Müller, K. M., K. M. Arndt, W. Strittmatter, and A. Plückthun. 1998. The first constant domain (C(H)1 and C(L)) of an antibody used as heterodimerization domain for bispecific miniantibodies. *FEBS Lett.* **422**:259–264.

Munro, S., and H. R. Pelham. 1986. An Hsp70-like protein in the ER: identity with the 78 kd glucose-regulated protein and immunoglobulin heavy chain binding protein. *Cell* **46**:291–300.

Nagai, K., and H. C. Thogersen. 1984. Generation of β-globin by sequence-specific proteolysis of a hybrid protein produced in *Escherichia coli*. *Nature* **309**:810–812.

Neu, H. C., and L. A. Heppel. 1965. The release of enzymes from *Escherichia coli* by osmotic shock and during formation of spheroplasts. *J. Biol. Chem.* **240**:3685–3692.

Nieba, L., A. Honegger, C. Krebber, and A. Plückthun. 1997. Disrupting the hydrophobic patches at the antibody variable/constant domain interface: improved *in vivo* folding and physical characterization of an engineered scFv fragment. *Protein Eng.* **10**:435–444.

Nielsen, H., J. Engelbrecht, S. Brunak, and G. Von Heijne. 1997. Identification of prokaryotic and eukaryotic signal peptides and prediction of their cleavage sites. *Protein Eng.* **10**:1–6.

Nilsson, B., T. Moks, B. Jansson, L. Abrahmsen, A. Elmblad, E. Holmgren, C. Henrichson, T. A. Jones, and M. Uhlen. 1987. A synthetic IgG-binding domain based on staphylococcal protein A. *Protein Eng.* **1**:107–113.

Nordström, K., and B. E. Uhlin. 1992. Runaway-replication plasmids as tools to produce large quantities of proteins from cloned genes in bacteria. *Biotechnology* **10**:661–666.

Oliver, D. B., and J. Beckwith. 1982. Regulation of a membrane component required for protein secretion in *Escherichia coli*. *Cell* **30**:311–319.

Ong, E., N. R. Gilkes, R. A. J. Warren, R. C. Miller, Jr., and D. G. Kilburn. 1989. Enzyme immobilization using the cellulose-binding domain of a *cellulomonas fimi* exoglucanase. *Biotechnology* **7**:604–607.

Ostermeier, M., K. De Sutter, and G. Georgiou. 1996. Eukaryotic protein disulfide isomerase complements *Escherichia coli* dsbA mutants and increases the yield of a heterologous secreted protein with disulfide bonds. *J. Biol. Chem.* **271**:10616–10622.

Pack, P., K. Müller, R. Zahn, and A. Plückthun. 1995. Tetravalent miniantibodies with high avidity assembling in *Escherichia coli*. *J. Mol. Biol.* **246**:28–34.

Pack, P., and A. Plückthun. 1992. Miniantibodies: use of amphipathic helices to produce functional, flexibly linked dimeric Fv fragments with high avidity in *Escherichia coli*. *Biochemistry* **31**:1579–1584.

Padlan, E. A. 1994. Anatomy of the antibody molecule. *Mol. Immunol.* **31**:169–217.

Park, S. J., G. Georgiou, and S. Y. Lee. 1999. Secretory production of recombinant protein by a high

cell density culture of a protease negative mutant *Escherichia coli* strain. *Biotechnol. Prog.* **15**:164–167.

Pérez-Pérez, J., J. L. Barbero, G. Márquez, and J. Gutiérrez. 1996. Different PrlA proteins increase the efficiency of periplasmic production of human interleukin-6 in *Escherichia coli*. *J. Biotechnol.* **49**:245–247.

Pérez-Pérez, J., G. Márquez, J. L. Barbero, and J. Gutiérrez. 1994. Increasing the efficiency of protein export in *Escherichia coli*. *Biotechnology* **12**:178–180.

Persson, M., M. G. Bergstrand, L. Bulow, and K. Mosbach. 1988. Enzyme purification by genetically attached polycysteine and polyphenylalanine affinity tails. *Anal. Chem.* **172**:330–337.

Pogliano, J., A. S. Lynch, D. Belin, E. C. C. Lin, and J. Beckwith. 1997. Regulation of *Escherichia coli* cell envelope proteins involved in protein folding and degradation by the Cpx two-component system. *Genes Dev.* **11**:1169–1182.

Poole, E. S., C. M. Brown, and W. P. Tate. 1995. The identity of the base following the stop codon determines the efficiency of an *in vivo* translational termination in *Escherichia coli*. *EMBO J.* **14**:151–158.

Prinz, W. A., F. Åslund, A. Holmgren, and J. Beckwith. 1997. The role of the thioredoxin and glutaredoxin pathways in reducing protein disulfide bonds in the *Escherichia coli* cytoplasm. *J. Biol. Chem.* **272**:15661–15667.

Puziss, J. W., R. J. Harvery, and P. J. Bassford. 1992. Alterations in the hydrophilic segment of the maltose-binding protein (MBP) signal peptide that affect either export or translation of MBP. *J. Bacteriol.* **174**:6488–6497.

Raffa, R. G., and T. L. Raivio. 2002. A third envelope stress signal transduction pathway in *Escherichia coli*. *Mol. Microbiol.* **45**:1599–1611.

Ramm, K., and A. Plückthun. 2000. The periplasmic *Escherichia coli* peptidylprolyl cis,trans-isomerase FkpA. *J. Biol. Chem.* **275**:17106–17113.

Reilly, D., and D. G. Yansura. 2004. Methods and composition for increasing antibody production. WO 2004 042017.

Richarme, G., and T. D. Caldas. 1997. Chaperone properties of the bacterial periplasmic substrate-binding proteins. *J. Biol. Chem.* **272**:15607–15612.

Richter, S. A., K. Stubenrauch, H. Lilie, and R. Rudolph. 2001. Polyionic fusion peptides function as specific dimerization motifs. *Protein Eng.* **14**:775–783.

Riechmann, L., and S. Muyldermans. 1999. Single domain antibodies: comparison of camel VH and camelised human VH domains. *J. Immunol. Methods* **231**:25–38.

Ringquist, S., S. Shinedling, D. Barrick, L. Green, J. Binkley, G. D. Stormo, and L. Gold. 1992. Translation initiation in *Escherichia coli*: sequences within the ribosome-binding site. *Mol. Microbiol.* **6:**1219–1229.

Rizzitello, A. E., J. R. Harper, and T. J. Silhavy. 2001. Genetic evidence for parallel pathways of chaperone activity in the periplasm of *Escherichia coli. J. Bacteriol.* **183:**679–680.

Robinson, C., and A. Bolhuis. 2001. Protein targeting by the twin-arginine translocation pathway. *Nat. Rev. Mol. Cell Biol.* **2:**350–356.

Rodrigues, M. L., M. R. Shalaby, W. Werther, L. Presta, and P. Carter. 1992. Engineering a humanized bispecific F(ab')$_2$ fragment for improved binding to T cells. *Intl J. Cancer Suppl.* **7:**45–50.

Rodrigues, M. L., B. Snedecor, C. Chen, W. L. T. Wong, S. Garg, G. S. Blank, D. Maneval, and P. Carter. 1993. Engineering Fab' fragments for efficient F(ab')$_2$ formation in *Escherichia coli* and for improved *in vivo* stability. *J. Immunol.* **151:**6954–6961.

Röthlishberger, D., A. Honegger, and A. Plückthun. 2005. Domain interactions in the Fab fragment: a comparative evaluation of the single-chain Fv and Fab format engineered with variable domains of different stability. *J. Mol. Biol.* **347:**773–789.

Sandee, D., S. Tungpradabkul, Y. Kurokawa, K. Fukui, and M. Takagi. 2005. Combination of Dsb coexpression and an addition of sorbitol markedly enhanced soluble expression of single-chain Fv in *Escherichia coli. Biotechnol. Bioeng.* **91:** 418–424.

Saul, F. A., J. P. Arié, B. Vulliez-Le Normand, R. Kahn, J. M. Betton, and G. A. Bentley. 2004. Structural and functional studies of FkpA from *Escherichia coli*, a cis/trans peptdyl-prolyl isomerase with chaperone activity. *J. Mol. Biol.* **335:** 595–608.

Schäfer, U., K. Beck, and M. Muller. 1999. Skp, a molecular chaperone of gram-negative bacteria, is required for the formation of soluble periplasmic intermediates of outer membrane proteins. *J. Biol. Chem.* **274:**24567–24575.

Schäffner, J., J. Winter, R. Rudolph, and E. Schwarz. 2001. Cosecretion of chaperones and low-molecular-size medium additives increases the yield of recombinant disulfide-bridged proteins. *Appl. Environ. Microbiol.* **67:**3994–4000.

Schmid, F. X. 2002. Prolyl isomerases. *Adv. Protein Chem.* **59:**243–282.

Schmidt, T. G. M., and A. Skerra. 1993. The random peptide library-assisted engineering of a C-terminal affinity peptide, useful for the detection and purification of a functional Ig Fv fragment. *Protein Eng.* **6:**109–122.

Schmidt, T. G. M., and A. Skerra. 1994. One-step affinity purifiaction of bacterially produced proteins by means of the "Strep tag" and immobilized recombinant core streptavidin. *J. Chromatogr.* **676:** 337–345.

Schmiedl, A., F. Breitling, and S. Dübel. 2000b. Expression of a bispecific dsFv-dsFv' antibody fragment in *Escherichia coli. Protein Eng.* **13:**725–734.

Schmiedl, A., F. Breitling, C. H. Winter, I. Queitsch, and S. Dübel. 2000a. Effects of unpaired cysteines on yield, solubility and activity of different recombinant antibody constructs expressed in *E. coli. J. Immunol. Meth.* **242:**101–114.

Shine, J., and L. Dalgarno. 1974. The 3' terminal sequence of *Escherichia coli* 16S ribosomal RNA: complementarity to nonesense triplets and ribosome binding sites. *Proc. Natl. Acad. Sci. USA* **71:** 1342–1346.

Simmons, L., L. Klimowski, D. E. Reilly, and D. G. Yansura. 2002b. Prokaryotically produced antibodies and uses thereof. WO 02/061090.

Simmons, L. C., D. Reilly, L. Klimowski, T. S. Raju, G. Meng, P. Sims, K. Hong, R. L. Shields, L. A. Damico, P. Rancatore, and D. G. Yansura. 2002a. Expression of full-length immunoglobulins in *Escherichia coli*: rapid and efficient production of aglycosylated antibodies. *J. Immunol. Methods* **263:**133–147.

Simmons, L. C., and D. G. Yansura. 1996. Translational level is a critical factor for the secretion of heterologous proteins in *Escherichia coli. Nat. Biotechnol.* **14:**629–634.

Sjöström, M., S. Wold, A. Wieslander, and L. Rilfors. 1987. Signal peptide amino acid sequences in *Escherichia coli* contain information related to final protein localization. A multivariate data analysis. *EMBO J.* **6:**823–831.

Skerra, A. 1994. Use of the tetracycline promoter for the tightly regulated production of a murine antibody fragment in *Escherichia coli. Genes* **151:**131–135.

Skerra, A., and A. Plückthun. 1988. Assembly of a functional immunoglobulin Fv fragment in *Escherichia coli. Science* **240:**1038–1041.

Skerra, A., and A. Plückthun. 1991. Secretion and *in vivo* folding of the Fab fragment of the antibody McPC603 in *Escherichia coli*: influence of disulphides and *cis*-prolines. *Protein Eng.* **4:**971–979.

Sletta, H., A. Nedal, T. E. V. Aune, H. Hellebust, S. Hakvåg, R. Aune, T. E. Ellingsen, S. Valla, and T. Brautaset. 2004. Broad-host-range plasmid pJB658 can be used for industrial-level production of a secreted host-toxic single-chain antibody fragment in *Escherichia coli. Appl. Environ. Microbiol.* **70:**7033–7039.

Sone, M., Y. Akiyama, and K. Ito. 1997. Differential *in vivo* roles played by DsbA and DsbC in the formation of protein disulfide bonds. *J. Biol. Chem.* **272:**10349–10352.

Spiess, C., A. Beil, and M. Ehrmann. 1999. A temperature-dependent switch from chaperone to protease in a widely conserved heat shcok protein. *Cell* **97:**339–347.

Stemmer, W. P. C., S. K. Morris, C. R. Kautzer, and B. S. Wilson. 1993. Increased antibody expression from *Escherichia coli* through wobble-base library mutagenesis by enzymatic inverse PCR. *Gene* **123**:1-7.

Stofko-Hahn, R. E., D. W. Carr, and J. D. Scott. 1992. A single step purification for recombinant proteins. Characterization of a microtubule associated protein (MAP2) fragment which associates with the type II cAMP-dependent protein kinase. *FEBS Lett.* **302**:274–278.

Studier, F. W. 1991. Use of bacteriophage T7 lysozyme to improve an inducible T7 expression system. *J. Mol. Biol.* **219**:37–44.

Studier, F. W., A. H. Rosenberg, J. J. Dunn, and J. W. Dubendorff. 1990. Use of T7 RNA polymerase to direct expression of cloned genes. *Methods Enzymol.* **185**:60–89.

Sydor, J. R., M. Mariano, S. Sideris, and S. Nock. 2002. Establishment of intein-mediated protein ligation under denaturing conditions: C-terminal labeling of a single-chain antibody for biochip screening. *Bioconjug. Chem.* **13**:707–712.

Tan, N. S., B. Ho, and J. L. Ding. 2002. Engineering of a novel secretion signal for cross-host recombinant protein expression. *Protein Eng.* **15**:337–345.

Thies, M. J. W., J. Mayer, J. G. Augustine, C. A. Frederick, H. Lilie, and J. Buchner. 1999. Folding and association of the antibody domain CH3: prolyl isomerization preceeds dimerization. *J. Mol. Biol.* **293**:67–79.

Todorovska, A., R. C. Roovers, O. Dolezul, A. A. Kortt, H. R. Hoogenboom, and P. J. Hudson. 2001. Design and application of diabodies, triabodies and tetrabodies for cancer targeting. *J. Immunol. Methods* **248**:47–66.

Torriani, A. 1990. From cell membrane to nucleotides: the phosphate regulon in *Escherichia coli.* *Bioessays* **12**:371–376.

Trepod, C. M., and J. E. Mott. 2002. A spontaneous runaway vector for production-scale expression of bovine somatotropin from *Escherichia coli.* *Appl. Microb. Biotechnol.* **58**:84–88.

Tziatios, C., D. Schubert, M. Lotz, D. Gundogan, H. Betz, H. Schägger, W. Haase, F. Duong, and I. Collinson. 2004. The bacterial protein-translocation complex: SecYEG dimers associate with one or two SecA molecules. *J. Mol. Biol.* **340**:513–524.

Vaara, M. 1992. Agents that increase the permeability of the outer membrane. *Microbiol. Rev.* **56**:395–411.

van Dijl, J. M., A. De Jong, H. Smith, S. Bron, and G. Venema. 1991. Signal peptidase I overproduction results in increased efficiencies of export and maturation of hybrid secretory proteins in *Escherichia coli. Mol. Gen. Genet.* **227**:40–48.

Völkel, T., T. Korn, M. Bach, R. Müller, and R. E. Kontermann. 2001. Optimized linker sequences for the expression of monomeric and dimeric bispecific single-chain diabodies. *Protein Eng.* **14**:815–823.

von Heijne, G. 1990. The signal peptide. *J. Membr. Biol.* **115**:195–201.

Voss, S., and A. Skerra. 1997. Mutagenesis of a flexible loop in streptavidin leads to higher affinity for the Strep-tag II peptide and improved performance in recombinant protein purification. *Protein Eng.* **10**:975–982.

Wada, K. N., Y. Wada, H. Doi, F. Ishibashi, T. Gojobori, and T. Ikemura. 1991. Codon usage tabulated from the GenBank genetic sequence data. *Nucleic Acids Res.* **19**:1981–1986.

Walker, P. A., L. E. C. Leong, P. W. P. Ng, S. H. Tan, S. Waller, D. Murphy, and A. G. Porter. 1994. Efficient and rapid affinity purification of proteins using recombinant fusion proteases. *Biotechnology* **12**:601–605.

Weir, A. N. C., and N. A. Bailey. 1997. Process for obtaining antibodies utilizing heat treatment. U.S. patent 5,665,866.

Wels, W., I. M. Harwerth, M. Zwickl, N. Hardman, B. Groner, and N. E. Hynes. 1992. Construction, bacterial expression and characterization of a bifunctional single-chain antibody-phosphatase fusion protein targeted to the human *erbB*-2 receptor. *Biotechnology* **10**:1128–1132.

Wenthzel, A. M. K., M. Stancek, and L. A. Isaksson. 1998. Growth phase dependent stop codon readthrough and shift of translation reading frame in *Escherichia coli. FEBS Lett.* **421**:237–242.

Wood, C. R., M. A. Boss, T. P. Patel, and J. S. Emtage. 1984. The influence of messenger RNA secondary structure on expression of an immunoglobulin heavy chain in *Escherichia coli. Nucleic Acids Res.* **12**:3937–3950.

Wörn, A., and A. Plückthun. 2001. Stability engineering of an antibody single-chain Fv fragments. *J. Mol. Biol.* **305**:989–1010.

Yi, K. S., J. Chung, K. H. Park, K. Kim, S. Y. Im, C. Y. Choi, M. J. Im, and U. H. Kim. 2004. Expression system for enhanced green fluorescent protein conjugated recombinant antibody fragment. *Hybrid. Hybridomics* **23**:279–286.

Zapata, G., J. B. B. Ridgway, J. Morden, G. Osaka, W. L. T. Wong, G. L. Bennett, and P. Carter. 1995. Engineering linear F(ab′)₂ fragments for efficient production in *Escherichia coli* and enhanced antiproliferative activity. *Protein Eng.* **8**:1057–1062.

Zhang, Y., D. R. Olsen, K. B. Nguyen, P. S. Olson, E. T. Rhodes, and D. Mascarenhas. 1998. Expression of eukaryotic proteins in soluble form in *Escherichia coli. Protein Expr. Purific.* **12**:159–165.

Zhang, Z., Z. H. Li, F. Wang, M. Fang, C. C. Yin, Z. Y. Zhou, Q. Ling, and H. L. Huang. 2002. Overexpression of DsbC and DsbG markedly improves soluble and functional expression of single-chain Fv antibodies in *Escherichia coli. Protein Expr. Purific.* **26:**218–228.

Zhang, Z., L. P. Song, M. Fang, F. Wang, D. He, R. Zhao, J. Liu, Z. Y. Zhou, C. C. Yin, Q. Lin, and H. L. Huang. 2003. Production of soluble and functional engineered antibodies in *Escherichia coli* improved by FkpA. *Biotechniques* **35:**1032–1042.

Zhu, Z., L. G. Presta, G. Zapata, and P. Carter. 1997. Remodeling domain interfaces to enhance heterodimer formation. *Protein Sci.* **6:**781–788.

Zhu, Z., G. Zapata, R. Shalaby, B. Snedcor, H. Chen, and P. Carter. 1996. High level secretion of a humanized bispecific diabody from *Escherichia coli. Biotechnology* **14:**192–196.

PROTEIN
COMPOSITION

METHODS FOR THE COMPUTATIONAL PREDICTION OF PERIPLASMIC PROTEINS

Jennifer L. Gardy and Fiona S. L. Brinkman

22

In this chapter, we address several different computational methods for the identification of periplasmic proteins from sequence information alone. These methods range from localization predictions tools like PSORTb and Proteome Analyst, which require little user intervention and provide high-quality predictions, to methods such as homology-based detection and signal peptide prediction, which can be helpful if predictive tools fail to provide adequate information. The benefits, pitfalls, and performance of the methods are discussed, and we present an approach for the optimal computational identification of periplasmic proteins from a sequenced genome.

A SHORT HISTORY OF METHODS FOR THE IDENTIFICATION OF PERIPLASMIC PROTEINS

For much of the twentieth century, the bulk of periplasmic localization experiments focused on either a single protein or a small selection of proteins. The most common technique for identifying these proteins involved fractioning the bacterial cell into its constituent compartments, isolating the periplasmic fraction, and purifying a protein or proteins of interest.

While such traditional methods can produce high-quality experimental data, they require both time and money and are small in scale.

The need for faster, more affordable, and larger-scale methods led to the development of early computational tools for localization analysis. One of the first approaches involved the prediction of signal peptides—N-terminal protein sequences directing the export of a protein out of the bacterial cytoplasm. Weight matrix-based analyses were frequently employed to identify these signal sequences. In this approach, frequency values reflecting each amino acid's occurrence in known signal peptides are assigned to each residue in a query sequence. A sliding window then moves down the sequence, summing the frequency scores. In this fashion, the region most likely to represent a signal peptide can easily be identified (von Heijne, 1986).

While signal peptide prediction allowed researchers to roughly classify a protein as cytoplasmic or noncytoplasmic, it did not provide further information about where a noncytoplasmic protein may reside, nor did it account for proteins with nontraditional export signals. Thus in 1991, PSORT I was released (Nakai and Kanehisa, 1991). PSORT I was the first true localization prediction method to be developed, capable of assigning a gram–negative

Jennifer L. Gardy and Fiona S. L. Brinkman, Department of Molecular Biology and Biochemistry, Simon Fraser University, Burnaby, BC V5A 1S6, Canada.

The Periplasm
Edited by Michael Ehrmann © 2007 ASM Press, Washington, D.C.

query protein to the cytoplasm, cytoplasmic membrane, periplasm, or outer membrane. The program employed a multicomponent approach to prediction: features influencing localization—including amino acid composition, signal peptides, and transmembrane helices—were identified in a query protein, and the resulting information was integrated to generate a final prediction.

Until 2003, PSORT I represented the only publicly available localization prediction method for prokaryotes. However, the program displayed low precision (Gardy et al., 2003) and did not predict secreted proteins. As a result, many secreted proteins were incorrectly predicted as periplasmic.

CURRENT COMPUTATIONAL PREDICTIVE METHODS FOR THE IDENTIFICATION OF PERIPLASMIC PROTEINS

Recognizing that PSORT I could be significantly improved, our group set out to develop a new method, PSORTb, for the prediction of protein subcellular localization in bacteria (Gardy et al., 2003). Subsequently, a number of research groups turned their attention to the problem of subcellular localization prediction, and as of September 2005, seven localization prediction tools relevant to the prediction of

periplasmic proteins from gram-negative bacteria have now been developed. While these seven methods vary with respect to the algorithms employed, the number of localization sites they are able to assign a protein to, and their predictive performance, they all offer an improvement in terms of both precision and coverage over PSORT I. The methods are described below, with a particular emphasis on their prediction of periplasmic proteins. A summary of the tools and their features is presented in Table 1.

A Word about Performance Evaluation

Before we describe the predictive methods, it is important to briefly address how the performance of a predictive tool is calculated and reported. Without an understanding of how terms such as "precision," "recall," and "MCC" are defined, it can be difficult to properly evaluate a method's performance.

When a predictive method is reported in the literature, the authors may use one or more of a variety of metrics to describe its performance. The choice of metrics depends primarily on the background of the authors and the aspect of their method that they wish to emphasize. All metrics, however, rely on four basic statistics—true positives (TP), false negatives

TABLE 1 A summary of the computational methods for bacterial protein localization prediction[a]

Program	Reference	Analytical method	Precision[b]	Recall	Open source?[c]	Forces predictions?[d]
PSORT[b]	Gardy et al., 2005	Multicomponent	High	Moderate	Yes	No
Proteome Analyst	Lu et al., 2004	Annotation keywords	High	High	No	No
SubLoc	Hua and Sun, 2003	Support vector machine	Moderate	Low	No	Yes
CELLO	Yu et al., 2004	Support vector machine	Low	Low	No	Yes
PSLpred	Bhasin et al., 2005	Support vector machine	Moderate	Low	No	Yes
LOCtree	Nair and Rost, 2005	Support vector machine	Moderate	Low	No	Yes
P-CLASSIFIER	Wang et al., 2005	Support vector machine	Moderate	Low	No	Yes

[a]Methods are listed in the order they appear in the chapter.

[b]For precision and recall, high refers to a value of 90% or greater, moderate to a value of 80 to 89%, and low to a value of less than 79%. These values were calculated using an independent test set of 144 proteins for all methods except SubLoc and LOCtree. Precision and recall values for these methods were taken from the manuscripts describing the programs. Note, recall of a method when applied to a whole genome or hypothetical proteins may be lower than the values reported here, which were calculated using a test set of well-characterized proteins.

[c]"Open source" indicates the authors of the program have made its source code available for download and modification by users.

[d]"Forces predictions" refers to methods that return a prediction for every protein submitted, even if the degree of confidence in the prediction is low.

(FN), false positives (FP), and true negatives (TN). Table 2 illustrates the matrix used to calculate these values.

Predictive methods developed by biologists tend to emphasize the importance of "quality," or correct predictions, over "quantity," or a high number of predictions. A barometer of a method's ability to generate correct predictions is the precision metric, also referred to as specificity. This is calculated as $\dfrac{TP}{TP + FP}$ and is reported as a percentage figure. A precision value of 95% indicates that for every 100 predicted periplasmic proteins, five of these will be false positives, or nonperiplasmic proteins.

Precision values are typically reported together with a method's recall. Recall, also referred to as sensitivity, is calculated as $\dfrac{TP}{TP + FP}$ and reflects a method's ability to identify all true positive cases. A recall value of 95% indicates that for every 100 periplasmic proteins in the test set, five of these will be false negatives—in other words, they will be predicted as nonperiplasmic when in fact they are periplasmic proteins.

Other metrics are frequently employed, primarily by computer science researchers developing predictive methods. Because in computing science the focus is on optimization of a method rather than erring on the side of precision/making a correct prediction, neither precision nor recall is emphasized at the expense of the other. Instead, a balance between the two is considered ideal. The Matthews Correlation Coefficient, or MCC, is a metric that rewards optimal combinations of precision and recall. Calculated as:

$$\frac{(TP)(TN) - (FN)(FP)}{\sqrt{(TP + FN)(TP + FP)(TN + FN)(TN + FP)}},$$

the MCC is reported as a value between 0 and 1. An MCC of 1 indicates a perfect predictive method, one that retrieves all periplasmic proteins in a test set and returns no false positives, while an MCC of 0 would indicate completely random prediction.

A small number of papers reporting subcellular localization prediction methods from the computer science domain report only a single metric—accuracy. To many, the use of the word accuracy implies a measure of quality, or $\dfrac{TP + TN}{TP + FP + TN + FN}$. In fact, some computer scientists' definition of accuracy is the same as our earlier definition of recall, or $\dfrac{TP}{TP + FN}$. This metric rewards methods that generate a large quantity of predictions and should not be used as an estimate of a method's false positive rate.

Therefore, when evaluating a predictive method it is important to be aware of which measure of accuracy is being reported by the authors. It is also critical to keep the requirements of your analysis in mind. In experiments where an incorrect prediction could cost time and money in expensive and time-consuming downstream laboratory analyses, a high-precision approach is often best and will ensure that laboratory study is focused on the correct protein and/or localization. However, in experiments where the downstream analysis is fast and cheap or in situations where one doesn't wish to miss a true positive and doesn't mind a few false positive results, a high recall approach may be preferred.

PSORTb—a Precise, Multicomponent Approach

PSORTb (Gardy et al., 2005), originally released in 2003 (Gardy et al., 2003), is the first,

TABLE 2 Statistics used in the calculation of performance[a]

Actual localization	Predicted localization	
	Periplasmic	Nonperiplasmic
Periplasmic	True positive	False negative
Nonperiplasmic	False positive	True negative

[a]Precision is calculated as $\dfrac{TP}{TP + FP}$. Recall is calculated as $\dfrac{TP}{TP + FN}$.

and to date only, method to follow in PSORT I's footsteps, offering a multicomponent approach to prediction that incorporates the detection of several sequence features known to influence subcellular localization. With a reported overall precision of 96%, it continues to be the most precise method presently available. It is also the most comprehensive method, capable of assigning a query protein to one or more of four gram-positive or five gram-negative localization sites. Most other methods do not make predictions for gram-positive bacteria at all, and the one method that does only sorts a protein to one of three potential sites.

When a query protein is submitted to PSORTb, it passes through a series of analytical steps. Each step, or module, is an independent piece of software that scans the protein for the presence or absence of a particular feature characteristic of a specific localization site. The modules return as output either a predicted localization site or, if the feature is not detected, a result of "unknown." The number of steps differs depending on whether a protein is gram negative or gram positive in origin—12 modules are used to analyze gram-negative proteins and 10 are used to analyze gram-positive proteins. Modules include SCL-BLAST for homology-based detection, the HMMTOP transmembrane helix prediction tool (Tusnady and Simon, 2001), a signal peptide prediction tool, a frequent subsequence-based support vector machine, and motif- and profile-matching modules. The predicted localization sites outputted by each module are then integrated by a Bayesian network into a final prediction. This comes in the form of a score distribution, showing the likelihood of the query protein being resident at each of the possible localization sites on a scale of 0 to 10.

Of the 12 modules used in the analysis of gram-negative proteins, four directly contribute to the identification of periplasmic proteins. The SCL-BLAST and SCL-BLASTe modules return near or exact matches, respectively, to a database of proteins of known localization. The motif module searches for the occurrence of localization-specific PROSITE motifs, includ-ing 13 periplasmic patterns. Finally, a support vector machine-based module scans a protein for the presence of 27,804 short sequences that are found frequently in periplasmic proteins and infrequently in proteins at other localization sites. When used in combination with support vector machines specific to the other four localization sites, this latter module is particularly effective at differentiating periplasmic proteins from those located in the outer membrane or extracellular space.

PSORTb was trained and tested with a set of 1,591 gram-negative and 576 gram-positive proteins obtained from the second release of our protein localization database ePSORTdb (Rey et al., 2005). The dataset includes 276 periplasmic proteins from 80 organisms. Note that the proteins in ePSORTdb were identified manually from the literature and represent a high-quality source of training data. Indeed, many of the other methods described in this chapter use ePSORTdb as a source of training and testing data. When PSORTb v2.0 (Gardy et al., 2005) was evaluated using this dataset, we reported that the method achieved 95.5% precision and 69.2% recall.

PSORTb offers several advantages over other prokaryotic localization prediction methods. It achieves the highest precision of any method to date due to both its robust multicomponent analysis and the fact that it does not force predictions—if not enough information is available to make a confident prediction, a result of "unknown" is returned. It is the only program presently available that flags proteins with potential dual localization sites, and is the only one to offer precomputed predictions for over 200 bacterial genomes, which are available in the cPSORTdb database at the PSORTdb site (http://db.psort.org). The principal drawback to the method, however, is that the emphasis on precision comes at the expense of recall. Predictions are only generated for approximately 60 to 70% of the proteins encoded in a given genome, and the recall for localization sites with small training datasets, such as the periplasm, is low. PSORTb accepts web-based submission of one or more sequences at http://

www.psort.org/psortb. The PSORTb source code is also freely available from the website under the GNU General Public License for Linux, Mac OSX, and Solaris operating systems, allowing users to run the program locally and customize it to suit their needs.

Proteome Analyst—an Accurate Annotation Keyword-Based Approach

Proteome Analyst's subcellular localization prediction server (Lu et al., 2004), released shortly after the first release of PSORTb, employs a drastically different approach to localization prediction. It too offers accurate classification, while improving on PSORTb's recall to generate more predictions.

Proteome Analyst employs an annotation keyword-based approach consisting of two steps. In the first step, the query protein is compared using BLAST to the SwissProt database, returning a set of homologous proteins with manually curated annotation. Keywords in the annotation that might be indicative of a particular localization site are extracted from the SwissProt records and, in the second step, are passed to a Naïve Bayes classifier. This classifier then uses the keywords to assign the query protein to one of three gram-positive or five gram-negative localization sites. The program is also capable of generating predictions for animal, plant, archaeal, and fungal sequences. Proteome Analyst's gram-negative classifier was trained and tested on the first release of the ePSORTdb dataset described above, and the authors report that it achieved precision of 89.8% and recall of 87.3%.

Proteome Analyst's unique approach and high recall make it an excellent complement to PSORTb, in particular, at sites such as the periplasm, for which PSORTb displays low recall. Like PSORTb, Proteome Analyst does not force predictions, resulting in high precision classification. A caveat to the method, however, is that to generate a prediction, a query protein must have homologs in the SwissProt database—hence not every protein encoded in a genome will return a result. The method may not perform as well when used to analyze bacterial sequences that are not closely related to well-studied or model bacteria.

Proteome Analyst can be accessed at http://www.cs.ualberta.ca/~bioinfo/PA/Sub/. While the tool is not available for download and thus cannot be used locally, the site does offer a selection of precomputed predictions for microbial and other genomes. The authors also note that they are willing to run predictions for specific genomes requested by users.

Amino Acid Composition Support Vector Machine Methods

While PSORTb and Proteome Analyst achieve the highest precision values of the available localization prediction tools by utilizing expert knowledge in the form of sequence features and annotations, they do not always return a result. If the quantity of predictions is of more importance than the quality, it is often necessary to turn to an amino acid composition-based support vector machine method. These methods exploit the differences in frequency of the 20 amino acids across different cellular compartments and are capable of making predictions when no prior information about a protein is available.

Support vector machine (Vapnik, 1995), or SVM, is a kernel learning algorithm trained to determine whether a query belongs or does not belong to a particular class. Using localization prediction as an example, an SVM might be used to determine if a protein is periplasmic or nonperiplasmic. By training a series of such binary SVMs, one for each localization site, multilocalization classification is possible.

SubLoc (Hua and Sun, 2001) was the first publicly available SVM-based localization tool to be released. Capable of sorting prokaryotic proteins to the cytoplasm, periplasm, or extracellular space, it generates its predictions using overall amino acid composition. When the method was evaluated on a dataset of 202 periplasmic proteins created by the SubLoc developers, it achieved a reported precision of 86.4% and a recall of 78.7%.

While the release of SubLoc marked an important milestone in the use of SVMs for localization prediction, the method itself carries two significant caveats. Because the program only sorts proteins to three compartments, known and suspected cytoplasmic membrane and outer membrane proteins must be removed prior to the analysis. Furthermore, the method does not distinguish between gram-positive and gram-negative proteins, thus proteins from a gram-positive organism can be mistakenly classified as periplasmic. SubLoc accepts web-based submissions of one or more sequences at http://www.bioinfo.tsinghua.edu.cn/SubLoc/.

A year after the release of SubLoc, CELLO was developed (Yu et al., 2004). CELLO also employs a composition-based SVM analytical approach, but it expands upon the technique used by SubLoc by implementing multiple composition-based SVMs. Five classes of SVMs are used: one that analyzes overall amino acid composition, one that incorporates sequence order information by analyzing dipeptide composition, and three that utilize modified composition analysis, in which amino acids are grouped according to their physiochemical properties. The predictions outputted by the five classes of SVM are integrated to generate a final prediction. The method is applicable only to gram-negative bacteria and can assign a protein to one of five localization sites. When evaluated on a set of 244 periplasmic proteins derived from the first release of the ePSORTdb database, the authors report that CELLO achieved 86.9% recall, with an MCC of 0.8. One or more FASTA format sequences can be submitted to the CELLO webserver at http://cello .life.nctu.edu.tw/cgi/main.cgi.

A third SVM-based method, PSLpred, was released in 2005 (Bhasin et al., 2005). Like CELLO, PSLpred uses different classes of SVMs to assign a query protein to one of five gram-negative localization sites. In addition to three SVMs that analyze overall composition, dipeptide composition, and composition incorporating physiochemical groupings, PSLpred also employs a PSI-BLAST module. Like PSORTb's SCL-BLAST module, this involves comparing the query to a database of proteins of known

localization. When PSLpred was evaluated using the 244 periplasmic proteins from the first release of the ePSORTdb dataset, it achieved a reported recall of 90.6% and an MCC of 0.84. PSLpred is available at http://www.imtech.res .in/raghava/pslpred/. Registration is required to use the program, and proteins must be submitted one at a time.

LOCtree (Nair and Rost, 2005) combines SVM analysis with a flowchart-style decision system designed to mimic cellular sorting. In the first step, an amino acid composition-based SVM determines whether the query protein is cytoplasmic or noncytoplasmic. Noncytoplasmic proteins are then passed to a second SVM, which determines whether the protein is periplasmic or extracellular. Like SubLoc, LOCtree only assigns a protein to one of three localizations, thus membrane proteins must be removed from the dataset prior to analysis. LOCtree was trained and tested using a dataset of 125 periplasmic proteins created by the program's authors. The authors report a precision of 86% and recall of 62%. The low recall is likely due to the small size of the training dataset, which contains both proteins known to be localized to the periplasm and predicted periplasmic proteins. Up to 100 sequences at a time can be submitted to LOCtree at http://cubic .bioc.columbia.edu/cgi-bin/var/nair/loctree/ query.

The most recent SVM-based method to be released is P-CLASSIFIER (Wang et al., 2005). Capable of sorting gram-negative proteins to one of five possible localizations, P-CLASSIFIER implements 15 SVMs in its analysis. The SVMs examine segments of length n in the sequence, where $n = 1$ to 4. Amino acids within a segment are further grouped according to physiochemical properties. P-CLASSIFIER was trained and tested on the ePSORTdb v.1.0 dataset of 244 periplasmic proteins, achieving a recall of 86.9% and an MCC of 0.81. P-CLASSIFIER accepts Web-based submissions of up to 100 sequences at a time at http://protein .bii.a-star.edu.sg/localization/gram-negative/ introduction.html.

A Comparison of Tools for Periplasmic Protein Prediction

In the paper reporting PSORTb v.2.0, Gardy et al. compared the performance of PSORTb, CELLO, and Proteome Analyst using a set of 144 novel proteins—32 of which are periplasmic—that are not contained in the training data of any of the methods (Gardy et al., 2005). Table 3 presents the results of this analysis for overall prediction and periplasmic prediction and extends the analysis to include PSLpred and P-CLASSIFIER. SubLoc and LOCtree were not included as these methods only predict three localization sites rather than five, and were trained using data that might overlap with the testing data used here.

A comparison of the five methods reveals that PSORTb achieves the highest precision, while Proteome Analyst achieves the highest recall, together with excellent precision. Both of these methods outperform the amino acid composition-based SVMs. Of the three SVM-based programs, PSLpred displays the highest performance with respect to precision and recall, both for periplasmic proteins and overall.

Note, however, that the 144 proteins used in the present evaluation are well characterized. Many of these are homologous to other proteins and are thus easily predicted by PSORTb and Proteome Analyst. This may result in a slight overestimation of these programs' recall—the actual predictive coverage when these methods are applied to whole genomes will be lower, though the precision will likely remain the same. The SVM-based methods will provide more predictions when applied to whole genome datasets, but they will still exhibit the low precision noted here.

ALTERNATIVE METHODS FOR THE IDENTIFICATION OF PERIPLASMIC PROTEINS

When attempting to predict a protein's localization site from sequence information alone, the automated methods described in the preceding section represent a critical first step. They are simple to use, provide results rapidly, require little to no user intervention, and, in the case of PSORTb and Proteome Analyst, can provide very accurate results. In certain cases, however, the results of an automated analysis may not be informative enough. PSORTb and Proteome Analyst may return a localization of "unknown" if a confident prediction cannot be made, while the lower precision SVM methods may return predictions requiring additional supporting evidence. In such cases, a researcher may turn to alternative methods for acquiring localization information.

TABLE 3 An independent comparison of the performance of five prokaryotic subcellular localization prediction methods

Localization	Method[a]	True positives	False positives	False negatives	Precision (%)	Recall (%)
Periplasm (32 proteins)	PSORTb	26	0	6	100.0	81.3
	Proteome Analyst	29	1	3	96.7	90.6
	CELLO	16	6	16	72.7	50.0
	PSLpred	20	4	12	83.3	62.5
	P-CLASSIFIER	13	3	19	81.3	40.6
All subcellular localizations (144 proteins)	PSORTb	122	3	22	97.6	84.7
	Proteome Analyst	126	14	18	90.0	87.5
	CELLO	103	41	41	71.5	71.5
	PSLpred	113	31	31	78.5	78.5
	P-CLASSIFIER	103	41	41	71.5	71.5

[a]Methods are listed in the order they appear in the chapter.

Identification of Periplasmic Proteins from Specialized Databases and Homology-Based Detection

With over 719 molecular biology databases available to the research community (Galperin, 2005), retrieving a specific subset of sequences has become a relatively trivial task. Several of these databases contain localization-related information, enabling the extraction of periplasmic proteins. Such data can then be used as a source of training/testing proteins in the development of a new predictive algorithm, as the input to a data-mining experiment to identify periplasm-specific patterns or trends, or as the basis for a local BLAST database. Because localization is an evolutionarily conserved trait, homologs tend to occur at the same site within a cell (Nair and Rost, 2002). Thus, if a query protein displays significant similarity to a known or confidently predicted periplasmic protein, chances are the query is also periplasmic. Note, however, that exceptions to this rule exist. For example, while the *Escherichia coli* trehalases TreF and TreA are homologous, TreF is found in the cytoplasm (Horlacher et al., 1996) and TreA is found in the periplasm (Boos et al., 1987).

UniProt

UniProt, the Universal Protein Resource, was established in 2002 through the union of Swiss-Prot, TrEMBL, and PIR (Apweiler et al., 2004). The SwissProt component of UniProt is the largest database of manually curated annotations of proteins currently available, spanning all kingdoms of life. The latest release of the full UniProt resource, which includes additional computationally derived annotations, contains over 790,000 bacterial proteins.

One of the annotation fields that may be found in a UniProt record is "SUBCELLULAR LOCATION." While common sense would dictate that searching for "periplasm" in this field would return all periplasmic proteins in UniProt, this is not the case. Because a controlled vocabulary is not utilized in this particular field, a single localization site may be described by many different text strings. For example, a recent search of the bacterial proteins with annotated localizations in UniProt returned 460 different types of annotation in this field! Nearly 100 of these may indicate a potential periplasmic localization or association (e.g., "membrane-anchored"), while 52 directly reference the periplasm.

With such a diverse set of annotations describing localization, it is clear that a traditional plain text search will only recover a small proportion of proteins of interest. Thus the ideal approach for retrieving the maximal number of periplasmic proteins is to download the complete UniProt database and manually filter the records, screening out nonperiplasmic proteins. Through the use of simple Perl scripts, a researcher can quickly remove any nonbacterial UniProt records and any records lacking an annotation in the "SUBCELLULAR LOCATION" field. By parsing the remaining records into an easy-to-manipulate format such as tab-delimited text format, researchers can then identify periplasmic proteins by either manually reviewing each annotated localization site or extracting any records with an instance of the word "periplasm." The former approach is slow, but has the advantage of allowing the researchers to incorporate their own expert knowledge into the review process. The latter approach is much faster, in particular, when using a database system such as MySQL, but it will miss potential periplasmic proteins, such as those that are simply annotated as "membrane-anchored."

If downloading the UniProt database and parsing out proteins of interest are not feasible, a second approach is to use one of the UniProt querying systems. Entry points to UniProt are provided by EBI, ExPASy, and PIR, with each entry point offering a different set of tools for querying the database. The SRS tool at the ExPASy site is the most user-friendly and powerful of these tools, offering excellent flexibility, an intuitive querying interface, and the ability to download up to 10,000 results at a time.

Once a subset of proteins of interest has been identified from UniProt, it can be used as a local database for a BLAST search. This first requires that the sequences of interest be in FASTA format. This is easily accomplished with a tool available through ExPASy, which returns such a file given a list of accession numbers. Next, the formatdb utility packaged with NCBI's standalone BLAST program is used to transform the FASTA file into a BLAST-ready database. Once the database has been formatted, a user can query these proteins locally using BLASTp.

Many of the annotations of periplasmic proteins in UniProt are based on similarity—not direct experimental study—and thus there may be a number of proteins incorrectly annotated as periplasmic in this database. For this reason, additional resources have been developed to address this issue as described below, including datasets of proteins that have been experimentally demonstrated to reside in the periplasm and other localization sites.

PSORTdb—a Database of Bacterial Protein Subcellular Localization

PSORTdb (Rey et al., 2005) is a specialized database designed to meet the needs of the bacterial protein localization community. It comprises ePSORTdb, containing proteins of experimentally verified localization, and cPSORTdb, containing proteins whose localization has been predicted using PSORTb. ePSORTdb contains 2,171 proteins and cPSORTdb contains 599,459 proteins as of September 2005.

ePSORTdb was created by first identifying bacterial proteins with an annotated localization site from SwissProt and then filtering this list to remove any proteins with localizations annotated as "probable," "potential," or "by similarity." The resulting list was then manually checked against PubMed to confirm the annotated localization sites. In addition, "localization annotation jamborees" were performed in our laboratory to identify additional proteins of experimentally known localization in other literature and in resources such as EcoSal (Neidhardt et al., 2004). ePSORTdb thus represents a small, high-quality dataset and as such has been used as training or testing data for

many of the automated localization prediction methods mentioned earlier in the chapter.

cPSORTdb stores the PSORTb predicted localization site for the proteins encoded in sequenced bacterial genomes available through NCBI. As of September 2005, it contains data from over 200 organisms. Because PSORTb is the most precise localization prediction method available, these predictions are made with a high level of confidence.

Both ePSORTdb and cPSORTdb can be accessed in several ways. For users new to the databases, the Browse option provides quick and easy access. At each stage of browsing, a user can choose to filter proteins according to localization, Gram stain, phylum, class, or organism. Any combination of these filters is possible, from "show me all periplasmic proteins" to "show me all the proteins from *Haemophilus influenzae*" to "show me all periplasmic proteins from gamma-proteobacteria."

For those users who wish to take a more advanced approach, the databases can be queried with a text search. In Simple Search mode, users enter a keyword, which is then searched against all fields of the database. In Advanced Search mode, users can select one or more fields to search against, combining them with Boolean operators. As a user selects a particular field from a dropdown list, a dynamic textbox displays a list of possible values for that field to assist the user in his or her query. Keywords can be directly imported from this dynamic textbox, or the user can enter his or her own values.

The output generated by browsing or text searching is extremely flexible. Records are displayed in table format. A user can show or hide different columns of the table and can sort on up to three columns in ascending or descending order. From the output page, the contents of the table displayed onscreen can be downloaded as tab-delimited text, or the FASTA format sequences of the individual proteins can be retrieved.

Through browsing and/or searching, a user can create a specific subset of proteins of interest which can be downloaded and used to form a BLAST database as described previously. However, if the user simply wishes to compare a query protein using BLAST to the entire ePSORTdb or cPSORTdb database, he or she can do so quickly and easily using a web-based BLAST tool available at the PSORTdb site. One or more FASTA-formatted protein sequences are subjected to a blastp search using NCBI defaults. The ePSORTdb- or cP-SORTdb-annotated localization site is returned as part of the BLAST results summary.

Other Localization Databases

In addition to PSORTdb, two other specialized protein localization databases are available. PA-GOSUB (Lu et al., 2005) stores the high-precision Proteome Analyst-predicted localization sites for a number of prokaryotic and eukaryotic genomes. As of September 2005 2005, localization data are available for 34 bacterial genomes. The data can be obtained as comma-separated value files for each proteome; however, the site does not offer the ability to browse or search the data or retrieve a subset of sequences. While users cannot search the database over the web using BLAST, it is possible to download a FASTA format file for use as a local BLAST database.

DBSubLoc (Guo et al., 2004) contains bacterial, eukaryotic, archaeal, and viral localization-related annotations extracted from Swiss-Prot and other literature sources. The data are stored in flat files similar to UniProt records, making it amenable to parsing with simple Perl scripts. While DBSubLoc can be downloaded in its entirety or as subsets corresponding to kingdom, the search utility on the website is limited in function, allowing users to search only by accession number, GO terms, or protein names. DBSubLoc offers a BLAST service over the web, in which a single protein sequence can be searched against the full database or a specific taxonomic subset.

Prediction of Signal Peptides

If localization prediction programs or BLAST-based detection fail to provide information regarding a protein's potential periplasmic localization, it may be necessary to turn to the identification of sequence features characteristic of periplasmic proteins. One of the most informative features is the presence of a signal peptide.

Signal peptides are short N-terminal stretches of sequence that cause a protein to be exported from the cytoplasm of a bacterial cell. After translocation across the cytoplasmic membrane, the signal peptide is cleaved from the mature protein. Predicting the presence of a signal peptide can rule out a cytoplasmic localization for that protein, although the inverse is not true: the absence of a signal peptide of course does not necessarily indicate a cytoplasmic localization.

When attempting to use signal peptide prediction for the discovery of periplasmic proteins, the dataset must first be screened to remove cytoplasmic membrane and outer membrane proteins. Several methods for the prediction of transmembrane helices, which are characteristic of cytoplasmic membrane proteins, are presented in Table 4, while Table 5 lists a number of recently developed β-barrel predictors, which can be used to identify outer membrane proteins. Any of these tools can be used to screen out membrane proteins from a dataset before submitting it to a signal peptide prediction program.

After screening for membrane proteins, the remaining sequences can be submitted to one or more signal peptide analysis tools. Several classes of signal peptide are known to exist, resulting in a number of different prediction methods, each best suited to a particular class. It is worthwhile to analyze a query sequence using all of the methods available, combining their predictions to form a final consensus. This approach will significantly reduce potential false positives, but false negatives—proteins with atypical signal peptides that are missed by one or more of the methods—are still possible.

Of all the methods developed for signal peptide prediction, the suite of tools developed at the Technical University of Denmark has consistently been ranked as the best by several independent evaluations. These programs in-

TABLE 4 Methods for the prediction of transmembrane α-helices

Method	Reference	URL
ConPredII	Arai et al., 2004	http://bioinfo.si.hirosaki-u.ac.jp/~ConPred2
DAS	Cserzo et al., 1997	http://www.sbc.su.se/~miklos/DAS
HMMTOP	Tusnady and Simon, 1998	http://www.enzim.hu/hmmtop
PHOBIUS	Kall et al., 2004	http://phobius.cgb.ki.se
SOSUI	Hirokawa et al., 1998	http://sosui.proteome.bio.tuat.ac.jp/sosuiframe0.html
TMAP	Milpetz et al., 1995	http://www.mbb.ki.se/tmap
TMHMM	Krogh et al., 2001	http://www.cbs.dtu.dk/services/TMHMM-2.0
TopPred 2	Claros and von Heijne, 1994	http://bioweb.pasteur.fr/seqanal/interfaces/toppred.html

clude SignalP, LipoP, and TatP, which are discussed individually below.

Note that while these are excellent methods for signal peptide prediction, some localization prediction methods such as PSORT I and PSORTb do contain signal peptide predictors as part of their algorithm. For PSORTb, signal peptide prediction results are available both in the program output and in PSORTdb. No localization prediction method, however, implements any of the SignalP, LipoP, and TatP algorithms themselves; thus, for optimal signal peptide prediction, it is recommended that users submit their query directly to these servers.

SignalP: Prediction of Type I Signal Peptides

Type I signal peptides are used to translocate proteins through the Sec machinery, after which they are cleaved by the integral membrane protein signal peptidase I (see Chapter 1). These are the most common type of signal peptides, and were the focus of the first signal peptide prediction methods, including SignalP. Originally developed in 1997, SignalP has undergone several updates and is currently at version 3.0 (Bendtsen et al., 2004). The program uses two machine-learning methods, a neural network and a hidden Markov model, to both identify the presence or absence of a signal peptide and predict the most probable cleavage site. The neural network offers the best performance, discriminating between signal peptides and nonsignal peptides with MCC = 0.95 and identifying cleavage sites with a measured precision of 92.5%.

SignalP is available over the web at http://www.cbs.dtu.dk/services/SignalP and can be run locally on an IRIX, Silicon Graphics, Solaris, Sun, OSF1, or Linux machine under a commercial license. Up to 2,000 sequences can be analyzed over the web in one submission. The output is available in a variety of formats, including a graphical format and a short text-only format ideal for parsing.

LipoP: Prediction of Type II Signal Peptides

Type II signal peptides are found at the N termini of lipoproteins, proteins tethered to the cytoplasmic or outer membrane via a lipid anchor (see Chapter 4). Although similar to type I signal peptides in their tripartite structure, type II signal peptides contain a different cleavage site mo-

TABLE 5 Methods for the prediction of β-barrel outer membrane proteins

Method	Reference	URL
BOMP	Berven et al., 2004	http://www.bioinfo.no/tools/bomp
Pred-TMBB	Bagos et al., 2004	http://bioinformatics.biol.uoa.gr/PRED-TMBB
Prof-TMB	Bigelow et al., 2004	http://www.rostlab.org/services/PROFtmb
TMBETA-NET	Gromiha et al., 2005	http://psfs.cbrc.jp/tmbeta-net
TMB-Hunt	Garrow et al., 2005	http://www.bioinformatics.leeds.ac.uk/betaBarrel

tif in the C region and are cleaved by a different signal peptidase. Because of these differences, they may not be recognized by SignalP.

In 2004, LipoP was developed to better recognize this important class of signal peptides (Juncker et al., 2003). It employs a hidden Markov model to predict whether the N terminus of a sequence is a type I or type II signal peptide, or whether the protein is cytoplasmic (lacking a predicted signal peptide) or an integral cytoplasmic membrane protein. The method is capable of distinguishing lipoprotein signal peptides from nonlipoprotein signal peptides with a reported MCC of 0.96.

LipoP is available over the web at http://www.cbs.dtu.dk/services/LipoP/. It is not yet available as a program that can be locally installed on a personal computer. Up to 4,000 sequences can be analyzed in one submission, and like SignalP, three output options are offered, including a graphical version and a short, text-only format. If LipoP predicts the presence of a type II signal peptide, the +2 residue of the mature protein is also reported. It is thought that the nature of the residue at this position determines the membrane to which the protein is anchored—aspartic acid (D) at the +2 position results in retention of the protein at the cytoplasmic membrane, while any other amino acid directs attachment of the protein to the outer membrane.

TatP: Prediction of Twin-Arginine Signal Peptides

A third class of signal peptides comprises the twin-arginine signal peptides, which direct proteins through the TAT transporter (see Chapter 2). While these signal peptides are similar to type I and type II signal peptides in that they consist of an N, H, and C region, they are considerably longer than Sec-directing signal peptides and contain a unique sequence motif at the border of the N and H regions, described as RRxFLK. Because of their unusual length, TAT signal peptides are often missed by methods like SignalP or LipoP. Thus, TatP was developed in 2005 with this group of peptides in mind (Bendtsen et al., 2005).

TatP utilizes a two-step prediction process. In the first step, query sequences are filtered using either the default regular expression of RR[FGAVML][LITMVF] or a user-submitted regular expression, which may be more or less stringent. In the second step, the filtered query proteins are passed to a series of two neural networks—one that recognizes cleavage sites, and one that discriminates between TAT and non-TAT signal peptides. On a test set of 35 TAT signal peptides, TatP correctly identified 91% of these and predicted the correct cleavage site in 84% of the cases. TatP is available over the web at http://www.cbs.dtu.dk/services/TatP-1.0/, and it will soon be available for local use on an IRIX, Silicon Graphics, Solaris, Sun, OSF1, or Linux machine under a commercial license. Up to 4,000 sequences can be analyzed in one submission, with three output options available.

Identification of Membrane-Anchored Periplasmic Proteins

Not all periplasmic proteins have a cleavable signal peptide. Rather, some are anchored in the cytoplasmic membrane via one or more transmembrane helices, with large, globular portions of the protein exposed to the periplasm. This class of proteins can be identified by using transmembrane helix topology prediction tools, methods that both identify the location of transmembrane helices within a protein as well as their orientation. By examining the orientation of predicted helices, a user can easily determine whether the non-helical regions are exposed to the cytoplasm or the periplasm.

Several transmembrane helix topology prediction methods are available; however, three of the most common and the most accurate are TMHMM (Krogh et al., 2001), PHOBIUS (Kall et al., 2004), and HMMTOP (Tusnády and Simon, 2001) (Table 4). Each of the methods employs a hidden Markov model to identify potential transmembrane helices, while topology assignments are made using the "positive-inside rule," which states that positively charged residues occur at a higher frequency on the cytoplasmic face of the membrane. PHOBIUS incorporates an additional step, in which transmembrane helices occurring in the N-terminal region of a protein are examined to determine whether they are, in fact, signal peptides, as the long H region of a signal peptide is frequently mistaken for a transmembrane helix by many predictors.

TMHMM is available over the web, while both PHOBIUS and HMMTOP can be accessed over the web or downloaded for local use under an academic license. The TMHMM and PHOBIUS web servers both support multiple sequence upload. Each of the three methods assigns each residue in a query sequence to one of three states: helix, inside (or cytoplasmic), or outside (or periplasmic). All three methods can output text-format predictions, in which each residue of the query sequence is noted as I, O, or H, while TMHMM and PHOBIUS also provide graphical output, with a color scheme allowing for easy visualization of a protein's predicted topology.

CONCLUSIONS

This chapter has presented an overview of a selection of methods for the computational identification of periplasmic proteins. Some of these methods, such as PSORTb and Proteome Analyst, display excellent predictive performance and require little in the way of user intervention. Other screening methods, such as signal peptide prediction, can be labor intensive but may return more results, although some periplasmic proteins, such as those lacking traditional signal peptides, can still be missed.

When attempting to predict the localization of a protein of interest, it is recommended that the user adopt a pipeline approach. This chapter has been organized in a similar fashion, with the methods presented in the order they might be used. In the first step, the protein of interest is submitted to one or more highly precise predictive methods, for example, PSORTb and Proteome Analyst. If a confident prediction of localization cannot be generated by one of these highly precise methods, the protein may then be submitted to a number of less precise methods. These "broad-spectrum" predictive tools include the amino acid composition-based SVM methods, which generate predictions and, in some cases, quality scores, for all proteins submitted. By combining the output of these more general methods and inspecting the associated quality scores, a user may feel confident enough to assign a localization site to the protein in question. If doubt remains, however, the protein can then be analyzed at the level of sequence features. Homology to a protein of known or confidently predicted localization can be used as a reliable indicator of localization in most cases, due to the high conservation of localization across diverse species (Nair and Rost, 2002). Information about the presence or absence of a signal peptide, transmembrane α-helices, or transmembrane β-barrel strands within the protein can also help elucidate whether a protein is likely periplasmic, or whether it has features consistent with other localizations.

While the analytical pipeline described above will identify a large proportion of periplasmic proteins with a moderate to high degree of confidence, the need still exists for improved localization prediction methods. Although the precision achieved by methods such as PSORTb and Proteome Analyst is excellent, an increase in the number of predictions generated by the programs would represent a significant improvement.

To increase these programs' ability to generate not just a greater number of predictions, but a greater number of highly confident predictions, more training data are required. Such data could be derived from the laboratory, where high-throughput analyses—such as the two-dimensional gel-based characterization of specific cellular compartments—produce large datasets. The data can also be derived computationally through the application of text-mining methods to the scientific literature. Through the use of machine-learning techniques, researchers can train a program to identify papers in PubMed that likely contain information about a localization experiment. By using text mining to reduce the literature space to a manageable workload for a manual curator, hundreds to thousands of abstracts and papers potentially containing localization information can be reviewed in a comparatively short time frame. This will become particularly useful in increasing the diversity of proteins in training datasets, expanding the taxonomic range of organisms represented in the data beyond what it is today.

Through the increase in the dataset of experimentally known periplasmic proteins, we anticipate that we will be able to learn more about the specific signals involved in transporting certain proteins to the periplasm. In particular, having a large dataset of periplasmic proteins may help uncover targeting features in proteins which presently have no discernible export signal.

As bioinformatics becomes an increasingly important aspect of molecular biology, the pace of discovery continues to quicken. New localization prediction methods are released on an almost regular basis, with as many as ten new methods being reported in a given year. Each new method brings a fresh approach to the problem and drives researchers to improve existing tools and to create even more powerful programs. Over the next several years, we expect that our ability to confidently predict periplasmic proteins computationally will continue to improve, providing the research community with a fast, simple, and reliable method for identifying this important and interesting class of proteins.

ACKNOWLEDGMENTS

We thank other members of the PSORT development community who have been involved in subcellular localization prediction research and who have provided helpful input. J.L.G. and F.S.L.B are a Michael Smith Foundation for Health Research Trainee and Scholar, respectively, as well as a Canada Graduate Scholarship holder and Canadian Institutes of Health Research New Investigator, respectively.

This work was funded by Natural Sciences and Engineering Research Council of Canada (NSERC).

REFERENCES

Apweiler, R., A. Bairoch, C. H. Wu, W. C. Barker, B. Boeckmann, S. Ferro, E. Gasteiger, H. Huang, R. Lopez, M. Magrane, M. J. Martin, D. A. Natale, C. O'Donovan, N. Redaschi, and L. S. Yeh. 2004. UniProt: the Universal Protein knowledgebase. *Nucleic Acids Res.* **32:**D115–D119.

Arai, M., H. Mitsuke, M. Ikeda, J. X. Xia, T. Kikuchi, M. Satake, and T. Shimizu. 2004. ConPred II: a consensus prediction method for obtaining transmembrane topology models with high reliability. *Nucleic Acids Res.* **32:**W390–W393.

Bagos, P. G., T. D. Liakopoulos, I. C. Spyropoulos, and S. J. Hamodrakas. 2004. PRED-TMBB: a web server for predicting the topology of beta-barrel outer membrane proteins. *Nucleic Acids Res.* **32:**W400–W404.

Bendtsen, J. D., H. Nielsen, G. von Heijne, and S. Brunak. 2004. Improved prediction of signal peptides: SignalP 3.0. *J. Mol. Biol.* **340:**783–795.

Bendtsen, J. D., H. Nielsen, D. Widdick, T. Palmer, and S. Brunak. 2005. Prediction of twin-arginine signal peptides. *BMC Bioinformatics* **6:**167–175.

Berven, F. S., K. Flikka, H. B. Jensen, and I. Eidhammer. 2004. BOMP: a program to predict integral beta-barrel outer membrane proteins encoded within genomes of Gram-negative bacteria. *Nucleic Acids Res.* **32:**W394–W399.

Bhasin, M., A. Garg, and G. P. Raghava. 2005. PSLpred: prediction of subcellular localization of bacterial proteins. *Bioinformatics* **21:**2522–2524.

Bigelow, H. R., D. S. Petrey, J. Liu, D. Przybylski, and B. Rost. 2004. Predicting transmembrane beta-barrels in proteomes. *Nucleic Acids Res.* **32:** 2566–2577.

Boos, W., U. Ehmann, E. Bremer, A. Middendorf, and P. Postma. 1987. Trehalase of Escherichia coli. Mapping and cloning of its structural gene and identification of the enzyme as a periplasmic protein induced under high osmolarity growth conditions. *J. Biol. Chem.* **262:**13212–13218.

Claros, M. G., and G. von Heijne. 1994. TopPred II: an improved software for membrane protein structure predictions. *Comput. Appl. Biosci.* **10:** 685–686.

Cserzo, M., E. Wallin, I. Simon, G. von Heijne, and A. Elofsson. 1997. Prediction of transmembrane alpha-helices in prokaryotic membrane proteins: the dense alignment surface method. *Protein Eng.* **10:**673–676.

Galperin, M. Y. 2005. The Molecular Biology Database Collection: 2005 update. *Nucleic Acids Res.* **33:** D5–D24.

Gardy, J. L., M. R. Laird, F. Chen, S. Rey, C. J. Walsh, M. Ester, and F. S. L. Brinkman. 2005. PSORTb v.2.0: expanded prediction of bacterial protein subcellular localization and insights gained from comparative proteome analysis. *Bioinformatics* **21:**617–623.

Gardy, J. L., C. Spencer, K. Wang, M. Ester, G. E. Tusnády, I. Simon, S. Hua, K. deFays, C. Lambert, K. Nakai, and F. S. L. Brinkman. 2003. PSORT-B: improving protein subcellular localization prediction for Gram-negative bacteria. *Nucleic Acids Res.* **31:**3613–3617.

Garrow, A. G., A. Agnew, and D. R. Westhead. 2005. TMB-Hunt: an amino acid composition based method to screen proteomes for beta-barrel transmembrane proteins. *BMC Bioinformatics* **6:**56–71.

Gromiha, M. M., S. Ahmad, and M. Suwa. 2005. TMBETA-NET: discrimination and prediction of membrane spanning beta-strands in outer membrane proteins. *Nucleic Acids Res.* **33:**W164–W167.

Guo, T., S. Hua, X. Ji, and Z. Sun. 2004. DB-SubLoc: database of protein subcellular localization. *Nucleic Acids Res.* **32:**D122–D124.

Hirokawa T., S. Boon-Chieng, and S. Mitaku. 1998. SOSUI: classification and secondary structure prediction system for membrane proteins. *Bioinformatics* **14:**378–379.

Horlacher, R., K. Uhland, W. Klein, M. Ehrmann, and W. Boos. 1996. Characterization of a cytoplasmic trehalase of Escherichia coli. *J. Bacteriol.* **178:**6250–6257.

Hua, S., and Z. Sun. 2001. Support vector machine approach for protein subcellular localization prediction. *Bioinformatics* **17:**721–728.

Juncker, A. S., H. Willenbrock, G. Von Heijne, S. Brunak, H. Nielsen, and A. Krogh. 2003. Prediction of lipoprotein signal peptides in Gram-negative bacteria. *Protein Sci.* **12:**1652–1662.

Kall, L., A. Krogh, and E. L. Sonnhammer. 2004. A combined transmembrane topology and signal peptide prediction method. *J. Mol. Biol.* **338:**1027–1036.

Krogh, A., B. Larsson, G. von Heijne, and E. L. Sonnhammer. 2001. Predicting transmembrane protein topology with a hidden Markov model: application to complete genomes. *J. Mol. Biol.* **305:**567–580.

Lu, Z., D. Szafron, R. Greiner, P. Lu, D. S. Wishart, B. Poulin, J. Anvik, C. Macdonell, and R. Eisner. 2004. Predicting subcellular local-ization of proteins using machine-learned classifiers. *Bioinformatics* **20:**547–556.

Lu, Z., D. Szafron, R. Greiner, D. S. Wishart, A. Fyshe, B. Pearcy, B. Poulin, R. Eisner, D. Ngo, and N. Lamb. 2005. PA-GOSUB: a searchable database of model organism protein sequences with their predicted Gene Ontology molecular function and subcellular localization. *Nucleic Acids Res.* **33:**D147–D153.

Milpetz, F., P. Argos, and B. Persson. 1995. TMAP: a new email and WWW service for membrane-protein structural predictions. *Trends Biochem. Sci.* **20:**204–205.

Nair, R., and B. Rost. 2002. Sequence conserved for subcellular localization. *Protein Sci.* **12:**2836–2847.

Nair, R., and B. Rost. 2005. Mimicking cellular sorting improves prediction of subcellular localization. *J. Mol. Biol.* **348:**85–100.

Nakai, K., and M. Kanehisa. 1991. Expert system for predicting protein localization sites in Gram-negative bacteria. *Proteins* **11:**95–110.

Neidhardt, F. C., R. Curtiss III, J. L. Ingraham, E. C. C. Lin, K. B. Low, B. Magasanik, W. S. Reznikoff, M. Riley, M. Schaechter, and H. E. Umbarger (ed.). 2004. Escherichia coli *and* Salmonella: Cellular and Molecular Biology. ASM Press, Washington, D.C.

Rey, S., M. Acab, J. L. Gardy, M. R. Laird, K. de-Fays, C. Lambert, and F. S. L. Brinkman. 2005. PSORTdb: a protein subcellular localization database for bacteria. *Nucleic Acids Res.* **33:**D164–D168.

Tusnády, G. E., and I. Simon. 2001. The HMM-TOP transmembrane topology prediction server. *Bioinformatics* **17:**849–850.

Vapnik, V. 1995. *The Nature of Statistical Learning Theory.* Springer, New York, N.Y.

von Heijne, G. 1986. A new method for predicting signal sequence cleavage sites. *Nucleic Acids Res.* **14:**4683–4690.

Wang, J., W. K. Sung, A. Krishnan, and K. B. Li. 2005. Protein subcellular localization prediction for Gram-negative bacteria using amino acid subalphabets and a combination of multiple support vector machines. *BMC Bioinformatics* **13:**174–183.

Yu, C., C. Lin, and J. Hwang. 2004. Predicting subcellular localization of proteins for Gram-negative bacteria by support vector machines based on n-peptide compositions. *Protein Sci.* **13:**1402–1406.

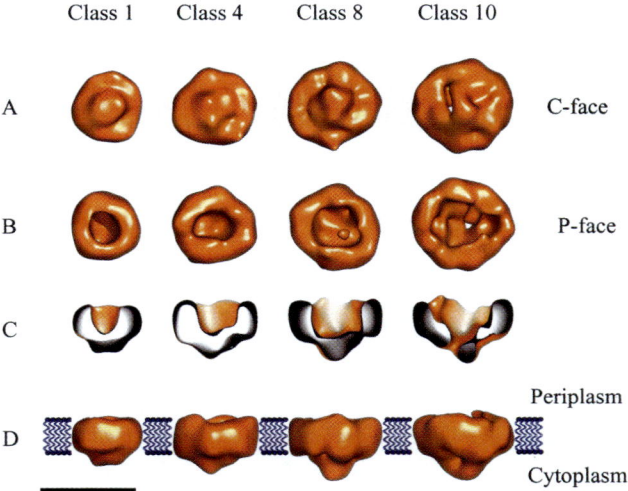

Class 1   Class 4   Class 8   Class 10

A   C-face

B   P-face

C

                                           Periplasm

D

                                           Cytoplasm

COLOR PLATE 1 (Chapter 2)   The structure of TatA from *E. coli*. TatA complexes were purified in detergent solution, negatively stained with uranyl acetate, and 3D structures obtained by single-particle electron microscopy and random conical tilt reconstruction. Shown are four size classes of TatA complexes with increasing diameter. The 3D maps are filtered between 150 Å and 25 Å and contoured at 4 standard deviations above the mean density. (A) TatA complexes viewed from the closed end of the channel, proposed to be at the cytoplasmic side of the membrane (C face). Density forming the lid domain can be clearly seen. (B) TatA complexes viewed from the open end of the channel, proposed to be at the periplasmic side of the membrane (P face). (C) Side views of TatA. The front half of each molecule has been cut away to reveal internal features. (D) Views of TatA parallel to the membrane plane. The proposed position of the lipid bilayer is indicated. (Scale bar, 100 Å.) The figure was taken from Gohlke et al. (2005).

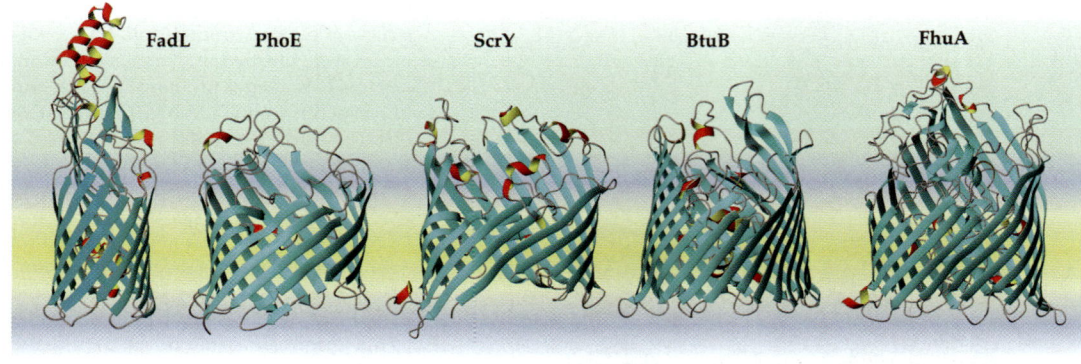

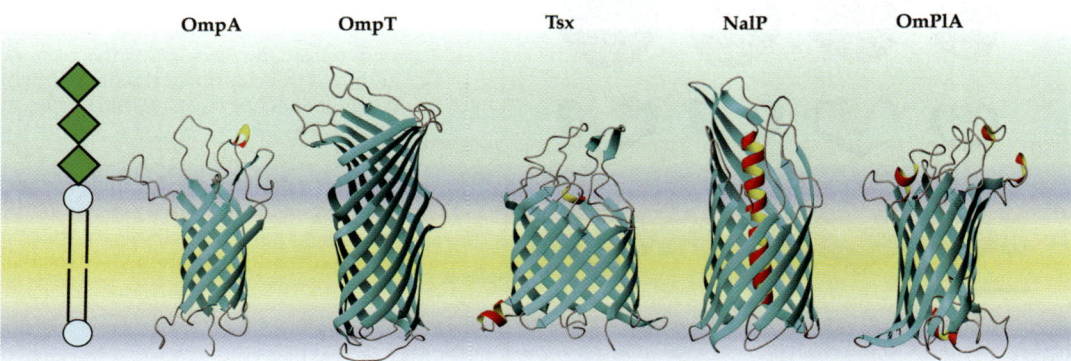

COLOR PLATE 2 (Chapter 3) Some representative crystal structures of β-barrel membrane proteins of the outer membranes of bacteria are shown. Transmembrane (TM) β-barrels have an even number of antiparallel TM strands, which is 8 for OmpA (shown here is the nuclear magnetic resonance [NMR] structure from Arora et al., 2001; for the crystal structure see Pautsch and Schulz, 1998, 2000), 10 for OmpT (Vandeputte-Rutten et al., 2001), 12 for Tsx (Ye and van den Berg, 2004), for NalP (Oomen et al., 2004), and OmPlA (Snijder et al., 1999), 14 for FadL (van den Berg et al., 2004), 16 for PhoE (Cowan et al., 1992), 18 for ScrY (Forst et al., 1998), and 22 for BtuB (Chimento et al., 2003b) and FhuA (Ferguson et al., 1998). OmpA is a small ion channel (Arora et al., 2000), OmpT is a protease, NalP is an autotransporter, FadL is a long-chain fatty acid transporter, PhoE is a diffusion pore, ScrY is a sucrose-specific porin, OmPlA is a phospholipase. BtuB and FhuA are active transporters for ferrichrome iron and vitamin B_{12} uptake, respectively. OMPs of mitochondria are predicted to form similar TM β-barrels. Examples are the VDAC channels, out of which more than a dozen have been sequenced (Heins et al., 1994). Protein structures were generated with MolMol (Koradi et al., 1996).

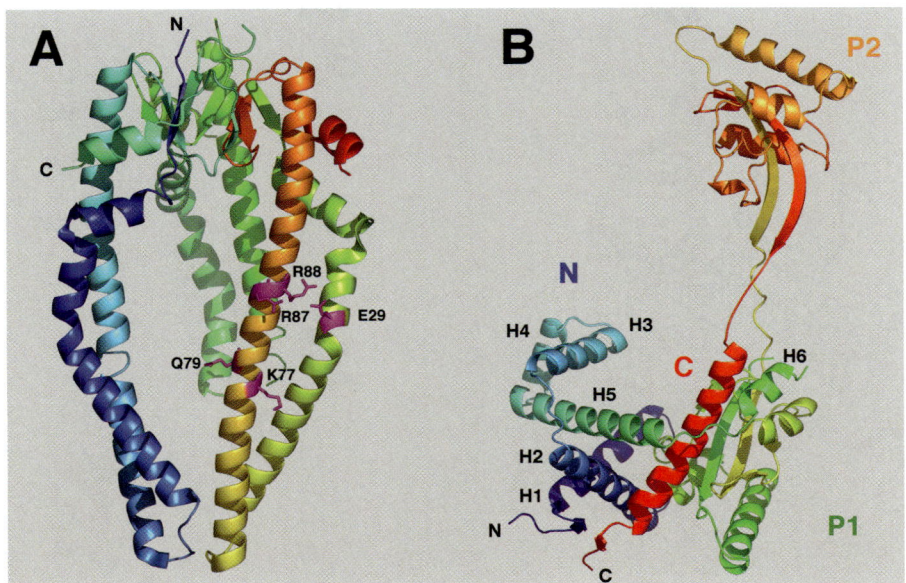

COLOR PLATE 3 (Chapter 3) (A) Crystal structure of the Skp trimer (PDB entry 1SG2; Korndörfer et al., 2004). The Skp trimer consists of a tightly packed 9-stranded β-barrel that is surrounded by C-terminal α-helices of the three subunits that point away from the barrel in the form of tentacles that are about 65 Å long. These tentacles form a cavity that may take up the unfolded OMP. The outside surface of the helical domain of Skp is highly basic. Each monomer of the trimeric Skp has a putative LPS binding site (Walton and Sousa, 2004) (Skp structure entry 1UM2 in the PDB). The LPS binding site was found by using a previously identified LPS binding motif (Ferguson et al., 2000) and consists of K77, R87, and R88. This motif matches the LPS binding motif in FhuA with residues K306, K351, and R382 and a root-mean-square (rms) deviation of 1.75 Å for the Cα – Cγ atoms was calculated (Walton and Sousa, 2004). Q99 in Skp may also form a hydrogen bond to an LPS phosphate, completing the four-residue LPS binding motif. (B) Crystal structure of Survival Factor A, SurA (PDB entry 1M5Y [Bitto and McKay, 2002]). The N-terminal domain (N) is composed of the α-helices H1 to H6 (residues 1 to 148) and connected to peptidyl-prolyl *cis/trans* isomerase (PPI) domain P1 (residues 149 to 260). The P2 domain (residues 261 to 369) connects the P1 domain to the C-terminal domain C (residues 370 to 428, colored in red). Thus, the N and C domains together constitute a compact core, which is traversed by a broad deep crevice of about 50 Å in length, suggesting a polypeptide binding site. The active PPIase domain 2 (P2) is tethered to this core by two extended peptide segments. It has been demonstrated that a mutant, SurAN(-Ct), which does not contain the two PPIase domains and is composed of the N and C domains only, functions like a chaperone (Behrens et al., 2001). This SurA "core domain" has been proposed to bind the tripeptide motif aromatic-random-aromatic, which is prevalent in the aromatic girdles of α-barrel membrane proteins (Bitto and McKay, 2003). Images of the structures were created with Pymol (Delano, 2002).

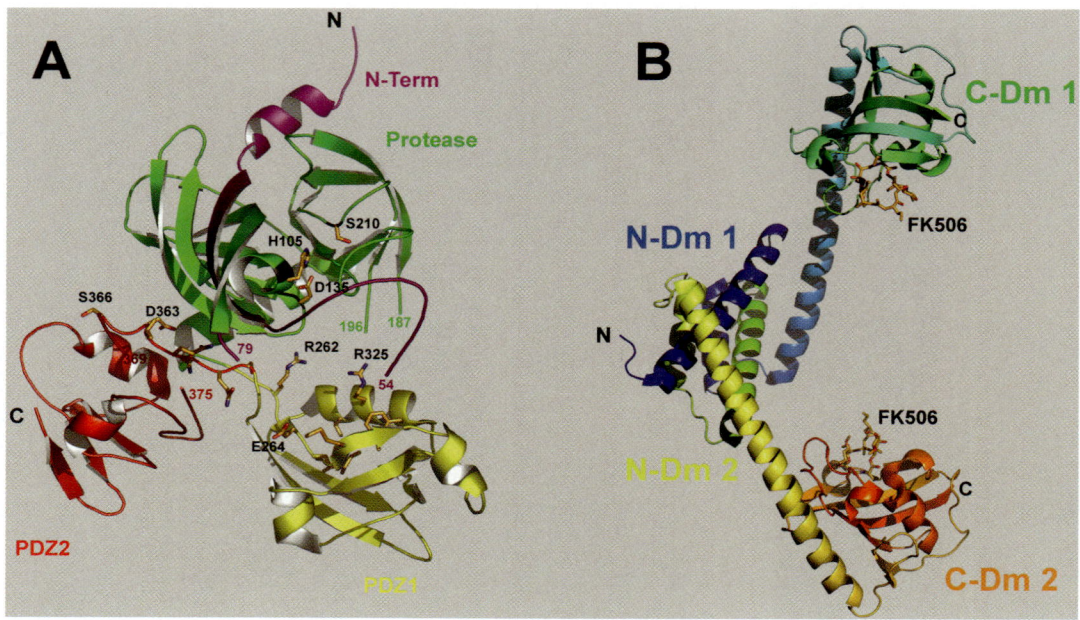

COLOR PLATE 4 (Chapter 3) (A) The crystal structure of DegP (Krojer et al., 2002) is shown (PDB entry 1KY9). DegP is a homo-hexamer that is composed of two stacked rings of three DegP subunits. Drawn here is one subunit of DegP. The subunit consists of three functionally distinct domains, the protease domain (residues 1 to 259, indicated in green and in purple) and two PDZ domains, PDZ1 (residues 260 to 358, yellow) and PDZ2 (residues 359 to 448, red). The catalytic triad of the protease domain, His-105, Asp-135, and Ser-210, is located in a crevice between two β-barrel lobes with a carboxy-terminal α-helix, similar to structures of other proteases of the trypsin family. In PDZ1, the residues Arg-262, Glu-264, Leu-265, Gly-266, Ile-267, Met-268, Phe-321, Arg-325, Leu-324, and Val-328 may constitute the peptide binding site. In PDZ2, the peptide binding site may consist of residues Ser-358, Gln-359, Asn-360, Gln-361, Val-362, Asp-363, Ser-366, Gly-370, Ile-371, Glu-372, Gly-373, and Ala-374. Some of these residues of PDZ1 and PDZ2 are shown. For further details on the hexameric organization of DegP and its two different conformational states, see Krojer et al. (2002) and the supplementary information on the journal website. (B) Crystal structure of the V-shaped FkpAΔCT-dimer, i.e., an FkpA mutant, which lacks the 21 C-terminal residues (Saul et al., 2004). FkpAΔCT is shown in complex with the immunosuppressant FK506 (PDB entry 1Q6I), which binds to the FKBP-type C domain (C-Dm). Each of the 2 monomers consists of the N domain (N-Dm, residues 15 to 114), which is composed of three α-helices (formed by residues 19 to 43, 51 to 62, and 70 to 111, respectively) and functions as a chaperone. For monomer 1, this domain is shown in blue, for monomer 2 it is shown in yellow and in light green. The helices of N-Dm 1 and N-Dm 2 are tightly interlaced and maintain FkpA in dimeric form. The C domains of the two monomers are indicated in dark green (C-Dm 1) and in orange (C-Dm 2) and contain the PPIase activity. The C domains belong to the FKBP family and, in the dimer, the bound FK506 molecules are separated by about 49 Å (Saul et al., 2004). Images of the structures were created with Pymol (Delano, 2002).

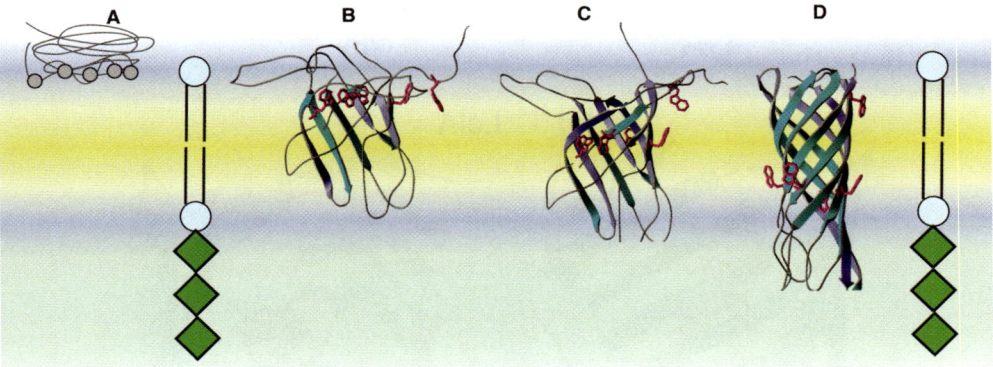

Locations of the Tryptophans of OmpA in Folding Intermediates identified by TDFQ:

Distance to the Center of the Lipid Bilayer

Tryptophan	I_{M1} (A) $\longrightarrow$	I_{M2} (B) $\longrightarrow$	I_{M3} (C) $\longrightarrow$	N (D)
⑦	~14-16 Å	~10 Å	~10 Å	~10 Å
15, 57, 102, 143		~10 Å	~0-5 Å	~10 Å

COLOR PLATE 5 (Chapter 3)   Folding model of OmpA. The kinetics of β–sheet secondary and β–barrel tertiary structure formation in OmpA have the same rate constants and are coupled to the insertion of OmpA into the lipid bilayer (Kleinschmidt et al., 1999a; Kleinschmidt and Tamm, 1999, 2002). The locations of the 5 tryptophans in the three identified membrane-bound folding intermediates and in the completely refolded state of OmpA (Kleinschmidt et al., 1999a; Kleinschmidt and Tamm, 1999) are shown. Additional details, such as the translocation of the long polar loops across the lipid bilayer, must still be determined. OmpA structures were generated with DeepView (Guex and Peitsch, 1997; Schwede et al., 2003).

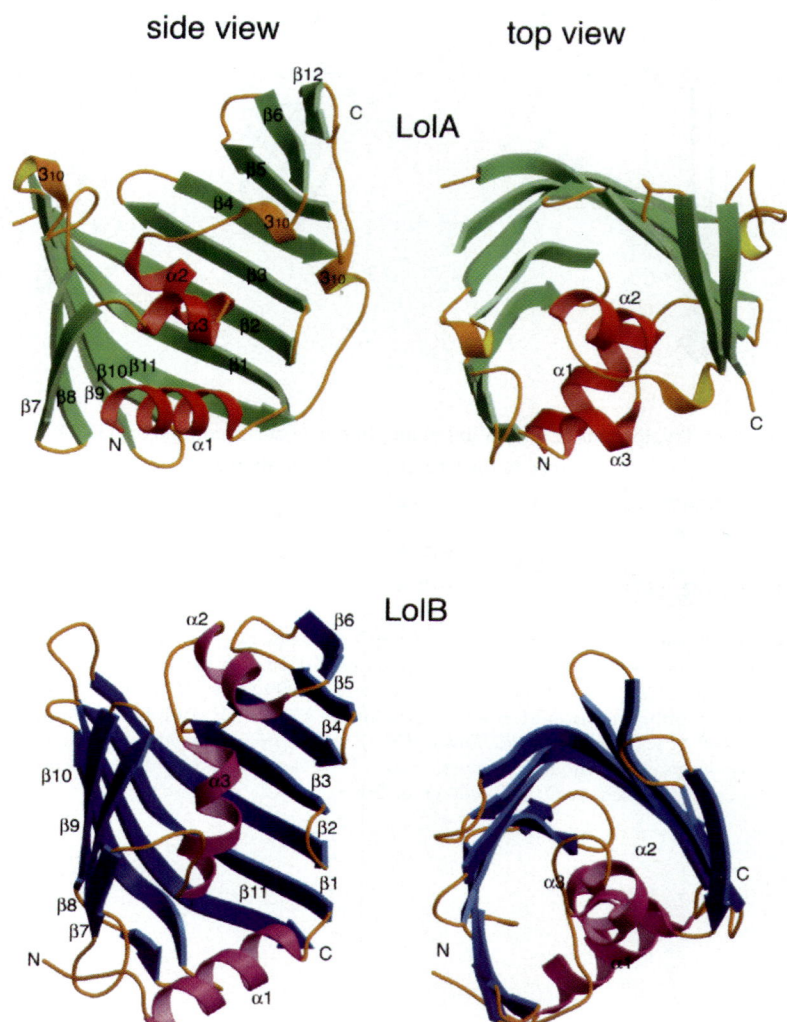

COLOR PLATE 6 (Chapter 4) Crystal structures of LolA and LolB. The LolA and LolB molecules are each shown as a ribbon model. The structural information on LolA (1UA8) and LolB (1IWM) was obtained from the RCSB protein data bank (http://pdb.protein.osaka-u.ac.jp/pdb/) and visualized with Molscript ver 2.1.2 (http://www.avatar.se/molscript/).

A)

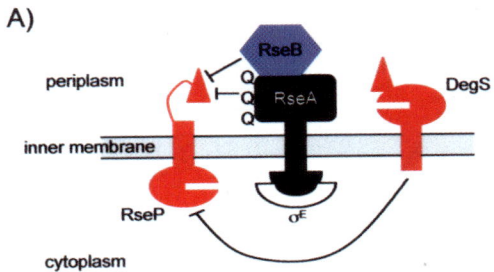

periplasm

inner membrane

cytoplasm

RseB

RseA

DegS

RseP

σ^E

B)

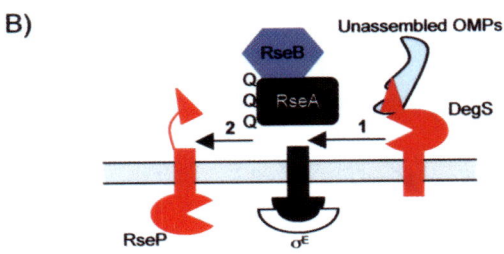

Unassembled OMPs

RseB

RseA

DegS

2 ← → 1

RseP

σ^E

C)

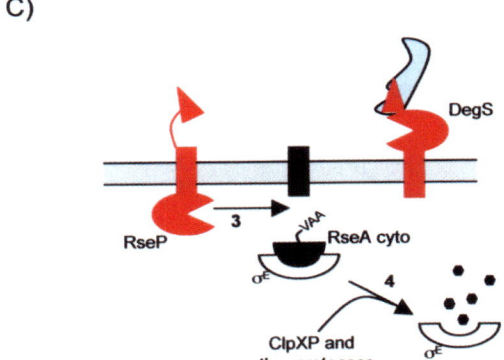

DegS

RseP

3 →

VAA

RseA cyto

σ^E

4

ClpXP and
other proteases

σ^E

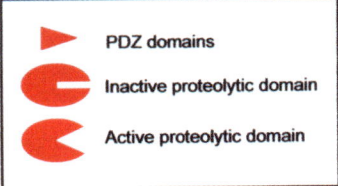

▶ PDZ domains

Inactive proteolytic domain

Active proteolytic domain

COLOR PLATE 7 (Chapter 6) Model of the proteolytic cascade that degrades RseA and releases σ^E. (A) Noninducing conditions: σ^E is tightly bound to RseA in the inner membrane; DegS protease is inactive and RseP protease is inhibited by interactions with its PDZ domain and RseB and the glutamine (Q)-rich periplasmic domain of RseA, and also by DegS. (B) Initiation of the proteolytic cascade by DegS cleavage: The C termini of unassembled outer membrane porins (OMPs) bind to the PDZ domain of DegS, triggering cleavage of RseA (1), which releases the RseA periplasmic domain and bound RseB, thereby relieving negative regulation of RseP (2). (C) Release of free σ^E: RseP cleavage of RseA (3) generates a membrane-free RseA with a VAA C terminus, which is an attractive substrate for cytoplasmic proteases. The released σ^E/RseA complex is bound by ClpXP and other proteases; RseA is degraded and free σ^E is released (4).

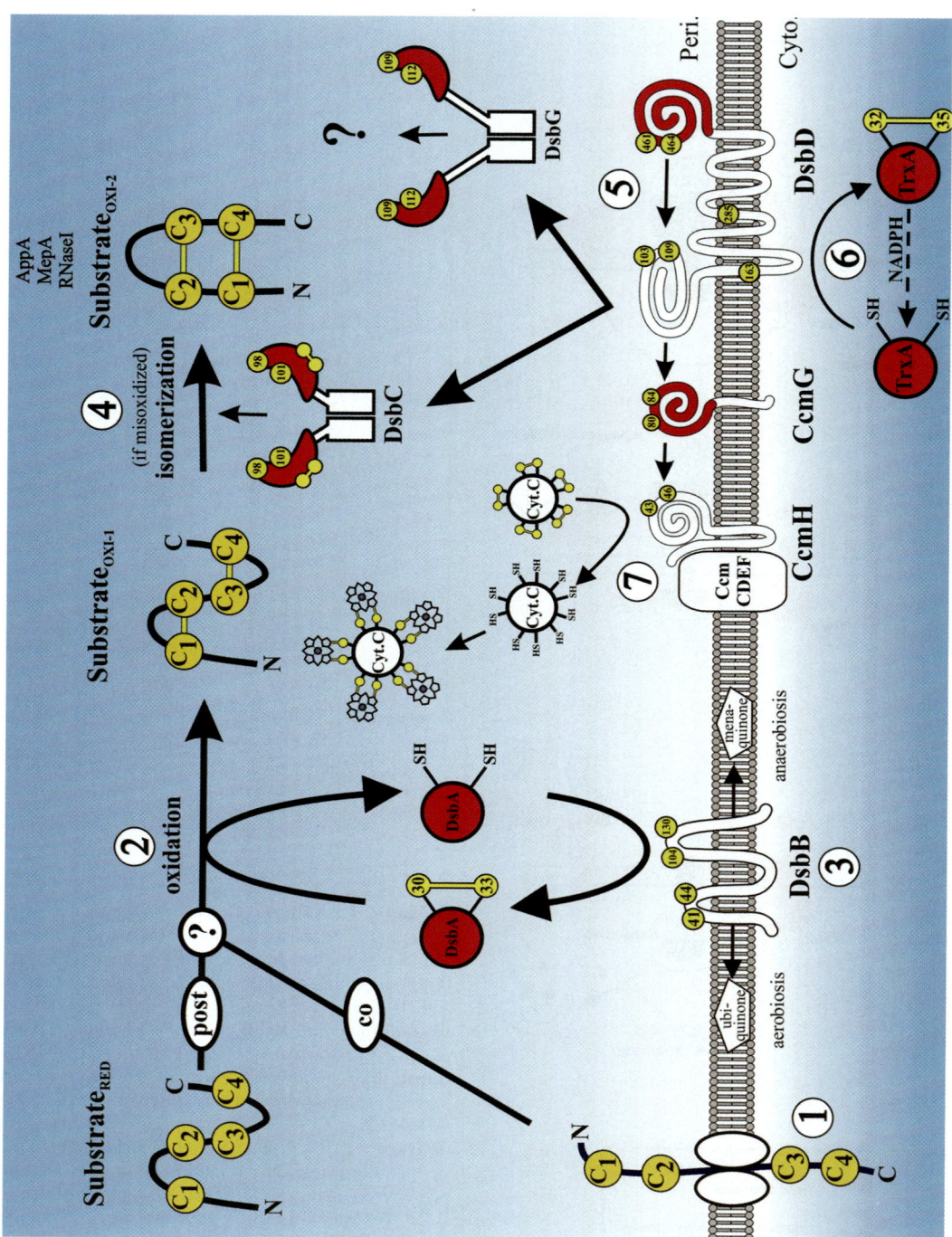

COLOR PLATE 8 (Chapter 7) Disulfide bond formation in the periplasm. A protein requiring disulfide bonds for its stability is translocated into the periplasm via the SecYEG translocon with their cysteines (arbitrarily labeled C1 to C4) in a reduced state (Substrate$_{RED}$) ①. Disulfide bond formation is catalyzed by DsbA, either during translocation, after translocation, or a ratio of both ②. DsbA is reoxidized back to its active oxidized state by DsbB ③. DsbB is oxidized by ubiquinone in aerobic conditions or by menaquinone in anaerobic conditions ③. If the substrate is misoxidized (Substrate$_{OXI-1}$) its disulfide bonds are isomerized to their native oxidized state (Substrate$_{OXI-2}$) by DsbC ④. DsbC, DsbG, and CcmG are maintained in their active reduced state by DsbD ⑤. DsbD in turn is reduced by the cytoplasmic thioredoxin TrxA, which receives its reducing potential ultimately from cytoplasmic pools of NADPH ⑥. CcmG maintains CcmH in a reduced state. Through the interaction of CcmH with CcmCDEF membrane complex, oxidized cytochrome c is reduced, enabling it to form thioether covalent bonds with its heme cofactor ⑦. Proteins with thioredoxin folds are in red and cysteines are in yellow. The amino acid residue numbers of the redox active cysteines are indicated.

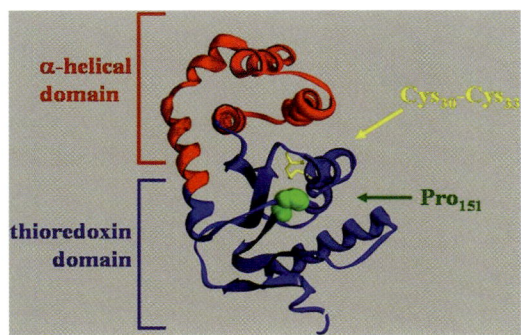

COLOR PLATE 9 (Chapter 7) Domain organization of DsbA. Crystal structure of oxidized DsbA (PDB ID: 1FVK). The thioredoxin domain is in blue and the α-helical domain is in red. The active site disulfide bond, along with the critical proline$_{151}$, is indicated.

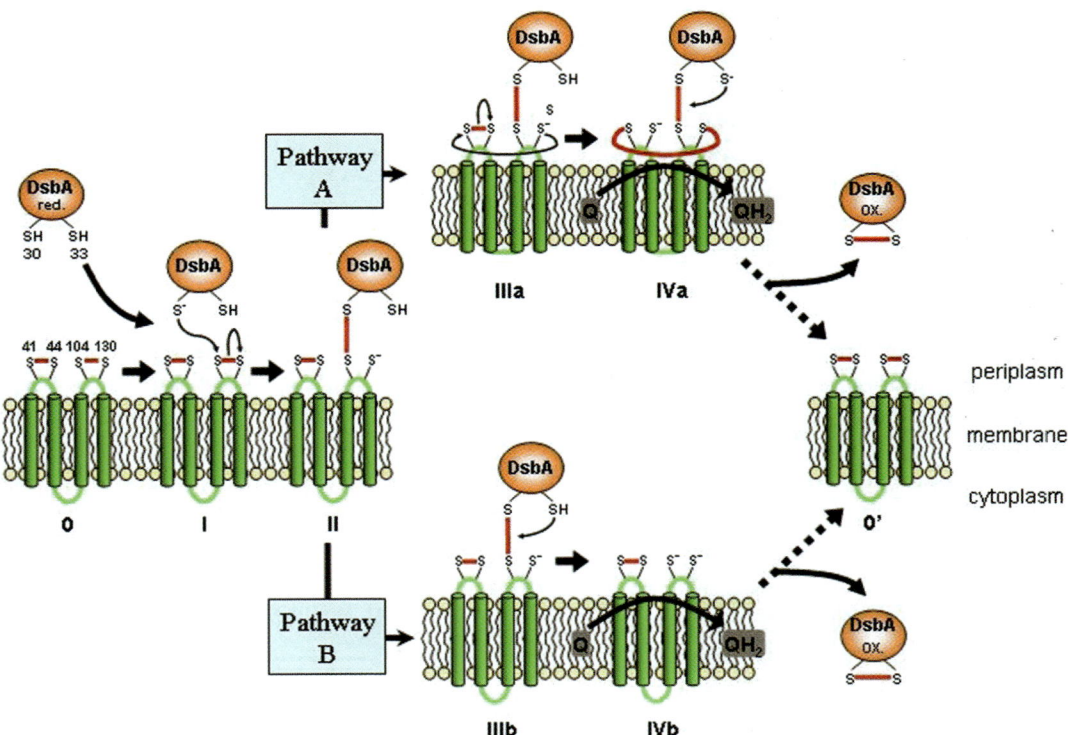

COLOR PLATE 10 (Chapter 7) The mechanism of DsbA reoxidation by DsbB. A reduced DsbA interacts with oxidized DsbB, resulting in the reoxidation of DsbA and reduction of DsbB. DsbA-DsbB complex is formed via a disulfide bond between C_{33} of DsbA and C_{104} of the second periplasmic loop of DsbB. The resolution of this complex is believed to occur through two pathways. In pathway A, a disulfide bond is formed between the first and second periplasmic loop, which is resolved by the oxidation of DsbB by quinones. In pathway B, DsbA-DsbB complex is resolved by quinones, without the interaction of the first periplasmic loop.

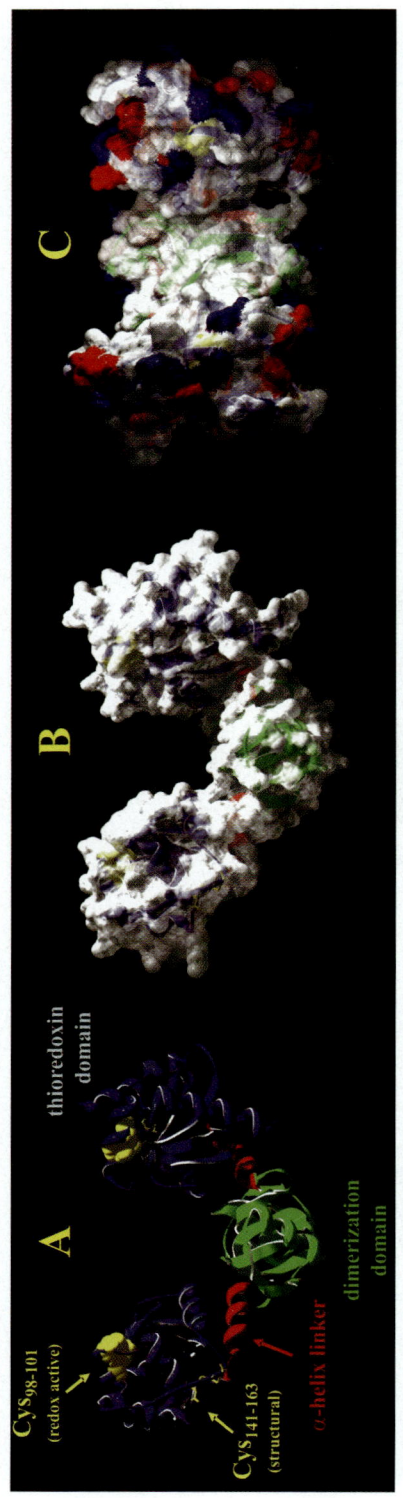

COLOR PLATE 11 (Chapter 7) Domain organization of DsbC. (A) The crystal structure of the homodimer DsbC showing the two domains (thioredoxin in blue and dimerization in green) separated by the short α-helix linker in red (PDB ID: 1EEJ). Redox active cysteines (C_{98}–C_{101}) are represented as yellow spheres and the structural disulfide bond (C_{141}–C_{163}) is indicated. (B) The molecular surface of DsbC is superimposed, visualizing the pocket formed by the dimerization of DsbC. (C) Top-down view of DsbC displaying the noncharged pocket devoid of acidic (red) and basic (blue) amino acid residues.

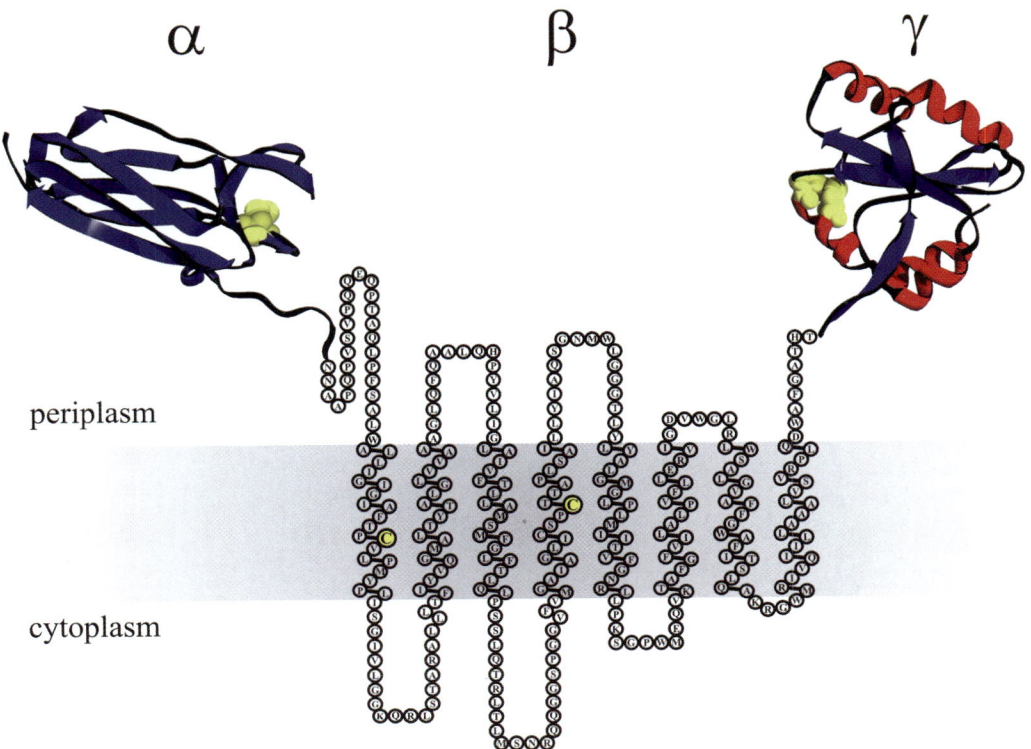

COLOR PLATE 12 (Chapter 7)   Domain organization of DsbD. The predicted membrane topology of DsbD from the web-based program PHOBIUS (www.phobius.binf.ku.dk). The immunoglobin-like α-domain is crystallized (PDB ID: 1L6P) devoid of its signal peptide from the amino acids Arg_8 to Asn_{125}. The amino acids of the β-domain from Asn_{126} to Thr_{422} are depicted as circles. The redox active cysteines (C_{163}–C_{285}) are highlighted as yellow circles. The thioredoxin-like γ-domain from Ala_{423} to Pro_{546} is crystallized (PDB ID: 1UC7). Active site cysteines in the crystal structures are shown as yellow spheres (α-domain C_{103}–C_{109} and β-domain C_{461}–C_{464}). The membrane is shaded grey.

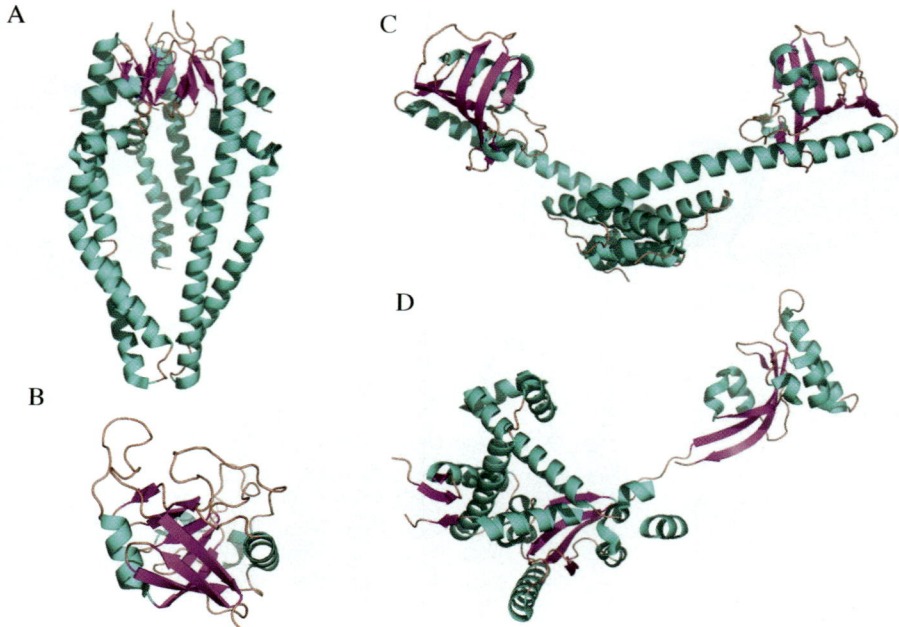

COLOR PLATE 13 (Chapter 8) Three-dimensional structure of periplasmic folding proteins. Ribbon drawing showing the overall structure of Skp (A), PpiA (B), FkpA (C), and SurA (D), with the secondary structure elements shown in cyan (α-helices) and magenta (β-strands). Structures are not shown to scale, and the Protein Data Bank identification numbers for these proteins are as follows: Skp, 1SG2; PpiA, 1J2A; FkpA, 1Q6H; SurA, 1M5Y.

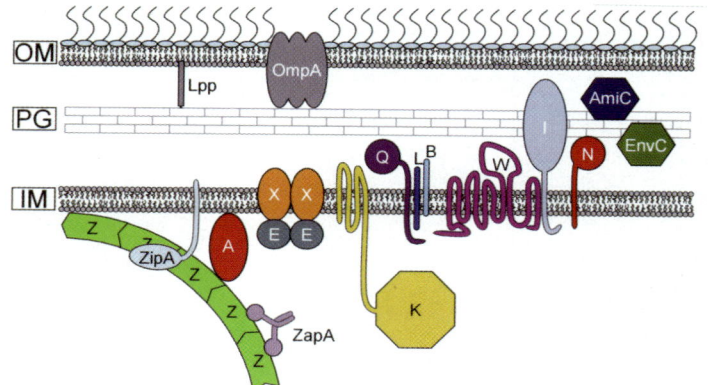

COLOR PLATE 14 (Chapter 10) Proteins that mediate septum assembly in *E. coli*. The septal ring contains at least 15 division proteins (in color). Those identified by single letters are Fts proteins. Braun's lipoprotein and OmpA, which probably mediate constriction of the outer membrane, are shown in gray because they are not considered to be components of the septal ring. OM, outer membrane; PG, peptidoglycan; CM, cytoplasmic membrane. (Adapted with permission from the *Journal of Bacteriology* [Goehring and Beckwith, 2005].)

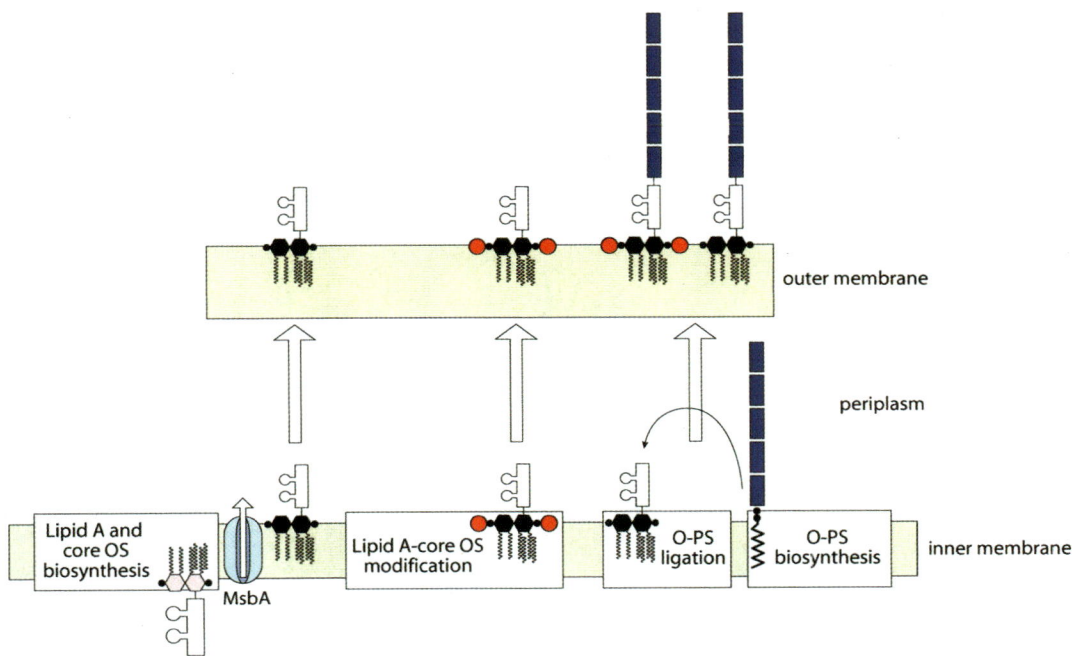

COLOR PLATE 15 (Chapter 12) Schematic overview of LPS biosynthesis and assembly. The composite lipid A–core oligosaccharide (lipid A–core OS) is synthesized in the cytoplasm by the activities of the Lpx★ and Waa★ enzymes. It is exported across the inner membrane by the ABC transporter, MsbA. From there, lipid A core can be translocated directly to the outer membrane. Alternatively, it may provide an acceptor for environmentally regulated modifications mediated by Pag★, Pmr★, and Arn★ enzymes, or for the repeat-unit O-polysaccharide (O-PS). O–Polysaccharide is assembled independently by one of three known mechanisms. All begin with the transfer of a hexose-1-P or acetamidohexose-1-P to the carrier lipid, undecaprenyl phosphate, and all terminate with undecaprenyl pyrophosphate-linked polymer at the periplasmic face of the inner membrane but the transmembrane processes occurring in between are different in each pathway. The LPS molecule is completed by a ligation step mediated by the WaaL protein. The fully modified LPS molecule is then translocated by a currently unknown process to the outer membrane.

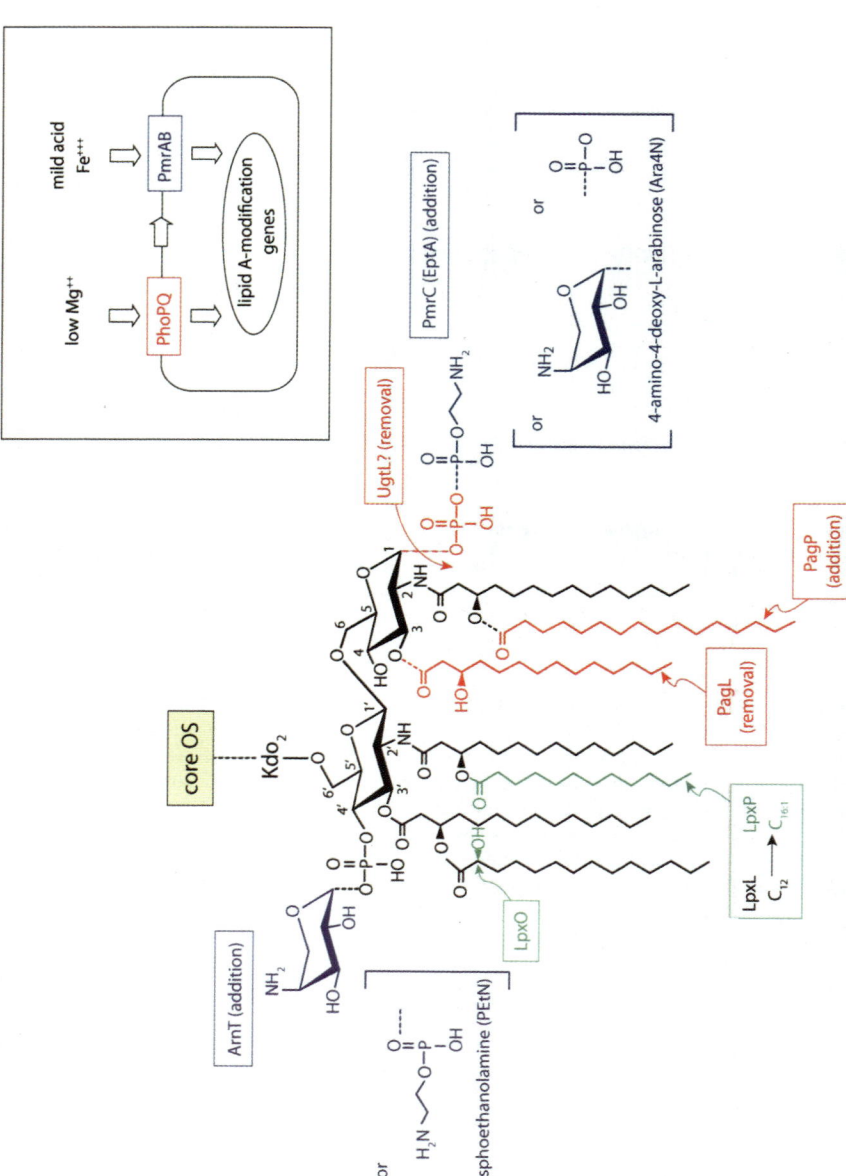

COLOR PLATE 16 (Chapter 12) Structure of lipid A and its environmentally regulated modifications in *Salmonella enterica*. The base structure of lipid A is modified by a series of enzymes that are active in the cytoplasm, the periplasm, or the outer membrane. Modifications include the nonstoichiometric addition of phospho-ethanolamine or 4-aminoarabinose to the 1 and 4' phosphates, removal of the 1-phosphate, and alterations of the acylation pattern. Many of these are regulated by the two-component regulatory systems PhoPQ (modifications in red) and PmrAB (in blue) in response to environmental signals that may reflect an intracellular lifestyle for the organism. These regulatory systems also interact. In *E. coli* K-12, many of these enzymes are cryptic but can be induced by growth in media containing metavanadate. In addition to enhancing resistance of the organism to host cationic antimicrobial peptides, these changes can influence LPS signaling and endotoxicity. Modifications in green are not dependent on PhoPQ or PmrAB.

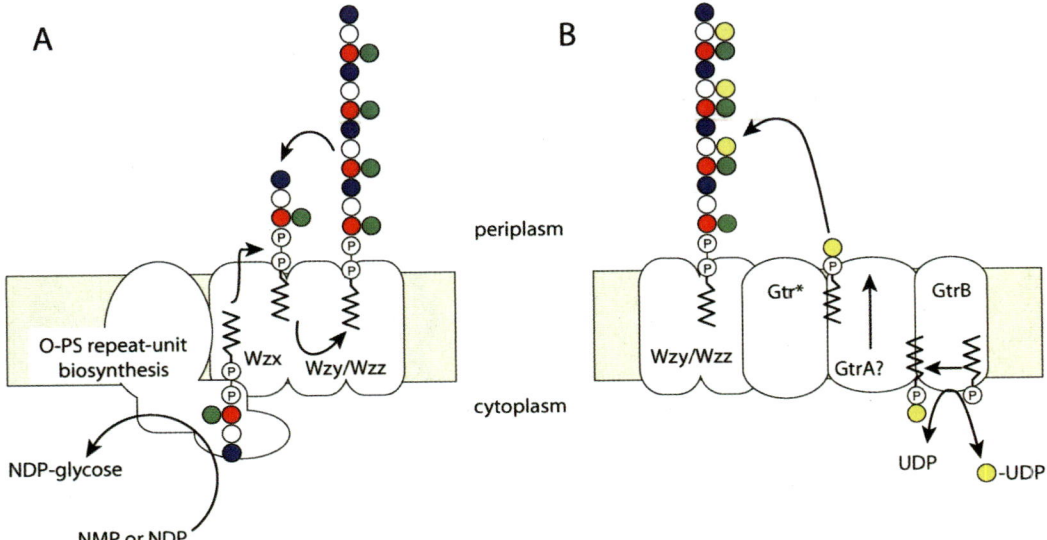

COLOR PLATE 17 (Chapter 12)   Schematic model of Wzy-dependent O-polysaccharide biosynthesis. (A) The formation of a hypothetical O-polysaccharide composed of a tetrasaccharide O-repeat unit (the four glycoses are identified by red, white, green, and blue filled circles). Undecaprenyl pyrophosphate-linked intermediates are formed at the cytoplasmic face of the membrane. These are initiated by transfer to undecaprenyl phosphate of hexose-1-P or acetamidohexose-1-P by homologs of the WbaP or WecA enzymes, respectively. The undecaprenyl pyrophosphate-linked O-repeat unit is flipped across the inner membrane by Wzx and polymerized at the periplasmic face by a blockwise process that requires Wzy and that is regulated by Wzz. (B) Modification of *Salmonella* or *Shigella* O-polysaccharide by lysogenic bacteriophage-encoded Gtr enzymes. The glucosyl donor is undecaprenyl phosphoryl-Glc and is synthesized by GtrB. The donor is transferred across the inner membrane by a mechanism that may involve GtrA. The serotype-specific Gtr* protein is then required for addition of glucosyl residues to the growing glycan. The reducing terminal O-repeat unit escapes modification.

WbdD	WbdA + WbdB	WbdCB	WecA
chain terminator	repeat-unit domain	adaptor	primer

methyl-O-P--[α-D-Manp-(1-3)-α-D-Man-(1-2)-α-D-Man-(1-2)-α-D-Man-(1-]$_N$-3)-α-D-Man-(1-3)-α-D-Man-(1-3)-β-D-GlcNAc-PP-und

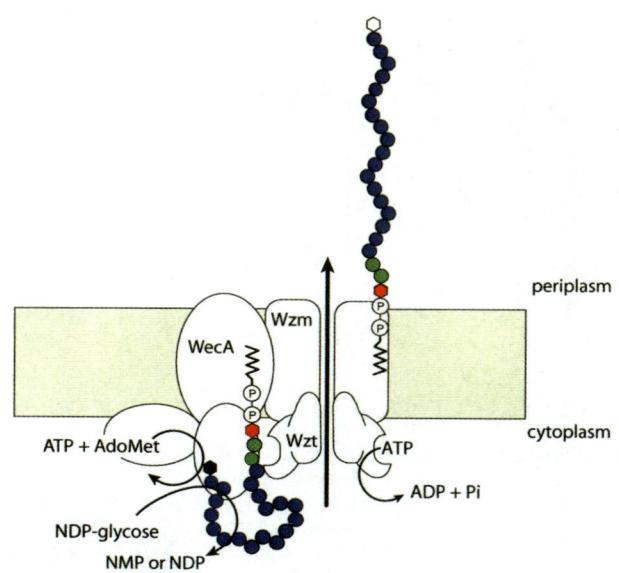

COLOR PLATE 18 (Chapter 12)   Schematic model of the biosynthesis of the *E. coli* O9a antigen by an ABC transporter–dependent pathway. The model is based on combined results from structural analysis of the LPS product and biochemical experiments. The polymer is formed by transfer of glycosyl residues to the nonreducing terminus of an undecaprenyl pyrophosphoryl–GlcNAc primer in the cytoplasm. The undecaprenyl pyrophosphate-linked intermediate (shown above) contains the identifiable primer, the adaptor domain, the repeat-unit domain, and a terminating phosphomethyl derivative whose precise linkage has not yet been resolved. The enzymes responsible for synthesis of each domain are indicated above the structure. The WbdD–mediated addition of the phosphomethyl terminator is essential for establishing modality and for recognition of the export substrate by the ABC transporter. (Modified from Raetz and Whitfield, 2002.)

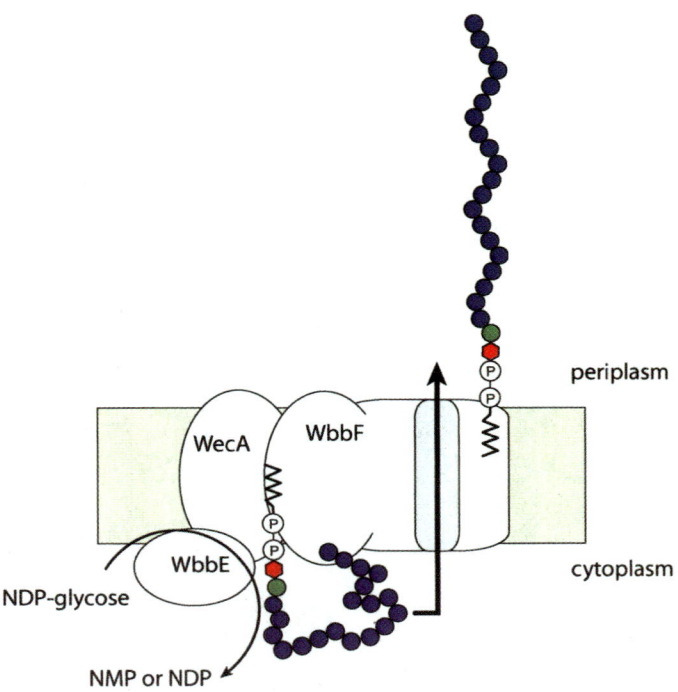

COLOR PLATE 19 (Chapter 12) Schematic model of the biosynthesis of the *S. enterica* O:54 antigen by a synthase-dependent pathway. The polymer is formed by transfer of glycosyl residues to the nonreducing terminus of an undecaprenyl pyrophosphoryl-GlcNAc primer in the cytoplasm. The undecaprenyl pyrophosphate–linked intermediate (shown above) contains the identifiable primer, the adaptor domain, the repeat-unit domain in an overall architecture resembling the *E. coli* polymannans formed by ABC transporter-dependent pathways (Color Plate 18). The synthase, WbbF, is required for chain extension, generating a repeat-unit domain with alternating β1,3- and β1,4-linkages. The currently available data suggest that this protein is sufficient for both chain extension and export of the nascent lipid-linked intermediate across the inner membrane. Details of the process and the mechanism by which chain termination is regulated have yet to be described. While chain extension and export are shown as separate processes, it is conceivable that they are temporally coupled in vivo. (Modified from Raetz and Whitfield, 2002.)

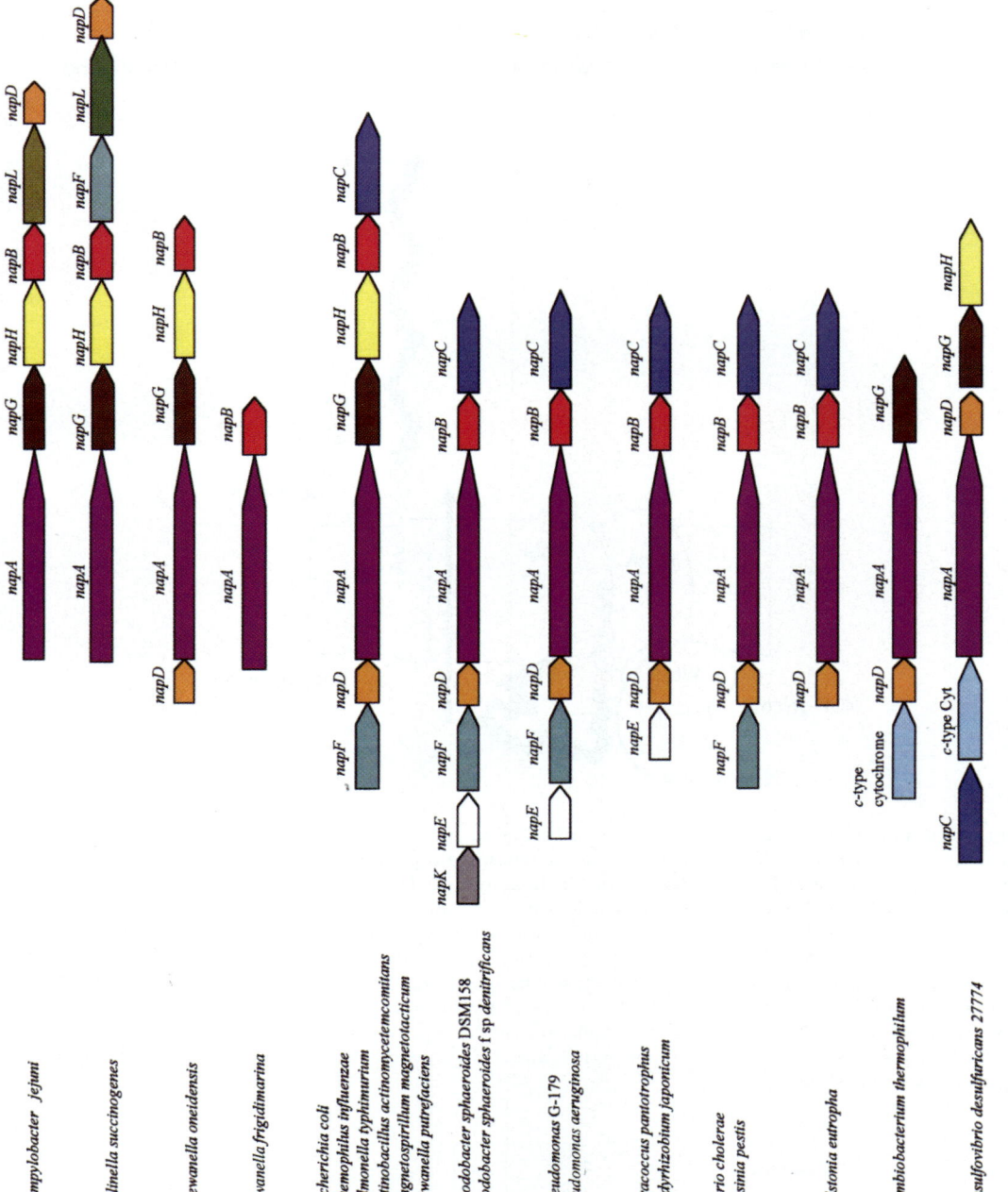

COLOR PLATE 20 (Chapter 14) Organization of *nap* gene clusters in different bacteria.

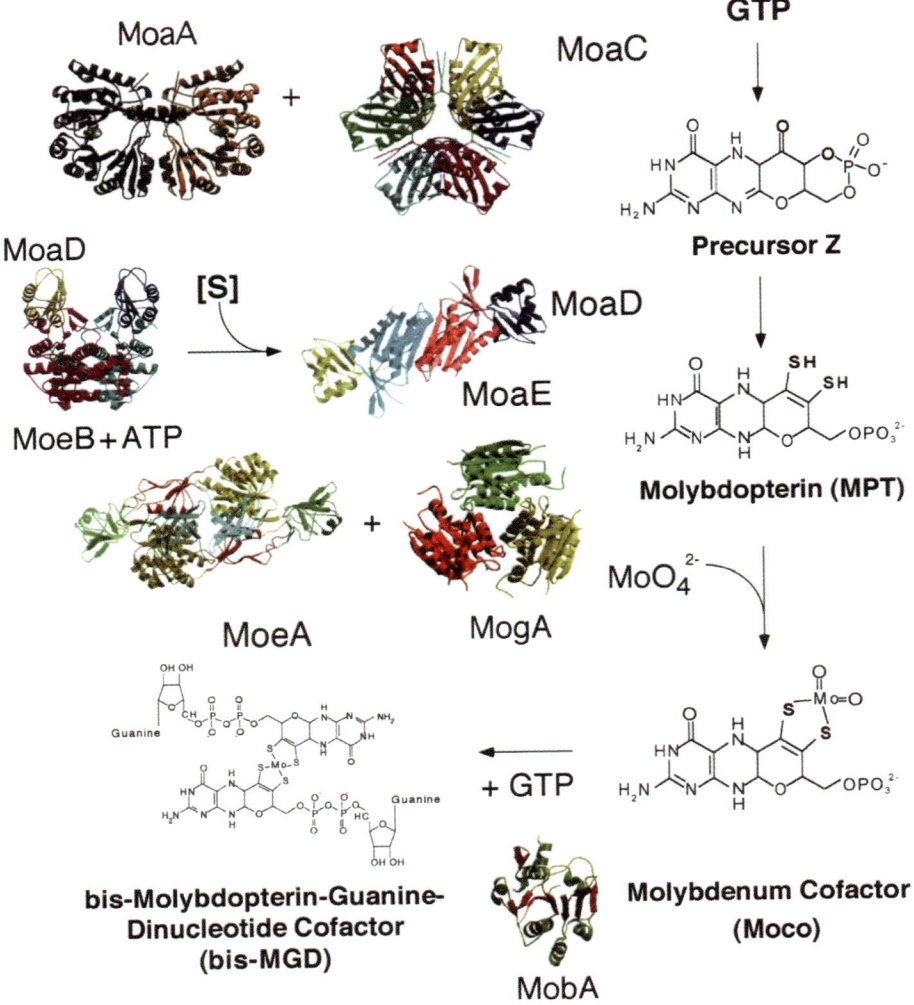

GTP

Precursor Z

Molybdopterin (MPT)

MoO_4^{2-}

MoaA

MoaC

MoaD

[S]

MoaD

MoaE

MoeB + ATP

MoeA

MogA

+ GTP

Molybdenum Cofactor
(Moco)

bis-Molybdopterin-Guanine-
Dinucleotide Cofactor
(bis-MGD)

MobA

Guanine

Guanine

COLOR PLATE 21 (Chapter 15)   The biosynthesis of Moco and bis-MGD in *E. coli*. Shown is a scheme of the biosynthetic pathway for Moco biosynthesis in *E. coli* and the proteins involved in these reactions. The crystal structures of MoaA from *S. aureus* (Hanzelmann and Schindelin, 2004) and MoaC (Wuebbens et al., 2000), MoaD/MoaE (Rudolph et al., 2001), MoeA (Xiang et al., 2001), MogA (Liu et al., 2000), and MobA (Lake et al., 2000; Stevenson et al., 2000) from *E. coli* have been solved. In *E. coli* Moco is further modified by the attachment of GMP, forming MGD, and two equivalents of MGD are bound to molybdenum, forming the so-called bis-MGD cofactor. The structures were modified after the published articles.

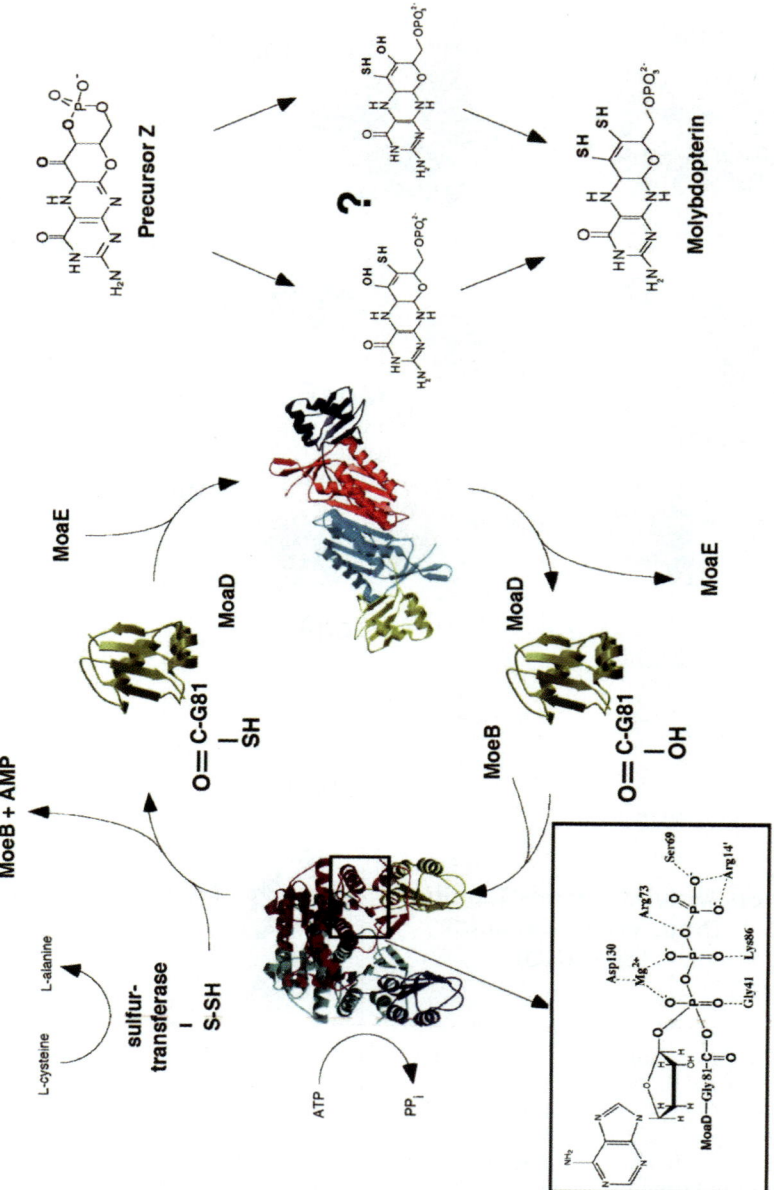

COLOR PLATE 22 (Chapter 15) Catalytic cycle of MoaD and its interactions with MoaE and MoeB. MoaD binds in its thiocarboxylated form to MoaE, forming active MPT synthase which converts precursor Z to MPT (Rudolph et al., 2001). The MoaD-carboxylate dissociates from the complex and interacts with dimeric MoeB, and in the presence of ATP an activated MoaD-adenylate is formed (Lake et al., 2001). MoaD-AMP is susceptible to sulfuration by a protein-bound persulfide group from a sulfurtransferase. The sulfur is most likely derived from L-cysteine (Leimkühler et al., 2001). After the formation of the thiocarboxylate group, MoaD dissociates from the MoeB dimer and reassociates with MoaE. Initial attack by the first MoaD thiocarboxylate could occur at either the C1′ or C2′ position of precursor Z to produce one of two hemisulfurated intermediates. For either intermediate structure, MPT formation would be completed by replacement of the remaining side-chain hydroxyl by the sulfhydryl from the second MoaD thiocarboxylate (Wuebbens and Rajagopalan, 2003). The structures were modified after the published articles.

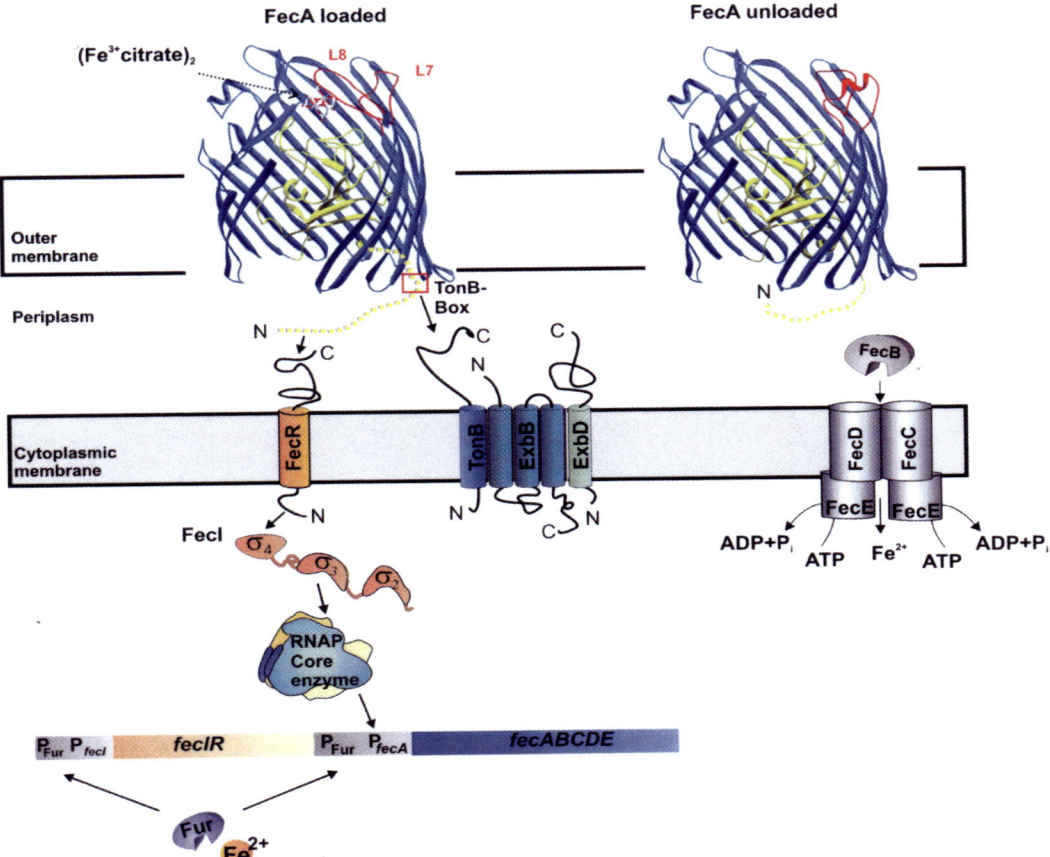

COLOR PLATE 23 (Chapter 16) The ferric citrate transport and regulatory system. The signaling pathway from FecA to FecI; the involvement of TonB, ExbB, and ExbD in signaling and transport; and transport of iron through the periplasmic FecB protein and the ABC transporter FecCDE proteins are shown. Fur repressor loaded with Fe^{2+} binds to the promoter upstream of *fecI* and *fecA* and dissociates from the promoter under low–iron conditions. Interactions between the FecA TonB box and TonB and between the FecA signaling domain (residues 1 to 79 of the mature protein) and FecR (see also Color Plate 25) are indicated. N indicates the N-terminal end, and C is the C-terminal end of the proteins. σ_2 and σ_4 indicate FecI domains involved in binding to FecR and DNA, respectively (see also Color Plate 26).

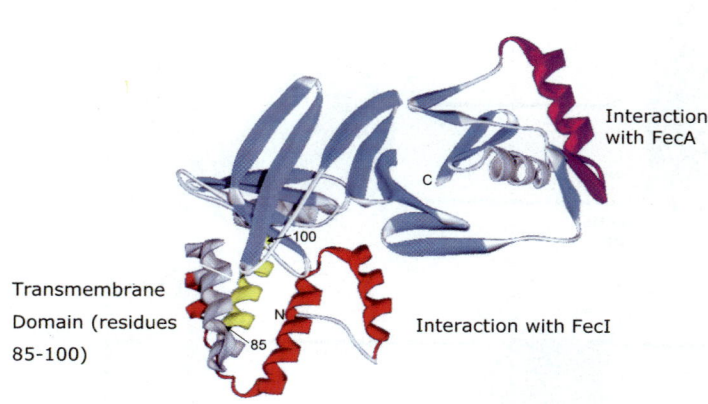

Transmembrane Domain (residues 85-100)

Interaction with FecA

Interaction with FecI

COLOR PLATE 24 (Chapter 16) Predicted FecR structure derived from known crystal structures of antisigma factors. The in silico-modeling was calculated by using the Rosetta-algorithm (http://www.bioinfo.rpi.edu/~bystrc/hmmstr/about.html and presented with Swiss-PDB-ViewerV3.7 and http://www.expasy.org/spdbv. and POV-Ray V3.5). The helical regions may be approximately predicted, but their relative orientation is less certain. The N-terminal FecR structure (residues 1 to 85) is similar to the N-terminal anti-σ RseA (residues 1 to 66) crystal structure composed of four short α-helices (Campbell et al., 2003).

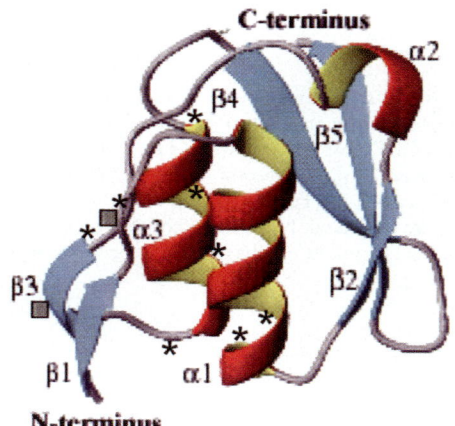

COLOR PLATE 25 (Chapter 16) Structure of mature FecA1–96 (Ferguson et al., 2002) deduced from NMR. The locations of suppressor mutations and single mutations that disrupt the interaction between FecA and FecR are indicated by arrows. Note that the mutations are allocated to α3, β1, and β3 (Eisenhauer et al., 2005; E. Breidenstein, S. Mahren, and V. Braun, unpublished results).

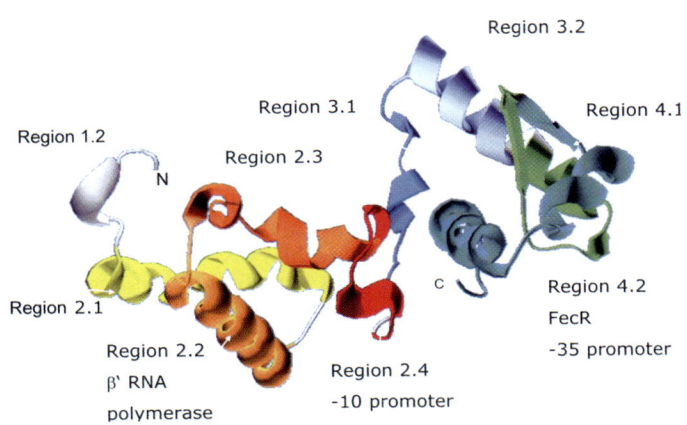

Region 1.2

Region 3.1

Region 3.2

Region 2.3

Region 4.1

N

Region 2.1

c

Region 2.2

β' RNA

polymerase

Region 2.4

-10 promoter

Region 4.2

FecR

-35 promoter

COLOR PLATE 26 (Chapter 16) Predicted FecI structure derived from known crystal structures of sigma factors. Note that all predicted sites of interaction with the −10 and −35 promoter regions, FecR, and the β′-subunit of the RNA polymerase are located on one side of the structure. The helical regions may be approximately predicted, but their relative orientation is less certain. The two functionally important regions 2 and 4 interconnected by region 3 are seen (see for comparison the crystal structure of σ^E in Campbell et al., 2003). The in silico-modeling was calculated by using the Rosetta-algorithm (http://www.bioinfo.rpi.edu/~bystrc/hmmstr/about.html and presented with Swiss-PDB-Viewer V3.7 and http://www.expasy.org/spdbv. and POV-Ray V3.5).

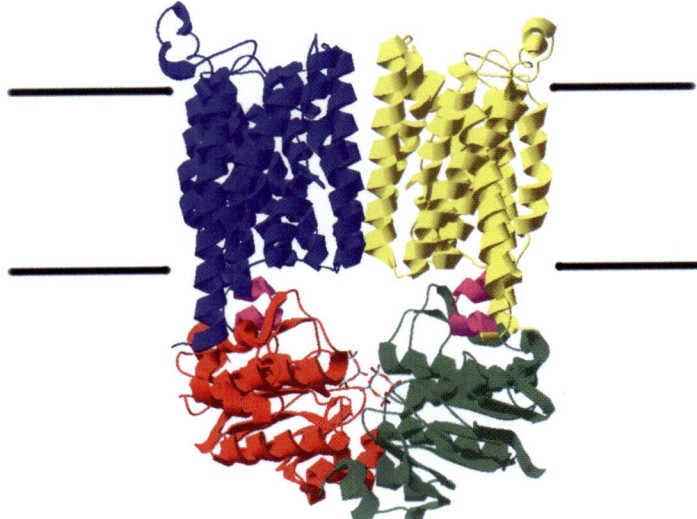

COLOR PLATE 27 (Chapter 17) Three-dimensional structure of the membrane-associated components BtuCD of the vitamin B$_{12}$ (cobalamin) ABC importer. The BtuCD dimer is represented from the side of the membrane, with BtuC helices (blue and yellow, respectively) probably embedded with the lipid bilayer, which was shown between thick horizontal lines. A vanadate molecule (green) is trapped within each cytoplasmic ABC ATPase domain (green and red, respectively). The so-called EAA loop (purple) intimately interacts with the helical domain of BtuD ATPase. The model is represented by using the SwissPDB_viewer software, from atomic coordinates deposited in the Protein Data Bank (Locher et al., 2002).

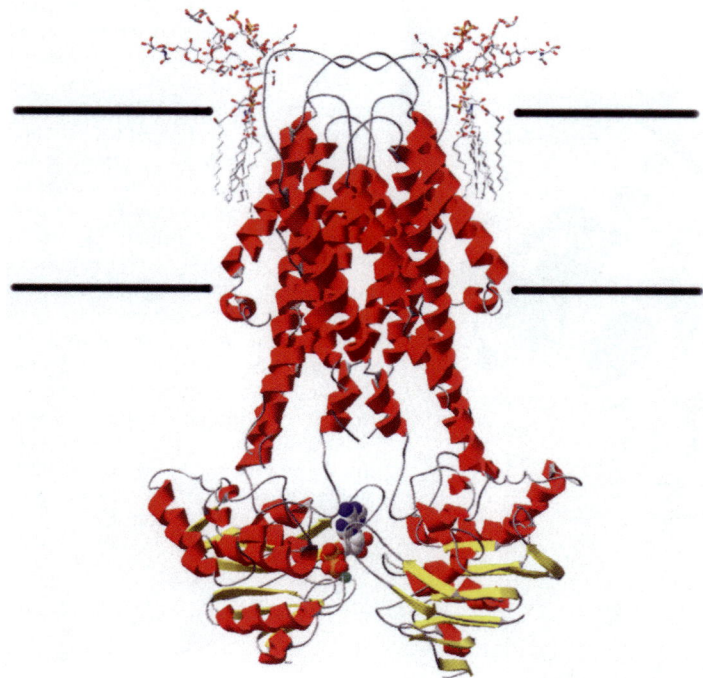

COLOR PLATE 28 (Chapter 17) Three-dimensional structure of the *S. enterica* serovar Typhimurium MsbA lipid A flippase, in complex with Mg^{2+} ADP-vanadate and rough LPS. The MsbA dimer is represented from the side of the membrane, with transmembrane helices probably embedded with the lipid bilayer, which was shown between thick horizontal lines. Two LPS molecules are tightly bound to a large periplasmic loop in the protein, with the lipid A portion buried in the membrane. The ADP-vanadate molecule is trapped within a composite binding site contributed by the two cytoplasmic ABC ATPase domains. The image model is drawn by using the Swiss PDB viewer software from atomic coordinates deposited in the Protein Data Bank (Reyes and Chang, 2005).

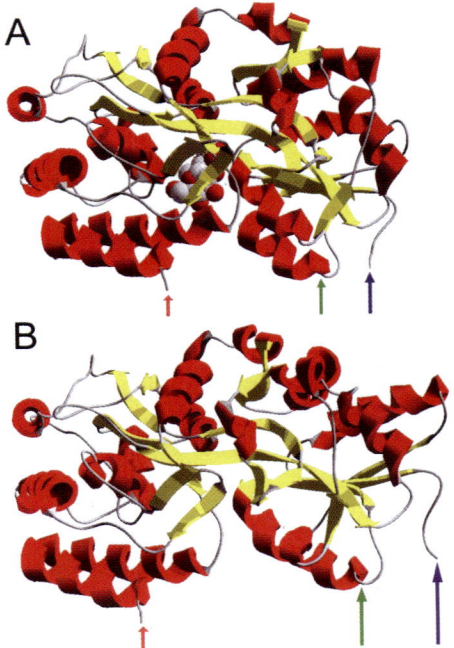

COLOR PLATE 29 (Chapter 17) Three-dimensional structures of maltose-bound closed form (A) and maltose-free open form (B) of *E. coli* maltose-binding protein. This figure illustrates the large substrate-induced conformational change (twist and rotation) in many periplasmic substrate-binding proteins. Compare the length and the relative distance of the three arrows, which point to the same atoms in the two structures. The image model is drawn by using the SwissPDB_viewer software from atomic coordinates deposited in the Protein Data Bank (Sharff et al., 1992).

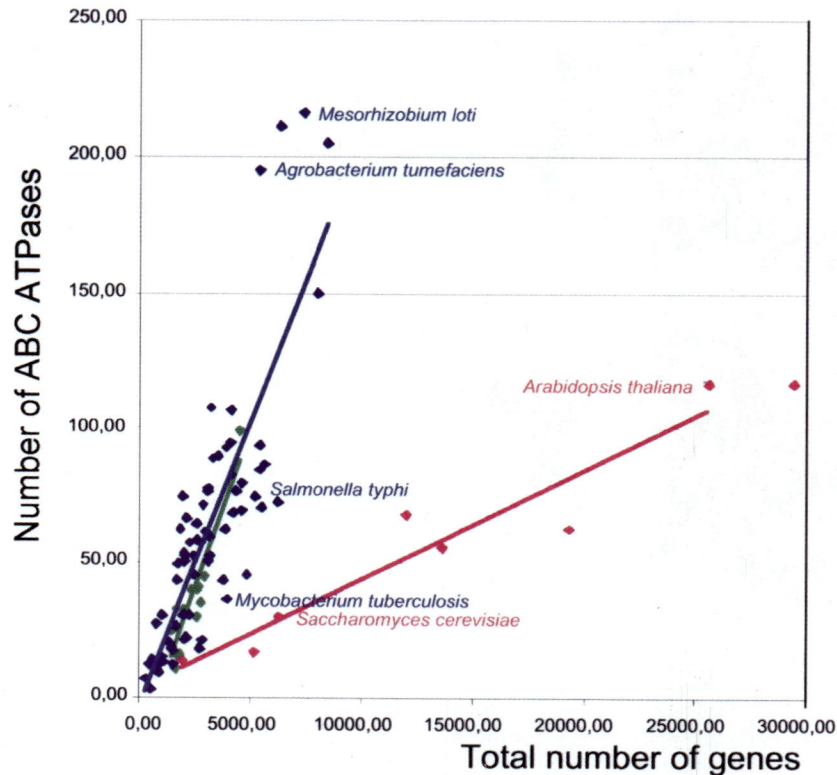

COLOR PLATE 30 (Chapter 17) Number of ABC ATPases versus number of genes in completely sequenced genomes. The number of ABC ATPases per genome (which roughly reflects the number of ABC systems) is plotted against the total number of genes (archaea, green dots; bacteria, blue dots; eukaryotes, purple dots). Selected genomes with exceptionally high or low ABC protein content are indicated on the graph.

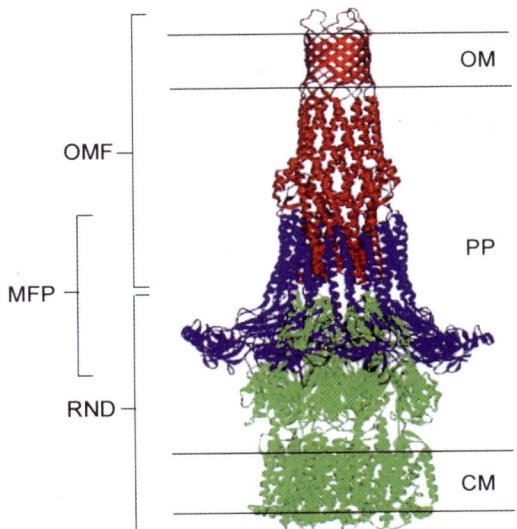

COLOR PLATE 31 (Chapter 18) Structure of a three-component RND–MFP–OMF efflux pump taken from the known crystal structures of AcrB (RND component [Murakami and Yamaguchi, 2003]), MexA (MFP [Akama et al., 2004a; Higgins et al., 2004]), and TolC (OMF [Koronakis et al., 2004]). (The figure was adapted from Eswaran et al. [2004] and is used with permission.)

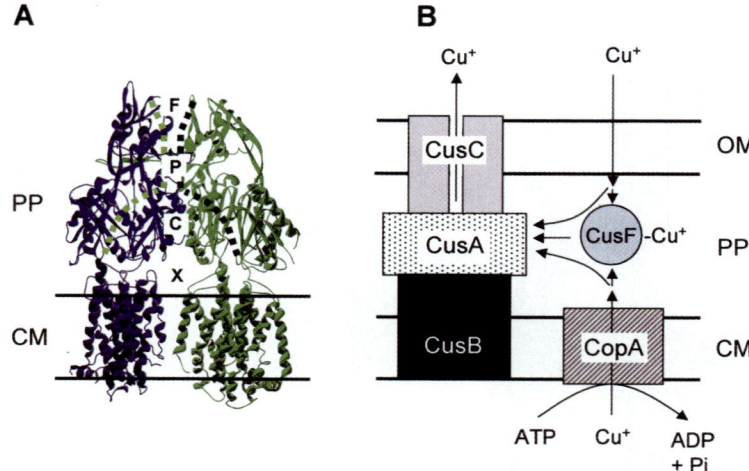

COLOR PLATE 32 (Chapter 18)   Structural (A) and functional (B) evidence that RND pumps export their substrates from the periplasm. (A) Side (cut-away) view of the MexB trimer (modeled on the AcrB crystal structure) with the monomer closest to the viewer removed to show the vestibule (x) between the two remaining monomers. Thus, the three vestibules in the trimer are likely portals of entry of export substrates from the periplasm into the central cavity (C), leading to the pore (P) and, finally, funnel (F) that exits the protein at the distal end of the trimer. CM, cytoplasmic membrane; PP, periplasm. (Adapted from Poole [2004c] and used with permission.) (B) Model of Cu^+ efflux by the CusCBA RND-type efflux system, highlighting its export of Cu^+ from the periplasm. CusF is a periplasmic Cu^+-binding protein that works with CusCBA to export Cu^+ from the cell, consistent with CusCBA being able to accommodate periplasmic Cu^+. CopA is a CM ATPase that exports cytoplasmic Cu^+, and mutants lacking this pump show a Cu^+-sensitive phenotype that is not compensated for by CusCBA, arguing that CusCBA is not an alternative pump for removal of cytoplasmic Cu^+. Thus, Cu^+ entering the periplasm from the extracellular milieu or from the cytoplasm (via CopA) is exported via CusCBA with or without CusF involvement.

INDEX

ABC binding protein-dependent uptake systems, prokaryote, 289–293
ABC importers, 293–297
ABC systems, ATPase of, genes encoding, 297, Color Plate 30
  high-affinity iron-uptake, 288
  inventory and classification of, 292–293
  organization and functions of, 288–289, Color Plates 27 and 28
ABC transporter-dependent pathway, 225
ABC transporters, biological role of, 287–288
  cytoplasmic membrane, to convey substrate into cytoplasm, 291–292, Color Plate 28
  periplasmic, 287–303
  substrates handled by, 289
Acetyl-CoA, 224–225
AcrB, 235–236
Agrobacterium tumefaciens, 297, 326
AhpC, 133
Alginate, 290
Alkaline phosphatase isozyme conversion protein (IAP), 162–163
α-Helices, transmembrane, prediction of, 400, 401
Amino acid composition support vector machine methods, 395–396
Amino acids, hydrophobic branched-chain, HAA family specific for, 295–296
  polar, PAO family specific for, 295
4-Aminoarabinose, addition to lipid A, 218–219
Amphiphiles, supramolecular assembly of, β-barrel structure formation and, 46
Anaeromyxobacter dehalogenans, 254–256
Antibody fragments, expression in periplasm, 362–368
  fusion partners and cleavage regimens for, 375, 377
  maximizing of, 368–379
  periplasmic expression of, 361–388

production of, 361–362
structure of, 363–364
types of, 364–368
uses of, 362
Antimicrobial resistance, 304–324
ATP-binding cassette transporters. *See* ABC transporters
ATPase subunit, 292

β-Barrel membrane proteins, 31
  folding of, into phospholipid bilayers, 32–33, 46
    lipid acyl chain length dependence and, 51–52
  genomic identification of, 34
  insertion and folding into lipid layers, 51–57
  insertion and folding of, electrophoresis and, 47–48
  structure formation of, 46
β-Lactamases, 315–316
Bacillus subtilis, 175, 177, 182, 183, 296, 297
Bae envelope stress response, Cpx and, 101
Binding protein-dependent transport systems. *See* ABC import systems
Biofilm, formation of, Cpx response influencing, 98
Bordetella avium, 283
Bordetella bronchiseptica, 283
Borrelia burgdorferi, 68, 76–77
Braun's lipoprotein, 201
Brominated phospholipids, quenching of tryptophan fluorescence by, 53–55
Bundle-forming pilus, 89

C-terminal processing protease-1 Prc. *See* Tsp
Campylobacter jejuni, 135, 237, 240, 250
Cardiolipin, 74
Catalase, periplasmic, 314
Caulobacter crescentus, 190
CcdA, 243

CcmA, 243
CcmB, 243
Cell division, 173–197
    ABC transporter and, 183–184
CELLO, 396
Cephalexin, 186
Chaperones, periplasmic, and peptidyl-prolyl
        isomerases, 141–149
Chloroflexus aurantiacus, 237
Chymotrypsin, catalytic triad of, 154
Chlamydomonas reinhardtii, 267
Clan PA(S), 154–156
Clan SE, with families S11, S12, and S13, 157
Clostridium perfringens, 248
Cold osmotic shock, for extraction of proteins from
        periplasm, 378–379
Conjugative pilus expression. *See* Cpx
Copper, 310–313
Copper proteins, 239–240
Cotranslational protein targeting, 6–9
Cpx, and Bae envelope stress responses, 101
    and σ^E envelope stress responses, 101
    envelope stress response, 83–106
        overlap with other stress responses, 100
    identification of locus of, independent genetic
        selection and, 83–85
    microbial surface interaction, 96–97
    overlap with EnvZ/OmpR regulon, 101–102
    response, and stationary phase, 102
        attachment to abiotic surfaces and, 97–98
        influencing biofilm formation, 98
    response activation, by general envelope
        perturbations, 88–89
        by overexpression of envelope proteins, 89–90
    signal transduction, 86–96
    two-component system, 86
Cpx-activating signals, 90–91
Cpx-inducing cues, and –signaling proteins, 88
Cpx-inducing signal, 87–90
Cpx pathway, 88
    induction of, bacterial attachment to surfaces and, 90
    pathogenesis and, 98
Cpx regulon, 91, 92–93
    influence on range of physiological processes and, 96
CpxA, 85, 87, 88, 90, 91
    and CpxR, two-component regulatory system,
        86–87
    sensing of envelope stress, 90–94
CpxP, 91–94
CpxR, 91, 94–95, 97
    and CpxA, two-component regulatory system,
        86–87
    as activator, 94–95
    as potentiator, 95–96
    as repressor, 94–95
CpxR ~ P, control of transcription by, 94–96
CpxR ~ P binding sites, 95, 96

Cupredoxins, 243–244
    electron delivery to periplasmic reductases from, 240
    electron transfer to periplasmic globular domains
        via, 240
Cysteine proteases, 163–164
Cytochrome bc_1 complex, 236–237
Cytochrome *c* peroxidases, 315
Cytochromes, *c*-type, electron delivery to periplasmic
        reductases from, 240
    electron transfer to periplasmic globular domains
        via, 240
    periplasmic, 242–244
Cytoplasmic membrane, energy transfer across
        periplasm into outer membrane, 278–280,
        Color Plate 23
Cytoplasmic protein turnover, conjugation/regulation
        of, 99–100

Daughter cells, separation of, 189–190
DBSubLoc, 400
DegP, 40–41, 70, 113, 155–156, 325
    functions of, 155
    structure of, 40, 155, Color Plate 4
DegP protease, 351
DegQ, 156
DegS, and activation of, 108–109
    in proteolytic pathway, 111–112
Desulfovibrio desulfuricans, 251, 253, 254, 256
Dimethyl sulfoxide (DMSO) reductase, 260, 262
Disulfide bonds, formation by oxidative process, 122
    formation in periplasm, 122–140, 345, Color Plate 8
        properties of, 123, 124
    formation of, DsbA catalyzes, 126
        history of, 123
    in protein, 362
    isomerization pathway, 129–133
    maturation of, 124–127
    three-dimensional structure of protein and, 122
DLM family, specific for methionine and methionine
        derivatives, 296–297
DNA translocation, 185
DsbA, 123, 124, 129, 132, 133, 134–135
    active site of, 126
    as member of thioredoxin family of proteins, 125
    catalyzation of formation of disulfide bonds by, 126
    domain organization of, 125, Color Plate 9
    exportation to periplasm, 126
    reoxidation by DsbB, 127–128, Color Plate 10
DsbB, 129
    color of, quinhydrone and, 128–129
    periplasmic cysteines in, 127
    reduced, oxidized by ubiquinone, 128
    reoxidation of DsbA by, 127–128, Color Plate 10
    topology of, 127, 128
DsbC, 129, 133
DsbC, domain organization of, 130–131, Color Plate 11
    isomerization by, 131, 132

DsbD, 132, 134, 243
   domain organization of, 134, Color Plate 12
DsbG, 133–134

Ecotin, 164–165
EFFLUX, 304–310
Efflux systems, gram-negative, periplasm-spanning, 305
   RND-type multidrug, 307
Electron transfer, by globular domains of membrane-anchored proteins, 240–241
Electron transfer proteins, periplasmic, assembly of, 242–244
Electron transfer system(s), cytoplasmic membrane, organizations of, 236–237
   pathways of, alternative and redundant, 237–238
   to periplasm, 244
Electron transport activities, cyclic, in purple bacteria, 238
   in periplasm, 235–246
Electrophoresis, kinetics of tertiary structure formation by (KTSE), 48, 50, 52
   to monitor OmpA folding, 55
   sodium dodecyl sulfate-polyacrylamide gel, 47, 222
Environmental signals, in Z-ring assembly, 177–178
EnvZ/OmpR regulon, Cpx overlap with, 101–102
EptB, 220–221
Ero1p, 129
Erv2p, 129
Escherichia coli, 326, 327, 329, 330, 331, 332, 336, 361, 362
   antibody fragments produced in, 364–367
   lipoproteins in, biochemically confirmed, 69
      structure, function, and transport of, 67–79
   murein segregation and morphogenesis of, 207–208
   protein secretion in, 347, 348
   proteins in cell division in, 173–174, 175, Color Plate 14
   signal recognition particle (SRP) system, 4
   SRP pathway, 8–9
   substrate proteins harboring Tat-targeting sequences, 17, 18–19
ExbB, 278, 279, 281
ExbD, 278, 279, 281
Extracytoplasmic proteins, biogenesis of, 141–142

Fab, 364
Fe^{3+}, 276, 277
Fe^{3+} siderophores, 276, 277
Fec genes, *E. coli* K-12, 282–283
Fec-type transcription regulation, occurrence of, 283–284
FecA, 277–278, 280–281
FecA1–96, structure of, 281, Color Plate 25
FecI, predicted structure of, 281, Color Plate 25
FecIR, and FecABCDE, regulation of transcription in cytoplasm, 281–282

FecR, structure of, 281, Color Plate 24
Ferric citrate transport and regulatory system, 277, 282–283, Color Plate 23
Ffh, 6
FhuA, 277, 278, 279, 280
FkpA, 144–145, Color Plate 13, 113
   homologues of, 42
   isomerase activity of, 41–42
   structure of, 41, Color Plate 4
Fluorescence quenching, folding intermediates by, 53–55
Folding factors, aggregation of periplasmic proteins and, 373–374
Formaldehyde, methanol oxidations to, 239–240
FptA, 277
FpvA, 277
FtsA, 173, 178, 180, 181–182
FtsB, 189
FtsEX, 183–184
FtsI, 185, 187–188, 208
FtsK, 185
FtsL, 189
FtsN, 189
FtsQLB complex, 178–180, 188–189
FtsW, 188
FtsY, 7–8
   ribosome-nascent-chain-SRP complex and, 8–9
FtsZ, 173, 174, 178, 180
FtsZ-binding proteins, 181–183

Genes, encoding ATPase of ABC systems, 297, Color Plate 30
Genome comparisons, 297–298, Color Plate 30
Glucans, osmoregulated periplasmic. *See* OPGs
Glucosylation, O-antigen, 223–224, Color Plate 17
GtrA, 223, 224
GtrB, 223, 224
GtrV, 224

HAA family, specific for hydrophobic branched-chain amino acids, 295–296
Heavy metals, efflux-mediated resistance to, 306, 307
HMMTOP, 401, 402, 403
HMW1B, 43
HtpX, transcription of, Cpx response in regulation of, 99, 100
HtrA. *See* DegP

IAP, 162–163
IgG1, 363
Imp, 44–45, 227
Integral membrane proteins, classification of, 31
   export through periplasm to outer membrane, 34, 35
   from periplasm, assembly into outer membrane, 30–66
   functions of, 30

Iron, transport and regulation of, energy and
    information transfer across periplasm in,
    276–286
  transport of, energy-coupled, across outer
    membrane, 276–280
Iron-siderophore uptake system family, for iron-
    siderophores, vitamin B_{12}, and hemin,
    293–294
Iron transporters, outer membrane, conformational
    changes in, 276–278

KdoII, 220
Klebsiella oxytoca, 76

LamB-LacZ-PhoA, 84, 89
Leaky hosts, use to produce recombinant proteins,
    353–354
Legionella pneumophila, 98, 99
Lipid A, -core, biosynthesis and export of, 215–221
  -core acceptor, ligation of O-polysaccharide to,
    227
  regulated modifications of, 217–220
  structure of, 217–218, Color Plate 16
Lipid layers, interaction of OmpA/Skp/LPS complex
    with, 50–51
LipoP, 401–402
Lipopolysaccharide(s), 113, 116–117
  acylation pattern, modification of, 219–220
  assembly of, periplasmic events in, 214–234
  biosynthesis of, 34
  in folding β-barrel OmpA, into lipid layers,
    49–51
  role of, 48–49
  synthesis of, 214–215, Color Plate 15
  trimerization of PhoE and, 49
  translocation to cell surface, 227–228
Lipoprotein box, 67
Lipoprotein-sorting signals, in vitro analysis of, 73–75
  in vivo analysis of, 70–71
Lipoproteins, biogenesis of, 67–68
  gene products, putative, analysis of, 67–70
  in *E. coli*, biochemically confirmed, 69
    structure, function, and transport of, 67–79
  sorting in gram-negative bacteria, 76–77
  sorting to outer membranes, mechanisms for, 75
  synthesis of, 67
LOCtree, 396
Lol, avoidance function, 74, 76
Lol pathway, 71–73
  sorting and localization of lipoproteins through, 75
LolA, crystal structure of, 71, Color Plate 6
  structure and function of, 71–72
LolB, crystal structure of, 71, Color Plate 6
  structure and function of, 71–72
LolC, 72
LolCDE, 72–73
LolCDE complex, 75

LolCDE system, 184
LolD, 73
LolE, 72

MacAB system, 184
MalE, 189
Maltoporin, 290
Membrane-anchored proteins, globular domains of,
    electron transfer and, 240–241
MepA, 162
Mercury, 310, 313
MET family, specific for metallic cations, 293
Metal trafficking, sequestration, and detoxification,
    310–313
Metalloproteases, 160–163
Methylamine, 239
Min system and nucleoid occlusion, in Z-ring
    assembly, 176–177
Mineral and organic ions, MOI family specific for, 294
MoaB, 264
MoaD, catalytic cycle of, 263–264, Color Plate 22
MoaE, 264
ModA, 262, 263, 265
ModC, 262, 263
MOI family, specific for mineral and organic ions, 294
Molecular chaperone, 142
Molybdate transfer, 262
Molybdenum, in biological systems, 260–262
  insertion into molybdopterin, 264–265
Molybdenum cofactor, 260
  biosynthesis in *E. coli*, 261, 262–267, Color Plate 21
  biosynthesis of, and incorporation into
    molybdoenzymes, 260–275
  carrier proteins, in eukaryotes, 267
  insertion into xanthine dehydrogenase, protein
    required for, 267
  modification of, and attachment of GMP,
    264–265
Molybdenum enzymes, mononuclear, active site
    structures of, 268
Molybdenum nitrogenase, 260–261
Molybdoenzymes, biosynthesis of molybdenum
    cofactor, and incorporation into, 260–275
  E. coli, insertion of bis-MGD into, 266–267
  localization in *E. coli*, 269–270
Molybdopterin, 260
Monosaccharides, MOS family specific for, 296
MOS family, specific for monosaccharides, 296
MsbA, 227
Murein. *See also* Peptidoglycan
  architecture of, 202–203
  chemical structure of, 199–201
  macromolecular properties of, 201–202
  segregation and morphogenesis of *E. coli* by,
    207–208
  synthesis and hydrolysis of, 203–207
Murein hydrolases, 190, 206–207

Murein lipoprotein, 201
Murein sacculus, growth of, 208–210
  in cell envelope, 203
  location of, 198, 199
  molecular architecture of, 202–203
  structure and biosynthesis of, 198–213
Murein synthase, 208
Mycobacterium tuberculosis, 297

N-terminal initiator methionine, 362
Nap, physiological roles of, 248
Nap complexes, organization of, in bacteria,
    Color Plate 20
Nap gene clusters, components of, and biochemical
    functions, 250–252, Color Plate 20
NapA, 237, 251, 252, 256
  electron transfer to, in different bacteria, 252, 253
NapB, 251, 252, 256
NapC, 237, 241, 251, 252, 254
NapD, 251, 254
NapG, 252, 254, 255
NapH, 252, 254
NapK, 251, 252
NapL, 251
NarL, 249
NarP, 249
NarQ, 249
NarX, 249
Nitrate reductases, periplasmic, discovery of,
    247–248
    distribution, regulation, and functions of,
      248–250
    evolution of, 254–256
Nitrate reduction, bacterial, 247–259
  periplasmic, 247–259
NlpB, 45
NlpC, putative cysteine protease, 164
NlpE, 70, 88, 90, 97
NrfA, 249
Nucleoid segregation, septum assembly and,
    184–185

O-Acetylation, 224–225
O-Acetyltransferases, 224–225
O-Antigens, biosynthesis of, 221
O-Polysaccharide(s), 223, 225–226
  biosynthesis of, Wzy-dependent, 221–222,
    Color Plate 17
  initiation reactions, 221
  ligation to lipid A-core acceptor, 227
OafA, 224
σ^E, and stress response of, discovery of, 107–108
  envelope stress response controlled by, regulation
    and function of, 107–121
  free, release of, 110–111, Color Plate 7
  signal transduction pathway regulating,
    components of, 108

  in maintenance of regulon core functions,
    117
  properties of, 111–112
σ^E regulon(s), and core functions of, and regulation
    of, 116–117
  core and extended, regulon functions of, 113, 114,
    115, 117
  extended, variable portion of, 116
  functions of, 112–116
  members in *E. coli*, functional classification of,
    112, 113
Oligosaccharides, core, postsynthetic modification of,
    220–221
  structure of, 215, 216
  OPN family specific for, 295
  OSP family specific for, 294
  periplasmic membrane-derived, 316
Omp85, 43–44
OmpA, 47–48
  β-barrel, folding and insertion by concerted
    mechanism, 55–57, Color Plate 5
  folding into lipid layers, 49–51
  folding kinetics into DOPC bilayers, 52–53
  Skp/LPS-assisted folding pathway of, 50–51
  structure formation in, 51–52
  folding kinetics of, 48
  folding model of, 55–56, Color Plate 5
  folding process of, 54–55
  insertion and folding into bilayers, 53
  interaction with lipid bilayer, 52
  native structure formation in, 48
OmpH. *See* Skp
OPG-defective mutants, pleiotropic phenotypes of,
    335
OPG substitution enzymes, 331
OpgG, 330
OpgH, 332
OPGs, and pathogenicity, 333–335
  and stress signaling, 335–336
  as molecular agents, 336–337
  biosynthesis of, genomic overview of, 328–329
    mechanisms of, 329–331
    regulation of, 331–333
  connection with trehalose synthesis, 333
  structures of, 325–328
OPN family, 295
Osmoregulation, in periplasm, 325–341
OSP family, 294
OTCN family, 294–295
Outer membrane, constriction of, 190
  electron transfer to and from, 242
  energy-coupled iron transport across, 276–280
  energy transfer from cytoplasmic membrane into,
    278–280, Color Plate 23
  integral membrane proteins from periplasm
    assembly into, 30–66
  substrates crossing through proteins, 289–290

Outer membrane proteins, as integral membrane
     proteins, 141
  assembly of, outer membrane proteins in, 43–45
     periplasmic proteins in, 34–43
     proteins in, 36, 37
  β-barrel, prediction of, 400, 401
  composition of, 30
  examples of, 32–33
  folding of, in vitro studies on, 45–51
     into detergent micelles, 45–46
  in gram-negative bacteria, 31
  in outer membrane protein assembly, 43–45
  in unfolded form, in periplasm, 30
  pathway from cytoplasm to outer membrane, 34, 35
  structure of, 31, Color Plate 2
Oxidative stress, resistance to, 313–315
OxyR, 133–134

P-CLASSIFIER, 396
PA-GOSUB, 400
PagL, 220
PagP palmitoyltransferase, 220
PAO family, specific for polar amino acids and opines,
     295
PapE, 89–90
PapG, 89
Paracoccus denitrificans, 238, 240, 241
Paracoccus pantotrophus, 238, 244, 248, 249
PBP3. See FtsI
PBP4, 157–158
PBP5, 158
PBP6, 159
PBP6B, 159
PBP7, 159–160
Penicillin-binding proteins, 157–160
  murein hydrolysis and, 204–205
Peptides, signal. See Signal peptides
Peptidoglycan, 157
  penicillin-insensitive synthesizing activity and, 186
  septal synthesis of, 185–189, 190
Peptidyl-prolyl isomerases, 144–146
  periplasmic chaperones and, 141–149
Periplasm, ABC transporters and, 287–303
  alternative and redundant electron transfer
     pathways in, 237–238
  as subcellular target for antibody fragment
     production, 361–362
  assembly of integral membrane proteins from, into
     outer membrane, 30–66
  augmentation of secretion apparatus and, 372–374
  bacteria in, 235
  chaperones in, 36–43
  compared to cytoplasm, 345
  control of resistance determinants by, 316–317, 318
  dimensions and physicochemical properties of,
     235–236
  disulfide bond formation in, 122–140, Color Plate 8

electron transport activities in, 235–246
  energy and information transfer across, in iron
     transport and regulation, 276–286
  expression of antibody fragments in, 361–388
     fusion partners and cleavage regimens for, 375, 377
     maximizing of, 368–379
  extraction methods, compared, 378
  folding environment in, 349–350
  folding factors in, 36
  osmoregulation in, 325–341
  oxidation reactions, and soluble c-type
     cytochromes, 239–240
  oxidation/reduction reactions catalyzed in, 238–239
  protein accumulation in, cytoplasmic events
     affecting, 347–349
     practical applications for, 345–360
  protein folding in, 142–143
     chemical strategies for improving, 354–355
  proteins in, computational prediction of, 391–405
     computational predictive methods for
        identification of, 392–397
     identification of, from homology-based
        detection, 398
     membrane-anchored, identification of, 402–403
     methods for identification of, 391–392
     tools for prediction of, compared, 397
  proteins in outer membrane protein assembly,
     34–43
  signal transduction in, 281, Color Plates 23, 24,
     and 25
  signaling across, 280
Periplasmic oxidation/reduction processes, 235
Periplasmic reductases, electron delivery from quinols
     to, 241–242
Phe, mutagenesis of, in LolA, 72
PHOBIUS, 401, 402, 403
PhoPQ, 217
1-Phosphate, removal from lipid A, 219
Phosphatidylethanolamine, 74
Phosphoethanolamine, 216, 220
  addition to lipid A, 219
Pitrilysin, 161–162
Plasmid factors, in antibody fragment expression in
     periplasm, 368
PmrAB, 217
Polyols, OSP family specific for, 294
Porin signaling, 117
Porin status, graded response to, 111–112
Posttranslational protein targeting, 4–6
PpiA, 144, Color Plate 13, 42
PpiD, 42, 146, 374
Precursor Z, formation of, 263
  formation of molybdopterin from, 263–264
Promoter choice, in antibody fragment expression in
     periplasm, 368–369
Protease inhibitors, 164–165
  periplasmic proteases and, 150–170

Proteases, of unknown classification, 164
  periplasmic, and protease inhibitors, 150–170
Protein(s), accumulation in periplasm, cytoplasmic
      events affecting, 346–349
  accumulation of, periplasmic, practical applications
      of, 345–360
  amino acid sequence of, ribonuclease A and, 123
  biotherapeutic, advantages for producing,
      353–354
  complex, 346–347
  exposure to extracellular chemical agents, 362
  extraction from periplasm, 375–379
  fates after secretion into periplasm, 349
  in cell division in *E. coli*, 173–174, 175, Color Plate 14
  multimeric, expression of, 372
  penicillin-binding, 157–160
  periplasmic, computational prediction of,
      391–405
    computational predictive methods for
        identification of, 392–397
    identification of, from homology-based
        detection, 398
      from specialized databases, 398
    membrane-anchored, identification of, 402–403
    methods for identification of, 391–392
    tools for prediction of, compared, 397
    translational fusion partners for, 375, 376
  periplasmic accumulation of, practical applications
      for, 345–360
  periplasmic folding, 142–143
    chemical strategies for improving, 354–355
  purification of, simplification of, 362
  recombinant, genetic selections and screens to
      improve yield of, 352–353
    leaky hosts and, 353–354
    production of, activities to achieve higher,
        350–352
  secretion in *E. coli*, 347, 348
  secretory, synthesis in cytoplasm, 5
  selection by SecB, 5
Protein disulfide bond isomerase, 123
Protein disulfide isomerase, 129, 130
Protein targeting, cotranslational, 6–9
  posttranslational, 4–6
Proteolysis, antibody fragments and, 374–375
Proteome Analyst, for prediction of periplasmic
      proteins, 391, 395, 397
Pseudoalteromonas haloplanktis, 135
Pseudomonas aeruginosa, 76, 220, 285, 305, 326, 334
Pseudomonas fluorescens, 283
PSLpred, 396
PSORT I, for prediction of periplasmic proteins,
      391–392
PSORTb, for prediction of periplasmic proteins, 391,
      392, 393–395, 397
PSORTdb, 399–400
Pyrococcus furiosus, 268

Quinhydrone, in DsbB function, 128–129
Quinols, 236
  membrane-bound, electron delivery to periplasmic
      reductases, 241–242
Quinones, 236
  membrane-bound, electron delivery from
      periplasm to, 241

Ralstonia solanacearum, 283
Resistance/selection marker, in antibody fragment
      expression in periplasm, 369–370
Rhodobacter capsulatus, 238, 249
Rhodobacter sphaeroides, 266
Rhodopseudomonas viridis, 238
Ribonucleotide reductase, 122
Ribosome-nascent-chain-SRP complex, membrane-
      associated receptor FtsY and, 8–9
Ribosomes, translating SRP substrates, 9
  4.5S RNA, 6–7
  7SL RNA, 6–7
RND pump(s), as multidrug exporters, 305–307
  periplasmic capture of substrate molecules by,
      309–310, Color Plate 32
  structure of, 307–308, Color Plates 31 and 32
RotA. *See* PpiA
Rotamase A. *See* PpiA
RseA, 117
  as negative regulator, 108, 109
  cleavage of, RseP and, 110
  degradation of, proteolytic cascade in, 110,
      Color Plate 7
  transduction of signal for, 111
RseB, as negative regulator, 108
RseP, and cleavage of RseA, 110

Salmonella enterica serovar Typhimurium, 98
SDS-polyacrylamide gel electrophoresis, 47
SecA, chaperoning newly synthesized nonsecretory
      proteins, 5
SecA protein, translocon-bound, targeted to
      SecB-bound preproteins, 9–10
SecB, protein selection by, 5
  targeting presecretory proteins, 5
SecB-bound preproteins, targeted to translocon-bound
      SecA protein, 9–10
Secretion apparatus, augmentation of, 372–373
SecYEG translocon, 3
  co- and posttranslational protein targeting to, 3–15
Septal ring, assembly of, 178–181
  and nucleoid segregation, 184–185
Serine peptidases, 150–160
  clan PA(S) of, 154–156
  clan SE with families S11, S12, and S13 of, 157
  penicillin-binding proteins and, 157–160
  putative protease YdgD of, 157
  tail-specific protease Prc/Tsp of, 156–157
Serine proteases, 150–160

Shewanella frigidimarina, 242, 251
Shewanella oneidensis, 250
Shigella sonnei, 9899
Signal peptides, description of, 400
    in antibody fragment expression in periplasm, 370–371
    prediction of, 400–401
    protein sequences of, 370, 371
    twin arginine, 402
    type I, 401
    type II, 401–402
Signal recognition particle (SRP) system, 3
    E. coli, 4
Signal recognition particle (SRP) system-mediated targeting, hydrophobic N-terminal polypeptide, 4
SignalP, 401
Skp, 36–39, 142
    as chaperone of OMP targeting to OM, 40
    as periplasmic chaperone, 143–144
    in folding β-barrel OmpA, into lipid layers, 49–51
    insertion into negatively charged lipids, 38–39
    structure of, 143–144, Color Plate 13, 38
Skp trimer, structure of, 36–38, Color Plate 3
SpoIIIE, 185
SRP pathway, *E. coli*, 8–9
SRP RNA, 7
SRP substrates, ribosomes translating, 9
Streptococcus pneumoniae, 182, 186, 187
SubLoc localization tool, 395–396
Substrate-binding proteins, substrates captured by, 290–291, Color Plate 29
Sucrose/EDTA, for extraction of proteins from periplasm, 378
Sulfate-reducing bacteria, 242
Sulfolobus acidocaldarius, 238
Superoxide dismutase, 314
Support vector machine, 395, 397
SurA, 39–40, 145–146
    as chaperone of OMP targeting to OM, 40
    crystal structure of, 39, 146, Color Plate 3, Color Plate 13
    N-terminal amino acids 21 to 133 of, 39–40
Synthase-dependent pathway, 226–227, Color Plates 18 and 19

Tail-specific protease. *See* Tsp
Tat complexes, and Tat protein transport cycle, 21–23, Color Plate 1
    structure of, transport channels and, 23
Tat-dependent cofactor, steps in preassembly of, 25–26
Tat-dependent protein chaperones, coordinating assembly and export, 18–19, 25–26
Tat-dependent proteome, 16–17
Tat export pathway, 16–29, 348
Tat pathway, chromosomal location and organization of genes encoding, 17–20
Tat protein export machinery, 17–21

Tat protein transport cycle, Tat complexes and, 21–23, Color Plate 1
Tat-signal peptide-bearing substrate proteins, in *E. coli*, 17, 18–19
Tat signal peptides, extended n-regions prior to twin-arginine motif, 18–19, 24–25
    twin-arginine motif of, 18–19, 24
Tat substrate, cofactor-containing, chaperone-mediated assembly of, 25–26
Tat targeting signal, 23–25
TatA proteins, 20, 21
TatB proteins, 20–21
TatC proteins, 20–21
TatE proteins, 20
TatP, 402
Thermotoga maritima, 182
Thiobacillus ferrooxidans, 239
Thioredoxin family of proteins, DsbA and, 125–126
    DsbC and, 130–131, Color Plate 11
TMHMM, 401, 402, 403
TolQ/TolR, 279–280
TonB, 278–279
TonB-dependent receptors, 290
TonB-dependent transporters, 280
TorA, 266–267
TorD, 266–267
TraJ, 99
TraM, 99, 100
Translocation process, Tat substrate and, 22–23
Translocon, post- and cotranslational pathways converge at, 9–10
Transmembrane proteins (TMPs). *See* Integral membrane proteins
TraY, 99
Trehalose, synthesis of, 333
Tris/EDTA, for extraction of proteins from periplasm, 378
Trp-7, movement toward bilayer center, 55, 56
Trps, translocation across bilayer, 56
Tryptophan fluorescence, quenching by brominated phospholipids, 53–55
Tsp, 156–157
Twin-arginine protein transport. *See* Tat
Typhimurium, *Salmonella enterica* serovar, 98

Ubiquinone, in DsbB oxidation, 128
UniProt (Universal Protein Resource), 398–399

Vector machine methods, amino acid composition support, 395–396
Vicia faba, 267

WaaL, 227
WbaP, 221–222
WbbF, 226, Color Plate 19
WecA, 221–222, 225
Wolinella succinogenes, 237, 251, 252, 254
Wzx homolog, 221–222

Wzy, 222
    -dependent pathways, O-antigen modifications
        linked to, 223–225
      polymerization in, 221–223, Color Plate 17
    specificity of, alteration of, 225
Wzz, 222

Xanthine dehydrogenase, molybdenum cofactor
        insertion into, protein required for, 267
Xanthine oxidase, 268

YaeT, 43, 44, 113
YafL, putative cysteine protease, 164
YdcP, putative protease, 164
YdgD, 157

YdhO, putative cysteine protease, 164
YebA, hypothetical metalloprotease, 163
YedY, 266
Yersinia enterocolitica, 98, 99
Yersinia pestis, 328
YfgC, hypothetical metalloprotease, 163
YfgL, 45
YfiO, 45
YhbU, putative protease, 164
YhjJ, hypothetical metalloprotease, 163

Z-ring, assembly of, regulation of, 175, 176–178
    constriction of, 176
    structure and construction of, 174–176
ZapA, 173, 178, 182–183
ZipA, 173, 178, 180, 181–182